Laurentia-Gondwana Connections before Pangea

Edited by

Victor A. Ramos
Laboratorio de Tectónica Andina
Departamento de Ciencias Geológicas
Universidad de Buenos Aires
1428 Buenos Aires
Argentina

and

J. Duncan Keppie
Instituto de Geología
Universidad Nacional Autónoma de México
04510 México D.F.
México

336

1999

Published by The Geological Society of America, Inc.
3300 Penrose Place, P.O. Box 9140, Boulder, Colorado 80301

Printed in U.S.A.

GSA Books Science Editor Abhijit Basu

Library of Congress Cataloging-in-Publication Data

Laurentia-Gondwana Connections before Pangea / edited by Victor A. Ramos and J. Duncan Keppie.

p. cm. -- (Special paper ; 336)

Includes bibliographical references and index.

ISBN 0-8137-2336-1

1. Geology, Stratigraphic--Paleozoic. 2. Geology--South America. 3. Laurentia (Geology) 4. Gondwana (Geology) I. Ramos, Victor A. II. Keppie, J. Duncan. III. Special papers (Geological Society of America) ; 336.

QE654 .L38 1999
551.7'2--dc21 99-047508

Cover: Shuttle 59 photo of the Highest Andes around Mount Aconcagua (33°S latitude) taken in April 1994 by astronauts Thomas Jones and Jerome Apt at the altitude of 121 nautical miles. Note the Precordillera fold and thrust belt and the Pie de Palo range to the east, two outstanding areas with Laurentian basement. The Calingasta-Uspallata valley, a tectonic depression located to the west of Precordillera, is along the suture between Chilenia and Cuyania terranes (courtesy of Patricia Dickerson, NASA, Johnson Space Center, Houston, Texas).

10 9 8 7 6 5 4 3 2 1

Contents

Preface

The first report of Borrello in 1963 about North American olenellid trilobites from near San Juan in the Precordillera of west-central Argentina prompted curiosity about possible connections between Laurentia and Gondwana. The first attempt to explain the presence of these olenellids in South America was made by Ross (1975), who explained it as larval transfer by oceanic currents.

Several years later, the striking coincidence in the subsidence curves of the Appalachian carbonates and the Precordillera carbonate platform caused Bond et al. (1984) to suggest that both regions were conjugate margins and shared a common rift-drift transition. The shared carbonate platformal history and faunal provinciality between the Precordillera and the Laurentian margin of the Appalachians led Ramos et al. (1986) to propose that the Precordillera was a far-traveled terrane derived from the northern Appalachians. This proposal was refined by Mpodozis and Ramos (1990), Comínguez and Ramos (1991), and Ramos (1992, 1993) within the tectonic framework of the early Paleozoic basement of the Andes.

In the Appalachians, debate about the provenance of its eastern margin began when Wilson (1966) asked the question "Did the Atlantic close and then reopen?" Faunal and paleomagnetic data suggested that much of the eastern margin of the Appalachians originated off Gondwana (e.g., Whittington and Hughes, 1972; Williams, 1973; Johnson and Van Der Voo, 1985, 1986). In this context, Schenk (1971), using sedimentologic data, proposed that the Meguma Terrane in Canada had been derived from Northwest Africa, although Keppie (1977) suggested a Central–South America provenance using tectonothermal criteria. Similarly, O'Brien et al. (1983) suggested the Avalon Terrane in southwest Canada and the northeast United States was derived from Northwest Africa. However, dating of detrital zircons in the Avalon suggested a provenance in South America rather than Africa (Krogh and Keppie, 1990). Subsequent detrital zircon studies in the neighboring Gander and Exploits Terranes suggested the same provenance (Keppie et al., 1996; Van Staal et al., 1996).

The challenging pre-Pangea Southwest U.S.–East Antarctic Connection (SWEAT) reconstruction of the Late Proterozoic continents proposed by Moores (1991) and complemented by Dalziel's 1991 Laurentian end-run, provided a context for the transfer of the Precordillera and Avalon. These varied from transfer during continent-continent collision to transfer of microcontinents. Thus, Dalziel (1991 and subsequent papers) proposed that Laurentia and Gondwana collided during the Ordovician, and, on separation, a segment of Laurentia derived from the Ouachita embayment—the Precordillera—was left on the Gondwanan margin. This hypothesis was complemented by the proposal of Dalla Salda and colleagues that the Famatinian orogen of central-western South America continued in the Taconian orogen of North America (Dalla Salda et al., 1992a, b, 1993).

On the other hand, Keppie (1991), Benedetto and Astini (1993), and Astini et al. (1995, 1996) suggested that Precordillera was a far-traveled exotic terrane derived from the Ouachita embayment. Similarly, models for the transfer of Avalon varied from rifting, drifting, and accretion of a microcontinent (e.g. Van der Voo, 1988) to transfer during transpressional collision between North and South America (Keppie, 1993). In this context, it became clear that many other terranes were involved in Laurentia-Gondwana interactions (Keppie et al., 1996).

These problems became the central focus of the 5-yr IGCP (International Geological Correlation Program) Project 376 of UNESCO (the United Nations Educational Scientific and Cultural Organization). A series of symposia and field meetings were organized in Canada (1994, Nova Scotia), Argentina (1995, San Juan and Jujuy; 1996, Buenos Aires), the United States (1996, Austin, Texas; 1997, Salt Lake City, Utah),

New Zealand (1997, Christchurch), and México (1998, Oaxaca). The GSA Penrose Conference held in 1995 in a piece of Laurentia, the San Juan Precordillera of Argentina, was a milestone in the understanding of the Laurentia-Gondwana connections (see conference results in Dalziel et al., 1996). Derived understandings of this conference were that isotopic and Nd model ages of the Precordilleran basement were similar to the ones of southern Appalachians and distinct from cratonic South America (Kay et al., 1996); that Precordillera was probably detached from the Ouachita embayment (Thomas and Astini, 1996); and that faunal links were strong in the Cambrian, but the Laurentian affinities vanished in the Ordovician, when they were replaced by typical Gondwanan faunas. Some new scenarios, such as the Texas plateau proposed by Dalziel (1997), suggest some sort of continent-continent collision during Ordovician times that fueled the initial dilemma.

The various chapters of this GSA Special Paper aim to shed new light on the nature of the Laurentian-Gondwanan connections. The first, discussed by Ricardo Astini and Bill Thomas (Chapter 1), considers the origin and evolution of the Precordillera terrane as a drifted Laurentian block. The precise paleontologic constraints in the successive positions of the Precordillera terrane as well as a comprehensive analysis of the affinities and dissimilarities of the early Paleozoic fauna, is presented by Luis Benedetto and his colleagues (Chapter 2). The latitudinal changes from Laurentian subequatorial warm faunas to cold subpolar Gondwana faunas recorded the nature and isolation of the terrane until its final docking with Gondwana.

The affinities between the Precordilleran carbonate platforms of Cambrian and Ordovician ages, and the equivalent Appalachian platforms, are enhanced through the facies and sequence analyses presented by Fernando Cañas (Chapter 3).

The arrival of Precordillera encompassed in the Cuyania composite terrane is reconstructed through the analysis of the early Paleozoic western Sierras Pampeanas magmatic belt presented by Sonia Quenardelle and Victor Ramos (Chapter 4). The overview of petrology, chemical composition, and ages of prekinematic, synkinematic, and postkinematic granitoids of the protomargin of Gondwana gives testimony of the subduction beneath western Gondwana. On the other hand, these independent data provide constraints on the time of docking and final collision of the Precordillera terrane. The time analysis of the magmatic rocks is complemented by new uranium-lead data on the felsic magmatic rocks of Sierras Pampeanas obtained by Peter Stuart-Smith and his colleagues (Chapter 5). These new time constraints permit reconstruction of the magmatic activity in this protomargin of Gondwana during early Paleozoic times.

The nature of the contact between the basement of the Main Andes, known as the Chilenia terrane, and western Precordillera, is discussed by Steve Davis and his colleagues (Chapter 6). The closing of the ocean between the Precordillera terrane and Chilenia during the early Devonian led to ophiolite emplacement and deformation in southwestern Precordillera.

The late Proterozoic paleogeography north of the Precordillera and Chilenia terranes indicates a different scenario for northwestern Argentina. Analysis of the Puncoviscana belt by Duncan Keppie and Heinrich Bahlburg (Chapter 7) shows the foreland basin nature of the Late Proterozoic–Early Cambrian deposits of this region. This segment of the Gondwana protomargin was under collision and compression during the Late Proterozoic as a part of the pampean or brasiliano orogeny, typical of other cratonic blocks of Gondwana at that time.

Analysis of the magmatic sources and tectonic setting of the Ordovician margin of Gondwana of northern Argentina and Chile is presented by Beatriz Coira and her coworkers (Chapter 8). The authors emphasize the different geologic history of this segment when compared with the Laurentian terranes farther south. Geologic evidence shows the para-authochthonous origin of the northern Argentina and Chile cratonic blocks.

The paleomagnetic data presented by Augusto Rapalini and his colleagues (Chapter 9) support the exotic nature of the Cuyania composite terrane. The paleolatitudes obtained for Early Cambrian Precordilleran sequences prove to be similar to the paleolatitude of the Ouachita embayment in Laurentia. Furthermore, paleomagnetic data from the Famatina and Puna terranes of northern Argentina contribute to the Ordovician paleogeography of western Gondwana.

Analysis of the Laurentia-Gondwana connections are extended to the Late Proterozoic–early Phanerozoic link between the basements of eastern México and northern South America by Joaquin Ruiz and his colleagues (Chapter 10). The new data yield constraints for the paleocontinental reconstructions of the Americas. The geology and P-T-t paths of the 1-Ga Precambrian basement of México and Central America are compared with Laurentian and South American correlatives by Duncan Keppie and Fernando Ortega (Chapter 11), and suggests that Oaxaquia-Chortis blocks originated in the Grenville ocean between North and South America.

An overview of the Paleozoic faunal affinities presented by José Sánchez and coworkers (Chapter 12), of the sedimentary cover of México, although dispersed and discontinuous shows the links between different terranes and Laurentia. The strata of Ciudad Victoria, México, are studied by John Stewart and his colleagues. (Chapter 13). These beds contain a non–North American early Paleozoic fauna with close affinities with northern South America, which poses important constraints to the Paleozoic Laurentia connections.

A review of the Avalon composite terrane by Brendan Murphy and coworkers (Chapter 14) suggests that 150 Ma of latest Precambrian subduction along the northwest margin of South America included subduction of a ridge. Based on the previous analyses, the final chapter by Duncan Keppie and Victor Ramos (Chapter 15) reconstructs a series of paleomagnetically constrained maps for the Paleozoic odyssey of the Iapetan and Rheic terranes.

We want to acknowledge to hundreds of International Geological Correlation Program Project 376 (Laurentia-Gondwana Connections before Pangea) participants who, through active and lively discussions in the different symposia, have directly contributed to the present knowledge of the Laurentian-Gondwana connections before Pangea. Special recognition is extended to Eldridge Moores and Ian Dalziel, whose initial ideas and challenging hypotheses provided much interest in the potential links between both continents.

Victor A. Ramos
Duncan Keppie

ACKNOWLEDGMENTS

The editors thank the reviewers of the papers for their many contributions, comments, and careful revisions, as much as for the efforts to improve the accuracy of English expression. The reviewers were: T. H. Anderson (University of Pittsburgh, U.S.A.), R. Ayuso (U.S. Geological Survey, Houston, U.S.A.), A. W. Bally (Rice University, Houston, U.S.A.), R. Butler (University of Arizona, Tucson, U.S.A.), Z. de Cserna (Universidad Nacional Autónoma de México, Mexico City), J. M. Cortés (Universidad de Buenos Aires, Argentina), C. M. González-León (Universidad Nacional Autónoma de México, Mexico City), P. Gromet (Brown University, Providence, U.S.A.). R. Harris (West Virginia University, Morgantown, U.S.A.), J. C. Hepburn (Boston College, U.S.A.), J. Hibbard (North Carolina State University, U.S.A.), N. P. James (Queen's University, Kingston, Ontario, Canada), T. Jordan (Cornell University, Ithaca, U.S.A.), P. Kraemer (Universidad Nacional de Córdoba, Códoba, Argentina), W. C. McClelland (University of Idaho, Moscow, U.S.A.), C. F. Miller (Vanderbilt University, Nashville, U.S.A.), C. A. Mpodozis (Servicio Nacional de Geología y Minería, Santiago, Chile), E. Nelson (Colorado School of Mines, Denver, Colorado, U.S.A.), R. B. Neuman (U.S. Geological Survey, Museum of Natural History, Washington, U.S.A.), R. Omarini (Universidad Nacional de Salta, Salta, Argentina), R. Palma (Universidad de Buenos Aires, Argentina), A. R. Palmer (Institute for Cambrian Studies, Boulder, Colorado, U.S.A.), S. Poma (Universidad de Buenos Aires, Argentina), J. E. Repetski (U.S. Geological Survey, Reston, U.S.A.), P. Schaaf (Universidad Nacional Autónoma de México, Mexico), R. Somoza (Universidad de Buenos Aires, Argentina), R. A. Strachman (Oxford University, U.K.), C. Van Staal (Geological Survey of Canada, Ottawa, Ontario), G. Viele (University of Missouri, Columbia U.S.A.), W. Von Gosen (Universität Erlangen, Erlangen, Germany), R. Weber (Universidad Nacional Autónoma de México), S. H. Williams (Memorial University, Canada), and J. L. Wilson (Rice University, Houston, U.S.A.).

REFERENCES CITED

Astini, R. A., Benedetto, J. L., and Vaccari, N. E., 1995, The early Paleozoic evolution of the Argentina Precordillera as a Laurentian rifted, drifted, and collided terrane: a geodynamic model: Geological Society of America Bulletin, v. 107, no. 3, p. 253–273.

Astini, R., Ramos, V. A., Benedetto, J. L., Vaccari, N. E., and Cañas, F. L. 1996, La Precordillera: un terreno exótico a Gondwana, in XIII Congreso Geológico Argentino y III Congreso Exploración de Hidrocarburos, Actas; Buenos Aires, v. 5, p. 293–324.

Benedetto, J. L., and Astini, R., 1993, A collisional model for the stratigraphic evolution of the Argentine Precordillera during the early Paleozoic, in 2nd Symposium on International Géodynamique Andine ISAG 93 (Oxford): Paris, p. 501–504.

Bond, G. C., Nickelson, P. A., and Kominz, M. A., 1984, Breakup of a supercontinent between 625 Ma and 55 Ma: new evidence and implications for continental histories: Earth and Planetary Science Letters, v. 70, p. 325–345.

Borrello, A. V., 1963, *Fremontella inopinata* n. sp. del Cámbrico de la Argentina: Ameghiniana (Buenos Aires), v. III(2), p. 51–55,

Comínguez, A. H., and Ramos, V. A., 1991, La estructura profunda entre la Precordillera y Sierras Pampeanas (Argentina): evidencias de la sísmica de reflexión profunda: Revista Geológica de Chile, v. 18, no. 1, p. 3–14.

Dalla Salda, L., Cingolani, C., and Varela, R., 1992a, Early Paleozoic orogenic belt of the Andes in southwestern South America: result of Laurentia-Gondwana collision?: Geology, v. 20, p. 617–620.

Dalla Salda, L., Dalziel, I. W. D, Cingolani, C. A., and Varela, R., 1992b, Did the Taconic Appalachians continue into southern South America?: Geology, v. 20, p. 1059–1062.

Dalla Salda, L., Varela, R. and Cingolani, C. A., 1993, Sobre la colisión de Laurentia-Sudamérica y el orógeno Famatiniano, *in* XII Congreso Geológico Argentino y II Congreso de Exploración de Hidrocarburos (Mendoza), Actas: Buenos Aires, v. 3, p. 358–366.

Dalziel, I. W. D., 1991, Pacific margins of Laurentia and east Antarctica–Australia as a conjugate rift pair: evidence and implications for an Eocambrian supercontinent: Geology, v. 19, p. 598–601.

Dalziel, I. W. D., 1992, Antarctica: a tale of two supercontinents: Annual Reviews of Earth and Planetary Science, v. 20, p. 501–526.

Dalziel, I. W. D., 1994, Precambrian Scotland as a Laurentia-Gondwana link: the origin and significance of cratonic promontories: Geology, v. 22, p. 589–592.

Dalziel, I. W. D., 1997, Neoproterozoic-Paleozoic geography and tectonics: review, hypothesis, environmental speculation: Geological Society of America Bulletin, v. 109, p. 16–42.

Dalziel, I. W. D., Dalla Salda, L., Cingolani, C., and Palmer, P., 1996, The Argentine Precordillera: a Laurentian terrane?: GSA Today, v. 6, p. 16–18.

Johnson, R. J., and Van Der Voo, R., 1985, Middle Cambrian paleomagnetism of the Avalon terrane in Cape Breton Island, Nova Scotia: Tectonics, v. 4, p. 629–651.

Johnson, R. J., and Van Der Voo, R., 1986, Paleomagnetism of the Late Precambrian Fourchu Group, Cape Breton Island, Nova Scotia: Canadian Journal of Earth Sciences, v. 23, p. 1673–1685.

Kay, S. M., Orrell, S., and Abbruzzi, J. M., 1996, Zircon and whole rock Nd-Pb isotopic evidence for a Grenville age and a Laurentian origin for the Precordillera terrane in Argentina: Journal of Geology, v. 104, p. 637–648.

Keppie, J. D., 1977, Plate tectonic interpretation of Paleozoic world maps (with emphasis on circum-Atlantic orogens and southern Nova Scotia): Nova Scotia Department of Mines Paper 77–3, 45 p.

Keppie, J. D., 1991, Avalon, an exotic Appalachian-Caledonide terrane of western South American provenance, *in* Proceedings, 5th Circum-Pacific Terrane Conference, Santiago: Resúmenes Expandidos, Comunicaciones, v. 42, Universidad de Chile, p. 109–111.

Keppie, J. D., 1993, Transfer of the northeastern Appalachians (Meguma, Avalon, Gander, and Exploits terranes) from Gondwana to Laurentia during middle Paleozoic continental collision, *in* Proceedings, 1st Circum-Pacific and Circum-Atlantic Terrane Conference, Guanajuato: Mexico City, Universidad Nacional Autónoma de México, Instituto de Geología, p. 71–73.

Keppie, J. D., Dostal, J., Murphy, J. B., and Nance, R. D. 1996, Terrane transfer between Laurentia and western Gondwana in the early Paleozoic; constraints on global reconstructions, *in* Nance, R. D., and Thompson, M. D., eds., Avalonian and related peri-Gondwanan terranes of the circum-North Atlantic, Geological Society of America Special Paper 304, p. 369–380.

Krogh, T. E., and Keppie, J. D., 1990, Age of detrital zircon and titanite in the Meguma Group, southern Nova Scotia: clues to the origin of the Meguma terrane: Tectonophysics, v. 177, p. 307–323.

Moores, E. M., 1991, Southwest U.S.–East Antarctic (SWEAT) connection: a hypothesis: Geology, v. 19, p. 425–428.

Mpodozis, C., and Ramos, V. A., 1990, The Andes of Chile and Argentina, *in* Ericksen, G. E., Cañas Pinochet, M. T., and Reinemud, J. A., eds., Geology of the Andes and its relation to hydrocarbon and mineral resources, Circum-Pacific Council for Energy and Mineral Resources: Earth Sciences Series, v. 11, p. 59–90.

O'Brien, S. J., Wardle, R. J., and King, A. F., 1983, The Avalon zone: a Pan-African terrane in the Appalachian orogen of Canada: Geological Journal, v. 18, p. 195–222.

Ramos, V. A., 1992, Laurentian affinities of the early Paleozoic Precordillera terrane of Argentina: 29th. International Geological Congress, Kyoto, Japan, Abstracts, v. 2, p. 254.

Ramos, V. A., 1993, Terranes of southern Gondwanaland and their control in the Andean Structure (30°E–33°E S Latitude), *in* Reutter, K.-J., Scheuber, E., and Wigger, P. J., eds., Tectonics of the southern central Andes: structure and evolution of an active continental margin: Springer-Verlag, Berlin, p. 249–261.

Ramos, V. A., Jordan, T. E., Allmendinger, R. W., Mpodozis, C., Kay, S., Cortés, J. M., and Palma, M. A., 1986, Paleozoic terranes of the central Argentine-Chilean Andes: Tectonics, v. 5, p. 855–880.

Ross, R. J., 1975, Early Paleozoic trilobites, sedimentary facies, lithospheric plates, and ocean currents: Fossil & Strata, v. 4, p. 307–329.

Schenk, P. E., 1971, Southeastern Atlantic Canada, northwestern Africa, and continental drift: Canadian Journal of Earth Sciences, v. 8, p. 1218–1251.

Thomas, W. A., and Astini, R. A. 1996, The Argentine Precordillera: a traveller from the Ouachita embayment of North American Laurentia: Science, v. 273, p. 752–757.

Van der Voo, R., 1988, Paleozoic paleogeography of North America, Gondwana, and intervening displaced terranes: comparisons of paleomagnetism with paleoclimatology and biogeographical patterns: Geological Society America Bulletin, v. 100, p. 311–324.

Van Staal, C. R., Sullivan, R. W., and Whalen, J. B., 1996, Provenance and tectonic history of the Gander zone in the Caledonian/Appalachian orogen: implications for the origin and assembly of Avalon, *in* Nance, R. D., and Thompson, M. D., eds., Avalonian and related peri-Gondwanan terranes of the circum-North Atlantic, Geological Society of America Special Paper 304, p. 347–367.

Whittington, H. B., and Hughes, C. P., 1972, Ordovician geography and faunal provinces deduced from trilobite distributions: Royal Society of London Philosophical Transactions, ser. B, v. 263, p. 235–278.

Williams, A., 1973, Distribution of Paleozoic assemblages in relation to Ordovician palaeogeography, *in* Hughes, N. F., ed., Organisms and continents through time: Palaeontological Association Special Papers in Palaeontology, v. 12, p. 241–269.

Wilson, J. T., 1966, Did the Atlantic Ocean close and reopen?: Nature, v. 211, p. 676–681.

Geological Society of America
Special Paper 336
1999

Origin and evolution of the Precordillera terrane of western Argentina: A drifted Laurentian orphan

Ricardo A. Astini
CONICET Estratigrafía y Geología Histórica, Facultad de Cs. Ex. Fís. y Nat., Universidad Nacional de Córdoba, Vélez Sarsfield 299, CC 395, 5000 Córdoba, Argentina
William A. Thomas
Department of Geological Sciences, University of Kentucky, Lexington, Kentucky 40506-0053

ABSTRACT

The Precordillera terrane of northwestern Argentina is a fragment of rifted Laurentian continental crust and passive-margin cover. Two separate episodes of extension, approximately 60–70 m.y. apart, and a contractional event are recorded in the early Paleozoic history of the Precordillera.

Crustal extension during the Early Cambrian (possibly starting in the latest Proterozoic) led to asymmetric continental rifting and separation of Precordillera from the Ouachita embayment of southern Laurentia. Synrift graben-fill successions of the Precordillera are overstepped by latest Early Cambrian carbonates indicating rift-to-drift transition and initiation of passive-margin deposition. Faunal evolution in the uppermost Lower Cambrian through Lower Ordovician passive-margin succession suggests isolated drifting of the Precordillera as a Laurentian orphan across the Iapetus Ocean during the Late Cambrian and Early Ordovician. Subsidence curves are typical of postrift thermal subsidence on rifted continental margins.

A contractional event, interpreted as the docking of the Precordillera with Gondwana, is documented by mylonitic fabrics, Ocloyic metamorphic ages (~464 Ma) imprinted on Grenville basement rocks, and west-directed thrusting of passive-margin limestones in the eastern Precordillera. The collision is bracketed to early Middle Ordovician time.

A second extension episode, during the Middle Ordovician starting in the Llanvirn, is evidenced by irregular distribution of sediments and hiatuses, abrupt changes in lithofacies, and local slope-scarp facies associated with block faults. Deep-seated extension of previously attenuated crust ultimately accounted for mafic sills and basaltic pillow lavas that interdigitate with hemipelagites. This postcollision extension is related to an unspecified reorientation of stresses generated within the orphaned plate, probably because of a fundamental change in plate motion immediately after the collision of Precordillera with Gondwana.

INTRODUCTION

The Argentine Precordillera is presently considered as one of the most convincingly documented allochthonous terranes in the literature. Its exotic nature to Gondwana and its derivation from Laurentia have been claimed by various authors and supported by different lines of evidence (Ramos et al., 1986; Dalla Salda et al., 1992a; Benedetto and Astini, 1993; Benedetto, 1993; Astini and Benedetto, 1993; Dalziel, 1993; Dalziel et al., 1994, Astini et al., 1995a, 1996; Benedetto et al., 1995; Thomas

Astini, R. A., and Thomas, W. A., 1999, Origin and evolution of the Precordillera terrane of western Argentina: A drifted Laurentian orphan, *in* Ramos, V. A., and Keppie, J. D., eds., Laurentia-Gondwana Connections before Pangea: Boulder, Colorado, Geological Society of America Special Paper 336.

and Astini, 1996a; Kay et al., 1996). Not only its allochthony is recognized, but also its provenance area is fairly well constrained (Thomas and Astini, 1996a), making it a particularly unique example in Earth history. Because of the importance of this exotic terrane to early Paleozoic paleogeography (cf. Torsvik et al., 1995, 1996; Dalziel et al., 1994; Dalziel, 1997), kinematic constraints for transfer of the Laurentian orphan to Gondwana must be strengthened. Consequently, the main focus has shifted to how this fairly small lithosphere fragment was rifted, and whether it drifted independently in the Iapetus Ocean before colliding with Gondwana. In this respect, two different hypotheses have been presented. One hypothesis is that the Precordillera is a far-traveled terrane (Ramos et al., 1986; Benedetto, 1993) with recognizable rift, drift, and collisional stages recorded in its early Paleozoic history. This "microcontinent model" or "raft idea," developed by Astini et al. (1995a, 1996) and Thomas and Astini (1996a), also has been called the "funeral ship model" by Dalziel et al. (1996) and Dalziel (1997). In this hypothesis, the Precordillera is considered to have been an independent microcontinent that separated from Laurentia and drifted through the Iapetus Ocean toward Gondwana, with which it finally collided around the Middle Ordovician. More specifically, the model proposes that the Precordillera terrane was the conjugate margin of the Ouachita embayment and Texas promontory (southern part of the pre-Appalachian Laurentian margin) and broke apart from Laurentia during opening of the Ouachita rift in Cambrian time (Thomas and Astini, 1996a). An alternate hypothesis proposes that a continent-continent collision between Laurentia and Gondwana during the Middle-Late Ordovician was followed by rifting that left the Precordillera fragment of Laurentia attached to Gondwana, thus remaining as a tectonic tracer of the collision (the "calling card model" of Dalla Salda et al., 1992b; Dalziel, 1993; Dalziel et al., 1994). A modified version of this hypothesis (Dalziel, 1997) suggests that, prior to collision, the Precordillera remained attached to Laurentia but was greatly extended outboard through stretched continental lithosphere that receives the name of "Texas plateau"; this could be called the "bridge hypothesis." Although the calling card or Texas plateau hypothesis denies Cambrian rifting and assigns real separation of Precordillera from Laurentia to Middle-Late Ordovician time (Dalziel, 1997), most interpretations hold that the Precordillera is the conjugate margin of the Texas promontory (Dalziel et al., 1996). This view is now strongly supported by new paleomagnetic evidence (Rapalini and Astini, 1998).

Although both alternatives are paleomagnetically permissible as suggested by Dalziel (1997), we maintain that the geologic evidence strongly favors the independent microcontinent alternative, which is more consistent with conventional paleogeography (Torsvik et al., 1995, 1996) for the early Paleozoic. Geologic models should not be forced to fit into paleogeographic schemes (model-driven paleogeography); rather, paleogeography should be based on geologic data. In this respect, although the microcontinent hypothesis does not allow precise relative positioning of Laurentia and Gondwana, indirect longitudinal positioning can be assumed, as discussed by Thomas and Astini (1996a).

In this chapter, we further develop the microcontinent hypothesis on the basis of strong evidence for the Early Cambrian rifting of the Precordillera terrane from Laurentia. Moreover, we suggest a preliminary kinematic model for separation of the Precordillera from the Ouachita embayment of Laurentia, considering differences in subsidence rates and inferred crustal structure. Furthermore, we add new evidence of a second rifting episode affecting the Precordillera more than 60 m.y. after its initial rifting. In the time between those two rifting episodes, the Precordillera drifted through the Iapetus Ocean. The second extension episode is interpreted to have occurred during and after docking of the Precordillera against Gondwana. Thus, we contend that, rather than a tectonic tracer of a major continent-continent collision, the Precordillera should be considered as a Laurentian orphan, which drifted alone to produce tectonic features that, with the exception of the initial rifting, should not be correlatable between Laurentia and Gondwana.

TECTONOSTRATIGRAPHY OF THE PRECORDILLERA TERRANE

The Argentine Precordillera basement rocks and lowermost Paleozoic cover strongly contrast with those of the surrounding geologic provinces in southern South America (Ramos et al., 1986; Astini et al., 1996). As defined by Astini et al. (1995a), the Precordillera terrane (Fig. 1) comprises the foothill region of the high Andes between the provinces of La Rioja and Mendoza in western Argentina. It includes the thin-skinned thrust belt that formed between 29° and 33°S as a response to Neogene flat-slab subduction (Jordan et al., 1983), as well as the outcrops south of Mendoza where Ordovician rocks lie nonconformably on Grenville basement. On the east, Grenville rocks, presently exposed in a discontinuous belt (western Pampean Ranges) in the Andean foreland, were considered by Astini et al. (1995a) to be like the probable basement beneath the Paleozoic succession in the Precordillera (see Kay et al., 1996) and, hence, are considered here as part of the Precordillera terrane (Cuyania terrane of Ramos, 1995). The Precordillera terrane is more than 200 km wide (east-west present-day coordinates) and more than 800 km long, although the northern and southern extremes remain to be strictly defined. The present width evidently is greatly shortened by Andean and pre-Andean orogenesis. The eastern and western boundaries are marked by the Bermejo and Rodeo-Calingasta-Uspallata alluviated valleys that probably reflect Andean reactivation of older structures (Ramos, 1988, 1993).

The tectonostratigraphic subdivisions defined by Astini and Benedetto (1996) and Astini et al. (1996) for the Precordillera (Fig. 2) include three main early and middle Paleozoic evolutionary stages on the basis of character and nature of the substrate, prevailing sedimentation, and deformational regime: a passive margin, a foreland basin I, and a foreland basin II. This

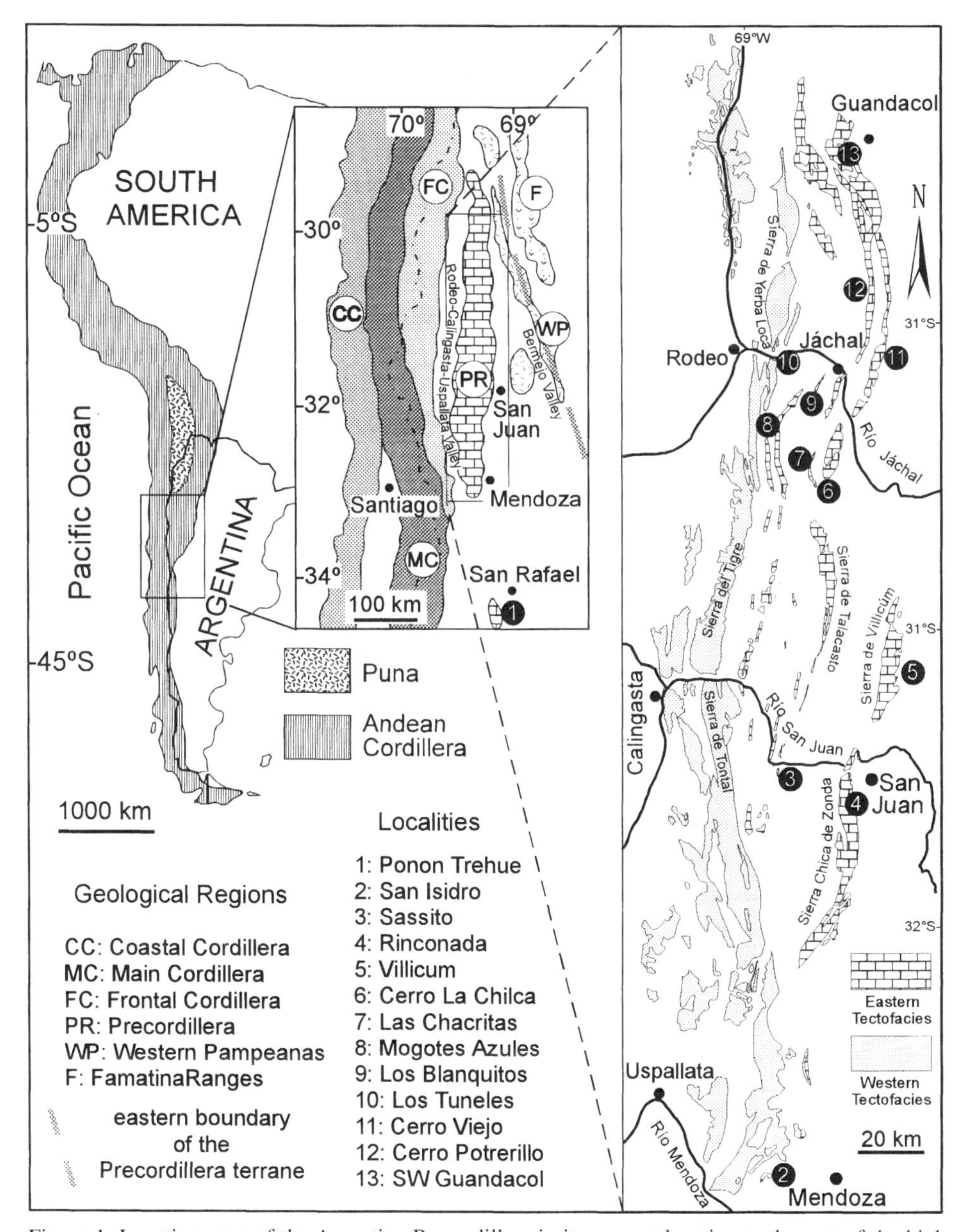

Figure 1. Location map of the Argentine Precordillera in its present location to the east of the high Andes, and detail of the eastern and western tectofacies recognized by Astini (1992) for the Ordovician. Note that part of the basement included in the western Pampeanas, as well as the outcrops to the south of San Rafael, are included in the Precordillera terrane. Numbered localities 1, 2, 3, 7, 8, 10, and 11 identify horsts on which carbonates nucleated in the Middle and Late Ordovician.

chapter focuses primarily on the first two stages because they encompass the history of rifting-drifting-collision of the Precordillera terrane (see Fig. 3). The foreland II stage includes the postcollisional middle Paleozoic history of the Precordillera and has been extensively treated in Astini (1996a) and Astini and Maretto (1996). This corresponds to the definitive Gondwanan part of Precordilleran history and originated as a response to contractional tectonics driven by the approach and collision of the Chilenia terrane (e.g., Ramos et al., 1986) to the present west of the Precordillera.

We are currently exploring in more detail the rifting of Precordillera from the Ouachita embayment (Thomas and Astini, 1996b). This is not an easy task because the original rifted margins of both Laurentia and the Precordillera have been deformed by later orogenies and partly covered by younger rocks that complicate and conceal the original geometry and paired rifted-margin paleogeography. The inherited geometry of the three-dimensional rift system has critical implications for the later contractional histories of both margins.

According to Astini et al. (1996), the passive-margin stage

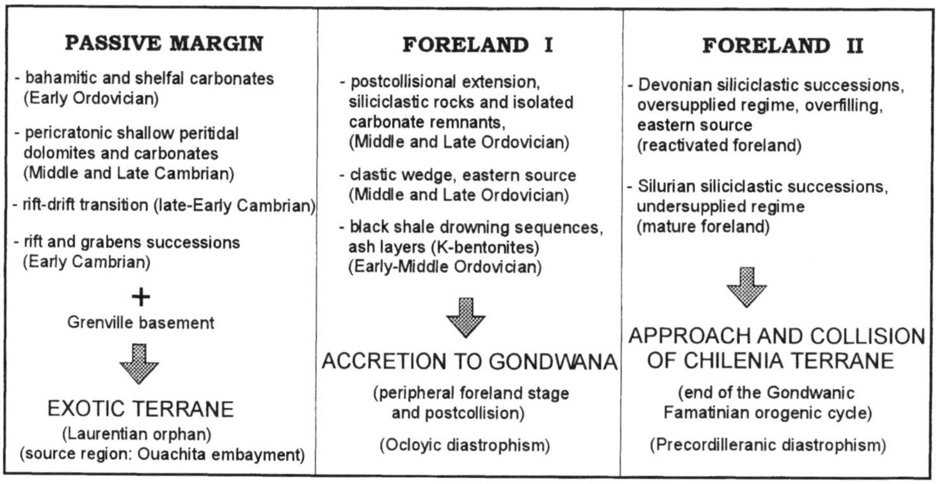

PASSIVE MARGIN	FORELAND I	FORELAND II
- bahamitic and shelfal carbonates (Early Ordovician) - pericratonic shallow peritidal dolomites and carbonates (Middle and Late Cambrian) - rift-drift transition (late-Early Cambrian) - rift and grabens successions (Early Cambrian) + Grenville basement ⇩ EXOTIC TERRANE (Laurentian orphan) (source region: Ouachita embayment)	- postcollisional extension, siliciclastic rocks and isolated carbonate remnants, (Middle and Late Ordovician) - clastic wedge, eastern source (Middle and Late Ordovician) - black shale drowning sequences, ash layers (K-bentonites) (Early-Middle Ordovician) ⇩ ACCRETION TO GONDWANA (peripheral foreland stage and postcollision) (Ocloyic diastrophism)	- Devonian siliciclastic successions, oversupplied regime, overfilling, eastern source (reactivated foreland) - Silurian siliciclastic successions, undersupplied regime (mature foreland) ⇩ APPROACH AND COLLISION OF CHILENIA TERRANE (end of the Gondwanic Famatinian orogenic cycle) (Precordilleranic diastrophism)

Figure 2. Synthetic chart showing the three main tectonostratigraphic subdivisions differentiated by Astini and Benedetto (1996) and Astini et al. (1996) in the early Paleozoic evolution of the Argentine Precordillera.

includes the complete carbonate-bank development, starting in the Early Cambrian with synrift facies and evolving to a mature-stage platform between the Middle Cambrian and the Early Ordovician (Figs. 2 and 3). The shelfal phase of the Precordilleran continental terrace constitutes the drifting stage of the Laurentian orphan (Astini et al., 1996). Black shales at the top of the passive-margin carbonates (Fig. 3) are interpreted to indicate initiation of flexural subsidence during the approach to Gondwana (see section on Docking). Younger Ordovician siliciclastic successions and related isolated carbonate deposits (Fig. 3) are included in the foreland I stage (see Fig. 2).

Although Middle Ordovician successions define the approach and ultimate collision of the Precordillera with Gondwana (Astini et al., 1995a), aside from faunal and sedimentologic criteria, the accretion event is not very clear. The timing of docking is bracketed (according to Astini et al., 1995a, 1996) between the Llanvirn and the Caradoc, but more precise timing is uncertain, especially considering that the Middle-Late Ordovician interval has cosmopolitan faunas (Benedetto et al., 1995) that are difficult to apply as straightforward criteria for exact timing. Moreover, this stratigraphic interval contains several hiatuses and laterally discontinuous units that yield few fossil collections. Although several authors (Borrello, 1969; Furque, 1972; Gosen, 1992; Gosen et al., 1995; Keller, 1995; Astini, 1994a, 1997a,b) have recognized indirect evidence of extensional tectonics through the Middle and Late Ordovician, none have suggested a clear kinematic context for this extension. Astini (1997c, 1998) shows that the mid-Ordovician extension could be related either to brittle subsidence induced by lithosphere flexure as the Precordillera approached Gondwana (cf. Bradley and Kidd, 1991), or, more likely, to postcollision extension as a response to a major change in plate motion after the docking of the Precordillera. Ordovician extension is separated by more than 60 m.y. from the initial rifting, suggesting a genetically different process.

RIFTING

As a supercontinent breaks up through rifting processes, some smaller fragments (microcontinents) surrounded by various rift branches and transfer (transform) faults are left between the larger continents, as implied by rift geometry models and mechanisms (e.g., Vink et al., 1984; Lister et al., 1986, 1991; Rosendahl, 1987). Following break-up, a microcontinent may remain near the parent continent and, during a subsequent continental collision, be welded back to the parent continent (e.g., Appalachian internal basement massifs of Thomas, 1977; Hatcher, 1989). Alternatively, a lithosphere fragment, a microcontinent, may be separated from the parent continent and drift independently to collide with a foreign continent (e.g., the Argentine Precordillera, as interpreted to be a far-traveled fragment of Laurentia) (see Ramos et al., 1986; Benedetto, 1993; Thomas and Astini, 1996a). Rift-related intracratonic fault systems (e.g., Mississippi Valley and Birmingham graben systems of southern Laurentia) indicate the width of brittle extension of the upper crust (Fig. 4). Extensional faults in continental crust, commonly recognized as "failed rifts" or "aulacogens" (Keller et al., 1983; Braile et al., 1986; Thomas, 1991), partially to completely outline basement blocks that represent prospective microcontinents arrested before completion of break-up of the parent continent.

A fragment of Grenville continental crust (now the Argentine Precordillera) was rifted from the Ouachita embayment of the Laurentian margin during the Early Cambrian (Thomas, 1991, 1993), as a result of diachronous rifting associated with the opening of the Iapetus Ocean (Thomas and Astini, 1996a,b). Both the Alabama-Oklahoma transform fault and the Ouachita rift, outlining the Ouachita embayment and the Texas promontory of the Laurentian margin (Fig. 4), are defined on the basis of deep wells and geophysical data (summarized in Thomas, 1991).

Contrasts in ages of latest synrift rocks and early passive-

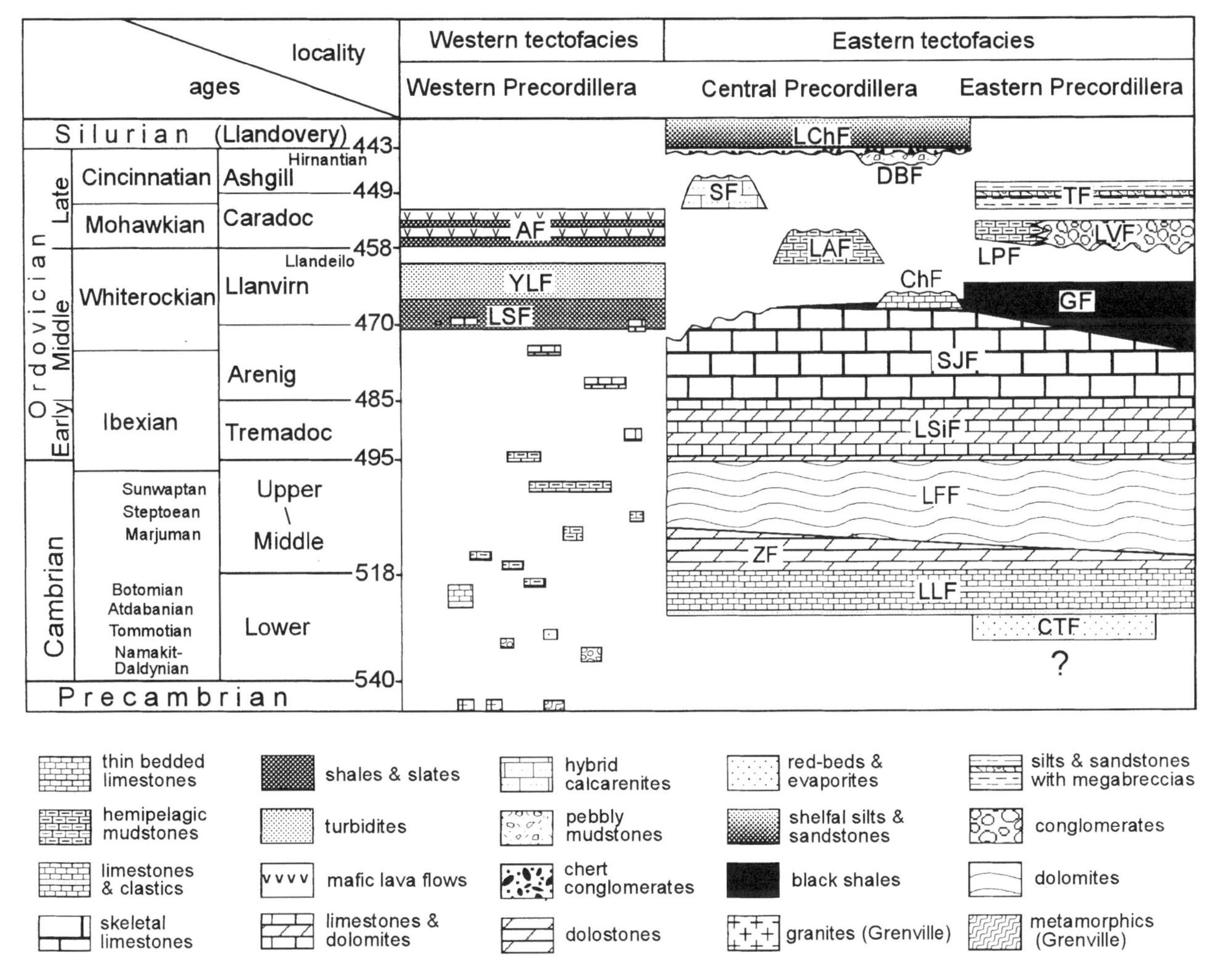

Figure 3. Generalized time-stratigraphic chart of the Cambrian and Ordovician across the Argentine Precordillera. Note contrasting stratigraphy of the eastern and western tectofacies (see distribution in Fig. 1). CTF = Cerro Totora Formation; LLF = La Laja Formation; ZF = Zonda Formation; LFF = La Flecha Formation; LSiF = La Silla Formation; SJF = San Juan Formation; GF = Gualcamayo Formation; ChF = Las Chacritas Formation; LAF = Las Aguaditas Formation; LPF = Las Plantas Formation; LVF = Las Vacas Formation (La Cantera Formation); SF = Sassito Formation; TF = Trapiche Formation; DBF = Don Braulio Formation; LChF = La Chilca Formation; LSF = Los Sombreros Formation; YLF = Yerba Loca Formation; AF = Alcaparrosa Formation. Time scales according to Tucker and McKerrow (1995) and Grotzinger et al. (1995). Olistoliths and basement boulders (granites and metamorphic rocks) present in the Los Sombreros Formation (western tectofacies) are represented according to their individual ages (Benedetto and Vaccari, 1992; Bordonaro and Banchig, 1996; R. A. Astini, unpublished data). (Sources for construction: Albanesi et al., 1990; Astini, 1991, 1994c; Astini and Brussa, 1997; Astini and Cañas, 1995; Astini and Vaccari, 1996; Benedetto and Herrera, 1987; Brussa, 1994; Keller et al., 1994; Lehnert, 1995; Ortega and Brussa, 1990; Ortega et al., 1991; Vaccari, 1994.)

margin deposits suggest that the spreading ridge shifted from the southern part of the Blue Ridge rift to the Ouachita rift at approximately 544 Ma, and propagation of the Alabama-Oklahoma transform fault transferred extension from the northern part of the Ouachita rift to a mid-Iapetus ridge outboard of the Blue Ridge rifted continental margin (Fig. 4) (Thomas, 1991). Along the Blue Ridge rift at the eastern margin of Laurentia (Fig. 4), the transition from thick synrift accumulations to passive-margin deposition is dated biostratigraphically as earliest Cambrian, ~544 Ma (Bond et al., 1984; Simpson and Sundberg, 1987). Extensive shelfal carbonate facies delimit the extent of the Early Cambrian–Early Ordovician passive margin along the Blue Ridge margin of Laurentia, and locally preserved east-facing Cambrian slope deposits locate the shelf edge (Pfeil and Read, 1980; Read, 1989). In contrast, adjacent to the Ouachita embayment, the Mississippi Valley and Birmingham graben systems (Fig. 4) are filled with Early and Middle Cambrian synrift successions, including fine-grained clastic rocks and evaporite interbeds (Mellen, 1977; Raymond, 1991; Thomas, 1991) in stratigraphic successions similar to those in the Cerro Totora graben in the northern Precordillera (Astini et al., 1995b; Astini and Vaccari, 1996). These graben systems indicate extension of continental crust parallel to the Alabama-Oklahoma transform fault. The Southern Oklahoma fault system parallels the

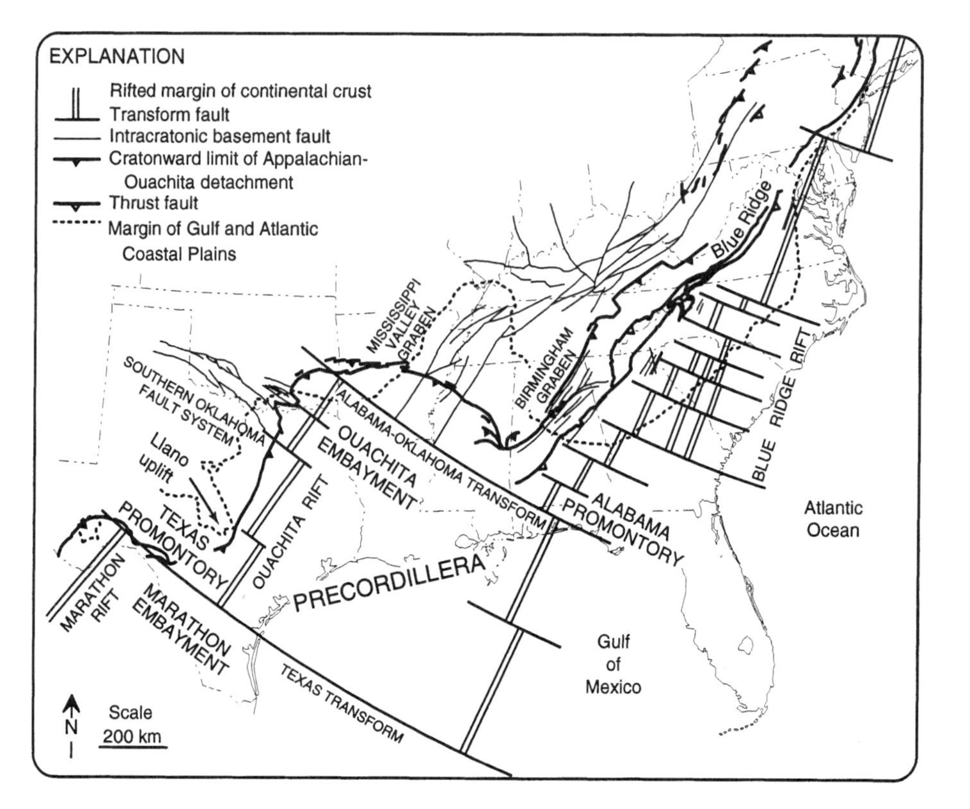

Figure 4. Map of the southeastern United States with location of the Ouachita embayment of Laurentia, the Argentine Precordillera block, and the general outline of the late Precambrian-Cambrian rifted margin and related structures (after Thomas and Astini, 1996a).

Alabama-Oklahoma transform fault (Fig. 4), and along the fault system, synrift igneous rocks (gabbro, granite, basalt, and rhyolite) range in age from 552 ± 7 to 525 ± 25 Ma (Gilbert, 1983; Lambert et al., 1988; Hogan et al., 1996). A transgressive passive-margin succession of middle Late Cambrian age overlaps the synrift sedimentary rocks of the Mississippi Valley and Birmingham grabens, as well as the synrift igneous rocks of the Southern Oklahoma fault system (Fig. 4).

No synrift rocks like those of the Mississippi Valley and Birmingham grabens have been recognized on the Texas promontory of Laurentia, where a transgressive Upper Cambrian–Lower Ordovician passive-margin carbonate-shelf succession is generally <1,000 m thick (Barnes, 1959; Denison, in Johnson et al., 1988). At the base of the passive-margin succession, uppermost Middle Cambrian sandstones rest directly on crystalline basement rocks with local paleotopographic relief of >200 m (Barnes et al., 1972). Passive-margin, deep-water, off-shelf deposits of Late Cambrian and younger age are well documented in outcrops and subsurface of the late Paleozoic Ouachita thrust belt (Fig. 4); the allochthonous off-shelf rocks were tectonically transported over the shelf edge and onto temporally equivalent shelf carbonates (Viele and Thomas, 1989).

The oldest part of the exposed stratigraphic succession in Andean thrust sheets of the Precordillera consists of Lower Cambrian red clastic rocks and evaporites of the Cerro Totora Formation (Fig. 5) (Astini et al., 1995a,b; Astini and Vaccari, 1996). The Lower Cambrian evaporites and clastic rocks suggest synrift graben-filling deposition, either in the "main" (Ouachita) rift or in a Precordilleran intracratonic synrift graben similar to the Mississippi Valley and Birmingham graben systems on Laurentia adjacent to the Alabama-Oklahoma transform margin. In the uppermost Lower Cambrian Series (Figs. 3 and 5), the red clastic rocks grade upward into the base of a thick Middle Cambrian–Lower Ordovician carbonate-shelf succession, recording cessation of rifting and establishment of a passive margin on the Precordillera (Astini et al., 1995a; Astini and Vaccari, 1996). West of the Precordillera carbonate shelf (Fig. 3), a west-facing slope facies of Ordovician mudstone turbidite contains slumped boulders of the shelf facies and olistoliths of the slope facies (Astini, 1988; Astini et al., 1995a), as well as blocks of conglomeratic sandstones, the composition of which indicates a provenance of continental basement rocks (Banchig et al., 1990; Astini, 1991, 1996b), suggesting synrift graben-filling rocks (Thomas and Astini, 1996b). The basal red clastic succession and the conglomeratic sandstone blocks in slope deposits in the Precordillera represent synrift sediment accumulations which possibly were associated with large-scale basement faults.

The Middle Cambrian–Lower Ordovician passive-margin-shelf succession in the Precordillera ranges from ~2,400 to ~3,100 m thick, approximately three times as thick as the passive-margin-shelf succession on the Texas promontory. However, the western margin of the Precordillera is interpreted to be the conjugate of the rifted margin of the Texas promontory (Fig. 4). The asymmetry of thickness of shelf deposits on opposing rift margins indicates an asymmetric structure of the rift, as well as asymmetry of postrift thermal subsidence. Backstripping of stratigraphic sec-

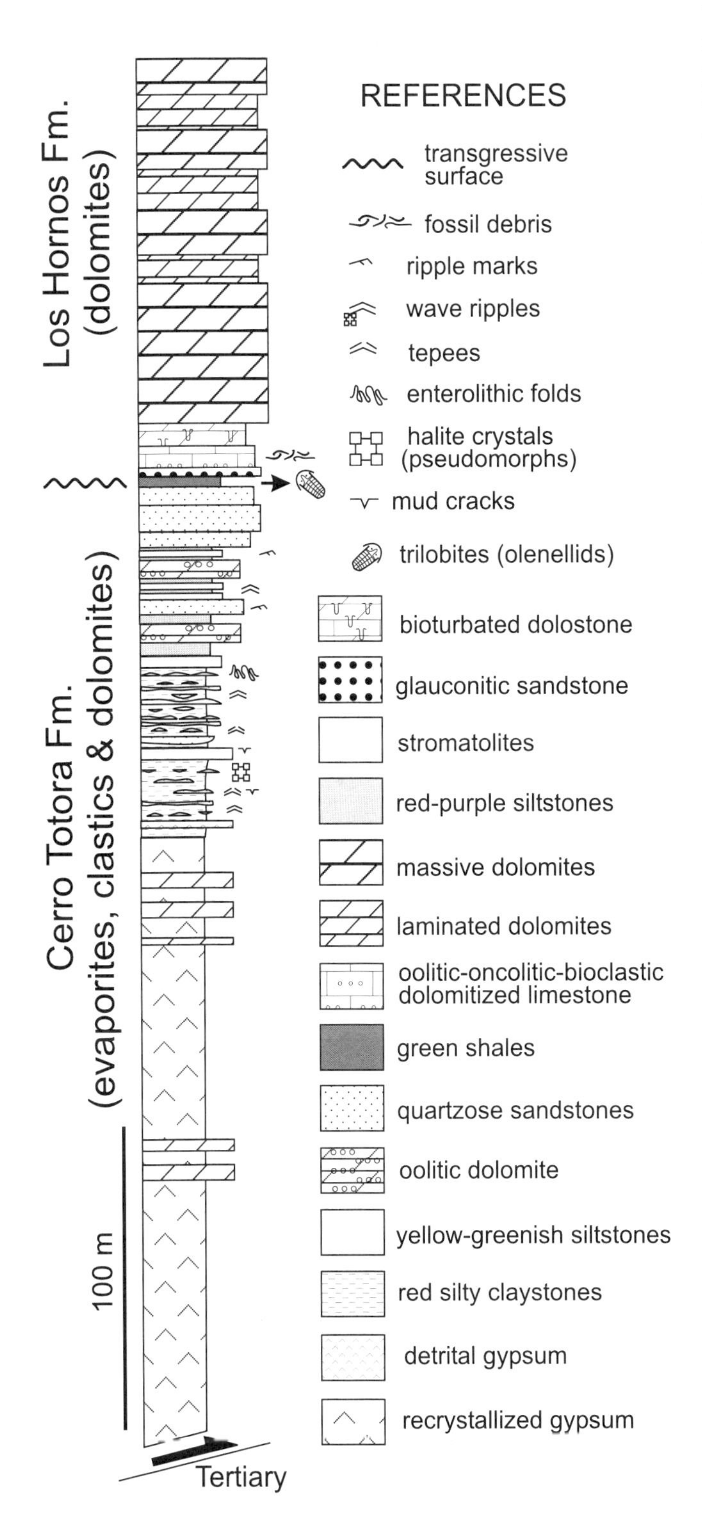

Figure 5. Detailed stratigraphic log of the synrift graben clastic and evaporitic fill succeeded by the transgressive surface representing the rift-drift transition in the northern Precordillera (modified from Astini and Vaccari, 1996).

tions from similar positions on the passive-margin shelves of the northern Precordillera and the Texas promontory of Laurentia yields subsidence profiles (Fig. 6) typical of passive-margin thermal subsidence; however, the profiles differ significantly in magnitude of subsidence and in time of initial subsidence. This asymmetry of postrift subsidence demonstrates asymmetry of the Ouachita rift and suggests a simple-shear, low-angle-detachment rift (Fig. 7) (as defined by Lister et al., 1986). Along the Ouachita rifted margin of the Texas promontory of Laurentia, the thin passive-margin-shelf succession resting with local paleotopographic relief directly on basement indicates slow and limited postrift subsidence of a broad uplift of the rift shoulder (Fig. 7), consistent with thermal buoyancy of the upper plate above a low-angle detachment (Buck et al., 1988). A narrow zone of transitional crust and the lack of synrift rocks further suggest an upper-plate structural configuration for the Ouachita rifted margin. The passive-margin-shelf succession of the Precordillera documents significantly greater subsidence and sediment accumulation than along the Ouachita rifted margin of Laurentia (Fig. 7). Passive-margin-shelf deposition around the Precordillera began in

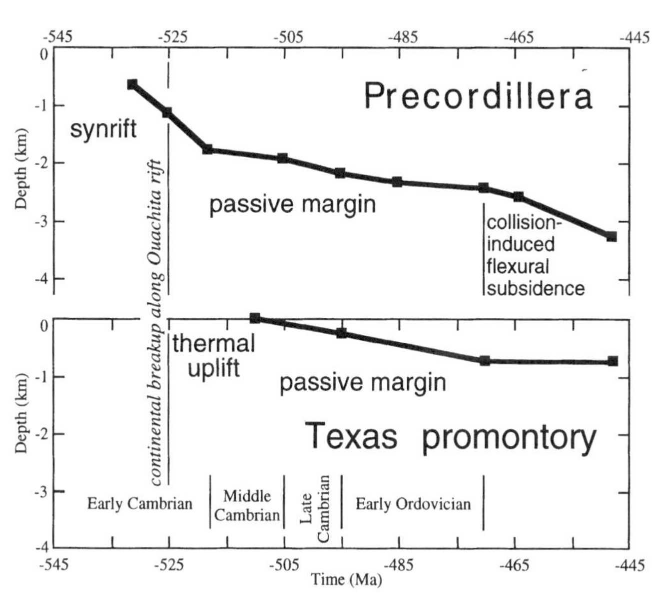

Figure 6. Tectonic subsidence curves for top of basement rocks computed from decompaction and backstripping of passive-margin successions from locations at similar distances from the shelf edges in the Precordillera and on the Texas promontory of Laurentia (program by Wilkerson and Hsui, 1989, using porosity/depth data from Sclater and Christie, 1980, and Schmoker and Halley, 1982). Stratigraphic data for the Precordillera from Baldis and Bordonaro (1981, 1984), Keller et al. (1994), and R. A. Astini (unpublished data). Stratigraphic data for the Texas promontory from Barnes (1959), Barnes et al. (1972), and Johnson et al. (1988). Time scale from Tucker and McKerrow (1995). Passive-margin sedimentary rocks rest unconformably on basement rocks on the Texas promontory, and the top of basement is defined at zero depth at the beginning of deposition. No contact of the synrift/passive-margin succession with basement is exposed in the Precordillera; therefore, the age of zero subsidence (initial deposition on basement) and the depth of tectonic subsidence of the top of basement at the time of deposition of the oldest exposed rocks can only be estimated.

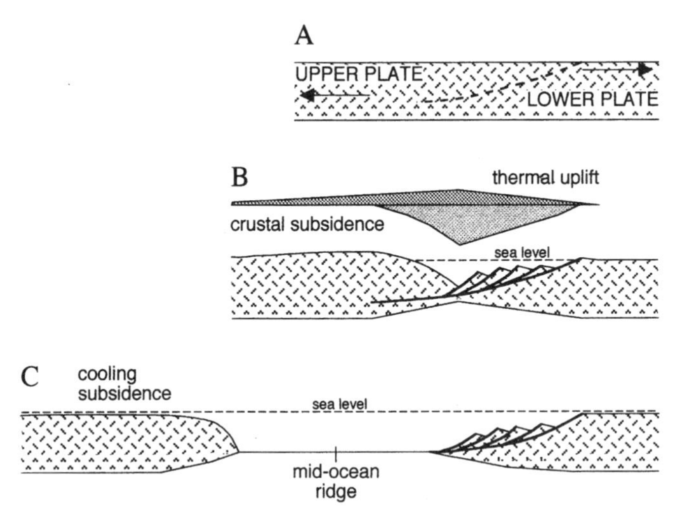

Figure 7. Sequence of diagrammatic cross sections (from Thomas and Astini, 1999) depicting the interaction of thermal uplift and isostatic crustal subsidence on opposite blocks of a low-angle detachment during continental rifting and break-up in simple shear (designed to illustrate concepts from Lister et al., 1986, 1991; Buck et al., 1988; Etheridge et al., 1989; Issler et al., 1989). A, Trajectory of low-angle detachment prior to initial stretching. B, Extended crust prior to break-up (geometry of faults from Lister et al., 1986); maximum heat flow is at the intersection of the detachment with the surface (Buck et al., 1988); subsidence of thinned crust counteracts thermal uplift so that maximum topographic elevation of thermal uplift is at the edge of full-thickness continental crust on the upper plate (proportions and distribution of thermal uplift, crustal subsidence, and topography from Buck et al., 1988); representative of Texas promontory (upper plate) and Precordillera (lower plate) at ~530 Ma. C, Following break-up and drift away from mid-ocean ridge; cooling and thermal-decay subsidence of upper plate, leading to transgression and initiation of passive-margin-shelf deposition on the upper plate (Texas promontory) at ~500 Ma; continued crustal subsidence and shelf deposition on lower plate (Precordillera).

latest Early Cambrian time, earlier than initiation of passive-margin-shelf deposition around the Ouachita embayment during late Middle Cambrian time. The earlier beginning and greater magnitude of postrift subsidence of the Precordillera are consistent with thinner, more extended continental crust and early separation from the synrift heat-flow maximum characteristic of a lower-plate margin (Buck et al., 1988). A lower-plate configuration of the western rifted margin of the Precordillera (Fig. 7) is conjugate with an upper-plate configuration on the Ouachita rifted margin of Laurentia (Thomas and Astini, 1996b; Astini, 1996b). Thermal doming of the upper-plate Ouachita margin prevented accumulation of synrift rocks and, as predicted by the model, retarded passive-margin-shelf sedimentation for almost 20 m.y. In contrast, "samples" of initial synrift sedimentary rocks are found in the western Precordillera, and deposition of a thick carbonate bank over locally important graben-fills (red sandstones and evaporites) began in the late Early to early Middle Cambrian, suggesting early thermal subsidence of a lower-plate margin.

A passive-margin carbonate rim surrounded much of Laurentia during Cambrian and Early Ordovician times, reflecting the general stability of the craton, the extensional nature of the margins, and the average equatorial position (Lochman-Balk, 1971; Alberstadt and Repetski, 1989; Witzke, 1990). Nevertheless, facies and passive-margin architecture are highly variable from margin to margin (Osleger and Read, 1991; Osleger, 1993) and along a same margin (Read, 1989). These variations apparently are related to the different behavior of separate segments of the rifted margin (inherited margin geometry) and the kinematics of the extended crustal blocks (e.g., Thomas, 1977, 1993; Lash, 1988), which in turn control thermal history and subsidence. Recent knowledge coming from the Argentine Precordillera allows reconsideration of the paleogeographic evolution and kinematics of the Cambrian margin of Laurentia, particularly that in the southeastern United States (Thomas and Astini, 1996a).

DRIFTING

Available data bearing on the drifting stage of the Laurentian orphan is detailed in Astini et al. (1996). The onset of the drifting stage is recorded in the thick carbonate passive-margin succession that characterizes the Precordillera terrane. The subsidence history of the Precordilleran platform is comparable to that of the Appalachian margin (Bond et al., 1984). Regional mapping shows that the Precordilleran carbonate platform developed on a fairly small continental block, and the Precordilleran subsidence curve (Fig. 6) illustrates an important component of thermal recovery after rifting. Subsidence driven by cooling of the lithosphere is the most typical pattern of drifting and is limited to the dimensions of the passive margin that followed mechanical stretching during rifting (Fig. 7) (Watts, 1982; Leeder, 1995). For the Precordillera, subsidence analysis (Fig. 6) allows reconstructing the carbonate bank history as a migrating isolated "raft" in the Iapetus Ocean (Astini et al., 1995a, 1996; Thomas and Astini, 1996a).

The rift-drift transition (Falvey, 1974; Royden and Keen, 1980; Scrutton, 1982) corresponds to a time of major change in basin tectonics from block faulting and graben formation during extension to regional subsidence controlled by lithospheric cooling and sediment loading at the onset of sea-floor spreading (Bond et al., 1995). Although identification of a postrift unconformity is not everywhere straightforward, in the Precordillera terrane, as well as at several sites in the southeastern United States, the transgressive contact representing rift-drift transition can be observed or rift-drift transition can be deduced from the overstep of synrift facies by Middle and Late Cambrian passive-margin facies (Thomas, 1991; Astini and Vaccari, 1996). This constitutes a very strong argument in favor of sea-floor spreading associated with the Ouachita rift and, hence, supports the hypothesis of complete rifting and separation of the Precordillera during the Cambrian.

The Precordilleran platform is very similar to the Appalachian carbonate passive margin (Ramos et al., 1986; Astini et al., 1995a). Studies by Cañas (this volume), Keller et al. (1993), and Lehnert et al. (1997) have greatly strengthened this view. Together with Astini et al. (1995b), these studies have favored the correlation of similar facies types and vertical successions with the southern Appalachians, where a wide-rimmed platform developed during the Late Cambrian (Read, 1989). By Late Cambrian time, similar to the central-southern Appalachian passive margin (cf. Demicco, 1985; Koerschner and Read, 1989; Osleger, 1993), the Precordillera had developed into an aggraded, flat-topped platform dominated by peritidal facies (Astini et al., 1995a; Cañas, 1995a,b; Armella, 1995). However, by the Middle Cambrian, east-west polarity with deep-water limestones to the west was clearly developing in the Precordillera (Palmer et al., 1996; Banchig and Bordonaro, 1997). Transition toward a distally steepened ramp characterizes the Early Ordovician Appalachian shelf margin (Alberstadt and Repetski, 1989). General similarities of the reef builders of the Precordillera to those of southern Laurentia (Cañas and Carrera, 1993; Keller and Flügel, 1996; Carrera and Cañas, 1996) suggest that, by the Early Ordovician, the Precordillera (like the southern Laurentian Midcontinent province) was still in the Pacific warm-water realm (Webby, 1984, 1992). These remarkable similarities indicate substrates and paleolatitudes like those of southern Laurentia. The combination of depositional setting and subsidence history indicates that the Precordillera passive margin is a low-latitude, lower-plate platform, in the context of low-angle detachment rifted margins. At present, studies of the internal architecture and lateral relationships between the Precordilleran units are still limited. Nevertheless, it is reasonable to suggest, on the basis of process-oriented facies analysis, that the Argentine Precordillera is somewhat incompletely preserved (Cañas, 1995a, this volume; Astini et al., 1996; Keller, 1996), and it might have been at least 400–500 km wide on the basis of comparisons with the Appalachian carbonate platform (Skehan, 1988; Read, 1989). Hence, what is preserved and exposed of the Precordilleran platform is only part of the original extent. In particular, the eastern Precordillera is not the original edge of the platform; and farther east in the subsurface beneath the Bermejo Valley, seismic lines show that the platform carbonates, although reduced in thickness, are well represented (Allmendinger et al., 1990; Zapata, 1996).

DOCKING

The foreland I stage (Fig. 2) (Astini et al., 1996) starts with the inception of anoxic depocenters that reflect progressive flexural subsidence caused by approach of the Precordillera to a subduction zone along the western Gondwana active margin. This stage includes siliciclastic mid-Ordovician units that prograde westward as a clastic wedge over a complex paleogeography. During this time, tectonism dominated over eustatic changes as the major accommodation control. A general shifting from carbonate- to siliciclastic-dominated environments is recorded. Carbonate deposition persisted in the central Precordillera (Las Chacritas) at the crest of an active peripheral forebulge (Astini, 1995; Astini et al., 1995a), whereas in the foredeep to the east, deposition of fine to coarse clastic sediment with a source to the east began.

Collision-induced west-vergent thrusting and metamorphism involved the basement rocks to the east of the Precordillera terrane (partly the source for the Ordovician detritus). Associated magmatism in the Pampeanas Ranges (inboard on the Gondwana margin) yields mostly late Llanvirn (Llandeilian) to early Caradoc ages (Pankurst et al., 1996). Well-developed mylonitic shear zones have been recorded at the surface and in the subsurface (Cominguez and Ramos, 1990, 1991; Zapata and Allmendinger, 1996; Ramos et al., 1996), and an Ocloyic metamorphic age (~464 Ma, in Ramos et al., 1996) imprints Grenville basement rocks in the eastern part of the Precordillera terrane. Kinematic indicators from ductile shear zones (Simpson et al., 1995; Martino and Astini, 1995; Ramos et al., 1996; Zapata and Allmendinger, 1996) show strong reverse faulting, with top-to-the-west movement in accordance with major east-west shortening. No apparent strike-slip tectonics have been demonstrated to suggest oblique convergence or escape tectonics associated with the collision, although diachronous along-strike variations in sedimentary patterns may suggest oblique convergence. Deformation in the leading edge of the Precordillera terrane (east of the present-day eastern Precordillera) would have led to uplifting and denudation of the sedimentary cover in the western Pampeanas Ranges (Fig. 1). Nevertheless, although not profuse throughout the Precordillera, west-directed thrust faults have also been recorded in the passive-margin limestones in the eastern Precordillera. These thrusts have been interpreted as related to pre-Carboniferous deformation (Gosen, 1992). However, they could be related to shortening that occurred during the docking of the Precordillera (Ocloyic diastrophism) or to the later accretion of Chilenia (Early Devonian diastrophism).

Starting by the mid-Arenig, diachronous widespread black shale deposition is evidence of increasing accommodation space partly related to regional subsidence, caused by lithospheric flexure (Fig. 6) as the Precordillera approached the subduction zone beneath the western margin of Gondwana (Astini, 1995; Thomas and Astini, 1996a). Highly altered volcanically derived ashes (K-bentonites) in mid-Arenig strata (Huff et al., 1995, 1998; Bergström et al., 1996) strongly support the interpretation that the Precordillera terrane was approaching a subduction-related volcanic arc. In this setting, brittle subsidence (cf. Bradley and Kidd, 1991) could have been responsible for the regional demise of the Early Ordovician carbonate platform and for the resulting facies associations observed in the foredeep and the forebulge. However, extensional tectonism continued into the Late Ordovician, a time interval too long to explain solely by a brittle component of load-induced subsidence.

MID–LATE ORDOVICIAN EXTENSION—SECOND RIFTING?

With detailed chronostratigraphically constrained mapping in the central Precordillera region, Astini (1997b) has shown that some unconformities, previously interpreted as related to sea-level fluctuations, can be better explained as tectonic drowning unconformities (yo-yo effect). In several localities, Caradocian black shales (e.g., Cerro La Chilca) and coarse conglomeratic successions (e.g., southwest of Guandacol) abruptly overlie early to middle Llanvirn (according to recent revision by Fortey et al., 1995) strata. Laterally adjacent to these relatively deep-water facies, isolated carbonate deposition in apparent topographic swells gave way to evenly stratified, black laminated mudstones (e.g., Las Aguaditas and Las Plantas Formations) interpreted as hemipelagites, mostly periplatform oozes. Associated widespread evidence of depositional slopes and by-pass margins is very common in this Caradoc interval. Slumped and contorted strata, debris flows, intraformational scars, and block falls are the most evident features that suggest internal highs in the basin, precluding open circulation and allowing development of small depocenters where differing fills reflect local sources. As was noted by Astini (1997a,c, 1998), depocenters on the east received mainly coarse sediment from basement sources farther east, whereas more isolated depocenters to the west (in present-day central Precordillera) were completely isolated from coarse extrabasinal siliciclastic dispersal and were filled with pelagic or hemipelagic sediments. The overall setting for the Caradoc can be visualized as a highly partitioned, block-faulted platform (Fig. 8). Thus, extensionally generated topography of horsts and grabens (Fig. 9) is suggested, indicating important lithosphere stretching. Substantial topographic relief is suggested in several localities (e.g., Sierra de La Invernada) by boulders of high-grade metamorphic rocks (presumably Grenville) within the intrabasinal carbonate debris flows.

One outstanding example of an upthrown block is in the Ponon Trehue area in the southern exposures of the Precordillera terrane (Fig. 1) (Astini et al., 1995a; Bordonaro et al., 1996; Heredia, 1996). There, a basal conglomerate, rich in feldspar sand and carbonate clasts, nonconformably overlies the Grenville basement (Astini et al., 1996; age ~1,100 Ma, according to V. Ramos, personal communication, 1996) and is succeeded by a deepening-upward sequence of bioclastic grainstones to deeper water dark limestones, progressively changing upward into dominantly fine siliciclastic beds (Fig. 10). *Pygodus serra* and *P. anserinus* elements indicate a prevailing late Llanvirn age for this unit (Bordonaro et al., 1996; Heredia, 1996) directly overlying the basement. Blocks (olistoliths) of Arenig and possibly Tremadoc limestones (see faunas in Bordonaro et al., 1996) in the basal olistostrome deposits, together with basement boulders, indicate substantial uplift and erosion before new submergence gave origin to the Ponon Trehue in situ limestones and overlying clastic rocks.

Deposition of carbonate sediments continued in some places of the central Precordillera, isolated from important detrital inflows. Not only in the Los Blanquitos Range (southwest of Jachal city) (Fig. 1), but also in the Sassito high (along the San Juan River) and in the topmost of the San Isidro section (to the west of Mendoza city), active carbonate fabric seems to have been able to catch-up and keep-up with relative sea level (Astini, 1997b). Nevertheless, slight changes in the carbonate fabrics and prevailing components suggest a progressive latitudinal change toward temperate carbonate environments (Astini, 1995; Astini and Cañas, 1995).

No beds overstepping faults have been seen in outcrop to mark the end of extensional movement. The Late Ordovician units (e.g., Trapiche-Empozada Formations) show very restricted areal distribution and contain conglomeratic limestone megabeds (Astini, 1994b), indicating continued block faulting and local exposure of limestones that provided the intrabasinal source for the megabeds. These beds, by analogy to Mesozoic Apennine and Pyrenees megabreccias, were previously interpreted as possible foreland-basin deposits (Astini, 1991, 1994b), but in light of the other evidence herein described, they can better be interpreted as products of the major extension episode that affected the Precordillera during the Middle and Late Ordovician. In most of the central Precordillera, uppermost Ordovician (post-Hirnantian) and Silurian successions overlap Ordovician substrates of various ages (Astini and Maretto, 1996), constituting the first clear overlapping of Late Ordovician extensional structures.

The Middle to Late Ordovician stretching probably reached a maximum when deep-seated extension generated the well-known basaltic pillow lavas of the western Precordillera. The origin of these mafic rocks interlayered with mid-Caradocian graptolitic shales (Blasco and Ramos, 1976; Ortega et al., 1991) has been a matter of debate for some time (Kay et al., 1984; Haller and Ramos, 1984, 1993; Ramos et al., 1986; Gosen, 1992; Dalla Salda et al., 1992a,b; Astini et al., 1996). In the light of the evidence presented so far (also see Astini, 1997c), it seems reasonable to consider the mafic rocks in the Alcaparrosa and Yerba Loca Formations ("Ofiolitas Famatinianas") as products

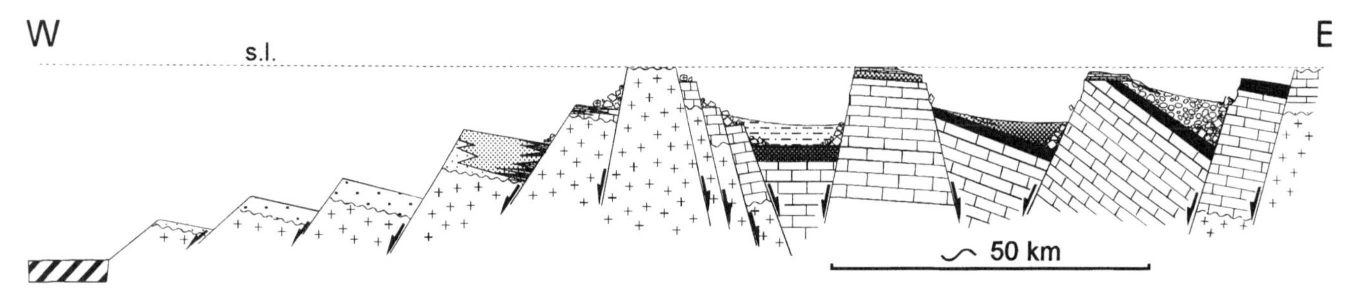

Figure 8. Schematic east-west sketch showing block faults in the Precordillera, herein suggested to be the main control on sedimentation and architecture during the Middle and Late Ordovician (after Astini, 1998).

Subsidence history has been used commonly to estimate the age of break-up at continental margins (e.g., Steckler and Watts, 1978; Bond et al., 1984) and, thus, to reconstruct paleogeography. However, potential variations in thermal history along a continental margin, implied in various rifting models, have not been considered until recently (e.g., Thomas and Astini, 1996b, 1999). Thus, thickness variations and age variations of paired-rift asymmetric margins may directly reflect the contrast in accommodation space caused by different thermally induced rates of change in subsidence (Buck et al., 1988; Issler et al., 1989). Furthermore, the kinematics and thermal evolution of continental transform margins are complex (e.g., Blarez and Mascle, 1988; Todd et al., 1988; Keen et al., 1990; Lorenzo and Vera, 1992; Bond et al., 1995) and are not well enough generalized to support the application of actualistic models based on modern analogs (see Dalziel, 1997). Preliminary facies mapping in the Precordillera suggests an initial approximately north-south polarity of subsidence that changed later, during the Middle Cambrian, to a distinct east-west polarity. This might be considered as evidence for an initial thermal doming decay orthogonally from the transform margin, followed by subsidence of the rift margin (Thomas and Astini, 1996b).

Evidence for an open ocean basin within the Ouachita embayment between the Laurentian margin and the Precordillera takes four forms. Pods of serpentinite and metagabbro within the Ouachita thrust belt/accretionary prism suggest oceanic crust beneath the Late Cambrian and younger slope-and-rise deposits of the Ouachita embayment of Laurentia (Nielsen et al., 1989; Viele and Thomas, 1989). Geophysical models of seismic-velocity and gravity data show thin oceanic-transitional crust within the Ouachita embayment south of the transform-defined margin of continental crust beneath the Ouachita thrust belt (Keller et al., 1989; Mickus and Keller, 1992). Even in the absence of an ophiolite suite in structurally complex areas, Fortey and Cocks (1986) have suggested that marginal faunal belts or platform-edge biofacies can be used, in principle, almost as the presence of ophiolites, to indicate the proximity of former ocean basins. Faunal assemblages diagnostic of shelf-margin biofacies, found in the Middle Cambrian olistoliths recovered from the western Precordillera (Vaccari, 1994; Palmer et al., 1996), reinforce the idea of a newly born oceanic basin between the Precordillera and the Texas promontory. Recently discovered Late Proterozoic–Early Cambrian mafic rocks (565 ± 45 Ma) representing oceanic lithosphere in the southwestern Precordillera (Davis et al., 1997, this volume) reinforce the existence of oeanic floor to the west of the Precordillera. The tectonic-lithologic, geophysical, and faunal interpretations are consistent in arguing for an open ocean between the Precordillera and Laurentia before the end of Cambrian time, thereby supporting the microcontinent model for the Precordillera in preference to the Texas plateau model.

Paleontologic criteria reveal that the Precordillera was isolated from Laurentia by the Early Ordovician (Astini et al., 1995a, 1996; Benedetto et al., 1995, this volume; Vaccari, 1995; Thomas and Astini, 1996a). Recent papers by Lehnert et al. (1997) and Albanesi and Barnes (1997) describe Early Ordovician endemism, allopatric speciation, and changeover from Laurentian to non-Laurentian conodonts. The main argument for isolation of the Precordillera during the earliest Ordovician is that the shelfal and cratonal faunas of limited migratory capacity (e.g., trilobites, brachiopods) started a gradual changeover in composition, departing from the 100% Laurentian domain in the latest Cambrian, and continuing in the Ordovician with the appearance in the Precordillera of specific genera that have not been recorded in Laurentia. *Annamitella*, for example (see Vaccari, 1993), is a typical non-Laurentian trilobite form of peri-Gondwanan/Avalonian affinity (R. Fortey, written communication, 1997). This has been used by Williams et al. (1995) as a strong argument to support the opposite drifting across the Iapetus Ocean (toward present-day north) of the Exploits arc, now located in the Newfoundland Central Mobile Belt (see van der Pluijm et al., 1995). The faunal evolution outlined by Benedetto (1993) and Astini et al. (1996) not only requires the separation of the Precordillera from Laurentia but also supports a connection with the Celtic and Baltic provinces, probably due to exchange with other intra-Iapetus islands or microcontinents through oceanic currents. An example of such an intra-Iapetus island is the Famatina-Puna island-arc system (Conti et al., 1996; Rapalini et al., this volume), with which the Precordillera terrane shares Arenig quasi-cosmopolitan and Celtic taxa (Vaccari, 1995; Benedetto and Sánchez, 1996). This reinforces the idea that the Precordillera, together with other different terranes and island arcs, makes a complex puzzle of drifting islands that moved independently within an open ocean similar to the present-day southwest Pacific (cf. Mac Niocaill et al., 1997). In this picture, whereas collision of successive northward-drifting terranes with Laurentia generated the protracted history of deformation that characterizes the Taconian orogeny, those drifting southeastward and colliding with Gondwana, like the Precordillera terrane, generated the Ocloyic orogeny. Hence, the Taconic and the Ocloyic orogenies have no spatial or kinematic connections, but overlap in time because of the complex and active plate motion that characterized the Iapetus Ocean.

Signs of isolation during the Early Ordovician come mainly from faunal arguments. With no more quantitative meaning than that suggested in Thomas and Astini (1996a), isolation means the existence of a barrier that prevented exchange of faunas and sediments between the main landmass of Laurentia and the Precordillera terrane. The argument that the Precordillera remained as part of Laurentia in a Falkland-Malvinas plateau analog (Texas plateau in Dalziel, 1997) counters the general theories of faunal provincialism and migrating capability of benthic faunas (Whittington and Hughes, 1972; Raup and Stanley, 1978; McKerrow and Cocks, 1986; Soper et al., 1991; Fortey and Mellish, 1992; Fortey and Cocks, 1992; Harper, 1992; Neuman and Harper, 1992; Williams et al., 1995). An epicratonal sea that separated the Precordillera from Laurentia in the Texas plateau model is invoked, as a geographic restriction and effective barrier to faunal exchange, to explain the faunal changes observed in the Precordillera (Dalziel, 1997). However, in the modern analog, apparent lack of endemism in the marine invertebrate fauna of the present-day Falkland-Malvinas islands (Magellan province) indicates no restriction of faunal exchange with continental southern South America (Benedetto, 1998).

From a compositional viewpoint, the Lower Ordovician limestone successions of the Precordillera, in contrast to those of the Upper Cambrian, have no detrital content (Astini et al., 1995a; Cañas, 1995a; Keller, 1996). This may be interpreted to indicate either major sea-level fluctuations (documented by Keller et al., 1994; Cañas, 1995a) or the existence of barriers between the Precordillera bank and continental Laurentia. Scarce detrital sediment in similar Appalachian settings is explained by extremely flat topography on a wide platform with intrabasinal depocenters (cf. Markello and Read, 1982; Demicco, 1985; Read, 1989). An equally viable explanation is that the Precordillera carbonate platform was completely isolated from the continent in a similar way to the present-day Bahamas, with which it has remarkable lithofacies similarities (Cañas, 1995a,b; Astini et al., 1995a, 1996).

In the Texas plateau hypothesis (Dalziel, 1997), the Precordillera at the leading edge of a major promontory should have functioned as the "bumper" that received the concentrated strain of a continent-continent collision. At a relatively small area of lithosphere, concentrated stress should be reflected in contractional strain, especially the inversion of older extensional faults in basement rocks. The Precordillera carbonate bank, however, hardly shows evidence of Ordovician contractional deformation. The Texas plateau model also includes a poorly explained basin west of the Precordillera, where syncollisional contraction and inversion would be predicted. Instead, the western Precordillera is characterized by extension. Contractional tectonics seem to have been important only at the eastern boundary of the Precordillera terrane and to have occurred approximately in the Middle Ordovician. Even there, deformation seems to be small for continent-continent collision.

Given the uncertainty of longitudinal control for paleomagnetic reconstructions, recent and ongoing stratigraphic and paleontologic research, as well as isotopic and rare earth geochemistry, provides the most valuable data for reliable paleogeographic reconstructions of the Iapetus Ocean. These data favor the conventional scheme of reconstruction in which most continents, terranes, and island arcs interacted within a "Northern Iapetus" open ocean (e.g., Neuman, 1984; Williams et al., 1995; Torsvik et al., 1996; Harper et al., 1996). In this context, the Precordillera terrane is considered as a separate Laurentian orphan (Fig. 11), which was rifted from Laurentia during Early Cambrian time, began to drift by Middle Cambrian time, and drifted independently through the Iapetus Ocean during Late Cambrian and Early Ordovician times, as suggested by Astini et al. (1995a, 1996), Thomas and Astini (1996a), and Astini (1997d).

Undoubtedly, future work must be focused on better constraints for timing, geometry, and kinematics of the two rifting events, as well as for the kinematics and timing of accretion. Better resolution of stratigraphic patterns will more precisely relate sedimentologic and biologic responses to these major events. Refining of paleontologic criteria, as well as new data on the polar wandering path of the Precordillera, will also be welcome to better define the time and route of the transfer of the Precordillera from Laurentia and Gondwana. A more systematic search for appropriate modern and ancient analogs likely will yield more actualistic kinematic models (e.g., Thomas, 1998). Nevertheless, presently available evidence greatly favors the rift-drift-collision (microcontinent) model, which explains the following: (1) the rifting and drifting history of the Precordillera in the context of stratigraphic and paleontologic successions of both the Precordillera and southeastern Laurentia; (2) the rift-drift transition for the Precordillera, suggested by exposed stratigraphic successions and subsidence analysis; (3) the east-west polarity of the Precordilleran platform as the conjugate margin of the Texas promontory; (4) the biologic evidence for oceanic isolation; (5) the volcanically derived K-bentonites in the Precordillera; and (6) the initial drowning-subsidence phase and postcollisional extension following accretion.

The subsequent early mid-Paleozoic history of the Precordillera bears directly on the nature and emplacement of the scarcely known Chilenia terrane, the present western neighbor of the Laurentian orphan. Thus, understanding the western tectofacies of the Precordillera is a clue to unravel its origin.

CONCLUSIONS

Rifting of the Precordillera terrane (the Laurentian orphan) from the Ouachita embayment and the brittle stretching of the continental lithosphere are suggested by synrift graben-fill deposits and by postrift subsidence history. The synrift facies indicate divergent margins in a low-latitude arid-climatic setting, both in southeast Laurentia and in the Precordillera. Contrasts in stratigraphic thickness, implied structures, and timing on opposite sides of the Ouachita rift demonstrate thermal and geometric differences in the paired-rift margins, consistent with the complementary asymmetry of a simple-shear low-angle-detachment model for continental rifting.

Thick platform carbonates ranging from late Early Cambrian to Early Ordovician transgressively overstep the synrift facies, indicating rift-drift transition and evolution of a passive margin on the Precordillera terrane. Dated latest Proterozoic gabbros and Middle Cambrian faunal assemblages diagnostic of shelf-margin biofacies in the western Precordillera reinforce the hypothesis of a spreading center opening between the Precordillera and the Texas promontory during the Cambrian. Isolated drifting of the Precordillera block during most of the Early Ordovician is suggested from biogeographic and lithologic considerations.

By the Early to Middle Ordovician, the approach of the Precordillera to subduction beneath western Gondwana is indicated by widespread drowning and diachronous deposition of black shales in response to flexurally induced deepening and progressive environmental restriction in an unstable foredeep. K-bentonite beds indicate proximity to active volcanic arcs of peri-Gondwanic position.

Evidence of contractile tectonics (Ocloyic event) caused by the collision of the Precordillera terrane and western Gondwana is stronger toward the eastern margin. Limestones of the eastern Precordillera were shortened by west-vergent thrusting, and the Grenville basement to the east was highly deformed and probably uplifted and denuded. Mylonite shear zones are associated with ~464-Ma metamorphic ages. These are considered as the strongest evidence for mid-Ordovician collision and accretion. Top-to-the-west reverse kinematic indicators show no evidence of major strike-slip related to the Ocloyic shortening.

Extensional tectonics affecting most of the Precordillera became the dominant control in the architecture of the sedimentary fill and intervening gaps during the Middle and Late Ordovician. Block faulting led to generation of horsts and grabens and to exposure of Grenville basement. Mass-wasting processes dominated in the slope break to the west, where deep-seated extension ultimately led to the development of an incipient spreading ridge. This extension episode is related to rifting that occurred after a fundamental change in plate motion after collision.

A two-stage rifting model (Cambrian and Middle Ordovician rifting) bracketing collision (sometime during the Llanvirn) encompasses the different tectonic stages of the Precordillera terrane, and brings new insights into the paleogeography of the southern Iapetus Ocean.

ACKNOWLEDGMENTS

We gratefully acknowledge informal discussion with many colleagues of some of the ideas presented in this chapter. Particularly, we thank Luis Benedetto, Stig Bergström, Edsel Brussa, Marcelo Carrera, Steven Davis, Bob Hatcher, Warren Huff, Sue Kay, Chuck Mitchell, Pete Palmer, Victor Ramos, Augusto Rapalini, and Sarah Roeske, for their stimulating thoughts. We especially thank Chris Schmidt for prompting our collaborative work, and Teresa Jordan and George Viele for extensive and helpful reviews. This work was supported by grants from the Consejo Nacional de Investigaciones Científicas y Técnicas, the Consejo de Investigaciones Científicas y Técnicas de la Provincia de Córdoba, the Agencia Nacional de Ciencia y Técnica, the Fundación Antorchas (to R.A.A.), and by National Science Foundation Grant EAR-9706735 (to W.A.T.). This is a contribution to the International Geological Correlation Program Project 376.

REFERENCES CITED

Albanesi, G. L., and Barnes, C. R., 1997, The Middle Ordovician conodont *Paroistodus horridus* in the Argentine Precordillera: 6th International Conodont Symposium, Warsaw, Poland, Abstracts, p. 1.

Albanesi, G. L., Hünicken, M., and Ortega, G., 1990, *Amorphognathus superbus* (Conodonta) from Trapiche Formation (Upper Ordovician), Cerro Potrerillos, Jáchal Department, San Juan Province, Argentina: 1st Latin American Conodont Symposium, Academia de Ciencias, Córdoba, Abstracts, p. 109–110.

Alberstadt, L., and Repetski, J. E., 1989, A Lower Ordovician sponge/algal facies in the southern United States and its counterparts elsewhere in North America: Palaios, v. 4, p. 225–242.

Allmendinger, R. W., Figueroa, D., Sydner, D., Beer, J., Mpodizis, C., and Isacks, B. L., 1990, Foreland shortening and crustal balancing in the Andes at 30°S latitude: Tectonics, v. 9, p. 789–809.

Armella, C., 1995, Thrombolitic-stromatolitic cycles of the Cambro-Ordovician boundary sequence, Precordillera oriental basin, western Argentina, *in* Bertrand Sarfati, J., and Monty, C., eds., Phanerozoic stromatolites: Dordrecht, Netherlands, Kluwer, v. 2, p. 421–441.

Astini, R. A., 1988, Yerba Loca Formation, an Ordovician clastic wedge in western Argentina: 5th International Symposium Ordovician System, Newfoundland, Canada, Abstracts, p. 3.

Astini, R. A., 1991, Paleoambientes sedimentarios y secuencias depositacionales del Ordovícico clástico de la Precordillera Argentina [Ph.D. thesis]: Córdoba, Argentina, Universidad Nacional de Córdoba, 851 p.

Astini, R. A., 1992, Tectofacies ordovícicas y evolución de la cuenca eopaleozoica de la Precordillera Argentina: Estudios Geológicos, v. 48, p. 315–327.

Astini, R. A., 1994a, Sucesiones calcáreo-silicoclásticas coetáneas del Ordovícico inferior de la Precordillera y su significado en la evolución de la cuenca: 5th Reunión Argentina de Sedimentología, San Miguel de Tucumán, Argentina, Tomo I, p. 113–118.

Astini, R. A., 1994b, Las megaturbiditas de la Formación Trapiche (Ordovícico superior de la Precordillera): procesos sedimentarios y marco geológico: 5th Reunión Argentina de Sedimentología, San Miguel de Tucumán, Argentina, Tomo I, p. 107–112.

Astini, R. A., 1994c, Significado estratigráfico del Miembro Superior de la Formación San Juan (Cordón de las Chacritas), Ordovícico medio de la Precordillera de San Juan: Revista de la Asociación Geológica Argentina, v. 49, p. 365–367.

Astini, R. A., 1995, Paleoclimates and paleogeographic paths of the Argentine Precordillera during the Ordovician: evidence from climatically sensitive lithofacies, *in* Cooper, J. D., Droser, M. L., and Finney, S. C., eds., Ordovician odyssey: SEPM Pacific Section, Book 77, p. 177–180.

Astini, R. A., 1996a, Las fases diastróficas del Paleozoico medio en la Precordillera del oeste argentino -evidencias estratigraficas-: 13th Congreso Geológico Argentino y 3rd Congreso de Exploración de Hidrocarburos, Buenos Aires, Argentina, Tomo V, p. 509–526.

Astini, R. A., 1996b, The Argentine Precordillera–lower plate of the Ouachita conjugate-pair rift: a suitable explanation for their stratigraphic differences: Geological Society of America Abstracts with Programs, v. 28, no. 1, p. 2.

Astini, R. A., 1997a, El "Conglomerado de Las Vacas" y el Grupo Trapiche de la Precordillera: tectónica distensiva en el Ordovícico Superior: Revista de la Asociación Geológica Argentina, v. 53, p. 489–503.

Astini, R. A., 1997b, Las unidades calcáreas del Ordovícico Medio y Superior de la Precordillera Argentina como indicadores de una etapa extensional: II Jornadas de Geología de Precordillera, San Juan, Argentina, Actas, p. 8–14.

Astini, R. A., 1997c, Stratigraphic evidence of two-stage rifting and collision in the Laurentian derived Precordillera terrane, south-central Andes: Geological Society of America Abstracts with Programs, v. 29, no. 6, A116.

Astini, R. A., 1997d, The Iapetus as viewed by southern sailors, the Precordillera and the Famatina-Puna oriental terranes: Geological Society of America Abstracts with Programs, v. 29, no. 6, A280.

Astini, R. A., 1998, Stratigraphical evidence supporting the rifting, drifting and collision of the Laurentian Precordillera terrane of western Argentina, *in* Pankhurst, R. J., and Rapela, C. W., eds., The Proto-Andean margin of Gondwana: Geological Society of London Special Publication 142, p. 11–33.

Astini, R. A., and Benedetto, J. L., 1993, Paleozoic evolution of the Argentine

Precordillera as a rifted, drifted and collided terrane: Geological Society of America Abstracts with Programs, v. 25, p. A232.

Astini, R. A., and Benedetto, J. L., 1996, Tectonostratigraphic development and history of an allochthonous terrane in the pre-Andean Gondwana margin: the Argentine Precordillera: 3rd International Symposium on Andean Geodynamics, St. Malo, France, Extended Abstracts, p. 759–762.

Astini, R. A., and Brussa, E. D., 1997, Dos nuevas localidades fosilíferas en el Conglomerado de Las Vacas (Caradociano) en la Precordillera Argentina: la importancia cronoestratigráfica: Reunión de Comunicaciones Paleontológicas, Ameghiniana, v. 43, p. 231.

Astini, R. A., and Cañas, F. L., 1995, La Formación Sassito, una nueva unidad calcárea en la Precordillera de San Juan: sedimentología y significado estratigráfico: Asociación Argentina de Sedimentología, Revista, v. 2, p. 19–37.

Astini, R. A., and Maretto, H. M., 1996, Análisis estratigráfico del Silúrico de la Precordillera Central de San Juan y consideraciones sobre la evolución de la cuenca: 13th Congreso Geológico Argentino y 3rd Congreso de Exploración de Hidrocarburos, Buenos Aires, Argentina, Tomo I, p. 351–368.

Astini, R. A., and Vaccari, N. E., 1996, Sucesión evaporítica del Cámbrico inferior de la Precordillera: significado geológico: Revista de la Asociación Geológica Argentina, v. 51, p. 97–106.

Astini, R. A., Benedetto, J. L., and Vaccari, N. E., 1995a, The early Paleozoic evolution of the Argentine Precordillera as a Laurentian rifted, drifted, and collided terrane: a geodynamic model: Geological Society of America Bulletin, v. 107, p. 253–273.

Astini, R. A., Vaccari, N. E., Thomas, W. A., Raymond, D. E., and Osborne, W. E., 1995b, Shared evolution of the Argentine Precordillerra and the southern Appalachians during the Early Cambrian: Geological Society of America Abstracts with Programs, v. 27, no. 6, p. A458.

Astini, R. A., Ramos, V. A., Benedetto, J. L., Vaccari, N. E., and Cañas, F. L., 1996, La Precordillera: un terreno exótico a Gondwana: 13th Congreso Geológico Argentino y 3rd Congreso de Exploración de Hidrocarburos, Buenos Aires, Argentina, Tomo V, p. 293–324.

Baldis, B. A., and Bordonaro, O., 1981, Evolución de las facies carbonáticas en la cuenca Cámbrica de la Precordillerra de San Juan: 8th Congreso Geológico Argentino, San Luis, Argentina, Tomo II, p. 385–397.

Baldis, B. A., and Bordonaro, O., 1984, Cámbrico y Ordovícico en la Sierra Chica de Zonda y cerro Pedernal, provincia de San Juan. Génesis del margen continental en la Precordillera: 9th Congreso Geológico Argentino, San Carlos de Bariloche, Argentina, Tomo 4, p. 190–207.

Banchig, A. L., and Bordonaro, O., 1997, Formación Alojamiento: una unidad carbonática—silicoclástica cámbrica de la Precordillera Mendocina: II Jornadas de Geología de la Precordillera, San Juan, Argentina, Actas, p. 16–21.

Banchig, A. L., Milana, J. P., and Bordonaro, O., 1990, Litofacies clásticas de la Formación Los Sombreros (Cámbrico Medio) en la Quebrada Ojos de Agua, Sa. del Tontal, San Juan: 3rd Reunión Argentina de Sedimentología, v. 1, p. 25–30.

Barnes, V. E., 1959, Stratigraphy of the pre-Simpson Paleozoic subsurface rocks of Texas and southeast New Mexico: University of Texas, Texas Bureau of Economic Geology Publication 5924, v. 1, p. 11–72.

Barnes, V. E., Bell, W. C., Clabaugh, S. E., Cloud, P. E., Jr., McGehee, R. V., Rodda, P. U., and Young, K., 1972, Geology of the Llano region and Austin area, field excursion, Guidebook No. 13: Austin, University of Texas, Texas Bureau of Economic Geology, 77 p.

Benedetto, J. L., 1993, La hipótesis de la aloctonía de la Precordillera Argentina: un test estratigráfico y biogeográfico: 12th Congreso Geológico Argentino y 2nd Congreso de Exploracion de Hidrocarburos, Mendoza, Argentina, Tomo III, p. 375–384.

Benedetto, J. L., 1998, Early Paleozoic brachiopods and associated shelly fauna from western Gondwana: their bearing on the geodynamic history of the pre-Andean margin, *in* Pankhurst, R. J., and Rapela, C. W., eds., The Proto-Andean margin of Gondwana: Geological Society of London Special Publication, v. 142, p. 57–83.

Benedetto, J. L., and Astini, R. A., 1993, A collisional model for the early Paleozoic stratigraphic evolution of the Argentine Precordillera: 2nd International Symposium on Andean Geodynamics, Oxford, Extended Abstracts, p. 501–504.

Benedetto, J. L., and Herrera, Z. A., 1987, Primer hallazgo de braquiópodos y trilobites de la Formación Trapiche (Ordovícico Tardío), Precordillera Argentina: 10th Congreso Geológico Argentino, San Miguel de Tucumán, Argentina, Actas 3, p. 73–76.

Benedetto, J. L., and Sánchez, T. M., 1996, Paleobiogeography of brachiopod and molluscan faunas along the South American margin of Gondwana during the Ordovician, *in* Baldis, B. A., and Aceñolaza, F. G., eds., Early Paleozoic evolution in NW Gondwana: San Miguel de Tucumán, Serie Correlación Geológica, v. 12, p. 23–38.

Benedetto, J. L., and Vaccari, N. E., 1992, Significado estratigráfico y tectónico de los complejos de bloques cambro-ordovícicos resedimentados de la Precordillera Occidental, Argentina: Estudios Geológicos, v. 48, p. 305–313.

Benedetto, J. L., Vaccari, N. E., Carrera, M. G., and Sánchez, T. M., 1995, The evolution of faunal provincialism in the Argentine Precordillera during the Ordovician: new evidence and paleontological implications, *in* Cooper, J. D., Droser, M. L., and Finney, S. C., eds., Ordovician odyssey: SEPM Pacific Section, Book 77, p. 181–184.

Bergström, S. M., Huff, W. D., Kolata, D. R., Krekeler, M. P. S., Cingolani, C., and Astini, R. A., 1996, Lower and Middle Ordovician K-bentonites in the Precordillera of Argentina: a preliminary review based on new data: 13th Congreso Geológico Argentino y 3rd Congreso de Exploración de Hidrocarburos, Buenos Aires, Argentina, Tomo V, p. 481–490.

Blarez, E., and Mascle, J., 1988, Shallow structures and evolution of the Ivory Coast and Ghana transform margins: Geology, v. 5, p. 54–64.

Blasco, G., and Ramos, V., 1976, Graptolitos caradocianos de la Formación Yerba Loca y del Cerro La Chilca, Departamento Jáchal, provincia de San Juan: Ameghiniana, v. 13, p. 312–329.

Bond, G. C., Nickeson, P. A., and Kominz, M. A., 1984, Breakup of a supercontinent between 625 Ma and 555 Ma: new evidence and implications for continental histories: Earth and Planetary Science Letters, v. 70, p. 325–345.

Bond, G. C., Kominz, M. A., and Sheridan, R. E., 1995, Continental terraces and rises, *in* Busby, C. J., and Ingersoll, R. V., eds., Tectonics of sedimentary basins: Oxford, United Kingdom, Blackwell Science, p. 149–178.

Bordonaro, O. L., and Banchig, A. L., 1996, Estratigrafía de los olistolitos cámbricos de la Precordillera Argentina: 13th Congreso Geológico Argentino y 3rd Congreso de Exploración de Hidrocarburos, Buenos Aires, Argentina, Tomo V, p. 471–479.

Bordonaro, O., Keller, M., and Lehnert, O., 1996, El Ordovícico de Ponón Trehue en la Provincia de Mendoza (Argentina): redefiniciones estratigráficas: 13th Congreso Geológico Argentino y 3rd Congreso de Exploración de Hidrocarburos, Buenos Aires, Argentina, Tomo I, p. 541–550.

Borrello, A. V., 1969, Embriotectónica y tectónica tensional. Su importancia en la evolución estructural de la Precordillera: Revista de la Asociación Geológica Argentina, v. 24, p. 5–13.

Bradley, D. C., and Kidd, W. S. F., 1991, Flexural extension of the upper continental crust in collisional foredeeps: Geological Society America Bulletin, v. 103, p. 1416–1438.

Braile, L. W., Hinze, W. J., Keller, G. R., Lidiak, E. G., and Sexton, J. L., 1986, Tectonic development of the New Madrid rift complex, Mississippi embayment, North America: Tectonophysics, v. 131, p. 1–21.

Brussa, E. D., 1994, Las graptofaunas ordovícicas del sector central de la Precordillera Occidental sanjuanina [Ph.D. thesis]: Córdoba, Argentina, Universidad Nacional de Córdoba, 323 p.

Buck, W. R., Martinez, F., Steckler, M. S., and Cochran, J. R., 1988, Thermal consequences of lithospheric extension: pure and simple: Tectonics, v. 7, p. 213–234.

Cañas, F. L., 1995a, Estratigrafía y evolución paleoambiental de las sucesiones carbonáticas del Cámbrico tardío y Ordovícico temprano de la Precordillera Septentrional, República Argentina [Ph.D. thesis]: Córdoba, Argentina, Universidad Nacional de Córdoba, 215 p.

Cañas, F. L., 1995b, Early Ordovician carbonate platform facies of the Argentine Precordillera: restricted shelf to open platform evolution, *in* Cooper, J. D., Droser, M. L., and Finney, S. C., eds., Ordovician odyssey: SEPM Pacific Section, Book 77, p. 221–224.

Cañas, F. L., and Carrera, M., 1993, Early Ordovician sponges—algal reef mound of the Precordillera basin, western Argentina: Facies (Erlangen), v. 29, p. 169–178.

Carrera, M. G., and Cañas, F. L., 1996, Los biohermos de la Formación San Juan (Ordovícico temprano, Precordillera Argentina): paleoecología y comparaciones: Asociación Argentina de Sedimentología Revista, v. 3, p. 85–104.

Cominguez, A. H., and Ramos, V. A., 1990, Sísmica de reflexión profunda entre Precordillera y Sierras Pampeanas: 11th Congreso Geológico Argentino, San Juan, Argentina, Tomo II, p. 311–314.

Cominguez, A. H., and Ramos, V. A., 1991, La estructura profunda entre Precordillera y Sierras Pampeanas de la Argentina: evidencias de la sísmica de reflexión profunda: Revista Geológica de Chile, v. 18, p. 3–14.

Conti, C. M., Rapalini, A. E., Coira, C., and Koukharsky, M., 1996, Paleomagnetic evidence of an early Paleozoic rotated terrane in northwest Argentina: a clue for Gondwana-Laurentia interaction?: Geology, v. 24, p. 953–956.

Dalla Salda, L., Cingolani, C. A., and Varela, R., 1992a, Early Paleozoic belt of the Andes and southwestern South America: result of Laurentia-Gondwana collision?: Geology, v. 20, p. 617–620.

Dalla Salda, L., Dalziel, I. W. D., Cingolani, C. A., and Varela, R., 1992b, Did the Taconic Appalachians continue into southern South America?: Geology, v. 20, p. 1059–1062.

Dalziel, I. W. D., 1993, Tectonic tracers and the origin of the proto-Andean margin: 12th Congreso Geológico Argentino y 2nd Congreso de Exploración de Hidrocarburos, Mendoza, Argentina, Tomo III, p. 367–374.

Dalziel, I. W. D., 1997, Neoproterozoic-Paleozoic geography and tectonics: review, hypothesis, environmental speculation: Geological Society of America Bulletin, v. 109, p. 16–42.

Dalziel, I. W. D., Dalla Salda, L. H., and Gahagan, L. M., 1994, Paleozoic Laurentia-Gondwana interaction and the origin of the Appalachian-Andean mountain system: Geological Society of America Bulletin, v. 106, p. 243–252.

Dalziel, I. W. D., Dalla Salda, L. H., Cingolani, C., and Palmer, A. R., 1996, The Argentine Precordillera: a Laurentian terrane? Penrose Conference Report: GSA Today, v. 6, no. 2, p. 16–18.

Davis, J. S., McClelland, W. C., Roeske, S. M., Gerbi, C., and Moores, E. M., 1997, The early to middle Paleozoic tectonic history of the SW Precordillera terrane and Laurentia-Gondwana interactions: Geological Society of America Abstracts with Programs, v. 29, no. 6, p. 380.

Demicco, R. V., 1985, Platform and off-platform carbonates of the Upper Cambrian of western Maryland, U.S.A.: Sedimentology, v. 32, p. 1–22.

Dewey, J. F., 1988, Extensional collapse of orogens: Tectonics, v. 7, p. 1123–1139.

Etheridge, M. A., Symonds, P. A., and Lister, G. S., 1989, Application of the detachment model to reconstruction of conjugate passive margins: American Association of Petroleum Geologists Memoir 46, p. 23–40.

Falvey, D. A., 1974, The development of continental margins in plate tectonic theory: Australian Petroleum Exploration Association Journal, v. 14, p. 95–106.

Fortey, R. A., and Cocks, L. R. M., 1986, Marginal faunal belts and their structural implications, with examples from the lower Palaeozoic: Journal of the Geological Society of London, v. 143, p. 151–160.

Fortey, R. A., and Cocks, L. R. M., 1992, The early Palaeozoic of the North Atlantic region as a test case for the use of fossils in continental reconstruction: Tectonophysics, v. 206, p. 147–158.

Fortey, R. A., and Mellish, C. J. T., 1992, Are some fossils better than others for inferring paleogeography? The early Ordovician of the North Atlantic region as an example: Terra Nova, v. 4, p. 210–216.

Fortey, R. A., Harper, D. A. T., Ingham, J. K., Owen, A. W., and Rushton, A., 1995, A revision of Ordovician series and stages from the historical type area: Geological Magazine, v. 132, p. 1–16.

Furque, G., 1972, Los movimientos Caledónicos en Argentina: Revista del Museo de La Plata, v. 8, p. 129–136.

Gilbert, M. C., 1983, Timing and chemistry of igneous events associated with the Southern Oklahoma aulacogen: Tectonophysics, v. 94, p. 439–455.

Gosen, W. von, 1992, Structural evolution of the Argentine Precordillera: the Río San Juan section: Journal of Structural Geology, v. 14, p. 643–667.

Gosen W. von, Buggisch, W., and Lenhert, O., 1995, Evolution of the early Paleozoic mélange at the eastern margin of the Argentine Precordillera: Journal of South American Earth Sciences, v. 8, p. 405–424.

Grotzinger, J. P., Bowring, S. A., Saylor, B. Z., and Kaufman, A. J., 1995, Biostratigraphic and geochronologic constraints on early animal evolution: Science, v. 270, p. 598–604.

Haller, M. J., and Ramos, V., 1984, Las ofiolitas famatinianas (Eopaleozoico) de las Provincias de San Juan y Mendoza: 9th Congreso Geológico Argentino, San Carlos de Bariloche, Argentina, v. 2, p. 66–83.

Haller, M. J., and Ramos, V. A., 1993, Las ofiolitas y otras rocas afines, *in* Ramos, V. A., ed., Geología y recursos naturales de Mendoza: 12th Congreso Geológico Argentino y 2nd Congreso de Exploración de Hidrocarburos, Mendoza, Argentina, Relatorio, p. 31–39.

Harper, D. A. T., 1992, Ordovician provincial signals from Appalachian-Caledonide terranes: Terra Nova, v. 4, p. 204–209.

Harper, D. A. T., Mac Niocaill, C., and Williams, S. H., 1996, The paleogeography of Early Ordovician Iapetus terranes: an integration of faunas and paleomagnetic constraints: Palaeogeography, Palaeoclimatology, Palaeoecology, v. 121, p. 297–312.

Hatcher, R. D., Jr., 1989, Tectonic synthesis of the U.S. Appalachians, *in* Hatcher, R. D., Jr., Thomas, W. A., and Viele, G. W., eds., The Appalachian-Ouachita orogen in the United States: Boulder, Colorado, Geological Society of America, Geology of North America, v. F-2, p. 511–535.

Heredia, S., 1996, El Ordovícico del Arroyo Ponon Trehue, sur de la provincia de Mendoza: 13th Congreso Geológico Argentino y 3rd Congreso de Exploración de Hidrocarburos, Buenos Aires, Argentina, Tomo I, p. 601–605.

Hogan, J. P., Gilbert, M. C., Price, J. D., Wright, J. E., Deggeller, M., and Hames, W. E., 1996, Magmatic evolution of the Southern Oklahoma aulacogen: Geological Society of America Abstracts with Programs, v. 28, no. 1, p. 19.

Huff, W. D., Bergström, S. M., Kolata, D. R., Cingolani, C., and Davis, D. W., 1995, Middle Ordovician K-bentonites discovered in the Precordillera of Argentina: geochemical and paleogeographical implications, *in* Cooper, J. D., Droser, M. L., and Finney, S. C., eds., Ordovician odyssey, SEPM Pacific Section, Book 77, p. 343–349.

Huff, W. D., Bergström, S. M., Kolata, D. R., Cingolani, C. A., and Astini, R. A., 1998, Ordovician K-bentonites in the Argentine Precordillera: relations to Gondwana margin evolution, *in* Pankhurst, R. J., and Rapela, C. W., eds., The Proto-Andean margin of Gondwana: Geological Society of London Special Publication 142, p. 107–126.

Issler, D., McQueen, H., and Beaumont, C., 1989, Thermal and isostatic consequences of simple shear extension of the continental lithosphere: Earth and Planetary Science Letters, v. 91, p. 341–358.

Johnson, K. S., Amsden, T. W., Denison, R. E., Dutton, S. P., Goldstein, A. G., Rascoe, B., Jr., Sutherland, P. K., and Thompson, D. M., 1988, Southern Midcontinent region, *in* Sloss, L. L., ed., Sedimentary cover—North American craton: U.S.: Boulder, Colorado, Geological Society of America, Geology of North America, v. D-2, p. 307–359.

Jordan, T. E., Isacks, B. L., Allmendinger, R. W., Brewer, J. A., Ramos, V. A., and Ando, C. J., 1983, Andean tectonics related to geometry of subducted Nazca plate: Geological Society of America Bulletin, v. 94, p. 341–361.

Kay, S. M., Ramos, V. A., and Kay, R. W., 1984, Elementos mayoritarios y trazas de las vulcanitas ordovícicas de la Precordillera Occidental: basaltos de rift oceánico temprano(?) próximos al margen continental: 9th Congreso Geológico Argentino, San Carlos de Bariloche, Argentina, v. 2, p. 48–65.

Kay, S. M., Orrell, S., and Abruzzi, J. M., 1996, Zircon and whole rock Nd-Pb isotopic evidence for a Grenville age and Laurentian origin for the basement of the Precordilleran terrane in Argentina: Journal of Geology, v. 104, p. 637–648.

Keen, C. E., Kay, W. A., and Roest, W. R., 1990, Crustal anatomy of a transform continental margin: Tectonophysics, v. 173, p. 527–544.

Keller, G. R., Lidiak, E. G., Hinze, W. J., and Braile, L. W., 1983, The role of rifting in the tectonic development of the Midcontinent, U.S.A.: Tectonophysics, v. 94, p. 391–412.

Keller, G. R., Braile, L. W., McMechan, G. A., Thomas, W. A., Harder, S. H., Chang, W.-F., and Jardine, W. G., 1989, Paleozoic continent-ocean transition in the Ouachita Mountains imaged from PASSCAL wide-angle seismic reflection-refraction data: Geology, v. 17, p. 119–122.

Keller, M., 1995, Continental slope deposits in the Argentina Precordillera: sediments and geotectonic significance, *in* Cooper, J. D., Droser, M. L., and Finney, S. C., eds., Ordovician Odyssey: SEPM Pacific Section, Book 77, p. 211–215.

Keller, M., 1996, Anatomy of the Precordillera (Argentina) during Cambro-Ordovician times: implications for the Laurentia-Gondwana transfer of the Cuyania Terrane: 3rd International Symposium on Andean Geodynamics, St. Malo, France, v. 1, p. 775–778.

Keller, M., and Flügel, E., 1996, Early Ordovician reefs from Argentina: stromatoporoid vs. stromatolite origin: Facies (Erlangen), v. 34, p. 177–192.

Keller, M., Bordonaro, O., and Cañas, F. L., 1993, Some plate tectonic aspects of Early Ordovician reefs, western Argentine Precordillera: 12th Congreso Geológico Argentino y 2nd Congreso de Exploracion de Hidrocarburos, Mendoza, Tomo III, p. 235–240.

Keller, M., Cañas, F. L., Lehnert, O., and Vaccari, N. E., 1994, The Upper Cambrian and Lower Ordovician of the Precordillera (Western Argentina): some stratigraphic reconsiderations: Newsletters on Stratigraphy, v. 31, p. 115–132.

Koerschner, W. F., III, and Read, J. F., 1989, Field and modelling studies of Cambrian carbonate cycles, Virginia Appalachians: Journal of Sedimentary Petrology, v. 59, p. 654–687.

Lambert, D. D., Unruh, D. M., and Gilbert, M. C., 1988, Rb-Sr and Sm-Nd isotopic study of the Glen Mountains layered complex: initiation of rifting within the Southern Oklahoma aulacogen: Geology, v. 16, p. 13–17.

Lash, G. G., 1988, Along-strike variations in foreland basin evolution: possible evidence for continental collision along an irregular margin: Basin Research, v. 1, p. 71–83.

Leeder, M. E., 1995, Continental rifts and proto-oceanic rift troughs, *in* Busby, C. J., and Ingersoll, R. V., eds., Tectonics of sedimentary basins: Oxford, United Kingdom, Blackwell Science, p. 119–148.

Lehnert, O., 1995, Ordovizische Conodonten aus der Präkordillere Westargentiniens: Ihre Bedeutung für Stratigraphie und Palaogeographie [Ph.D. thesis]: Erlangen, Germany, Erlanger Geologische Abhandlungen, v. 125, 193 p.

Lehnert, O., Miller, J. L., and Repetski, J. E., 1997, Paleogeographic significance of *Clavohamulus hintzei* Miller (Conodonta) and other Ibexian conodonts in an early Paleozoic carbonate platform facies of the Argentine Precordillera: Geological Society of America Bulletin, v. 109, p. 429–443.

Lister, G. S., Ethridge, M. A., and Symonds, P. A., 1986, Detachment faulting and the evolution of passive continental margins: Geology, v. 14, p. 246–250.

Lister, G. S., Ethridge, M. A., and Symonds, P. A., 1991, Detachment models for the formation of passive continental margins: Tectonics, v. 10, p. 1038–1064.

Lochman-Balk, C., 1971, The Cambrian of the craton of the United States, *in* Holland, E. R., ed., Cambrian of the New World: London, Wiley-Interscience, p. 79–167.

Lorenzo, J. M., and Vera, E. E., 1992, Thermal uplift and erosion across the continent-ocean transform boundary of the southern Exmouth Plateau: Earth and Planetary Science Letters, v. 108, p. 79–92.

Mac Niocaill, C., van der Pluijm, B. A., and Van der Voo, R., 1997, Ordovician paleogeography and the evolution of the Iapetus ocean: Geology, v. 25, p. 159–162.

Markello, J. R., and Read, J. F., 1982, Upper Cambrian intrashelf basin, Nolichucky Formation, southwest Virginia Appalachians: American Association of Petroleum Geologists Bulletin, v. 66, p. 860–878.

Martino, R. D., and Astini, R. A., 1995, The Cerro Salinas–Angaco Belt and its relationship with western Sierras Pampeanas: Laurentian-Gondwanan connections before Pangea, Field Conference, San Salvador de Jujuy, Abstracts with Programs, p. 27–28.

McKerrow, W. S., and Cocks, L. R. M., 1986, Oceans, island arcs and olistostromes: the use of fossils in distinguishing sutures, terranes and environments around the Iapetus Ocean: Journal of the Geological Society of London, v. 143, p. 185–191.

Mellen, F. F., 1977, Cambrian System in Black Warrior basin: American Association of Petroleum Geologists Bulletin, v. 61, p. 1897–1900.

Mickus, K. L., and Keller, G. R., 1992, Lithospheric structure of the south-central United States: Geology, v. 20, p. 335–338.

Neuman, R. B., 1984, Geology and paleobiology of islands in the Ordovician Iapetus Ocean: review and implications: Geological Society of America Bulletin, v. 95, p. 1188–1201.

Neuman, R. B., and Harper, D. A. T., 1992, Paleogeographic significance of Arenig-Llanvirn Toquima–Table Head and Celtic brachiopod assemblages, *in* Webby, B. D., and Lurie, J. R., eds., Global perspectives on Ordovician geology: Rotterdam, Netherlands, Balkema, p. 241–254.

Nielsen, K. C., Viele, G. W., and Zimmerman, J., 1989, Structural setting of the Benton-Broken Bow uplifts, *in* Hatcher, R. D., Jr., Thomas, W. A., and Viele, G. W., eds., The Appalachian-Ouachita orogen in the United States: Boulder, Colorado, Geological Society of America, Geology of North America, v. F-2, p. 635–660.

Ortega, G., and Brusa, E. D., 1990, La subzona de *Climacograptus bicornis* (Caradociano temprano) en la Formación Las Plantas en su localidad tipo, Precordillera de San Juan, Argentina: Ameghiniana, v. 27, p. 281–288.

Ortega, G., Brussa, E. D., and Astini, R. A., 1991, Nuevos hallazgos de graptolitos en la Formación Yerba Loca y su implicancia estratigráfica (Precordillera de San Juan, Argentina): Ameghiniana, v. 28, p. 163–178.

Osleger, D. A., 1993, Cyclostratigraphy of Late Cambrian carbonate sequences: an interbasinal comparison of the Cordilleran and Appalachian passive margins, *in* Cooper, J. D., and Stevens, C. H., eds., Paleozoic paleogeography of the western United States: SEPM Pacific Section, Book 67, p. 801–828.

Osleger, D. A., and Read, J. F., 1991, Relation of eustasy to stacking patterns of meter-scale carbonate cycles, Late Cambrian, U.S.A.: Journal of Sedimentary Petrology, v. 61, p. 1225–1252.

Palmer, A. R., Vaccari, N. E., Astini, R. A., Cañas, F. L., and Benedetto, J. L., 1996, Cambrian history of the Argentine Precordillera: Adios Laurentia!: Geological Society of America Abstracts with Programs, v. 28, no. 1, p. 57.

Pankurst, R. J., Rapela, C. W., Saavedra, J., Baldo, E., Dahlquist, J., and Pascua, I., 1996, Sierras de Los Llanos, Malanzán and Chepes: Ordovician I and S-type granitic magmatism in the Famatinian orogen: 13th Congreso Geológico Argentino y 3rd Congreso de Exploración de Hidrocarburos, Buenos Aires, Argentina, Tomo V, p. 415.

Pfiel, R. W., and Read, J. F., 1980, Cambrian carbonate platform margin facies, Shady Dolomite, southwestern Virginia, U.S.A.: Journal of Sedimentary Petrology, v. 50, p. 91–116.

Ramos, V. A., 1988, The tectonics of the central Andes, 30 to 33° S latitude, *in* Clark, S., and Burchfiel, D., eds., Processes in continental lithospheric deformation: Geological Society of America Special Paper 218, p. 31–54.

Ramos, V. A., 1993, Interpretación tectónica, *in* Ramos, V. A., ed., Geología y recursos naturales de Mendoza: 12th Congreso Geológico Argentino y 2nd Congreso de Exploración de Hidrocarburos, Mendoza, Argentina, Relatorio, p. 257–266.

Ramos, V. A., 1995, Sudamérica: un mosaico de continentes y océanos: Ciencia Hoy (Buenos Aires), v. 6, no. 32, p. 24–29.

Ramos, V. A., Jordan, T. E., Allmendinger, R. W., Mpodozis, C., Kay, S. M., Cortés, J. M., and Palma, M. A., 1986, Paleozoic terranes of the central Argentine-Chilean Andes: Tectonics, v. 5, p. 855–880.

Ramos, V. A., Vujovich, G. I., and Dallmeyer, D., 1996, Klippes y ventanas tectónicas de la Sierra de Pie de Palo: sus relaciones con la colisión de Precordillera: 13th Congreso Geológico Argentino y 3rd Congreso de Exploración de Hidrocarburos, Buenos Aires, Argentina, Tomo V, p. 377–391.

Rapalini, A. E., and Astini, R. A., 1998, Paleomagnetic confirmation of the Laurentian origin of the Argentine Precordillera: Earth and Planetary Science Letters, v. 155, p. 1–14.

Raup, D. M., and Stanley, S. M., 1978, Principles of paleontology (second edition): San Francisco, W. H. Freeman, 481 p.

Raymond, D. E., 1991, New subsurface information on Paleozoic stratigraphy of the Alabama fold and thrust belt and the Black Warrior basin: Alabama Geological Survey Bulletin 143, 185 p.

Read, J. F., 1989, Controls on evolution of Cambro-Ordovician passive margin U.S. Appalachians, *in* Crevello, P. D., Wilson, J. L., Sarg, J. F., and Read, J. F., eds., Controls on carbonate platform and basin development: Society of Economic Paleontologists and Mineralogists Special Publication 44, p. 147–165.

Rosendahl, B. R., 1987, Architecture of continental rifts with special reference to East Africa: Annual Review of Earth and Planetary Sciences, v. 15, p. 445–503.

Royden, L. H., 1993a, The tectonic expression slab pull at continental convergent boundaries: Tectonics, v. 12, p. 303–325.

Royden, L. H., 1993b, Evolution of retreating subduction boundaries formed during continental collision: Tectonics, v. 12, p. 629–638.

Royden, L. H., and Keen, C. E., 1980, Rifting process and thermal evolution of the continental margin of eastern Canada determined from subsidence curves: Earth and Planetary Science Letters, v. 51, p. 343–361.

Schmoker, J. W., and Halley, R. B., 1982, Carbonate porosity versus depth: a predictable relation for south Florida: American Association of Petroleum Geologists Bulletin, v. 66, p. 2561–2570.

Sclater, J. C., and Christie, P. A. F., 1980, Continental stretching: an explanation of the post-mid-Cretaceous subsidence of the central North Sea basin: Journal of Geophysical Research, v. 85, p. 3711–3739.

Scrutton, R. A., 1982, Passive continental margins: a review of observations and mechanisms, *in* Scrutton, R. A., ed., Dynamics of passive margins: American Geophysical Union and Geological Society of America Geodynamics Series, v. 6, p. 5–11.

Simpson, C., Law, R. D., and Martino, R. D., 1995, Dip-slip tectonic events in basement rocks of the western and eastern Pampean Ranges, central Argentina: Geological Society of America Abstracts with Programs, v. 27, no. 6, p. A–125.

Simpson, E. L., and Sundberg, F. A., 1987, Early Cambrian age for synrift deposits of the Chilhowee Group of southwestern Virginia: Geology, v. 12, p. 123–126.

Skehan, J. W., 1988, Evolution of the Iapetus Ocean and its borders in pre-Arenig times: a synthesis, *in* Harris, A. L., and Fettes, D. J., eds., The Caledonian-Appalachian orogen: Geological Society of London Special Publication 38, p. 185–229.

Soper, N. J., Gibbons, W., and Mc Kerrow, W. S., 1991, Displaced terranes in Britain and Ireland: Journal of the Geological Society of London, v. 146, p. 365–367.

Steckler, M. S., and Watts, A. B., 1978, Subsidence of the Atlantic-type continental margin off New York: Earth and Planetary Science Letters, v. 41, p. 1–13.

Thomas, W. A., 1977, Evolution of Appalachian-Ouachita salients and recesses from reentrants and promontories in the continental margin: American Journal of Science, v. 277, p. 1233–1278.

Thomas, W. A., 1991, The Appalachian-Ouachita rifted margin of southeastern North America: Geological Society of America Bulletin, v. 103, p. 415–431.

Thomas, W. A., 1993, Low-angle detachment geometry of the Late Precambrian–Cambrian Appalachian-Ouachita rifted margin of southeastern North America: Geology, v. 21, p. 921–924.

Thomas, W. A., 1998, Neoproterozoic-Paleozoic geography and tectonics: review, hypothesis, environmental speculation: Discussion: Geological Society of America Bulletin, v. 110, p. 1617–1618.

Thomas, W. A., and Astini, R. A., 1996a, The Argentine Precordillera: a traveler from the Ouachita embayment of North American Laurentia: Science, v. 273, p. 752–757.

Thomas, W. A., and Astini, R. A., 1996b, Asymmetric conjugate rift margins of the Ouachita rift and the Argentine Precordillera: Geological Society of America Abstracts with Programs, v. 28, no. 1, p. 66.

Thomas, W. A., and Astini, R. A., 1999, Simple-shear conjugate rift margins of the Argentine Precordillera and the Ouachita embayment of Laurentia: Geological Society of America Bulletin, v. 111, p. 1069–1079.

Todd, B. J., Reid, I., and Keen, C. E., 1988, Crustal structure across the southwest Newfoundland transform margin: Canadian Journal of Earth Sciences, v. 25, p. 744–759.

Torsvik, T. H., Tait, J., Moralev, V. M., McKerrow, W. S., Sturt, B. A., and Roberts, D., 1995, Ordovician paleogeography of Siberia and adjacent continents: Journal of the Geological Society of London, v. 152, p. 279–287.

Torsvik, T. H., Smethurst, M. A., Meert, J. G., Van der Voo, R., McKerrow, W. S., Brasier, M. D., Sturt, B. A., and Walderhaug, H. J., 1996, Continental break-up and collision in the Neoproterozoic and Palaeozoic—a tale of Baltica and Laurentia: Earth Science Review, v. 40, p. 229–258.

Tucker, R. D., and McKerrow, W. S., 1995, Early Paleozoic chronology: a review in light of new U-Pb zircon ages from Newfoundland and Britain: Canadian Journal of Earth Sciences, v. 32, p. 368–379.

Vaccari, N. E., 1993, El género *Annamitella* Mansuy, 1920 (Trilobita, Leiostegiidae) en el Ordovícico de la Precordillera Argentina: Ameghiniana, v. 30, p. 395–405.

Vaccari, N. E., 1994, Las faunas de trilobites de las sucesiones carbonáticas del Cámbrico y Ordovícico temprano de la Precordillera Septentrional, República Argentina [Ph.D. thesis]: Córdoba, Argentina, Universidad Nacional de Córdoba, 271 p.

Vaccari, N. E., 1995, Early Ordovician trilobite biogeography of Precordillera and Famatina, western Argentina: preliminary results, *in* Cooper, J. D., Droser, M. L., and Finney, S. C., eds., Ordovician odyssey: SEPM Pacific Section, Book 77, p. 193–196.

Vaccari, N. E., and Bordonaro, O., 1993, Trilobites en los olistolitos cámbricos de la Formación Los Sombreros (Ordovícico), Precordillera de San Juan, Argentina: Ameghiniana, v. 30, p. 383–393.

van der Pluijm, B. A., Van der Voo, R., and Torsvik, T. H., 1995, Convergence and subduction at the Ordovician margin of Laurentia, *in* Hibbard, J. P., van Staal, C. R., and Cawood, P. A., eds., Current perspectives in the Appalachian-Caledonian orogen: Geological Association of Canada Special Paper 41, p. 127–136.

Viele, G. W., and Thomas, W. A., 1989, Tectonic synthesis of the Ouachita orogenic belt, *in* Hatcher, R. D., Jr., Thomas, W. A., and Viele, G. W., eds., The Appalachian-Ouachita orogen in the United States: Boulder, Colorado, Geological Society of America, Geology of North America, v. F-2, p. 695–728.

Vink, G. E., Morgan, W. J., and Zhao, W.-L., 1984, Preferential rifting of continents: a source of displaced terranes: Journal of Geophysical Research, v. 89, no. B12, p. 10072–10076.

Watts, A. B., 1982, Tectonic subsidence, flexure, and global changes of sea level: Nature, v. 297, p. 469–474.

Webby, B. D., 1984, Ordovician reefs and climate: a review, *in* Bruton, D. L., ed., Aspects of the Ordovician System: Oslo, Norway, Universitetsforlaget, p. 87–98.

Webby, B. D., 1992, Global biogeography of Ordovician corals and stromatoporoids, *in* Webby, B. D., and Lurie, J. R., eds., Global perspectives on Ordovician geology: Rotterdam, Netherlands, Balkema, p. 261–276.

Whittington, H. B., and Hughes, C. G., 1972, Ordovician geography and faunal provinces deduced from trilobite distributions: Royal Society of London Philosophical Transactions, v. B263, p. 235–278.

Wilkerson, M. S., and Hsui, A. T.-K., 1989, Application of sediment backstripping corrections for basin analysis using microcomputers: Journal of Geological Education, v. 37, p. 337–340.

Williams, S. H., Harper, D. A. T., Neuman, R. B., Boyce, W. D., and Mac Niocaill, C., 1995, Lower Paleozoic fossils from Newfoundland and their importance in understanding the history of the Iapetus Ocean, *in* Hibbard, J. P., van Staal, C. R., and Cawood, P. A., eds., Current perspectives in the Appalachian-Caledonian orogen: Geological Association of Canada Special Paper 41, p. 115–126.

Witzke, B. J., 1990, Paleoclimatic constraints for Palaeozoic palaeolatitudes of Laurentia and Euramerica, *in* McKerrow, W. S., and Scotese, C. R., eds., Paleozoic paleogeography and biogeography: Geological Society of London Memoir 12, p. 57–73.

Zapata, T. R., 1996, Crustal evolution of the Precordillera thrust belt-Bermejo basin, Argentina [Ph.D. thesis]: Ithaca, New York, Cornell University, 243 p.

Zapata, T. R., and Allmendinger, R. W., 1996, La estructura cortical de la Precordillera oriental y valle del Bermejo a los 30° de latitud sur: 13th Congreso Geológico Argentino y 3rd Congreso de Exploración de Hidrocarburos, Buenos Aires, Argentina, Tomo II, p. 211–224.

Manuscript Accepted by the Society September 4, 1998

Geological Society of America
Special Paper 336
1999

Paleontological constraints on successive paleogeographic positions of Precordillera terrane during the early Paleozoic

Juan L. Benedetto, Teresa M. Sánchez, Marcelo G. Carrera, Edsel D. Brussa, and María J. Salas
Cátedra de Estratigrafía y Geología Histórica, Facultad de Ciencias Exactas, Fisicas y Naturales, Universidad Nacional de Córdoba. Av. Velez Sarsfield 299, 5000 Córdoba, Argentina

ABSTRACT

Faunal data from Cambrian and Ordovician rocks of the Precordillera terrane are examined in order to refine their biogeographic relationships with Laurentia and Gondwana. This study is based principally on benthic organisms (sponges, bryozoans, brachiopods, bivalves, and ostracods), but data from graptolites were also considered. The data lead us to recognize four successive stages of Precordilleran biogeographic evolution: (1) Laurentian stage (Cambrian-Tremadoc), (2) isolation stage (Arenig–early Llanvirn), (3) pre-accretion stage (Llanvirn-Caradoc) and (4) Gondwanan stage (Hirnantian-Silurian). Each denotes its specific position during the rifting-drifting-collision sequence and each reflects a different pattern of faunal exchange. During the Laurentian stage, the nearly complete identity with Appalachian faunas supports a close geographic connection between Precordillera and Laurentia. The isolation stage begins when taxa that have not been recorded in Laurentia appear for the first time in the Precordillera basin. Through this stage the Laurentian faunal influence decreases and the number of endemic Baltic-Avalonian genera correlatively increases. The pre-accretion stage is characterized by its paucity of Laurentian forms, incoming of Gondwanan taxa, and an unusually high level of endemicity. The latter may reflect a degree of geographic isolation, and may in part be due to biologic factors related to dispersal mechanisms. Finally, the Gondwanan stage starts after accretion of the Precordillera terrane at the end of the Ordovician. Faunal data reviewed here support the "far traveled microplate" hypothesis that is consistent, in general, with the geological evidence.

INTRODUCTION

Attempts to reconstruct southern Iapetus paleogeography have provided several different paleogeographic maps depending on the nature of information on which each was based. In the Cambrian and Ordovician, paleomagnetically controlled maps (e.g., Scotese and McKerrow, 1991; Torsvik and Trench, 1991) of eastern Laurentia and the Andean margin of Gondwana are depicted with a longitudinal separation of ~110°. Authors such as Bond et al. (1984), Herrera and Benedetto (1990), Benedetto (1993), and Dalziel (1991, 1992) have argued for an opposite positioning of both continents, and proposed different kind of interactions between them through the early Paleozoic. The Argentine Precordillera played a key role in debates about the evolution of the proto-Andean margin and in establishing paleogeographic reconstructions.

The Precordillera mountain belt was first considered to be a suspect terrane by Ramos et al. (1986), but its probable Laurentian connection had earlier been deduced from paleontological evidence (Bond et al., 1984). It has long been known that its thick and richly fossiliferous Cambrian carbonate rocks contain olenellid trilobites very similar to those of North America

Benedetto, J. L., Sánchez, T. M., Carrera, M. G., Brussa, E. D., and Salas, M. J., 1999, Paleontological constraints on successive paleogeographic positions of Precordillera terrane during the early Paleozoic, *in* Ramos, V. A., and Keppie, J. D., eds., Laurentia-Gondwana Connections before Pangea: Boulder, Colorado, Geological Society of America Special Paper 336.

(Poulsen, 1958; Borrello, 1971). Later studies by us and by Argentine colleagues working on Cambrian and Ordovician trilobites (Vaccari, 1994, 1995), Ordovician brachiopods (Herrera and Benedetto, 1990, and references therein; Benedetto, 1995, 1998c), sponges and bryozoans (Carrera, 1985, 1994a,b, 1995, 1996a,b; Beresi and Rigby, 1993) lead us to adopt a model in which the Precordillera is considered to be a microplate that had rifted from the Ouachita embayment and afterward accreted to the pre-Andean margin of Gondwana (Benedetto, 1993; Benedetto and Astini, 1993; Astini et al., 1995; Benedetto et al., 1995, Thomas and Astini, 1996). Others (Dalla Salda, 1992a,b; Dalziel et al., 1994; Dalziel, 1997), while accepting the general idea that the Precordillera is a Laurentia-derived terrane, suggested a contrasting scenario in which the Precordillera is transferred to Gondwana as a result of a Laurentia/Gondwana continental collision followed by a rifting event. This hypothesis differs substantially from our interpretation because the Precordillera, a part of the "Texas plateau" of Dalziel (1997) remains attached to Laurentia until the mid-Ordovician, and that by Llandeilo–early Caradocian times, Laurentia and Gondwana had been assembled into a megacontinent. This implies not only a different geographic configuration for evolution and migration of benthic organisms but a different tectono-stratigraphic evolution of the pre-Andean margin.

This chapter presents our current interpretations of Cambrian and Ordovician paleogeographic relations of the Precordillera, Famatina, and Central Andean basins (Fig. 1), based on an updated review of their fossils, elucidating a previous summary (Benedetto et al., 1995; see also Benedetto 1998a–c). Here we present new paleontological information that supports the "far traveled microplate" hypothesis, including a more detailed analysis of drifting and collision of the Precordillera terrane. We also attempt to establish some relationships between biogeographic patterns and southern Iapetus paleo-oceanography, and possible interactions between paleogeographic scenarios and faunal provincialism.

It has long been observed that different taxonomic groups (i.e., brachiopods, bivalves, trilobites, ostracods) exhibit substantial differences in provinciality. Fortey and Mellisch (1992) drew attention to the variation of paleogeographic configurations of different groups of organisms. To minimize such uncertainty, we have based our biogeographic analysis principally on five groups of benthic organisms: sponges, bryozoans, brachiopods, bivalves, and ostracods. Data from graptolites were also included, even though they were plancktonic and their provincialism seems not to be strictly controlled by geographic proximity. Distribution patterns of graptolite faunas along the pre-Andean margin, however, would be helpful to recognize latitudinal thermal gradients in surface waters. Trilobites have been considered especially for the Cambrian, and are included to complement information for the Ordovician times. They were not included in the tables because most of them are actually under taxonomic revision by other colleagues.

GEOLOGIC AND BIOSTRATIGRAPHIC FRAMEWORK

Precordillera region

The Argentine Precordillera is a high-level thrust and fold belt that involved a thick Paleozoic sedimentary succession of Cambrian to Devonian marine rocks unconformably overlain by mainly continental Carboniferous-Triassic deposits. Younger strata include Tertiary continental clastics deposited in intermontane basins. Crystalline basement rocks are nowhere exposed in the Precordillera so that their nature is virtually unknown, although xenoliths of metamorphic rocks in Miocene andesites were interpreted by Abbruzzi et al. (1993) as basement rocks of Grenville affinity. The original stratigraphic relationships between basement and sedimentary cover are preserved at the southern edge of the Cuyania terrane, where Ordovician limestones overlie metamorphic rocks of Grenvillian age (Keller and Dickerson, 1996).

From a tectono-stratigraphic viewpoint, early Paleozoic strata were divided into western and eastern tectofacies (Fig. 2) (Astini, 1992). The eastern belt consists mainly of carbonates and shallow marine clastics of shelfal and restricted environments, whereas the western tectofacies includes a thick wedge of slope to basin olistostromes, turbidites, and basinal black shales, the latter interbedded with mafic lava flows (some pillowed) and intruded by basic dikes. The overall western belt succession is affected by a very low-grade regional metamorphism.

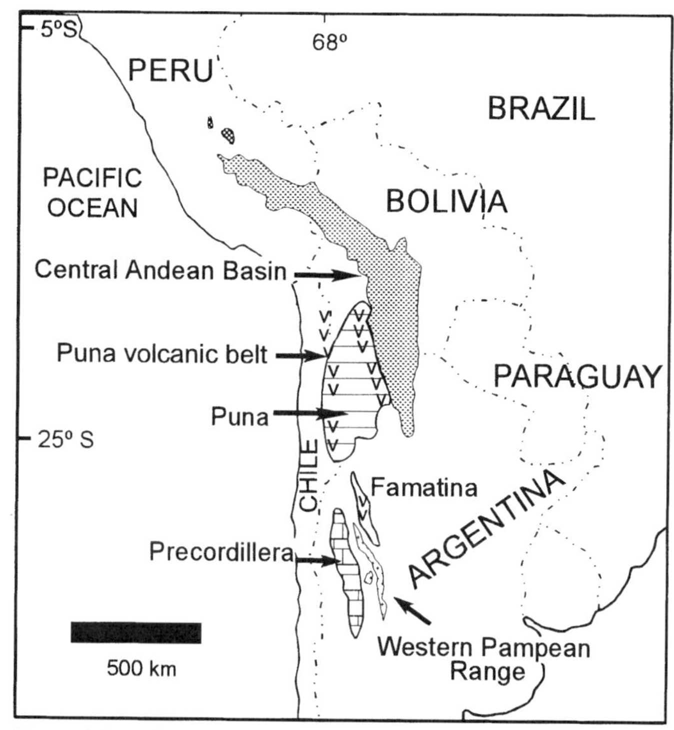

Figure 1. Location map showing basins and geologic regions mentioned in the text.

Cambrian and Early Ordovician carbonates reflect the transition from marginal marine environments probably related to initial rifting stages (Cerro Totora Formation in Fig. 2), to open carbonate platform/ramp (San Juan Formation) via restricted subtidal and peritidal platforms (La Laja, Zonda, and La Flecha formations) and isolated rimmed platform (La Silla Formation). According to Cañas (1995a,b and this volume), the absence of quartz detritus suggests that the Tremadocian La Silla Formation is the first carbonate unit clearly isolated from nearby land masses. Development near the Tremadoc-Arenig transition of widespread muddy fossiliferous carbonates with metazoan-microbial reef-mounds (Carrera, 1991; Cañas and Carrera, 1993) occurring on an open platform (San Juan Formation) represents an important change in the basin environment. The onset of this major transgressive event was related to the cumulative effects of eustatic sea-level rise and extensional tectonics (Cañas, 1995a,b). The mid-Arenig fossiliferous wackestones and grainstones (*Niquivilia* and *Huacoella* zones) are capped by stromatoporoid patch reefs and stromatoporoid-lithistid-algal reef-mounds (Cañas and Keller, 1993). Carbonate sedimentation culminates with nodular wackestones and packstones of early Llanvirn age (*Ahtiella* zone) containing high-diversity open-platform faunas. Paleoecologic analysis of these faunas suggests a regional slope to the north (Sánchez et al., 1996).

The transition between the carbonate platform and basin deposits is recorded along the boundary between eastern and western tectono-stratigraphic belts. Typical slope deposits are represented by olistostromes containing huge (kilometer-scale) olistoliths of Cambrian and Early Ordovician limestones (Los Sombreros Formation) embedded in late Arenig–early Llanvirn shales (Benedetto and Vaccari, 1992; Keller, 1995).

Deposition changes significantly after the early Llanvirn (but locally as early as the late Arenig), when the drowning of the platform stopped carbonate production followed by widespread

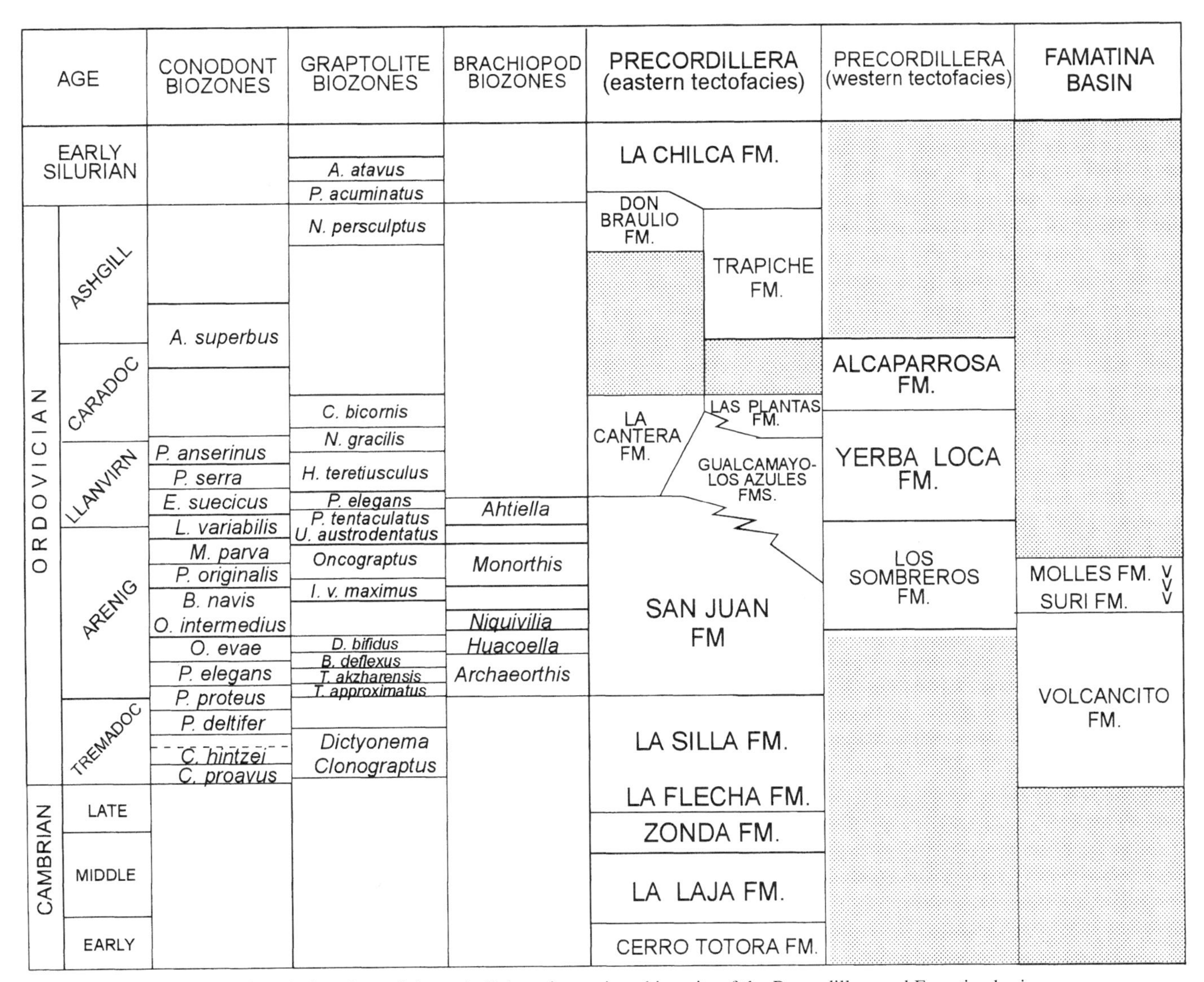

Figure 2. Correlation chart of the early Paleozoic stratigraphic units of the Precordillera and Famatina basins.

deposition of graptolitic black shales (Los Azules and Gualcamayo Formations). Toward the upper part of the San Juan Limestone and in the lower part of black shale units (late Arenig–early Llanvirn), numerous K-bentonites occur at several localities (Huff et al., 1995; Bergström et al., 1996). Associated with this change in the sedimentary regime, siliciclastic sedimentation continued from late Llanvirn to Ashgill time, including huge intrabasinal limestone olistoliths mixed with extrabasinal resedimented conglomerates (Las Vacas, La Cantera, and Trapiche Formations) (Astini, 1996a). In the central Precordillera, foramol carbonate remnants (Sassito Formation) developed on structural highs (Astini and Cañas, 1995). Ordovician deposition concluded with late Ashgill glacial diamictites followed by transgressive fossiliferous mudstone (Don Braulio Formation) (Peralta and Carter, 1990; Buggish and Astini, 1993).

Silurian and Devonian mud- to sand-dominated platform successions unconformably overlie different levels of the Ordovician succession.

The main deformation event in the Precordillera affecting Paleozoic sedimentary rocks, of Late Silurian–Early Devonian age (Precordilleran orogeny *sensu* Astini, 1996b), was responsible for low-grade regional metamorphism, penetrative cleavage, and folding of the flysch-like Ordovician successions of the western Precordilleran belt (Buggisch et al., 1994), and the strong angular unconformity between Silurian and Carboniferous formations in the eastern belt (Von Gosen et al., 1995). The existence of previous tectonic disturbances can be detected mainly by stratigraphic and sedimentologic evidence. For instance, the "Guandacol orogeny" is recognized by widespread deposition of coarse-grained sediments during the late Llanvirn, and the "Ocloyic orogeny" is reflected by the local erosional unconformity between early–mid Ordovician and Silurian beds. The rock-preservation curve constructed by Astini (1996b) shows a main noneustatic interruption in the sedimentary record encompassing the Llandeilo-Caradoc time interval.

Complete reviews of the Paleozoic stratigraphy of the Precordillera basin were given by Baldis et al. (1982), Ramos et al. (1986), and Astini et al. (1995, 1996). Biostratigraphy of Cambrian units was established by Bordonaro (1980, 1992), Bordonaro and Liñán (1994), Vaccari (1994), and Vaccari and Bordonaro (1993). Ordovician biostratigraphic schemes are largely based on conodonts (Albanesi et al., 1995a; Lehnert, 1993; 1995), graptolites (Ortega et al., 1993, 1995 and references therein; Brussa, 1994 and references therein), brachiopods (Herrera and Benedetto, 1990; Benedetto, 1995, 1998c), and trilobites (Vaccari, 1994; Baldis, 1995; Baldis and Pöthe de Baldis, 1995).

Famatina Range

The latitude of the northern segment of the Precordillera region overlaps that of the southern half of the Famatina Range (Fig. 1). This geologic province is characterized by early Paleozoic magmatic, volcaniclastic, and epiclastic rocks. The accumulation of thick (~1,000 m), volcaniclastic rocks in the Suri and Los Molles Formations has been interpreted as marking an active margin in a back-arc setting (Aceñolaza and Toselli, 1988; Rapela et al., 1992; Mannheim, 1993). Petrologic and geochemical evidence suggests that these volcanic rocks represent oceanic island-arc magmatism with collisional-arc magmatic features (Mannheim, 1993). Astini and Benedetto (1996) recognized a first stage of rapid thermal subsidence characterized by deposition of graptolitic black shales followed by a shallowing-upward succession of siltstones and sandstones. Deposition culminates with marginal marine reddish siltstones interbedded with prograding volcaniclastic wedges (Mangano and Buatois, 1994). The age of this sequence, according to the conodonts (Albanesi and Vaccari, 1994), graptolites (Toro, 1997; Toro and Brussa, 1997), brachiopods (Benedetto, 1994), and trilobites (Vaccari et al., 1993; Vaccari and Waisfeld, 1994), ranges from early to middle Arenig.

FACTORS CONTROLLING DISPERSAL OF BENTHIC FAUNAS AND THEIR BIOGEOGRAPHIC SIGNIFICANCE

Geographic dispersion of marine benthic organisms is controlled by several physical and biotic factors. Among the latter, one of the most important is the type of larval development. At the end of the larval stage, the success of settling in a new area depends on the availability of a niche to colonize, the absence of competitors, and an adequate substrate-level environment. Influence of such factors depends, in turn, on the opportunistic or highly selective character of the species. Abiotic factors include direction and velocity of marine currents, water salinity, oxygenation, and temperature.

Some planktotrophic larvae can survive several weeks before settlement. Depending on marine currents they can be transported as much as 500 km in a single breeding period (Thorson, 1950; Scheltema, 1977). Nonplanktotrophic larvae survive no more than a few days. Organisms that have nonfeeding larvae or that lack larval stages may also attain wide geographic dispersion by means of rafting (Jokiel, 1990). Research on drifting of modern reef corals demonstrates the high possibilities of long dispersal by means of rafting on pumice and other floating debris. Large quantities of pumice have probably been common in oceans throughout time, but especially during the Ordovician, when widespread volcanic activity forming volcanic islands and arc-volcanic systems has been documented in many localities of the world, particularly within the Iapetus ocean (Neuman, 1984; Jokiel, 1990).

On the other hand, larval type is related to speciation rates and endemism. High capability of dispersal produces widespread species, while species having nonplanktotrophic larvae occupy more reduced geographic areas and may have higher rates of extinction and speciation (Jablonski and Lutz, 1980).

Dispersal also depends on pattern of marine currents. Transport along the same latitudinal belt does not prevent dispersion, but currents perpendicular to latitude could inhibit dispersal of stenothermic species (Valentine, 1971). Based on the bipolar or antitropical pattern of distribution of some modern faunas and their origins in late Cenozoic times, Lindberg (1991) stated that if temperature and currents are appropriate, the dispersal of marine species, with or without pelagic larvae, would be effective and could appear instantaneously in geologic time. The strong relationship between current pattern and dispersal is evident, for example, in a recent species of the brachiopod *Terebratulina*, which migrated up to 4,000 km, carried by the North Atlantic current (Curry and Endo, 1991). Studies on recent brachiopods of the South African platform show that species distribution is controlled by two different currents, the warm Agulhas current in the east coast, and the cold Benguela Upwelling system in the west coast. Only some tolerant taxa live in both coasts (Hiller, 1994).

Successful dispersal of organisms among wide geographic areas depends on a conjunction of factors that prevail at each place at a given time. In a paleobiogeographic study one should ideally take into account the whole of these variants for each group of organisms. If not, we risk overestimating the role of geographic separation. For example, the occurrence of the same restricted taxa in two now-distant regions is important and can infer closer geographic relationships between them in the past. Conversely, the absence of shared taxa in two compared regions may not be the result of geographic separation but could indicate such conditions as low capability of these species for dispersal, absence of suitable substrate for settlement, unsuitable water temperature, or unfavorable current directions.

Life habits, reproductive cycle, and dispersal mechanisms of Paleozoic organisms are, in general, unknown, especially in those of extinct groups (trilobites, graptolites, and many kind of brachiopods and bryozoans). More accurate inferences can be made on fossil organisms phylogenetically related to living groups. Some biologic aspects of those groups considered in our biogeographic analysis are briefly reviewed below.

Sponges are sessile benthic organisms, mainly concentrated in shallow marine environments. Most of the species have a short larval period and their settlement is rapid. These characteristics point out their potential importance in paleobiogeography. Several additional factors could have been important to sponge dispersal, such as flotation of colonies and rafting. Biogeographic distribution of present-day sponges shows remarkable differences between interoceanic populations (Wilkinson, 1987). During the Ordovician, some genera were widespread while others are endemic or have more restricted distribution. Sponges have been considered as a very conservative group (long stratigraphic ranges of genera and families), which implies slow but persistent migration rates and, conversely, low rates of speciation.

Most of the species of bryozoans have short-lived, lecithotrophic larvae that can move only short distances after settling (McKinney and Jackson, 1991). In some species of Cheilostomata (e.g., *Membranipora*) however, cyphonautes planktic larvae permit long-distance dispersal (Scheltema, 1977); such a larval type may have been common among Paleozoic bryozoans (Taylor and Cope, 1987). Additionally, many authors accept a rafting mechanism for bryozoan dispersal that could allow them a wide geographic range.

Recent brachiopods are survivors of a large Paleozoic group. Fossils reveal that these animals attained a great morpho-functional diversification and inhabited nearly all marine environments. In modern articulate brachiopods the timing of larval pelagic development is very short, rarely exceeding 2 days (Scheltema, 1977). However, evidence from brachiopod distribution in the past seems to indicate the contrary. As Rong et al. (1995) pointed out, distribution of Silurian brachiopods and the abundance of cosmopolitan genera suggest the existence of teleplanic larvae in many of the species. Water temperature may be the primary factor controlling brachiopod provincialism. This conclusion is based on comparison between biogeographic patterns and paleogeographic reconstructions based on independent criteria such as paleomagnetism and sedimentary regime (climatically sensitive lithologies). Provincial differentiation within a realm, however, may depend on factors other than climate, such as barriers and connections between basins, rates of exchange between these basins, and the open ocean and local paleoenvironments (Benedetto and Sánchez, 1996b). The Ordovician was a time of high provincialism, which has long been related to the splitting of continental plates (Williams, 1973). Development of intraoceanic islands and pericontinental volcanic arcs also could have played an important role in Ordovician diversification (Neuman, 1984).

Larval development of recent species of molluscs (including gastropods and bivalves) is well known (Jablonski and Lutz, 1980), but much uncertainty exists about the types of larvae of Paleozoic species, especially those belonging to extinct clades. Studies of modern larvae in eastern North Atlantic waters demonstrate that bivalve larvae can be transported long distances across the ocean, especially if there are islands or subsurface elevations such as guyots that can be used as standing points (Thiede, 1974).

Some 65–70% of living marine bivalves have planktotrophic larvae (Jablonski and Lutz, 1980), a condition that depends on several factors. Generalized "r-strategist" species produce large numbers of small planktotrophic larvae, whereas the more specializad "k-strategist" species produce fewer nonplanktotrophic larvae. Under certain conditions, however, the opposite situation may occur. High latitude species (normally viewed as r-strategists) have nonplanktotrophic larvae, probably due to environmental fluctuations, especially the timing of primary production that restricts the timing of life of the planktotrophic stage (Jablonski and Lutz, 1980).

Ostracods have benthic habitats; most of them are restricted to shallow waters. Deep-water barriers apparently separated lower Paleozoic ostracod associations, because none are found in deep-water environments. Their dimorphic features suggest that they had no pelagic larvae (Siveter, 1984; Schallreuter and Siveter, 1985). In addition, the ontogeny of modern species, indicates that they and their predecessors live and reproduce in a single habitat (Siveter, 1984). In general, early Paleozoic ostracods tended to differentiate into biogeographic regions, each characterized by a relatively high percentage of endemic genera (Fortey and Mellisch, 1992).

Trilobites inhabited shallow to deep waters, so that a succession of more or less continuous biofacies can be recognized across the marine bathymetric profile (Fortey, 1975). It is known that the widespread olenid biofacies, for instance, includes deep-water trilobites that inhabit sites that are marginal to continents, from high to low latitudes (Cocks and Fortey, 1990). Epipelagic trilobites are dispersed over long distances but are restricted to climatic belts. For these reasons, use of that group of trilobites in paleobiogeography is limited. Conversely, trilobites inhabiting shallow- water platforms, such as the warm-water bathyurid biofacies or the cool to cold-water *Neseuretus* biofacies, include a characteristic suite of endemic genera useful for paleogeographic discrimination.

Because of their pelagic mode of life, many graptolites exhibit a global distribution. However, during the Ordovician, a low-latitude Pacific province and a high-latitude Atlantic province are commonly recognized. Although it has been considered that during the late Arenig–Llanvirn graptolite provincialism attains a maximum, a recent evaluation by Maletz and Mitchell (1995) suggested that Llanvirn endemism was not as marked as supposed. In graptolites, provincialism was relatively independent of geographic barriers, and development of different graptolite biofacies may reflect ecologic rather than biogeographic controls. Cooper et al. (1991) have suggested that the principal factors that contribute to the spatial differentiation of graptolites are the oceanic currents and water mass specificity, latitudinal surface-water temperature gradients, depth distribution, and partitioning of shelf oceanic water masses, producing contrast of salinity, turbidity, and phytoplankton distribution.

FAUNAL AFFINITIES AND INFERRED DISPLACEMENTS OF THE PRECORDILLERA TERRANE

The biogeographic history of the Precordillera is the key to inferring its origin, trajectory, and accretion site. The first ideas about its Laurentian origin arose from the strong affinities between the Cambrian trilobite faunas from the North American and Precordilleran carbonate belts (Bond et al., 1984). Likewise, the first attempts to establish the timing of detachment of Precordillera from Laurentia and its subsequent accretion to Gondwana have been based on biogeographic evidence (Benedetto, 1993). We present here a more accurate picture of the successive positions of the Precordillera terrane based on a more complete compilation of paleontological data than has been available until now. Faunal evidence supports the microcontinental or far-traveled microplate hypothesis of Benedetto (1993), later refined and amplified (Benedetto and Astini, 1993; Astini et al., 1995, 1996; Benedetto et al., 1995; Thomas and Astini, 1996; Benedetto, 1998c).

Faunal evidence allows us to recognize four successive stages in the biogeographic evolution of the Precordillera terrane: Laurentian stage (Cambrian-Tremadoc), isolation stage (Arenig–early Llanvirn), pre-accretion stage (Llanvirn-Caradoc), and Gondwanan stage (Hirnantian?-Silurian). The paleontological signature of each of them results of its specific paleogeographic position during the rifting-drifting-collision cycle, and reflects different sources and rates of faunal exchange.

Laurentian stage (Cambrian-Tremadoc)

This stage started when a rift system developed along the southeastern edge of the Laurentian plate, marking its separation of a crustal block, the future Precordillera terrane. The first faunal evidence of a seaway penetrating between them is the record of olenellid trilobite faunas in some early Cambrian formations (La Laja, Cerro Totora). Coeval shallow- to deep-water trilobite associations were also encountered in redeposited limestone blocks within the Los Sombreros Formation (Benedetto and Vaccari, 1992). These associations, belonging to the *Bonnia-Olenellus* zone, are typical of Laurentia (Vaccari, 1988, 1994). From the overlying Late Cambrian La Flecha Formation, Vaccari (1994) has described several species characteristic of the Dresbachian carbonatic rocks of southern and central Appalachians, such as *Madarocephalus laetus, Komaspidella laevis,* and *Dytremacephaalus strictus*. All of the trilobite taxa recorded in the Precordillera are typical Laurentian forms and no inner-platform trilobites belonging to a different biogeographic province have been recorded to date (Vaccari, 1994; Palmer et al., 1996; Astini et al., 1995).

Macrofaunas from uppermost Cambrian–lowermost Ordovician rocks are very scarce because of their restricted depositional environment (e.g., La Silla Formation). From the base of this unit Vaccari (1994) reported the Laurentian trilobite *Plethometopus obtusus*. Overlying limestone yielded *Clavohamulus hintzei,* a typical conodont of the warm-water North American Mid Continent faunal province. Outside Laurentia, it has been recorded only in the Georgina basin of Australia (Lehnert et al., 1997).

Microbe-sponge bioherms occur at the base of the San Juan Formation (Cañas and Carrera, 1993), near the Tremadoc/Arenig boundary. They share many features with early Ordovician buildups that occur in a wide belt along the southern margin of North America. Both include the *Archaeoscyphia-Calathium* association and stromatoporoid-like organisms (Keller and Flugel, 1996). The presence of *Archaeoscyphia-Calathium-Pulchrilamina* reef-mounds in China (Rigby et al., 1995) however, suggests that this association represents a world-wide episode of tropical to subtropical areas without any other paleogeographic significance.

Few Tremadoc–early Arenig graptolite faunas are known in the Precordillera. Two associations were recorded in the Empozada Formation, at the southern Precordillera (Bordonaro and Peralta, 1987). This fauna contains many pandemic species, like *Pendeograptus fruticosus, P. pendens,* and *Tetragraptus bigsbyi,* as well as the Pacific province *T. approximatus*.

In summary, there is abundant faunal evidence to support a close geographic connection between Laurentia and the Precordillera until earliest Ordovician times. We infer that, by the end of the Tremadoc, the width of the seaway between the Precordillera terrane and Laurentia was some hundreds of kilometers, not enough to prevent a nearly complete exchange of benthic faunas.

Isolation stage (Arenig– early Llanvirn)

We recognize the onset of the isolation stage as gradual, marked by the first appearance in the Precordillera terrane of non-Laurentian taxa that is interpreted to indicate the effects of the Precordillera drift toward a higher latitude (Fig. 3). Biogeographic affinities of fossils involved in our analysis are summarized in the histogram shown in Figure 3.

Precordilleran Arenig-age sponges include Appalachian genera, such as *Allosacus, Eospongia, Rhopalocoelia, Psarodictyum,* and *Hudsonospongia*. Endemic taxa include the megamorinid *Nexospongia*, which belongs to the endemic family Nexospongiidae (Carrera, 1996a), and isolated spicules of the genus *Chilcaia* (Table 1). In the early Llanvirn, the number of endemic genera increases (*Protachileum, Talacastonia, Rugospongia,* and *Chilcaia*) (Table 3). The genera *Incrassospongia* and *Aulacopium* are known only from younger beds of Baltica and Laurentia, so they can be considered as endemic taxa of the Precordillera in the early Ordovician. In summary, the occurrence in Arenig limestones of the Precordillera basin of sponges recorded elsewhere in the Appalachians confirms that their origin and early history were strongly influenced by the Laurentian margin faunas, although differentiation is marked by an incipient endemism. By the Early Llanvirn these similarities diminished, due to the incorporation of forms not known in the Appalachians and to the increase of endemic taxa. This increase in endemicity is significant and reinforces the idea of the isolation of the Precordillera after middle Arenig time.

Brachiopods from the early Arenig age (*Archaeorthis* zone) lower San Juan Limestone belong to the Laurentian genera *Syntrophia, Orthidium,* and *Notorthis*. Brachiopod diversity gradually increases in the mid-Arenig (*O.evae* and *O. intermedius* conodont zones), recording the first appearance of nonpustulose *Platystrophia* like those of coeval beds of the Baltic and New World Island, Newfoundland (Neuman, 1976). Associated brachiopods of Laurentian type include *Leptella (Petroria), Leptella (Leptella)* and *Cuparius*. The genus *Niquivilia* was probably also present in coeval beds of Tasmania (Benedetto and Sánchez, 1996a). The genus *Acanthotoechia* has been reported only from the Tourmakeady Limestone of Ireland (Williams and Curry, 1985). *Monorthis*, a Celtic assemblage genus, appears in the late Arenig part of the formation (*Monorthis* zone), together with the Laurentian *Oligorthis* and the cosmopolitan *Aporthohyla* (Table 2).

The brachiopod assemblage of the upper San Juan Limestone (*Ahtiella* zone) is characterized by an increasingly diverse

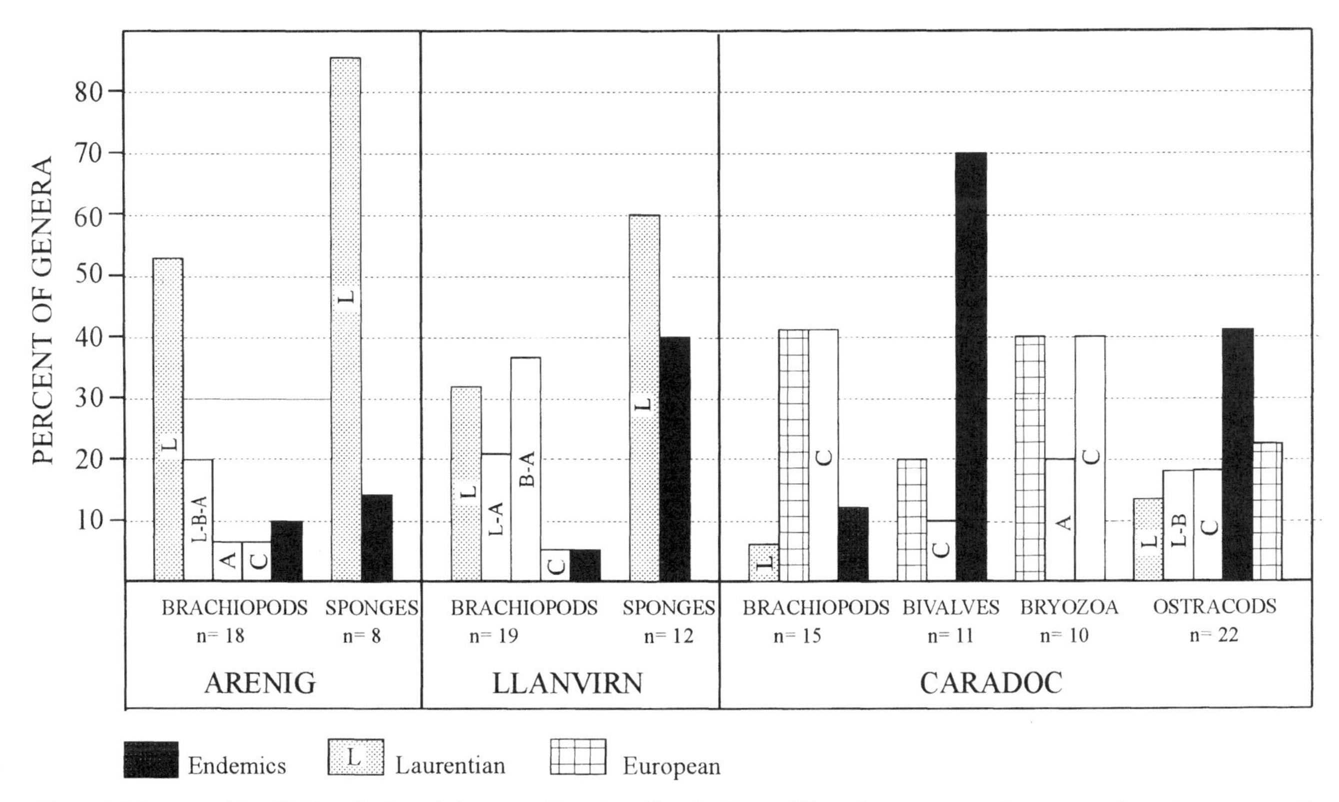

Figure 3. Biogeographic affinities of selected elements of Arenig to Caradoc Precordilleran faunas expressed as percent of taxa in common with selected geographic areas. European indicates "Mediterranean" (Bohemia, southern Europe, and North Africa) and Baltic genera. A = Avalonia; C = cosmopolitan genera; L = Laurentia; L-A = genera common to Laurentia and Avalonia; L-B = genera common to Laurentia and Baltica; L-B-A = genera common to Laurentia, Baltica, and Avalonia.

TABLE 1. ARENIGIAN GENERA OF THE PRECORDILLERA, FAMATINA, AND CENTRAL ANDEAN BASIN, AND OTHER SELECTED AREAS (SPONGES, BIVALVES, ROSTROCONCHIA)

Taxa	Regions									
	Precordillera	Famatina Basin	Central Andean Basin	Laurentia	Baltica	Avalonia	Armorica	Perunica	North Africa	Australia
Sponges										
Allosacus	X			X						
Archaeoscyphia	X			X						
*Chilcaia**	X									
*Calicocoelia**	X			X						
Eospongia	X			X						
Hudsonospongia	X			X						
*Nexospongia**	X									
Rhopalocoelia	X			X						
Bivalves										
*Catamarcaia**		X								
Goniophorina		X	X	X	X					
Modiolopsis			X	X	X					
*Natasia**			X							
Redonia		X	X				X		X	
*Suria**		X								
Rostroconchids										
Ribeiria	X	X	X							

*Endemic taxa.
Note: Data from: Carrera, 1985, 1994a, b, 1996a, b; Gutierrez-Marco and Aceñolaza, 1991; Harrington, 1938; Sánchez and Babin, 1997a, 1998b; Sánchez and Babin, 1993, 1994.

mixture of Celtic, Baltic, and Laurentian genera, together with a few that are endemic and several that are cosmopolitan (Benedetto, 1985; Herrera and Benedetto, 1990; Benedetto et al., 1995). Distinctive forms are smooth-shelled species of *Platystrophia, Productorthis, Skenidioides, Ahtiella, Inversella* and *Rugostrophia,* genera known only from Avalonia and Baltica. Based on statistical analysis, Neuman and Harper (1992) considered the Precordillera association as a Celtic assemblage with strong Toquima–Table Head influences. Figure 3 shows graphically faunal changes spanning the Arenig–Llanvirn boundary.

In comparison with the coeval volcano-sedimentary rocks of the Famatina Range, Precordilleran faunas have in common the Celtic genus *Monorthis* and some widespread genera such as *Tritoechia, Paralenorthis,* and *Hesperonomia*. Note that the clitambonitacean *Tritoechia* has not yet been recorded from either the Central Andean basin or the other autochthonous Gondwanan basins. The Celtic affinities of the Famatina fauna have been noted in the work of several researchers (Benedetto, 1994; Benedetto and Sánchez, 1996a; Benedetto et al., 1995).

Arenig–early Llanvirn bivalves are unknown in the Precordillera. Among the Rostroconchia, the genus *Ribeiria* occurs in Arenig-age rocks in both the Precordillera and Famatina basins (Sánchez, 1998). The species from the San Juan Formation most closely resembles *R. compressa* Whitington from Vermont (USA) and is clearly different from the typically Gondwanan *R. spinosa* Babin and Branisa (1987) from the Acoite Formation of the Cordillera Oriental of northwestern Argentina and the Suri Formation of the Famatina Range. Llanvirn rostroconchids from the San Juan Formation include the endemic *Talacastella,* and *Tolmachovia,* a widespread genus known in Europe, Siberia, Australia, Tasmania, and northwestern Argentina (in Arenigian beds) (Tables 2 and 3).

The succession of trilobite assemblages in the Precordillera records the transition from Laurentian to peri-Gondwanan affinities (Vaccari, 1994, 1995). The assemblage of early Arenig age consists of genera of the warm-water Bathyurid province such as *Uromystrum*, restricted to Laurentia, and *Peltabelia*, recorded in North America, Siberia, and northeastern China. Lower Llanvirnian limestone have yielded some widespread taxa of the Illaenid-Cheirurid biofacies and the endemic genera *Waisfeldaspis* and *Mendolaspis* (Vaccari, 1995).

Trilobites from the coeval clastic rocks (Suri Formation) of the Famatina basin belong to the peri-Gondwanan *Neseuretus* fauna (Vaccari et al., 1993). Vaccari (1995) showed that Famatinian trilobites exhibit a mixture of eastern (*Hungioides, Gogoella*) and western Gondwana (*Illaenopsis, Merlinia*) genera, the last two suggesting a faunal exchange with the tropical *Asaphopsis* province of Australia. Vaccari concluded that such a distribution pattern cannot be explained if the Precordillera was adjacent to Famatina during the Arenig.

Because early and middle Arenig graptolites are unknown in the Precordillera basin, no comparisons can be made with other

TABLE 2. ARENIGIAN BRACHIOPODS OF THE PRECORDILLERA, FAMATINA, CENTRAL ANDEAN BASIN, AND OTHER SELECTED AREAS

	Regions									
Taxa	Precordillera	Famatina Basin	Central Andean Basin	Laurentia	Baltica	Avalonia	Armorica	Perunica	North Africa	Australia
Brachiopods										
*Acanthotoechia**	X			X						
*Aporthophyla**	X			X			X			X
Archaeorthis	X			X						X
Camerella		X	X	X						
*Cuparius**	X			X						
Desmorthis			X							
Euorthisina			X			X		X	X	
Famatinorthis		X				X				
Glyptorthis			X			X				
Hesperonomia	X		X	X			X			
Hesperonomiella		X	X	X						X
Huacoella†	X									
Incorthis		X	X					X	X	
Leptella (Leptella)	X			X						X
Leptella (Petroria)	X			X						
Monorthis	X	X				X				
*Nanorthis**	X			X						X
Niquivilia†	X									
*Notorthis**	X			X	X					
*Oligorthis**	X			X						
Orthidium	X			X		X				
Paralenorthis	X	X	X	X	X	X	X			
Platystrophia	X				X	?				
Skenidioides		X				X				
*Syntrophia**	X			X						?
Tritoechia	X	X		X		X				X

*Unpublished data.
†Endemic taxa.
Note: Data from: Benedetto, 1987a, 1994, 1997a; Benedetto and Herrera, 1993; Herrera and Benedetto, 1989, 1990.

regions. Graptolite genera of this age from both Famatina and northwestern Argentina have mixed Baltic-Atlantic-Pacific affinities, suggesting that these regions could have been located in the transitional zone of intermediate latitudes (Toro and Brussa, 1997; Cooper et al., 1991). The distribution of taxa indicates that the majority of them consists of pandemic forms from the epipelagic biotope *Pendeograptus fruticosus, Dichograptus octobrachiatus, Didymograptus (E.) similis, D. (s.l.) simulans, Tetragraptus reclinatus* cf. *reclinatus,* and *Phyllograptus* sp. Some species are common in Baltoscandia (*Baltograptus vacillans, B. kurki,* and *B.* cf. *geometricus*). Associated forms are the Pacific species *D. bifidus* and the Atlantic *B. deflexus.*

Graptolites from the late Arenig–age *Isograptus victoriae maximus, Oncograptus,* and *Paraglossograptus tentaculatus* zones indicate that faunas of the Precordillera belong to the diverse isograptid biofacies of the Pacific province. Principal component taxa of this Pacific province are *Holmograptus spinosus,* isograptids of the *Isograptus victoriae* group, and species of *Paraglossograptus, Oncograptus, Exigraptus, Pseudisograptus,* and *Zygograptus* (Ortega et al., 1993; Brussa, 1994, 1995, 1997a,c,d). During the early Llanvirn, some Atlantic province faunas appear, for example, *Hustedograptus* (Da2) and *Pseudoclimacograptus angulatus* (Da3-Da4) (Mitchell et al., 1997). Such exchange with Atlantic faunas could be explained if during the Llanvirn the Precordillera terrane drifted into higher latitudes than in the Arenig.

The main conclusion resulting from the above discussion and the data plotted in Figure 3 is that Laurentian faunal influence in the Precordillera decreased through the Arenig-Llanvirn time interval. Taking into account the total fauna, 77% of Arenigian genera were also recorded in different Laurentian localities, but this percentage reduced to 57% in the early Llanvirn. The reciprocal increase consists of Avalonian taxa (in particular, among brachiopods) and, to a lesser extent, from the Baltic region (brachiopods, trilobites), accompanied by an increase of endemic forms (Fig. 3). The appearance of endemism at familial level among the sponges, which constitute a very conservative group, is noteworthy.

One particularly interesting feature is that, although the Precordillera received Laurentian immigrants, Precordilleran

TABLE 3. LOWER LLANVIRNIAN GENERA OF THE PRECORDILLERA AND RECORDS IN OTHER SELECTED AREAS

Taxa	Regions					
	Precordillera	Laurentia	Baltica	Avalonia	Armorica	Australia
Brachiopods						
Ahtiella	X		X	X		
*Camerella**	X	X		?		
*Idiostrophia**	X	X				
Inversella (Reinversella)	X			X		
Leptella (Petroria)	X	X				
Leptellina	X	X				X
*Orthidiella**	X	X				
Orthidium	X	X		X		
Paralenorthis	X	X	X	X	X	
Paurorthis	X		X	X	x	
Pharagmorthis	X	X				
Platystrophia	X		X	X		
Pomatotrema	X	X				
Productorthis	X		X	X		
*Rugostrophia**	X			X		
Sanjuanella†	X					
*Skenidioides**	X			X		
Taffia	X	X		?		
Tritoechia	X	X		X		X
Sponges						
Allosacus	X	X				
Anthaspidella	X	X				
Aulocopium†	X					
Calicocoelia	X	X				
Chilcaia†	X					
Hudsonospongia	X	X				
Icarassospongia†	X					
Patellispongia	X	X				
Protachileum†	X					
Psarodictyum	X	X				
Rhopalocoelia	X	X				
Rugospongia†	X					
Talacastonia†	X					
Rostroconchids						
Talacastella†	X					
Tolmachovia•	X			X		X

*Unpublished data.
†Endemic taxa.
Note: Data from: Benedetto and Herrera, 1986, 1987a, b, 1993; Beresi and Rigby, 1993; Carrera, 1985, 1994a, b; Herrera and Benedetto, 1987, 1989, 1990; Sánchez, 1986a.

endemics or associated Celtic forms never came into Laurentia. Likewise, some Precordilleran forms, such as the North American pentamerid *Camerella*, colonized the Famatina basin and penetrated into Gondwana as far as the Bolivian basin, but at present there are no known taxa that followed the inverse route (Gondwana to Precordillera). *Monorthis* is perhaps another good example in that it has been recorded in Arenig beds of Precordillera and Famatina and in younger horizons of the northwestern basin of Argentina (upper member of Sepulturas Formation) (Benedetto, 1997a). We infer that this unidirectional migration reflects an anticlockwise gyre of surface oceanic currents comparable to the present-day zonal wind circulation pattern. This implies a westerly directed warm equatorial current and a cool eastward flowing current at about 30°–40° south latitude (Fig. 4).

The gyre of the equatorial warm current could have been responsible for the arrival from Australia of some taxa recorded from the western margin of Gondwana. A similar pattern was proposed (Jell et al., 1984) to explain the presence in Australia and northwestern Argentina of the gastropod *Peelerophon* and the

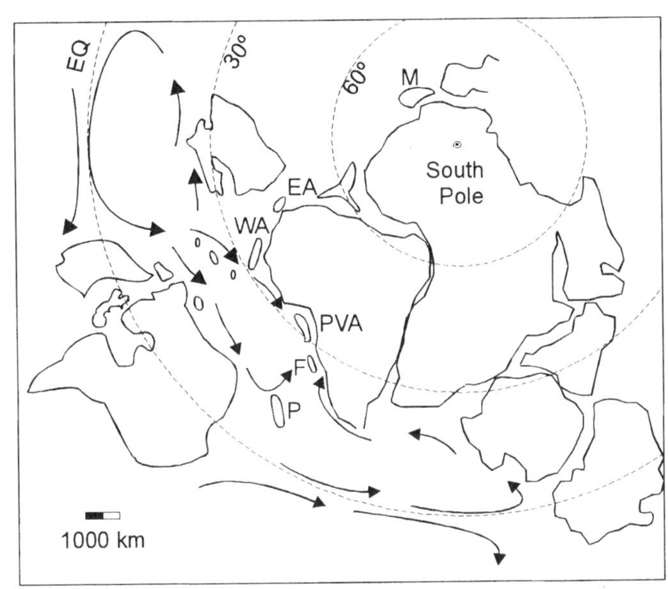

Figure 4. Simplified paleogeography of the early Ordovician (late Arenig–early Llanvirn). Arrows show inferred principal oceanic currents. EA = eastern Avalonia; F = Famatina arc system; M = Meguma terrane; P = Precordillera terrane; PVA = Puna volcanic arc; WA = western Avalonia; EQ = equator.

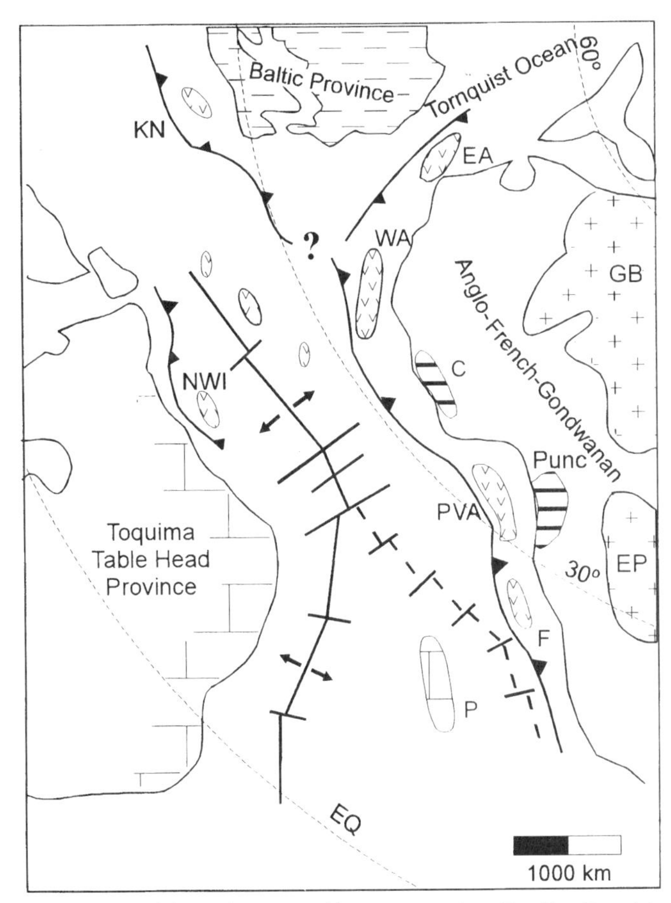

Figure 5. Arenigian paleogeographic reconstruction. C = Carolina slate belt; EA = eastern Avalonia; EP= eastern Pampean Range; F = Famatina arc system; G-B = Guyana-Brazilian craton; KN = Kola nappe; NWI = northwestern Ireland terrane (includes Tourmakeady Lms.); P = Precordillera terrane; Punc = Puncoviscana basin; WA = western Avalonia (after Benedetto, 1998c).

associated trilobite *Kayseraspis*. Such a current also might explain the presence of the Australian trilobite genera *Gogoella* and *Hungioides* in the Suri Formation of Famatina basin. Such a process may have been initiated earlier in the Ordovician, as is suggested by the occurrence of three early Tremadocian species common to northwestern Argentina and Australia (Webby, 1987). The absence of eastern Gondwanan trilobites (*Asaphopsis* province) in the Precordillera basin and their occurrence in the nearby Famatina basin implies that at this time the Precordillera and Famatina basins were not adjacent (Vaccari, 1995).

We assume that an important faunal exchange was established between waters of the Precordillera basin and those of the Iapetus ocean, particularly around islands inhabited by Celtic brachiopods. These islands, according to one reconstruction (Fig. 5) (Benedetto, 1998c) may have constituted a discontinuous peri-Gondwanan volcanic-arc system termed the Famatina-Puna-Avalonia volcanic-arc system (Benedetto, 1998c). The record of *Famatinorthis* in the Famatina basin and in coeval volcanic rocks from Maine (Neuman, 1997) is also consistent with an oceanic current flowing almost parallel to the peri-Gondwanan volcanic-arc system. The equatorial current bordering northeastern Laurentia (Fig. 4) could have been deflected by interposed islands and land masses within Iapetus, resulting in a complex circulation pattern.

Pre-accretion stage (Llanvirn-Caradoc)

This stage represents the final approach of the Precordillera terrane to the Gondwana margin. Its onset is marked by the first appearance in the Precordillera of typical Gondwanan taxa. Unlike the preceding stages, faunal evidence from early vertebrates, bivalves, and ostracods is significant. No sponges have yet been found. Intensive study of trilobite faunas was begun a few years ago; most of the information is still unpublished. Concerning the bryozoans (Table 5), we note that by the Early Caradoc (Las Aguaditas Formation) most of them are widespread forms known from coeval rocks in south-central Europe, Baltica, Siberia, Laurentia, and North Africa regions. However, 2 of the 10 genera recorded in the Precordillera are found only in Laurentia (Tuckey, 1990). Younger bryozoans occur in Lower Ashgill beds (Sassito Formation) of the central Precordillera. Similar-appearing faunas are found also in the Late Ordovician of North Africa (Lybia and Algeria), Spain, France, and Germany (N. Spjeldnaes, written com.).

Late Llanvirn brachiopods are very scarce because the stratigraphic record is incomplete, but 15 genera have been identified from the Early Caradoc Las Plantas and La Cantera Formations (Benedetto, 1995, 1998b). Of these, only *Campylorthis* is restricted to Laurentia. The remainder of the fauna includes a sig-

TABLE 4. CARADOCIAN GENERA OF THE PRECORDILLERA AND RECORDS IN OTHER SELECTED AREAS (BRACHIOPODS AND BIVALVES)

Taxa	Regions							
	Pre-cordillera	Central Andean Basin	Laurentia	Baltica	Avalonia	Mediter-ranean Region*	Australia	Other Areas
Brachiopods								
Aegiromena	X	X				X		
Anchoramena†	X							
Anoptambonites	X		X					X
Bicuspina	X				X	X		
Camerella	X		X					
Campylorthis	X		X					
Dinorthis	X	?	X		X			X
Drabovia	?				X	X		
Drabovinella		X			X	X		
Hesperorthis	X		X					X
Howellites	X				X	X		
Oanduporella	X	X		X	X			
Oepikoides†	X							
Ptychoglyptus	X		X	X	X			
S. (Sowerbyella)	X		X		X		X	X
Tissintia	X	X			X	X		
Bivalves								
Cadomia		X				X		
Cardiolaria		X				X\		
Concavodonta	X		?			X		
Concavoleda†	X							
Cuyopsis†	X							
Cycloconchid new gen†	X							
Emiliania†	X							
Hemiconcavodonta†	X							
Modiolopsis	X		X			X	X	
Protomya?	X		X					
Praenucula	X					X		
Trigonoconcha†	X							
Villicumia†	X							

*South-central Europe and North Africa.
†Endemic taxa.
Note: Data from: Benedetto, 1995, 1997b; Sánchez, 1986b, 1990, 1999.

nificantly high proportion (44%) of European realm genera, some of which have Anglo-Welsh affinities (Table 4). The presence of *Aegiromena, Tissintia,* and probably of *Drabovia*, indicates a strong link with the Berounian *Aegiromena-Drabovia* fauna that characterizes the Mediterranean province (Havlíček, 1989) and represents the first appearance of typically Gondwanan taxa in the Precordillera. Similar faunas are now known from numerous localities of South America. *Aegiromena corolla* Havlíček and *Tissintia canalifera* Havlíček, for instance, were recorded from Llanvirn(?) and early Caradoc beds of Bolivia, respectively (Havlíček and Branisa, 1980), and *Tissintia simplex* was reported from the Contaya Formation (late Llanvirn) of Peru (Hughes et al., 1980). Also significant is the presence in the Precordillera of two endemic plectambonitaceans, *Oepikoides* and *Anchoramena*. In a previous study (Benedetto and Sánchez, 1996a), we noted that the relatively high proportion of strophomenids and the record of genera restricted to temperate or warm waters (i.e., *Oanduporella, Campylorthis*) contrasts with the cold-water heterorthid-draboviid associations of Gondwana. The location of the Precordillera at a latitude lower than that of Central Andean basin, probably within the subtropical belt, is consistent with the relative abundance of carbonate rocks.

Early Caradoc bivalve faunas of the Precordillera (Table 4) are unusual not only by their relatively high diversity but also by their high level of endemicity. Of 12 described genera, 7 (58%) are endemics, 2 are European genera, and 3 are widespread forms (Sánchez, 1999).

Late Llanvirnian and Caradocian trilobites also show high levels of endemicity. Endemic genera are *Guandacolithus* from the Las Plantas Formation and *Hunickenolithus, Bancroftolithus,* and *Raphioampyx* from the Las Aguaditas Formation (Baldis and Pöthe de Baldis, 1995). The recently described species of *Ceratocara* and *Stenoblepharum* are closely related to Laurentian species (Chatterton et al., 1997; Edgecombe et al., 1997). *Incaia*, in contrast, is a typical Gondwanan taxa (Astini et al., 1996).

Caradoc ostracod fauna includes 37% of endemic genera. The remainder consists of a mixture of genera that show several affinities—Baltic, peri-Gondwanan, Laurentian, and Australian (Table 5).

All the graptolites from the *Nemagraptus gracilis* and *Climacograptus bicornis* zones, reported from different units of the Precordillera (Las Aguaditas, Las Plantas, Los Azules, Las Vacas Formations) are cosmopolitan (Ortega, 1995; Ortega and Brussa, 1990; Brussa, 1996, 1997b). Late Caradoc–early Ashgill rocks of Laurentia contain several endemic species and genera that have not been found in the Precordillera. Moreover, the diversity of coeval faunas from the Precordillera is less than that of Laurentia (Mitchell et al., 1997).

In summary, the biogeographic relationships of groups analyzed here have in common the following aspects: (1) almost all contain few genera related to Laurentian faunas; (2) all contain a varied number of typically Gondwanan genera, and (3) in all endemicity is unusually high. This pattern (Fig. 3) supports a model that implies that during the pre-accretion stage the Precordillera terrane gradually moved away from Laurentia and correspondingly approached to Gondwana. We infer that during the

TABLE 5. CARADOCIAN GENERA OF THE PRECORDILLERA AND RECORDS IN OTHER SELECTED AREAS (BRYOZOAN AND OSTRACODS)

Taxa	Regions					
	Precordillera	Laurentia	Baltica	Mediterranean Region*	Australia	Other Areas
Briozoans						
Amplexopora	X	X	X	X	X	X
Carinopylloporina	X	X				
Enallopora	X	X	X	X		
Hallopora	X	X	X			X
Nematopora	X	X	X	X		
Phyllodyctia	X	X	X			
Phylloporina	X	X	X	X		
Prasopora	X	X	X	X		X
Pseudostictoporella	X	X				
Ulrichostylus	X	X	X			X
Ostracods						
Acanthoscapha	X	X				
Baldiscella[†]	X					
Berdanellina	X		X			
Brevidorsa	X	?	X	X		
Cribobolbina[†]	X					
Cryptophylus	X	X	X			
Ectoprimitoides	X	X				
Garciana[†]	X					
Inversibolbina[†]	X					
Jachalipisthia[†]	X					
Kosuriscapha	X			X		
Lodesia[†]	X					
Medianella	X		X	X		
Ningulella	X	X	X			
Parenthatia	X	X				
Pilla	X				X	
Pseudulrichia	X	X	X	X		X
Revisyithere	X		X			
Sohniella[†]	X					
Steusloffina	X	X	X			
Trispinatia[†]	X					
Velapezoides	X		X		X	X

*South-central Europe and north Africa.
[†]Endemic taxa.
Note: Data from: Carrera, 1988; Schallreuter, 1996.

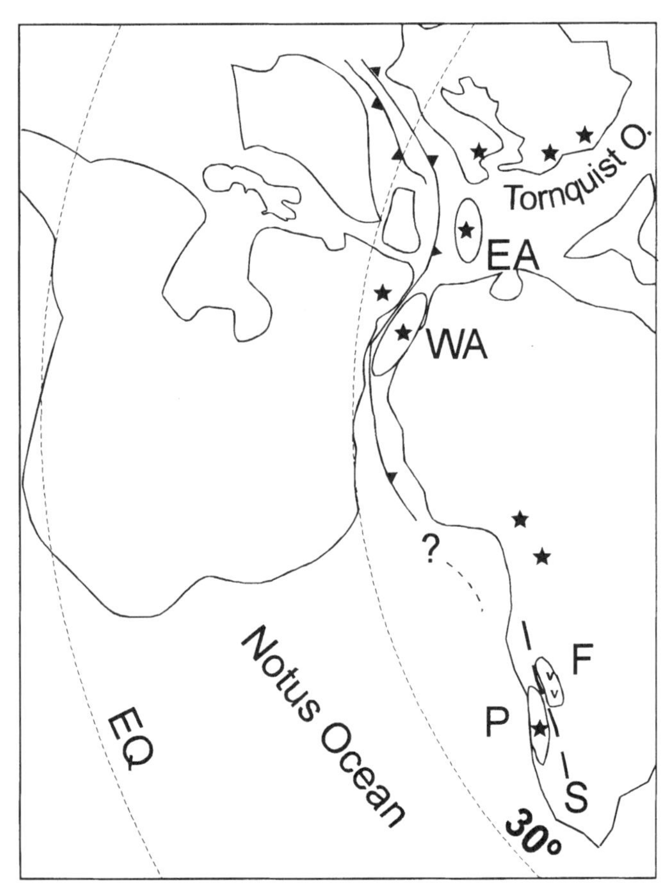

Figure 6. Paleogeographic reconstruction of the upper Ordovician (Caradoc-Ashgill). EA = eastern Avalonia; F = Famatina arc system; P = Precordillera terrane; S = suture; WA = western Avalonia; EQ = equator. Stars indicate location of *Hirnantia* fauna of the Kosov province (modified from Benedetto, 1998c).

pre-accretion stage the ocean between the Precordillera terrane and Gondwana was less a barrier to faunal exchange than it had been in Llanvirn time (Fig. 6). This is supported by the record of the jawless fish *Sacabambaspis* in mid-Ordovician rocks of the Precordillera and Bolivia (Gagnier et al., 1986; Albanesi et al., 1995b). Like other primitive vertebrates, this genus was unable to migrate across oceans, and was restricted to Gondwanan distribution (Gagnier et al., 1986).

Such a breakdown of faunal isolation, however, seems to be in contrast to the increasing levels of endemism notedly among the bivalves, trilobites, and ostracods, and to a lesser extent, among the brachiopods. The importance of isolated areas like island arcs in the origin of new endemic taxa has been long noted (cf. Neuman, 1984; Bruton and Harper, 1985; Webby, 1992). Thus, at first sight, the simultaneous appearance of several endemic genera in the Caradoc could be considered as strong evidence for a relative geographic (and therefore genetic) isolation of the Precordillera. Analysis of some biologic aspects of each group is necessary to clarify this biogeographic problem.

About half of the Precordilleran Caradocian ostracod genera appear to be endemic, the other half have congeners elsewhere; some are cosmopolitan (Table 5) and do not support a model of complete geographic isolation.

The bivalves are noteworthy because, although most are widely distributed genera, most of the Caradocian taxa of the Precordillera are endemic. Colonization of young islands is initiated by species that have larvae that are pelagic for various lengths of time. However, most benthic species of older islands have nonplanktotrophic larval stages (Jablonski and Lutz, 1980). The reduction of the duration of the planktonic larval stage could be advantageous because it reduces the loss of larvae from shallow-water habitats to deeper waters offshore. That five of the seven Caradocian endemic genera belong to the same family suggests that they arose from closely related predecessors that have acquired nonplanktonic larvae during their long history of isolation. The restricted geographic distribution of Precordilleran bivalves also might have been controlled, at least in part, by ecologic factors, but biofacies have not been studied enough to corroborate this possibility.

The biogeographic pattern of trilobites is similar to that of the other groups, but in addition to the relatively high number of endemics (six genera from a total of about 15 taxa) and Gondwanan forms, there are two taxa phylogenetically related to Laurentian species: *Ceratocara argentina* (Chatterton et al., 1997) and *Stenoblepharum astinii* (Edgecombe et al., 1997). *Ceratocara argentina* was a benthic form with a nonpelagic protaspid stage, and consequently it was unable to cross deep, wide oceans by mean of oceanic currents. On this basis, Chatterton et al. (1997) inferred a geographic proximity between Laurentia and the Precordillera, or at least some form of corridor between them during the Caradoc. Another possibility suggested by these authors is that both Laurentian and Precordilleran species differentiated earlier in the Ordovician from a common ancestor that inhabited the Precordillera terrane when it was closer to Laurentia. In light of the new paleogeographic hypothesis proposed recently by the senior author, in which eastern Laurentia is placed at or near the northwest corner of South America (Fig. 6), a migration route from Laurentia to Gondwana or vice versa could have been established along their continental margins during the early Caradoc (Benedetto, 1998c).

In summary, we consider that the strong endemism of the Precordilleran faunas may be due not only to a certain degree of geographic isolation, as is discussed below, but also to biologic factors related to the dispersal mechanisms and, to a lesser extent, environmental barriers. The presence of numerous Gondwanan taxa, seems better explained by means of a geographic proximity. Certainly, many more data will be needed to understand these complex biogeographic patterns.

Gondwana stage (Ashgill-Silurian)

During this stage, the benthic fauna of the Precordillera basin became indistinguishable from those of autochthonous Gondwanan basins. They include the *Dalamanitina* and *Hirnantia* fauna of late Ashgill (Hirnantian) age, recorded in the Don Braulio and Trapiche Formations of the Precordillera basin

(Benedetto, 1986, 1990; Benedetto and Herrera, 1987c; Astini and Benedetto, 1992; Sánchez et al, 1991; Sánchez, 1990), and in northwestern Argentina (Zapla Formation) (Monaldi and Boso, 1987) and Bolivia (Toro et al., 1991; Benedetto et al., 1992). The brachiopod assemblage from the Precordillera (Table 6) can be attributed to the subtropical to temperate Kosov province of Rong and Harper (1988). Plotting of the typical *Hirnantia* fauna localities on a global reconstruction (Fig. 6) shows that all of them fall within the temperate belt (more than 30° south latitude). The higher paleolatitude of Hirnantian assemblages from the Precordillera and Central Andean basins is based on their association with glaciogenic rocks (Peralta and Carter, 1990; Sánchez et al., 1991; Buggish and Astini, 1993).

After the latest Ordovician global extinction event, a cosmopolitan early Llandovery fauna developed (Sánchez et al., 1993). Brachiopod assemblages dominated by *Cryptothyrella* are present in both the La Chilca Formation of the Precordillera and the Salar del Rincón Formation of the Puna region (northwestern Argentina). Gondwanan Silurian faunas attain their distinctiveness through the Llandovery time, with the appearance of endemic taxa such as *Heterorthella* and *Anabaia* in the La Chilca Formation. Typical, low-diverse Afro–South American (= "Malvinokaffric") faunas are characteristic of the intracratonic Brazilian and Paraguayan basins, while in the marginal Precordilleran and Central Andean basins mixed Llandoverian brachiopod faunas consist of several North Silurian realm elements and a few endemics (Benedetto and Sánchez, 1996b). The *Clarkeia* fauna became widespread only in Early Wenlock and later time, when Precordilleran faunas attain their definitive Gondwanan character, indistinguishable from other South American faunas.

TABLE 6. ASHGILL FAUNA FROM THE PRECORDILLERA

Brachiopods
Anisopleurella cf. *A. gracilis* (Jones)
Cliftonia oxoplecioides Wright
Dalmanella testudinaria (Dalman)
Drabovia undulata Astini and Benedetto
Eostropheodonta aff. *E. hirnantensis* (M'Coy)
Hirnantia sagittifera (M'Coy)
Paromalomena polonica (Temple)
Plectatrypa sp.
Plectothyrella crassicosta (Dalman)
Reuschella sp.
Trematis sp.

Bivalves
Costaledopsis fuertensis Sánchez
Modiolopsis cuyana Sánchez
Palaeoneilo sp.
Whiteavesia sp.

Trilobites
Dalmanitina
Eohomalonotus

Note: Data from: Astini and Benedetto, 1992; Benedetto, 1986, 1987b, 1990; Benedetto and Herrera, 1987c; Sánchez, 1990.

GEOLOGIC EVIDENCE, BIOTIC CONSTRAINTS, AND PALEOGEOGRAPHIC HYPOTHESIS

There is general consensus that the counterpart of the Precambrian eastern Laurentian margin is the proto-Andean margin of Gondwana (Bond et al., 1984; Dalziel 1991, 1997 and references therein). This implies a rifting event that was followed by an ocean opening in late Proterozoic time leading to progressive separation of the two margins. The near-absence of Precambrian fossils prohibits paleontologic support of this hypothesis. Along the proto-Andean margin, widespread flysch-like facies, includ-

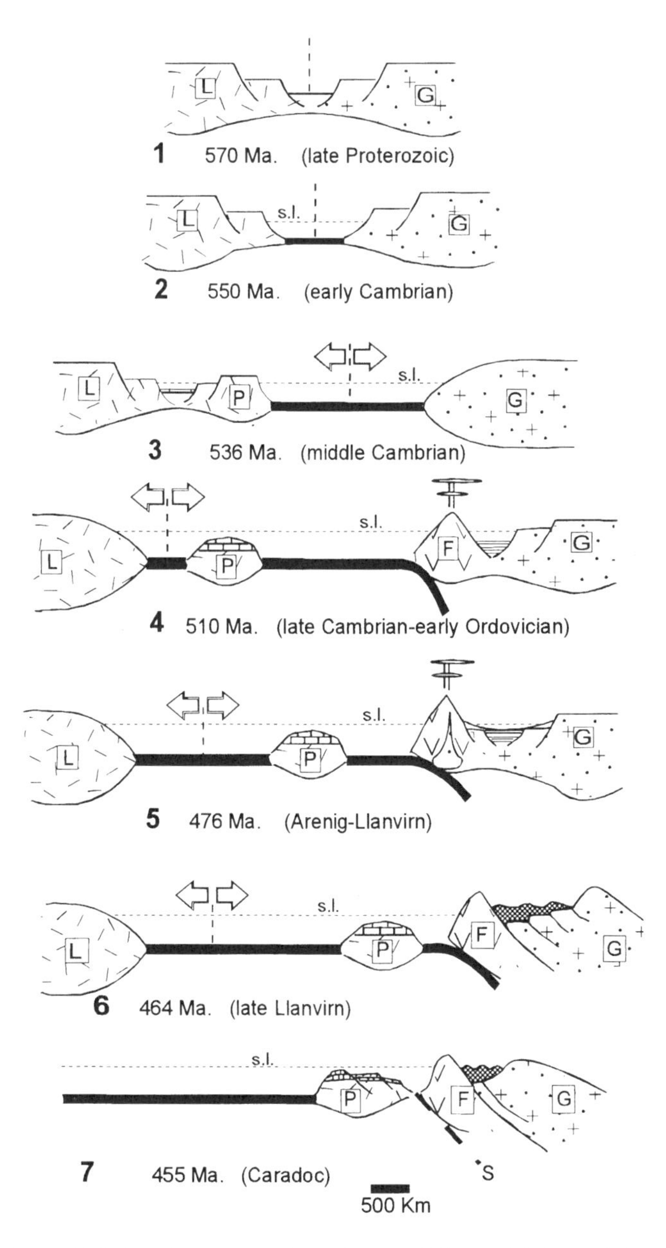

Figure 7. Schematic cross sections showing rifting and drifting of the Precordillera terrane, and the evolution of the proto-Andean margin from the late Proterozoic to Caradoc. L = Laurentia; G = Gondwana; F = Famatina arc system; P = Precordillera; s.l. = sea level.

ing *Oldhamia*-bearing slates of Vendian to Middle Cambrian age, and associated with passive-margin magmatic activity (Puncoviscana Formation), have been interpreted as evidence of this rifting event (Ramos et al., 1986). The nearly complete absence of early Cambrian platform carbonates in the proto-Andean basins suggests that at this time Gondwana was at a higher latitude than the Appalachian rifted margin.

During the Laurentia-Gondwana continental break-up and the subsequent drift phase, active crustal stretching along the eastern Laurentia plate might have caused an incipient rift system that culminated in the separation of a crustal block, the Precordillera or Cuyania terrane. Actualistic models show that the earliest rifting stage predates by some ten million years the effective continental separation, so that the initial crustal thinning could have occurred near the Precambrian/Cambrian boundary. The first evidence of a seaway interposed between the Precordillera terrane and Laurentia is the record of shallow-water sandstones, shales, and carbonates containing late Early Cambrian olenellid trilobites (Vaccari, 1988). These fossiliferous rocks underlain a succession of red beds and evaporites interpreted as syn-rift facies (Astini and Vaccari, 1996). These and other clastic-carbonate Cambrian rocks of the Argentine Precordillera may originally have deposited on the Appalachian margin and may represent postrift deposits related to the Laurentia-Gondwana break-up (Cañas, this volume). This suggests the presence of two sets of rifting deposits in the Precordillera basin, the older one (Cambrian) related to the initial rifting of Laurentia from Gondwana, and the younger one (Ordovician) reflecting the separation of the Precordillera microplate. It should be noted that none of them contradict the Cambrian biogeographic evidence.

We interpret the newly created margins to have established a shelf and slope along the western (present-day coordinates) side of the Precordillera (cf. Astini et al., 1995). Possible confirmation of its presence may be the deep-water hemipelagites and slope deposits that contain upper Middle Cambrian agnostid trilobites, which are preserved in huge olistoliths of Early Ordovician, east-derived olistostromes of the Los Sombreros Formation (Benedetto and Vaccari, 1992; Banchig and Bordonaro, 1994). The outer platform edge may have foundered along normal faults beginning in the Late Cambrian (Keller and Dickerson, 1996), but shallow shelf-lagoon and peritidal carbonate rocks of Early Ordovician (Tremadoc) age show the establishment of an isolated platform (Cañas, 1995b) that characterizes the end of the Laurentian stage when the Precordillera was separated from Laurentia by a narrow (~300–400 km) proto-Iapetus sea, which was not a barrier to migration of benthic organisms, not even for the shallow-water, low-dispersal species (Fig. 7).

Faunal changes that characterize the "isolation stage," which begin in the mid-Arenig, are synchronous with or slightly postdate a major change in the sedimentary regime represented by the transition from bahamitic carbonates to open marine deposits (San Juan Formation). This transgressive event is now thought to be related to a combination of regional extensional tectonics and a long-term eustatic rise (Cañas, 1995, this volume). Continued drifting gradually transported the Precordillera block to an intraoceanic position. If a conservative sea-floor spreading rate of 3 cm/yr is assumed, we estimate that near the Arenig/Llanvirn boundary it was separated from Laurentia by an ocean width greater than 1,500 km (Fig. 7). At this time, the extent of migrations appears to have been controlled by oceanic currents, which selectively transported several taxa onto the proto-Precordillera block from both Laurentia and the "northern" Iapetus volcanic islands and the Baltic plate (Fig. 4).

The continued sea-floor spreading probably was compensated by subduction of Iapetus oceanic crust beneath the Gondwana plate. As a result, subduction-related volcanic arcs and associated back-arc basins developed along the Gondwana margin (Mannheim, 1993). Taking into account that subduction progressively reduced the ocean located to the (present-day) east of the Precordillera, we infer that by the late Arenig it was nearer to the Famatina volcanic arc than to the Laurentian margin. K-Bentonites recorded throughout the Precordillera basin may represent volcanic ash derived from the coeval Famatina magmatic arc (Astini and Benedetto, 1996). Stratigraphic evidence from the Famatina basin shows that, by the end of the Arenig, sedimentation was interrupted and volcanic activity ceased. Until recently we interpreted this to have been result of the collision with the Precordillera terrane (Benedetto and Astini, 1993; Astini et al., 1995). This juxtaposition, however, seems to be contradicted by the faunal differences between coeval rocks of the Famatina and Precordillera successions. From the data reviewed here, a more likely possibility is that closing and deformation of the Famatina basin was not linked to the Precordillera accretion but may have resulted from collision of the island-arc complex with the Gondwanan continental margin. The Ordovician basin of northwest Argentina, in particular the Puna and Cordillera Oriental regions, had broadly similar sedimentary, magmatic, and tectonic history (Bahlburg et al., 1996). It appears that events affecting both basins are of regional extent, perhaps representing a collision of the peri-Gondwanan volcanic-arc system with an extinct mid-oceanic ridge (Fig. 5) or a change in the subduction rate.

If, as faunal evidence suggests, by the early Llanvirn the Precordillera terrane was still separated from Gondwana by a relatively narrow seaway (Figs. 5 and 7), then the drowning of the carbonate platform and the onset of widespread graptolitic black shales (Gualcamayo–Los Azules Formations) could have been related to a global sea-level rise combined with block-faulting of the platform in an extensional regime (Keller, 1995; Astini, 1995). This scenario, however, is also compatible with the gradual approach of the "eastern" Precordillera edge toward Gondwana, leading to the initial stages of a peripheral foreland and the upwarping of the platform edge (Astini et al., 1995; Astini, 1995).

The more striking change in the sedimentary regime of the Precordillera basin is reflected by the influx of east-derived turbidites, conglomerates, and olistostromes of late Llanvirn to early Caradoc age. Variations in thickness and facies indicate considerable topographic relief (Guandacol Orogeny), which has been related to crustal stretching (Keller, 1995) or a postcollisional relaxation (Astini et al., 1995). Paleobiogeographic evidence points to an increasing geographic connection between the autochthonous Gondwanan basins and the Precordillera since the late Llanvirn–early Caradoc ("pre-accretion stage"). From a geodynamic perspective, we believe that at this time the accretion of the Precordillera began, which is consistent with the intrusion in the Famatinian belt of large amounts of collisional granitoids near the Ordovician/Silurian boundary (Rapela et al., 1992; Toselli et al., 1993).

To account for the northward regional slope of the carbonate platform as well as the diachronism of early Ordovician black shales, we postulate an oblique collision of the Precordillera terrane with the Gondwana margin (Sánchez et al., 1996). If such a diachronous collision occurred, the history of the Precordillera (or Cuyania) terrane amalgamation should have been extremely complex, involving a progressive closure of the remanent ocean basin and repeated pulses of loading, normal faulting, subsidence, and sedimentation. Moreover, as Miall (1995) has stated, oblique collisions of small terranes with larger continents may lead to the development of peripheral basins dominated by strike-slip faulting. An oblique-collisional scenario of relatively isolated basins developed near the orogenic front, between or on top of faulted blocks, helps to explain, together the biologic aspects discussed above, the anomalously high endemism displayed by the Caradocian faunas. Although analysis of stratigraphic evidence supporting the oblique-collision model is beyond the scope of this chapter, it is important to note that it accounts for the large volume of latest Ordovician clastic deposits (Trapiche Formation) in the northern part of the basin, the subsequent migration of the depocenter to the south in the Silurian and the formation of a highly subsiding, block-faulted, strike-slip basin to the east (Rinconada–Mogotes Negros trough).

The onset of typically Gondwanan faunas during mid-Silurian times is the first conclusive evidence of the complete amalgamation of the Precordillera, even if compressional stresses could have continued during the entire Silurian and the early Devonian. However, sedimentary deposition across the suture zone did not begin until the early Carboniferous.

SOME OUTSTANDING QUESTIONS

Biogeographic patterns of benthic faunas strongly suggest that the Precordillera is a Laurentian-derived terrane that was accreted to the Gondwana margin during the early Paleozoic. Based on paleontological evidence we visualize a sequence of events that is generally consistent with geologic evidence. Important details, however, remain to be evaluated, particularly the specific site of the Precordillera on the Laurentian margin and the timing of its separation and accretion.

Current interpretations concerning the pre-rifting site of the Precordillera are based mainly on structural or stratigraphic comparisons (Thomas and Astini, 1996; Keller and Dickerson, 1996; Cañas, this volume). Paleontological comparisons are compatible but emphasize different relations; the late Cambrian trilobite faunas, for example, have greater affinities with those of the south-central Appalachians than with those from western United States (Vaccari, 1994; Astini et al., 1995); furthermore, Precordilleran earliest Ordovician brachipods are most similar to coeval Laurentian assemblages of western Newfoundland, perhaps because the stratigraphic (and paleontologic) record is most nearly complete there.

The other question concerns the timing of separation. According to Dalla Salda et al. (1992a,b), Appalachian and proto-Andean margins separated during a post-Taconic/Ocloyic (Late Ordovician) rift event. Dalziel et al. (1994, p.248), however, based on paleoclimatic indicators, suggested that ". . . a major ocean basin may have opened between the two continental masses after the Taconic-Ocloyic collision." Keller (1996) and Keller and Dickerson (1996) assumed a fully extensional scenario in which Llandeilo-Caradocian basic extrusions (pillow lavas) of western Precordillera tectofacies are interpreted as a result of rifting which marks the final separation of the Precordillera from Laurentia. Recently, Dalziel (1997) considered that the Precordillera may have been part of a plateau-like promontory located outboard the Ouachita embayment (the "Texas plateau") that was detached from Laurentia in a rift event ca. 455 Ma (mid-Caradoc), following its collision with Gondwana.

Differences between our microcontinental model and all of these hypotheses originate in the fact that they were based on nonpaleontologic criteria. Biogeographic analysis, however, is an excellent tool to discriminate continental separations and, as Fortey and Mellisch (1992) stated, certain fossils can "see" the existence of oceans of different magnitude that cannot be detected by other criteria. In this respect, evidence from sponges, bryozoans, brachiopods, and trilobites is consistent in pointing to a geographic separation between the Precordillera and Laurentia since the mid-late Arenig. Such a separation attains its maximum by the end of the Ordovician, when the accretion of Cuyania to Gondwana occurred. In an effort to make his "Texas plateau" hypothesis compatible with paleontological evidence, Dalziel (1997) explained the presence of deep-water trilobites in the Cambrian rocks of western Precordillera. This point, however, is not in dispute because in both the far-travelled plate and the Texas plateau scenarios a deep sea opened to the "west" of the Precordillera. The Dalziel 1997 model, although compatible with the trend toward increasing Gondwanan faunal affinities of the Precordillera seen through the Ordovician, does not account for the post-Arenig breakdown of Precordilleran faunal affinities with Laurentia. From an actualistic perspective, such faunal differentiation cannot be explained by a plateau model. Biogeographic studies in the possibly modern analog, the Malvinas plateau, show that the benthic fauna of the plateau and nearby islands (belonging to the Magellanic province) is identical to that of the adjacent Patagonian continental margin (Benedetto, 1998c).

Nevertheless, we accept that a Texas plateau-like configuration could have been achieved during the initial stages of the Precordillera terrane by crustal stretching (Fig. 7). However, the subsequent evolution in our model is substantially different. The opening of a true oceanic basin between them is largely inferred on biogeographic evidence. The well-preserved suite of Cambrian–Early Ordovician rift, slope, and deep basin facies in the Precordillera fit well with both, a rift basin developed on continental crust and a proto-oceanic basin that evolved in an ocean basin. The resolution of much of this geologic controversy depends on finding the pre-Caradoc ophiolites in the Precordillera. Such evidence, together with additional studies on the tectonic behavior (extensional vs. compressional), the structural response to the collision, and the inferred suture zone, will be needed to produce a more satisfactory model.

ACKNOWLEDGMENTS

We thank the volume editors, Victor Ramos and Duncan Keppie, for their invitation to participate in this volume. We are especially indebted to Robert Neuman, National Museum of Natural History, and S. H. Williams, Memorial University of Newfoundland, for their helpful comments and revision of the English translation. Paleontologic studies carried out in the last 5 years were made possible through support of the Consejo de Investigaciones Científicas y Tecnológicas de la Provincia de Córdoba (CONICOR), Secretaria de Ciencia y Técnica of the Córdoba University, and Consejo Nacional de Investigaciones Científicas y Técnicas (CONICET), Grant PMT-PICT 0029.

This is a contribution to International Geological Correlation Program Projects 376 and 410.

REFERENCES CITED

Abbruzi, J., Mahlburg, S., and Bickford, M. E., 1993, Implications for the nature of the Precordillera basement from Precambrian xenoliths in Miocene volcanic rocks, San Juan Province, Argentina: 12th Congreso Geológico Argentino and 2nd Congreso de Exploración de Hidrocarburos, v. 3, p. 331–339.

Aceñolaza, G. F., and Toselli, A. J., 1988, El sistema de Famatina; Su interpretación como orógeno de margen activo: 5th Congreso Geológico Chileno, Santiago, v. 1, p. 55–67.

Albanesi, G. L., and Vaccari, N. E., 1994, Conodontes del Arenig en la Formación Suri, Sistema del Famatina, Argentina: Revista Española de Micropaleontología, v. 26, p. 125–146.

Albanesi, G. L., Hünicken, M. A., and Ortega, G., 1995a, Review of Ordovician conodont-graptolite biostratigraphy of the Argentine Precordillera, *in* Cooper, J. D., Droser, M. L., and Finney, S. C., eds., Ordovician odyssey: SEPM, Pacific Section, Book 77, p. 31–36.

Albanesi, G. L., Benedetto, J. L., and Gagnier, P.-Y., 1995b, *Sacabambaspis janvieri* (Vertebrata) y conodontes del Llandeiliano temprano en la Formación La Cantera, Precordillera de San Juan, Argentina: Boletín de la Academia Nacional de Ciencias, t. 60, p. 519–543.

Astini, R. A., 1992, Tectofacies Ordovícicas y evolución de la cuenca Eopaleozoica de la Precordillera Argentina: Museo Nacional de Ciencias Naturales, Estudios Geológicos, v. 48, p. 315–327.

Astini, R. A., 1995, Geologic meaning of Arenig-Llanvirn diachronous black shales (Gualcamayo Aloformation) in the Argentine Precordillera, tectonic or eustatic?, *in* Cooper, J. D., Droser, M. L., and Finney, S. C., eds., Ordovician odyssey: SEPM, Pacific Section, Book 77, p. 217–220.

Astini, R. A., 1996a, Estratigrafía de secuencias del Paleozoico inferior en Precordillera. Secuencias depositacionales, secuencias genéticas o aloestratigrafía?: 6° Reunión Argentina de Sedimentología, p. 89–96.

Astini, R. A., 1996b, Las fases diastróficas del Paleozoico medio en la Precordillera del oeste argentino—evidencias estratigráficas: 13° Congreso Geológico Argentino y 3° Congreso de Exploración de Hidrocarburos, v. 5, p. 509–526.

Astini, R. A., and Benedetto, J. L., 1992, El Ashgilliano tardío (Hirnantiano) del Cerro La Chilca, Precordillera de San Juan, Argentina: Ameghiniana, v. 29, p. 249–264.

Astini, R. A., and Benedetto, J. L., 1996, Paleoenvironmental features and basin evolution of a complex volcanic arc region in the proto-Andean western Gondwana: 3rd International Symposium on Andean Geodynamics, St. Malo, p. 755–758.

Astini, R. A., and Cañas, F. L., 1995, La Formación Sassito, una nueva unidad calcárea en la Precordillera de San Juan: sedimentología y significado estratigráfico: Revista de la Asociación Argentina de Sedimentología, v. 2, p. 19–37.

Astini, R. A., and Vaccari, N. E., 1996, Sucesión evaporítica del Cámbrico inferior de la Precordillera: Significado geológico: Revista de la Asociación Geológica Argentina, v. 51, p. 97–106.

Astini, R. A., Benedetto, J. L., and Vaccari, N. E., 1995, The early Paleozoic evolution of the Argentine Precordillera as a Laurentian rifted, drifted, and collided terrane: a geodynamic model: Geological Society of America, Bulletin 107, p. 253–273.

Astini, R. A., Ramos, V., Benedetto, J. L., Cañas, F. L., and Vaccari, N. E., 1996, La Precordillera: Un terreno exótico a Gondwana: 13° Congreso Geológico Argentino y 3° Congreso de Exploración de Hidrocarburos, v. 5, p. 293–324.

Balburg, H., Pankhurst, R. J., Hervé, F., Goettert, M., and Zimmermann, U., 1996, The Ordovician basin in the southern Puna: new data on basin evolution and the Paleozoic terranes in NW-Argentina and N-Chile [abs.]: 13° Congreso Geológico Argentino y 3º Congreso de Exploración de Hidrocarburos, v. 5, p. 427.

Baldis, B. A., 1995, Ordovician trilobite zonation in western Argentina, *in* Cooper, J. D., Droser, M. L., and Finnry, S. C., eds., Ordovician odyssey: SEPM, Pacific Section, Book 77, p. 27–30.

Baldis, B. A., and Pöthe de Baldis, E. D., 1995, Trilobites Ordovícicos de la Formación Las Aguaditas (San Juan, Argentina) y consideraciones estratigráficas: Boletín de la Academia Nacional de Ciencias, Córdoba, t. 60, p. 409–448.

Baldis, B. A, Beresi, M., Bordonaro, O., and Vaca, A., 1982, Síntesis evolutiva de la Precordillera Argentina: 5° Congreso Latinoamericano de Geología, v. 4, p. 399–445.

Banchig, A. L., and Bordonaro, O., 1994, Reinterpretación de la Formación Los Sapitos: Secuencia olistostrómica de talud, Precordillera Argentina: 5 Reunión Argentina de Sedimentología, v. 2, p. 283–288.

Benedetto, J. L., 1985, Early Ordovician brachiopods of Argentine Precordillera and their palaeogeographic significance [abs.]: International Congress on Brachiopods I, Brest, p. 14.

Benedetto, J. L., 1986, The first typical *Hirnantia* fauna from South America (San Juan province, Argentine Precordillera), *in* Racheboeuf, P. R., and Emig, C. C., eds., Les brachiopodes fossiles et actuels: Biostratigraphie du Paléozoïque, v. 4, p. 439–447.

Benedetto, J. L., 1987a, Braquiópodos Clitambonitaceos de la Formación San Juan (Ordovícico temprano), Precordillera Argentina: Ameghiniana, v. 24, p. 95–108.

Benedetto, J. L., 1987b, Braquiópodos neo-Ordovícicos del flanco occidental del Cerro del Fuerte, Provincia de San Juan: Ameghiniana, v. 24, p. 169–174.

Benedetto, J. L., 1990, Los géneros *Cliftonia y Paromalomena* (Brachiopoda) en el Ashgilliano tardío de la Sierra de Villicum, Precordillera de San Juan, Argentina: Ameghiniana, v. 27, p. 151–159.

Benedetto, J. L., 1993, La hipótesis de la aloctonía de la Precordillera Argentina: Un test estratigráfico y biogeográfico: 12° Congreso Geológico Argentino and 2º Congreso de Exploración de Hidrocarburos, v. 3, p. 375–384.

Benedetto, J. L., 1994, Braquiópodos ordovícicos (Arenigiano) de la Formación Suri en la región del Río Chaschuil, Sistema de Famatina, Argentina: Ameghiniana, v. 31, p. 221–238.

Benedetto, J. L., 1995, La fauna de braquiópodos de la Formación Las Plantas (Ordovícico tardío, Caradoc), Precordillera Argentina: Revista Española de Paleontología, v. 10, p. 239–258.

Benedetto, J. L., 1998a, Early Ordovician (Arenig) brachiopods from the Acoite and Sepulturas formations, Cordillera Oriental, northwestern Argentina: Geologica et Palaeontologica, v. 32, p. 7–27.

Benedetto, J. L., 1998b, Braquiópodos caradocianos en los bloques de la diamictita glacigénica de la Formación Don Braulio (Ashgilliano), Sierra de Villicum, Precordillera Argentina: Ameghiniana, v. 35, p. 243–254.

Benedetto, J. L., 1998c, Early Palaeozoic brachiopods and associated shelly faunas from western Gondwana: its bearing on the geodynamic history of the pre-Andean margin, *in* Rapela, C. W., and Pankhurst, R., eds., The Proto-Andean margin of Gondwana: Geological Society of London, Special Publications, v. 142, p 57–83.

Benedetto, J. L., and Astini, R. A., 1993, A collisional model for the stratigraphic evolution of the Argentine Precordillera during the early Palaeozoic: International Symposium on Andean Geodynamics 2, Oxford, p. 501–504.

Benedetto, J. L., and Herrera, Z., 1986, Braquiópodos del Suborden Strophomenidina de la Formación San Juan (Ordovícico temprano), Precordillera Argentina: 4° Congreso Argentino de Paleontología y Bioestratigrafía, Mendoza, v. 1, p. 113–123.

Benedetto, J. L., and Herrera, Z., 1987a, *Sanjuanella*, un nuevo género de la Subfamilia Ahtiellinae (Brachiopoda, Plectambonitacea) del Ordovícico de la Precordillera Argentina: 4° Congreso Latinoamericano de Paleontología, Bolivia, v. 1, p. 97–109.

Benedetto, J. L., and Herrera, Z., 1987b, El género *Platystrophia* King (Brachiopoda) en la Formación San Juan de la Precordillera Argentina: Ameghiniana, v. 24, p. 51–59.

Benedetto, J. L., and Herrera, Z., 1987c, Primer hallazgo de braquiópodos y trilobites en la Formación Trapiche (Ordovícico tardío), Precordillera Argentina: 10° Congreso Geológico Argentino, Tucumán, v. 3, p. 73–76.

Benedetto, J. L., and Herrera, Z., 1993, New early Ordovician Leptellinae (Brachiopoda) from the San Juan Formation, west-central Argentina: Ameghiniana, v. 30, p. 39–58.

Benedetto, J. L. and Sánchez, T. M., 1996a, Paleogeography of brachiopod and molluscan faunas along the South American margin of Gondwana during the Ordovician, *in* Baldis, B., and Aceñolaza, G. F., eds., Early Paleozoic evolution of NW Gondwana: Serie Correlación Geológica, n. 12, p. 23–38.

Benedetto, J. L., and Sánchez, T. M., 1996b, The "Afro-South American Realm," and Silurian "Clarkeia Fauna," *in* Copper, P., and Jisuo Jin, eds., International Brachiopod Congess 3, Sudbury, Canada: Rotterdam, Netherlands, Balkema, p. 29–33.

Benedetto, J. L., and Vaccari, N. E., 1992, Significado estratigráfico y tectónico de los complejos de bloques cambro-ordovícicos resedimentados de la Precordillera Occidental Argentina: Museo Nacional de Ciencias Naturales, Estudios Geológicos, v. 48, p. 305–313.

Benedetto, J. L., Sánchez, T. M., and Brussa, E., 1992, Las cuencas silúricas de América Latina, *in* Gutierrez-Marco, J. C., Saavedra, J., and Rábano, I., eds., Paleozoico Inferior de Ibero-América: Mérida, Spain, Universidad de Extremadura, p. 119–148.

Benedetto, J. L., Vaccari, N. E., Carrera, M., and Sánchez, T. M., 1995, The evolution of faunal provincialism in the Argentine Precordillera during the Ordovician: new evidence and paleogeographic implications, *in* Cooper, J. D., Droser, M. L., and Finney, S. C., eds., Ordovician odyssey: SEPM, Pacific Section, Book 77, p. 181–184.

Beresi, M. S., and Rigby, J. K., 1993, The Lower Ordovician sponges of San Juan, Argentina: Brigham Young University Geology Studies, v. 39, p. 1–64.

Bergström, S. M., Huff, W. D., Kolata, D. R., Krekeler, M. P. S., Cingolani, C., and Astini, R. A., 1996, Lower and middle Ordovician K-bentonites in the Precordillera of Argentina: a progress report: 13° Congreso Geológico Argentino y 3er Congreso de Exploración de Hidrocarburos, v. 5, p. 481–490.

Bond, G. C., Nickeson, P. A., and Kominz, M. A., 1984, Breakup of a supercontinent between 625 Ma and 555 Ma: new evidence and implications for continental histories: Earth and Planetary Science Letters, v. 7, p. 325–345.

Bordonaro, O., 1980, El Cámbrico de la Quebrada de Zonda: Revista de la Asociación Geológica Argentina, v. 35, p. 26–40.

Bordonaro, O., 1992, El Cámbrico de Sudamérica, *in* Gutierrez Marco, J. C., Saavedra, J., and Rábano, I., eds., Paleozoico inferior de Iberoamérica: Mérida, Spain, Universidad de Extremadura, p. 69–84.

Bordonaro, O., and Liñán, E., 1994, Some Middle Cambrian agnostids from the Precordillera Argentina: Revista Española de Paleontología, v. 9, p. 105–114.

Bordonaro, O., and Peralta, S., 1987, El Arenigiano Inferior de la Formación Empozada en la localidad de San Isidro, Mendoza, Argentina: X Congreso Geológico Argentino, v. 3, p. 81–84.

Borrello, A. V., 1971, The Cambrian of South America, *in* Holland, C. H., ed., Cambrian of the New World: New York, Wiley-Interscience, p. 385–438.

Brussa, E. D., 1994, Las graptofaunas ordovícicas del sector central de la Precordillera Occidental sanjuanina, Argentina [Ph.D. thesis]: Córdoba, Argentina, Universidad Nacional de Córdoba, 323 p.

Brussa, E. D., 1995a, Preliminary analysis of the distribution of Early Ordovician graptolites from the Argentine Precordillera, *in* Cooper, J. D., Droser, M. L., and Finney, S. C., eds., Ordovician odyssey: SEPM, Pacific Section, Book 77, p. 185–188.

Brussa, E. D., 1995b, Upper Middle Ordovician–Lower Upper Ordovician graptolites from the western Precordillera, Argentina: Graptolites News, v. 8, p. 15–16.

Brussa, E. D., 1996, Las graptofaunas ordovícicas de la Formación Las Aguaditas, Precordillera de San Juan, Argentina. Parte I: Familias Thamnograptidae, Dichograptidae, Abrograptidae y Glossograptidae: Ameghiniana, v. 33, p. 421–434.

Brussa, E. D., 1997a, La Biozona de *Paraglossograptus tentaculatus* (Graptolithina) en la Formación Sierra de La Invernada, Precordillera Occidental, Argentina: Geobios, v. 30, p. 15–29.

Brussa, E. D., 1997b, Las graptofaunas ordovícicas de la Formación Las Aguaditas, Precordillera de San Juan, Argentina. Parte II: Familias Cryptograptidae, Dicranograptidae, Diplograptidae y Orthograptidae: Ameghiniana, v. 34, p. 93–105.

Brussa, E. D., 1997c, Graptolitos del Arenigiano tardío–Llanvirniano de la Formación Sierra de La Invernada, Precordillera Occidental sanjuanina, Argentina. Parte 1: Ameghiniana, v. 34, p. 357–372.

Brussa, E. D., 1997d, Graptolitos del Arenigiano tardío-Llanvirniano de la Formación Sierra de La Invernada, Precordillera Occidental sanjuanina, Argentina. Parte 2: Ameghiniana, v. 34, p. 373–383.

Bruton, D. L., and Harper, D. A. T., 1985, Early Ordovician (Arenig-Llanvirn) faunas from oceanic islands in the Appalachian-Caledonide orogen, *in* Gee, D. G., and Sturt, B. A., eds., The Caledonide orogen—Scandinavia and related areas: Chichester, United Kingdom, John Wiley, p. 359–368.

Buggish, W., and Astini, R. A., 1993, The Late Ordovician ice age: new evidence from the Argentine Precordillera, *in* Findlay, R. H., Unrug, R., Banks, M. R., and Veevers, J. J., eds., Gondwana Eight: Assembly, evolution and dispersal: Rotterdam, Netherlands, Balkema, p. 439–447.

Buggish, W., Von Gosen, W., Heinjes-Kunts, F., and Krumm, S., 1994, The age of early Paleozoic deformation and metamorphism in the Argentine Precordillera—evidence from K-Ar data: Zentralblatt für Geologie und Paläontologie, v. 1, p. 275–286.

Cañas, F. L., 1995a, Estratigrafía y evolución paleoambiental de las rocas carbonáticas cambro-ordovícicas de la Precordillera septentrional [Ph.D. thesis]: Córdoba, Argentina, Universidad Nacional de Córdoba, 300 p.

Cañas, F. L., 1995b, Early Ordovician Carbonate platform facies of the Argentine Precordillera: restricted shelf to open platform evolution, *in* Cooper J. D., Droser, M. L., and Finney, S. C., eds., Ordovician odyssey: SEPM, Pacific Section, Book 77, p. 221–224.

Cañas, F. L., and Carrera, M. G., 1993, Early Ordovician microbial-sponge-receptaculitid bioherms of the Precordillera, western Argentina: Facies, v. 29, p. 169–178.

Cañas, F. L., and Keller, M., 1993, Arrecifes y montículos arrecifales en la Formación San Juan (Precordillera Sanjuanina Argentina); Los arrecifes más antiguos de Sudamérica: Real Sociedad Española de Historia Natural (Geología), v. 88, p. 127–136.

Carrera, M. G., 1985, Descripción de algunos poríferos de la Formación San Juan (Ordovícico), [abs.]: Reunión de Comunicaciones Paleontológicas, Asociación Paleontológica, Delegación San Juan, p. 51–53.

Carrera, M. G., 1991, Los géneros *Selenoides* Owen y *Calathium* Billings (Receptaculitaceae) en el Ordovícico de la Precordillera de San Juan, Argentina: Ameghiniana, v. 28, p. 375–380.

Carrera, M. G., 1994a, Taxonomía, bioestratigrafía y significado paleoambiental de las faunas de poríferos y briozoos del Ordovícico de la Precordillera Argentina [Ph.D. thesis]: Córdoba, Argentina, Universidad Nacional de Córdoba, 197 p.

Carrera, M. G., 1994b, An Ordovician sponge fauna from San Juan Formation, Precordillera basin, western Argentina: Neues Jahrbüch für Geologie und Paläontologie (Abhandlungen), v. 191, p. 201–220.

Carrera, M. G., 1995, El género *Nicholsonella* (Bryozoa) en el Ordovícico de la Precordillera Argentina, su significado paleoecológico y paleobiogeográfico: Ameghiniana, v. 32, p. 181–190.

Carrera, M. G., 1996a, Ordovician megamorinid demosponges from San Juan Formation, Precordillera, western Argentina: Geobios, v. 29, p. 643–650.

Carrera, M. G., 1996b, Nuevos poríferos de la Formación San Juan (Ordovícico), Precordillera Argentina: Ameghiniana, v. 33, p. 335–342.

Chatterton, B. D. E., Edgecombe, G. D., Vaccari, N. E., and Waisfeld, B. G., 1997, Ontogeny and relationship of the Ordovician odontopleurid trilobite *Ceratocara*, with new species from Argentina and New York: Journal of Paleontology, v. 71, p. 108–125.

Cocks, L. R. M., and Fortey, R. A., 1990, Biogeography of Ordovician and Silurian faunas, *in* McKerrow, W. S., and Scotese, C. R., eds., Palaeozoic palaeogeography and biogeography: Geological Society of London Memoir 12, p. 97–194.

Cooper, R. A., Fortey, R., and Lindholm, K., 1991, Latitudinal and depth zonation of early Ordovician graptolites: Lethaia, v. 24, p. 199–218.

Curry, G. B., and Endo, K., 1991, Migration of brachiopod species in the North Atlantic in response to Holocene climatic change: Geology, v. 19, p. 1101–1103.

Dalla Salda, L. H., Cingolani, C. A., and Varela, R., 1992a, El orógeno colisional Paleozoico de Argentina: Serie Correlación Geológica, no. 9, p. 165–178.

Dalla Salda, L. H., Cingolani, C. A., and Varela, R., 1992b, Early Paleozoic orogenic belt of the Andes in southwestern South America: result of Laurentia-Gondwana collision?: Geology, v. 20, p. 617–620.

Dalziel, I. W. D., 1991, Pacific margins of Laurentia and East Antarctica–Australia as a conjugate rift pair: evidence and implication for an Eocambrian supercontinent: Geology, v. 19, p. 598–601.

Dalziel, I. W. D., 1992, On the organization of American plates in the Neoproterozoic and the breakout of Laurentia: Geology Today, v. 2, p. 237–241.

Dalziel, I. W. D., 1997, Neoproterozoic-Paleozoic geography and tectonics: Review, hypothesis, environmental speculation: Geological Society of America Bulletin, v. 109, p. 16–42.

Dalziel, I. W. D., Dalla Salda, L. H., and Gahagan, L. M., 1994, Paleozoic Laurentian-Gondwana interaction and the origin of the Appalachian-Andean mountain system: Geological Society of America Bulletin, v. 106, p. 243–252.

Edgecombe, G. D., Chatterton, B. D. E., Vaccari, N. E., and Waisfeld, B. G., 1997, Ontogeny of the proetoid trilobite *Stenoblepharum*, and relationships of a new species from the upper Ordovician of Argentina: Journal of Paleontology, v. 71, p. 419–433.

Fortey, R. A., 1975, Early Ordovician trilobite communities: Fossils and Strata, v. 4, p. 331–352.

Fortey, R. A., and Mellish, C. J. T., 1992, Are some fossils better than others for inferring palaeogeography? The early Ordovician of the North Atlantic region as an example: Terra Nova, v. 4, p. 210–216.

Gagnier, P. Y., Blieck, A. R. M., and Rodrigo S. G., 1986, First Ordovician vertebrate from South America: Geobios, v. 19, p. 629–634.

Gutierrez-Marco, J. C., and Aceñolaza, G. F., 1991, *Ribeiria y Tolmachovia* (Mollusca, Rostroconchia) en el Ordovícico inferior de la Cordillera Oriental Argentina: Zentralblatt für Geologie und Paläontologie, v. 1, p. 1799–1814.

Harrington, H. J., 1938, Sobre las faunas del Ordoviciano inferior del Norte Argentino: Revista del Museo de La Plata (Nueva Serie), 1, Sección Paleontología, p. 109–289.

Havlíček, V., 1989, Climatic changes and development of benthic communities through the Mediterranean Ordovician: Sbornik geologickych ved Geologie, v. 44, p. 79–110.

Havlíček, V., and Branisa, L., 1980, Ordovician brachiopods of Bolivia: Rozpravy Ceskoslovenske Akademie Ved, v. 90, p. 1–54.

Herrera, Z., and Benedetto, J. L., 1987, El género *Reinversella* (Brachiopoda) en el Ordovícico temprano de la Precordillera Argentina y su significado paleobiogeográfico: 10º Congreso Geológico Argentino, Tucumán, v. 3, p. 77–80.

Herrera, Z., and Benedetto, J. L., 1989, Braquiópodos del Suborden Orthidina de la Formación San Juan (Ordovícico temprano), en el área de Huaco–Cerro Viejo, Precordillera Argentina: Ameghiniana, v. 26, p. 3–22.

Herrera, Z., and Benedetto, J. L., 1990, Early Ordovician brachiopod faunas of the Precordillera basin, western Argentina: biostratigraphy and paleobiogeographical affinities, *in* MacKinnon, D. I., Lee, D. E., and Campbell, J. D., eds., Brachiopods through time. Rotterdam, Netherlands, Balkema, p. 283–301.

Hiller, N., 1994, The environment, biogeography, and origin of the southern African recent brachiopod fauna: Journal of Paleontology, v. 68, p. 776–786.

Huff, W. D., Bergström, S. M., Kolata, D. R., Cingolani C., and Davis, D. W., 1995, Middle ordovician K-bentonites discovered in the Precordillera of Argentina: Geochemical and paleogeographical implications, *in* Cooper, J. D., Droser, M. L., and Finney, S. C., eds., Ordovician odyssey: SEPM, Pacific Section, Book 77, p. 343–350.

Hughes, C. P., Rickards, R. B., and Williams, A., 1980, The Ordovician fauna from the Contaya Formation of eastern Perú: Geological Magazine, v. 117, p. 1–21.

Jablonski, D., and Lutz, R. A., 1980, Molluscan larval shell morphology. Ecological and paleontological applications, *in* Rhoads, D. C., and Lutz, R. A., eds., Skeletal growth of aquatic organisms, Chapter 9: New York, Plenum Publishing, p. 323–377.

Jell, P. A., Burret, C. F., Stait, B., and Yochelson, E. L., 1984, The Early Ordovician bellerophontoid *Peelerophon oehlerti* (Bergeron) from Argentina, Australia and Thailand: Alcheringa, v. 8, p. 169–176.

Jokiel, P. L., 1990, Long-distance dispersal by rafting: reemergence of an old hypothesis: Endeavour, n.s., v. 14, p. 66–73.

Keller, M., 1995, Continental slope deposits in the Argentine Precordillera: sediments and geotectonic significance, *in* Cooper, J. D., Droser, M. L., and Finney, S. C., eds., Ordovician odyssey: SEPM, Pacific Section, Book 77, p. 211–216.

Keller, M., 1996, Anatomy of the Precordillera (Argentina) during Cambro-Ordovician times: implications for the Laurentia-Gondwana transfer of the Cuyania terrane: 3rd International Symposium on Andean Geodynamics, S. Malo, p. 775–778.

Keller, M., and Dickerson, P. W., 1996, The missing continent of Llanoria—Was it the Argentine Precordillera?: 13° Congreso Geológico Argentino y 3° Congreso de Exploración de Hidrocarburos, v. 5, p. 355–367.

Keller, M., and Flügel, E., 1996, Early Ordovician reefs from Argentina Stromatoporoid vs Stromatolite origin: Facies, v. 34, p. 177–192.

Lehnert, O., 1993, Bioestratigrafía de los conodontes arenigianos de la Formación San Juan en la localidad de Niquivil (Precordillera Sanjuanina, Argentina) y su correlación intercontinental: Revista Española de Paleontología, v. 8, p. 153–164.

Lehnert, O., 1995, The Tremadoc/Arenig transition in the Argentine Precordillera, *in* Cooper, J. D., Droser, M. L., and Finney, S. C., eds., Ordovician odyssey: SEPM, Pacific Section, Book 77, p. 145–148.

Lehnert, O., Miller, J. F., and Repetski, J. E., 1997, Paleogeographic significance of *Clavohamulus hintzei* Miller (Conodonta) and other Ibexian conodonts in an early Paleozoic carbonate platform facies of the Argentine Precordillera: Geological Society of America Bulletin 109, p. 429–443.

Lindberg, D. E., 1991, Marine biotic interchange between the Northern and Southern Hemispheres: Paleobiology, v. 17, p. 308–324.

Maletz, J., and Mitchell, C. E., 1995, Atlantic versus Pacific province graptolite faunas in the Llanvirn; where is the big difference?: Graptolite News, v. 8, p. 43–45.

Mangano, M. G., and Buatois, L. B., 1994, Historia deposicional de las secuencias Ordovícicas marinas del Sistema de Famatina en el Noroeste de la Sierra de Narváez, Catamarca, Argentina: 4º Reunión Argentina de Sedimentología, v. 2, p. 215–222.

Mannheim, R., 1993, Geodynamic evolution of the Gondwana margin in western Argentina (Famatina System): International Symposium on Andean Geodynamics 2, Oxford, p. 531–534.

McKinney, F. K., and Jackson, B. C., 1991, Bryozoan evolution: Chicago, University of Chicago Press, 238 p.

Miall, A. D., 1995, Remnant ocean basins, in Busby, C. J., and Ingersoll, R.V., eds., Tectonics of sedimentary basins: Oxford, United Kingdom, Blackwell Science, p. 363–392.

Mitchell, C. E., Brussa, E. D., and Astini, R. A., 1997, Biogeography of Middle and Upper Ordovician graptolites, Precordilleran terrane, Argentina: Plate tectonic implications [abs.]: Geological Society of America Abstracts with Programs, v. 29, p. A379.

Monaldi, R., and Boso, M. A., 1987, *Dalmanitina (Dalmanitina) subandina* n. sp. (Trilobita) en la Formación Zapla del Norte Argentino: 4° Congreso Latinoamericano de Paleontología, Bolivia, v. 1, p. 149–158.

Neuman, R. B., 1976, Early Ordovician (late Arenig) brachiopods from Virgin Arm, New World Island, Newfoundland: Geological Survey of Canada Bulletin 261, p. 11–61.

Neuman, R. B., 1984, Geology and paleobiology of islands in the Iapetus ocean: review and implications: Geological Society of America Bulletin, v. 95, p. 1188–1201.

Neuman, R. B., 1997, *Famatinorthis* cf. *F. turneri* Levy and Nullo 1973 (Brachiopoda, Orthida) from the Shin Brook Formation (Ordovician, Arenig) in Maine: Journal of Paleontology, v. 71, p. 812–815.

Neuman, R. B., and Harper, D. A. T., 1992, Paleogeographic significance of Arenig–Llanvirn Toquina–Table Head and Celtic brachiopod assemblages, *in* Webby, B. D., and Laurie, J. R., eds., Global perspectives on Ordovician geology: Rotterdam, Netherlands, Balkema, p. 241–254.

Ortega, G., 1995, Graptolite zones of the Los Azules Formation (Middle Ordovician) from Precordillera, western Argentina: Graptolite News, v. 8, p. 57–59.

Ortega, G., and Brussa, E. D., 1990, La subzona de *Climacograptus bicornis* (Caradociano temprano) en la Formación Las Plantas en su localidad tipo, Precordillera de San Juan, Argentina: Ameghiniana, v. 27, p. 281–288.

Ortega, G., Toro, B., and Brussa, E. D., 1993, Las zonas de graptolitos de la Formación Gualcamayo (Arenigiano tardío–Llanvirniano temprano) en el norte de la Precordillera (Provincias de La Rioja y San Juan), Argentina: Revista Española de Paleontología, v. 8, p. 207–219.

Ortega, G., Albanesi, G., and Hünicken, M., 1995, Bioestratigrafía en base a conodontes y graptolitos de las formaciones San Juan (techo) y Gualcamayo (Arenigiano-Llanvirniano) en el cerro Potrerillo, Precordillera de San Juan, Argentina: Boletín de la Academia Nacional de Ciencias, Córdoba, v. 60, p. 317–364.

Palmer, A. R., Vaccari, N. E., Astini, R. A., Cañas, F. J., and Benedetto, J. L., 1996, Cambrian history of the Argentine Precordillera: adiós Laurentia!: Geological Society of America Abstracts with Programs, South-Central Section, v. 28, p. A57.

Peralta, S. H., and Carter, C., 1990, La glaciación gondwánica del Ordovícico tardío: evidencia en fangolitas guijarrosas de la Precordillera de San Juan, Argentina: XI Congreso Geológico Argentino, v. 2, p. 181–185.

Poulsen, V., 1958, Contribution to the middle Cambrian paleontology and stratigraphy of Argentina: Kongelige Danske Videnskabernes Selskab, Matematisk-fysiske Meddelelser, v. 31, p. 1–22.

Ramos, V. A., Jordan, T. E., Allmendinger, R. W., Mpodozis, C., Kay, S. M., Cortés, J. M., and Palma, M. A., 1986, Paleozoic terranes of the central Argentine–Chilean Andes: Tectonics, v. 5, p. 855–880.

Rapela, C. W., Coira, B., Toselli, A., and Saavedra, J., 1992, El magmatismo del Paleozoico Inferior en el sudoeste de Gondwana, *in* Gutierrez-Marco, J. L., Saavedra, J., and Rábano, I., eds., Paleozoico Inferior de Ibero-América, Mérida, Spain, Universidad de Extremadura, p. 21–68.

Rigby, J. K., Nitecki, M. H., Zhu Z., Liu B., and Jiang Y., 1995, Lower Ordovician reefs of Hubei, China, and the western United States, *in* Cooper, J., Droser, M. L. and Finney, S. C., eds., Ordovician odyssey: SEPM, Pacific Section, Book 77, p. 423–426.

Rong, J.-Y., and Harper, D. A. T., 1988, A global synthesis of the latest Ordovician Hirnantian brachiopod faunas: Transactions of the Royal Society of Edinburgh: Earth Sciences, v. 79, p. 383–402.

Rong, J.-Y., Boucot, A. J., Su, Y.-Z., and Strusz, D.L., 1995, Biogeographical analysis of Late Silurian brachiopod faunas, chiefly from Asia and Australia: Lethaia, v. 28, p. 39–60.

Sánchez, T. M., 1986a, La presencia de la Familia Eopteriidae (Mollusca, Rostroconchia) en la Formación San Juan (Ordovícico inferior de la Precordillera Argentina): 4° Congreso Argentino de Paleontología y Bioestratigrafía, Mendoza, v. 1, p. 123–131.

Sánchez, T. M., 1986b, Una fauna de bivalvos de la Formación Santa Gertrudis (Ordovícico) de la Provincia de Salta (Argentina): Ameghiniana, v. 23, p. 131–139.

Sánchez, T. M., 1990, Bivalvos del Ordovícico Medio-Tardío de la Precordillera de San Juan (Argentina): Ameghiniana, v. 27, p. 251–261.

Sánchez, T. M., 1997, Additional Mollusca (Bivalvia and Rostroconchia) from the Suri Formation, Early Ordovician (Arenig), western Argentina: Journal of Paleontology, v. 71, p. 1046–1054.

Sánchez, T. M., 1998, Rostroconchia (Mollusca, Diasoma) en la Formación San Juan (Ordovícico temprano) Precordillera Argentina: Ameghiniana, v. 34, p. 345–347.

Sánchez, T. M., 1999, New late Ordovician (Caradoc) bivalves from the Sierra de Villicum (Precordillera Argentina): Journal of Paleontology, v. 73, p. 66–76.

Sánchez, T. M., and Babin, C., 1993, Un insolite mollusque bivalve, *Catamarcaia* n.g. de l'Arenig (Ordovicien Inférieur) d'Argentine: Comptes Rendus de l'Académie des Sciences, v. 316, sér. II, p. 265–271.

Sánchez, T. M., and Babin, C., 1994, Los géneros *Redonia y Catamarcaia* (Mollusca, Bivalvia) de la Formación Suri (Ordovícico temprano, Oeste de Argentina) y su interés paleobiogeográfico: Revista Española de Paleontología, v. 9, p. 81–90.

Sánchez, T. M., Benedetto, J. L., and Brussa, E. D., 1991, Late Ordovician stratigraphy, paleoecology, and sea level changes in the Argentine Precordillera, *in* Barnes, C. R., and Williams, S. H., eds., Advances in Ordovician geology: Geological Survey of Canada Paper 90-9, p. 245–258.

Sánchez, T. M., Benedetto, J. L., and Astini, R. A., 1993, Eventos de recambio faunístico en las secuencias depositacionales del Paleozoico temprano de la Precordillera Argentina: 12 Congreso Geológico Argentino, v. 2, p. 281–288.

Sánchez, T. M., Carrera, M. G., and Benedetto, J. L., 1996, Variaciones faunísticas en el techo de la Formación San Juan (Ordovícico temprano, Precordillera Argentina): Significado paleoambiental: Ameghiniana, v. 33, p. 185–200.

Scotese, C. R., and McKerrow, W. S., 1991, Ordovician plate tectonics reconstructions, *in* Barnes, C. R., and Williams, S. H., eds., Advances in Ordovician geology: Geological Survey of Canada Paper 90-9, p. 271–282.

Schallreuter, R., 1996, Ordovizische Ostracoden Argentiniens II, Mitt. Geol.-Paläontologie Inst. Univ. Hamburg, v. 79, p. 139–169.

Schallreuter, R., and Siveter, D. J., 1985, Ostracodes across the Iapetus ocean. Palaeontology, v. 28, p. 577–598.

Scheltema, R. S., 1977, Dispersal of marine invertebrate organisms: paleobiogeographic and biostratigraphic implications, *in* Kauffman, E. G., and Hazel, J. E., eds., Concepts and methods of biostratigraphy: Stroudsburg, Pennsylvania, Dowden, Hutchinson, and Ross, p. 73–108.

Siveter, D. J., 1984, Habitats and mode of life of Silurian ostracodes: Palaeontological Association Special Papers in Palaeontology, v. 32, p. 71–85.

Taylor, P. D., and Cope, J. C., 1987, A trepostome bryozoan from the Lower Arenig of South Wales: implications of the oldest described bryozoan: Geological Magazine, v. 124, p. 367–371.

Thiede, J., 1974, Marine bivalves: distribution of mero-planktonic shell-bearing larvae in eastern North Atlantic surface waters: Palaeogeography, Palaeoclimatology, Palaeoecology, v. 15, p. 267–290.

Thomas, W. A., and Astini, R. A., 1996, The Argentine Precordillera: a traveler from the Ouachita embayment of North American Laurentia: Science, v. 273, p. 752–757.

Thorson, G., 1950, Reproductive and larval ecology of marine bottom invertebrates: Biological Review, v. 25, p. 1–25.

Toro, B., 1997, Asociaciones de graptolitos del Arenig de la localidad tipo de la Formación La Alumbrera, Sistema de Famatina, Argentina: Revista Española de Paleontología, v. 12, p. 43–51.

Toro, B., and Brussa, E. D., 1997, Nuevos hallazgos de graptolitos ordovícicos en la Puna Oriental, Argentina: Ameghiniana, v. 34, p. 126.

Toro, M., Varga, C., and Birhuet, R., 1991, Los trilobites ashgillianos de la Formación Cancañiri en el área de Milluni, Cordillera Real, Departamento de La Paz, Bolivia: 4° Reunión Internacional, Proyecto 271, IUGS, Resúmenes, p. 27.

Torsvik, T. H., and Trench, A., 1991, The Ordovician history of the Iapetus ocean in Britain: new palaeomagnetic constraints: Journal of the Geological Society of London, v. 148, p. 423–425.

Tuckey, M., 1990, Biogeography of Ordovician bryozoans: Palaeogeography, Palaeoclimatology, Palaeoecology, v. 77, p. 91–126.

Vaccari, N. E., 1988, Primer hallazgo de trilobites del Cámbrico inferior de la Provincia de La Rioja (Precordillera Septentrional): Revista de la Asociación Geológica Argentina, v. 43, p. 558–561.

Vaccari, N. E., 1994, Las faunas de trilobites de las sucesiones carbonáticas del Cámbrico y Ordovícico temprano de la Precordillera Septentrional, República Argentina [Ph.D. thesis]: Córdoba, Argentina, Universidad Nacional de Córdoba, 271 p.

Vaccari, N. E., 1995, Early Ordovician trilobite biogeography of Precordillera and Famatina, Western Argentina: preliminary results, *in* Cooper, J. D., Droser, M. L., and Finney, S. C.,eds., Ordovician odyssey: SEPM, Pacific Section, Book 77, p. 193–196.

Vaccari, N. E., and Bordonaro, O., 1993, Trilobites en los olistolitos cámbricos de la Formación Los Sombreros (Ordovícico), Precordillera de San Juan, Argentina: Ameghiniana, v. 30, p. 383–393.

Vaccari, N. E., and Waisfeld, B. G., 1994, Nuevos trilobites de la Formación Suri (Ordovícico temprano) en la región de Chaschuil, Provincia de Catamarca. Implicancias bioestratigráficas: Ameghiniana, v. 31, p. 73–86.

Vaccari, N. E., Benedetto, J. L., Waisfeld, B. G., and Sánchez, T. M., 1993, La fauna de *Neseuretus* en la Formación Suri (Oeste de Argentina): edad y relaciones paleobiogeográficas: Revista Española de Paleontología, v. 8, p. 185–190.

Valentine, J. W., 1971, Plate tectonics and shallow marine diversity and endemism, an actualistic model: Systematic Zoology, v. 20, p. 253–264.

Von Gosen, W., Buggisch, W., and Lehnert, O., 1994, Evolution of the Early Paleozoic mélange at the eastern margin of the Argentine Precordillera: Journal of South American Earth Sciences, v. 8, p. 405–424.

Webby, B. D., 1987, Biogeographic significance of some Ordovician faunas in relation to east Australian Tasmanide suspect terranes, *in* Leitch, E. C., and Scheibner, E., eds., Terrane accretion and orogenic belts: American Geophysical Union Geodynamics Series, v. 19, p. 103–116.

Webby, B. D., 1992, Ordovician island biotas: New South Wales record and global implications: Journal and Proceedings, Royal Society of New South Wales, no. 125, p. 51–77.

Wilkinson, C. R., 1987, Interocean differences in size and nutrition of coral reef sponge populations: Science, v. 236, p. 1654–1657.

Williams, A., 1973, Distribution of brachiopod assemblages in relation to Ordovician palaeogeography, *in* Hughes, N. F., ed., Organisms and continents through time: Palaeontological Association Special Papers in Palaeontology, v. 12, p. 241–269.

Williams, A., and Curry, G. B., 1985, Lower Ordovician Brachiopoda from the Tourmakeady Limestone, Co. Mayo, Ireland: British Museum (Natural History) Bulletin of Geology, v. 38, p. 183–269.

Manuscript Accepted by the Society September 4, 1998

Printed in U.S.A.

Geological Society of America
Special Paper 336
1999

Facies and sequences of the Late Cambrian–Early Ordovician carbonates of the Argentine Precordillera: A stratigraphic comparison with Laurentian platforms

Fernando L. Cañas
Departamento de Geología, Universidad Nacional de Río Cuarto, Ruta Nacional 36 Km 601, (5800) Río Cuarto, Argentina

ABSTRACT

The Cambrian to Early Ordovician platform carbonates of the Argentine Precordillera are part of the Cuyania terrane, accreted to Gondwana during the Ordovician. The interplay between thermo-tectonic driven subsidence and eustasy is reflected in changing platform styles that occurred within the different stages of the miogeocline evolution. Late Cambrian carbonates represent a regional peritidal shelf. That shelf was replaced by an extensive restricted carbonate platform close to the Cambrian-Ordovician boundary, which in turn evolved into an open shelf toward the Arenigian. Lithostratigraphy of the platform succession reflects these distinct platform configurations, developed within the mature stage of platform growth characterized by decaying rates of subsidence.

Detailed facies analysis allowed the recognition of third-order scale depositional sequences, providing a framework for comparing the Cambro-Ordovician sedimentary record of the Argentine Precordillera with "autochthonous" Laurentian sequences. Comparison reveals a parallel evolution of the Precordilleran and the Appalachian miogeocline successions until the Arenigian. This result, together with current paleontologic knowledge, suggests that the Argentine Precordillera was part of the early Paleozoic eastern margin of Laurentia prior to the detachment of Cuyania in the Early Ordovician.

INTRODUCTION

As in many other regions of the world, one of the most outstanding features of the early Paleozoic rocks exposed in the Argentine Precordillera (Fig. 1) is the development of a thick succession (2.0–2.5 km) of shallow water, Cambro-Ordovician platform carbonates. These rocks were recognized to have been deposited on a passive continental margin (e.g., González Bonorino, 1975; Baldis et al., 1982; Ramos et al., 1986), whereas their thickness was shown to fit with post-rifting, thermo-tectonic driven subsidence, comparable in rates and timing with that of most Laurentian miogeoclinal sequences (Bond et al., 1984). This fact, together with the presence of Laurentian olenellid trilobites in Cambrian rocks of the Argentine Precordillera (Borrello, 1965; Ross, 1975), led Bond et al. (1984) to suggest a probable Laurentian origin for the latter (cf. Ramos et al., 1984, 1986). Recent detailed paleontologic work has confirmed the allochthonous nature of the Argentine Precordillera relative to most of South America in the Cambro-Ordovician (Herrera and Benedetto, 1991; Benedetto, 1993; Vaccari, 1994; see Benedetto et al., this volume, for a review), now recognized as part of a larger terrane encompassing a

Cañas, F. L., 1999, Facies and sequences of the Late Cambrian–Early Ordovician carbonates of the Argentine Precordillera: A stratigraphic comparison with Laurentian platforms, *in* Ramos, V. A., and Keppie, J. D., eds., Laurentia-Gowana Connections before Pangea: Boulder, Colorado, Geological Society of America Special Paper 336.

Grenvillian basement exposed to the east, named the "Cuyania" terrane by Ramos (1995). Different attempts to explain how the Cuyania terrane reached its present location include: (1) a Middle Ordovician collision of eastern Laurentia and western Gondwana (e.g. Dalla Salda et al., 1992a,b; Dalziel et al., 1994), (2) a microcontinent detached from Laurentia in Early to Middle Cambrian times (e.g., Astini et al., 1995, 1996), and (3) a hypothetical promontory of Laurentia, the "Texas plateau" (Dalziel, 1997).

This chapter summarizes physical stratigraphic aspects of the Late Cambrian–Early Ordovician platform carbonates of the Argentine Precordillera. Detailed stratigraphy, sedimentology, and paleontology were integrated to reconstruct the evolution of the mature stage of this passive-margin platform succession, allowing a better comparison with the "autochthonous" Laurentian stratigraphy.

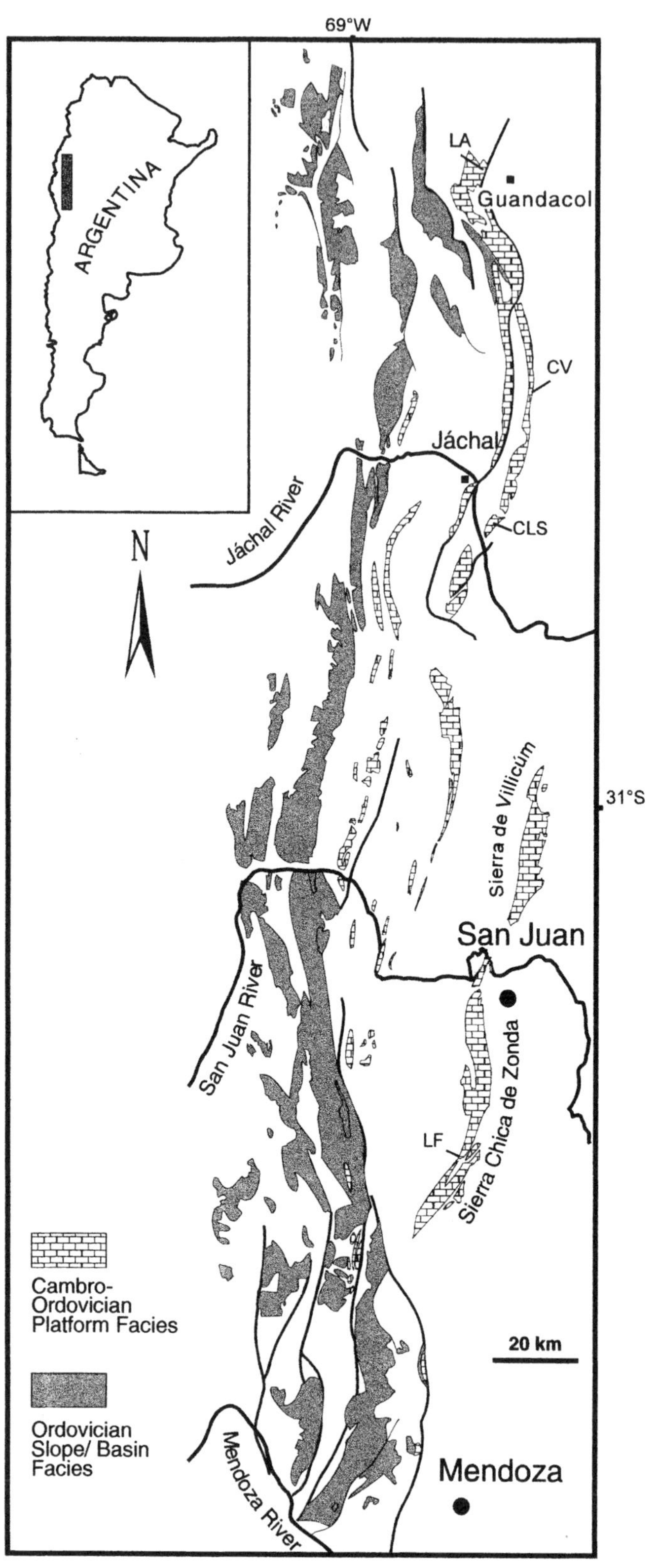

Figure 1. Geologic map of the Precordillera of western Argentina, illustrating the present distribution of the Cambrian and Lower Ordovician carbonate platform facies, and the Ordovician slope-to-basin facies. Sections mentioned in the text: LA = Quebrada La Angostura; CV = Cerro Viejo; CLS = Cerro La Silla; LF = Quebrada La Flecha.

STRATIGRAPHY, FACIES, AND ENVIRONMENTS OF LATE CAMBRIAN–EARLY ORDOVICIAN CARBONATES OF THE ARGENTINE PRECORDILLERA

The Late Cambrian–Early Ordovician succession is represented by the La Flecha, La Silla, and San Juan Formations (for a revision of the lithostratigraphy of these units, see Keller et al., 1994) (Fig. 2). These three lithostratigraphic units present different facies associations and internal architecture recognizable over all the basin, denoting different stages of the platform evolution. A major reorganization of the platform took place above each formation boundary, which are here taken as supersequence boundaries.

Regional peritidal carbonate shelf: La Flecha Formation

The La Flecha Formation (ca. 700 m) (Baldis et al., 1981) consists mostly of peritidal dolostones and limestones, with abundant chert (Fig. 3). A striking feature of the La Flecha is its diversity of cryptomicrobial sediments, ranging from laminites to small thrombolitic build-ups. The earliest beds assigned to the La Flecha Formation are Dresbachian (*Crepicephalus* zone), whereas the top is Trempealeauan (*Saukia* zone) (Vaccari, 1994; Keller et al., 1994), but no continuous section of the unit has been recorded. The Zonda Formation (Bordonaro, 1980) is a 300–400-m-thick succession of dolostones, mostly cryptomicrobial laminites and mud-cracked (dolo-) mudstones, which underlies the La Flecha Formation at its type section. There, a Steptoean (late Franconian) age was assigned to the base of the La Flecha Formation (Vaccari, 1994), implying a Marjuman age for the Zonda Formation. Thus, the Zonda Formation interfingers northward with the lower part of the La Flecha Formation (Keller et al., 1994).

Although lithologies as well as their vertical arrangement in different sections are varied, nine recurrent lithofacies are widely discernible (Glossary 1). Facies recurrence ("cyclicity") is an outstanding feature of these rocks (Cañas, 1995a).

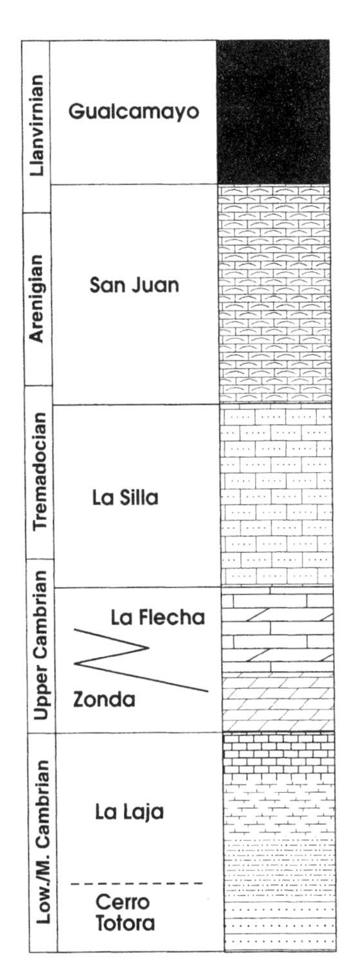

Figure 2. Sketch illustrating principal rock units in the Cambrian and Lower Ordovician platform succession of the Argentine Precordillera (after Keller et al., 1994).

Peritidal lithofacies are widespread, making up the Upper Cambrian sections in the Argentine Precordillera arranged as small-scale sequences. Most sequences are 1–6 m thick (Fig. 4), analogous to the "muddy," shallowing-upward sequences of James (1984). The base of each sequence is a subtidal unit that may differ in character. Low-energy, restricted subtidal settings are represented by the mudstone lithofacies, in which abnormal salinity is suggested by the low diversity and scarcity of fauna (trilobites and lingulid brachiopods). Indications of anoxic conditions such as bioturbation-free, bituminous, pyritic calcilutites are present in some cycles. Small (0.3–1.2 m), mound-shaped microbial bioherms (microbialites) were developed on this lithotope, probably profiting from the general environmental restriction. These lithofacies are interpreted to have been deposited in a relatively deep lagoonal environment.

The mudstone lithofacies may grade into or be replaced at the base of sequences by peloidal wacke- to packstones, mottled limestones or ribbon limestones, that represent shallow subtidal to intertidal settings. In this part of the sequence, the general decreasing-upward content of carbonate mud mixed with grains through burrowing, and the increase of preserved tractive sedimentary structures resembles that of modern subtidal carbonate sand flats. The ribbon limestones share many features in common with subtidal-intertidal, mixed lime sand-mud flats interpreted elsewhere from the rock record (e.g., Demicco, 1985; Pratt and James, 1986). Sequences are capped by thick laminite or cryptomicrobial laminites that usually show evidence of subaerial exposure.

Shoaling-up "grainy sequences" (James, 1984) are also present. In this case, the middle part of the sequence is composed of peloidal or ooid grainstones. These sequences may also present stromatolitic or laminite caps.

A different kind of sequence is developed in some sections, as in the type locality of the La Flecha Formation (Fig. 5), where most of the section is Franconian-Trempealeauan. There, mound- to fan-shaped thrombolitic bioherms (0.4–1.5 m) are a conspicuous feature of the succession (Baldis et al., 1981), and calcimicrobial limestones constitute an important part of the total thickness (Fig. 5). Keller et al. (1989) regarded the succession as a stacking of small-scale shallowing-upward cycles.

Depositional environments: The Upper Cambrian record of paleoenvironments of the Argentine Precordillera includes a variety of related lithotopes. A mosaic of low-energy muddy tidal flats, shoals, sand flats, and shallow subtidal settings covered most of the platform (Cañas, 1995a). To render a precise paleogeographic reconstruction of the Upper Cambrian platform is difficult because rocks of this age crop out in a relatively narrow belt, without clear indications of the actual polarity of platform during that time. Terrigenous clastic grains occur only as traces and are restricted to the lower and upper parts of third-order, transgressive-regressive sequences, indicating that the succession was formed far from any shoreline. Nevertheless, the presence of detrital quartz denotes that some connection with emerged land still existed during Late Cambrian time. Moreover, thick bank-margin shoal deposits or organic build-ups are also absent, suggesting that a shelf interior was the most likely setting for the La Flecha Formation. Many Upper Cambrian stacked peritidal carbonate successions have been described around North America, formed on extensive tidal flat complexes at the lee of the regional shelf's bank edges, frequently facing intrashelf basins (e.g,. Aitken, 1978; Mazzullo et al., 1978; Markello and Read, 1982; Demicco, 1985; Chow and James, 1987). All these examples share some common features, such as laminite-dominated supratidal environments, ribbon carbonates, thrombolites, and restricted subtidal facies. However, Chow and James (1987) pointed out important differences in the arrangement and distribution of lithologies in western North America, the northern Appalachians (western Newfoundland), and the southern Appalachians. Sections in western North America appear to lack the overall peritidal nature of the Appalachian successions, with frequent interfingering of platform-edge carbonate shoals and intrashelf basin siliciclastics (Aitken, 1966, 1978). A probable cause of this difference is the steepened ramp morphology of the Late Cambrian Cordilleran platforms (Osleger and Read, 1993).

Siliciclastic muds also constitute an important part of peritidal facies in the northern Appalachian Petit Jardin Formation (Chow and James, 1987). In this respect, the Upper Cambrian record of the Argentine Precordillera resembles more closely the thick peritidal seaward belt of carbonates developed in the southern Appalachians (Read, 1989). Small-scale facies sequences of the La Flecha Formation are remarkably similar to those illustrated by Markello and Read (1982) for the Honaker/Elbrook Formations, and by Demicco (1985) for the Conococheague Formation.

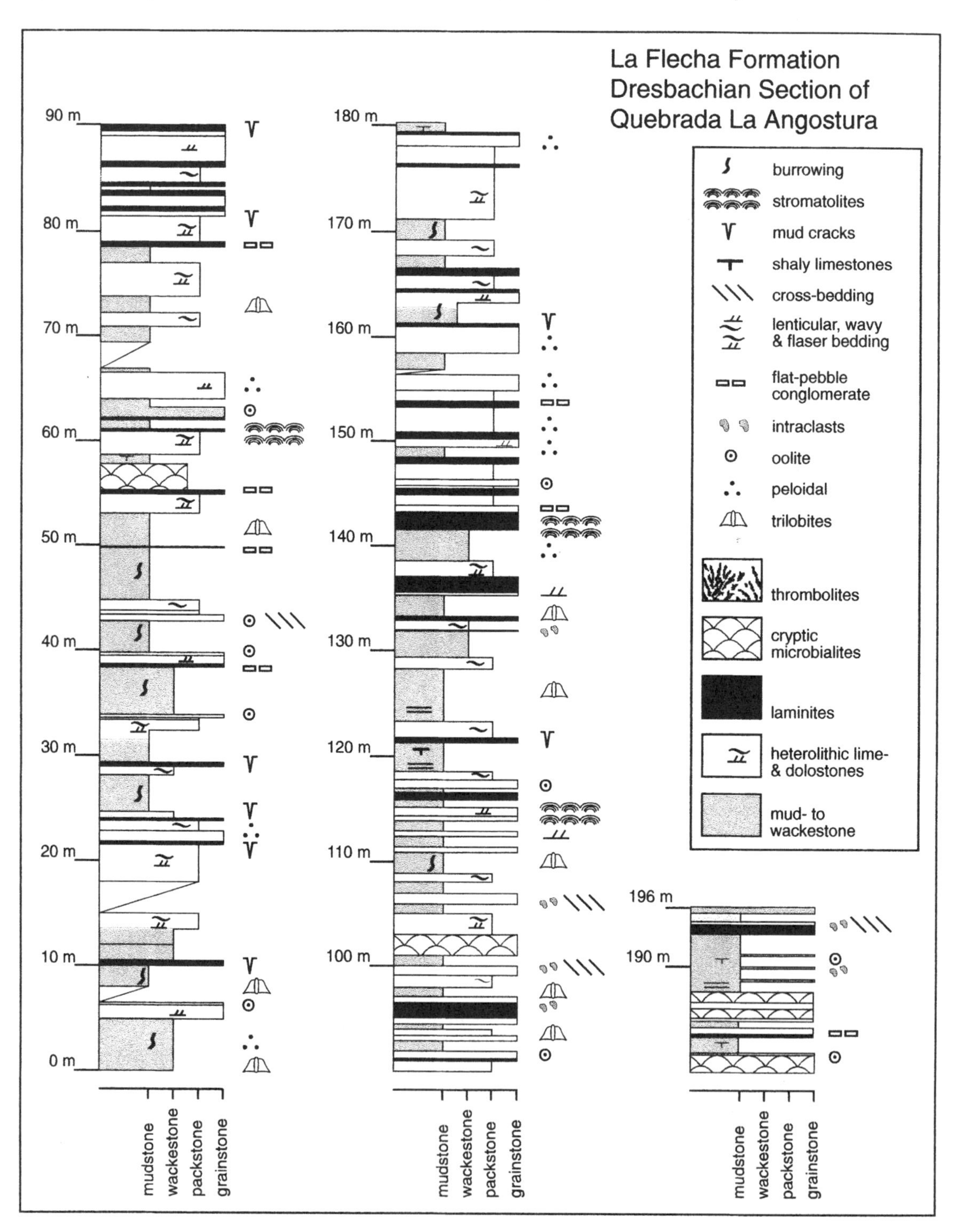

Figure 3. Detailed stratigraphic column of the lower portion (Dresbachian) of the La Flecha Formation at Quebrada La Angostura, west of Guandacol in Figure 1. Facies (see Glossary 1) arranged in small-scale peritidal sequences are representative of the Upper Cambrian in the northern Precordillera.

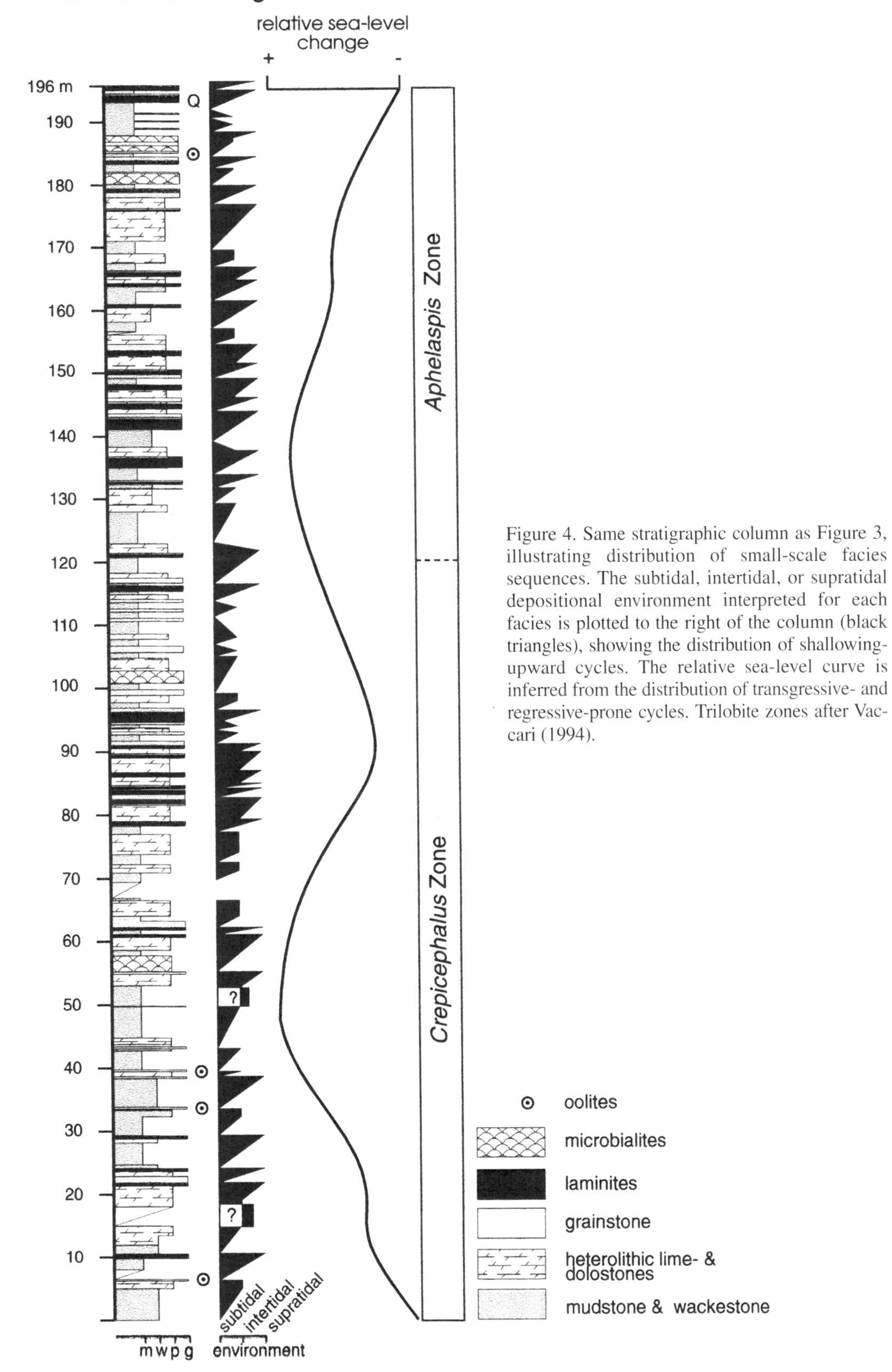

Figure 4. Same stratigraphic column as Figure 3, illustrating distribution of small-scale facies sequences. The subtidal, intertidal, or supratidal depositional environment interpreted for each facies is plotted to the right of the column (black triangles), showing the distribution of shallowing-upward cycles. The relative sea-level curve is inferred from the distribution of transgressive- and regressive-prone cycles. Trilobite zones after Vaccari (1994).

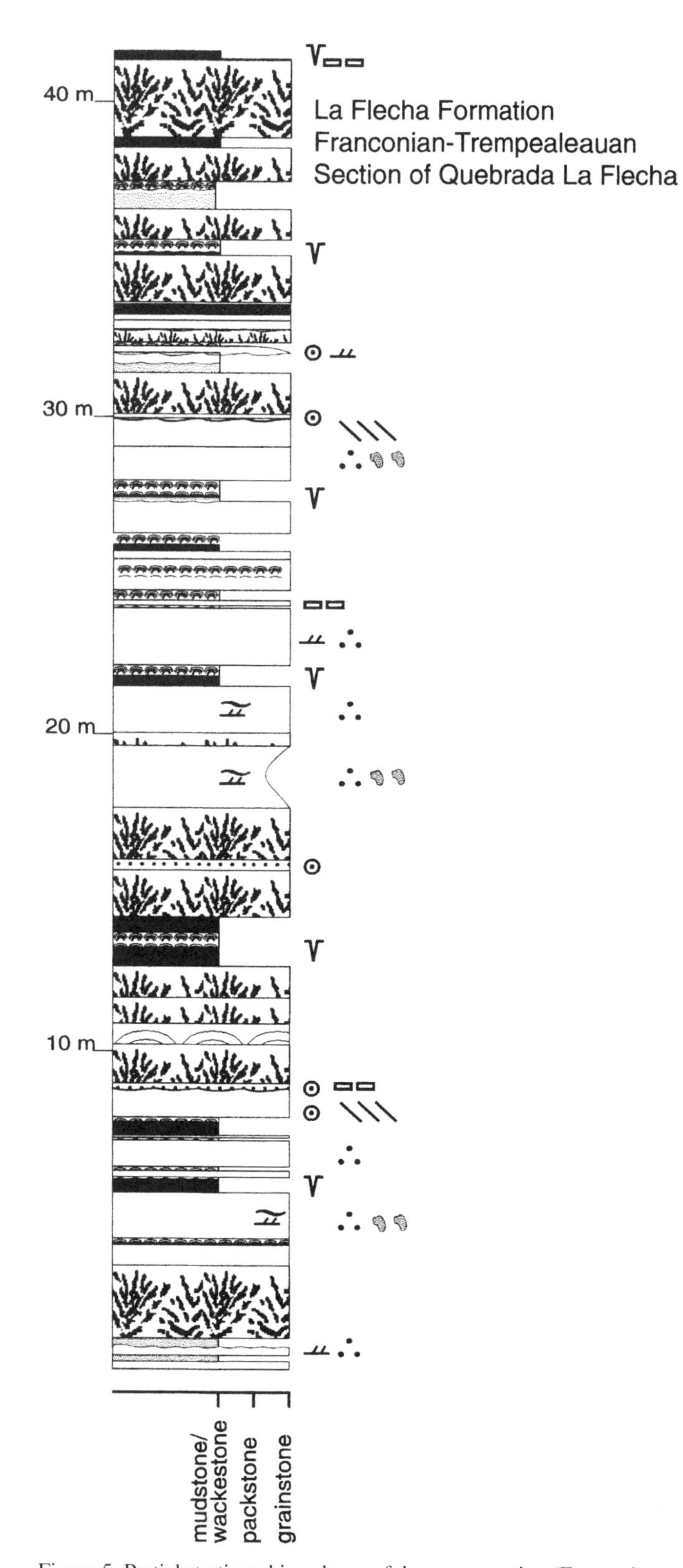

Figure 5. Partial stratigraphic column of the upper portion (Franconian-Trempeauleauan) of the La Flecha Formation at the type section of Quebrada La Flecha, Sierra Chica de Zonda, in Figure 1. Note the abundance of thrombolites and biolaminated deposits (cf. with Fig. 3). See Figure 3 for explanation of symbols.

Restricted carbonate platform: La Silla Formation

The La Silla Formation (300–400 m) (Keller et al., 1994) lies (para)conformably on the La Flecha Formation. The unit is composed chiefly of medium- to thick-bedded peloidal and ooid limestones and lime mudstones, with subordinate amounts of laminated and stromatolitic dolostones and chert (Fig. 6). The lower contact with the La Flecha Formation was placed at the lowest occurrence of thick-bedded limestones, which at the type section of Cerro La Silla are filling an erosive relief on top of the La Flecha dolostones. Sparse trilobites and conodonts (Keller et al., 1994) have proven an Ibexian age for the La Silla Formation, ranging from the *Missisquoia* zone into the *Paltodus deltifer* zone.

Five recurring lithofacies have been recognized in the La Silla Formation based on lithology, bed thickness, sedimentary structures, and faunal content (Glossary 2).

Peloidal and intraclast grain- to packstone, and peloidal wackestone constitute the dominant rock types in the formation. They are chiefly calcarenites (grain- to lime-mud supported), composed of peloids, carbonate "grains" (sensu Halley et al., 1983), aggregate grains, micritized ooids, and rounded intraclasts. Bioclasts are very scarce and include calcareous algae?, gastropods, trilobites, and nautiloid fragments. The bedding is thick to very thick, generally massive due to pervasive burrowing. Silicified *Thalassinoides*-like burrows are preserved in some beds. Tabular or trough cross-bedding is rare. Some layers present internally planar dissolution and/or erosion surfaces that might be interpreted as hardgrounds or as surfaces of subaerial exposure; it is difficult to assign a precise origin.

These lithofacies are analogous to the "oolitic and grapestone" and the "mud and pellet mud" lithofacies of modern Bahamian environments (Purdy, 1963; Bathurst, 1975), representing mostly platform interior to platform margin settings. Allochems present (peloids, micritic grains, lumps, and micritized ooids and bioclasts) are formed today in shallow subtidal to intertidal environments. A subtidal lithotope under near-normal marine conditions is indicated by the absence of evidence of subaerial exposure, ubiquitous burrowing, and fossil content.

The oolite lithofacies occurs interbedded with the previously described rock types, composed of normal and composite, radial-concentric ooids, reworked oolitic intraclasts, and rare bioclasts. Thick beds (0.3–1.0 m) with uni- or bidirectional, medium-scale (0.3–0.6 m) cross-bed sets formed by clean, well-sorted ooid grainstone have sharp bases. Erosive bases are present in medium beds (0.1–0.3 m) of oolitic-lithoclast lime conglomerates. Gradational contacts with underlying peloidal packstone and wackestone also are present where burrowing has mixed both lithofacies.

Figure 6. Stratigraphic column of the La Silla Formation at the type section of Cerro La Silla (see Fig. 1).

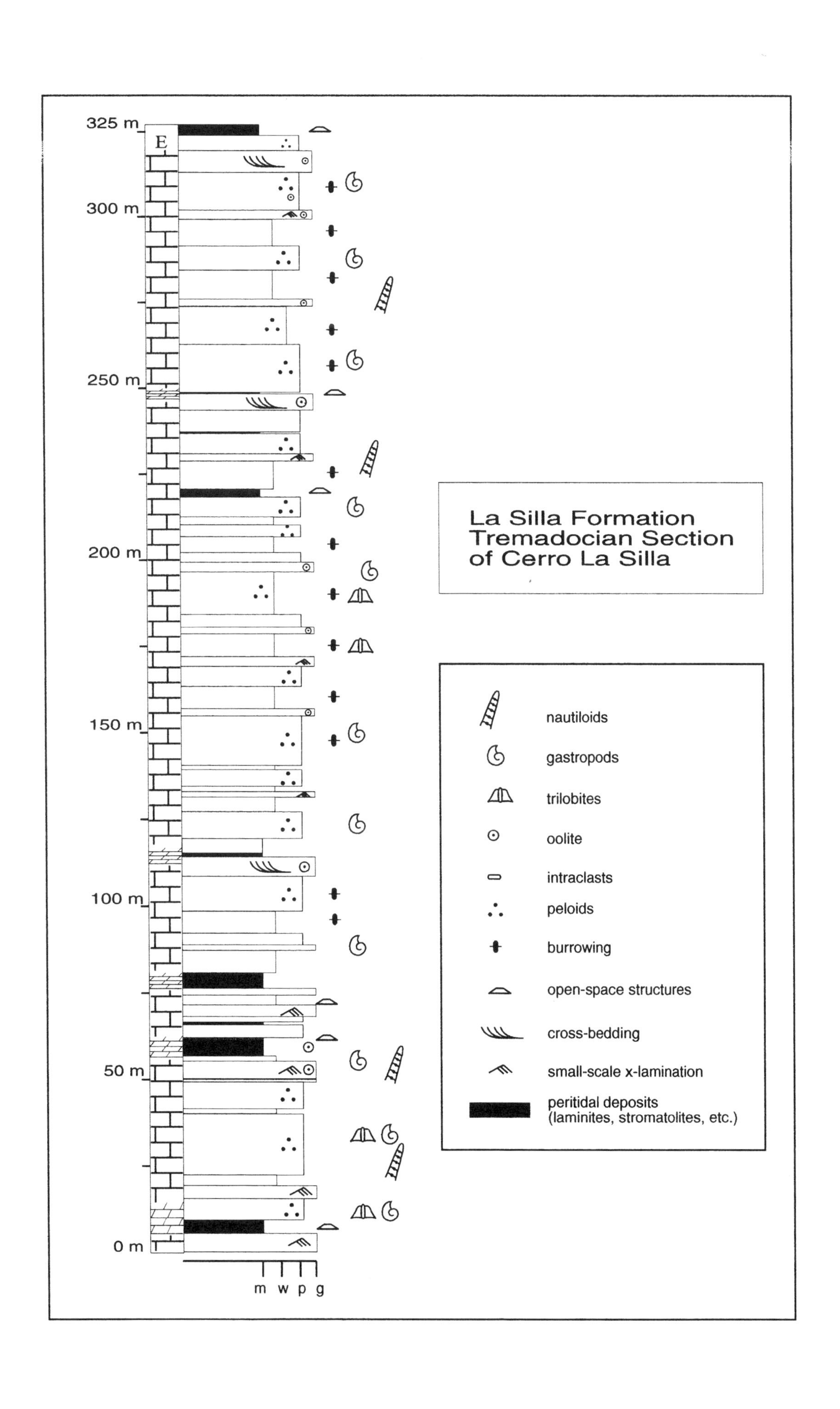

325 m
E
300 m
250 m
200 m
150 m
100 m
50 m
0 m
m
w
p
g
La Silla Formation
Tremadocian Section
of Cerro La Silla
nautiloids
gastropods
trilobites
oolite
intraclasts
peloids
burrowing
open-space structures
cross-bedding
small-scale x-lamination
peritidal deposits
(laminites, stromatolites, etc.)

The textural and compositional maturity (sensu Smosna, 1987) indicates moderate and persistent currents, whereas cross-bedding suggests relatively high-energy shallow subtidal conditions. "Shoaling" sequences up to 10 m thick were observed, in which burrowed peloidal-bioclastic wackestone/packstone with mixed ooids grades vertically into peloidal-oolitic grainstone/ packstone, which in turn is sharply covered by clean oolitic grainstone. This sequence is comparable to the one described by Harris (1979) in a modern Bahamian shoal complex, where peloids and ooids are churned through burrowing with muddy sediments of the platform interior in mixed sand-mud flats developed at the lee side of the shoals.

Erosive-based oolitic-lithoclasts seem to represent storm-wave reworking of hardgrounds, as suggested by edge-wise arranged flat clasts.

Cryptomicrobial laminites and stromatolites occur throughout the section, but are not frequent lithologies. Included in this lithofacies are fenestral peloidal mudstone, flat pebble and dissolution breccias, and irregular erosional surfaces accounting for subaerial exposure.

Although shallowing-upward sequences do occur in the La Silla Formation, the unit is predominantly composed of thick-bedded, shallow subtidal lithofacies, strongly contrasting with the cyclic appearance of the stacked peritidal facies of the underlying La Flecha Formation.

Depositional environments: Facies associated in the La Silla Formation represent different platform interior lithotopes, and probably platform-margin–related subenvironments (Cañas, 1995a,b). This is supported not only by the lithofacies that are comparable to modern and ancient "shelf-lagoon" sediments (Mazzullo and Friedman, 1975), but also by the exclusion of both land-derived siliciclastics and open marine deposits. Oolites punctuating the sequence are interpreted as expansions of carbonate sand bodies of the bank margin, crossing over the platform interior (cf. Beach and Ginsburg, 1980; Demicco, 1985).

Similar facies relationships were coevally established along the eastern margin of Laurentia. Mazzullo and Friedman (1975) postulated there the development of a shallow shelf-lagoon with a gastropod- and nautiloid-dominated fauna during Gasconadian times. Inter- to supratidal flats were built at the platform-side oolite shoals that formed the bank-margin, stretching along most of the Iapetus-facing border (Sternbach and Friedman, 1984).

The belt of Cambro-Ordovician carbonates that crop out in the Argentine Precordillera is narrow (ca. 50 km across the inferred depositional strike) in comparison to the dimensions of known platforms of that age. For example, Chow and James (1987) estimated a minimum width of 200 km for platforms developed along the Appalachian margin, and 1,100 km for those of the Canadian Cordillera, which gives an idea of how much could be absent at the Argentine Precordillera.

Open shelf: San Juan Formation

The topmost unit of the platform carbonates succession is represented by the San Juan Formation (300–380 m, as defined by Keller et al., 1994). At the type-section of Cerro La Silla (Fig. 7), the unit conformably overlies rocks of the La Silla Formation.

A meter-scale parasequence capped by peritidal fenestral limestones occurs near the base of the formation, exhibiting a subaerial erosional relief and evidence of subsequent marine erosion. After this event, subtidal lithotopes prevailed during deposition of the San Juan Formation.

These limestones are distinguished by a conspicuous pseudobedding showing alternating massive, gray horizons and "nodular," yellow to tan weathered levels that concentrate dolomite and argillaceous material in solution seams and compacted burrows. The earliest conodonts recorded from this unit demonstrate that the lower levels still belong to the uppermost Tremadocian (*P. deltifer* zone; Keller et al., 1994). The top of the San Juan Formation was dated at several localities. In the northern Argentine Precordillera conodonts and graptolites gave a late Arenigian age for the transitional beds between the San Juan Formation and the black shales of the Gualcamayo Formation (Ortega et al. 1983), whereas elsewhere conodonts have demonstrated an early Llanvirnian age (*Eoplacognathus suecicus* zone) for the top of the limestones (e.g., Hünicken and Ortega, 1987; Sarmiento, 1986; Lehnert, 1993).

Lithologies of the San Juan Formation are grouped into five lithofacies associations (Glossary 3), all representing fully marine, open platform deposits (Cañas, 1995a,b) (Fig. 7). These shallow water carbonates are tectonicaly juxtaposed with slope and deep basinal rocks to the west (Western Tectofacies of Astini, 1992), whereas by late Arenigian time they graded northward into deeper, outer ramp to intrashelf-basinal deposits (Fig. 7).

Reef-mound facies association. This association (Glossary 3) comprises a series of small coalesced mound-shaped bioherms and patch-reefs of varied composition, flanking deposits, and coarse channel-fills. The earliest bioherms arose at late Tremadocian time near the base of the San Juan Formation, and were described in detail by Cañas and Carrera (1993). They consist of *Girvanella*-rich calcimicrobial biolithite, lithistid sponges, and the receptaculitid *Calathium* as main builders, forming banks of coalesced mounds up to 5 m in height. The fauna contained in the bioherms and related beds is indicative of an open marine environment, while the associated channel and

Figure 7. Stratigraphic columns of the San Juan Formation illustrating shallow-to-deeper ramp transition: (a) composite of the Cerro La Silla–Cerro Viejo sections; (b) composite column of the Guandacol area (see locations in Fig. 1); (c) detail of the transitional beds (parted limestone and shale facies association) into the black shales of the Gualcamayo Formation (GU). Brachiopod biozones after Herrera and Benedetto (1991).

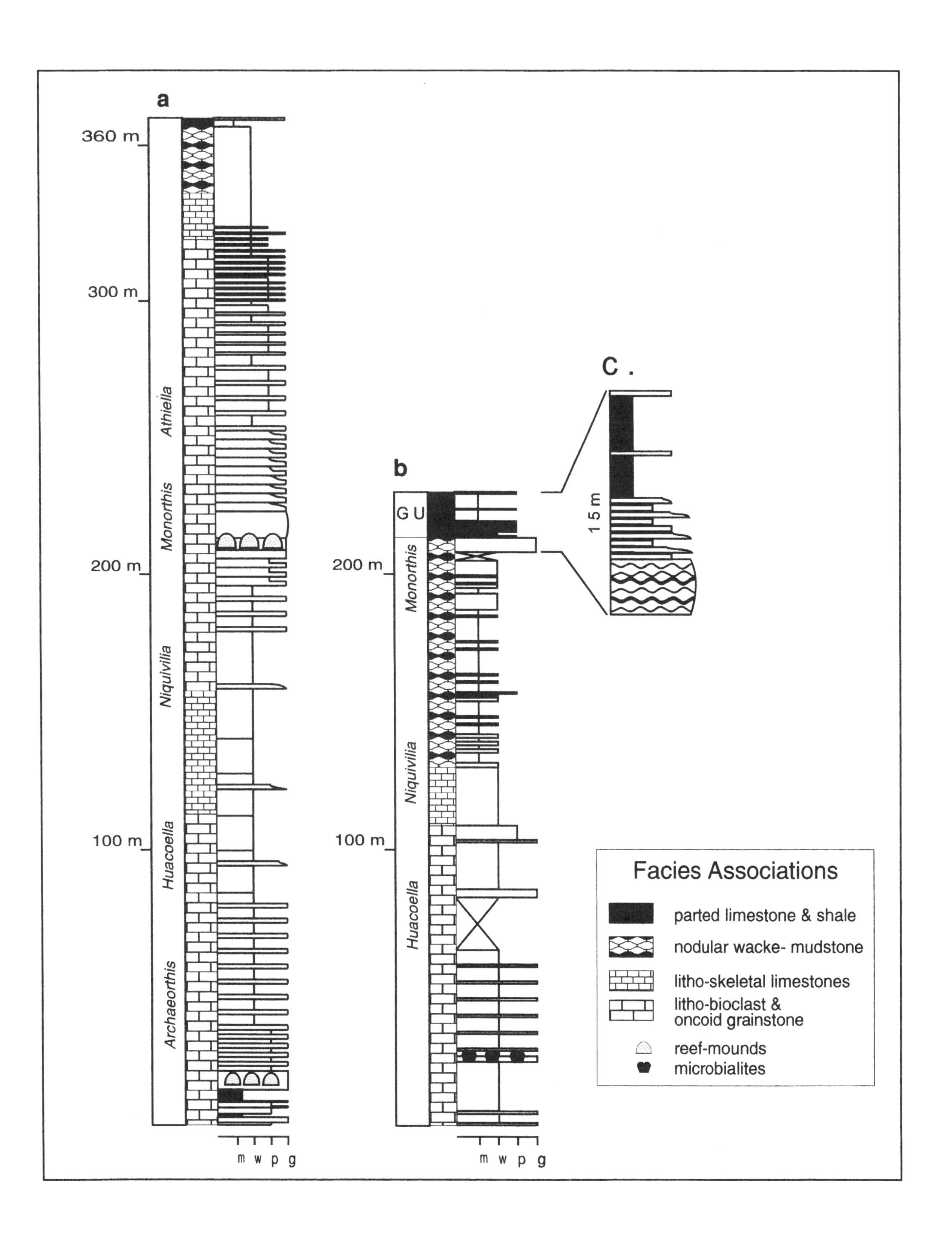
a
360 m
300 m
200 m
100 m
Athiella
Monorthis
Niquivilia
Huacoella
Archaeorthis
m w p g
b
G U
Monorthis
Niquivilia
Huacoella
m w p g
C .
1 5 m
Facies Associations
parted limestone & shale
nodular wacke- mudstone
litho-skeletal limestones
litho-bioclast &
oncoid grainstone
reef-mounds
microbialites

intermound facies reflect a high-energy agitated setting. These bioherms rest directly on a subaerial exposure surface with a decimeter-scale erosional relief cut into supratidal fenestral mudstones.

A second episode of reef-mound development took place in beds of late Arenigian age (Cañas and Keller, 1993; Lehnert and Keller, 1993). The stomatoporoid-like organism *Zondarella* (Keller and Flügel, 1996) was added to the reef-mound building community as an accessory, together with calcimicrobes, lithistids, and receptaculitids, or as the main frame builder. In both cases the bioherms are associated with coarse-grained, cross-bedded shoal, intermound, and tidal channel deposits, indicating that growth occurred in shallow water agitated settings.

Lithoclastic-skeletal facies association. This association (Glossary 3) comprises burrowed, whole-fossil wackestones interbedded with thin lithoclastic-bioclastic packstones and grainstones, which are the most representative rock types of the San Juan Formation. At the type-section of Cerro La Silla this association onlaps the reef-mound association, elsewhere it rests (para)conformably on top of the La Silla Formation.

Burrowed skeletal wackestone-packstone is the prevailing lithology. Whole fossils and abundant bioclasts represent a diverse fauna that included brachiopods, trilobites, pelmatozoans, sponges, receptaculitids, bryozoans, gastropods, ostracods, nautiloids, conodonts, *Girvanella*, and algae such as *Nuia, Halysis, Sphaerocodium,* and forms comparable to the *Solenopora-Hedstroemia* group. Peloids are abundant in this facies together with lithoclasts. These latter are mostly derived from underlying beds of the same lithology, and may show evidence for multiple generation indicating consecutive events of cementation (hardgrounds), reworking, and redeposition. Frequent centimeter- to decimeter-thick, graded lithoclastic-bioclastic grainstone beds interpreted as storm layers, usually have sharp, irregular erosive bases developed atop underlying wackestone.

A diverse marine fauna and ubiquitous burrowing, along with the lack of structures indicative of shallow-water traction processes or of subaerial exposure, suggest a low-energy, open marine environment, located within the photic zone as indicated by calcareous algae. Hardgrounds and reworked lithoclasts are indicative of reduced sedimentation rates. The frequent graded grainstone beds punctuating the section are interpreted as storm deposits (cf. Kreisa, 1981; Markello and Read, 1981).

Intraclastic-bioclastic and oncoid packstone-grainstone facies association. This association (Glossary 3) forms successions up to 50 m thick of medium- to thick-bedded, massive or more rarely cross-bedded strata. Internal erosion-solution surfaces within beds are frequent, showing that early cementation was common (firm- to hardgrounds).

Bioclastic, peloidal-lithoclastic grainstones are moderately to well sorted, composed of fragments of brachiopods, pelmatozoans, algae (*Nuia*), ostracods, gastropods, trilobites, and bryozoans. Bioclasts are mostly disarticulated and reworked, and may be completely micritized. Lithoclasts in this lithofacies are rounded, derived from micritic rocks, or are reworked clasts from similar grainstone. Lithoclastic-bioclastic grainstones are usually more coarsely grained and are poorly sorted. Oncolitic (litho-pel–bioclastic) grainstone-packstones are variably sorted, with *Girvanella* oncoids up to 1–2 cm in diameter.

Lithofacies of this association are interpreted as carbonate shoal or sand flat deposits. Both textures and the extent of allochem reworking indicate that relatively high-energy shallow subtidal conditions prevailed. Micritization and biogenic coating of grains, and the formation of firm- and hardgrounds might have taken place under lowered rates of sedimentation in intershoal areas.

Nodular wackestones and mudstone association. This association (Glossary 3) attains its thickest development at a local depocenter in the northern extreme of the Argentine Precordillera (Guandacol area) (Figs. 7 and 8). This association is absent at the type section of Cerro La Silla, where it correlates with the upper part of the lithoclastic-skeletal facies association (upper *Huacoella* to lower *Niquivilia* biozones (Herrera and Benedetto, 1991)).

Lithologies consist mostly of dark gray to black burrowed wackestone and mudstone. Layers are thin to medium thickness, characteristically exhibiting a stylo-nodular to wavy bedding with dolomitic partings. Peloidal mud and spar filled horizontal burrows are frequent between bedding planes.

Wackestones include scarce peloids and bioclasts as well as whole fossils of brachiopods, trilobites, lithistid sponges and spicules, calcispheres, and ramose bryozoans. Mudstone and fine peloidal packstone layers that contain disseminated pyrite may show a diffuse parallel, slightly undulatory lamination.

The association is interpreted as deposited in an outer ramp environment, below the storm wave-base for most of the time. This is implied by the lack of evidence for shallow water deposition, as well as by the stratigraphic position of the association between the underlying lithoclastic-skeletal limestones, and the basinal limestones and shale association (see below). Algal-related forms such as *Nuia, Halysis, Sphaerocodium,* or receptaculitids are notably absent in these rocks, indicating deposition below the photic zone, suggesting that sediments mostly represent periplatform muds derived from nearby shallow-water settings.

Parted limestone and shale facies association. This association (Glossary 3) occurs on top of the San Juan Formation in most sections of the Argentine Precordillera as a thin "transitional" package (10–35 m) between the limestones and black shales of the Gualcamayo and Los Azules Formations. In the northern area (west of Guandacol) deposition of these rocks took place during the late Arenigian, whereas south of Río Jáchal platform carbonates continued being deposited until early Llanvirnian times. In the Guandacol section (Fig. 7), the basal 5 m are formed by thin, irregularly bedded bioclastic grainstone and packstone. Bioclasts include trilobites, brachiopods, pelmatozoan and bryozoan fragments, calcispheres, sponge spicules and algae (e.g., *Nuia, Halysis*). Normal grading, perched

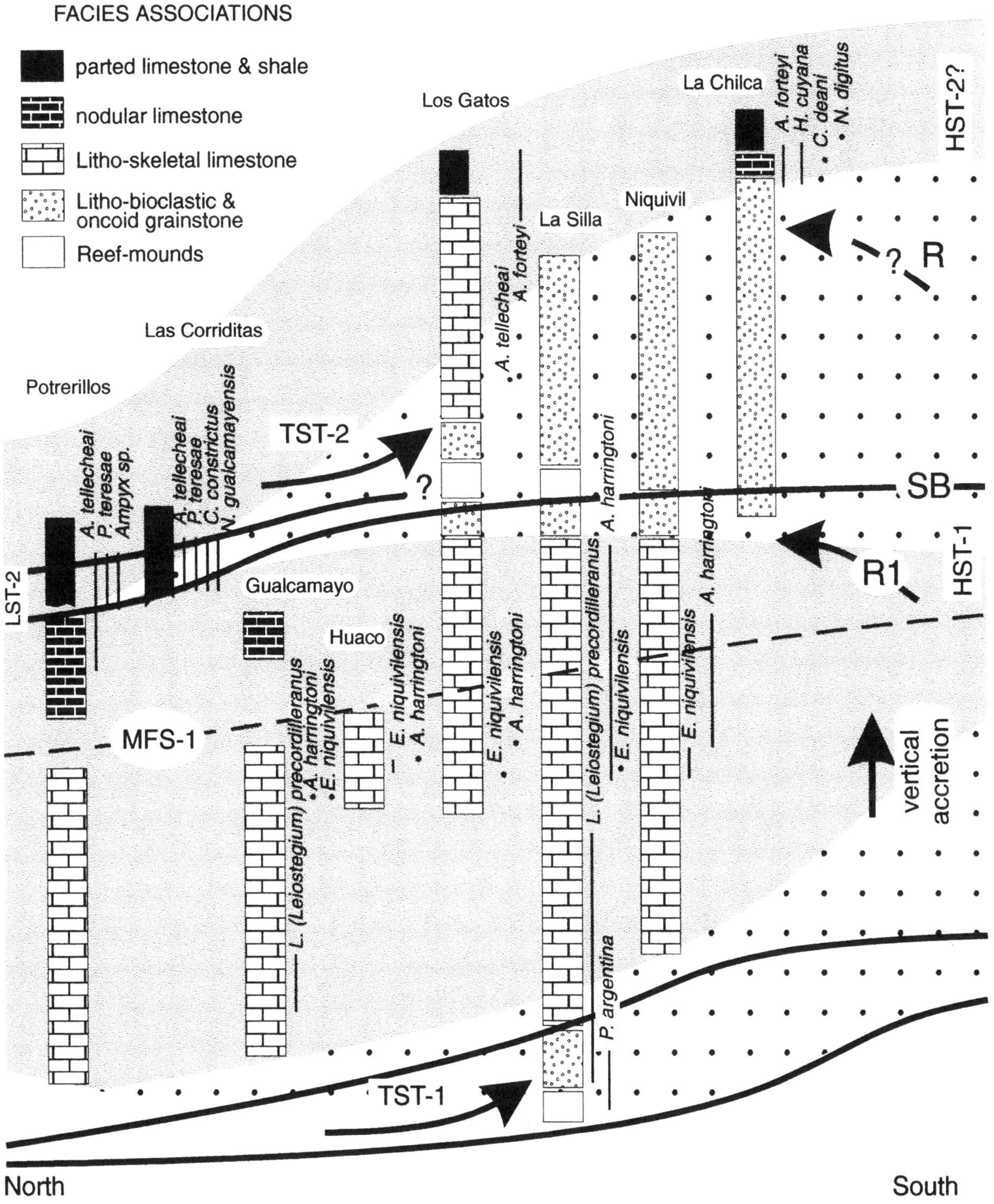

Figure 8. Regional cross section of the San Juan Formation showing the ramp-to-basin transition; the section is approximately north-south, with the north to the left. Sections Potrerillos to Gualcamayo are located in the Guandacol area; sections Huaco to La Chilca are located south of the Jáchal River. Regional facies relationships define two third-order sequences that can be traced from the shallow platform area south of Jáchal into the depocenter located west of Guandacol (see explanation in the text). Vertical distribution of trilobites after Vaccari (1994).

fabrics, spar-occluded shelter porosity and millimeter-thick shale partings are common features of these rocks, supporting an interpretation of them as graded tempestites deposited between the fair weather wave-base and the storm wave-base. Most bioclasts in this lithology probably represent parautochthonous elements, but the association of calcareous algae with agnostid and trinucleid trilobites suggests that some components were derived from a shallower source.

A sharp boundary, in places determined by a hardground, separates these grainstones from an overlying deepening-upward sequence formed by an alternation of bioclastic micro-graded wackestone, rare fine beds of grainstone, and finely laminated pyritic mudstone and shale. The dominant process was deposition of argillaceous carbonate and terrigenous mud from suspension, while the thin carbonate beds are interpreted as distal tempestites deposited from clouds of sediment put into suspension during storms, or from dilute turbidites (Rogers, 1984). The environment may have been low in oxygen, as indicated by disseminated pyrite and the paucity of bioturbation. These rocks grade upward into the graptolitic black shales of the Gualcamayo Formation. Thus, the succession records a deepening event, from a middle to outer ramp setting, and finally to an intrashelf-basinal environment.

Depositional Environments. The San Juan Formation was deposited on top of an open (unrimmed) carbonate shelf, bounded to the west by continental slope and oceanic basin deposits. Despite this long-recognized relationship, no significant changes in facies associations are recorded in this direction in the shallow platform carbonates. Although microbial-sponge reef mounds and stromatoporoid patch reefs, as well as thick oncoid grainstone, are restricted to the eastern sections, no facies association formed below the storm wave-base was recorded at the westernmost sections. This indicates that, on palinspastic restoration, inclination in that direction was negligible.

At least from the late Arenigian onward, a homoclinal ramp was developed in association with a local depocenter in the northern Guandacol area. The inner ramp was defined by the fair weather wave-base, a high-energy zone where *Girvanella*-oncoids banks, lime sand shoals, and small bioherms were formed. Remarkably similar facies associations were developed in coeval (*Isograptus victoriae* zone) open platform settings of western North America as part of regressive oncoid-shoal systems that included lithistid sponges–*Calathium* bioherms.

Skeletal-lithoclastic limestones aggraded over most of the open platform and middle ramp environments, which was a low-energy setting periodically affected by storms, whereas the outer ramp was the site of accumulation of nodular wackestones and mudstones, probably as periplatform hemipelagic muds derived from the shallower platform, together with autochthonous skeletal material. This association occurs onto the platform as outer shelf deposits before complete drowning in the early Llanvirnian.

Rocks of the San Juan Formation may be considered as part of the sponge/algal facies that was widespread around the North American continental edge during the Early Ordovician (Alberstadt and Repetski, 1989). This regional facies is characterized by wackestones and mudstones with a distinct biotic assemblage formed by sponges (especially *Archaeoscyphia*), and *Calathium, Nuia, Girvanella,* and *Sphaerocodium.* These rocks hosted small reef-mounds (Church, 1974; Toomey and Nitecki, 1979) that are almost identical to those of the San Juan Formation, and were interpreted to have been deposited in a normal, open marine environment, away from the more restricted shallow shelf settings of the continental interior (Alberstadt and Repetski, 1989).

DEPOSITIONAL SEQUENCES

Despite the close lithologic and stratigraphic correspondence found between many Appalachian and Precordilleran lithostratigraphic units, no individual section is directly comparable (cf. Astini et al., 1995). However, a sequence-stratigraphic–oriented analysis, which affords a framework for comparing the miogeocline development, suggests that both areas shared a common accommodation history.

Upper Cambrian. The overall flat-topped nature of the Upper Cambrian platform and the lack of outcropping platform-margin facies makes difficult the identification of third-order sequences within the La Flecha Formation. In this situation, third-order sequence boundaries are less obvious (Goldhammer et al., 1993). Nevertheless, analysis of stacking patterns of peritidal sequences in parts of the unit, together with available biostratigraphic information, allows recognition of transgressive-regressive trends in the third-order scale.

A detailed section was measured at Quebrada La Angostura in the northern Argentine Precordillera (Fig. 4), that yielded several paleontologic control points (trilobites) ranging from the *Crepicephalus* zone to the *Aphelaspis* zone (Vaccari, 1994). Third-order sequences may be recognized on the basis of small-scale cycle distribution. Figure 4 shows that meter-scale cycles at the lower *Crepicephalus* zone possess transgressive-prone facies successions with thicker subtidal portions, and are systematically thinner up-section. Amalgamation of intertidal-supratidal facies and subaerial exposure surfaces toward the upper *Crepicephalus* zone indicates slower rates of accommodation change. This part of the succession is interpreted to contain a sequence boundary. A similar transgressive-regressive trend was developed from the top of the *Crepicephalus* zone to the *Aphelaspis* zone. The sea-level fall at the sequence boundary is further indicated by the occurrence of detrital quartz grains in lithoclastic grainstone beds. This sequence boundary is time-equivalent to the top of the Sauk II subsequence (term supersequence in this chapter) of Palmer (1983), which is recognized everywhere in North America (James et al., 1989). A comparison of the inferred curve of relative sea-level change of the Dresbachian section at Quebrada La Angostura with similar ones obtained for Cordilleran and Appalachian successions (Osleger and Read,

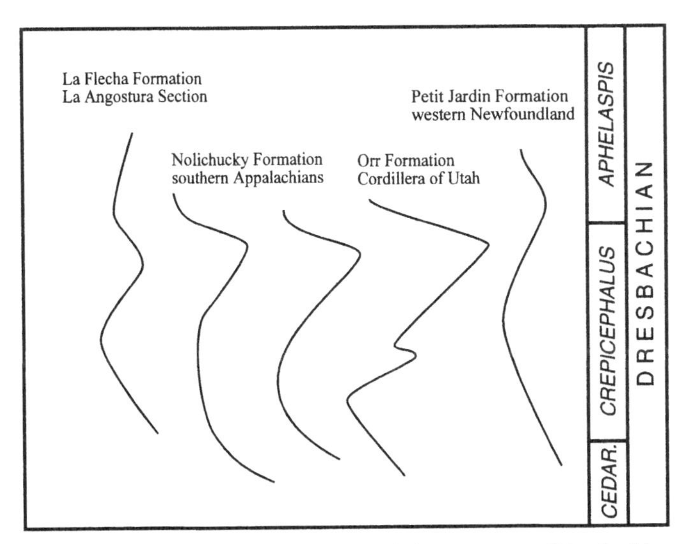

Figure 9. Comparison of relative sea-level change curves of Dresbachian sections of the Argentine Precordillera (see Fig. 4), the southern Appalachians and the Cordillera in North America (after Osleger and Read, 1993), and from western Newfoundland in the northern Appalachians (after James et al., 1989).

1993), along with a curve of western Newfoundland (James et al., 1989), shows a remarkable correlation of the third-order sequences (Fig. 9).

The lack of a continuous section and poor biostratigraphic control hindered a similar analysis for the Franconian-Trempealeauan portion of the La Flecha Formation. However, the transition from the laminite-dominated dolostones of the Zonda Formation to the thrombolitic La Flecha Formation reflects a long-term transgressive event corresponding to this time interval, that ends with renewed flooding at the base of the overlying La Silla Formation. The contact between these units is a prominent subaerial exposure surface that occurs within the *Saukia* zone (Vaccari, 1994), below the Cambro-Ordovician boundary, and is interpreted to represent a sea-level fall. This event was recorded on most continents (Miller, 1984; Fortey, 1984; James et al., 1989), suggesting an eustatic origin. The Franconian-Trempeauleauan sequence thus defined appears to be equivalent to Grand Cycle C distinguished in the northern Appalachians by Chow and James (1987), and to the single, long-term Franconian-Trempeauleauan accommodation event recognized by Osleger and Read (1993) in Cordilleran and southern Appalachian successions.

Tremadocian. The La Silla Formation is separated from the underlying La Flecha Formation by a subaerial break followed by extensive flooding of the shelf that changed its overall peritidal nature. Erosion-dissolution surfaces are also present at the top of the uppermost beds of the La Silla Formation, as well as in the base of the overlying San Juan Formation accompanying initial transgression during the *P. deltifer* conodont zone. Thus, the La Silla Formation represents a single third-order sequence roughly coinciding with the Tremadocian epoch (Fig. 10), at which time rapid vertical accretion took place on top of a wide, and probably rimmed, shelf. Transgression at the base of the formation started within the *Missisquoia depressa* trilobite zone (Vaccari, 1994) below the *Cordylodus intermedius* conodont zone (Keller et al., 1994). This episode corresponds to the Lange Ranch Eustatic Event recognized by Miller (1984), and its stratigraphic expression in the Argentine Precordillera is remarkably similar to many Appalachian sections. In Pennsylvania, peritidal facies of the Conococheague Formation identical to those of the La Flecha Formation are succeeded by predominantly shallow-subtidal, "noncyclic" Stonehenge Formation. Transgression was interpreted to occur in the *C. angulatus* conodont zone (Taylor et al., 1992).

Thus, the sequence is equivalent to a third-order sequence widely recognized in North America, like the Chepultepec interval of the southern Appalachians (Bova and Read, 1987), the Event 6 identified by James et al. (1989) in the northern Appalachians, and to the division (supersequence) C-1 of the Sauk, as distinguished by Goldhammer et al. (1993) in southern North America (Fig. 10).

Arenigian. The study of sedimentary facies and their vertical evolution in the San Juan Formation, together with available biostratigrafic control, led to the recognition of deepening and shallowing sequences correlatable over areas of different paleogeographic settings. This permitted establishment of a stratigraphic scheme based on outcrop data, using the definitions and terminology of Van Wagoner et al. (1988).

A late Tremadocian transgressive event began at the base of the San Juan Formation (*P. deltifer* zone) is indicated by the first occurrence of open marine faunas onto the platform. The rate of relative sea-level change initially was slow enough to allow the formation of two parasequences (4–5 m each) capped by laminites or subaerial exposure surfaces. This relationship is observed only at the type section of Cerro La Silla where the lower contact of the formation is well exposed. This interval may be regarded as a lowstand systems tract or as the base of a transgressive systems tract (TST) (Fig. 8). However, the lack of regional data on stratal geometry prevents this resolution, and it is here included within the TST-1. Calcimicrobial-sponge reef-mounds were developed on the last subaerial surface identified in the San Juan Formation, and are in turn covered by a thick sequence of skeletal limestones of the middle ramp or open platform that aggraded, "keeping up" with the rising sea level. The top of the TST-1 is best expressed in deeper, outer ramp settings (Guandacol area) by hardgrounds in addition to glauconitic and phosphatized horizons, interpreted as a maximum flooding surface (MFS-1) (Fig. 8). Sedimentation resumed in this area with outer ramp nodular mudstones and wackestones, which are correlated onto the shallow platform areas to the south with a large-scale shallowing sequence ending with thick grainstone shoal deposits, interpreted as the highstand systems tract (HST-1 in Fig. 8).

In the outer ramp succession of the Guandacol area, mudstones and wackestones are conformably overlain by proximal tempestites of the base of the parted limestones and shale facies

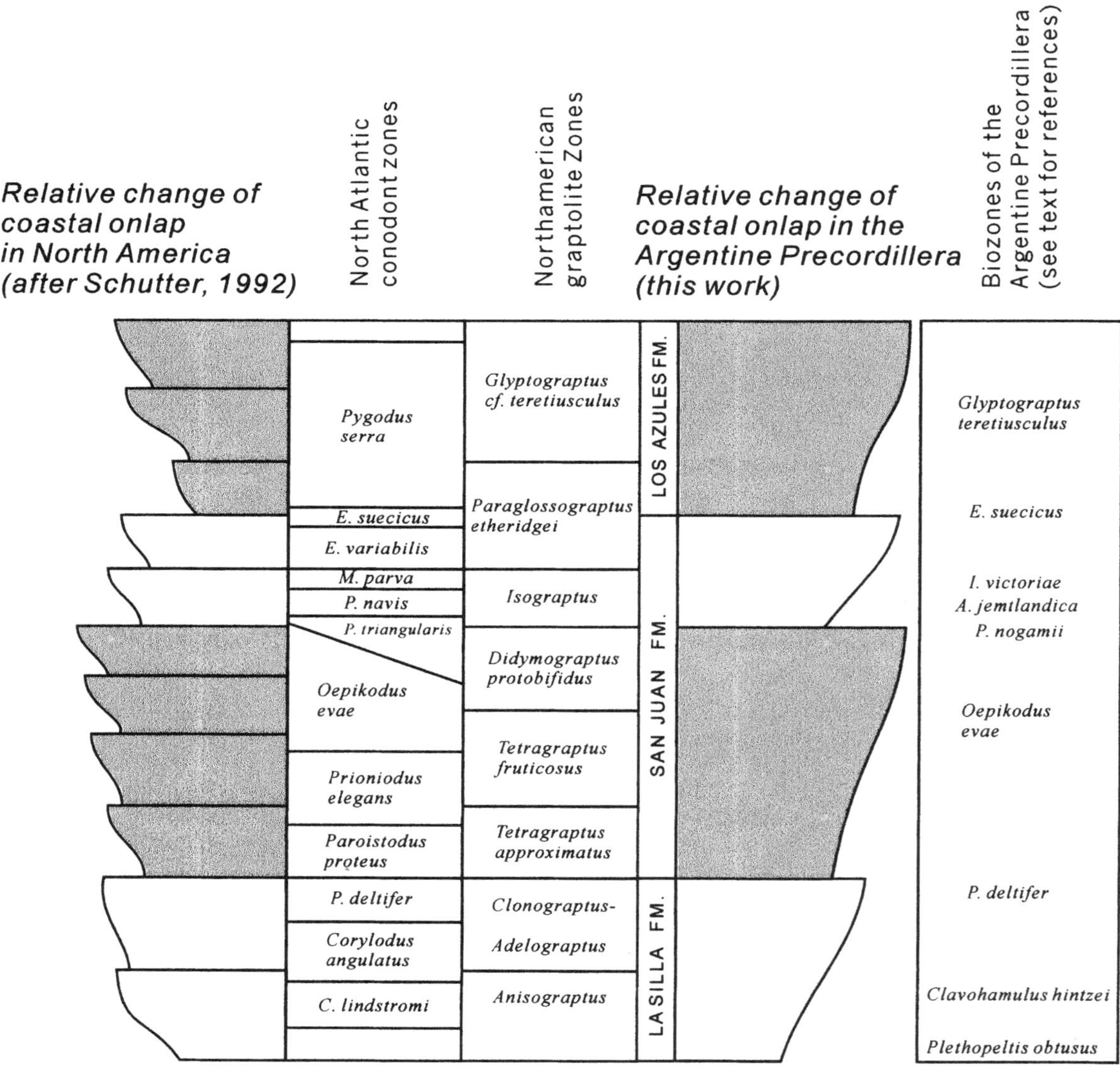

Figure 10. Correlation of supersequences of the Ordovician of the Argentine Precordillera and a sequence stratigraphic scheme of North America after Schutter (1992).

association. These rocks appear as an "out-of-sequence" deposit (Burchette and Wright, 1992), probably indicating the development of a lowstand prograding wedge, corresponding to the LST-2, below which the sequence boundary (SB) has to be located (Fig. 8). The age of the SB in this area is well constrained by conodonts of the *O. intermedius* zone (Lehnert and Keller, 1993) occurring below the reefs, and graptolites of the *Isograptus victoriae* zone (Ortega et al., 1983) from the overlying shales. The deepening-up sequence of distal tempestites to basinal muds that follows indicates the onset of higher rates of relative sea-level change that resulted in drowning of this part of the shelf (TST-2). A reduced rate of carbonate production in the shallow shelf areas during the time elapsed between the LST-2 and the TST-2 (Fig. 8) reduced drastically lime mud supply to periplatform, outer ramp settings (cf. Droxler and Schlager, 1985) at the time an unidentified area started to supply terrigenous mud.

Beds of the shallow platform area (south of Río Jáchal) correlatable with the SB and LST-2 are represented by thick intraclast-bioclast grainstones that rest on open shelf or middle ramp deposits, indicating a strong progradational system. No evidence of subaerial exposure has been recorded, and the distal portion of the highstand inner ramp may be amalgamated with the proximal part of the lowstand inner ramp tract, making difficult the identification of a SB. Nevertheless, according to conodont data of Lehnert and Keller (1993) reef-mound development in this part of the shelf started above the *B. navis/B. triangularis* conodont zone, an age corresponding to the TST-2 in the northern outer ramp area. Thus, it is reasonable to place the SB below the reef horizons and to include them in the TST-2.

Shelf carbonates continued to aggrade until the early Llanvirnian (*E. suecicus* zone). A consistent deepening-upward succession of facies associations has been found at some sections (e.g., Cerro Viejo del Fuerte, Talacasto), whereas it is not clear if a shallowing event took place before the early Llanvirnian drowning of the platform.

The first transgressive-regressive sequence of the San Juan Formation lasted from the late Tremadocian to the late Arenigian, following a pattern recognized in other platform areas of the world (Leggett et al., 1981; Fortey, 1984; James et al., 1989), suggesting a eustatic control (Fig. 10).

In western Newfoundland a regression equivalent to the HST-1 and SB of the San Juan Formation is represented by the Aguathuna Formation, formed of peritidal carbonates that overlie open marine limestones of the Catoche Formation. Barnes (1984) identified a regressive event in cratonic Cordilleran and Appalachian sections of Canada, immediately below the *Isograptus victoriae* zone. The following transgressive-regressive sequence spanning from the *I. victoriae* zone to the *E. suecicus* zone is limited at the top by renewed transgression at the *E. suecicus* zone (Barnes, 1984; Schutter, 1992). In the Argentine Precordillera, the *I. victoriae* zone corresponds to the TST-2 of the San Juan Formation and to the start of black shale deposition in depocenters (Gualcamayo Formation), whereas transgression at the *E. suecicus* zone (Sarmiento, 1986; Hünicken and Ortega, 1987) is marked by drowning of the carbonate platform.

PARALLEL PLATFORM GROWTH STAGES

Previous works have compared in more or less detail the Cambrian to Lower Ordovician stratigraphy of the Precordillera and Laurentia, pointing out remarkable similarities (e.g., Baldis and Bordonaro, 1982; Ramos et al., 1986; Dalla Salda et al., 1992a; Benedetto, 1993; Astini et al., 1995; Thomas and Astini, 1996), especially with the Appalachian successions that also present the closest paleobiogeographic affinities (Vaccari, 1994; Lehnert et al., 1997; see Benedetto et al., this volume, and references therein). Carbonate platforms of the Cordilleran and Appalachian miogeoclines overlie a wedge of siliciclastics that accumulated during the early postrift stages of the passive margins, the transition from these to carbonate sedimentation occurring within the *Bonnia-Olenellus* trilobite zone (Bond et al., 1984, 1989; James et al., 1989; Read, 1989). Similar red siltstones and evaporites crop out in the northern Argentine Precordillera in the same stratigraphic position (Astini and Vaccari, 1996), i.e., underlying late Early Cambrian peritidal carbonates of the *Bonnia-Olenellus* zone (Cañas, 1988; Vaccari, 1988). Although these rocks were implied to represent initial rift-stage facies (Astini et al., 1995; Thomas and Astini, 1996), no evidence for active rifting has been recorded. Instead, their lithologic similarity and stratigraphic position at the base of a passive-margin succession that mirrors the Appalachian margins suggest that these rocks correspond most likely to the postrift transgressive cover represented there by such units as the upper Chilhowee and equivalents in the U.S. Appalachians and the Forteau Formation of Canadian western Newfoundland (Bond et al., 1984).

A long-term rise of relative sea level corresponding to a second-order oscillation, coupled to the thermal contraction that followed rifting of the margins of Laurentia, allowed the deposition of the thick Cambo-Ordovician platform carbonate package known as the Sauk Sequence (Sloss, 1963; Palmer, 1983). Initiation of carbonate platforms was synchronous in the Appalachian, Cordilleran, and Great Basin areas by early Middle Cambrian time, characterized by high and rapidly falling rates of net subsidence reflecting exponential decay of thermally controlled subsidence, and an increase of flexural bending of the cratonic edge (Bond et al., 1989). This resulted in a progressive retreat of the cratonic siliciclastic shorelines and a cratonward expansion of the carbonate belt through the Middle to Upper Cambrian.

The earliest stage of platform development in the Argentine Precordillera is represented by the Early to Middle Cambrian La Laja Formation (Borrello, 1962; Bordonaro, 1980), composed of alternating siliciclastic, mixed, and carbonate facies of open marine settings (Berkowski et al., 1990), ordered in sequences comparable to North American Grand Cycles (Baldis and Bordonaro, 1982).

The next stage of platform growth around Laurentia, comprising the Upper Cambrian and Lower Ordovician successions, was characterized by slowly declining rates of net subsidence (Bond et al., 1989). As in the Laurentian successions, the cratonward onlap of carbonate platforms in the Argentine Precordillera is reflected by the decreasing influx of siliciclastic sediments until total elimination at Early Ordovician times. Estimates of net subsidence for the Argentine Precordillera based on uncorrected thickness and available biostratigraphic data yield values of over 70 m/m.y. for the early Late Cambrian to 30 m/m.y for the Early Ordovician (Cañas, 1995a), equating net subsidence values of most Laurentian miogeoclines. The tectonic subsidence curve of the lower Paleozoic carbonate succession of the Argentine Precordillera has been shown to fit with curves of the Appalachians and western Newfoundland, implying the same age of break-up and miogeocline initiation from about 575 m.y. (Bond et al., 1984) to close to the Precambrian-Cambrian boundary (545–540 m.y.) (Williams and Hiscott, 1987; Dalziel, 1997).

DISCUSSION

A combination of thermo-tectonic subsidence with multiple-order eustasy was the main control on the lithostratigraphic framework of the carbonate platforms. Correlation of third-order sequences, and the overall lithologic similarity within individual sequences shown in previous sections of this work, suggest that the Precordilleran carbonate platform and the Laurentian miogeoclines had the same accommodation histories. This holds especially for the southern and central Appalachians, where sedimentary facies, thickness of lithostratigraphic units, and faunal assemblages resemble more closely those of the Argentine Precordillera (Astini et al., 1995).

A parallel platform evolution involving a Laurentian origin of the Argentine Precordillera has been explained in three ways. Bond et al. (1984) suggested that the two miogeoclines could have been paired margins, rifting away the western Gondwana margin from eastern Laurentia in the Late Proterozoic. As Benedetto (1993) pointed out, this model would not explain significant faunal differences of the Argentine Precordillera with other Andean Cambro-Ordovician basins, and the lack of carbonate successions in the southern ranges of Buenos Aires and South Africa. Moreover, the Laurentian faunal affinities of the Argentine Precordillera persisted through Early Ordovician times (Benedetto et al., this volume).

A second hypothesis holds that the rifting event that separated the Argentine Precordillera from Laurentia took place in the Late Ordovician, following Middle Ordovician collision with Gondwana (Dalla Salda et al., 1992a,b; Dalziel et al., 1994, 1996). In this case the Cambrian to Lower Ordovician carbonates of the Argentine Precordillera would be a relict part of the Appalachian passive-margin, or of a hypothetical promontory of Laurentia (Dalziel, 1997). Nevertheless, the paleontologic evidence does not support the implicit Ordovician South America–Laurentia connection (Benedetto, 1993; Astini et al., 1995).

The third hypothesis favored a conjugate rift-pair origin of the Precordilleran and the Appalachian carbonate platforms, considering the former as a microcontinent detached from Laurentia (Benedetto, 1993). Recent papers have argued that the Argentine Precordillera was derived from the Ouachita embayment in an Early to Middle Cambrian rifting event (Astini et al., 1995, 1996; Keller and Dickerson, 1996; Thomas and Astini, 1996), as part of the larger Cuyania terrane (Ramos, 1995) that includes Grenvillian rocks of the western Pampeanas Ranges. This proposal was sustained by Thomas's prediction of a missing fragment of the Laurentian platform from the Ouachita embayment (Thomas, 1991), by the Late Cambrian trilobite faunas, which show closest affinities to southern-central Appalachian faunas (Vaccari, 1994), by resemblance of sedimentary histories, and by the size recognized for the Cuyania terrane, which matches that of the Ouachita embayment (Astini et al., 1996). However, in reconstructing the early Paleozoic southeastern margin of North America, Thomas (1991) concluded that the Ouachita rifted margin is younger (Early Cambrian) than the Appalachian (Late Proterozoic) margin. As stated above, the Argentine Precordillera appears to have shared a common accommodation history with the Appalachian miogeocline from the Early Cambrian through the Early Ordovician, suggesting that both arose from the same rifting event (Bond et al., 1984). Thus, the paleontologic evidence indicating that Laurentian conodont, trilobite, and brachiopod faunas were present in the Argentine Precordillera until the lower Arenigian (Lehnert et al. 1997; Benedetto et al., this volume), along with evidence for a source of siliciclastics which persisted at least into Late Cambrian times (Cañas, 1995a), suggest that the Argentine Precordillera remained adjacent to the eastern Laurentian margin as a southern continuation of the Appalachian regional shelf until the Early Ordovician. Evidence for large-scale extension at the Argentine Precordillera from the Arenigian onward includes the development of local subsiding depocenters onto the shelf (Cañas, 1995a) and a regional slope in the western Argentine Precordillera (Cuerda et al., 1983; Astini, 1992; Benedetto and Vaccari, 1992; Keller, 1995). The onset of this extensional tectonic regime during the Arenigian was related to the separation of the Cuyania terrane from southern Laurentia (Cañas, 1995a; Keller and Dickerson, 1996), which is consistent with the onset of diverging faunas and sedimentary histories (Benedetto, 1993).

CONCLUSIONS

Detailed facies analysis combined with available biostratigraphic data provide a sequence stratigraphic framework for the Upper Cambrian and Lower Ordovician of the Argentine Precordillera. Transgressive-regressive facies relationships define third-order scale depositional sequences that can be correlated with sequences recognized in Laurentian platforms.

Stacking pattern analysis shows two sequences in the Dresbachian, the boundary between them occurring in the *Crepicephalus* zone. The top of the second sequence occurs above the *Aphelaspis* zone, recording an important drop of relative sea level when quartz sand was shed onto the platform. The third sequence in the Upper Cambrian spans to near the Cambrian-Ordovician boundary, within the *Saukia* zone.

Transgression at the *Missisquoia depressa* zone determined a drastic change of platform configuration, with expansion of shallow-subtidal and shoal depositional settings replacing the regional peritidal shelf of the Late Cambrian. The top of this succession reaches the *P. deltifer* zone, defining a sequence approximately coincident with the Tremadocian. A new transgressive-regressive sequence extending from the Upper Tremadocian to the Upper Arenig can be traced regionally across a ramp to intrashelf basin transition. The last transgressive event on the carbonate platform spans through the Upper Arenig, whereas it is not clear if regression took place before the final drowning of the platform recorded in the Lower Llanvirnian.

Comparison of platform growth stages, paleoenvironments, subsidence history, and third-order depositional sequences confirm a parallel evolution of the Precordilleran passive-margin succession of western Argentina and the Appalachian margin of Laurentia, began after a rifting event near the Precambrian-Cambrian boundary.

The most suitable explanation consistent with current paleobiogeographic knowledge (see Benedetto et al., this volume) is that the Argentine Precordillera terrane was part of the eastern North American margin. Divergent depositional histories and faunal composition from the Arenigian onward, together with evidence for extensional tectonics, suggest that the Cuyania terrane was detached from Laurentia during the Early Ordovician.

ACKNOWLEDGMENTS

This chapter represents part of my Ph.D. dissertation in the Department of Geology at the Universidad Nacional de Córdoba, carried out under the supervision of J. L. Benedetto (Córdoba, Argentina), and W. Buggisch (Erlangen, Germany). Colleagues at Córdoba—T. Sanchez, E. Vaccari, R. Astini, M. Carrera, and E. Brussa, and at Erlangen—M. Keller, O. Lehnert, and W. von Gossen—are gratefully acknowledged for shared data and ideas, years of cooperative field work, and countless discussions. I especially thank N. P. James, R. M. Palma, and J. E. Repetski for thorough reviews and constructive suggestions that contributed greatly toward improvement of this manuscript. The study was supported by doctoral fellowships from the Consejo Nacional de Investigaciones Científicas y Técnicas (CONICET–Argentina) and the Deutscher Akademischer Austauschdienst (DAAD–Germany). This is a contribution to International Geological Correlation Program Project 376.

REFERENCES CITED

Aitken, J. D., 1966, Middle Cambrian to Middle Ordovician cyclic sedimentation, southern Rocky Mountains of Alberta, Canada: Bulletin of Canadian Petroleum Geology, v. 14, p. 405–441.

Aitken, J. D., 1978, Revised models for depositional Grand Cycles, Cambrian of the southern Rocky Mountains, Canada: Bulletin of Canadian Petroleum Geology, v. 26, p. 515–542.

Alberstadt, L., and Repetski, J. E., 1989, A Lower Ordovician sponge/algal facies in the southern United States and its counterparts elsewhere in North America: Palaios, v. 4, p. 225–242.

Astini, R. A., 1992, Tectofacies ordovícicas y evolución de la cuenca eopaleozoica de la Precordillera Argentina: Estudios Geológicos, v. 48, p. 315–327.

Astini, R. A., and Vaccari, N. E., 1996, Sucesión evaporítica del Cámbrico Inferior de la Precordillera: significado geológico: Revista de la Asociación Geológica Argentina, v. 51, p. 97–106.

Astini, R. A., Benedetto, J. L., and Vaccari, N. E., 1995, The early Paleozoic evolution of the Argentine Precordillera as a Laurentian rifted, drifted, and collided terrane: a geodynamic model: Geological Society of America Bulletin, v. 107, p. 253-273.

Astini, R. A., Ramos, V. A., Benedetto, J. L., Vaccari, N. E., and Cañas, F. L., 1996, La Precordillera: un terreno exótico a Gondwana: Buenos Aires, 13° Congreso Geológico Argentino y 3° Congreso de Exploración de Hidrocarburos, v. 5, p. 293–324.

Baldis, B. A., and Bordonaro, O. L., 1982, Comparación entre el Cámbrico del Great Basin norteamericano y la Precordillera Argentina. Su implicancia intercontinental: Buenos Aires, 5° Congreso Latinoamericano de Geología, v. 1, p. 97–108.

Baldis, B. A., Bordonaro, O. L., Beresi, M., and Uliarte, E., 1981, Zona de dispersión estromatolítica en la secuencia calcáreo dolomítica del Paleozoico inferior de San Juan: San Luis, 8° Congreso Geológico Argentino, v. 2, p. 419–434.

Baldis, B. A., Beresi, M., Bordonaro, O. L., and Vaca, A., 1982, Síntesis evolutiva de la Precordillera Argentina: Buenos Aires, 5° Congreso Latinoamericano de Geología, v. 4, p. 399–445.

Barnes, C. R., 1984, Early Ordovician eustatic events in Canada, *in* Bruton, D. L., ed., Aspects of the Ordovician system: Paleontological Contributions of the University of Oslo, Universitetsforlaget, v. 295, p. 51–63.

Bathurst, R. G., 1975, Carbonate sediments and their diagenesis: Amsterdam, Elsevier, Developments in Sedimentology, v. 12, 658 p.

Beach, D. K., and Ginsburg, R. N., 1980, Facies successions of Pliocene-Pleistocene carbonates, northwestern Great Bahama Bank: American Association of Petroleum Geologists Bulletin, v. 64, p. 1634–1642.

Benedetto, J. L., 1993, La hipótesis de la aloctonía de la Precordillera Argentina: un test estratigráfico y bioestratigráfico: Mendoza, XII° Congreso Geológico Argentino, v. 3, p. 375–384.

Benedetto, J. L., and Vaccari, N. E., 1992, Significado estratigráfico y tectónico de los complejos de bloques resedimentados Cambro-Ordovícicos de la Precordillera Occidental, Argentina: Estudios Geológicos, v. 48, p. 305–313.

Berkowski, F., Keller, M., and Bordonaro, O. L., 1990, Litofacies de la Formación La Laja (Cámbrico) en la Sierra Chica de Zonda, Precordillera sanjuanina, Argentina: San Juan, 3° Reunión Argentina de Sedimentología, p. 31–36.

Bond, G. C., Nickeson, P. A., and Kominz, M. A., 1984, Breakup of a supercontinent between 625 Ma and 555 Ma: new evidence and implications for continental histories: Earth and Planetary Science Letters, v. 70, p. 325–345.

Bond, G. C., Kominz, M. A., Steckler, M. S., and Grotzinger, J. P., 1989, Role of thermal subsidence, flexure and eustasy in evolution of Early Paleozoic passive-margin carbonate platforms, *in* Crevello, P. D., Wilson, J. L., Sarg, J. F., and Read, J. F., eds., Controls on carbonate platform and basin development: Society of Economic Paleontologists and Mineralogists Special Publication 44, p. 39–61.

Bordonaro, O. L., 1980, El Cámbrico en la Quebrada de Zonda: Revista de la Asociación Geológica Argentina, v. 35, p. 26–40.
Borrello, A., 1962, Caliza La Laja (Cámbrico Medio de San Juan): Noticias y Comunicaciones de Investigaciones Científicas, v. 2, p. 3–8.
Borrello, A., 1965, Sobre la presencia del Cámbrico inferior del olenellidiano en la Sierra de Zonda, Precordillera de San Juan: Ameghiniana, v. 3, no. 10, p. 313–318.
Bova, J. A., and Read, J. F., 1987, Incipiently drowned facies within a cyclic pertidal ramp sequence, Early Ordovician Chepultepec interval, Virginia Appalachians: Geological Society of America Bulletin, v. 98, p. 714–727.
Burchette, T. P., and Wright, V. P., 1992, Carbonate ramp depositional systems: Sedimentary Geology, v. 79, p. 3–57.
Cañas, F. L., 1988, Facies perimareales del Cámbrico inferior en el área de Guandacol, La Rioja: Buenos Aires, II° Reunión Argentina de Sedimentología, p. 46–50.
Cañas, F. L., 1995a, Estratigrafía y evolución paleoambiental de las sucesiones carbonáticas del Cámbrico tardío y Ordovícico temprano de la Precordillera Septentrional, República Argentina [Ph.D. thesis]: Córdoba, Argentina, Universidad Nacional de Córdoba, 215 p.
Cañas, F. L., 1995b, Early Ordovician carbonate platform facies of the Argentine Precordillera: Restricted shelf to open platform evolution, *in* Cooper, J. D., Droser, M. L., and Finney, S. C., eds., The Ordovician odyssey: SEPM, Pacific Section, v. 77, p. 221–224.
Cañas, F. L., and Carrera, M. G., 1993, Early Ordovician microbial-sponge-receptaculitid bioherms of the Precordillera, western Argentina: Facies, v. 29, p. 169–178.
Cañas, F. L., and Keller, M., 1993, Arrecifes y montículos arrecifales en la Formación San Juan (Precordillera Sanjuanina, Argentina): Los arrecifes más antiguos de Sudamérica: Boletín de la Real Sociedad Española de Historia Natural (Geología), v. 88, p. 127–136.
Chow, N., and James, N. P., 1987, Cambrian grand cycles: a north Appalachian perspective: Geological Society of America Bulletin, v. 98, p. 418–429.
Church, S. B., 1974, Lower Ordovician patch reefs in western Utah: Brigham Young University Geology Studies, v. 21, p. 41–62.
Cuerda, A. J., Cingolani, C. A., and Varela, R., 1983, Las graptofaunas de la Formación Los Sombreros, Ordovícico inferior, de la vertiente oriental de la Sierra del Tontal, Precordillera de San Juan: Ameghiniana, v. 20, p. 239–260.
Dalla Salda, L. H., Cingolani, C. A., and Varela, R., 1992a, Early Paleozoic orogenic belt on the Andes in South America: results of Laurentia-Gondwana collision?: Geology, v. 20, p. 617–620.
Dalla Salda, L. H., Dalziel, I. W. D., Cingolani, C. A., and Varela, R., 1992b, Did the Taconic Appalachians continue into southern South America?: Geology, v. 20, p. 1059–1062.
Dalziel, I. W. D., 1997, Neoproterozoic-Paleozoic geography and tectonics: review, hypothesis, environmental speculation: Geological Society of America Bulletin, v. 109, p. 16–42.
Dalziel, I. W. D., Dalla Salda, L. H., and Gahagan, L. M., 1994, Paleozoic Laurentia-Gondwana interaction and the origin of the Appalachian-Andean mountain system: Geological Society of America Bulletin, v. 106, p. 243–252.
Dalziel, I. W. D., Dalla Salda, L. H., Cingolani, C. A., and Palmer, A. R., 1996, The Argentine Precordillera: a Laurentian terrane?: GSA Today, v. 6, p. 16–18.
Demicco, R. V., 1985, Platform and off-platform carbonates of the Upper Cambrian of western Maryland, U.S.A.: Sedimentology, v. 32, p. 1–22.
Droxler, A., and Schlager, W., 1985, Glacial versus interglacial sedimentation rates and turbidite frequency in the Bahamas: Geology, v. 13, p. 799–802.
Fortey, R. A., 1984, Global earlier Ordovician transgressions and regressions and their biological implications, *in* Bruton, D. L., ed., Aspects of the Ordovician system: Paleontological Contributions of the University of Oslo, Universitetsforlaget, v. 295, p. 37–50.
Goldhammer, R. K., Lehmann, P. J., and Dunn, P. A., 1993, The origin of high-frequency platform carbonate cycles and third-order sequences (Lower Ordovician El Paso group, west Texas): constraints from outcrop data and stratigraphic modeling: Journal of Sedimentary Petrology, v. 63, p. 318–359.
González Bonorino, G., 1975, Sedimentología de la Formación Punta Negra y algunas consideraciones sobre la geología regional de la Precordillera de San Juan y Mendoza: Revista de la Asociación Geológica Argentina, v. 30, p. 223–246.
Halley, R., Harris, P., and Hine, A., 1983, Bank margin environment, *in* Scholle, P.A., Bebout, D. G., and Moore, C. H., eds., Carbonate depositional environments: American Association of Petroleum Geologists Memoir 33, p. 464–506.
Harris, P. M., 1979, Facies anatomy of a Bahamian ooid shoal: Sedimenta, The Comparative Sedimentology Laboratory, University of Miami, v. VII, 163 p.
Herrera, Z. A., and Benedetto, J. L., 1991, Early Ordovician bachiopod faunas of the Precordillera basin, western Argentina: biostratigraphy and paleobiogeographical affinities, *in* McKinnon, D. I., Lee, D. E., and Campbell, J. D., eds., Brachiopods through time: Rotterdam, Netherlands, Balkema, p. 283–301.
Hünicken, M. A., and Ortega, G., 1987, Lower Llanvirn–Lower Caradoc (Ordovician) conodonts and graptolites from the Argentine central Precordillera: Nottingham, England, 4th European Conodont Symposium, p. 136–145.
James, N. P., 1984, Shallowing-upward sequences in carbonates, *in* Walker, R. G., ed., Facies models: Geological Association of Canada Geoscience Canada Reprint Series 1, p. 213–228.
James, N. P., Stevens, R. K., Barnes, C. R., and Knight, I., 1989, Evolution of a lower Paleozoic continental-margin carbonate platform, northern Canadian Appalachians, *in* Crevello, P. D., Wilson, J. L., Sarg, J. F., and Read, J. F., eds., Controls on carbonate platform and basin development: Society of Economic Paleontologists and Mineralogists Special Publication 44, p. 123–146.
Keller, M., 1995, Continental slope deposits in the Argentine Precordillera: sediments and geotectonic significance, *in* Cooper, J. D., Droser, M. L., and Finney, S. C., eds., The Ordovician odyssey: SEPM, Pacific Section, v. 77, p. 211–215.
Keller, M., and Dickerson, P. W., 1996, The missing continent of Llanoria. Was it the Argentine Precordillera?: Buenos Aires, 13° Congreso Geológico Argentino y 3° Congreso de Exploración de Hidrocarburos, v. 5, p. 355–367.
Keller, M., and Flügel, E., 1996, Early Ordovician reefs from Argentina: stromatoporoid vs. stromatolite origin: Facies, v. 34, p. 177–192.
Keller, M., Buggisch, W., and Bercowski, F., 1989, Facies and sedimentology of Upper Cambrian shallowing-upward cycles in the La Flecha Formation (Argentine Precordillera): Zentralblatt für Geologie und Paläontologie, v. 1, p. 999–1011.
Keller, M., Cañas, F. L., Lehnert, O., and Vaccari, N. E., 1994, The Upper Cambrian and Lower Ordovician of the Precordillera (western Argentina): some stratigraphic reconsiderations: Newsletters on Stratigraphy, v. 31, p. 115–132.
Kreisa, R. D., 1981, Storm-generated sedimentary structures in subtidal marine facies with examples from the middle Ordovician of southwestern Virginia: Journal of Sedimentary Petrology, v. 51, p. 823–848.
Leggett, J. K., McKerrow, W. S., Cocks, L. R. M., and Richards, R. B., 1981, Periodicity in the early Paleozoic marine realm: Journal of the Geological Society of London, v. 138, p. 167–176.
Lehnert, O., 1993, Bioestratigrafía de los conodontes arenigianos de la Formación San Juan en la localidad de Niquivil (Precordillera Sanjuanina, Argentina) y su correlación intercontinental: Revista Española de Paleontología, v. 8, p. 153–164.
Lehnert, O., and Keller, M., 1993, Posición estratigráfica de los arrecifes arenigianos en la Precordillera Argentina: Document du Laboratoire Géologique du l'Université de Lyon, v. 125, p. 263–275.
Lehnert, O., Miller, J. F., and Repetski, J. E., 1997, Paleogeographic significance of *Clavohamulus hintzei* Miller (Conodonta) and other Ibexian

conodonts in an early Paleozoic carbonate platform facies of the Argentine Precordillera: Geological Society of America Bulletin, v. 109, p. 429–443.

Markello, J. R., and Read, J. F., 1981, Carbonate ramp-to-deeper shale shelf transitions of an Upper Cambrian intrashelf basin, Nolichucky Formation, southwest Virginia Appalachians: Sedimentology, v. 28, p. 573–597.

Markello, J. R., and Read, J. F., 1982, Upper Cambrian intrashelf basin, Nolichucky Formation, southwest Virginia Appalachians: American Association of Petroleum Geologists Bulletin, v. 66, p. 860–878.

Mazzullo, S. J., and Friedman, G. M., 1975, Conceptual model of tidally influenced deposition on margins of epeiric seas: Lower Ordovician (Canadian) of eastern New York and southwestern Vermont: American Association of Petroleum Geologists Bulletin, v. 59, p. 2123–2141.

Mazzullo, S. J., Agostino, P., Seitz, J., and Fisher, D., 1978, Stratigraphy and depositional environments of the Upper Cambrian–Lower Ordovician sequence, Saratoga Springs, New York: Journal of Sedimentary Petrology, v. 48, p. 99–116.

Miller, J. F., 1984, Cambrian and earliest Ordovician conodont evolution, biofacies and provincialism, *in* Clark, D. L., ed., Conodont biofacies and provincialism: Geological Society of America Special Paper 196, p. 43–68.

Ortega, G., Cañas, F. L., and Hünicken, M. A., 1983, Sobre la presencia de *Isograptus victoriae* Harris en la Formación Gualcamayo, La Rioja, Argentina: Revista Técnica de Yacimientos Petrolíferos Fiscales de Bolivia, v. 9, p. 215–221.

Osleger, D., and Read, J. F., 1993, Comparative analysis of methods used to define eustatic variations in outcrop: Late Cambrian interbasinal sequence development: American Journal of Science, v. 293, p. 157–216.

Palmer, A. R., 1983, Subdivision of the Sauk sequence, *in* Taylor, M. E., ed., 2nd International Symposium on the Cambrian System: U.S. Geological Survey Open-File Report 81-473, p. 160–162.

Pratt, B. R., and James, N. P., 1986, The St. George group (Lower Ordovician) of western Newfoundland: tidal flat island model for carbonate sedimentation in shallow epeiric seas: Sedimentology, v. 33, p. 313–343.

Purdy, E. G., 1963, Recent calcium carbonate facies of the Great Bahama Bank. Pt. 2, Sedimentary facies: Journal of Geology, v. 71, p. 472–497.

Ramos, V. A., 1995, Sudamérica: un mosaico de continentes y océanos: Ciencia hoy, v. 6, p. 24–29.

Ramos, V.A., Jordan, T., Allmendinger, R., Kay, S. M., Cortés, J. M., and Palma, M. A., 1984, Chilenia: Un terreno alóctono en la evolución paleozoica de los Andes Centrales: Buenos Aires, IX° Congreso Geológico Argentino, v. 2, p. 84–106.

Ramos, V. A., Jordan, T., Allmendinger, R., Mpodozis, C., Kay, S. M., Cortés, J. M., and Palma, M., 1986, Paleozoic terranes of the central Argentine–Chilean Andes: Tectonics, v. 5, p. 855–880.

Read, J. F., 1989, Controls on evolution of Cambrian-Ordovician passive margin, U.S. Appalachians, *in* Crevello, P. D., Wilson, J. L., Sarg, J. F., and Read, J. F., eds., Controls on carbonate platform and basin development: Society of Economic Paleontologists and Mineralogists Special Publication 44, p. 147–165.

Read, J. F., and Goldhammer, R.K., 1988, Use of Fischer plots to define third-order sea-level curves in Ordovician peritidal cyclic carbonates, Appalachians: Geology, v. 16, p. 895–899.

Rogers, J. C., 1984, Depositional environments and paleoecology of two quarry sites in the Middle Cambrian Marjum and Wheeler Formations, House Range, Utah: Brigham Young University Geology Studies, v. 31, p. 97–115.

Ross, R. J., 1975, Early Paleozoic trilobites, sedimentary facies, lithospheric plates and ocean currents, *in* Martinson, A., ed., Evolution and morphology of Trilobita, Trilobitoidea and Merostomata: Fossils and Strata, v. 4, p. 307–329.

Sarmiento, G. N., 1986, La biozona de *Amorphognathus variabilis–Eoplacognathus pseudoplanus* (Conodonta), Llanvirniano Inferior, en el flanco oriental de la Sierra de Villicúm: San Juan, Primera Jornada Geológica de la Precordillera, p. 119–123.

Schutter, S. R., 1992, Ordovician hydrocarbon distribution in North America and its relationship to eustatic cycles, *in* Webby, B. D., and Laurie, J. R., eds., Global perspectives on Ordovician Geology: Rotterdam, Netherlands, Balkema, p. 421–432.

Sloss, L. L., 1963, Sequences in the cratonic interior of North America: Geological Society of America Bulletin, v. 74, p. 93–114.

Smosna, R., 1987, Compositional maturity of limestones: a review: Sedimentary Geology, v. 51, p. 137–146.

Sternbach, L. R., and Friedman, G. M., 1984, Deposition of ooid shoals marginal to the Late Cambrian Proto-Atlantic (Iapetus) ocean in New York and Alabama; influence on the interior shelf: Society of Economic Paleontologists and Mineralogists, Core Workshop 5, p. 2–19.

Taylor, J. F., Repetski, J. E., and Orndorff, R. C., 1992, The Stonehenge transgression: a rapid submergence of the central Appalachian platform in the Early Ordovician, *in* Webby, B. D., and Laurie, J. R., eds., Global perspectives on Ordovician geology: Rotterdam, Netherlands, Balkema, p. 409–418.

Thomas, W. A., 1991, The Appalachian-Ouachita rifted margin of southeastern North America: Geological Society of America Bulletin, v. 103, p. 415–431.

Thomas, W. A., and Astini, R. A., 1996, The Argentine Precordillera: a traveler from the Ouachita embayment of North American Laurentia: Science, v. 283, p. 752–757.

Toomey, D. F., and Nitecki, M. H., 1979, Organic buildups in the Lower Ordovician (Canadian) of Texas and Oklahoma: Fieldiana (New Series), v. 2, 181 p.

Vaccari, N. E., 1988, Primer hallazgo de trilobites del Cámbrico inferior en la Provincia de La Rioja (Precordillera septentrional): Revista de la Asociación Geológica Argentina, v. 43, p. 558–561.

Vaccari, N. E., 1994, Las faunas de trilobites de las sucesiones carbonáticas del Cámbrico tardío y Ordovícico temprano de la Precordillera Septentrional, República Argentina [Ph.D. thesis]: Córdoba, Argentina, Universidad Nacional de Córdoba, 271 p.

Van Wagoner, J. C., Posamentier, H. W., Mitchum, R. M., Vail, P.R., Sarg, J. F., Loutit, T. S., and Hardenbol, J., 1988, An overview of the fundamentals of sequence stratigraphy and key definitions, *in* Wilgus, C. K., Hastings, B. S., Posamentier, H. W., Ross, C. A., and Kendall, G. C. St. C., eds., Sea-level changes: an integrated approach: Society of Economic Paleontologists and Mineralogists Special Publication 42, p. 39–46.

Williams, H., and Hiscott, R. N., 1987, Definition of the Iapetus rift-drift transition in western Newfoundland: Geology, v. 15, p. 1044–1047.

GLOSSARY 1. LITHOFACIES OF THE LA FLECHA FORMATION: ROCK TYPES, FEATURES, BIOTA, AND INFERRED DEPOSITIONAL ENVIRONMENTS

Cryptomicrobial laminites

Boundstone; millimeter-scale alternation of micritic to microsparitic dolomite and lime mudstone; peloidal; pale brown to tan, thin to medium beds; millimeter-scale lamination (cryptomicrobial), mud cracks, centimeter-scale tepees, fenestrae, "rip-up" clasts; no fossils, inferred microbial benthic communities. Supratidal to high intertidal mud and sand flats

Thick laminites

Dolomite; carbonate mud, peloidal; microsparitic dolomite; light gray to pale brown, thin to medium beds with thick lamination to very fine bedding; cut and fill structures, mud cracks, "sausage" features; no fossils. Supratidal flats

Stromatolites

Boundstone; millimeter-scale alternation of micritic to microsparitic dolomite and lime mudstone; peloidal; pale brown to very dark gray, medium beds: laterally linked hemispheroidal, stacked hemispheroidal, and digitate stromatolites; no fossils, inferred microbial bentic communities. Intertidal to shallow subtidal

Flat-Pebble rudstone

Sand- and clast-supported conglomerates; rounded to angular clasts of lime mudstone, laminites, ooid and peloid grainstone; rare reworked shell debris; color depends on clasts and matrix, medium to thick, tabular beds with erosive bases; frequent clast imbrication; rare grading; finer textures might show cross-bedding. Supratidal to shallow subtidal

Ribbon limestone

Centimeter-scale interbedded gray limestone, usually peloidal-intraclast grainstone, and pale brown dolomite; thin to medium beds; heterolithic, wavy to flaser bedding; cut and fill structures; rare trilobites. Intertidal to shallow subtidal

Mottled limestone

Carbonate mud- to wackestone, peloidal grainstone and crystalline textures patchilly distributed within beds; pale brown with dark gray mottles (burrowing) or vice-versa; medium to thick beds; scarce trilobites. Mostly shallow subtidal

Ooid and peloid grainstone

Oolitic, ooid-intraclast and peloid intraclast grainstones; light to very dark gray, medium to thick beds; tabular or trough cross-bedding, wave-ripples, planar lamination; scarce trilobites. Shallow subtidal to intertidal sand flats and shoals

Cryptic and thrombolitic microbialites

Microbialitic boundstones; texturally clotted carbonate mud dominates; peloidal textures occur in patches; usually dark gray, small (meter-scale), mound-shaped build-ups; thrombolite-bioherms may be tabular-, mound-, or fan-shaped. Inferred microbial communities; *Renalcis*-like microbial rests, trilobites. Mostly subtidal, occasionally develop into intertidal environment

Mudstone

Burrowed mudstones to pel-bioclastic wackestone and pyritic calcilutites; dark gray to black; burrowed mud- to wackestones are medium to thick bedded; calcilutites are thin bedded; trilobites and lingulid brachiopods. Restricted subtidal

GLOSSARY 2. LITHOFACIES OF THE LA SILLA FORMATION: ROCK TYPES, FEATURES, BIOTA, AND INFERRED DEPOSITIONAL ENVIRONMENTS

Cryptomicrobial laminites

Dolomite; millimeter-scale alternation of carbonate, mud, and peloidal laminae; light gray to tan; thin to medium beds with cryptomicrobial lamination; fenestral fabrics, mud cracks; laterally linked hemispheroidal/stacked hemispheroidal stromatolites; no fossils, inferred microbial mats. Supratidal to intertidal flats

Peloidal grainstone

Peloidal (intraclast) grainstone/ packstone; micritized bioclasts, carbonate "grains" and lumps; light to medium gray; medium to thick beds; usually burrowed, rare planar lamination; *Thallasinoides*-like traces; hardgrounds; calcareous algae, scarce trilobites, gastropods. Shallow subtidal and restricted sand flats

Oolite

Well-sorted oolitic grainstone; millimeter-scale alternation of ooid and peloid grainstone and carbonate mud; oolitic-clasts rudstone; whitish to dark gray; medium to thick beds; tabular and trough cross-bedding; massive beds; millimeter-thick heterolithic bedding; scarce trilobites and brachiopods, nautiloids, and gastropods. Shoals and shallow subtidal sand flats

Intraclast grainstone

Grainstone to fine rudstone of micritic intraclasts, peloids and scarce bioclasts; medium gray; fine to medium massive beds; erosive bases; frequently rests on hardgrounds; nautiloids and gastropods. Shallow subtidal, tidal channels, and lag deposits

Pel-bioclastic wackestone

Burrow-mottled peloidal wacke- to packstone; scarce bioclasts; light to dark gray; medium to thick, burrow mottled beds; frequent firm- or hardgrounds; gastropods and nautiloids; scarce trilobites and brachiopods; conodonts. Shallow subtidal, restricted platform interior

GLOSSARY 3. LITHOFACIES ASSOCIATIONS OF THE SAN JUAN FORMATION: ROCK TYPES, FEATURES, BIOTA, AND INFERRED DEPOSITIONAL ENVIRONMENT

Reef-mounds and patch-reefs

Wacke- to packstone; microbial boundstone; grain- to rudstone as channel-fills; stromatoporoid framestone; massive, mound-shaped bioherms (1–4 m high) and small patch-reefs; cross-bedding in channels and intermound sands; inferred bacterial mats; lithistid sponges, *Calathium*, stromatoporoids, *Girvanella*, pelmatozoans, gastropods, trilobites, brachiopods, nautiloids, calcareous algae. Shallow subtidal, probably just below mean level wave base.

Lithoclast-bioclast grainstone

Bio-intraclastic and peloidal grainstone; medium to thick beds, massive (burrowed); tabular and trough cross-bedding; pelmatozoan fragments, *Nuia*, brachiopods, *Girvanella*, receptaculitids, trilobites, nautiloids, ostracods. Shoals and shallow subtidal sand flats

Oncoidal grainstone

Oncolitic, intra-bioclastic coarse grainstone; fine to medium beds; frequent erosive bases; pervasive bioturbation; pelmatozoan fragments, *Nuia*, brachiopods, *Girvanella*, receptaculitids, trilobites, nautiloids, ostracods. Shoals and shallow subtidal (? to intertidal) sand flats

Litho-skeletal limestones

Skeletal wacke-to packstone; lithoclast grainstone; fine pel-bioclastic grainstone; micro-bioclastic wackestone; medium to thick, burrow-homogenized skeletal wackestone beds; frequent hardgrounds; grainstones are thin, graded, planar, or ripple cross-laminated layers; brachiopods, pelmatozoans, trilobites, calcareous algae, receptaculitids, gastropods, bryozoans, nautiloids, ostracods. Shallow open platform, between fair-weather and storm wave-base

Nodular mudstone and wackestone

Mudstone and pel-bioclastic wackestone; dolomitic shale partings; thin to medium, wavy to nodular (stylonodular) beds; burrow-homogenized wackestones; rare laminated lensoid mudstones with clay seams; trilobites, brachiopods, sponge spicules, calcispheres. Open platform to deep ramp, below storm wave-base

Parted limestone and shale

Lime mudstone, wackestone, pel-bio–intraclast calcarenite with shale partings; minor shale; thin to medium beds of massive to nodular limestones; graded calcarenites; frequent clay seams and shale partings; trilobites, brachiopods, pelmatozoans, bryozoans, calcispheres, *Nuia*. Basin margin to restricted intrashelf basin

Manuscript Accepted by the Society September 4, 1998

Printed in U.S.A.

Geological Society of America
Special Paper 336
1999

Ordovician western Sierras Pampeanas magmatic belt: Record of Precordillera accretion in Argentina

Sonia M. Quenardelle and Victor A. Ramos
Departamento de Ciencias Geológicas, Universidad de Buenos Aires, 1428 Buenos Aires, Argentina

ABSTRACT

Accretion of the Precordillera terrane, as part of a composite terrane of Laurentia derivation, is well founded, mainly based on paleontologic grounds, sedimentologic similarities, and isotopic and paleomagnetic evidence. However, the active proto-margin of Gondwana has not been analyzed to verify if the distribution, age, and composition of the early Paleozoic magmatic rocks are consistent with accretion of the Precordillera.

This chapter presents an overview of the volcanic rocks and granitoids exposed in the western Sierras Pampeanas. The analysis of the composition, structure, timing, and other relevant characteristics help to define a series of pretectonic subduction-related granodiorites, tonalites, and other granitoids associated with volcanic rocks, ranging in age between 515 and 470 Ma. Scarce and volumetrically less important small bodies of syntectonic granitoids range in age from 470 to 450 Ma. Large post-tectonic batholiths, mainly in subcircular shapes of monzo- to syenogranitic composition, were emplaced from 450 to 360 Ma. A magmatic arc along the western Sierras Pampeanas is possible to reconstruct based on the available data. The time of emplacement of the subduction-related igneous rocks, as well as the collisional and postcollisional granitoids, matches the time of accretion and final amalgamation of the Precordillera terrane inferred from the sedimentary record. The time constraints of both areas, Precordillera and western Sierras Pampeanas, show a coherent pattern that reinforces the interpretation of Precordillera as a far-traveled terrane exotic to Gondwana.

INTRODUCTION

The identification of exotic terranes amalgamated to the protomargin of Gondwana during early Paleozoic times, led to the identification of an active margin in western Sierras Pampeanas. This active margin is represented by a series of calc-alkaline granitoids and volcanic rocks of Cambro-Ordovician age developed over 1,300 km (Fig. 1) along the western border of the Sierras Pampeanas (Ramos et al., 1984, 1986). Although these early hypotheses were challenged by other tectonic models, it was not until recent times that the allochthonous nature of the Precordillera was generally accepted (Astini et al., 1995; Dalziel et al., 1996).

The new biostratigraphic, sedimentologic, and paleomagnetic data tightly constrained the time of rifting and drifting of the Precordillera from Laurentia, as well as the beginning of docking and amalgamation of this terrane to the southwestern part of South American protomargin of Gondwana (Thomas and Astini, 1996; Ramos et al., 1996). This chapter views in an independent way the evidence of magmatic activity of the South American margin in order to verify if the period of Precordilleran's drift combined with the sedimentologic and metamorphic evidence for its accretion is compatible with the tectonic setting and age of magmatism along the western Sierras Pampeanas.

EARLY PALEOZOIC TECTONIC SETTING

Two different models have been proposed to explain the accretion of the Precordillera to the protomargin of Gondwana. One of the models proposed an independent microcontinent or

Quenardelle, S., and Ramos, V. A., 1999, Ordovician western Sierras Pampeanas magmatic belt: Record of Precordillera accretion in Argentina, *in* Ramos, V. A., and Keppie, J. D., eds., Laurentia-Gondwana Connections before Pangea: Boulder, Colorado, Geological Society of America Special Paper 336.

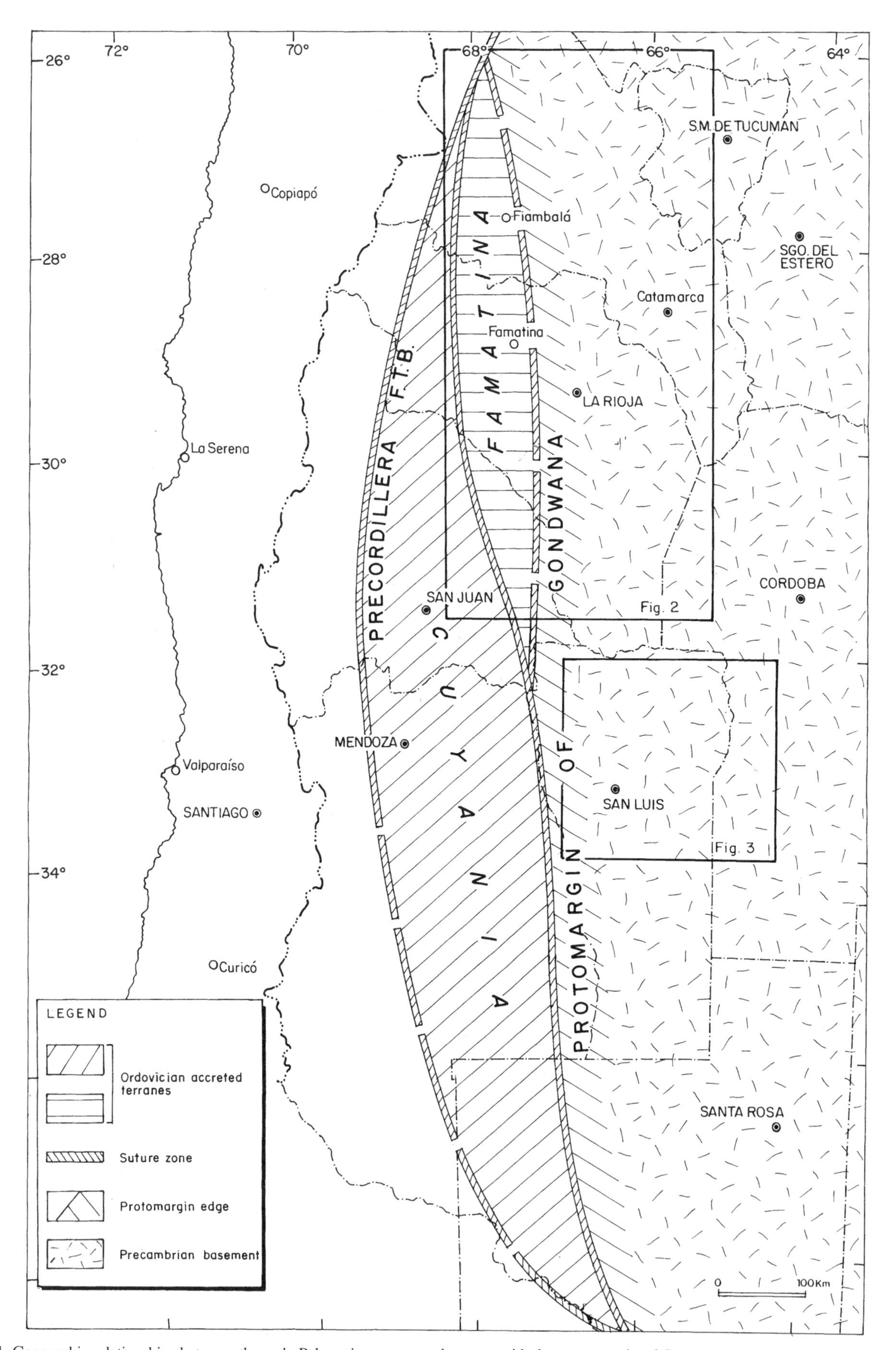

Figure 1. Geographic relationships between the early Paleozoic terranes and sutures with the protomargin of Gondwana (modified from Ramos, 1989).

microplate, detached from Laurentia during early Cambrian time, that collided against Gondwana during Middle to Late Ordovician time (Ramos et al., 1986; Benedetto and Astini, 1993; Astini et al., 1995, 1996). The second model proposed a continent-to-continent collision between Laurentia and Gondwana during Early to Middle Ordovician time, from 487 to 467 Ma. Subsequently, during the Late Ordovician the detachment of the two continents left behind the Precordillera terrane on the Gondwanan side, with the opening of an ocean on the western side of Precordillera (Dalla Salda et al., 1992a,b: Dalziel, 1992, 1993, 1997; Dalziel et al., 1994, 1996).

Both proposals required an active early Paleozoic margin of the Sierras Pampeanas, and explained the magmatism and the Ordovician deformation as a result of a collisional orogeny known as the Ocloyic deformation (Ramos, 1986; Dalla Salda et al., 1992a,b).

The microcontinent hypothesis required a second early Paleozoic accretion to close the ocean that bounded the western Precordillera. The accretion of the Chilenia microcontinent produced a second foreland basin and a shifting of the magmatic activity to the Pacific margin (Ramos et al., 1984). This collision occurred either in Late Devonian time (Ramos et al., 1986) or in Early Devonian time (Astini, 1996). Sedimentologic studies of these foreland basins and geochronologic data of peak metamorphism and associated deformation suggest an Early Devonian age for the beginning of the accretion of Chilenia.

Recently, petrologic studies and dating performed in the basement of Precordillera and adjacent terranes (Kay et al., 1996; Tosdal, 1996) have demonstrated that Precordillera was part of a composite terrane of Grenvillian age, named Cuyania (see Fig. 1), which was accreted as a whole to the Gondwana margin (Ramos et al., 1996).

MAGMATIC ACTIVITY OF SIERRAS PAMPEANAS

Although the two contrasting models require an active margin in the Sierras Pampeanas, the authors of the continent-to-continent collision model between Laurentia and Gondwana grouped all the magmatic activity in a single orogenic belt: the Famatinian orogen (Dalla Salda et al., 1992a,b). These workers recognized a subduction-related magmatism (Dalla Salda et al., 1995a,b; López de Luchi and Dalla Salda, 1995, 1996) followed by a generalized collisional magmatism widely distributed across the western and eastern Sierras Pampeanas. According to this model, most of the ductile deformation seen in the Sierras Pampeanas was due to this Famatinian deformation.

On the other hand, based on the limited geochronologic data available in the Sierras Pampeanas, the paleogeographic distribution of the magmatism, as well as the time of uplift of the eastern and western Sierras Pampeanas, Ramos (1988) has recognized two different orogens. He identified a Brasiliano belt developed along eastern Sierras Pampeanas over 1,000 km long with main activity during Late Proterozoic times (Ramos, 1989). Recent geochemical studies in the eastern belt proposed the subduction-related nature of part of this magmatism (Lira et al., 1997). This proposal was also based on the uplift time dated by the angular unconformity between the granitoids of the eastern belt and the rhyolitic plateau of Oncán, located in the northern sector of eastern Sierras Pampeanas. These rhyolites have a radiometric age of 494 ± 11 Ma (Rb/Sr) (Rapela et al., 1991) and overlie metamorphic and igneous rocks that evolved at mid-crustal levels, recording an exhumation of at least 20 km.

The second belt was concentrated along the western border of Sierras Pampeanas, and is represented by Late Cambrian and Ordovician magmatic (Ramos, 1988) and metamorphic rocks (Rapela et al., 1990). Although Rapela et al. (1990) recognized the two belts, they were considered as nearly coeval and interpreted as inner and outer arcs, produced during the same subduction regime.

Recent studies by Kraemer et al. (1995) demonstrated that the eastern arc was located in a different terrane, and that the two belts were separated during Late Proterozoic times by an ocean. Recent, preliminary radiometric dates on the late granitic intrusives of the eastern magmatic belt indicates that magmatism extended until 530–520 Ma (Early Cambrian) (Rapela and Pankhurst, 1996; Stuart Smith et al., this volume).

The new data base available from the Sierras Pampeanas belt shows that it contains two magmatic belts developed in two distinct orogens involving different subduction regimes. A suture occurs in the basement of the Sierras Pampeanas, separating the two magmatic belts (Escayola et al., 1996).

As the western Sierras Pampeanas belt was approximately coeval with the period of accretion of the Precordillera, it is described below in order to evaluate whether these data are consistent with the evidence collected by researchers working on the sedimentary strata and low-grade metamorphic rocks of the Precordillera.

CAMBRO-ORDOVICIAN WESTERN SIERRAS PAMPEANAS BELT

In order to describe the magmatic activity of this belt, it has been divided in a series of discrete segments. Knowledge of these segments is not homogeneous, and therefore the amount of information and the available geochronologic petrologic data are uneven.

Sierra de San Luis granitoids

These granitoids belong to the southern sector of Sierras Pampeanas (see Figs. 1 and 2), where several units have been identified from the early Paleozoic to the Carboniferous. A metamorphic complex is the country rock of these intrusives. Several north-south–trending subparallel metamorphic belts have been recognized. One of the belts is composed of gneiss, migmatite, and amphibolite with polyphase deformation and amphibolite-grade metamorphism. This belt has a sharp contact with mica-quartz schists that have a low-pressure greenschist to lower

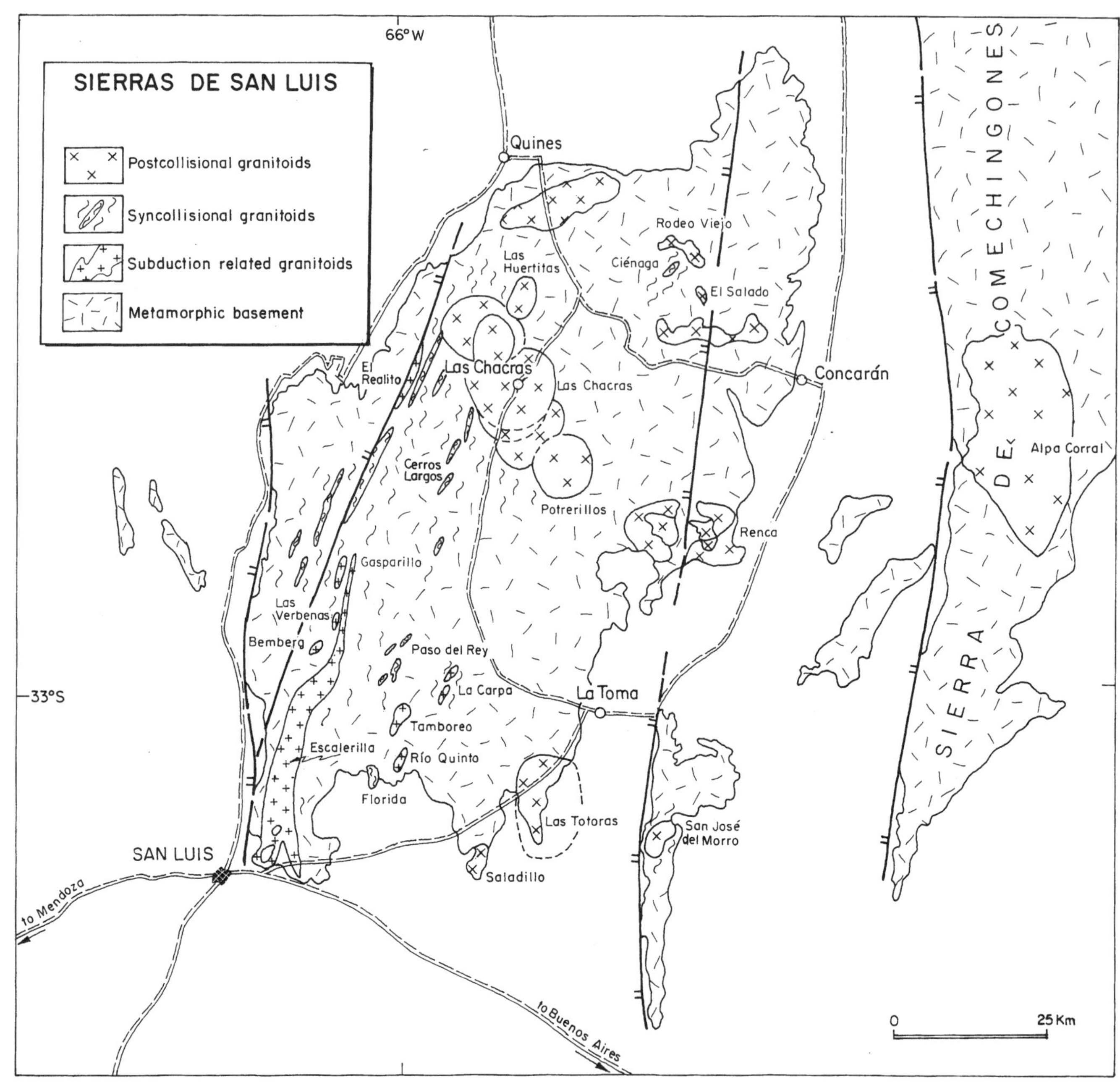

Figure 2. Granitoids in the Sierra de San Luis (based on Ortiz Suárez et al., 1992; Llambías et al., 1996; Ortiz Suárez, 1996; Sato et al., 1996). Location in Figure 1.

amphibolite grade, and that record a polyphase deformation history. This last belt transitionally grades to the San Luis Formation of Prozzi and Ramos (1988), which is composed of slate, phyllite, metaconglomerate, and acidic metavolcanic rocks. This last unit reached greenschist grade, but still preserves original bedding and is affected only by the last deformational episode (Ortiz Suárez et al., 1992). Recent U/Pb dating in zircons from acidic tuffs yielded ages of 529 ± 12 Ma, indicating an Early Cambrian age for the San Luis Formation (Söllner et al., 1998). Based on the degree of deformation and the relationship with the metamorphic country rock, prekinematic, synkinematic, and postkinematic granitoids have been recognized in both belts by Ortiz Suárez et al. (1992).

Prekinematic granitoids. La Escalerilla, Las Verbenas, Tamboreo, Bemberg, Río Quinto, and El Realito are the prekinematic granitoids (Fig. 2). These granitoids, as well as the metamorphic aureoles in the country rocks, are characterized by foliation with a regional northwest trend similar to the main schistosity of the metamorphic rocks. These granitoids have a sharp contact with the wall rock, fine-grained margins, and an aureole of contact metamorphism, suggesting their emplacement in shallow crustal levels. These rocks are emplaced in the slate and phyllite belt as well as in the adjacent schist (Ortiz Suárez et al., 1992). Tonalite is dominant (Tamboreo, Bemberg, Las Verbenas, and Gasparillo) over the granite (La Escalerilla, Río Quinto). Within the El Realito stock, some tonalites have also been identified, and interpreted as prekinematic granitoids (Brogioni et al., 1994). The superimposed metamorphic foliation of these granitoids has been ascribed to the Ocloyic deformational phase by Sato et al. (1996) that occurred in middle–late Ordovician time (Ocloyic collision of Ramos, 1986; D2 deformational episode of Rapela et al., 1992).

The Tamboreo Tonalite is a north-northeast elongated stock of 14.5 km^2 with aligned mafic inclusions parallel to the main foliation of the stock. Oblique shear zones deflected the main foliation and some aplites of granitic composition are present with a north-south trend (Sato et al., 1996).

Like the Tamboreo tonalite, the Las Verbenas Tonalite is a north-south–trending elongated stock of 10 km^2. It also has a thermal metamorphic aureole and is in contact with the La Escalerilla Granite along the eastern margin. The inner part of the stock has a strong foliation that produced a gneissic texture. It also has been overprinting by ductile shear zones, and contains small, elongated enclaves of fine-grained mafic rocks (Sato et al., 1996).

The Bemberg Tonalite is composed of several northeast-trending stocks, with a combined exposure of 4.9 km^2. It is composed of three different facies: tonalite, melanotonalite, and quartz gabbro. The dominant tonalitic facies consists of quartz, plagioclase, biotite, and hornblende; apatite and zircon are the main accessory phases, occasionally with muscovite, clinozoisite, and opaque minerals. In the other facies the hornblende content increases as the quartz decreases. The contact among different facies is sharp to transitional, indicating coeval emplacement (Sánchez et al., 1996). In the previous bodies, they have also fine-grained mafic enclaves and xenoliths of schistose metamorphic rocks. They are crossed by tonalitic, granitic, and pegmatitic dikes (Sato et al., 1996).

The Gasparillo Tonalite has a rhomboidal shape and is emplaced between gneiss and mica-schist. The western contact with the gneiss is a ductile shear zone that affects the tonalite as well as the wall rock, with variably developed cataclastic and mylonitic textures. The thermal contact aureole is difficult to recognize due to the superimposed metamorphic foliation. Tonalite is dominant with subordinate melanocratic facies that consist of plagioclase (andesine), quartz, biotite, and K-feldspar in accessory proportions. Amphibole is again dominant in the melanocratic facies, plagioclase is more calcic, and the texture is ophitic. There are frequent biotite gneiss roof-pendants along the western contact, where original contacts are visible due to the minimum deformation of the stock (Llambías et al., 1996).

The La Escalerilla granite is conspicuous due to its axial ratio (52 km long in continuous outcrop and 2–3 km wide), and a strike bend, across 33°S lat. (see Fig. 2) that may indicate some synplutonic deformation (Ortiz Suárez et al., 1992). The eastern border of the pluton also has a ductile shear zone (Llano et al., 1987) along the contact with the mica-schist. To the southwest, it grades to gneiss and migmatite (Costa, 1983), and to the northwest it has either an intrusive or sheared contact with the Las Verbenas Tonalite. Monzogranitic to granodioritic facies are present, partially porphyritic, with microcline megacrysts, and biotite schist enclaves. The pluton is crossed by aplitic dikes, as well as the country rocks (Sato et al., 1996).

The Río Quinto Granite is a set of minor bodies, smaller than 1 km^2, emplaced in slate and phyllite of the San Luis Formation. Epizonal monzogranite with chilled borders and porphyritic facies is strongly deformed by ductile shear that transformed the granitic rocks in protomylonite (Carugno Durán et al., 1992).

The El Realito stock, an elongated body 23 km long, is located in the northern sector of Sierra de San Luis. It consists of tonalite with subordinates granodiorite, and abundant mafic enclaves. Monzogranite is emplaced in the core of the stock. The tonalite is emplaced in mica schists, where no metamorphic aureole was recognized. The tonalitic unit shows metamorphic recrystallization in greenschist facies (Brogioni et al., 1994).

The geochemical data of this suite of prekinematic granitoids show a calc-alkaline trend, where tonalite is the most mafic facies with an SiO_2 content varying from 48 to 67%, the granite varying from 68 to 76% (Fig. 3).

The alumina ratio (A/CNK) ranges from metaluminous to slightly peraluminous in tonalite, and slightly peraluminous in granite. Among the tonalites, only the Bemberg evolved from low-K to high-K while the remaining stocks are high-K. The rare earth element (REE) patterns are similar to the Bemberg and Gasparillo tonalites, as well as for the La Escalerilla Granite: LILE relative enrichment, and depleted HFS elements, typical of subduction-related environments (Fig. 4). The Rb, Sr, and Ba vary between the tonalitic to granitic units as the tonalites have

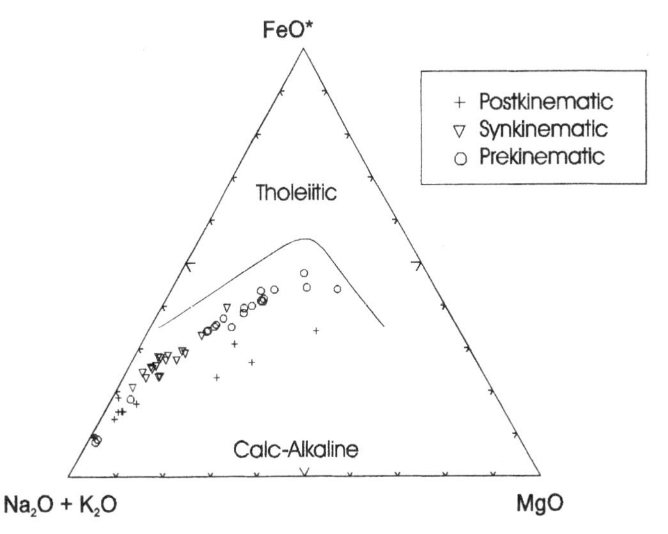

Figure 3. AFM diagram of San Luis granitoids. Note that all available data fall in the calc-alkaline field (data from Quenardelle, 1995; Llambías et al., 1996, and Sato et al., 1996).

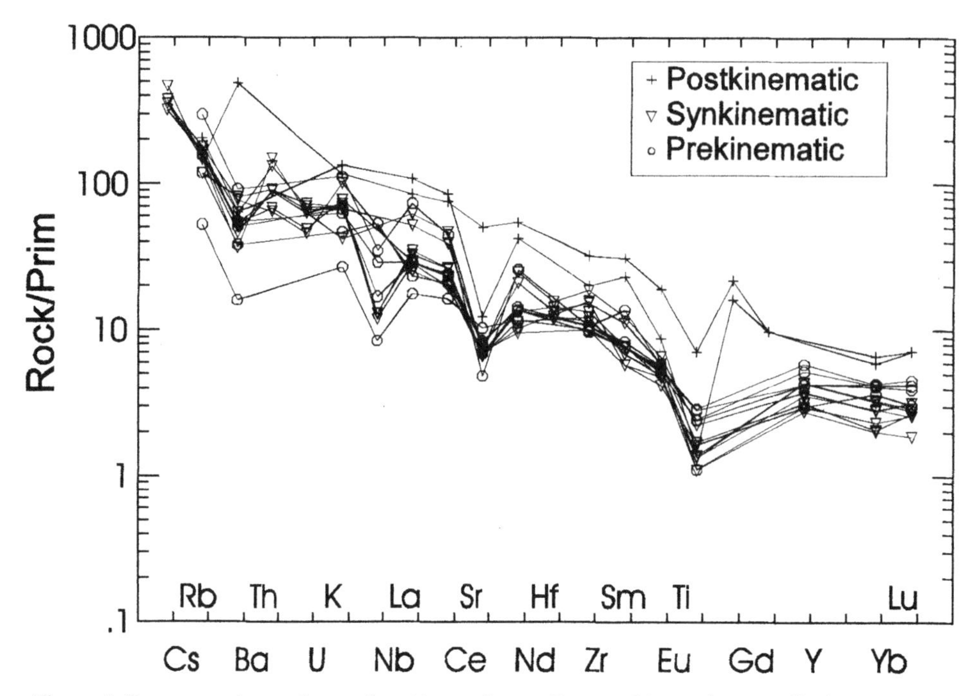

Figure 4. Representative analyses of prekinematic tonalites, synkinematic granodiorites, and postkinematic granites normalized to the primordial mantle (data from Quenardelle, 1995; Llambías et al., 1996; Sato et al., 1996).

smaller proportion of fractionated plagioclase, and do not have K-feldspar. The granites have a larger plagioclase fractionation, and also have K-feldspar. REE patterns are compatible with orogenic environments (Fig. 5), with LREE enrichment, small negative Eu anomalies, and gentle HREE slope. Most of the tonalites have lower total REE contents (REE <43 ppm), lower La/Yb ratios, and larger negative Eu anomaly. The La Escalerilla Granite is the unit with larger total REE contents (>202 ppm), and higher enrichment of LREE, and depleted HREE (Sato et al., 1996).

The chemical characteristics, as well as the geologic setting indicate that these are arc granitoids, emplaced previously to the Ocloyic phase that took place at about 454 Ma. The Bemberg stock yielded 513 Ma by Rb/Sr method although the error is considerable (Sato et al., 1996). No reliable geochronologic data were available in these intrusives until this present volume. New U/Pb dating in zircons in part of these granitoids confirm pre-Ocloyic ages (Stuart Smith et al., this volume).

Synkinematic granitoids. A suite of synkinematic stocks, emplaced along the Ocloyic deformation (Llambías et al., 1991) are located in the Sierra de San Luis. These stocks are emplaced in a wide belt within medium grade schist, and are closely linked with pegmatite swarms. There are several bodies up to 1.4 km^2 such as Río de La Carpa, Cruz de Caña, Cerros Largos, Paso del Rey Norte, and Sur (Fig. 2). Most of them have lensoid shape, although some are more elongated with folded borders, conformable with the wall rocks and elongated in a northeast trend, with axial ratios between 3.7 and 18.0 (Llambías et al., 1996). Small bodies have uniform composition, but the larger ones have a K-enrichment in the inner parts. They do not have chilled margins. Granodiorite is the dominant composition with subordinate monzogranite and tonalite. They consist of quartz, plagioclase, microcline, biotite, and muscovite, with apatite, zircon, garnet, and fibrolite as accessory phases. The modal quartz content is high, between 33 and 50%, and they are neither texturally nor mineralogically similar to arc granodiorites (Llambías et al., 1996). They have sharp contacts with the wall rocks and no contact aureoles, indicating a low thermal contrast. The inner part of these bodies has a rudimentary foliation due to the alignment of the micas, which do not form continuous bands.

The La Ciénaga and La Represa stocks are included as synkinematic by Ortiz Suárez (1996). La Ciénaga is interpreted as a phacolith of 4 km^2 and northeast trend. They are monzogranite and leucogranite with quartz, microcline, muscovite, and occasional garnet. These stocks also have pegmatoid differentiates and quartz veins in the wall rock.

The geochemical characteristics of the synkinematic granitoids indicate a meta- to peraluminous composition (A/CNK between 1.0 and 1.2), with a peraluminous modal mineralogy, high-K, and calc-alkaline trend, although they are not a calc-alkaline suite, with SiO_2 ranging from 66 to 75%. Trace elements indicate high LILE values, and relatively low high field strength (HFS) elements. The REE patterns show slight LREE enrichment and moderate La/Sm ratio, with a small negative Eu anomaly and flat HREE pattern with gentle Sm/Yb slope. The REE patterns are dominantly flat for the different bodies (Fig. 5).

The age of these bodies has been established at 454 ± 12 Ma by whole-rock Rb/Sr isochron, with an $^{87}Sr/^{86}Sr$ initial ratio of 0.712, while the K/Ar biotite ages obtained by Varela et al. (1994)

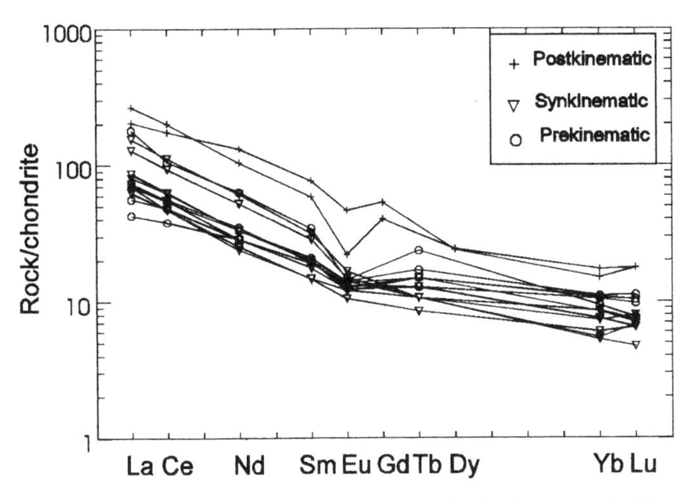

Figure 5. Rare earth element (REE) pattern of San Luis granitoids. Note the similar trend of synkinematic and prekinematic granitoids, less enriched in REE than the postkinematic granitoids (data from Quenardelle, 1995; Llambías et al., 1996; Sato et al., 1996). Symbols as in Figure 3.

are 391–372 Ma. It is inferred that the pluton cooled slowly, approximately from 500° to 300°C, and that after deformation the granitoids recovered by recrystallization, eliminating the internal strain of the minerals (Llambías et al., 1996).

The synkinematic magma evolution was confined to the crust, as inferred from the high $^{87}Sr/^{86}Sr$ initial ratio and the trace element and RRE patterns. This magma falls in the syncollisional granite field in the diagram of Batchelor and Bowden (1985).

Postkinematic granitoids. The post-tectonic stocks have a striking discordance with the metamorphic wall rock (Ortiz Suárez et al., 1992). They have subcircular outlines, well-developed contact aureoles, and roof pendants of deformed metamorphic rocks. The Las Chacras–Piedras Coloradas, and the Renca batholiths, as well as San José del Morro, Paso del Tigre, Saladillo, and Las Totoras stocks are postkinematic plutons (Fig. 2). Some authors have distinguished early postkinematic intrusives such as Rodeo Viejo and El Salado with a north-northwest trend, emphasizing that post-tectonic intrusives have a defined northwest trend (see Ortiz Suárez, 1996).

The Las Chacras–Piedras Coloradas, the largest batholith with an exposed surface of approximately 500 km², is composed of four stocks: Potrerillos, Las Chacras, La Mesilla, and Las Huertitas. There are distinctive facies and textures in each stock, and the contacts among them are transitional. They consist of microcline megacrysts, plagioclase, quartz, biotite, and hornblende; apatite, sphene, and opaque minerals are accessory phases. Rock types are monzogranite with syenite, and quartz-monzonite enclaves, granitic and metamorphic xenoliths, and aplitic, pegmatitic, and microgranitic dikes.

Las Huertitas stock has high and anomalous U-Th and REE contents. Wall rocks of these intrusives consist of mica-quartz schist that grades to gneiss, which in sectors has sericitic nodules with relic sillimanite that may represent some contact metamorphism. Miarolithic structures are frequent, indicating the epizonal character of the intrusives (Brogioni, 1987, 1991, 1992, 1993).

The chemical characteristics as illustrated in Table 1 indicate a meta- to slightly peraluminous composition (A/CNK between 0.84 and 1.2), high-K calc-alkaline trend, and relatively high field strength elements contents. The age of these rocks has been established by the Rb/Sr method at 412 ± 30 Ma with an initial ratio of 0.70348 for the equigranular La Mesilla stock, while the cooling age of the same stock was 322 ± 9 Ma. Las Chacras stock in the amphibole-biotite facies is dated at 317 ± 54 Ma, while in the facies without amphibole is 307 ± 72 Ma (Brogioni, 1993). These ages agree with the ages obtained by K/Ar in amphibole (336 ± 17 Ma), and in biotite (321 ± 16 Ma) according to Brogioni (1987). The Potrerillos stock has a K/Ar age in biotite of 335 ± 17 Ma, equivalent to the other stocks.

The Renca batholith of 270 km² has a main monzogranitic pluton with porphyritic and equigranular facies and a series of minor intrusives. These granitoids have microcline megacrysts, plagioclase, quartz, biotite, and amphibole with apatite and sphene as accessory phases. Wall rock consists of mica-quartz schist, with sharp contacts with the intrusive rocks denoting a diapiric emplacement (López de Luchi, 1987a). The intrusive has not produced neoblasts during contact metamorphism, with the only exception of the most pelitic levels where sillimanite and andalusite have been observed associated with biotite. Metamorphic foliation near the contact with the adjacent schists is lost and is superimposed by development of a granoblastic fabric (López de Luchi, 1993). Fine-grained mafic enclaves with microcline and plagioclase megacrysts are common, as well as equigranular mafic and metamorphic rock enclaves in the porphyritic facies. There are also schlieren in the same facies (López de Luchi, 1987b).

The geochemistry indicates a metaluminous to slightly peraluminous composition, with a calc-alkaline trend. The Rb/Sr age is 415 Ma (Halpern et al., 1970), although the isochron included samples of Renca and Las Chacras stocks.

The San José del Morro stock, exposed in the southeast sector of Sierra de San Luis, has a north-northeast trend and a 66-km² area. There are two dominant facies: a porphyritic monzogranite and a monzonite with high mafic contents. The monzogranite has perthitic microcline megacrysts, plagioclase, quartz, and biotite with sphene and apatite as accessory phases. The monzonite has equigranular textures, with a tendency to develop amphibole cummulates, and consists of plagioclase, microcline, amphibole and biotite, with quartz, sphene, apatite, pyrite, magnetite, and ilmenite as accessory phases. The contact between the two facies is transitional to sharp, showing a low thermal contrast. A hybrid facies is observed in some sectors, with evidence of mingling and ocellar quartz. Both facies have mafic enclaves, with pertithic microcline megacrysts. Wall rocks of this stock are mica schist and strongly foliated gneiss (Quenardelle, 1993). There are abundant pegmatitic dikes in the stock, as well as in the country rocks, of syntectonic and post-tectonic emplacement (Llambías and Malvicini, 1982). The con-

TABLE 1. REPRESENTATIVE ANALYSES FROM GRANITOIDS OF SIERRA DE SAN LUIS

	Postkinematic		Synkinematic			Prekinematic		
	Sig69	Sig86	L94-7	L94-2	L94-3	S174	SI67	19413
SiO_2	49.77	72.16	69.04	68.36	66.68	52.66	72.27	63.01
TiO_2	1.71	0.36	0.33	0.42	0.55	0.68	0.27	0.58
Al_2O_3	13.49	13.90	16.26	16.36	17.06	17.40	14.50	15.30
Fe_2O_3	8.53	1.89	3.31	3.67	4.62	n.d.	n.d.	n.d.
FeO	n.d.	n.d.	n.d.	n.d.	n.d.	7.92	1.76	5.66
MnO	0.13	0.05	0.09	0.07	0.07	0.18	0.05	
MgO	8.03	0.41	0.96	1.25	1.47	6.40	0.40	2.50
CaO	8.51	1.20	3.76	3.87	4.54	8.40	1.68	5.56
Na_2O	2.17	3.48	3.16	3.03	3.17	2.66	2.90	2.30
K_2O	4.72	5.53	3.05	2.82	1.74	1.10	4.61	2.56
P_2O_5	0.64	0.05	0.14	0.17	0.19	0.13	0.06	0.25
H_2O	1.49	0.54	n.d.	n.d.	n.d.	n.d.	n.d.	n.d.
LOI	n.d.	n.d.	0.60	0.74	0.79	1.68	0.93	1.23
Total	99.19	99.57	100.70	100.96	100.88	99.21	99.43	98.95
Rb	102.00	146.00	112.00	116.00	105.00	37.00	210.00	110.00
Ba	3824.00	513.00	424.00	518.00	295.00	127.00	734.00	438.00
Sr	1196.00	301.00	164.00	179.00	182.00	247.00	117.00	166.00
Nb	n.d.	n.d.	11.00	n.d.	n.d.	7.00	29.00	24.00
Zr	396.00	254.00	146.00	196.00	239.00	125.00	134.00	132.00
Y	n.d.	n.d.	22.00	16.00	19.00	21.00	15.00	21.00
La	67.50	87.60	26.60	42.40	51.00	14.10	59.10	23.10
Ce	151.00	174.00	54.00	81.00	97.00	33.00	92.00	47.00
Nd	82.80	65.60	21.00	33.00	39.00	18.00	40.00	21.00
Sm	15.60	11.90	4.06	5.80	6.41	3.87	6.98	4.24
Eu	3.60	1.70	0.92	1.15	1.29	1.02	1.11	1.08
Gd	14.70	11.10	n.d.	n.d.	n.d.	n.d.	n.d.	n.d.
Tb	n.d.	n.d.	0.70	0.50	0.50	0.60	1.00	0.80
Dy	8.20	8.30	n.d.	n.d.	n.d.	n.d.	n.d.	n.d.
Yb	3.30	3.80	1.86	1.20	1.61	2.30	2.07	2.42
Lu	0.60	0.60	0.24	0.23	0.27	0.33	0.26	0.38

Note: Data from Quenardelle, 1995; Llambías et al., 1996; and Sato et al., 1996.
n.d. = no data.

tact between the granites and the wall rock is sharp, indicating a high thermal contrast. There are no mineral changes in the wall rocks, although the granite has roof pendants with conspicuous variations in the foliation of the metamorphic rocks.

The monzogranitic pluton has a calc-alkaline affinity, metaluminous to slightly peraluminous with high alkali contents, dominance of K_2O over Na_2O, and low CaO, with enrichment of Sr, Ba, and LREE, small negative Eu anomaly, and general moderate to high REE pattern slope. The magma was high temperature and H_2O-poor, with high K_2O, in both facies (Quenardelle, 1995). This body gave a whole-rock Rb/Sr age of 382 ± 4 Ma, with an $^{87}Sr/^{86}Sr$ initial ratio of 0.70602 (Varela et al., 1994), and a K/Ar age of 390 ± 10 Ma in biotite and of 360 ± 20 Ma in amphibole (Lema, 1980).

The early postkinematic granites of Rodeo Viejo and El Salado stocks have exposed areas of 30 and 3.5 km^2, respectively (Ortiz Suárez, 1996). These intrusives have sharp contacts, partially conformable or interfingered with metamorphic rocks, with frequent metamorphic enclaves. The tonalite, the dominant facies, with subordinate quartz-diorite, consist of plagioclase, quartz, variable amount of biotite and amphibole, with muscovite, sphene, apatite, occasional tourmaline, and garnet as accessory minerals. The intrusive has produced a conspicuous metamorphic aureole in the mica schist wall rock, with scarce garnet or sillimanite porphyroblasts, belonging to hornblende hornfels facies with pressure higher than 2.5 kb (Ortiz Suárez, 1996).

The Paso del Tigre stock is a small subcircular body emplaced in the El Salado stock. It consists of equigranular granite and leucogranite, with quartz microcline, plagioclase, and biotite. It is associated with the late stages of Las Chacras–Piedras Coloradas batholith (Ortiz Suárez, 1996).

Famatina granitoids and volcanic rocks

The Famatina system is exposed in a series of ranges located between eastern and western Sierras Pampeanas (Fig. 6). This system comprises a well-developed Phanerozoic sequence deposited on a crystalline basement of Precambrian to early

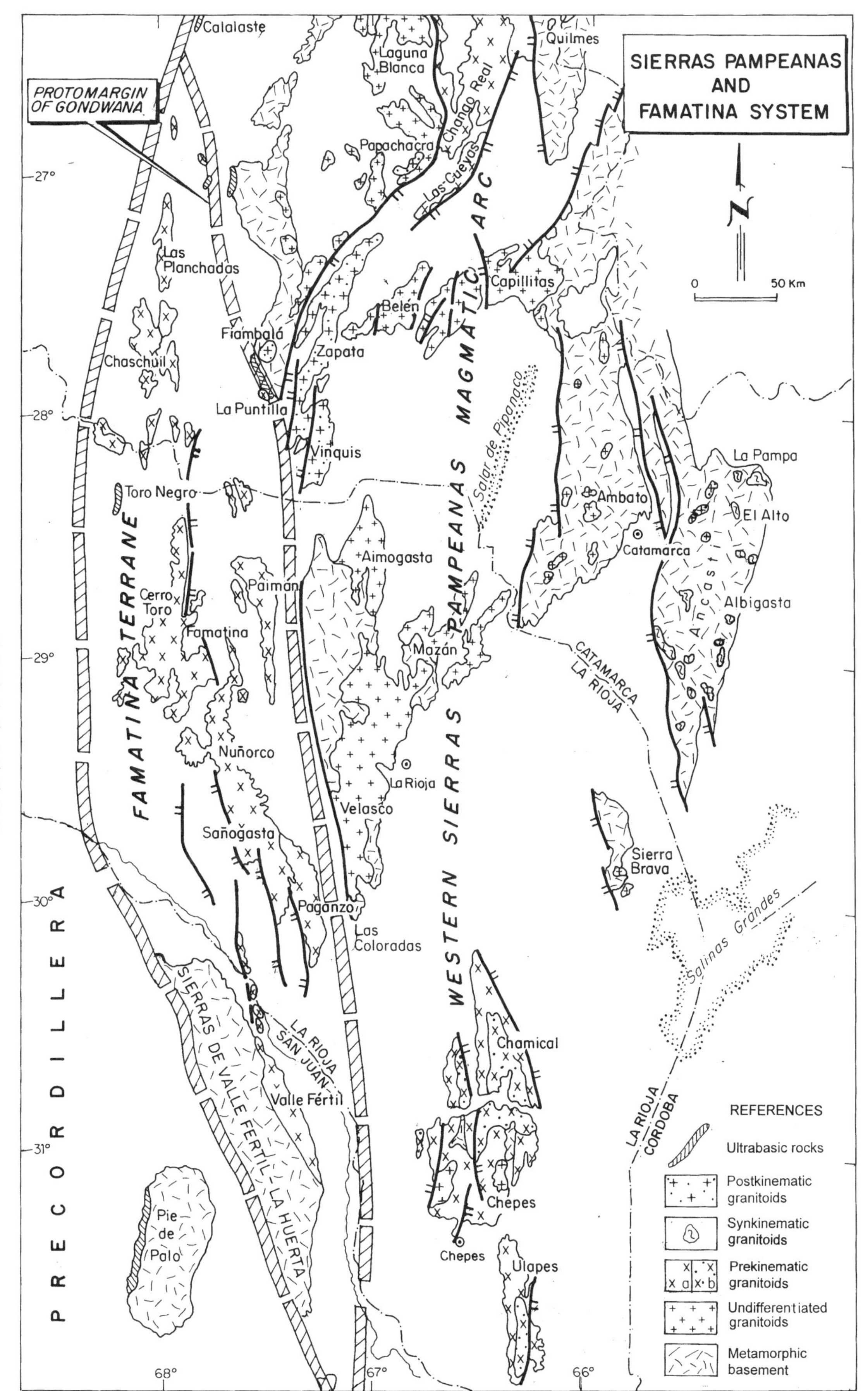

Figure 6. Ordovician magmatic belt of western Sierras Pampeanas and Famatina between Chango Real in the north and Ulapes in the south, with names of the main plutons (see location in Fig. 1). Prekinematic a: mostly tonalites; b: mostly granodiorites.

Paleozoic age. One of its distinctive characteristic is the presence of granitoids with volcanic and sedimentary sequences, all of Ordovician age (Toselli et al., 1987).

Early Paleozoic volcanic rocks. Both slopes of the Sierra de Famatina, as well as the northern continuation in the Sierras de Las Planchadas and Narváez, expose a thick sequence of synsedimentary volcanic rocks associated with pyroclastic rocks of Tremadoc to Llanvirn age (Aceñolaza et al., 1996). The pyroclastic rocks are dominantly rhyolitic in composition, and lava flows include calc-alkaline dacites, andesites, and basalts. Sandstones and shales containing abundant trilobites, brachiopods, and graptolites of Arenig age are interbedded with the volcanic rocks (Aceñolaza and Toselli, 1984).

Sedimentologic studies in these sequences have demonstrated that both the subaqueous volcanic rocks, as well as the sedimentary sequences, were deposited in an extensional regime controlled by faults typical of an intra-arc rift setting (Mángano, 1993). The petrography of the pyroclastic rocks, associated with the fauna-bearing levels, indicates that the subaqueous eruptions were produced in relative shallow water environment not far from an emergent continent.

Chemical analyses of lava flows and pyroclastic rocks (see Table 2) indicate that the dominant group is composed of andesites and dacites, followed by rhyolites and basalts (Fig. 7A). This first group consists of porphyritic rocks with zoned plagioclase phenocrysts, and biotite, amphibole, and pyroxene, generally heavily altered to chlorite and actinolite. The alteration minerals occur in assemblages typical of very low-grade metamorphism. Basalts are also deeply altered with clinopyroxene and amphibole relics. Rhyolitic rocks are also porphyritic with quartz and plagioclase phenocrysts in a fine matrix. Pyroclastic rocks correspond to tuffs, ignimbrites, and breccia with abundant lithoclasts (Toselli, 1992).

The synsedimentary volcanic rocks have typical calc-alkaline geochemical characteristics (Fig. 7B). Basalts are enriched in LILE (Fig. 8), when normalized to mid ocean ridge basalt (MORB), with negative anomalies of Nb, Zr, and Ti. These characteristics indicate a subduction component, probably the addition of hydrated fluids to the asthenospheric wedge. The high $^{87}Sr/^{86}Sr$ ratio (0.707–0.7095) of the basalts has been interpreted as evidence of high Sr-ratio sediment supply to the mantle source instead of crustal contamination processes during the rise of magma to the surface (Mannheim, 1993b).

Famatina basalts are typical island arc tholeiites (Fig. 9) according to Mannheim (1993b). However, as the basalts are minor components in the arc, and the bulk composition of the magmatic rocks of Famatina are granodiorites and granites, it is interpreted that the arc was developed in continental crust. The arc tholeiites are related to the intra-arc extensional regime as proposed by Mángano (1993).

The chemical characteristics of intermediate to acidic volcanic rocks indicate peraluminous composition, high Rb and high $^{87}Sr/^{86}Sr$ ratio (0.720), compatible with a syncollisional trend (Mannheim, 1993b).

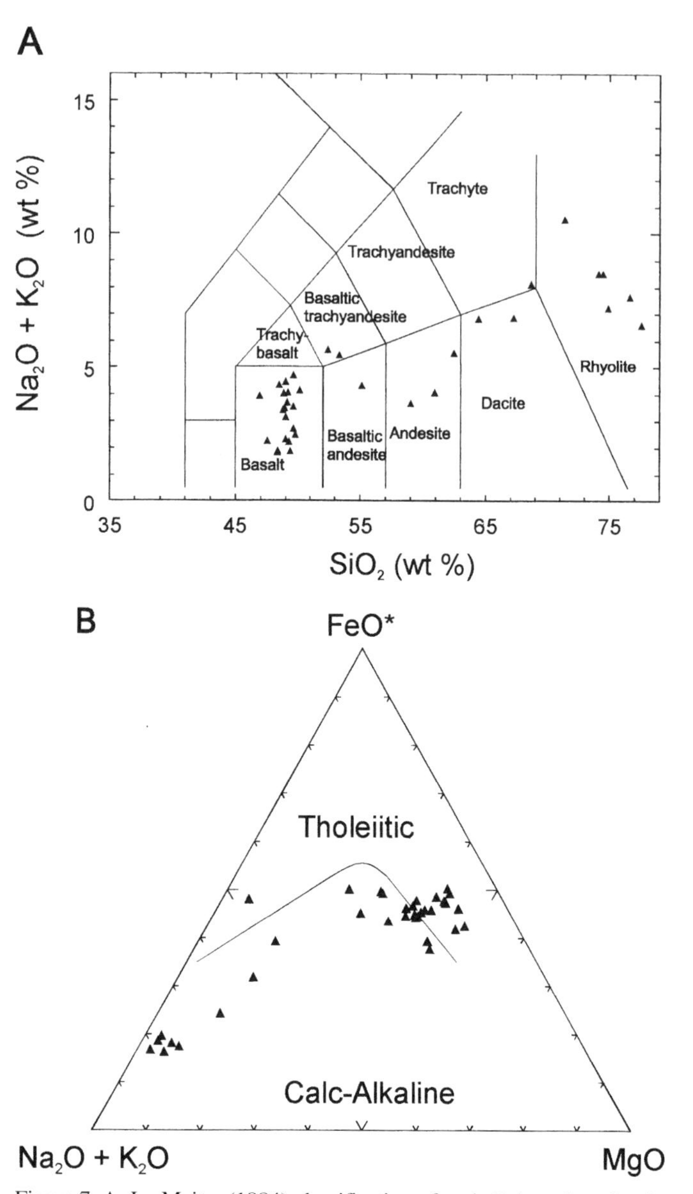

Figure 7. A, Le Maitre (1984) classification of early Paleozoic volcanic rocks of Famatina; note the dominance of basalt and rhyolite. Data from Mannheim (1993a). B, Early Paleozoic volcanic rocks from Famatina; note the tholeiitic to calc-alkaline trend of the volcanic rocks (data from Mannheim, 1993a).

Early Paleozoic granitoids. The intrusive rocks of the Famatina system correspond to a variety of granitoids that have received a series of local names (e.g., Paimán, Ñuñorco-Sañogasta, Cerro Toro) based on the different ranges where they have been identified. These varying petrologic and geochemical characteristics justify their description from east to west.

Paimán Granites. These rocks are exposed in the Sierra de Paimán (Fig. 6) and extend north to the Sierra de Fiambalá. They are composed of granodiorite, tonalite, and granite, with amphibole and biotite as the main mafic minerals. Small bodies of gabbro and quartz-diorite with olivine-orthopyroxene and clinopyroxene-amphibole-biotite (Durand et al., 1990) are widely exposed.

Mafic igneous enclaves are common in the granitic rocks, with rounded shape outlines, K-feldspar megacrysts, and chilled margins. Metamorphic enclaves, on the other hand, are rare (Toselli et al., 1993). There is also evidence of hybrid rocks formed by mingling of basic and acidic magmas. In such rocks amphibole, biotite, and ocellar quartz are common. The country rocks to the intrusive suite are low-grade metamorphic, with a superimposed contact metamorphism, characterized by cordierite and andalusite. The arc granitoids and the mafic rocks are intruded by the Potrerillos granite stock and its associated dike swarms. This is an A-type granite and has a monzogranitic to syenogranitic composition, with garnet, muscovite, and fluorite. They have granophiric textures that indicate a shallow emplacement level and no evidence of deformation (Pérez and Kawashita, 1992).

The age of the granitoids varies from 459 to 379 Ma (K/Ar and Rb/Sr), with initial ratios of $^{87}Sr/^{86}Sr$ from 0.704 to 0.711 (Pérez and Kawashita, 1992). These ages are concordant with the ones obtained by McBride et al. (1976) and González et al. (1985a). The Potrerillos stock has an age of 39 ± 15 Ma, with an initial ratio of 0.711 (Pérez and Kawashita, 1992).

The geochemical data of these granitoids (see Table 2) show a calc-alkaline trend (Fig. 10) with normal K and a highly variable A/CNK ratio between 0.7 and 1.4 (Toselli et al., 1996a).

A mylonitic belt is developed on the eastern side of the Sierra de Paimán that affects all the described rocks with the only exception of the Potrerillos Granite. The deformed rocks are unconformably covered by Carboniferous sedimentary rocks (Marcos, 1984).

Paganzo Granitoids. These granitoids are exposed south of the Sierra de Paimán and consist of monzogranite, granodiorite, and tonalite. They have frequent igneous and metamorphic enclaves, as well as evidence of mingling of basic and acidic magmas (Saal, 1988). These bodies have been emplaced in biotite–muscovite–cordierite–sillimanite–K-feldspar gneisses, which also form roof pendants in the intrusives.

The granitoids, as well as the enclaves, have a calc-alkaline trend and tholeiitic mafic rocks. The A/CNK ratio varies between 0.7 and 1.2 (see Table 2), similar to the Paimán granitoids (Toselli, 1992). The high-temperature mylonitic belt that gradually passes to magmatic foliation may indicate its synkinematic emplacement in ductile shear zones, according to Saal (1993).

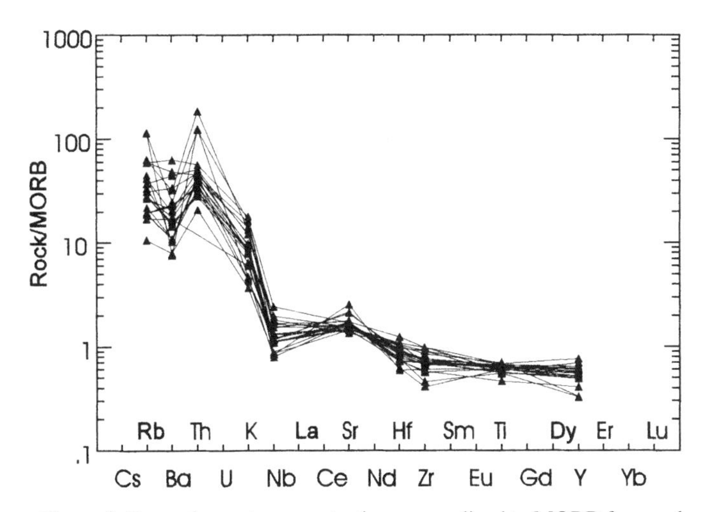

Figure 8. Trace elemente concentrations normalized to MORB from volcanic rocks of Famatina. Note the enrichment of incompatible elements in relation to more compatible elements (data from Mannheim, 1993a).

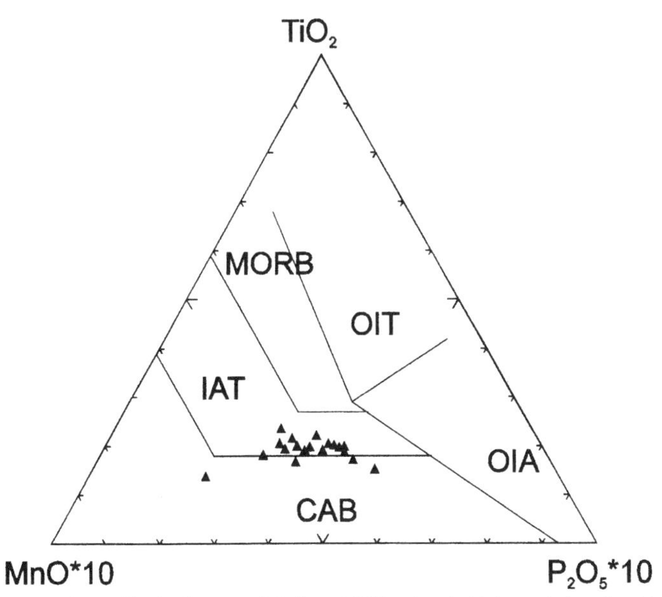

Figure 9. Mullen's diagram (Mullen, 1983) of early Paleozoic basalts of Famatina. Most of the samples falls in the IAT field (data from Mannheim, 1993a).

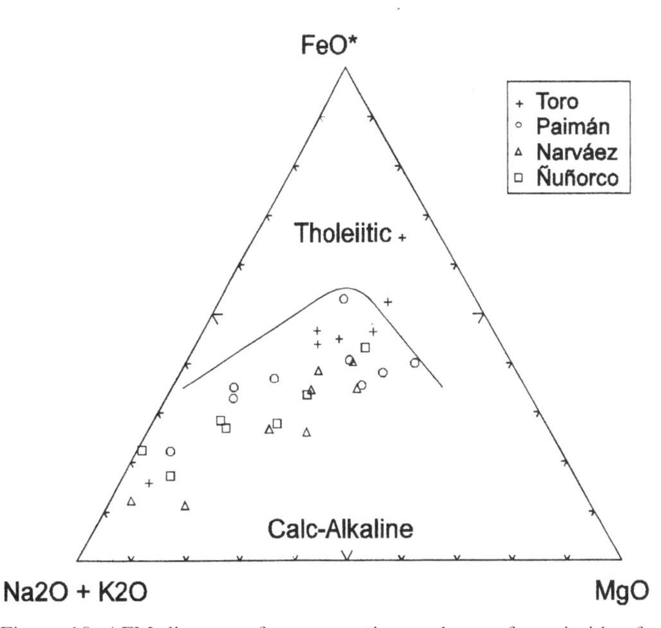

Figure 10. AFM diagram of representative analyses of granitoids of Sierra de Famatina (data from Toselli, 1992).

TABLE 2. REPRESENTATIVE ANALYSES FROM THE FAMATINA SYSTEM

	Granitoids									Volcanic Rocks				
	Fn5191	Fn5181	Ft4562	Ft4575	Ft4580	Ft4581	Ft4763	Ft4782	F3812	Fhc21	Fhb10216	Fmh6015	Fm0401	Fmc10
SiO_2	75.43	60.43	74.92	64.64	54.64	68.42	49.69	56.75	65.20	52.92	67.91	76.46	45.67	49.23
TiO_2	0.45	0.89	0.20	0.71	1.30	0.63	4.62	1.21	0.44	1.01	0.36	0.11	1.03	0.96
Al_2O_3	15.03	13.55	12.91	16.46	16.88	15.17	13.97	15.27	12.62	17.04	16.38	13.03	18.78	16.65
Fe_2O_3	1.15	7.31	1.74	0.93	2.62	1.58	16.20	1.95	n.d.	9.86	3.40	1.69	11.57	11.26
FeO	n.d.	n.d.	n.d.	4.55	6.76	3.62	n.d.	7.59	5.37	n.d.	n.d.	n.d.	n.d.	n.d.
MnO	0.03	0.03	0.08	0.08	0.33	0.07	0.29	0.44	0.09	0.28	0.13	0.03	0.22	0.18
MgO	0.33	6.52	0.54	2.78	6.25	2.29	6.14	5.61	4.11	5.34	1.45	0.64	8.03	8.52
CaO	0.55	5.26	1.75	5.61	6.74	4.63	7.18	8.14	5.49	7.11	1.14	0.09	7.81	10.83
Na_2O	3.36	4.55	2.10	2.47	2.05	1.63	0.86	1.67	2.66	3.94	3.13	3.17	1.98	1.38
K_2O	3.67	1.21	5.66	1.67	2.26	1.87	0.71	1.22	3.89	1.51	4.91	3.35	1.86	0.50
P_2O_5	n.d.	0.24	0.09	0.11	0.16	0.08	0.36	0.16	0.12	0.22	0.11	0.04	0.35	0.19
H_2O	0.48	2.43	0.71	0.95	1.14	0.90	0.47	1.38	0.73	n.d.	n.d.	n.d.	n.d.	n.d.
Total	100.48	102.42	100.70	100.96	101.13	100.89	100.49	101.39	99.99	99.23	98.92	98.61	97.30	99.70
Rb	143.73	86.97	176.04	75.79	n.d.	104.22	n.d.	n.d.	104.33	71.00	229.00	251.00	125.00	21.00
Ba	1173.29	354.98	849.01	334.90	n.d.	184.94	n.d.	n.d.	265.93	461.00	525.00	457.00	346.00	156.00
Sr	73.39	111.25	156.93	185.37	n.d.	184.94	n.d.	n.d.	177.97	212.00	98.00	21.00	190.00	181.00
Nb	7.14	6.07	15.09	16.39	n.d.	11.24	n.d.	n.d.	n.d.	6.00	9.40	8.20	5.80	4.70
Zr	134.56	94.05	146.87	124.95	n.d.	138.96	n.d.	n.d.	117.62	88.00	92.00	83.00	74.00	64.00
Y	26.50	14.16	19.11	22.53	n.d.	8.17	n.d.	n.d.	17.39	26.00	4.00	1.00	11.00	21.00
La	n.d.	n.d.	10.06	10.24	8.21	19.41	29.23	28.37	n.d.	n.d.	n.d.	n.d.	n.d.	n.d.
Ce	n.d.	n.d.	24.14	26.63	23.61	36.78	68.53	83.10	n.d.	n.d.	n.d.	n.d.	n.d.	n.d.
Nd	n.d.	n.d.	10.06	25.60	22.58	19.41	40.31	59.79	n.d.	n.d.	n.d.	n.d.	n.d.	n.d.
Sm	n.d.	n.d.	n.d.	5.12	10.26	n.d.	11.09	15.20	n.d.	n.d.	n.d.	n.d.	n.d.	n.d.
Eu	n.d.	n.d.	n.d.	1.02	1.03	1.02	3.02	2.03	n.d.	n.d.	n.d.	n.d.	n.d.	n.d.
Gd	n.d.	n.d.	2.01	7.17	7.19	2.04	12.09	15.20	n.d.	n.d.	n.d.	n.d.	n.d.	n.d.
Dy	n.d.	n.d.	n.d.	6.15	6.16	1.02	11.09	14.19	n.d.	n.d.	n.d.	n.d.	n.d.	n.d.
Er	n.d.	n.d.	2.01	4.10	5.13	n.d.	6.05	8.11	n.d.	n.d.	n.d.	n.d.	n.d.	n.d.
Yb	n.d.	n.d.	2.01	3.07	5.13	1.02	5.04	8.11	n.d.	n.d.	n.d.	n.d.	n.d.	n.d.
La	n.d.	n.d.	n.d.	n.d.	1.03	n.d.	1.01	1.01	n.d.	n.d.	n.d.	n.d.	n.d.	n.d.

Note: Data from Toselli, 1992; Cisterna, 1994; and Mannheim, 1993a.
n.d. = no data; Fn = Sierra de Narvaez; Ft = Cerro Toro; F = Sierra de Ñuñorco; Fhc, Fnb, Fmh, Fm, and Fmc = Sierra de Famatina.

The age, based on Rb/Sr isochrons, yields 456 ± 9 Ma, with $^{87}Sr/^{86}Sr$ initial ratios of 0.706–0.709 (Saal, 1993).

Ñuñorco-Sañogasta Granitoids. The central core of the Famatina system is composed of the Ñuñorco-Sañogasta granitoids. Petrographic varieties range from dominant granodiorite and monzogranite to subordinate tonalite. They have small igneous basic enclaves, with biotite more abundant than hornblende, plagioclase, microcline, and quartz. Some sectors of the Famatina range show strong foliation with a north-northwest trend, accompanied by the development of quartz subgrains, rotation of feldspar grains, and fusiform biotite (Toselli, 1992).

The emplacement of these granitoids into the low-grade metamorphic rocks of Negro Peinado Formation produced cordierite-biotite hornfels. These epizonal assemblages indicate pressures of 2.0–3.5 kb, according to Turner (1981). A shallow emplacement, in the order of 2 km, is supported by the presence of high-level granophyres in close association with granitoids in the Sierra de Narváez, the northern extension of the Sierra de Famatina (Rossi de Toselli et al., 1987).

The geochemical trend of the granitoids corresponds to the most evolved magmas of a calc-alkaline series. The alumina ratio indicates that the granitoids are peraluminous (see Table 2). The granitoids of Sierra de Narváez, as well as the Ñuñorco granitoids (Fig. 11A,B) present high Rb and K, and low Nb, Ti, and relatively low Sr (Cisterna, 1994).

The U/Pb age of the Ñuñorco granite is dated at 459 Ma, obtained in igneous zircons (Loske in Toselli et al., 1996a), coherent with previous K/Ar ages of 451–395 Ma obtained by González et al. (1985a,b) and Linares and González (1990).

Cerro Toro Granitoids. The Cerro Toro granitoids are exposed along the western side of the Sierra de Famatina and the Sierra de Umango. They are dominantly tonalite with monzogranite, granodiorite, and subordinate gabbro. They frequently contains enclaves and display magma mingling (Toselli et al., 1988). Country rocks of these granitoids are schist, gneiss, and migmatite characterized by kyanite-garnet mineral assemblage. These minerals, as well as the pressure values obtained from alumina contents of hornblende from

Cerro Toro tonalites, suggest a pressure of approximately 6 kb (Rossi de Toselli et al., 1991).

The granitoids show a calc-alkaline trend in the AFM diagram (Fig. 10). The A/CNK alumina index varies from 0.7 to 1.2 (see Table 2). These granitoids have been interpreted as syn- to late tectonic, and emplaced at mesozonal levels (Toselli et al., 1996a). The spiderdiagrams of these granitoids show a great variability, typical of combined processes of crystal fractionation, magma mixing, and crustal assimilation. The REE have a flat pattern, with a small LREE enrichment and a weak Eu negative anomaly (Fig. 12). In general, there is not a unique trend with parallel patterns, corroborating more than one petrogenetic process. These bodies yielded Rb/Sr ages of 456 ± 14 Ma, with a 0.70967 $^{87}Sr/^{86}Sr$ initial ratio (Saavedra et al., 1992).

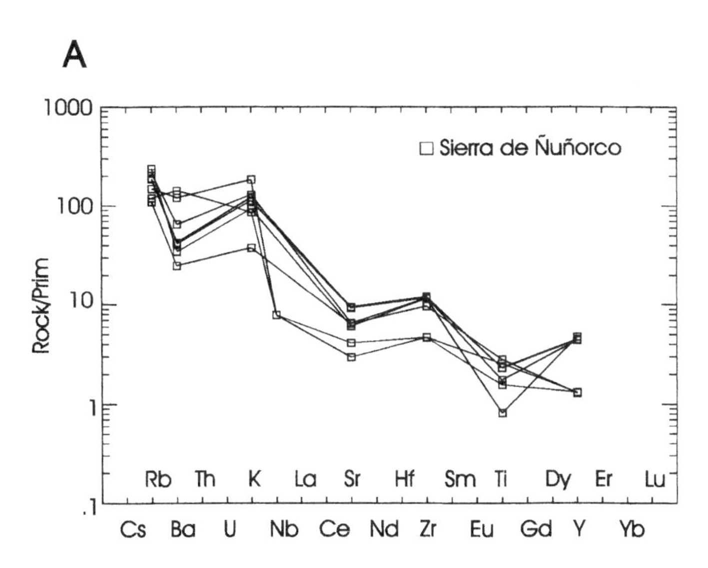

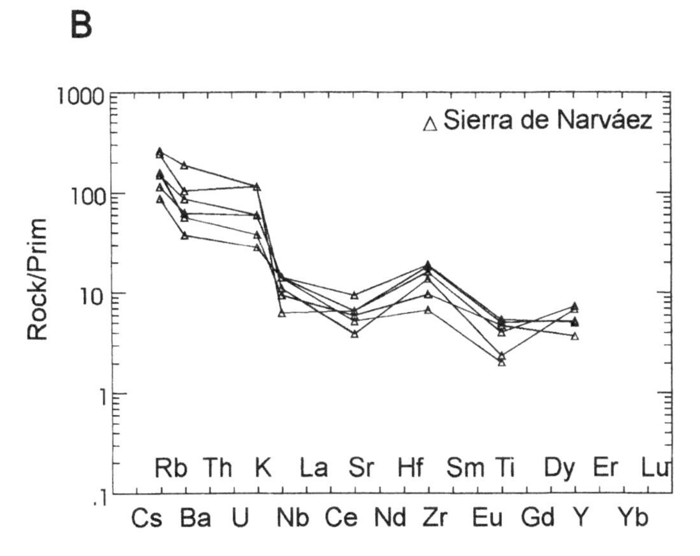

Figure 11. Spiderdiagrams of Ordovician granitoids. A, Ñuñorco granitoids. Note the relative enrichment of some incompatible elements and depletion in Nb and Sr (data after Toselli, 1992). B, Narváez granitoids; note similar trend as in Ñuñorco (data from Cisterna, 1994).

Western Sierras Pampeanas granitoids

A magmatic arc coeval with the magmatic arc described in the Famatina terrane can be identified along different ranges of the western Sierras Pampeanas (Fig. 6).

Sierra Valle Fértil–La Huerta. The magmatic rocks in western Sierras Pampeanas are exposed in the Sierra de Valle Fértil and La Huerta. Granodiorite with tonalite and diorite are dominantly present in the eastern flank and display a large variety of textures ranging from strongly foliated to equigranular. The foliated granitoids contain a tectonic layering up to 1 cm wide. The microlithons consist of quartz, oligoclase-sodic andesine, and K-feldspar, with scarce hornblende and biotite. The melanocratic foliae dominantly consist of hornblende and biotite. The transition between the foliated and the equigranular facies, which have the same composition, is gradual. A strong tectonic ductile deformation repeated, in a west-to-east section, the different facies (Mirré, 1976).

The basement rocks mainly consist of amphibolite and lenses of garnet and sillimanite-bearing paragneiss. They occur most frequently in the central part and are associated with some subordinate basic and ultrabasic bodies.

Preliminary U/Pb zircon ages indicate an age of ca. 467 Ma (Vujovich et al., 1996). These ages are consistent with the 456 ± 15- and 430 ± 15-Ma K/Ar biotite and amphibole ages obtained by González and Toselli (1974) and the 484 ± 15-Ma K/Ar whole-rock age (Toubes Spinelli, 1983).

Northern Sierras Pampeanas granitoids

Chango Real. Two different types of granitoids are exposed in the Sierras de Chango Real and Papachacra, in the northwestern extreme of Sierras Pampeanas, close to the boundary with the Puna. The oldest granitoid corresponds to a biotite

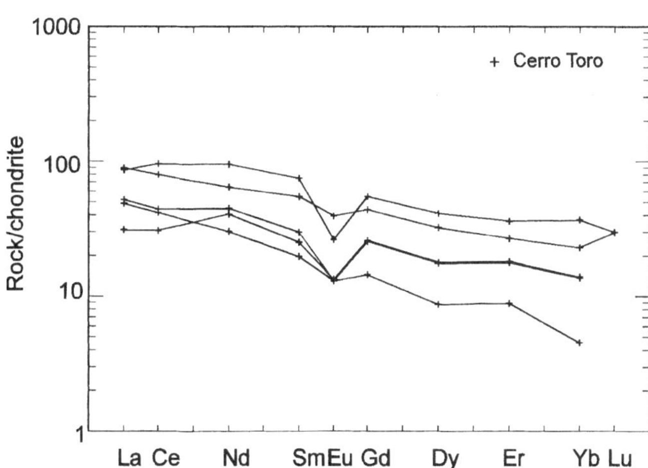

Figure 12. REE pattern of Cerro Toro granitoids normalized to the Nakamura (1974) chondrite. Note the flat pattern and the lack of LREE enrichment (data from Toselli, 1992).

orthogneiss batholith, known as the Chango Real Granite (Turner, 1962). It is emplaced in a sequence of phyllite and schist that interfingers with metabasite and marble of Precambrian age. The batholith of Chango Real consists of strongly foliated granodiorite and monzogranite. Petrography shows deformational textures such as quartz subgrains, polygonal recrystallization of feldspars, myrmekites, flexured biotite, and kink bands. The accessory minerals are composed of magmatic epidote and occasional tourmaline, amphibole, and sphene. Metasedimentary enclaves are also frequent (Lazarte, 1987, 1991, and 1992).

On geochemical grounds this body is peraluminous (A/CNK between 0.94 and 1.66), with high K and normal calc-alkaline trend (Fig. 13) and REE patterns (see Table 3). Magma formation was under 680°C temperature and water-saturated pressures of the order of 4 kb (14–15-km depth), according to Lazarte (1992).

These granitoids yielded K/Ar-biotite ages of 505 ± 20, 441 ± 15, 430 ± 15, and 408 ± 15 Ma (González et al., 1985b), consistent with the 498 ± 15–474 ± 15-Ma K/Ar-biotites ages obtained by García and Rossello (1984). The younger 441–430-Ma K/Ar data may be indicating ages of uplift and deformation, whereas the 505- and 498-Ma ages are thought to date the time of crystallization. These K/Ar ages are slightly younger than the pretectonic tonalite ages, dated at 510–515 Ma by U/Pb from zircons of the La Puntilla Orthogneiss in the Sierra de Fiambalá (Grissom, 1991).

The Papachacra is a younger batholith (García et al., 1981) and includes the Altohuasi and El Portezuelo stocks. These stocks have sharp contacts with the Chango Real Orthogneiss and have a syenogranitic to monzogranitic composition, with strong evidence of volatile rich fluids (Lazarte, 1991).

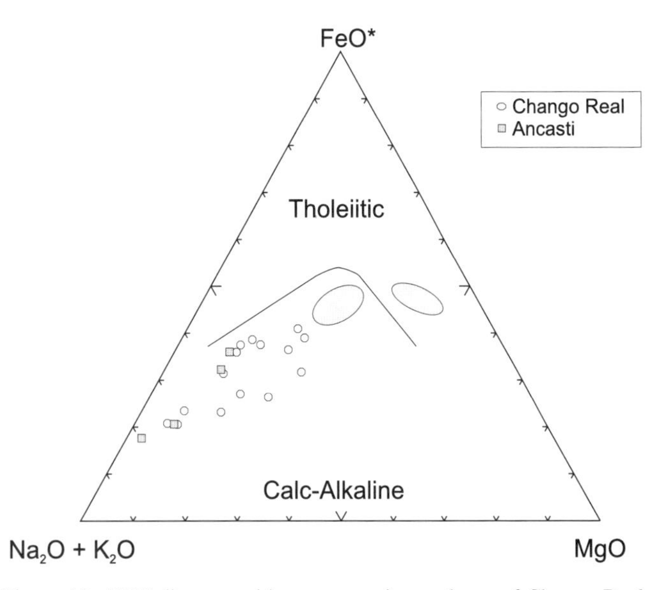

Figure 13. AFM diagram with representative analyses of Chango Real and Ancasti granitoids. Shaded area in the tholeiitic field represents gabbro and quartz-diorite; shaded area in the calc-alkaline field corresponds to tonalite of Ancasti (data from Lazarte, 1992; Lottner and Miller, 1986).

The geochemistry indicates a calc-alkaline trend, with peraluminous, high-silica–potassium compositions. There is an alkaline trend in relation to the alumina in the last stages of crystallization. The emplacement conditions correspond to ca. 670°C and 2 kb, with depths shallower than 7 km and high fluorine and boron contents, which controlled the formation of topaz and tourmaline, and the release of Na in the last magmatic stages, which produced a relative enrichment of K (Lazarte, 1995).

The Papachacra granitoids are post-tectonic, associated with a late-stage greisen with occurrences of Au, W-Sn, Cu-Pb-Zn mineralization. The age is not well established, but it was considered as pre–early Carboniferous (Lazarte, 1995).

Sierra de Fiambalá. The oldest magmatic rocks are represented by the strongly foliated La Puntilla Orthogneiss. This gneiss has a mainly tonalitic composition: locally it has variations to granitic and granodioritic compositions with equigranular to porphyritic textures. The orthogneiss is composed of quartz, plagioclase, microcline, biotite, and muscovite. This gneiss is in tectonic contact with other rocks of the Sierra de Fiambalá. The intrusion of the tonalitic pluton (La Puntilla Orthogneiss) predated the main tectonic foliation and metamorphism in the Sierra de Fiambalá (Grissom et al., 1991). The prekinematic body has a U/Pb-age of ca. 515–510 Ma (Grissom, 1991).

The magmatic sequence continues with the Fiambalá Gabbro-norite, composed of homogeneous fine-grained clinopyroxene, orthopyroxene, plagioclase, biotite, epidote, and tourmaline. The rocks show some variable composition of plagioclase, hornblende, quartz, and clinopyroxene, often transformed to epidote, chlorite, and limonite. Cumulate levels, partially serpentinized, are linked to this intrusive. The emplacement of this body was coeval with the metamorphism that accompanied the deformation (Grissom et al., 1991). The geochemistry indicates a subduction-related setting, based on the LILE enrichment, deflected HFS relative to MORB, which is partially corroborated by the REE pattern. The LREE are slightly enriched with a small positive Eu anomaly (DeBari, 1994). The age of the gabbronorite, based on Nd/Sm analyses, is 501 ± 21 Ma (Grissom, 1991).

The latest intrusive in the Sierra de Fiambalá corresponds to the Los Ratones Granite, emplaced either in La Puntilla Orthogneiss or the Fiambalá Gabbronorite. The granite is equigranular, with porphyritic margins, associated with hydrothermal alteration. Primary minerals are quartz, microcline, plagioclase, amphibole, and biotite. The age of this body is assumed to be Carboniferous (Arrospide, 1974).

Sierras de Vinquis and Zapata. The granitoids of this region (Fig. 6) have porphyritic textures. Monzogranite is dominant in the Sierra de Vinquis and composed of quartz, plagioclase, megacrysts of perthitic microcline, biotite, and primary muscovite, with minor accessory minerals like fibrolitic and prismatic sillimanite, apatite, and tourmaline. The country rocks are amphibolite facies quartz-mica schists. There are frequent metasedimentary enclaves, with conspicuous schistosity, and with muscovite, sillimanite, cordierite, tourmaline, and apatite (Toselli et al., 1992).

TABLE 3. REPRESENTATIVE ANALYSES FROM NORTHWESTERN SIERRAS PAMPEANAS

	Chango Real Orthogneiss				Sierra de Ancasti			
	Chr100	Chr338	Chr95	Chr80	An6270	An6263	An6201	An6244
SiO_2	75.08	66.93	64.06	68.86	69.95	75.1	71.41	67.16
TiO_2	0.27	0.35	0.48	0.49	0.44	0.15	0.62	0.62
Al_2O_3	11.31	18.3	16.71	16.21	14.45	14.04	14.49	17.19
Fe_2O_3	1.73	1.98	2.66	1.56	2.91	1.73	3.63	3.37
FeO	2.35	2.68	3.61	2.12	n.d.	n.d.	n.d.	n.d.
MnO	0.15	0.23	0.19	0.08	0.05	0.09	0.08	0.11
MgO	2.33	1.43	3.13	1.32	0.97	0.25	0.96	1.01
CaO	2.01	1.83	3.2	2.9	2.01	1.44	3.76	4.98
Na_2O	1.71	2.13	2.42	2.87	4.11	3.22	2.99	3.42
K_2O	2.77	3.89	3.13	3.52	5.01	3.79	1.85	1.92
P_2O_5	0.29	0.24	0.4	0.07	0.09	0.19	0.22	0.22
H_2O	1.04	2.04	1.13	1.04	1.5	1.42	1.27	1.24
Total	101.04	102.03	101.12	101.04	101.49	101.42	101.28	101.24
Rb	n.d.	n.d.	n.d.	n.d.	177.11	184.98	61.61	62.29
Sr	n.d.	n.d.	n.d.	n.d.	122.07	101.54	203.01	266.25

Note: Data from Lazarte, 1992, and Lottner and Miller, 1986.
n.d. = no data.

The monzogranite is also dominant in Sierra de Zapata where the country rocks consist of fine-grained mica-quartz schists with lower metamorphic grade in comparison with the Sierra de Vinquis metamorphic rocks. The monzogranite are porphyritic with large perthitic microcline megacrysts, quartz, plagioclase, and biotite and muscovite. Accessory minerals are fluorite, cordierite, andalusite, and allanite, which are different from the accessory phase of the Vinquis granites (Toselli et al., 1992).

Although both granites have conspicuous calc-alkaline trends (Fig. 14), with high silica, peraluminous (A/CNK between 1 and 1.4), and high K (see Table 4), there are significant differences between them. The Zapata granite had fluorine in the volatile phases, as evidenced by fluorite and andalusite, and the feldspars show textures of a hypersolvus crystallization. The Zapata granite has also a higher content of HFS elements, total RRE (300–600 ppm), Th (60–90 ppm), and U (8–15 ppm), according to Rapela et al. (1996). These characteristics imply a crystallization at shallower depths in the order of 650–600°C and 1.5–2 kb for the Zapata granite, and of 650°C and 2 kb for the Vinquis granite, according to Toselli et al. (1992). It may be concluded that the Zapata granite has more affinities with the post-tectonic granitoids emplaced in a way similar to the Papachacra and part of the San Luis granites.

Capillitas granite. This large batholith extends from the Sierras de Capillitas to Belén, Zapata, eastern flank of Fiambalá, and to the southwestern Sierra de Aconquija, according to González Bonorino (1951a). Several facies have been recognized in the batholith, with older equigranular facies intruded by later porphyritic facies (Toselli and Indri, 1984; Indri, 1986). Monzogranite is the dominant composition, with subordinate leucogranite, consisting of perthitic microcline, oligoclase, quartz, biotite, and muscovite. Accessory minerals are cordierite, andalusite, sillimanite, tourmaline, and apatite. The abundance of aluminosilicates, albeit localized, is a distinctive characteristic of these granitoids. Cordierite is particularly common in the equigranular facies and reaches up to 9% modal content. Andalusite, cordierite, and sillimanite only locally occur together; the most frequent assemblages are cordierite and/or andalusite and andalusite-fibrolite. Biotite-rich enclaves are common with abundant andalusite, sillimanite, and cordierite (Toselli et al., 1996b).

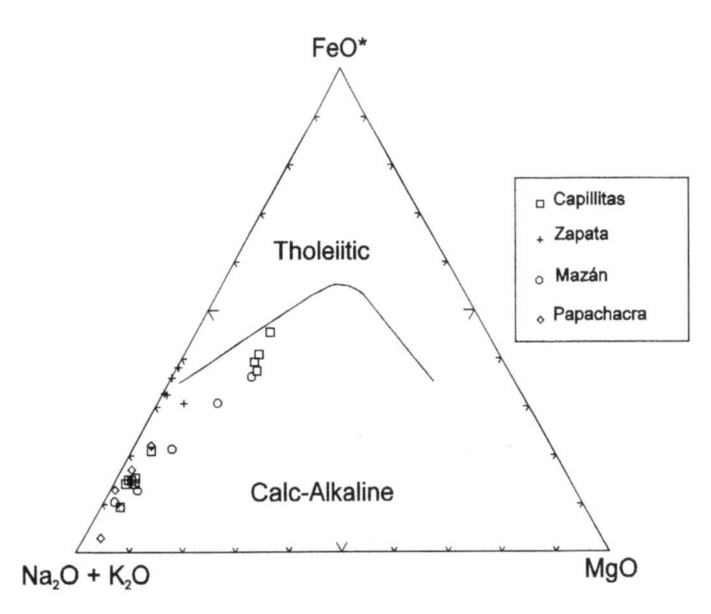

Figure 14. AFM diagram of Northern Sierras Pampeanas granitoids. Note the marked MgO depletion of the postkinematic Zapata granitoids (data from Rapela and Heaman, 1982; Toselli et al., 1991; Toselli 1992).

TABLE 4. REPRESENTATIVE ANALYSES FROM POSTKINEMATIC GRANITOIDS FROM NORTHERN SIERRAS PAMPEANAS

	Mazán		Zapata		Capillitas		
	Mz9	Mz15	Z564	Z565	Cap51	Cap49	Catu6
SiO_2	68.87	76.71	76.00	75.00	66.10	74.16	70.02
TiO_2	0.81	0.12	0.08	0.05	0.63	0.09	0.18
Al_2O_3	13.86	13.71	1.53	11.70	13.95	14.42	14.64
Fe_2O_3	4.80	0.97	4.56	4.28	4.98	0.96	1.64
MnO	0.02	0.06	0.02	0.03	0.10	0.05	0.05
MgO	1.79	0.35	0.04	0.02	1.85	0.25	0.30
CaO	1.36	0.85	0.42	0.28	1.11	0.79	0.86
Na_2O	2.02	2.95	2.00	2.15	1.34	1.76	3.29
K_2O	3.76	2.61	4.60	4.65	4.28	3.21	4.85
P_2O_5	0.05	0.05	0.05	0.05	0.18	0.30	0.45
H_2O	1.04	0.94	0.17	0.22	n.d.	n.d.	n.d.
Total	98.38	99.32	89.47	98.43	94.52	95.99	96.28
Rb	n.d.	n.d.	502.00	502.00	217.00	277.00	452.00
Ba	n.d.	n.d.	45.00	45.00	264.00	66.00	54.00
Sr	n.d.	n.d.	33.00	33.00	78.00	41.00	47.00
Nb	n.d.	n.d.	47.00	47.00	28.00	31.00	34.00
Zr	n.d.	n.d.	160.00	160.00	195.00	64.00	120.00
Y	n.d.	n.d.	81.00	81.00	40.00	18.00	27.00
La	n.d.	n.d.	n.d.	n.d.	26.00	6.00	13.00
Ce	n.d.	n.d.	n.d.	n.d.	58.00	10.00	22.00
Nd	n.d.	n.d.	n.d.	n.d.	35.00	9.00	15.00

Note: Data from Toselli et al., 1992, and Rapela et al., 1996.
n.d. = no data.

These granitoids are intruded discordantly in schist and phyllite in greenschist facies that locally have developed some contact aureoles with biotite-cordierite-muscovite hornfels. The sharply discordant contacts indicate an emplacement at epizonal levels, while the mineral paragenesis suggests conditions near 4 kb of pressure and near-solidus temperatures, under-water saturated conditions that stabilized the aluminosilicates (Toselli et al., 1996a,b).

On geochemical grounds (see Table 4), these granitoids are high silica, with normal calc-alkaline trend, and peraluminuous (A/CNK > 1.2), with K_2O/Na_2O ratios > 1, high values of LILE (Fig. 15), with a high isotopic $^{87}Sr/^{86}Sr$ initial ratio (0.7146) that corroborates their crustal derivation. The REE patterns are flat, with a pronounced negative Eu anomaly (Rapela et al., 1996). Based on K/Ar dating on biotite and muscovite, the age of the Capillitas Granite varies from 471 to 404 Ma (McBride et al., 1976; Linares and González, 1990).

Sierra de Ancasti. Different stocks have been recognized in the Sierra de Ancasti that show variable compositions and characteristics. A suite of hornblendite, gabbro, and diorite is exposed in the Ramblones–La Majada area, in the southern sector of the range. These rocks are associated with tonalite and trondhjemite, granodiorite, and biotite and muscovite granite (Lottner and Miller, 1986). The igneous bodies are emplaced in a metamorphic basement and are spatially associated with northwest-trending shear zones.

Several stocks have been identified, such as the La Majada, La Pampa–Unquillo, El Alto, Albigasta, and El Taco. La Majada is the largest body that is composed of the basic rocks mentioned and tonalite and granodiorite. La Pampa–Unquillo stock is of granodioritic to granitic composition. El Alto stock is a muscovite-granite with garnet as an accessory phase. The Albigasta

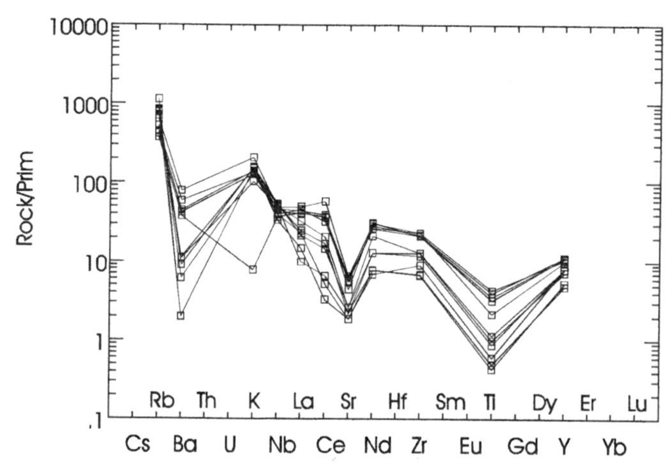

Figure 15. Spiderdiagram of trace elements of Capillitas granitoids normalized to primordial mantle. Note the relative enrichments of LIL elements (data from Rapela and Heaman, 1982).

stock is a monzogranite that varies to a quartz-monzonite with K-feldspar megacrysts and abundant biotite-schist enclaves. The El Taco stock is a two-mica granite with perthitic microcline megacrysts and schlieren-type enclaves (Toselli et al., 1983).

The geochemical characteristics (see Table 3) show a calc-alkaline trend differentiated from a tholeiitic initial stage (Fig. 13). The rocks have evolved from initial partial melting of a peridotite by crystal fractionation, according to Lottner and Miller (1986). These authors recognized in this subduction-related suite two parental magmas, one that evolved to tonalite and trondhjemite, and a second one more completely evolved to granite. Some of the granites, such as the Albigasta and El Alto, have a relatively low initial $^{87}Sr/^{86}Sr$ ratio, which indicates a mantle source. Others, like La Pampa, one of the older granites with ages of 496 ± 26 Ma (Rb/Sr), have $^{87}Sr/^{86}Sr$ initial ratios of 0.70767. The latest granites, such as Las Cañadas, with ages of 435 ± 23 Ma (Rb/Sr) have an $^{87}Sr/^{86}Sr$ initial ratio of 0.7101 (Knüver, 1983).

The magmatic evolution in the Sierra de Ancasti ends with the Sauce Guacho and Santa Rosa Granites. These are two-mica granites with microcline megacrysts. The first granite has an orbicular texture with sillimanite, and both granites are related to a fluorite mineralization (Toselli et al., 1983). The ages of these granites vary from 373 ± 10 (Santa Rosa, K/Ar, Linares, 1977) to 334–10 Ma (Sauce Guacho, Rb/Sr, Knüver, 1983), with an $^{87}Sr/^{86}Sr$ initial ratio of 0.715. Both granites are LILE-enriched, and have a crustal derivation typical of post-orogenic granitoids.

Sierra de Velasco–Mazán. A large batholith is exposed between the locality of Aimogasta in the north and Los Colorados to the south (Fig. 6). The northern part presents equigranular and porphyritic granitoids, while toward the south the porphyritic facies are dominant. In Pampa de Los Altos, orbicular textures have been described in the porphyritic granites (Quartino and Villar Fabre, 1963). These orbicular granites have microperthitic microcline megacrysts, surrounded by biotite and an outer layer of oligoclase. The geochronologic data indicate for the granitoids, ages between 458 and 425 Ma (K/Ar) (Stipanicic and Linares, 1975; Linares, 1977; Linares and Quartino, 1979) and 332 ± 16 Ma by Rb/Sr with an $^{87}Sr/^{86}Sr$ initial ratio of 0.7109 (Rapela et al., 1982).

Porphyritic granites, some of them with foliated textures are exposed in the eastern sector of the Sierra de Velasco and in the Sierra de Mazán (Fig. 6). The feldspars have hypersolvus characteristics and sometimes contain magmatic cordierite. Primary minerals consist of K-feldspar quartz, plagioclase, and biotite. The available geochemical data indicate that the granitoids are peraluminous (A/CNK between 1.1 and 1.5), high K and, based on the trace elements, two groups can be distinguished (see Table 4). A group with high Rb and low Sr and Ba, from another group with low Rb and high Sr and Ba. The first group corresponds to the hypersolvus variety with primary muscovite and cordierite. These granitoids have scarce enclaves of metasedimentary rocks with peraluminous assemblages and other small enclaves of igneous rocks (Toselli et al., 1991). These granitoids yielded ages between 475 and 345 Ma (K/Ar method) Linares and González (1990).

Sierra de Los Llanos. The batholith of Sierra de Los Llanos, exposed between Malanzán and Chepes, is composed of five different magmatic units (Pankhurst et al., 1996). The main and more extensive unit is the Chepes Granodiorite, that locally grades to tonalitic and monzogranitic compositions.

These granitoids have hornblende and biotite; epidote, sphene, and allanite are accessory phases. The Chepes Granodiorite has fine-grained mafic enclaves, aligned with a penetrative foliation with northeast to northwest trends. The contact with the country rock is marked by a shear zone that was developed during the magmatic emplacement (Dahlquist and Baldo, 1996).

The second unit is a porphyritic granodiorite, with a similar composition and foliation as the Chepes Granodiorite. The main characteristics are the K-feldspar megacrysts (Ramos, 1982), mafic enclaves, and xenoliths of cordierite gneiss. These granitoids gave an Rb/Sr isochron age of 471 ± 10 Ma ($^{87}Sr/^{86}Sr$ initial ratio of 0.7089) and 452 ± 9 Ma (initial ratio of 0.7111), according to Pankhurst et al. (1996). Similar isotopic initial ratios are present in the Los Llanos Diorite, although no age is available for this diorite. These diorites show mingling with the regional granodiorite and are better exposed in the northwest sector of the region. The three described units represent metaluminous granitoids.

The third unit in regional extension is the Tuani Granite, which is emplaced in the granodiorite. This is a two-mica granite with tourmaline and is associated with small bodies of cordierite-bearing monzogranites, sometimes also containing sillimanite. The Rb/Sr age of these bodies is 464 ± 5 Ma, with an $^{87}Sr/^{86}Sr$ initial ratio of 0.7089 that increases to 0.7155 in the cordierite-monzogranite.

The last intrusive event in the region is the El Elefante Granite. This is a small stock emplaced in the granodiorite that grades to a high SiO_2 syenogranite (74–77% SiO_2 and in the aplites up to 78% of SiO_2), with scarce biotite and muscovite. The Rb/Sr age varies from 469 ± 5 to 456 ± 3 Ma, with $^{87}Sr/^{86}Sr$ initial ratios of 0.7088–0.7105 (Pankhurst et al., 1996). New U/Pb ages in these granitoids are consistent with the previous ages (Stuart Smith et al., this volume).

DISCUSSION

Since the early description and classification of granitoids by González Bonorino (1951b), mainly based on the northern sector of Sierras Pampeanas, tectonic, syntectonic, and post-tectonic granites have been recognized. A similar history was presented for the southern Sierras Pampeanas of San Luis by Llambías et al. (1991). In recent years, the studies of Dalla Salda et al. (1995a,b) and López de Luchi and Dalla Salda (1995, 1996) of several granitic suites of Sierras Pampeanas correlated the pre-, syn-, and post-tectonic groups with subduction-related, syncollisional, and postcollisional granites, respectively, based on their compositional differences (Fig. 16).

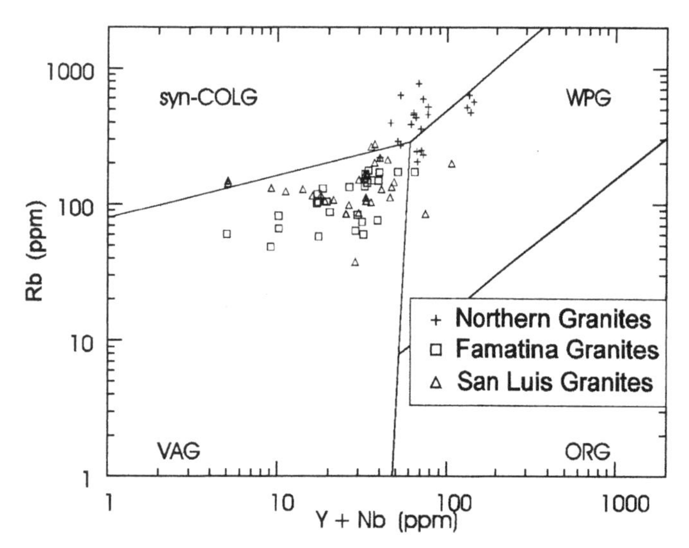

Figure 16: Rb vs. (Y + Nb) diagram of Pearce et al. (1984) summarizing the tectonic settings of the different granitoids of western Sierras Pampeanas and Famatina.

Although, there are some geochemical data available from several specific areas, there is no uniform coverage. On the other hand, there is a high abundance of radiometric ages, mainly determined by the K/Ar method in whole rock and in specific minerals, but with an uneven distribution.

This database allowed recognition along the whole length of the western Sierras Pampeanas of orthogneiss and foliated tonalites and granodiorites, which correspond to the pretectonic suites. These granitoids have intrusive contacts with the metamorphic country rock, developed contact aureoles, and planar foliation. The petrographic composition varies, including granodiorite, tonalite, and monzogranite. There are significant variations along strike. For example, tonalites are dominant in San Luis, while in the northern sector, the porphyritic textures are more common, with large megacrysts of microcline. These northern granitoids have common metasedimentary and mafic microgranular enclaves.

The Famatina area exposes volcanic rocks as well as granitoid plutons. These granitoids are also porphyritic, but dominantly monzogranitic in composition. The abundance of mafic enclaves show evidence of mingling.

These granitoids are interpreted as subduction-related arc magmas based on their geochemical composition. They are metaluminous or peraluminous (Rapela et al., 1992; Pankhurst et al., 1996), with typical calc-alkaline trends. Some of them are linked to small volumes of initial tholeiitic composition. Trace elements indicate an arc developed on continental, perhaps locally attenuated, crust. Conversely, in the northern sector there is evidence of sedimentation interbedded with volcanics under a synextensional regime (Mángano, 1993).

These subduction-related granitoid and volcanic rocks were emplaced in the western Sierras Pampeanas belt between 510 and ca. 470 Ma (Linares and González, 1990), although ages within the 490–470-Ma interval are abundant in the northern sector. The La Puntilla Orthogneiss described by Grissom et al. (1991) in the Fiambalá area, with an U/Pb-age of 515–510 Ma, may represent one of the oldest pretectonic rocks. The last emplacement of Famatina granitoids seems to be slightly younger, reaching common values of 459–450 Ma. Some of these ages are confirmed by scarce U/Pb determinations in zircons (Loske and Miller, 1996).

The syncollisional granites, identified mainly in the Sierras de San Luis and Ancasti, are small bodies concordant with regional foliation and have ages varying from 460 to 435 Ma (Linares and González, 1990). Granite contacts are sharp and are not accompanied by contact aureoles. Plutons are emplaced in migmatite and medium- to high-grade schist, associated with pegmatoid facies. Compositions vary from leucogranodioritic to monzo- and syenogranitic. In general, these granitoids are peraluminous, high-K, LILE-enriched, and have clear crustal signatures.

The postcollisional granitoids are well represented in the entire belt, although volumetrically they seem to be better exposed in the San Luis region. They constitute stocks and large batholiths, in general with circular shapes. Locally, these bodies are associated with structural lineaments. Their compositions vary from monzo- to syenogranitic, with porphyritic textures; peraluminous rocks are dominant, but metaluminous are also common. Some of them have an alkaline trend, are evolved with high-K, with high temperature and low water content. They have high HFS elements, REE patterns with moderate to high slopes, and small Eu anomalies. These characteristics point out to a thickened crust, with slight arc influence, that changed in time to an intraplate setting.

Available ages from these rocks vary from 440 to 360 Ma, although locally it is possible to recognize several pulses. The Rb vs. (Y + Nb) diagram of Pearce et al. (1984) summarizes the regional trend and tectonic settings of the different granitoids of western Sierras Pampeanas and Famatina (Fig. 16). The analyses of the northern sector are mainly located in the syncollisional field, with a trend toward within-plate granitoids. The Famatina granitoids fell in the volcanic arc granites with a trend toward the triple point of VAG-synCOLG and WPG. The granitoids of San Luis vary from VAG to WPG.

These chemical characteristics are consistent with postcollisional granites originated and evolved in a continental crust (Pearce, 1996). The granitoids have more affinities with upper crustal granites, as indicated by the low Nb and Y contents, controlled by the influence of previous subduction processes.

CONCLUDING REMARKS

The Cambro-Ordovician magmatic belt of western Sierras Pampeanas records a series of episodes that can be matched to the sedimentary evolution of the Precordillera early Paleozoic sequences. A coherent evolution can be proposed when both systems are integrated in a single model (see Fig. 17).

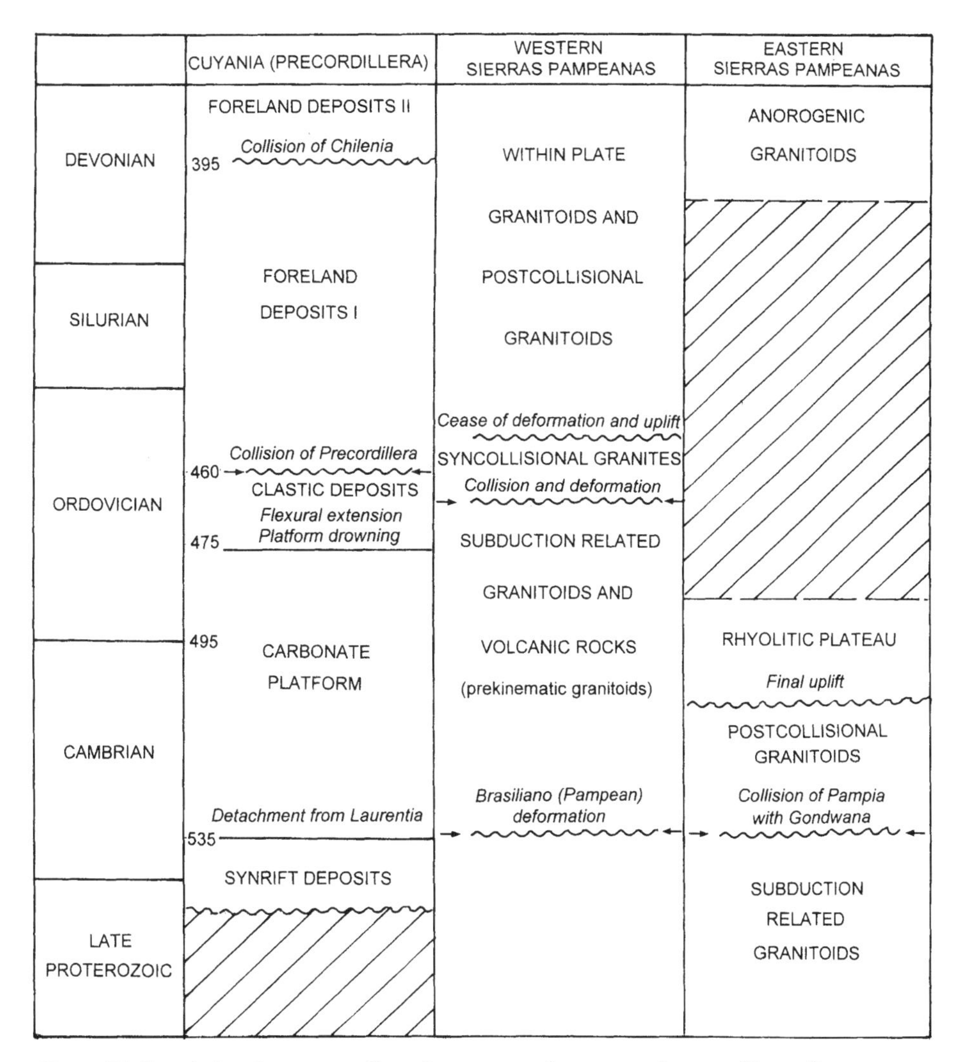

Figure 17. Correlation chart among Cuyania terrane, and western and eastern Sierras Pampeanas.

Early Cambrian (545–518 Ma)

In this period there is no evidence of subduction-related granitoids or volcanic rocks. It could be a period of collision, as proposed by Aceñolaza and Toselli (1984), or a period without subduction. This period coincides with detachment of Cuyania composite terrane from Laurentia, which started its drift toward Gondwana in the Middle to Late Cambrian (see Astini et al., 1995, 1996).

Middle Cambrian to Early Ordovician (515–470 Ma)

Start-up of subduction and arc magmatism along the protomargin of Gondwana, evidence of which is preserved in a series of volcanic and granitic rocks that were emplaced in the western Sierras Pampeanas. This happened while the carbonate platform of Precordillera was thermally subsiding and isolated from Laurentia (Astini et al., 1996).

Middle to Late Ordovician (470–450 Ma)

In this period syncollisional granitoids were emplaced synchronously with drowning of the Precordilleran carbonate platform, and subsequent flexural extension. Deformation by ductile thrusting at depth was followed by general uplift of the areas close to the present suture (Ramos et al., 1996).

The northern Sierras Pampeanas may have had a more complex history. Here a collision of the Famatina terrane against the protomargin of Gondwana was followed by the final collision of Precordillera with the Famatina terrane. This could explain the delay of subduction-related magmatism in comparison with the Ancasti area and other northern batholiths.

If correct, the Famatina terrane was an independent terrane sutured to the protomargin of Gondwana ca. 470–460 Ma, as proposed on paleomagnetic grounds by Conti et al. (1996). We favor this alternative (Fig. 18) and disregard the hypothesis of Toselli et al. (1996a), which proposed that Famatina was separated only by a backarc basin from the Sierras Pampeanas.

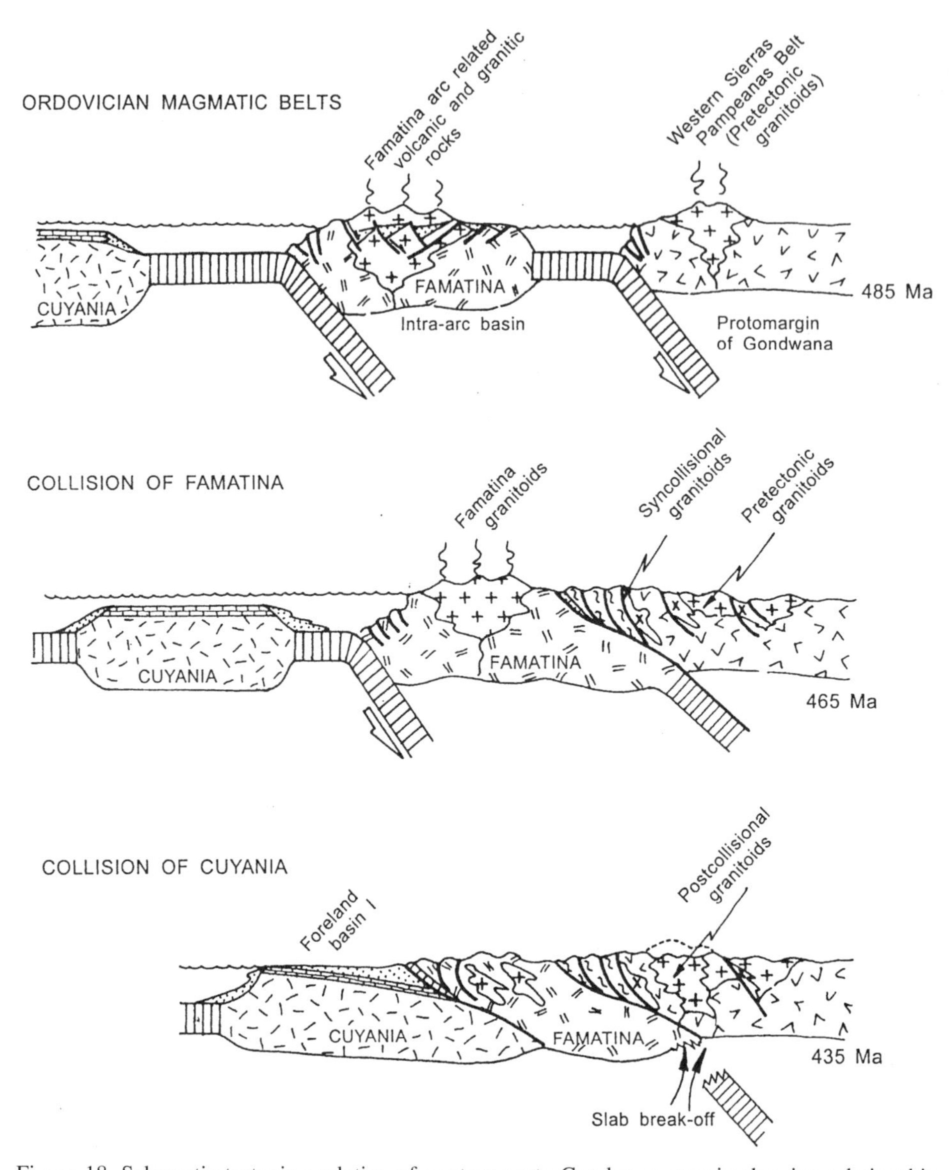

Figure 18. Schematic tectonic evolution of western proto-Gondwana margin showing relationship between Cuyania and Famatina terranes. Crosscutting relationships between those terranes required that Famatina magmatic arc collided prior to docking of Cuyania. Precordillera is included as part of Cuyania terrane.

Latest Ordovician to Late Devonian

In this period, intense anorogenic magmatism occurred in most of Sierras Pampeanas. This cratonization of the Sierras Pampeanas basement was followed in the Early Carboniferous by the beginning of rifting and sedimentation of the conglomerate and red beds of the Paganzo group.

ACKNOWLEDGMENTS

This work was partially funded by Universidad de Buenos Aires (UBACYT TW87) and Consejo Nacional de Investigaciones (CONICET PIP 4162/97). It is a contribution to International Geological Correlation Program Project 376 (Laurentia-Gondwana Connection before Pangea). The authors wish to acknolwedge Cees Van Staal, Geological Survey of Canada; Eric Nelson, Colorado School of Mines; Stella Poma, Universidad de Buenos Aires; and Suzanne M. Kay, Cornell University, for their critical review of an early version of the manuscript.

REFERENCES CITED

Aceñolaza, F. G., and Toselli, A. J., 1984, Lower Ordovician volcanism in northwest Argentina, *in* Bruton, D. L., ed., Aspects of the Ordovician system: Oslo, Norway, University of Oslo, v. 295, p. 203–209.

Aceñolaza, F. G., Miller, H., and Toselli, A. J., 1996, Geología del Sistema de Famatina: Münchner Geologische Heste, A 19, p. 1–410.

Arrospide, A., 1974, Petrografía, estructura y génesis de los depósitos de "greisen," del sector comprendido entre las quebradas "Los Arboles" y "Los Ratones," Sierra de Fiambalá, Catamarca [Ph.D. thesis]: La Plata, Argentina, Universidad de La Plata, 268 p.

Astini, R. A., 1996, Las fases diastróficas del Paleozoico medio en la Precordillera del oeste argentino: 13° Congreso Geológico Argentino y 3° Congreso Exploración de Hidrocarburos, Buenos Aires, v. 5, p. 509–526.

Astini, R. A., Benedetto, J. L., and Vaccari, N. E., 1995, The early Paleozoic evolution of the Argentina Precordillera as a Laurentian rifted, drifted, and collided terrane: a geodynamic model. Geological Society of America Bulletin, v. 107, p. 253–273.

Astini, R., Ramos, V. A., Benedetto, J. L., Vaccari, N. E., and Cañas, F. L., 1996, La Precordillera: un terreno exótico a Gondwana: 13° Congreso Geológico Argentino y 3° Congreso Exploración de Hidrocarburos, Buenos Aires, v. 5, p. 293–324.

Batchelor, R., A. and Bowden, P., 1985, Petrogenetic interpretation of granitic rock series using multicationic parameters: Chemical Geology, v. 48, p. 43–55.

Benedetto, J. L., and Astini, R., 1993, A collisional model for the stratigraphic evolution of the Argentine Precordillera during the early Paleozoic: 2nd Symposium International Geodynamique Andine ISAG 93, Oxford, United Kingdom, p. 501–504.

Brogioni, N., 1987, El Batolito Las Chacras–Piedras Coloradas, provincia de San Luis: Geología y edad: 10° Congreso Geológico Argentino, S. M. de Tucumán, v. 4, p. 115–118.

Brogioni, N., 1991, Caracterización petrográfica y geoquímica del Batolito de Las Chacras–Piedras Coloradas, San Luis, Argentina: 6° Congreso Geológico Chileno, Viña del Mar, v. 1, p. 766–770.

Brogioni, N., 1992, Geología del Batolito Las Chacras–Piedras Coloradas, provincia de San Luis: Revista del Museo de La Plata (Nueva Serie), sección Geología, v. 11, p. 1–16.

Brogioni, N., 1993, El Batolito de Las Chacras–Piedras Coloradas, provincia de San Luis: Geocronología Rb/Sr y ambiente tectónico: 12° Congreso Geológico Argentino y 2° Congreso de Exploración de Hidrocarburos, Mendoza, v. 4, p. 54–60.

Brogioni, N., Parrini, P., and Pecchioni, E., 1994, Magmatismo pre y sin-colisional en el cordón de El Realito, Sierra de San Luis, Argentina: 7° Congreso Geológico Chileno, Concepción, v. 2, p. 962–966.

Carugno Durán, A., Ortiz Suárez, A., and Prozzi, C., 1992, Deformación en granitoides del Río V, provincia de San Luis: 8° Reunión de Microtectónica (S.C. de Bariloche), v. 1, p. 83–86.

Cisterna, C. E., 1994, Contribución a la petrología de los granitoides del extremo norte de la Sierra de Narváez, Sistema de Famatina, provincia de Catamarca [Ph.D. thesis]: Salta, Argentina, Universidad Nacional de Salta, 228 p.

Conti, C. M., Rapalini, A. E., Coira, B., and Koukharsky, M., 1996, Paleomagnetic evidence of an early Paleozoic rotated terrane in northwest Argentina: a clue from Gondwana-Laurentia interaction?: Geology, v. 24, p. 953–956.

Costa, C., 1983, Geología del perfil El Durazno–Suyuque Viejo (Sierra Grande de San Luis), República Argentina: Revista de la Asociación Argentina de Mineralogía, Petrografía y Sedimentología, v. 14, p. 70–79.

Dahlquist, J., and Baldo, E., 1996, Metamorfismo y deformación famatinianos en la sierra de Chepes, La Rioja, Argentina: 13° Congreso Geológico Argentino y 3° Congreso de Exploración de Hidrocarburos, Mendoza, v. 5, p. 393–409.

Dalla Salda, L., Cingolani, C., and Varela, R., 1992a, Early Paleozoic orogenic belt of the Andes in southwestern South America: result of Laurentia-Gondwana collision?: Geology, v. 20, p. 617–620.

Dalla Salda, L., Dalziel, I. W. D, Cingolani, C. A., and Varela, R., 1992b, Did the Taconic Appalachians continue into southern South America?: Geology, v. 20, p. 1059–1062.

Dalla Salda, L., Cingolani C., and López de Luchi, M., 1995a, South American pre- and collisional Taconian granitoids: Geological Society of America Abstracts with Programs, v. 27, p. A6.

Dalla Salda, L., Cingolani, C., Varela, R., and López de Luchi, M. 1995b, The Famatinian orogenic belt in south-western South America, granites and metamorphism: an Appalachian similitude: Caracas, Venezuela, 11° Congreso Geológico Latinoamericano (electronic version).

Dalziel, I. W. D., 1992, Antarctica: a tale of two supercontinents: Annual Reviews of Earth and Planetary Science, v. 20, p. 501–526.

Dalziel, I. W. D., 1993, Tectonic tracers and the origen of the proto-Andean margin: 12° Congreso Geológico Argentino y 2° Congreso de Exploración de Hidrocarburos, Mendoza, v. 3, p. 367–374.

Dalziel, I. W. D., 1997, Neoproterozoic-Paleozoic geography and tectonics: review, hypothesis, environmental speculation. Geological Society of America Bulletin, v. 109, p. 16–42.

Dalziel, I. W. D., Dalla Salda, L. H., and Gahagan, L. M., 1994, Paleozoic Laurentia-Gondwana interaction and the origin of the Appalachian-Andean mountain system: Geological Society of America Bulletin, v. 106, p. 243–252.

Dalziel, I. W. D., Dalla Salda, L., Cingolani, C., and Palmer, P., 1996, The Argentine Precordillera: a Laurentian terrane?: GSA Today, v. 6, p. 16–18.

DeBari, S., 1994, Petrogenesis of the Fiambalá Gabbroic intrusion, northwestern Argentina, a deep crustal syntectonic pluton in a continental magmatic arc: Journal of Petrology, v. 35, no. 3, p. 679–713.

Durand, F., Toselli, A. J., Aceñolaza, F. G., Lech, R., Pérez, W., and Lencina, R., 1990, Geología de la Sierra de Paimán, provincia de La Rioja, Argentina: 10° Congreso Geológico Argentino, San Juan, v. 2, p. 15–18.

Escayola, M. P., Ramé, G. A., and Kraemer, P. E., 1996, Caracterización y significado geotectónico de las fajas ultramáficas de Córdoba: 13° Congreso Geológico Argentino y III° Congreso Exploración de Hidrocarburos, Buenos Aires, Actas, v. 3, p. 421–438.

García, H., and Rosello, E., 1984, Geología y yacimientos minerales de Papachacra, Departmento Belén, Catamarca: 10° Congreso Geológico Argentino, San Juan, v. 7, p. 245–259.

García, H., Massabie, A., and Rosello, E., 1981, Contribución a la Geología de La Cuesta, Belén, Catamarca: 8° Congreso Geológico Argentino, San Luis, v. 4, p. 833–865.

González, R., Omil, M., and Ruiz, D., 1985a, Observaciones y edades potasio-argón de formaciones de la Sierra de Paimán, provincia de La Rioja: Acta Geológica Lilloana, v. 16, no. 2, p. 281–287.

González, R., Cabrera, M., Bortolotti, P., Castellote, P., Cuenya, P., Omil, M., Moyano, R., and Ojeda, J., 1985b, La actividad eruptiva de Sierras Pampeanas. Esquematización gráfica y temporal: Acta Geológica Lilloana, v. 16, no. 2, p. 289–318.

González, R. R., and Toselli, A. 1974, Radiometric dating of igneous rocks from Sierras Pampeanas, Argentina: Revista Brasileira de Geociencias, v. 4, p. 137–141.

González Bonorino, F., 1951a, Descripción geológica de la Hoja 12e Aconquija (Catamarca-Tucumán): Dirección Nacional de Minería Boletín (Buenos Aires), v. 75, p. 1–49.

González Bonorino, F., 1951b, Granitos y migmatitas de la falda occidental de la Sierra de Aconquija: Asociación Geológica Argentina, Revista, v. 6, no. 3, p. 137–186.

Grissom, G. C., 1991, Thermal processes in the deep crust of magmatic arcs, Sierra de Fiambalá, northwestern Argentina [Ph.D. thesis]: Palo Alto, California, Stanford University, 262 p.

Grissom, G., DeBari, S., Page, S., Page, R., Villar, L., Coleman, R., and Ramirez, M. V., 1991, The deep crust of an early Paleozoic arc; the Sierra de Fiambalá, northwestern Argentina: Geological Society of America Special Paper 265, p. 189–200.

Halpern, M., Linares, E., and Latorre, C., 1970, Estudio preliminar por el método Rb/Sr de rocas metamórficas y graníticas de la provincia de San Luis, República Argentina: Asociación Geológica Argentina Revista, v. 25, no. 3, p. 293–261.

Indri, D., 1986, Rasgos geológicos de la Cuesta Mina Capillitas, provincia de

Catamarca, Argentina: Revista del Instituto de Geología y Minería, v. 6, p. 191–209.

Kay, S. M., Orrell, S., and Abbruzzi, J. M., 1996, Zircon and whole rock Nd-Pb isotopic evidence for a Grenville age and a Laurentian origin for the Precordillera terrane in Argentina: Journal of Geology, v. 104, p. 637–648.

Knüver, M., 1983, Dataciones radimé tricas de rocas plutónicas y metamórficas, *in* Aceñolaza, F., Miller, H., and Toselli, A., ed., Geología de la Sierra de Ancasti: Münstersche Forschungen zur Geologie und Paläontologie, H59, p. 201–218.

Kraemer, P. E., Escayola M. P., and Martino, R. D., 1995, Hipótesis sobre la evolución tectónica neoproterozoica de las Sierras Pampeanas de Córdoba (30°40′–32°40′), Argentina: Asociación Geológica Argentina, Revista, v. 50, p. 47–59.

Lazarte, J. E., 1987, Contribución a la petrología de los granitoides de la sierra de Papachacra, Catamarca, República Argentina: 10° Congreso Geológico Argentino, S. M. de Tucumán, v. 4, p. 69–72.

Lazarte, J. E., 1991, Estudio petrológico y geoquímico de los granitoides de las sierras de Papachacra y Culampajá. Relaciones metalogenéticas [Ph.D. thesis]: Tucumán, Argentina, Universidad Nacional de Tucumán, 270 p.

Lazarte, J. E., 1992, La Formación Chango Real (NW de Sierras Pampeanas, República Argentina), ejemplo del magmatismo paleozoico (Cámbrico?). Diferencias geoquímicas con batolitos ordovícicos: Estudios Geológicos, v. 48, p. 257–267.

Lazarte, J. E., 1995, Geología y geoquímica del Granito Papachacra (Carbonífero?), Sierras Pampeanas, Catamarca: Asociación Geológica Argentina, Revista, v. 49, p. 337–352.

Lema, H., 1980, Geología de los afloramientos del arroyo Peñas Blancas, Sierra de Yulto, provincia de San Luis: Asociación Geológica Argentina, Revista, v. 35, p. 147–150.

Le Maitre, R. W., 1984, A proposal by the IUGS Subcomission on the Systematics of Igneous Rocks for a chemical classification of volcanic rocks based on the total alkali silica (TAS) diagram: Australian Journal of Earth Science, v. 31, p. 243–255.

Linares, E., 1977, Catálogo de edades radimétricas para la República Argentina: Asociación Geológica Argentina, serie B 4, 35 p.

Linares, E., and González, R., 1990, Catálogo de edades radimétricas de la República Argentina 1957–1987: Publicación Especial Asociación Geológica Argentina. Serie B (Didáctica y Complementaria) v. 19, 628 p.

Linares, E., and Quartino, B., 1979, Nuevas aportaciones a la génesis de las rocas orbiculares de La Rioja y el control recíproco de datos K/Ar e interpretaciones petrogenéticas: 7° Congreso Geológico Argentino, Neuquén, v. 2, p. 585–594.

Lira, R., Millone, H. A., Kirschbaum, A. M., and Moreno, R. S., 1997, Magmatic-arc calc-alkaline granitoid activity in the Sierra Norte–Ambargasta Ranges, central Argentina: Journal of South American Earth Sciences, v. 10, p. 157–178.

Llambías, E., and Malvicini, L., 1982, Geología y génesis de los yacimientos de tungsteno de las sierras del Morro, Los Morrillos y Yulto, provincia de San Luis: Asociación Geológica Argentina, Revista, v. 37, p. 100–143.

Llambías, E., Cingolani, C., Varela, R., Prozzi, C., Ortiz Suárez, A., Toselli, A., and Saavedra, J., 1991, Leucogranodioritas sin-cinemáticas ordovícicas en la Sierra de San Luis: 6° Congreso Geológico Chileno, Viña del Mar, v. 1, p. 187–191.

Llambías, E., Quenardelle, S., Ortiz Suárez, A., and Prozzi, C., 1996, Granitoides sincinemáticos de la Sierra Central de San Luis: 13° Congreso Geológico Argentino and 3° Congreso de Exploración de Hidrocarburos, Buenos Aires, v. 3, p. 487–496.

Llano, J., Castro de Machuca, B., Rossa, N., and Vaca, A., 1987, Las rocas cataclásticas en el perfil Valle de Pancanta, Paso del Rey, Sierra de San Luis, República Argentina: 10° Congreso Geológico Argentino, S. M. de Tucumán, v. 3, p. 31–34.

López de Luchi, M., 1987a, Caracterización geológica y geoquímica del plutón La Tapera y del batolito de Renca, provincia de San Luis: 10° Congreso Geológico Argentino, S.M. de Tucumán, v. 4, p. 84–87.

López de Luchi, M., 1987b, Enclaves en el Batolito de Renca: 10° Congreso Geológico Argentino, S.M. de Tucumán, v. 4, p. 89–91.

López de Luchi, M., 1993, Caracterización geológica y emplazamiento del Batolito de Renca: 12° Congreso Geológico Argentino y 2° Congreso de Exploración de Hidrocarburos, Mendoza, v. 4, p. 42–53.

López de Luchi, M. G., and Dalla Salda, L. H., 1995, Some Pampean pre-tectonic (Famatinian) granites of southwestern South America, *in* Brown, M., and Piccoli, P. M., eds., The origin of granites and related rocks: College Park, Maryland, 3rd Hutton Symposium, Abstracts, p. 91.

López de Luchi, M. G., and Dalla Salda, L., 1996, South America Famatinian granitoids and the evolution of southern Iapetus, 30th International Geological Congress, Beijing, Abstracts, v. 1, p. 205.

Loske, W., and Miller, H. 1996, Sistemática U/Pb de circones del granito de Ñuñorco-Sañogasta, *in* Aceñolaza, F. G., Miller, H., and Toselli, A. J., eds., Geología del Sistema de Famatina: Münchner Geologische Heste A 19, p. 221–228.

Lottner, U. S., and Miller, H., 1986, The Sierra de Ancasti as an example of the structurally controlled magmatic evolution in the lower Paleozoic basement of the NW Argentine Andes: Zentralblatt für Geologie Palöntologie Teil Y, H9/10, p. 1269–1281.

Mángano, M. A., 1993, Dinámica sedimentaria y tafonomía en las secuencias ordovícicas volcaniclásticas de la Formacián Suri, Sistema de Famatina [Ph.D. thesis]: Buenos Aires, Argentina, Universidad de Buenos Aires, 260 p.

Mannheim, R., 1993a, Genese der vulkanite und subvulkanite des altpal ozischen Famatina-systems, NW-Argentinien, und seine geodynamische entwicklung. Münchner Geologische Hefte 9, 130 p.

Mannheim, R., 1993b, Génesis de las vulcanitas eopaleozoicas del Sistema de Famatina, noroeste de Argentina: 12° Congreso Geológico Argentino y 2° Congreso de Exploración de Hidrocarburos, Mendoza, v. 4, p. 147–155.

Marcos, O., 1984, Geología y prospección de la zona del Río Oro, Sierra de Famatina, provincia de La Rioja: 9° Congreso Geológico Argentino, S.C. de Bariloche, v. 7, p. 191–206.

McBride, S. L., Caelles, J. C., Clark, A. H., and Farrar, E., 1976, Palaeozic radiometric age provinces in the Andean basement, latitudes 25°–30°: Earth and Planetary Science Letters, v. 29, p. 373–383.

Mirré, J. C., 1976, Descripción geológica de la Hoja 19e, Valle Fértil, provincias de San Juan y La Rioja: Carta geológico-económica de la República Argentina, Servicio Geológico Nacional, Boletín 147, 70 p.

Mullen, E. D., 1983, $MnO/TiO_2/P_2O_5$: a minor element discriminant for basaltic rocks of oceanic environments and its implications for petrogenesis: Earth Planetary and Science Letters, v. 62, p. 53–62.

Nakamura, K., 1974, Determination of REE, Ba, Fe, Mg, Na and K in carbonaceous and ordinary chondrites: Geochimica Cosmochimica Acta, v. 38, p. 757–775.

Ortiz Suárez, A., 1996, Geología y petrografía de los intrusivos de Las Aguadas, provincia de San Luis: Asociación Geológica Argentina, Revista, v. 51, p. 321–330.

Ortiz Suárez, A., Prozzi, C., and Llambías, E., 1992, Geología de la parte sur de la Sierra de San Luis y granitoides asociados, Argentina: Estudios Geológicos, v. 48, p. 269–277.

Pankhurst, R., Rapela, C., Saavedra, J., Baldo, E., Dahlquist, J., and Pascua, I., 1996, Sierra de Los Llanos, Malanzán, and Chepes: Ordovician I and S-type granitic magmatism in the Famatinuan orogen: 13° Congreso Geológico Argentino y 3° Congreso de Exploración de Hidrocarburos, Buenos Aires, v. 5, p. 415.

Pearce, J. A., 1996, Source and settings of granitic rocks: Episodes, v. 19(4), p. 120–125.

Pearce, J. A., Harris, N. B. W., and Tindle, A. G., 1984, Trace element discrimination diagrams for the tectonic interpretation of granitic rocks: Journal of Petrology, v. 24, p. 956–983.

Pérez, W., and Kawashita, K., 1992, K/Ar and Rb/Sr geochronology of igneous rocks from the Sierra de Paimán, northwestern Argentina: Journal of South American Earth Sciences, v. 5, no. 3–4, p. 251–264.

Prozzi, C., and Ramos, G., 1988, La Formación San Luis: 1° Jornadas de trabajo de Sierras Pampeanas, San Luis, Abstracts, p. 1.

Quartino, B., and Villar Fabre, J., 1963, El cuerpo granítico orbicular precámbrico de la Pampa de los Altos, Sierra de Velasco (provincia de La Rioja): Asociación Geológica Argentina, Revista, v. 28, p. 11–42.

Quenardelle, S., 1993, Caracterización geológico-petrológica del granito San José del Morro, provincia de San Luis: 12° Congreso Geológico Argentino y 2° de Exploración de Hidrocarburos, Mendoza, v. 4, p. 61–67.

Quenardelle, S., 1995, Petrografía y geoquímica del plutón San José del Morro, provincia de San Luis: Asociación Geológica Argentina, Revista, v. 50, p. 229–236.

Ramos, V. A., 1982, Descripción Geológica de la Hoja 20 f Chepes, provincia de La Rioja: Servicio Geológico Nacional Boletín 184, p. 152.

Ramos, V. A., 1986, El diastrofismo oclóyico: un ejemplo de tectónica de colisión durante el Eopaleozoico en el noroeste Argentino: Revista Instituto Ciencias Geológicas, v. 6, p. 13–28.

Ramos, V. A., 1988, Tectonics of the late Proterozoic–early Paleozoic: a collisional history of southern South America: Episodes, v. 11, p. 168–174.

Ramos, V. A., 1989, Southern South America: an active margin for the past 700 Ma: 28th International Geological Congress, Washington, Abstracts, v. 2, p. 664.

Ramos, V. A., Jordan, T., Allmendinger, R. W., Kay, S. M., Cortés, J. M., and Palma, M. A., 1984, Chilenia un terreno alóctono en la evolución paleozoica de los Andes Centrales: 4° Congreso Geológico Argentino, S.C. Bariloche, v. 2, p. 84–106.

Ramos, V. A., Jordan, T. E., Allmendinger, R. W., Mpodozis, C., Kay, S. M., Cortés, J. M., and Palma, M. A., 1986, Paleozoic terranes of the central Argentine-Chilean Andes: Tectonics, v. 5, p. 855–880.

Ramos, V. A., Vujovich, G. I., and Dallmeyer, R. D., 1996, Los klippes y ventanas tectónicas de la estructura preándica de la Sierra de Pie de Palo (San Juan); edad e implicaciones tectónicas: 13° Congreso Geológico Argentino y 3° Congreso Exploración de Hidrocarburos, Buenos Aires, v. 5, p. 377–392.

Rapela, C. W., and Heaman, L., 1982, Composición química de granitos batolíticos de las Sierras Pampeanas: Revista del Museo de la Plata, Sección Geología, v. 9(75), p. 89–96.

Rapela, C. W., and Pankhurst, R., 1996, The Cambrian plutonism of the Sierras de Córdoba: pre-Famatinian subduction? and crustal melting: 13° Congreso Geológico Argentino y 3° Congreso de Exploración de Hidrocarburos, Buenos Aires, Actas, v. 5, p. 491.

Rapela, C. W., Heaman, L., and McNutt, R., 1982, Rb-Sr geochronology of granitoid rocks from the Pampean Ranges, Argentina: Journal of Geology, v. 90, p. 574–582.

Rapela, C. W., Toselli, A. J., Heaman, L., and Saavedra, J., 1990, Granite plutonism of Sierras Pampeanas; an Inner Cordilleran Paleozoic arc in the southern Andes, *in* Kay, S. M., and Rapela, C. W., eds., Plutonism from Antarctica to Alaska: Geological Society of America Special Paper 241, p. 77-90.

Rapela, C. W., Pankhurst, R. J., and Bonalumi, A. A., 1991, Edad geoquímica del pórfido granítico de Oncán, Sierra Norte de Córdoba, Sierras Pampeanas, Argentina: 6° Congreso Geológico Chileno, Viña del Mar, v. 1, p. 19–22.

Rapela, C., Coira, B., Toselli, A., and Saavedra, J., 1992, El magmatismo del Paleozoico inferior en el sudoeste de Gondwana, Paleozoico Inferior de Ibero-América, *in* Gutiérrez Marco, J. C., Saavedra, J., and Rábano, I., eds., Mérida, Spain, Universidad de Extremadura, p. 21–68.

Rapela, C., Saavedra, J., Toselli, A., and Pellitero, E., 1996, Eventos magmáticos fuertemente peraluminosos en las Sierras Pampeanas: 13° Congreso Geológico Argentino y 3° Congreso de Exploración de Hidrocarburos, Buenos Aires, v. 5, p. 337–353.

Rossi de Toselli, J., Toselli, A. J., Medina, M., and Saal, A., 1987, Los stocks granofíricos de Chaschuil, Sierra de Narváez, Catamarca: 10° Congreso Geológico Argentino, S.M. de Tucumán, v. 4, p. 151–153.

Rossi de Toselli, J., Toselli, A., and Wagner, A., 1991, Geobarometría de hornblendas en granitoides calcoalcalinos, Sistema de Famatina, Argentina: 6° Congreso Geológico Chileno, Viña del Mar, v. 6, p. 244–247.

Saal, A., 1988, Los granitoides de la sierra de Paganzo, La Rioja, Argentina: 5° Congreso Geológico Chileno, Santiago, v. 3, p. I1–I15.

Saal, A., 1993, El basamento cristalino de la sierra de Paganzo, provincia de La Rioja, Argentina [Ph.D. thesis]: Córdoba, Argentina, Universidad Nacional de Córdoba, 345 p.

Saavedra, J., Pellitero-Pascual, E., Rossi, J. N., and Toselli, A. J., 1992, Magmatic evolution of the Cerro Toro Granite, a complex Ordovician pluton of northwestern Argentina: Journal of South American Earth Science, v. 5, p. 21–32.

Sánchez, V., Ortiz Suárez, A., and Prozzi, C., 1996, Geología y petrografía de la tonalita precinemática Bemberg, provincia de San Luis: 13° Congreso Geológico Argentino y 3° Congreso de Exploración de Hidrocarburos, Buenos Aires, v. 3, p. 669–677.

Sato, A. M., Ortiz Suárez, A., Llambías, E. J., Cavarozzi, C. E., Sánchez, V., Varela, R., and Prozzi, C., 1996, Los plutones preoclóyicos del sur de la Sierra de San Luis: arco magmático al inicio del ciclo famatiniano: 13° Congreso Geológico Argentino y 3° Congreso Exploración de Hidrocarburos, Buenos Aires, v. 5, p. 259–272.

Söllner, F., de Brodtkorb, M. K., Miller, H., Pezzutti, N., and Fernández, R., 1998, Early Cambrian metavolcanic rocks from the Sierra de San Luis, Argentina: evidence from U-Pb age determinations on zircons: 10° Congreso Latinoamericano de Geología, Buenos Aires, Actas, v. 2, p. 387.

Stipanicic, P., and Linares, E., 1975, Catálogo de edades radimétricas determinadas para la República Argentina: Asociación Geológica Argentina, Publicaciones Especiales, Serie B, v. 3, p. 1–42.

Thomas, W. A., and Astini, R. A., 1996, The Argentine Precordillera: a traveller from the Ouachita embayment of North American Laurentia: Science, v. 273, p. 752–757.

Tosdal, R. M., 1996, The Amazon-Laurentian connection as viewed from the Middle Proterozoic rocks in the central Andes, western Bolivia and northern Chile: Tectonics, v. 15, p. 827–842.

Toselli, A., and Indri, D., 1984, Consideraciones sobre los silicatos de aluminio en el granito Capillitas, Catamarca: 9° Congreso Geológico Argentino, Buenos Aires, v. 3, p. 205–215.

Toselli, A., Reissinger, M., Durand, F., and Bazán, C., 1983, Rocas graníticas, *in* Aceñolaza, F., Miller, H., and Toselli, A., eds., Geología de la Sierra de Ancasti: Münstersche Forschungen zur Geologie und Paläontologie, H59, p. 79–99.

Toselli, A., Rossi de Toselli, J., Saavedra, J., and Medina, M., 1987, Granitoides del Famatina, La Rioja, Argentina: Algunos aspectos geológicos y geoquímicos: 10° Congreso Geológico Argentino, Tucumán, Argentina, v. 4, p. 147–150.

Toselli, A., Rossi de Toselli, J., Saavedra, J., Pellitero, E., and Medina, M., 1988, Aspectos petrológicos y geoquímicos de los granitoides del entorno de Villa Castelli, Sierras Pampeanas Occidentales–Sistema de Famatina, Argentina: 5° Congreso Geológico Chileno, Santiago; v. 3, p. I17–I28.

Toselli, A., Rossi de Toselli, J., Pellitero, E., and Saavedra, J., 1993, El arco magmático granítico del Paleozoico Inferior en el Sistema de Famatina, Argentina: 12° Congreso Geológico Argentino and 2° Exploración de Hidrocarburos, Mendoza, v. 4, p. 7–15.

Toselli, A. J., 1992, El magmatismo del noroeste argentino. Reseña sistemática e interpretación: Serie Correlación Geológica 8, Instituto Superior de Correlación Geológica, Facultad de Ciencias Naturales e Instituto Miguel Lillo, Universidad Nacional de Tucumán, 243 p.

Toselli, A. J., Durand, F., Rossi de Toselli, J., and Saavedra, J., 1996a, Esquema de evolución geotectónica y magmática eopaleozoica del Sistema de Famatina y sectores de Sierras Pampeanas: 13° Congreso Geológico Argentino y 3° Congreso de Exploración de Hidrocarburos, Buenos Aires, v. 5, p. 443–462.

Toselli, A. J., Sial, A. N., Saavedra, J., Rossi de Toselli, J., and Ferreira, V., 1996b, The Famatinian peraluminous Capillitas batholith, Argentina: genesis by collision-related crustal anatexis: 13° Congreso Geológico Argentino y 3° Congreso de Exploración de Hidrocarburos, Buenos Aires, v. 5, p. 463.

Toselli, G., Saavedra, J., Córdoba, G., and Medina, M. E., 1991, Petrología y geoquímica de los granitos de la zona Carrizal-Mazán, La Rioja y Catamarca: Asociación Geológica Argentina, Revista, v. 46, p. 36–50.

Toselli, G., Saavedra, J., Córdoba, G., and Medina, M. E., 1992, Los granitos peraluminosos de las Sierras de Vinquis, Cerro Negro y Zapata (Sierras Pampeanas), provincia de Catamarca, Argentina: Estudios Geológicos, v. 48, p. 247–256.

Toubes Spinelli, R., 1983, Edades potasio-argón de algunas rocas de la sierra de Valle Fértil, provincia de San Juan: Revista de la Asociación Geológica Argentina, v. 38, p. 405–411.

Turner, F. J., 1981, Metamorphic petrology. Mineralogical, field and tectonic aspects: New York, McGraw-Hill, 524 p.

Turner, J. C., 1962, Estratigrafía de la región al naciente de la Laguna Blanca, Catamarca: Asociación Geológica Argentina, Revista, v. 27, p. 11–46.

Varela R., Llambías, E., Cingolani, C., and Sato, A., 1994, Datación de algunos granitoides de la Sierra de San Luis (Argentina) e interpretación evolutiva: 7° Congreso Geológico Chileno, Concepción, v. 2, p. 1249–1253.

Vujovich, G., Godeas, M., Marín, G., and Pezzutti, N., 1996, El complejo magmático metamórfico de la Sierra de la Huerta, provincia de San Juan: 13° Congreso Geológico Argentino y 3° Congreso Exploración de Hidrocarburos, Buenos Aires, v. 2, p. 465–475.

Manuscript Accepted by the Society September 4, 1998

Printed in U.S.A.

Geological Society of America
Special Paper 336
1999

Uranium-lead dating of felsic magmatic cycles in the southern Sierras Pampeanas, Argentina: Implications for the tectonic development of the proto-Andean Gondwana margin

Peter G. Stuart-Smith*, Alfredo Camacho*, John P. Sims*, Roger G. Skirrow, Patrick Lyons, Peter E. Pieters*, Lance P. Black
Australian Geological Survey Organization, P.O. Box 378, Canberra, ACT 2601 Australia
Robert Miró
Subsecretaría de Minería, Delegación Córdoba, Av. Poeta Lugones 161, B° Nueva Córdoba, 5000 Córdoba, Argentina

ABSTRACT

Uranium-lead ion probe dating of zircons from granites of the southern Sierras Pampeanas, Argentina, has provided new constraints on the timing of major felsic magmatic events and tectonic development of the proto-Andean Gondwana margin.

The zircon data indicate that there were three separate episodes of granite emplacement in the southern Sierras Pampeanas. The cycles, occurring during the Cambrian (ca. 530–510 Ma) , Early Ordovician (ca. 490–470 Ma), and Devonian (ca. 403–382 Ma), are referred to as the Pampean, Famatinian, and Achalian Cycles, respectively. There is no evidence to support Late Ordovician or Silurian granite intrusion, as indicated in previous Rb-Sr and K-Ar isotopic age data. The revised Early Ordovician age of the Famatinian Cycle indicates that final amalgamation of the Precordillera had taken place prior to 470 Ma, rather than by about 450 Ma as previously suggested.

INTRODUCTION

The southern Sierras Pampeanas, lying to the east of the Precordillera in northwest Argentina, comprise a series of basement ranges of early Paleozoic metamorphic rocks and Paleozoic granitoids, separated by intermontane Mesozoic and Cainozoic sediments (Fig. 1). The basement rocks form a series of north-trending lithologic and structural domains separated by major midcrustal shear zones. These domains have been variously interpreted to form part of an ensialic mobile belt (e.g., Dalla Salda, 1987) or as terranes that either accreted or developed on a western convergent margin of the Río Plata craton (e.g., Ramos 1988; Demange et al., 1993; Escayola et al., 1996; Kraemer et al., 1995).

Despite numerous age determinations of igneous rocks in the region mainly by K-Ar and Rb-Sr methods (e.g., Linares and Gonzalez, 1990), there has been, until recently, a lack of critical data to constrain the timing and correlation of major deformational events and granite intrusion in different parts of the Sierras Pampeanas. This has been a major limitation on previous tectonic models and histories. Consequently, the relationship of the development of the Sierras Pampeanas and the history of the proto-Andean Gondwana margin and its interaction with Laurentia has not been clearly understood.

The results of new geochronologic studies of granites in the southern Sierras Pampeanas, utilizing SHRIMP U-Pb zircon age determinations, are presented here. The study was part of a collaborative regional geoscientific mapping program between the Aus-

*Present address: Stuart-Smith, SRK Consulting, Suite 7, Deakin House, P.O. Box 250, Deakin West, A.C.T. Australia; Camacho, Research School of Earth Sciences, Australian National University, Canberra, A.C.T. 2600, Australia; Sims, Geoverde Pty Ltd., P.O. Box 479, Jamison Centre, A.C.T. 2614, Australia; Pieters, 13 Arbana Street, Aranda, A.C.T. 2614, Australia.

Stuart-Smith, P. G., Miró, R., Sims, J. P., Pieters, P. E., Lyons, P., Camacho, A., Skirrow, R. G., and Black, L. P., 1999, Uranium-lead dating of felsic magmatic cycles in the southern Sierras Pampeanas, Argentina: Implications for the tectonic development of the proto-Andean Gondwana margin, *in* Ramos, V. A., and Keppie, J. D., eds., Laurentia-Gondwana Connections before Pangea: Boulder, Colorado, Geological Society of America Special Paper 336.

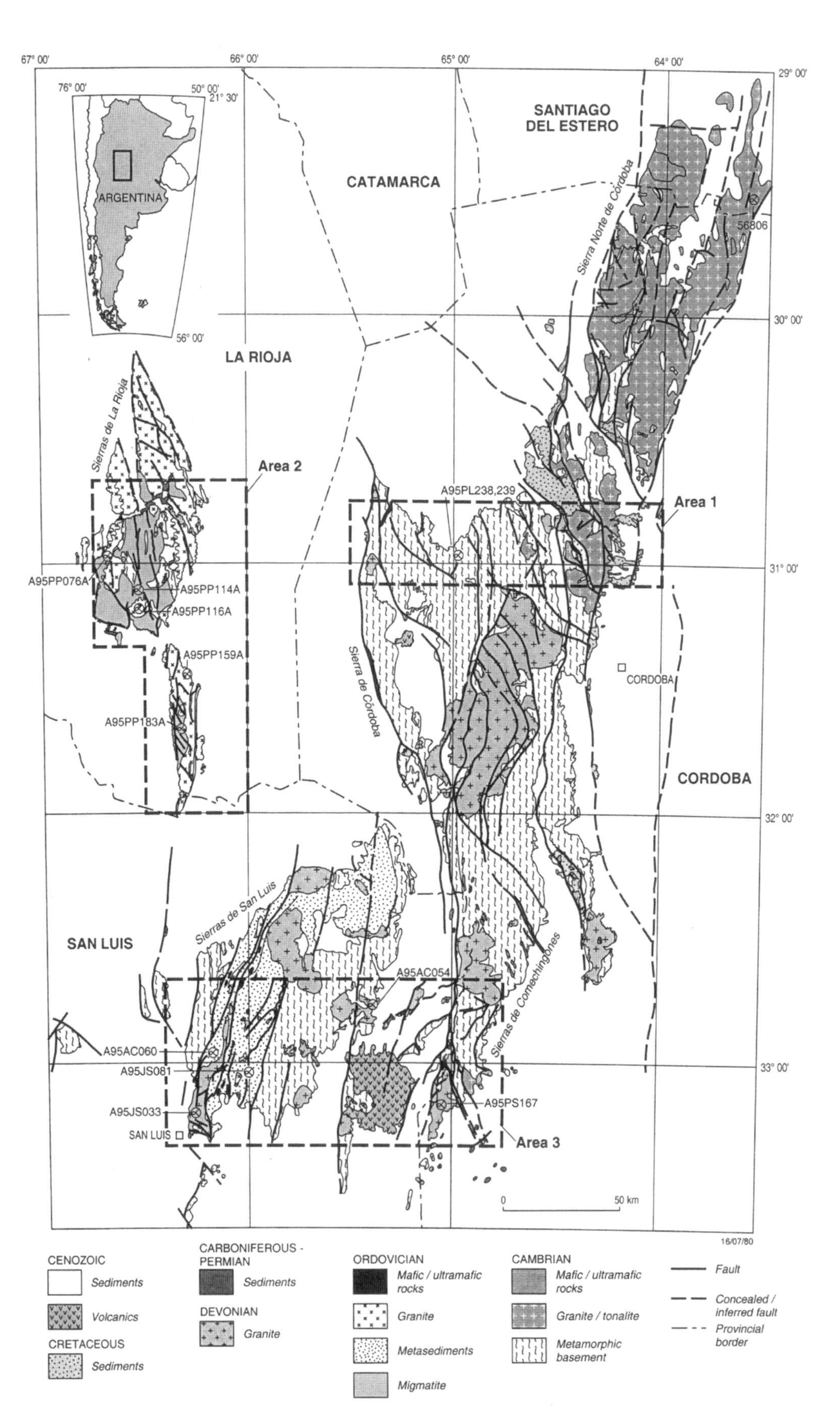

Figure 1. Simplified geology of the southern Sierras Pampeanas and location of geochronology samples of the present study. The three areas of the AGSO-SEGEMAR collaborative mapping project are indicated. Geology based on AGSO-SEGEMAR mapping (Sims et al., 1997; Pieters et al., 1997; Lyons et al., 1997), and maps of the Secretaría de Minería (1995), scale 1:500, 000.

tralian Geological Survey Organisation (AGSO) and the Servício Geológico Minero Argentino (SEGEMAR). The aim of the geochronology program was to provide key data to constrain the timing of igneous rock crystallization, major metamorphic/deformation episodes, and mineralizing events in the region. Mapping in three areas (Fig. 1), totaling about 27,000 km^2 and representing cross sections through the principal tectono-stratigraphic packages of the southern Sierras Pampeanas, was supported by airborne magnetic and radiometric surveys.

Major outcomes of the work include a refinement of the age of the early Paleozoic Pampean and Famatinian tectono-magmatic cycles recognized by previous workers (e.g., Aceñolaza and Toselli, 1976, Dalla Salda, 1987, Toselli et al., 1992) and the delineation of a third magmatic event, during the Devonian. These regional tectono-magmatic cycles are related to the development of the Gondwana margin and its interaction with Laurentia.

REGIONAL GEOLOGIC SETTING

The geology of the southern Sierras Pampeanas is shown in Figure 1. The province consists of a Neoproterozoic-Paleozoic crystalline basement comprising two principal metamorphic/igneous domains: an older Cambrian "Pampean" domain mainly in the east (Sierras de Córdoba, Norte de Córdoba, and Comechingones), and a mostly younger Ordovician "Famatinian" domain to the west (Sierras de San Luis and La Rioja). Several low-grade Ordovician meta sedimentary belts structurally overlie both metamorphic domains (Sims et al., 1998). Both crystalline basement domains share a common geologic history since early Ordovician time.

Polydeformed Neoproterozoic-Cambrian rocks, forming the "Pampean" domain, include pelitic, psammitic, and carbonate-rich meta sedimentary belts and metamorphic and igneous complexes, such as the Pichanas Complex in the Sierras de Córdoba (Lyons et al. 1997). Peraluminous felsic intrusions, common in the metamorphic complexes, are interpreted as partial melts that were emplaced during the oldest medium- to high-grade metamorphic event in the region (Rapela et al., 1996; Lyons et al., 1997). Several calc-alkaline granitic plutons and composite batholiths intrude the metamorphic complexes. Relationships show that the latter granites intruded late in the Cambrian deformation/metamorphic event (Rapela and Pankhurst, 1996; Lyons et al., 1997; Sims et al., 1998). The largest of these late-stage intrusions is the Ascochinga Igneous Complex, an extensive batholith that extends throughout the Sierra Norte de Córdoba in the provinces of Córdoba and Santiago del Estero.

The "Famatinian" domain, exposed mainly in the Sierras de San Luis and La Rioja, comprises Cambrian-Ordovician quartz-rich pelitic and psammitic meta sedimentary rocks and granitic complexes. These rocks were metamorphosed at low to high grades during a major deformation event dominated by westward-directed thrusting at about 480 Ma (Sims et al., 1998). Numerous bodies of granite, tonalite, and mafic and ultramafic rocks intrude the metasediments and show intrusive relationships indicative of syn- and postcompressive deformation emplacement. The Chepes Igneous Complex (including the Asperezas Granite) is the dominant basement unit in the Sierras de La Rioja (Sierras de Chepes, Las Minas, Ulapes, and Los Llanos) and is comprosed of granite, monzogranite, granodiorite, tonalite, and migmatite (Pieters et al., 1997). These intrusions were interpreted by Pankhurst et al. (1996), Dahlquist and Baldo (1996), and Pieters et al., (1997) as part of a magmatic belt with collisional characteristics. Similarly, Sato et al. (1996) interpreted that some felsic intrusions in the Sierras de San Luis formed part of a Famatinian magmatic arc.

Voluminous mid-Paleozoic granites intrude older rocks throughout the region. They are exposed in the Sierras de Córdoba and Sierras de San Luis (e.g., Sims et al., 1997) and are interpreted to be present in the subsurface, intruding basement rocks in the La Rioja Province (Pieters et al., 1997). Examples of these granites include the Renca Granite (Sierras de San Luis), the Achala and Cerro Aspero Batholiths (Sierras de Córdoba), and the Achiras Igneous Complex in the Sierras de Comechingones. The latter complex consists of multiply injected subconcordant granite and pegmatite intrusions within a long-lived mylonite zone that was active during and after intrusion (Sims et al., 1998). Most of the mid-Paleozoic granites form subcircular, zoned, and fractionated plutons that are commonly coalesced, forming batholiths. The granites cross-cut high-grade metamorphic fabrics in both Pampean and Famatinian basement domains, intruding either during or after a regional retrogressive deformation event (Sims et al., 1998).

The crystalline basement rocks are locally capped by Tertiary volcanics and overlain by relatively undeformed Carboniferous-Permian, Cretaceous and Quaternary continental deposits.

TECTONO-MAGMATIC CYCLES: PREVIOUS WORK

Two major tectono-magmatic cycles have long been recognized in the southern Sierras Pampeanas. Granitic magmatism accompanied medium- to high-grade metamorphism and deformation during the Cambrian Pampean Cycle, and the mid-Paleozoic Famatinian Cycle. A younger post-Famatinian magmatic event was also recognized by Rapela et al. (1992).

Pampean cycle

The Pampean Cycle is the oldest tectono-magmatic event recognized in the Sierras Pampeanas and is well developed in the Sierras de Córdoba. The event includes both the D1 and part of the D2 deformation domains of Dalla Salda (1987)[1] and has been previously termed the "Ciclo orogénico Pampeano" (Aceñolaza and Toselli, 1976) or "Ciclo Pampeano" (Dalla Salda, 1987; Toselli et al., 1992). During this event Cambrian

[1]Dalla Salda (1987) considered all of the D2 domains to be part of the Famatinian Cycle.

sediments and mafic dikes were deformed at midcrustal levels by a compressive event (D1) and metamorphosed at mostly upper amphibolite facies, and locally granulite-facies (e.g., Martino et al., 1995).

During the closing stages of the Pampean Cycle, an extensive phase of felsic magmatism is evident by widespread subconcordant intrusion of tonalite, granodiorite, and granite (e.g., Ascochinga Igneous Complex) (Lyons et al., 1997). These granitoids are mostly medium-K calc-alkaline (Pérez et al., 1996) varieties, interpreted to indicate a continental magmatic arc setting (Pérez et al., 1996; Lira et al., 1996). Local anatexis of paragneiss during regional metamorphism and migmatization is also inferred to be the origin for some late-stage strongly peraluminous granites (Baldo and Casquet, 1996).

Several of the peraluminous granites occur within the Pichanas Metamorphic Complex (Lyons et al., 1997) in the Sierras de Córdoba. Rubidium-strontium (Rb-Sr) and Sm-Nd isotopic studies (Rapela et al., 1995) indicate that the granites were coeval and were products of partial melting during high-T metamorphism of pelitic metasediments during the Pampean Cycle. The granites show both concordant and discordant relations with D1 high-grade metamorphic fabrics in enclosing gneiss. One of the bodies, the El Pilón Granite, a cordierite- and sillimanite-bearing porphyritic biotite granite, has yielded an Rb-Sr age of 520 ± 5 Ma (Rapela et al., 1995).

Famatinian Cycle

During the mid-Paleozoic, a widespread deformational, metamorphic and magmatic event known as the "Ciclo orogénico Famatiniano" (Aceñolaza and Toselli 1976), Famatinian Orogen (e.g., Dalla Salda et al., 1992) or "Ciclo Famatiniano" (Dalla Salda, 1987) affected the southern Sierras Pampeanas. Dalla Salda (1987) and Toselli et al. (1992) included their D2 and D3 deformation domains within this cycle. Intrusion of numerous granitic to tonalitic bodies (G2 granites of Rapela et al., 1992) accompanied the orogenic event (syn-, late-, and post-D2). K-Ar and Rb-Sr isotopic dating of these rocks indicates Ordovician to Devonian ages of 490–399 Ma (Linares and Gonzales, 1990; Llambías et al., 1991; Rapela et al., 1992; Pankhurst et al., 1996).

Peraluminous to slightly peralkaline felsic melts, which intruded the metamorphics after late Famatinian retrogressive low-grade metamorphism and mylonite development (D3), were interpreted by Rapela et al. (1992) as forming during a third cycle of felsic magmatism (G3) generated from partial melting of MgO-depleted crustal rocks (Dalla Salda et al., 1995). The age of this magmatism is interpreted as Devonian to Carboniferous, based on K-Ar and Rb-Sr isotopic dating (e.g., Rapela et al., 1982, 1992). The G3 granites include the Capillitas syenogranite, Cerro Amarillo granite (Rapela et al., 1992), and parts of the Achala Batholith.

METHODOLOGY

Samples

The location of granite samples is shown in Figure 1 and details are listed in Table 1. Petrography and whole-rock geochemistry are discussed by Sims et al. (1997), Lyons et al. (1997), and Pieters et al. (1997).

Three "Pampean" granites were selected for U-Pb isotopic age determination. Two granite samples (A95PL238 and A95PL239) from the Pichanas Metamorphic Complex (Lyons et al., 1997) were selected in order to verify Rb-Sr ages and to better constrain the timing of Pampean felsic magmatism and deformation. The Ojo de Agua granite (a northern extension of the Ascochinga Igneous Complex) from the Sierra Norte de Córdoba, an example of calc-alkaline intrusion emplaced during the closing stages of the Pampean Cycle, was selected to provide a minimum age of felsic magmatism in the cycle.

Ten "Famatinian" granites were selected for U-Pb isotopic age determination on the basis of their known field relationships. These comprise five granites from the Chepes Igneous Complex in La Rioja, four granites from the Sierra de San Luis (Bemberg Tonalite, Tamborero Granodiorite, Escalerilla Granite, Renca Granite), and a granite from the Achiras Igneous Complex of the Sierra de Comechingones. Previous age determinations of the Chepes Igneous Complex by Rb-Sr isotopic methods indicate an Ordovician age for the complex (471–456 Ma) (Pankhurst et al., 1996).

In the Sierra de San Luis, the Tamboreo Granodiorite and the Bemberg Tonalite form a suite of intermediate to mafic intrusions, and intrude low-grade metasediments of the San Luis Formation (Sims et al., 1997). The granodiorite and tonalite were selected for age determination to provide a minimum age for "Famatinian" deformation and metamorphism.

The Escalerilla Granite in the Sierras de San Luis and the Achiras Igneous Complex (includes the Los Nogales Granite) in the southern Sierras de Comechingones intruded medium-grade metamorphic rocks and are examples of subconcordant intrusions emplaced during localized retrogressive shearing (Sims et al., 1997). The granites were selected for age determination to constrain the timing of this low-grade deformation event. There are no previous isotopic data on the Achiras Igneous Complex.

The Renca Granite (Sierra de San Luis) is representative of a number of undeformed, subcircular, zoned, and fractionated plutons, that postdate all regional deformation and metamorphic events in the southern Sierras Pampeanas (Sims et al., 1997). An Rb-Sr whole-rock isotopic age of 415 ± 25 Ma has been obtained for the Renca Granite (Halpern et al., 1970).

SHRIMP U-Pb analyses

Uranium-lead isotopic analyses of zircons separated from the granites were carried out using the SHRIMP I (Sensitive High Resolution Ion MicroProbe) at the Australian National Univer-

TABLE 1. SUMMARY OF U-Pb ZIRCON AGES FOR GRANITES OF THE SOUTHERN SIERRAS PAMPEANAS

Sample	Latitude (S)	Longitude (W)	Unit (Rock type samples)	Age (Ma)
Cambrian				
A95-PL238	30.96323	64.98086	Pichanas Metamorphic Complex (El Pilón Granite, porphyritic)	Ca. 527, ca. 548
A95-PL239	30.96419	64.98095	Pichanas Metamorphic Complex (El Pilón Granite, equigranular)	Ca. 480, ca. 514
56806	29.5167	63.5833	Ojo de Agua granite	515 ± 4
Ordovician				
A95-PP076A	30.96634	66.67940	Chepes Igneous Complex (biotite-hornblende granodiorite)	491 ± 6
A95-PP159A	31.44962	66.2915	Chepes Igneous Complex (Asperezas Granite, biotite monzogranite)	490 ± 7
A95-PP116A	31.18457	66.52772	Chepes Igneous Complex (biotite monzogranite)	485 ± 7
A95-PP183A	31.67874	66.32181	Chepes Igneous Complex (epidote-bearing biotite-hornblende tonalite)	480 ± 6
A95-PP114A	31.10643	66.53083	Chepes Igneous Complex (biotite granodiorite)	477 ± 7
A95-JS081	33.05848	66.99416	Tamboreo Granodiorite	470 ± 5
A95-AC060	32.72854	65.61784	Bemberg Tonalite	468 ± 6
Devonian				
A95-JS033	33.19602	66.25120	Escalerilla Granite	403 ± 6
A95-AC054	32.76876	65.38741	Renca Granite	393 ± 5
A95-PS167	33.17095	65.04894	Achiras Igneous Complex (Los Nogales Granite)	382 ± 6

Notes: Errors quoted are 2σ. Geographical coordinates determined by GPS. Ca. = generally implies a ± 10 Ma uncertainty.

sity. The advantage of SHRIMP dating over conventional U-Pb zircon dating is that petrographically selected areas of complexly zoned zircons can be analyzed in situ, thus enabling the determination of the crystallization age of a rock and the identification, and dating, of inherited components within magmatic crystals (e.g., Williams, 1992). In comparison to conventional analysis, the precision of SHRIMP analyses is much lower, simply because of the volume of sample consumed in the analysis (only 2 ng of zircon is typically consumed for a spot analysis). Standard statistical pooling can be used to combine analyses when the range is consistent with the cited errors. However, some samples remain complicated even at the microscale.

Zircons were separated, mounted in epoxy, and polished to reveal midsections. Cathodoluminescence (CL) imaging enabling detection of internal structures not evident in the optical images (Fig. 2) and were used as a guide for ion probe U-Th-Pb analysis. The procedures for U-Th-Pb analyses are similar to those used by Muir et al. (1996), although some variances in the peaks analyzed were employed according to the types of analyses involved. For granitic zircons, where a high proportion of inheritance is expected (based on the cathodoluminescence imaging), ^{204}Pb and a background measurement were included so that a common Pb correction could be made from the ^{204}Pb/^{206}Pb ratio and thus age information could be obtained from both ^{207}Pb/^{206}Pb and ^{206}Pb/^{238}U. The "age" of any grain is derived from the weighted mean ^{206}Pb/^{238}U and ^{207}Pb/^{206}Pb ages; in this way the best estimate for the age is assessed without any user bias. The ages for discordant analyses were taken from the ^{207}Pb/^{206}Pb age. Common Pb compositions are taken as the Cumming and Richards (1975) model Pb age for the inferred ^{206}Pb/^{238}U age of each zircon.

For granites with low degrees of inheritance, ^{204}Pb and the background were generally omitted because the time resolution afforded by ^{207}Pb/^{206}Pb is not sufficient to assess discordance. Rather, the time spent on the ^{204}Pb/^{206}Pb measurement is better used for more ^{206}Pb/^{238}U scans, so that the reliability of that ratio can be better constrained, and/or on more analyses of different grains, so that the reliability of the pooled ^{206}Pb/^{238}U ages can be better constrained. In this case, the reliability of the age is assessed by the reproducibility of the ^{206}Pb/^{238}U ages after a common Pb correction based on the linear extrapolation from the assumed common Pb composition (model Pb based on inferred mean age of population) through the datum to concordia. Thus, Tera-Wasserburg diagrams of the rocks, with only single age components, are plotted as the uncorrected Pb isotopic ratios and the deviation from the line (drawn through common Pb and the mean concordant age) can be assessed. If the data are excessively scattered, then outliers are sought. Such rejection can be nonunique in terms of whether a younger (attributable to Pb loss), or older (attributable to xenocrysts) point is rejected. In a robust data set, the exclusion of a few points on either side of the mean has little effect on the mean itself.

Additionally, any systematic effect, such as a distributed Pb loss over the entire range of data, will produce a data distribution that is non-Gaussian and is evident in the statistical treatment used. For data such as these with an apparently more mixed parentage, mixture unmixing using the MIX program of Sambridge and Compston (1994) was employed. However, caution must be used in interpreting these unmixed age spectra because of the finite numbers of data contributing to the data sets and the nonunique selection of finite numbers of components to be modelled.

The final ages reported are at the 2σ level of uncertainty. The cited error is based on the sum of the variances of the mean of the pooled analyses and the mean of the standard analyses, since the standard calibration can be only as good as the age determination on the standard.

RESULTS

The uranium-lead isotopic analyses and sample details are summarized in Table 1. Analytical data sets are presented in Appendix A.

Pichanas Metamorphic Complex, Sierras de Córdoba

El Pilón Granite, porphyritic phase (sample A95-PL238). Zircons from the peraluminous El Pilón Granite are pale brown and dominated by doubly terminated crystals. Cathodoluminescence imaging shows that the majority of the grains consist of rounded zircons surrounded in whole or in part by a thin, zoned overgrowth. Petrographically, the cores are interpreted as inherited grains from the source, whereas the zoned rims are interpreted to have formed from a melt. Most analyses were made on the rims to better constrain the magmatic age.

All U-Pb data from PL238 are shown in Figure 3a. The dominance of the ~500-Ma peak is largely a function of rim selection rather than provenance of the cores. Figure 3b shows

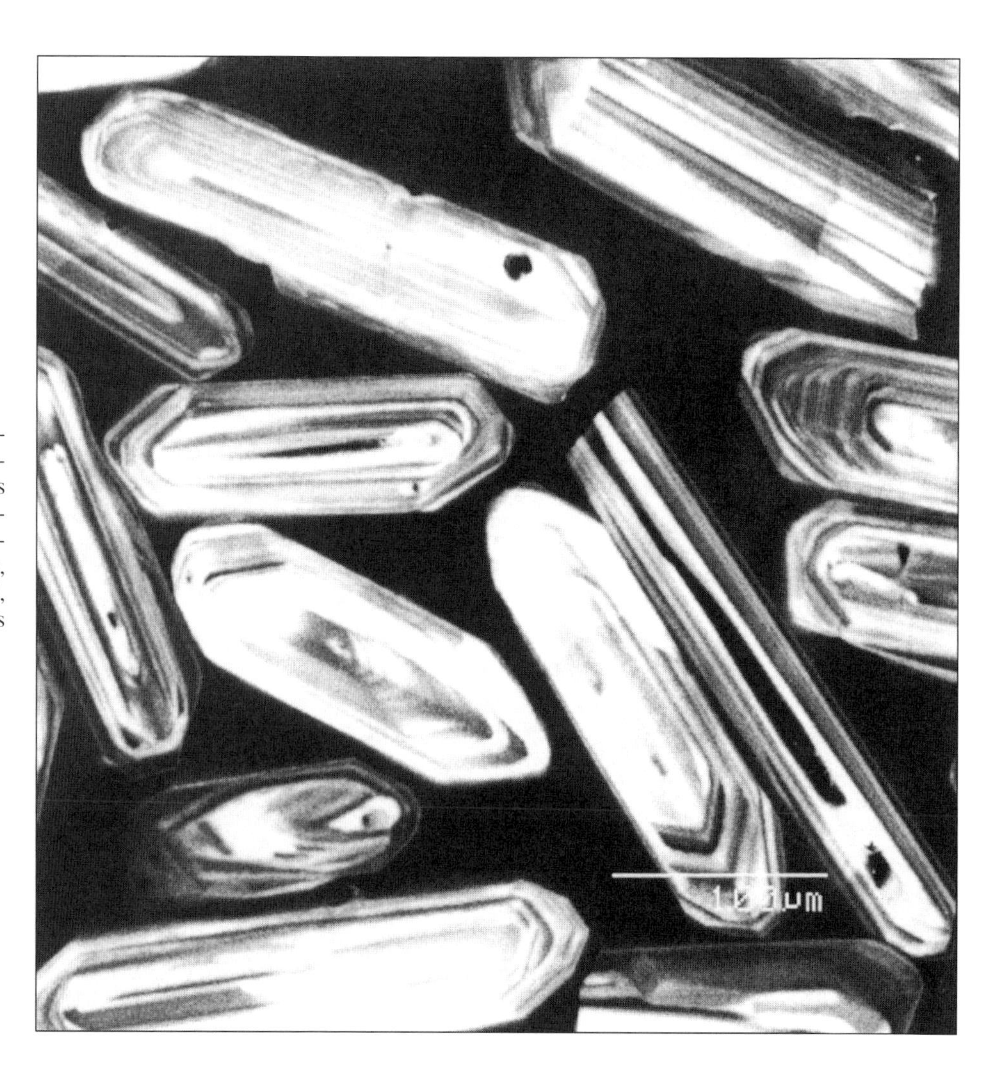

Figure 2. An example of a scanning electron microprobe (SEM)-based cathodoluminescence (CL) image showing zircons from A95PP114. The image shows typical euhedral magmatic zircons, displaying ocillatory zoning. The images, routinely taken of all samples analyzed, enabled detection of internal structures not evident in the optical images.

an expanded view of the main peak. After rejecting analyses not forming part of this peak, significant scatter remains. The peak was deconvolved with the MIX program and the results shown in the figure inset. The best fit was for three components with ages of 510.0 ± 4.2, 526.9 ± 4.0, and 547.5 ± 3.5 Ma (all $1\sigma_m$) with subequal proportions of the two older peaks and a minor contribution from the younger. A conclusive age for this granite is difficult to obtain. The rims in PL238 analyzed in sessions 2 and 3 have extremely high U concentrations (as high as 1.4 wt%) and consistently low Th/U ratios. While Pb loss is possibly responsible for the youngest peak, there is little to indicate which of the 527- or 548-Ma peaks are magmatic. However, the younger population is similar to the 523 ± 2-Ma U-Pb zircon age obtained by Rapela et al. (1998) and is the preferred age of crystallization of the granite.

El Pilón Granite, equigranular phase (sample A95-PL239). Zircons from this sample are petrographically and morphologically similar to PL238 with the inherited cores surrounded by zoned rims interpreted to have formed from a melt. Most analyses were made on the rims to better constrain the magmatic age. The expanded scale of the main peak (Figs. 3c, d) reveals more structure than was apparent in sample PL238. The two youngest peaks are ca. 480 and 514 Ma, distinctly younger than the ages derived for sample PL238. Both of these peaks are constrained by five zircon analyses each.

Like PL238, the rims analyzed in session 2 also have very high U and very low Th/U. Despite the chemical and petrographic similarities, the main age peaks for PL239 are decidedly younger than those for PL238. It should be noted that zircon with such high U concentrations would be rather susceptible to resetting and the PL239 zircons may be disturbed. The widespread Ordovician granitic magmatism in the La Rioja and San Luis regions could be a cause for Pb loss in the PL239 sample. The alternate explanation that both intrusions might be the younger age (ca. 480 Ma) belies the absence of young rims on PL238 zircons. Further work is required to elucidate the ages of samples A95-PL238 and A95-PL239.

Ojo de Agua Granite, Sierra Norte de Córdoba

Sample 56806. The zircon grains from this sample are typically euhedral, prismatic grains with aspect ratios generally around 3:1 and ocasionally up to 8:1. The zircons are generally 50–100 μm wide and contain inclusions that are mostly clear, colorless blebs and rods. Opaque inclusions are scarce whereas cracks occur in most grains and are severe in some. Iron staining due to weathering is also common. The zircons contain some rounded inherited cores. Figure 3e shows the analyses for this sample. Uranium concentrations are variable and range between 122 and 736 ppm (with both from the same grain), with Th/U ratios ranging from 0.21 to 1.59. Apart from a single outlier, 21 ages define a single population with a weighted $^{238}U/^{206}Pb$ age of 515 ± 4 Ma (2σ).

Chepes Igneous Complex, Sierras de La Rioja

Biotite-hornblende granodiorite (sample A95-PP076A). Sample PP076A, a granodiorite from the Sierra de Chepes, contains zircons that are light brown to pink and dominated by elongate and equant prismatic crystals with pyramidal terminations. The grains are mostly clear and contain abundant fine apatite needles and possible fluid inclusions. Grains are zoned with some cores apparent. Uranium contents range between 115 and 396 ppm, with a range of Th/U ratios of 0.48–0.97. Zircon U-Pb ages form a single population with only one outlier indicative of inheritance (Fig. 4a). Sixteen analyses define a population with a weighted mean $^{238}U/^{206}Pb$ age of 491 ± 6 Ma (2σ; MSWD = 0.94).

Biotite monzogranite (Asperezas Granite) (sample A95-PP159A). Zircons from the Asperezas Granite are colorless to light brown to pink and are dominated by elongate to equant prismatic crystals with pyramidal terminations. The grains are mostly clear, containing some fine apatite needles and possible fluid inclusions. Most grains are zoned with no apparent inherited cores. Cathodoluminescence imaging shows predominantly oscillatory zoning and few cores. Uranium contents range between 124 and 566 ppm, with a range of Th/U ratios of 0.50–0.98. All 15 zircon U-Pb analyses combine to give a single population with an age of 490 ± 7 Ma (2σ; MSWD = 1.46) (Fig. 4b).

Biotite monzogranite (sample A95-PP116a). Sample PP116a represents the porphyritic phase of the Chepes monzogranite. Zircons are colorless to light brown to pink, and are dominated by elongate prismatic crystals with pyramidal terminations, some of which are slightly rounded. The grains are mostly clear, containing some fine apatite needles and possible fluid inclusions. Most grains are zoned with no apparent inherited cores. Uranium concentrations range between 134 and 510 ppm, with a range of Th/U ratios of 0.63–1.13. Fifteen U-Pb analyses define a single population of age of 485 ± 7 Ma (2σ; MSWD = 0.99) (Fig. 4c) with the exclusion of one older (670 Ma) grain.

Epidote-bearing biotite-hornblende tonalite (sample A95-PP183A). Sample PP183A represents an epidote-bearing granodiorite from the Sierra de Las Minas. Zircons are colorless to light brown to pink and are dominated by elongate, prismatic crystals with pyramidal terminations, some of which are slightly rounded. The grains are mostly clear, containing fine apatite needles and possible fluid inclusions. The grains show sector and oscillatory zoning with no apparent inherited cores. Uranium concentrations show a relatively narrow range between 140 and 305 ppm, with Th/U ratios ranging from 0.37 to 0.88. After exclusion of one U-Pb analysis marginally older than the remainder, zircon U-Pb data show a main peak with a $^{238}U/^{206}Pb$ age of 480 ± 6 Ma (2σ; MSWD = 0.67) (Fig. 4d).

Biotite granodiorite (sample A95-PP114A). Sample PP114A represents a biotite-bearing granodiorite from the Sierra de Chepes. Zircons are light brown to pink and are dominated by elongate prismatic crystals with pyramidal terminations. The grains are mostly clear, containing fine apatite needles and "tubes" that may have contained fluid inclusions.

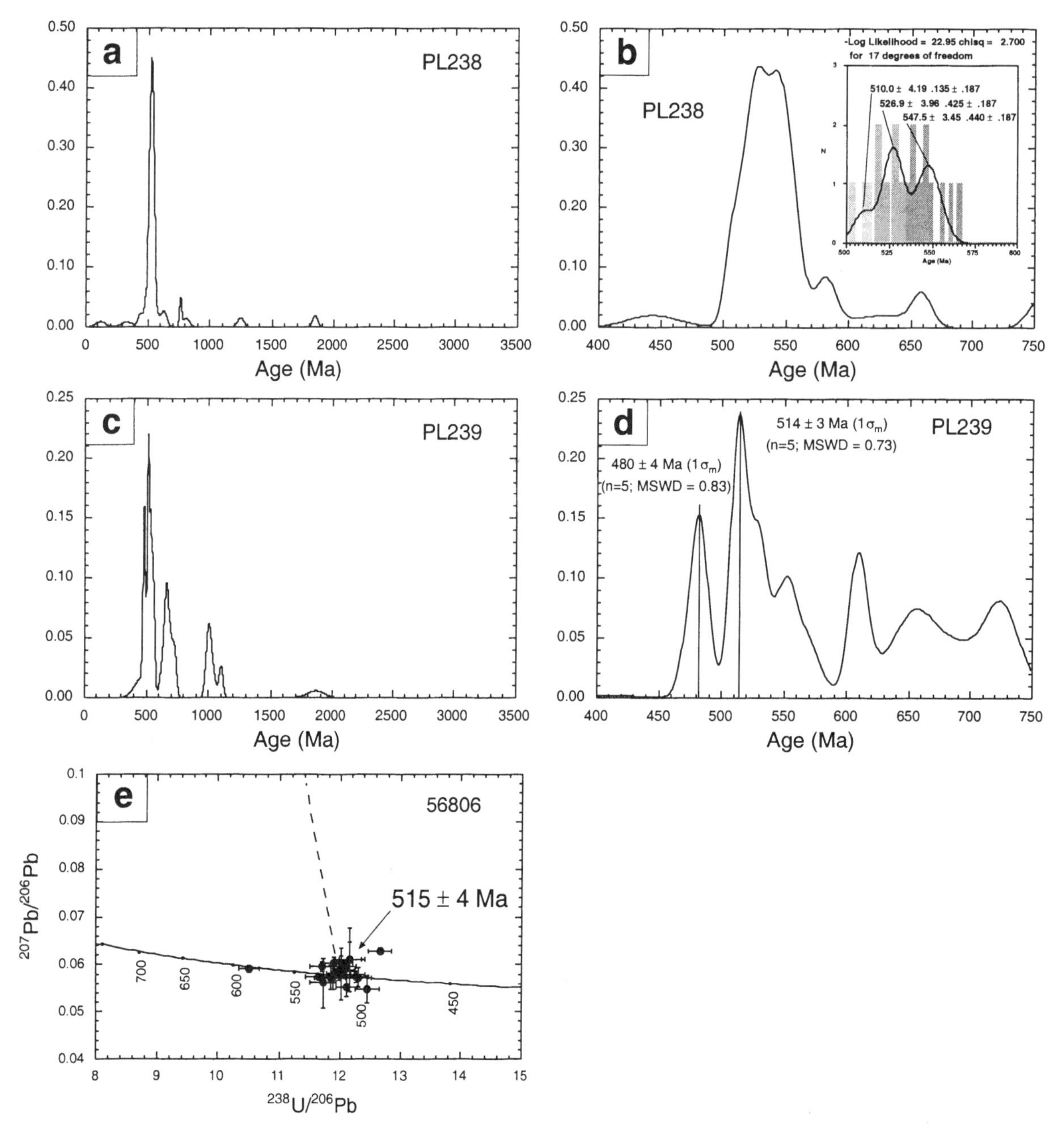

Figure 3. a, Cumulative probability diagram of weighted mean $^{206}Pb^*/^{238}U$ and $^{207}Pb^*/^{206}Pb^*$ ages for zircons from A95PL238. The analyses are dominated by rims in an attempt to constrain the magmatic age and so the relative proportions do not indicate the inheritance components in terms of abundance. b, The cumulative probability diagram for the youngest zircons (rims) from A95PL238. The data are not consistent with a single mean and were deconvolved to give ages of ca. 510, ca. 527, and ca. 548 Ma (2σ error approximately 10 Ma on each). While the data can be satisfactorily modeled using these three age components, it remains unclear as to which if any reflects the magmatic age. c, Cumulative probability diagram for zircons from A95PL239. Similar to A95PL238, most analyses were of rims in an attempt to constrain magmatic age. d, Cumulative probability diagram for the youngest zircons from A95PL239. The youngest zircon ages form coherent peaks at ca. 480 and ca. 514 Ma (2σ error approximately 10 Ma on each). The ages for PL238 and PL239 therefore differ in detail. e, Tera-Wasserburg plot showing common Pb corrected data for zircons from sample 56806. Unlike PL238 and PL239, little inheritance is found in this granite and a mean age of 515 ± 4 Ma is well constrained. Dashed line represents locus of common Pb correction. Error bars are 1σ.

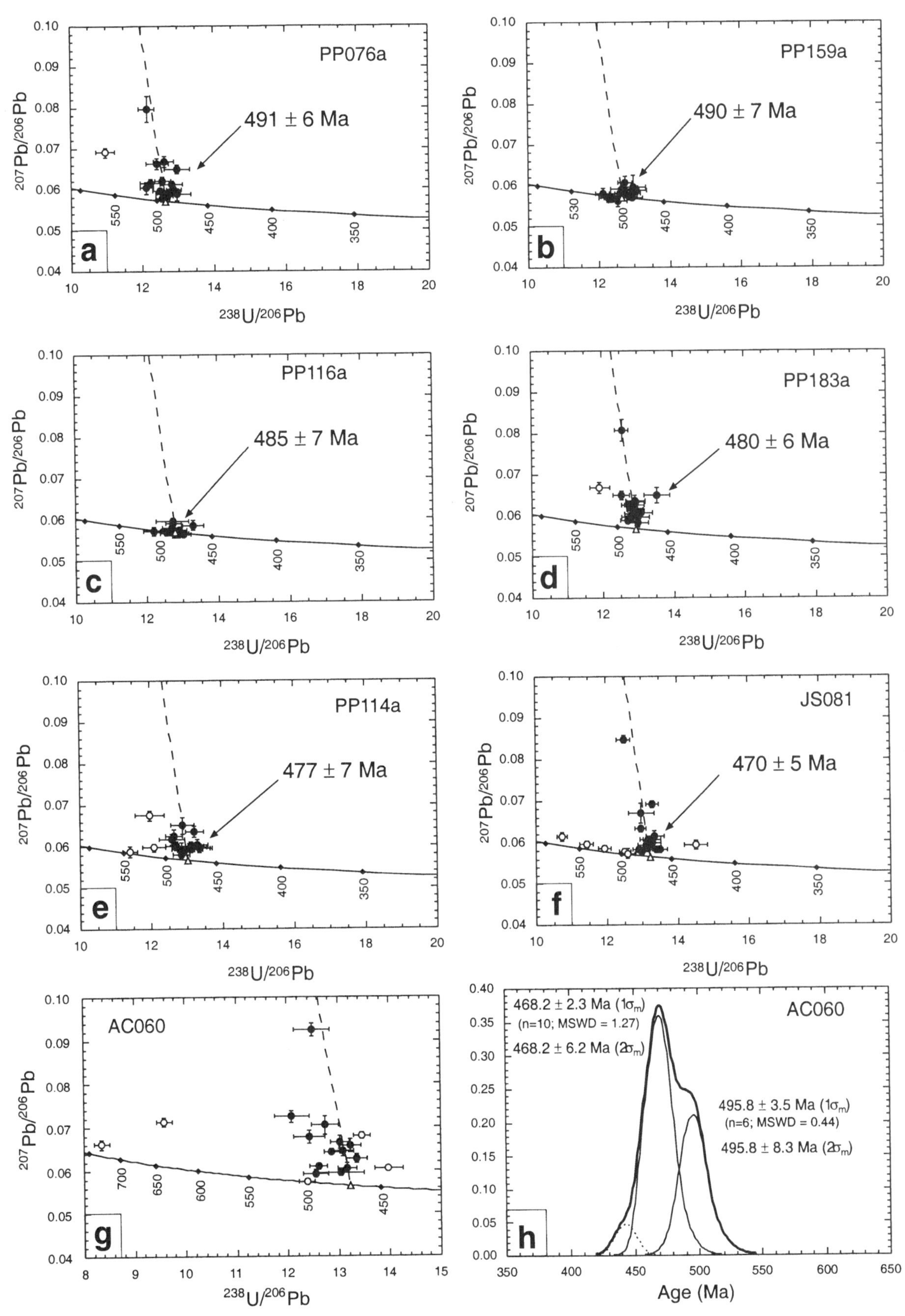

Figure 4. a–g, Tera-Wasserburg diagrams showing uncorrected Pb compositions. The dashed line represents the common Pb mixing curve between the inferred radiogenic age and the common Pb composition inferred from the model Pb composition of Cumming and Richards (1975). Data lying within error of the line are consistent with the mean age. A magmatic age is forthcoming from samples shown in (a) to (f) (see text for details). Sample AC060 (g) however, shows excess scatter. This can be modeled most simply as a two-component mixture as shown using the cumulative probability plot (h) with components around 468 ± 6 and 496 ± 8 Ma. Alternat scenarios involving distributed Pb loss are also possible. Error bars on Tera-Wasserburg diagrams are 1σ.

Most grains show broad oscillatory zones with no apparent inherited cores. Uranium concentrations range between 61 and 269 ppm, with Th/U ratios ranging from 0.20 to 1.32. There is some scatter in the U-Pb analyses (Fig. 4e). However, after excluding three older analyses (with ages ranging up to 550 Ma), the $^{238}U/^{206}Pb$ age from the remaining 14 analyses gives a weighted mean of 477 ± 7 Ma (2σ; MSWD = 0.68).

Sierras de San Luis

Tamboreo Granodiorite (sample A95-JS081). Sample JS081, from the Tamboreo Granodiorite, contains zircons that are pale pink and dominated by elongate prismatic crystals with pyramidal terminations. The grains are mostly clear and contain abundant fine apatite needles and possible fluid inclusions. Most grains are zoned with no apparent inherited cores. Cathodoluminescence images reveal cores but it is not clear whether these are inherited or early formed nuclei. Uranium concentrations are variable and range between 118 and 1,532 ppm, and Th/U ratios are also variable, ranging from 0.19 to 1.17. Zircon U-Pb data are shown in Figure 4f. Of the 20 U-Pb analyses, 6 must be excluded to obtain a satisfactory mean. Five of these are older, indicating inheritance, while one analysis is younger, indicating Pb loss. Despite the rejection of these outliers, the main population is well defined by 14 analyses that give a weighted mean of 470 ± 5 Ma (2σ; MSWD = 0.99).

Bemberg Tonalite (sample A95-AC060). The Bemberg Tonalite contains zircons that are light brown and dominated by elongate and equant prismatic crystals with pyramidal terminations. The grains are mostly clear, containing abundant fine apatite needles and possible fluid inclusions. Most grains are zoned with no apparent inherited cores. Cathodoluminescence images reveal dominantly oscillatory zoning around cores. Uranium concentrations are variable over an order of magnitude between 96 and 922 ppm, but with a small range of Th/U ratios from 0.22 to 0.69. Zircon U-Pb data are plotted in Figure 4g. In assessing the zircon population, two clear outliers older than 700 Ma were removed. The remaining analyses are not consistent with a single population. The main peak was unmixed and the data are consistent with two components (Fig. 4h), although one analysis was excluded and appears to have lost Pb. The two peaks are 468 ± 6 Ma (10 analyses) and 496 ± 8 Ma (6 analyses) (2σ). Without any geologic evidence otherwise, the younger group of $^{206}Pb/^{238}U$ ages is interpreted as representing the emplacement age for the Bemberg Tonalite and the older as inheritance. Alternatively, the older age may represent the emplacement age, with the younger analyses having been affected by varying degrees of Pb loss.

Escalerilla Granite (sample A95-JS033). The zircons from this porphyritic phase of the Escalerilla Granite are dominated by elongate and equant prismatic crystals with pyramidal terminations. The grains are mostly clear, pale brown, and zoned. Many of the zircons contain "tubes" that may have contained fluid inclusions. Cathodoluminescence images show oscillatory zoning around some cores. Uranium concentrations show a large range between 128 and 3,257 ppm, with a large range of Th/U ratios from 0.27 to 2.13. With the exclusion of three of the grains that have clearly lost Pb (Fig. 5a),16 analyses define a population that gives a weighted mean age of 403 ± 6 Ma (2σ; MSWD = 1.81).

Renca Granite (sample A95-AC054). Sample AC054, from the porphyritic phase of the Renca Batholith, contains zircons dominated by brown, elongate prismatic crystals with pyramidal terminations. The grains are clear and zoned with no apparent inherited cores, and may contain fluid inclusions. Cathodoluminescence imaging shows fine oscillatory zoning around more uniform centers. Uranium concentrations are highly variable, between 139 and 1,194 ppm, with Th/U ratios ranging from 0.30 to 1.40. Zircon U-Pb data are plotted in Figure 5b. Three of the U-Pb analyses appear to have lost Pb with respect to the main population, which defines a $^{238}U/^{206}Pb$ age of 393 ± 5 Ma (2σ; MSWD = 1.69).

Sierras de Comechingones

Achiras Igneous Complex (sample A95-PS167). The zircons from sample PS167, of the Los Nogales Granite (Achiras Igneous Complex), are pale brown in color and are dominated by equant, prismatic crystals. Elongate crystals are also present and generally have rounded pyramidal terminations. The grains are mostly clear, zoned, and contain abundant fine apatite needles and possible fluid inclusions. Cathodoluminescence imaging shows dominant sector zoning with superposed oscillatory zoning. Uranium concentrations range between 158 and 610 ppm, and the zircons have Th/U ratios ranging from 0.06 to 0.89. Zircon U-Pb data (Fig. 5c) define a single population with a $^{238}U/^{206}Pb$ age of 382 ± 6 Ma (2σ; MSWD = 0.79).

DISCUSSION

Uranium-lead isotopic dating of igneous zircons confirms that there are three significant episodes of granite emplacement in the southern Sierras Pampeanas. These occurred during the Cambrian, Early Ordovician, and Devonian (Fig. 6). Although there is some agreement with the threefold classification of Rapela et al. (1992), there is no direct correspondence of some granites, particularly those classified as G2 or G3. Furthermore, there is no evidence to support Late Ordovician, or Silurian granite intrusion as indicated by previous Rb-Sr and K-Ar isotopic age data. The Cambrian and Early Ordovician magmatic events correspond to the Pampean and Famatinian Cycles, respectively. The third period of granite magmatism corresponds to a newly defined tectonic cycle in the southern Sierras Pampeanas, known as the Achalian orogeny (Sims et al., 1998).

Pampean Cycle

Two phases of the peraluminous El Pilón granite from the Pichanas Metamorphic Complex crystallized during the Pampean Cycle. Zircons from the porphyritic phase of the El Pilón granite do not yield a well-constrained age with two main age populations at ca. 527 and ca. 548 Ma (2σ error on mean ages ca. 10 Ma in both cases), the latter of which is older than the Rb-Sr age of 520 ± 5 Ma obtained by Rapela et al. (1995) for the same granite. The younger U-Pb zircon age is similar to the 523 ± 2 Ma U-Pb zircon age obtained by Rapela et al. (1998). The equigranular El Pilón granite sample yields two ages of ca. 480 and ca. 514 Ma (error ca. 10 Ma 2σ) . Noticeably, zircon rims in this sample have extremely high U concentrations (up to 1.4 wt%), making them susceptible to Pb loss in subsequent events. The younger ages may therefore reflect Pb loss during a later event such as the Famatinian tectono-magmatic Cycle. Alternatively, the 480-Ma age may represent zircon growth during the Famatinian Cycle. The isotopic analyses also show that both granites have an inherited component, indicating either contamination from the surrounding country rocks during magma genesis or assimilation during intrusion.

The Ojo de Agua granite from the Sierra Norte de Córdoba, emplaced late in the Pampean deformation, yields an age of 515 ± 4 Ma, which is a possible minimum age of felsic magmatism associated with the Pampean Cycle. The zircon data confirm that granitic melts of the Pampean Cycle crystallized in the Cambrian most likely between ca. 510 and 530-Ma, consistent with the interpreted age of ca. 530 Ma for peak metamorphism (Sims et al., 1998).

Famatinian Cycle

The zircon data indicate that the Famatinian granites from the Chepes Igneous Complex (Sierras de La Rioja) crystallized over a narrow time range bracketed between 491 ± 6 and 477 ± 7 Ma (Table 1). The U-Pb dates are older than K-Ar and Rb-Sr isotopic ages of 450–475 Ma obtained on these rocks by other workers (e.g., Linares and Gonzales, 1990; Llambías et al., 1991; Pankhurst et al., 1996).

In the Sierra de San Luis, zircons dated from the Tamboreo Granodiorite and the Bemberg Tonalite produced crystallization ages (ca. 470 Ma) that are within error of the younger ages in the sierras of southern La Rioja province.

Achalian Cycle

A new tectonic and magmatic cycle in the Devonian is defined in the southern Sierras Pampeanas by regional mapping and U-Pb zircon dating. The Achalian Cycle derives its name from the Achala Batholith, the largest of the Devonian granite bodies in the southern Sierras Pampeanas, that intruded the metamorphic rocks discontinuously during and after a widespread deformation characterized by major ductile shear zones with intensive greenschist facies retrogressive fabrics (Sims

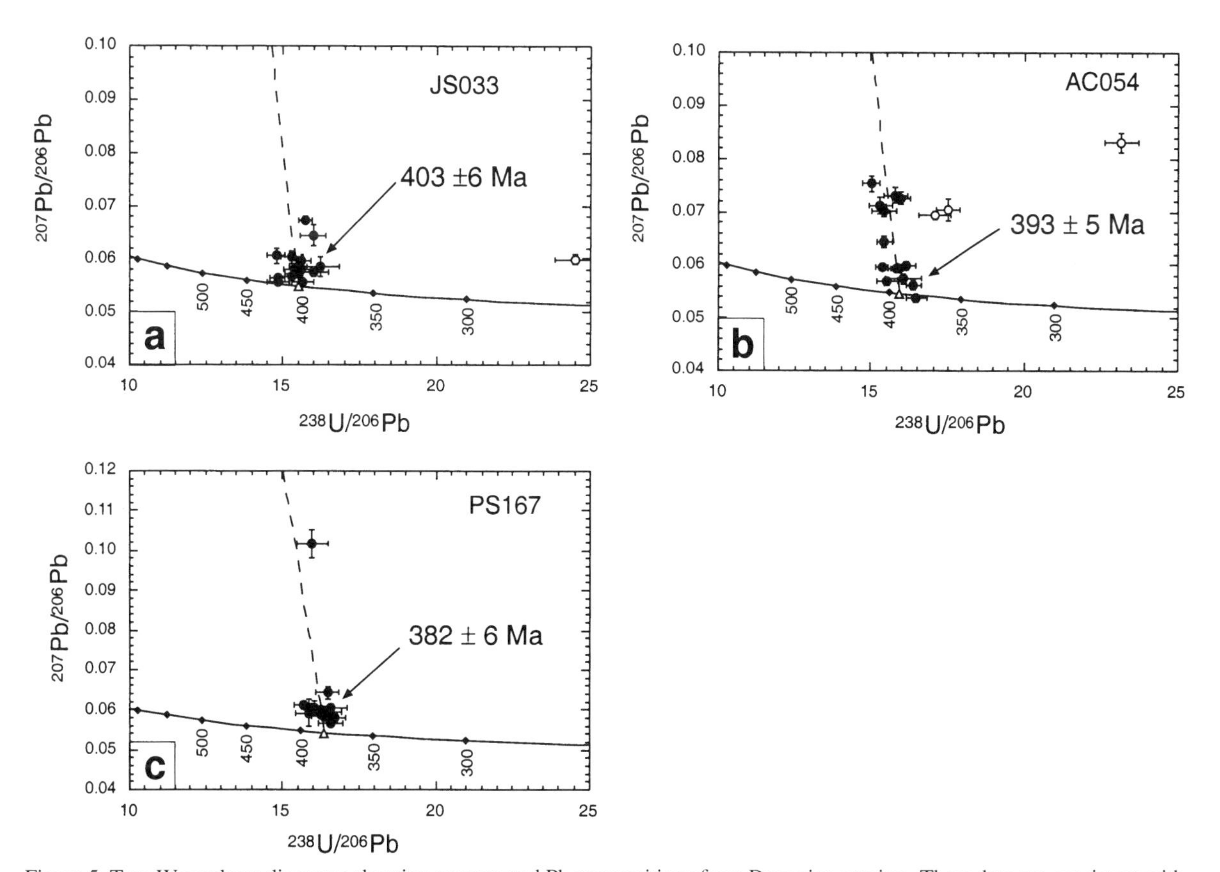

Figure 5. Tera-Wasserburg diagrams showing uncorrected Pb compositions from Devonian granites. These data are consistent with simple magmatic ages with a small degree of Pb loss apparent in AC054. Error bars are 1σ. Dashed line as described in Figure 4.

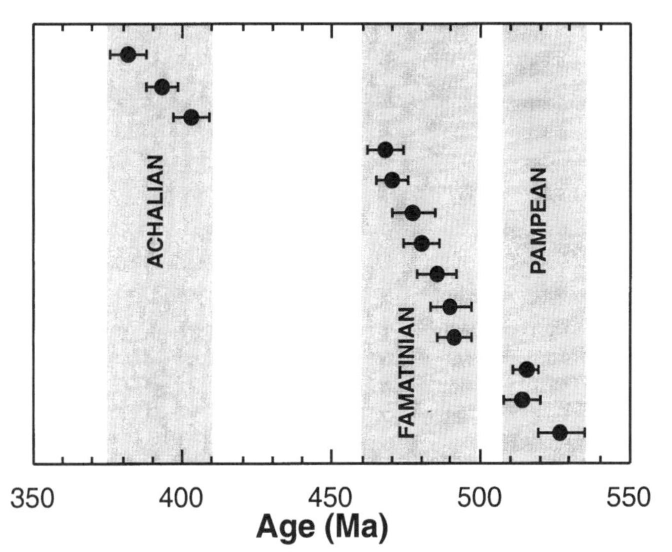

Figure 6. Summary of U-Pb age determinations from the present study for granitic rocks from the southern Sierras Pampeanas. The shaded bands schematically indicate the duration, based on the data presented herein, of the three magmatic events within the Pampean, Famatinian, and Achalian tectonic cycles.

et al., 1998). Dalla Salda (1987) defined this deformation as D3, placing it in the "Ciclo Famatiniano." Some Achalian granites were emplaced as sheets subparallel to the greenschist facies shear fabrics (e.g., Achiras Igneous Complex and Escalerilla Granite), whereas others form subcircular, zoned, and fractionated plutons (e.g., Renca Granite, Inti Huasi Granite, Comechingones Batholith). Limited U-Pb zircon dating of the Los Nogales, Escalerilla, and Renca granites suggests crystallization of the felsic magmas may have occurred over a ca. 20-m.y. period between ca. 403 and ca. 382 Ma. However, K-Ar isotopic ages for similar intrusions elsewhere in the southern Sierras Pampeanas range to as young as the Carboniferous (Linares and Gonzalez, 1990). The Achalian granites were identified previously as either Famatinian (G2) or post-Famatinian (G3) by Rapela et al. (1992).

Tectonic development of the proto-Andean Gondwana margin

The U-Pb zircon data for the granites presented here provide new limits on the timing of the major magmatic events in the Sierras Pampeanas, and place timing constraints on the tectonic development of the proto-Andean Gondwana margin. The following tectonic history is an attempt to reconcile these new data with previous tectonic models applied to the region.

The oldest rocks in the region form a structurally thick sequence of pelitic and lesser psammitic gneiss interlayered with a series of fault-bounded semicontinuous belts of marbles and amphibolite exposed in the Sierras de Córdoba and San Luis. These metasediments are interpreted as being deposited on a passive margin, developed during intracontinental rifting and the break-up of Laurentia from Gondwana in Early Cambrian time at about 540 Ma (Dalziel et al., 1994) or in the late Neoproterozoic (e.g., Sims et al., 1998). Lithologic similarities and comparable ages indicate that the metasediments may be correlatives of the Early Cambrian (Aceñolaza and Toselli, 1981) Puncoviscana Formation in the northern Sierras Pampeanas, as first postulated by Willner and Miller (1986) and more recently by Willner (1990) and Lork et al. (1989). This formation was interpreted by Dalla Salda et al. (1992) to be related to the rift-drift transition during postcollisional Gondwana-Laurentia break-up.

Cambrian convergence. The Pampean deformation is interpreted as the first in a series of deformational events associated with convergence on the newly created Pacific Gondwana margin (e.g., Dalziel et al., 1994). At the closing stages of the Pampean Cycle, an extensive phase of felsic magmatism is evident by widespread subconcordant intrusion of tonalite, granodiorite, and granite, which may be indicative of magmatic arc formation (Rapela and Pankhurst 1996). Alternatively, these granites may have been generated during postcollisional extension (Sims et al., 1998). The U-Pb zircon data presented here indicate that this magmatism persisted until the end of the Early Cambrian.

Ordovician collision. Closure of the Iapetus ocean and collision of the Precordillera with the Pampean margin of the Gondwana craton (Dalla Salda et al., 1992, 1996; Dalziel et al., 1996) occurred during the Famatinian Cycle, a widespread deformational, metamorphic, and magmatic event.

The petrography and the geochemistry of granites associated with the Famatinian Cycle are consistent with a calc-alkaline magmatic arc setting associated with subduction early in the collisional event (e.g., Toselli et al., 1996; Sato et al., 1996). In the southern Sierras de La Rioja Province, occurrences of some strongly peraluminous granites also indicate minor crustal reworking or fusion of short-lived mafic rocks underplated during the Famatinian Cycle (Rapela et al., 1996). Both granite types intruded at depths of less than 10 km (Dahlquist and Baldo, 1996; Rapela et al., 1996). Uranium-lead dating of zircons from the granites in both the Sierra de San Luis and the southern Sierras de La Rioja Province constrains this magmatic event to between 491 ± 6 and 468 ± 6 Ma (Table 1).

In most tectonic interpretations of the Sierras Pampeanas, other authors have suggested that the final amalgamation of the Precordillera, with Gondwana took place at about 450 Ma during the "Oclóyic" orogeny (e.g., Ramos et al., 1986; Martino et al., 1994; Astini et al., 1996; Ramos et al., 1996; Toselli et al., 1996) contemporaneous with the Taconic Orogeny (Dalla Salda et al. 1995) in North America (Dalziel, 1991). Paleomagnetic reconstructions of the mid-Ordovician (Van der Voo, 1993) place Laurentia and the western Gondwana margin continents in close proximity at about 450 Ma (e.g., Dalla Salda et al., 1992). However, the proximity of both continents during the mid-Ordovician is also consistent with a slightly earlier age for the Famatinian collisional event. A simple explanation is that collision and later initiation of the second and final separation of Laurentia and Gondwana began some 20 Ma earlier. There is no U-Pb isotopic or structural evidence of a mid-Ordovician magmatic event in the southern Sierras Pampeanas. Owing to the widespread Devonian deformation and greenschist facies metamorphism the mid-Ordovician ages obtained by K-Ar and Rb-Sr methods must be treated with caution and interpreted as minimum ages only.

An Early Ordovician connection between the Precordillera and the Famatinian orogen is also indicated by the non-Laurentia source of Precordilleran Arenigian-Llanvirnian K-bentonites interpreted by Keller and Dickerson (1996) and Bergstrom et al. (1996) to be derived from a continental margin volcanic arc source to the east of the Precordillera: a source consistent with an Early Ordovician Famatinian magmatic arc.

Devonian convergence. Mid-Paleozoic resumption of convergence on the western margin of Gondwana is evidenced by a widespread compressive deformation in the Sierras Pampeanas and the development of an Early Devonian magmatic arc. The deformation was dominated by orthogonal westerly directed thrusting and the development of regionally extensive ductile shear zones with intensive greenshist facies retrogressive fabrics (Sims et al., 1997). Dalla Salda (1987) defined this deformation as D3, placing it within the "Ciclo Famatiniano." Felsic melts generated during the cycle intruded the metamorphic rocks as either syn-tectonic sheets within shear zones or as large fractionated plutons. Emplacement of these magmas had commenced by the Early Devonian, as suggested by the age of the Escalerilla Granite (403 ± 6 Ma) (Table 1). The cycle probably corresponds to the "Fase Precordilleránica" of Astini (1996) in the Precordillera west of the Sierras Pampeanas where it is related to the amalgamation of the Chilenia terrane.

CONCLUSIONS

The U-Pb isotopic dating of zircons provides new limits on the timing of the major felsic magmatic events in the southern Sierras Pampeanas. These zircon data confirm that there are three separate episodes of granite emplacement in the southern Sierras Pampeanas. The cycles occurred during the Cambrian (ca. 530–510 Ma), Early Ordovician (ca. 490–470 Ma), and Devonian (ca. 403–382 Ma), and are referred to as the Pampean, Famatinian, and Achalian Cycles, respectively. Although the Pampean and Famatinian Cycles were widely documented, the latter was previously considered to span most of the mid-Paleozoic. The Achalian Cycle corresponds to a regional tectono-magmatic event previously placed within the Famatinian Cycle. Although there is some broad agreement with the threefold classification of Rapela et al. (1992), there is no direct correspondence of some granites, particularly those classified as G2 or G3. Neither is there evidence to support Late Ordovician or Silurian granite intrusion, as indicated in previous Rb-Sr and K-Ar isotopic age data.

The isotopic data on magmatic and deformational cycles within the southern Sierras Pampeanas constrain the time of Paleozoic plate convergence and magmatic arc development on the proto-Andean Gondwana margin. The revised Early Ordovician age of the Famatinian Cycle indicates that final amalgamation of the Precordillera had taken place prior to 470 Ma, rather than about 450 Ma as previously suggested.

ACKNOWLEDGEMENTS

We thank all those who participated in the field programs and contributed to discussion and the development of the ideas presented herein. Trevor Ireland is thanked for comments on analytical methodology and critique of the data. Constructive reviews by L. P. Gromet, W. McClelland, G. Gibson, and K. Cassidy are gratefully acknowledged. The name Achalian Cycle was suggested by the late Roberto Caminos during discussions on the results presented in this chapter. Chris Fouldoulis and Tas Armstrong are thanked for their assistance in sample preparation. This study was conducted under the auspices of the Geoscientific Mapping of the Sierras Pampeanas Cooperative Project between the Australian Geological Survey Organisation (AGSO) and the Servico Geológico Minero Argentino (SEGEMAR). Fieldwork was conducted in Argentina between November 1994 and December 1996. The work is published with the permission of the Directors of AGSO and SEGEMAR.

APPENDIX A

PL238

Label	U ppm	Th ppm	Th/U	#f206Pb %	207Pb/206Pb	238U/206Pb	Age, Ma
Session 1							
12.2	126	71	0.566 ± 0.009	1.14	0.05941 ± 0.00527	10.98 ± 0.33	562.1 ± 16.0
13.2	161	70	0.434 ± 0.007	0.52	0.08296 ± 0.00328	4.69 ± 0.12	1249.6 ± 27.2
14.2	328	223	0.680 ± 0.014	0.31	0.11321 ± 0.00146	3.08 ± 0.10	1844.4 ± 21.3
15.2	217	264	1.218 ± 0.022	4.35	0.06500 ± 0.00877	7.47 ± 0.28	809.5 ± 28.0
16.2	357	135	0.378 ± 0.005	0.67	0.05622 ± 0.00386	11.53 ± 0.26	535.9 ± 11.4
17.2	21	10	0.450 ± 0.013	-0.78	0.10436 ± 0.03307	10.28 ± 0.68	601.4 ± 37.9
18.2	353	22	0.062 ± 0.002	7.26	0.05711 ± 0.01370	14.02 ± 0.66	444.3 ± 20.2
19.2	379	209	0.552 ± 0.010	3.86	0.06392 ± 0.00684	9.69 ± 0.35	633.8 ± 21.5
Session 2							
14.3	1452	7	0.005 ± 0.000	0.44	0.05465 ± 0.00115	12.15 ± 0.14	508.5 ± 5.7
19.3	1350	22	0.017 ± 0.000	1.09	0.05076 ± 0.00220	11.98 ± 0.18	515.3 ± 7.6
15.1	1086	15	0.013 ± 0.000	0.82	0.05174 ± 0.00238	11.63 ± 0.19	530.4 ± 8.5
30.1	3785	39	0.010 ± 0.000	0.17	0.05521 ± 0.00104	11.81 ± 0.14	522.2 ± 5.8
20.3	3699	34	0.009 ± 0.000	0.72	0.05541 ± 0.00146	11.34 ± 0.13	543.7 ± 6.0
18.3	1069	12	0.011 ± 0.001	1.03	0.05250 ± 0.00218	85.59 ± 54.34	119.3 ± 42.7
22.3	1948	14	0.007 ± 0.000	0.84	0.05300 ± 0.00112	11.93 ± 0.14	328.8 ± 48.6
25.3	1858	36	0.019 ± 0.000	0.11	0.05686 ± 0.00130	12.05 ± 0.18	513.3 ± 7.1
24.3	1095	33	0.030 ± 0.001	0.02	0.05688 ± 0.00194	11.26 ± 0.19	547.8 ± 8.8
26.3	14262	79	0.006 ± 0.000	0.18	0.05684 ± 0.00027	9.35 ± 0.12	485.2 ± 10.6
31.1	1377	82	0.059 ± 0.002	0.02	0.06011 ± 0.00128	11.06 ± 0.12	558.8 ± 5.6
27.3	1937	84	0.043 ± 0.000	0.30	0.05609 ± 0.00105	11.64 ± 0.18	528.8 ± 7.6
Session 3							
32.1	1348	15	0.011 ± 0.000	0.09	0.05715 ± 0.00119	11.90 ± 0.21	519.2 ± 8.6
33.1	1384	23	0.017 ± 0.000	0.33	0.05550 ± 0.00113	11.37 ± 0.18	539.9 ± 7.9
34.1	1345	11	0.009 ± 0.001	0.25	0.05750 ± 0.00117	52.86 ± 28.01	510.7 ± 45.2
35.1	881	26	0.030 ± 0.000	0.37	0.05517 ± 0.00169	11.50 ± 0.19	535.9 ± 8.5
36.1	1898	268	0.141 ± 0.004	1.51	0.05862 ± 0.00155	11.29 ± 0.16	547.1 ± 7.4
37.1	1054	56	0.053 ± 0.001	0.34	0.06704 ± 0.00090	8.04 ± 0.10	764.1 ± 8.8
38.1	5066	13	0.003 ± 0.000	0.15	0.05742 ± 0.00034	10.60 ± 0.14	507.7 ± 13.3
39.1	1743	56	0.032 ± 0.008	0.18	0.05747 ± 0.00090	11.68 ± 0.18	528.8 ± 7.7
40.1	2845	65	0.023 ± 0.000	0.16	0.05672 ± 0.00054	11.44 ± 0.21	530.3 ± 8.7
41.1	1399	10	0.007 ± 0.000	0.31	0.05565 ± 0.00089	11.68 ± 0.20	524.5 ± 8.5
42.1	3326	29	0.009 ± 0.000	0.78	0.05642 ± 0.00067	12.29 ± 0.13	502.9 ± 5.2
43.1	1804	73	0.040 ± 0.000	0.44	0.05558 ± 0.00173	11.40 ± 0.50	531.9 ± 21.7

Notes:

Corrected for common Pb. All errors in table are 1σ. #f206Pb% is percentage common ^{206}Pb in the total measured ^{206}Pb. Common Pb: $^{204}Pb/^{206}Pb$ correction. Age: Weighted Mean $^{206}Pb/^{238}U$ & $^{207}Pb/^{206}Pb$ age.

Standards

Session 1 (with PL147 and PL239)
See PL147. Error in mean is 0.73%.

Session 2
Six SL13 standards give an MSWD of 1.02. No rejects. Error in mean is 3.2 Ma which is 0.56%.

Session 3
Seven AS3 standards give an MSWD of 1.95. No data rejected. Error in mean is 6.8 Ma which is 0.62%.

PL239

Label	U ppm	Th ppm	Th/U	#f206Pb %	$^{207}Pb/^{206}Pb$	$^{238}U/^{206}Pb$	Age, Ma
Session 1							
2.2	384	183	0.476 ± 0.008	0.70	0.06904 ± 0.00402	5.94 ± 0.12	1000.6 ± 18.0
12.2	307	129	0.421 ± 0.012	2.25	0.06042 ± 0.00450	9.33 ± 0.50	654.9 ± 33.1
3.2	224	53	0.236 ± 0.003	1.63	0.11403 ± 0.00433	4.56 ± 0.12	1864.6 ± 70.2
13.2	454	154	0.340 ± 0.004	0.80	0.05883 ± 0.00262	9.18 ± 0.24	663.7 ± 16.2
14.2	39	34	0.879 ± 0.013	2.60	0.08160 ± 0.00852	5.84 ± 0.16	1021.4 ± 26.2
15.2	190	206	1.085 ± 0.035	1.74	0.05862 ± 0.00605	9.68 ± 0.32	633.4 ± 19.7
8.2	568	193	0.339 ± 0.006	0.89	0.06095 ± 0.00201	8.54 ± 0.26	707.9 ± 19.8
10.2	986	70	0.071 ± 0.001	1.49	0.06329 ± 0.00305	8.31 ± 0.18	732.0 ± 14.8
16.2	35	86	2.459 ± 0.044	4.32	0.07643 ± 0.01686	5.77 ± 0.20	1030.5 ± 32.3
11.2	638	109	0.170 ± 0.003	0.42	0.07053 ± 0.00241	5.91 ± 0.14	1002.3 ± 21.1
17.2	334	100	0.301 ± 0.002	1.28	0.06939 ± 0.00458	5.34 ± 0.08	1104.7 ± 15.6
18.2	729	222	0.304 ± 0.004	2.29	0.05994 ± 0.00221	8.84 ± 0.26	685.7 ± 19.1
19.2	342	154	0.449 ± 0.006	0.37	0.06521 ± 0.00212	9.34 ± 0.20	660.1 ± 13.0
Session 2							
2.3	2485	3	0.001 ± 0.000	0.27	0.05848 ± 0.00131	12.75 ± 0.23	488.4 ± 8.2
3.3	895	31	0.034 ± 0.000	1.61	0.05145 ± 0.00360	11.31 ± 0.25	544.7 ± 11.6
1.3	4187	8	0.002 ± 0.000	2.42	0.05332 ± 0.00186	10.90 ± 0.21	562.1 ± 10.1
4.3	3882	687	0.177 ± 0.009	0.18	0.05726 ± 0.00030	8.49 ± 0.15	501.7 ± 11.5
6.3	8055	26	0.003 ± 0.000	1.57	0.05657 ± 0.00095	10.10 ± 0.13	474.7 ± 37.4
5.3	3496	44	0.013 ± 0.000	3.71	0.05425 ± 0.00263	13.11 ± 0.20	473.7 ± 6.9
7.3	2784	32	0.011 ± 0.001	1.05	0.05667 ± 0.00143	20.45 ± 5.69	425.1 ± 47.1
8.3	3518	9	0.003 ± 0.000	0.27	0.05605 ± 0.00077	12.02 ± 0.18	511.7 ± 7.4
10.3	2435	3	0.001 ± 0.000	1.20	0.06203 ± 0.00088	12.05 ± 0.13	675.2 ± 30.7
9.3	9480	60	0.006 ± 0.000	0.27	0.05872 ± 0.00037	10.14 ± 0.11	556.8 ± 13.9
34.3	4724	17	0.004 ± 0.000	0.71	0.05546 ± 0.00077	11.64 ± 0.21	523.5 ± 8.7
26.3	5075	18	0.004 ± 0.000	1.14	0.05830 ± 0.00042	11.67 ± 0.13	531.4 ± 5.4
11.3	2309	39	0.017 ± 0.001	0.87	0.05556 ± 0.00126	12.88 ± 0.15	481.5 ± 5.5
35.1	2516	7	0.003 ± 0.000	0.94	0.05761 ± 0.00075	11.94 ± 0.16	518.1 ± 6.6
27.3	1122	61	0.054 ± 0.001	2.63	0.04590 ± 0.00318	11.36 ± 0.17	543.8 ± 7.6
16.1	2755	5	0.002 ± 0.000	0.63	0.05694 ± 0.00109	12.08 ± 0.15	512.1 ± 5.9

Notes:
Corrected for common Pb.
All errors in table are 1σ.
#f206Pb% is percentage common ^{206}Pb in the total measured ^{206}Pb.
Common Pb: $^{204}Pb/^{206}Pb$ correction.
Age: Weighted Mean $^{206}Pb/^{238}U$ & $^{207}Pb/^{206}Pb$ age.

Standards
Session 1 (with PL147 and PL239) See PL147. Error in mean is 0.73%.

Session 2
Twelve standards give an MSWD of . Remaining points all within 2σ. MSWD is 1.68 and error in mean for 11 standards is 4.4 Ma (0.40%).

56806

Label	U ppm	Th ppm	Th/U	#f206Pb %	$^{207}Pb/^{206}Pb$	$^{238}U/^{206}Pb$	Age, Ma
1.1	302	274	0.906±0.008	0.01	0.05523±0.00195	12.12±0.18	511.2±7.3
2.1	268	276	1.028±0.033	0.00	0.05798±0.00544	12.02±0.40	515.2±16.6
3.1	264	160	0.609±0.004	0.00	0.05799±0.00126	11.99±0.19	516.4±8.0
4.1	182	88	0.485±0.011	0.01	0.05727±0.00258	11.84±0.26	522.8±11.2
4.2	148	127	0.859±0.006	0.01	0.05729±0.00269	11.87±0.19	521.5±7.9
5.1	316	200	0.634±0.004	0.00	0.05740±0.00114	12.27±0.20	504.9±7.9
6.1	262	408	1.559±0.032	0.00	0.06099±0.00680	12.17±0.25	509.1±10.1
7.1	122	48	0.399±0.003	0.01	0.05952±0.00152	12.09±0.19	512.4±7.8
7.2	736	278	0.378±0.002	0.01	0.06292±0.00072	12.67±0.19	489.7±6.9
8.1	413	128	0.311±0.002	0.00	0.05734±0.00066	11.64±0.23	531.2±9.9
9.1	283	110	0.388±0.003	0.00	0.05961±0.00103	11.69±0.20	529.1±8.6
10.1	278	128	0.462±0.002	0.01	0.05857±0.00115	12.04±0.20	514.4±8.1
11.1	260	414	1.589±0.008	0.02	0.06100±0.00364	12.17±0.19	509.0±7.8
12.1	144	105	0.732±0.004	0.01	0.05796±0.00179	11.89±0.19	520.5±7.8
13.1	181	195	1.077±0.014	0.02	0.05607±0.00525	11.71±0.21	528.2±9.3
14.1	212	291	1.375±0.006	0.01	0.05467±0.00296	12.46±0.19	497.7±7.5
15.1	504	336	0.666±0.002	0.00	0.05885±0.00084	12.07±0.18	512.9±7.4
16.1	348	312	0.896±0.005	0.00	0.06031±0.00157	12.03±0.19	514.9±7.9
17.1	178	135	0.759±0.005	0.00	0.05726±0.00208	12.31±0.21	503.5±8.4
18.1	245	207	0.845±0.005	0.00	0.06032±0.00120	11.89±0.19	520.6±8.1
19.1	696	216	0.311±0.001	0.00	0.05817±0.00093	11.94±0.17	518.4±7.3
20.1	509	109	0.214±0.002	0.01	0.05910±0.00062	10.50±0.17	586.5±8.8

Notes:
Corrected for common Pb.
All errors in table are 1σ.
#f206Pb% is percentage common ^{206}Pb in the total measured ^{206}Pb.
Common Pb: $^{204}Pb/^{206}Pb$ correction.
Age: Weighted Mean $^{206}Pb/^{238}U$ & $^{207}Pb/^{206}Pb$ age.

PP076A

Label	U ppm	Th ppm	Th/U	#f206Pb %	207Pb/206Pb	238U/206Pb	Age, Ma
1.1	274	214	0.780 ± 0.010	0.49	0.06141 ± 0.00109	12.23 ± 0.30	504.3 ± 11.9
2.1	115	69	0.599 ± 0.007	1.09	0.06606 ± 0.00131	12.42 ± 0.31	493.8 ± 11.9
3.1	271	224	0.825 ± 0.007	0.34	0.06029 ± 0.00151	12.13 ± 0.19	508.9 ± 7.9
4.1	141	100	0.706 ± 0.009	1.01	0.06490 ± 0.00113	12.97 ± 0.34	474.1 ± 12.1
5.1	211	100	0.474 ± 0.005	1.25	0.06906 ± 0.00117	10.98 ± 0.26	555.2 ± 12.8 •
6.1	302	207	0.684 ± 0.004	0.09	0.05781 ± 0.00100	12.57 ± 0.18	493.0 ± 6.9
7.1	384	281	0.730 ± 0.007	0.37	0.05976 ± 0.00078	12.89 ± 0.21	480.0 ± 7.5
8.1	328	279	0.850 ± 0.014	0.31	0.05935 ± 0.00084	12.79 ± 0.28	484.0 ± 10.2
9.1	262	160	0.609 ± 0.010	0.53	0.06115 ± 0.00089	12.81 ± 0.31	482.0 ± 11.3
10.1	115	52	0.448 ± 0.005	1.20	0.06678 ± 0.00137	12.61 ± 0.29	486.2 ± 10.7
11.1	313	194	0.620 ± 0.005	0.21	0.05875 ± 0.00075	12.60 ± 0.22	491.3 ± 8.2
12.1	275	193	0.700 ± 0.005	0.61	0.06203 ± 0.00106	12.56 ± 0.22	490.8 ± 8.4
13.1	255	178	0.696 ± 0.008	0.25	0.05865 ± 0.00157	12.99 ± 0.36	477.1 ± 12.7
14.1	242	134	0.554 ± 0.005	2.73	0.07969 ± 0.00325	12.14 ± 0.22	496.7 ± 8.8
15.1	289	182	0.632 ± 0.005	0.30	0.05952 ± 0.00070	12.50 ± 0.21	494.7 ± 8.1
16.1	396	382	0.966 ± 0.004	0.18	0.05838 ± 0.00054	12.71 ± 0.16	487.5 ± 5.9
17.1	337	260	0.771 ± 0.005	0.24	0.05901 ± 0.00062	12.59 ± 0.19	491.5 ± 7.2

$^{206}Pb/^{238}U$ AGE
Weighted Mean: (n=16/17; MSWD = 0.94) 490.6 ± 2.2 (1σ)
Error in standard: 0.34% (1σ)
Final age: 490.6 ± 5.5 Ma (2σ)

Notes:
Uncorrected for common Pb.
All errors in table are 1σ.
#f206Pb% is percentage common ^{206}Pb in the total measured ^{206}Pb.
Common Pb: $^{204}Pb/^{206}Pb$ correction.
Age: Weighted Mean $^{206}Pb/^{238}U$ & $^{207}Pb/^{206}Pb$ age.

Standards
(with PP183A)
The weighted mean of all 18 standards has an MSWD of 0.76. The most deviant point is –2.9σ and was not rejected. The error of the weighted mean for 18 standards is 3.7 Ma which is 0.34%.

PP159A

Label	U ppm	Th ppm	Th/U	#f206Pb %	$^{207}Pb/^{206}Pb$	$^{238}U/^{206}Pb$	Age, Ma
1.1	233	130	0.557 ± 0.006	0.23	0.05900 ± 0.00078	12.64 ± 0.31	489.7 ± 11.6
2.1	158	130	0.821 ± 0.010	-0.14	0.05610 ± 0.00125	12.53 ± 0.27	495.8 ± 10.2
3.1	294	246	0.838 ± 0.008	0.06	0.05758 ± 0.00077	12.62 ± 0.25	491.5 ± 9.6
4.1	506	390	0.771 ± 0.005	0.16	0.05810 ± 0.00058	12.95 ± 0.22	478.8 ± 7.9
5.1	170	86	0.504 ± 0.006	0.06	0.05809 ± 0.00113	12.12 ± 0.21	510.8 ± 8.5
6.1	141	85	0.603 ± 0.005	-0.08	0.05686 ± 0.00101	12.26 ± 0.25	505.8 ± 9.9
7.1	566	301	0.533 ± 0.004	0.00	0.05686 ± 0.00049	12.93 ± 0.32	480.1 ± 11.4
8.1	212	126	0.597 ± 0.007	-0.02	0.05723 ± 0.00084	12.34 ± 0.27	502.6 ± 10.5
9.1	159	156	0.981 ± 0.008	0.12	0.05797 ± 0.00088	12.73 ± 0.28	486.8 ± 10.2
10.1	237	145	0.613 ± 0.010	0.04	0.05724 ± 0.00084	12.86 ± 0.22	482.5 ± 8.0
11.1	160	106	0.665 ± 0.005	0.46	0.06075 ± 0.00147	12.74 ± 0.21	485.1 ± 7.7
12.1	181	109	0.599 ± 0.005	-0.05	0.05698 ± 0.00105	12.35 ± 0.22	502.0 ± 8.5
13.1	124	77	0.625 ± 0.009	0.34	0.05957 ± 0.00278	12.99 ± 0.34	476.6 ± 12.2
14.1	176	98	0.554 ± 0.006	0.09	0.05790 ± 0.00134	12.59 ± 0.30	492.1 ± 11.2
15.1	201	128	0.640 ± 0.006	0.25	0.05868 ± 0.00105	13.07 ± 0.28	474.1 ± 9.7

$^{206}Pb/^{238}U$ AGE
Weighted Mean: (n=15/15; MSWD = 1.46) 490.4 ± 2.5 (1σ)
Error in standard: 0.47% (1σ)
Final age: 490.4 ± 6.9 Ma (2σ)

Notes:
Uncorrected for common Pb.
All errors in table are 1σ.
#f206Pb% is percentage common ^{206}Pb in the total measured ^{206}Pb.
Common Pb: $^{204}Pb/^{206}Pb$ correction.
Age: $^{206}Pb/^{238}U$.
Outliers •

Standards
The weighted mean of all 14 AS3 standards has a MSWD of 3.7. The most deviant point is -3.8σ and was rejected; MSWD =2.57. Next deviant is -2.8σ; reject and MSWD falls to 2.02. Next is -2.6σ; reject and MSWD falls to 1.47. The three most deviant points in successive iterations are all low and indicative of Pb loss; two of the points come from the same AS3 crystal (grain 6), the other from grain 2 is the only analysis from that grain. One low precision analysis which is low is also rejected. There is no suggestion of a systematic drift during the day. The mean changes by -1.4 % with the rejection of the four low points.Final MSWD is 1.24 and the error of the weighted mean for 10 standards is 5.2 Ma which is 0.47%.

PP116A

Label	U ppm	Th ppm	Th/U	#f206Pb %	$^{207}Pb/^{206}Pb$	$^{238}U/^{206}Pb$	Age, Ma
5.1	292	248	0.850 ± 0.008	0.02	0.05736 ± 0.00056	12.92 ± 0.19	480.5 ± 6.8
6.1	161	242	1.499 ± 0.026	0.28	0.05951 ± 0.00085	12.73 ± 0.46	486.1 ± 17.0
7.1	231	233	1.008 ± 0.022	-0.01	0.05713 ± 0.00110	12.20 ± 0.28	507.9 ± 11.1
8.1	168	170	1.009 ± 0.010	0.16	0.05854 ± 0.00120	13.30 ± 0.28	466.7 ± 9.6
9.1	225	246	1.095 ± 0.008	0.02	0.05740 ± 0.00103	12.69 ± 0.21	488.9 ± 7.9
10.1	161	162	1.006 ± 0.005	0.02	0.05739 ± 0.00120	12.68 ± 0.20	489.2 ± 7.4
11.1	254	288	1.135 ± 0.010	-0.07	0.05665 ± 0.00097	12.88 ± 0.29	482.5 ± 10.6
12.1	172	138	0.801 ± 0.011	0.10	0.05804 ± 0.00124	12.72 ± 0.26	487.4 ± 9.8
13.1	146	141	0.968 ± 0.011	-0.03	0.05695 ± 0.00073	12.56 ± 0.28	494.2 ± 10.7
14.1	176	173	0.985 ± 0.006	-0.09	0.05647 ± 0.00072	13.03 ± 0.21	477.1 ± 7.4
15.1	150	159	1.061 ± 0.029	-0.08	0.05658 ± 0.00106	12.78 ± 0.39	486.2 ± 14.3
16.1	161	144	0.899 ± 0.008	0.22	0.05905 ± 0.00113	12.75 ± 0.23	485.6 ± 8.4
17.1	90	57	0.640 ± 0.027	1.23	0.06718 ± 0.00247	9.03 ± 0.28	669.2 ± 20.2 •

$^{206}Pb/^{238}U$ AGE
Weighted Mean: (n=12/13; MSWD = 0.99) 484.9 ± 2.6 (1σ)
Error in standard: 0.40% (1σ)
Final age: 484.9 ± 6.5 Ma (2σ)

Notes:
Uncorrected for common Pb.
All errors in table are 1σ.
#f206Pb% is percentage common ^{206}Pb in the total measured ^{206}Pb.
Common Pb: $^{204}Pb/^{206}Pb$ correction.
Age: $^{206}Pb/^{238}U$.
Outliers •

Standards
See JS129c. Sixteen AS3 standards give a MSWD of 0.95. No rejections were made. The error of the mean is 4.5 Ma which is 0.40 %.

PP183A

Label	U ppm	Th ppm	Th/U	#f206Pb %	$^{207}Pb/^{206}Pb$	$^{238}U/^{206}Pb$	Age, Ma
1.1	251	130	0.521 ± 0.007	0.52	0.06076 ± 0.00092	13.09 ± 0.31	472.1 ± 10.7
2.1	208	165	0.795 ± 0.006	0.44	0.06032 ± 0.00091	12.85 ± 0.23	481.0 ± 8.3
3.1	273	155	0.568 ± 0.009	0.56	0.06125 ± 0.00085	12.95 ± 0.24	476.8 ± 8.4
4.1	150	73	0.483 ± 0.005	0.96	0.06487 ± 0.00102	12.53 ± 0.23	490.5 ± 8.9
5.1	260	100	0.386 ± 0.004	0.83	0.06349 ± 0.00138	12.92 ± 0.27	476.5 ± 9.6
6.1	186	95	0.511 ± 0.006	0.36	0.05955 ± 0.00089	13.01 ± 0.31	475.8 ± 10.9
7.1	182	112	0.617 ± 0.005	0.57	0.06134 ± 0.00093	12.91 ± 0.24	478.3 ± 8.8
8.1	275	102	0.371 ± 0.004	0.70	0.06244 ± 0.00123	12.90 ± 0.27	478.0 ± 9.6
9.1	256	224	0.875 ± 0.010	0.20	0.05828 ± 0.00104	12.99 ± 0.29	477.0 ± 10.2
10.1	305	175	0.575 ± 0.004	0.23	0.05881 ± 0.00094	12.70 ± 0.19	487.6 ± 7.0
11.1	247	157	0.637 ± 0.004	2.91	0.08069 ± 0.00257	12.56 ± 0.19	480.1 ± 7.2
12.1	201	125	0.624 ± 0.008	1.07	0.06487 ± 0.00174	13.53 ± 0.34	454.9 ± 11.2
13.1	218	121	0.553 ± 0.006	0.34	0.05963 ± 0.00100	12.80 ± 0.27	483.2 ± 9.8
14.1	220	144	0.654 ± 0.008	0.81	0.06326 ± 0.00080	12.95 ± 0.26	475.8 ± 9.3
15.1	264	166	0.627 ± 0.006	0.69	0.06254 ± 0.00091	12.72 ± 0.20	484.5 ± 7.4
16.1	140	77	0.552 ± 0.007	1.10	0.06665 ± 0.00129	11.92 ± 0.27	513.7 ± 11.3 •

$^{206}Pb/^{238}U$ AGE
Weighted Mean: (n=15/16; MSWD = 0.67) 479.5 ± 2.3 (1σ)
Error in standard: 0.34% (1σ)
Final age: 479.5 ± 5.6 Ma (2σ)

Notes:
Uncorrected for common Pb.
All errors in table are 1σ.
#f206Pb% is percentage common ^{206}Pb in the total measured ^{206}Pb.
Common Pb: $^{204}Pb/^{206}Pb$ correction.
Age: $^{206}Pb/^{238}U$.
Outliers •

Standards
(with PP076A)
The weighted mean of all 18 standards has an MSWD of 0.76. The most deviant point is –2.9σ and was not rejected. The error of the weighted mean for 18 standards is 3.7 Ma which is 0.34%.

PP114A

Label	U ppm	Th ppm	Th/U	#f206Pb %	207Pb/206Pb	238U/206Pb	Age, Ma
1.1	133	175	1.317 ± 0.013	0.89	0.06365 ± 0.00133	13.21 ± 0.25	466.3 ± 8.4
2.1	186	199	1.071 ± 0.017	0.57	0.06168 ± 0.00130	12.59 ± 0.36	490.0 ± 13.7
3.1	191	160	0.836 ± 0.011	0.40	0.06013 ± 0.00109	12.69 ± 0.28	487.0 ± 10.3
4.1	269	301	1.120 ± 0.017	0.34	0.05932 ± 0.00089	13.11 ± 0.34	472.4 ± 11.7
5.1	211	43	0.203 ± 0.002	0.27	0.05972 ± 0.00099	12.09 ± 0.32	511.0 ± 12.9 •
6.1	165	214	1.298 ± 0.023	0.41	0.05982 ± 0.00112	13.18 ± 0.38	469.5 ± 13.0
7.1	146	91	0.625 ± 0.006	0.31	0.05930 ± 0.00170	12.85 ± 0.27	481.8 ± 9.7
8.1	241	98	0.407 ± 0.009	1.20	0.06742 ± 0.00112	11.97 ± 0.42	511.2 ± 17.3 •
9.1	145	181	1.252 ± 0.034	0.37	0.05953 ± 0.00102	13.13 ± 0.53	471.6 ± 18.4
10.1	145	149	1.030 ± 0.011	0.50	0.06041 ± 0.00115	13.30 ± 0.28	465.0 ± 9.6
11.1	61	55	0.905 ± 0.012	1.02	0.06502 ± 0.00197	12.89 ± 0.33	477.1 ± 11.9
12.1	244	234	0.961 ± 0.009	0.15	0.05796 ± 0.00111	12.86 ± 0.24	482.1 ± 8.5
13.1	191	182	0.952 ± 0.015	0.30	0.05911 ± 0.00094	12.93 ± 0.21	478.7 ± 7.4
14.1	190	160	0.839 ± 0.011	0.39	0.05951 ± 0.00109	13.35 ± 0.36	464.0 ± 12.0
15.1	170	81	0.475 ± 0.005	0.06	0.05881 ± 0.00126	11.42 ± 0.21	540.6 ± 9.4 •
16.1	101	79	0.780 ± 0.007	0.68	0.06246 ± 0.00150	12.65 ± 0.25	487.2 ± 9.1
17.1	152	127	0.836 ± 0.011	0.42	0.05997 ± 0.00112	13.12 ± 0.35	471.6 ± 12.1

$^{206}Pb/^{238}U$ AGE
Weighted Mean: (n=14/17; MSWD = 0.68) 476.5 ± 2.7 (1σ)
Error in standard: 0.47% (1σ)
Final age: 476.5 ± 7.0 Ma (2σ)

Notes:
Uncorrected for common Pb.
All errors in table are 1σ.
#f206Pb% is percentage common ^{206}Pb in the total measured ^{206}Pb.
Common Pb: $^{204}Pb/^{206}Pb$ correction.
Age: $^{206}Pb/^{238}U$.
Outliers •

Standards
The weighted mean of all 14 standards has an MSWD of 0.49. The most deviant point is +1.6σ and was not rejected. The error of the weighted mean for 14 standards is 5.2 Ma which is 0.47%.

JS081

Label	U ppm	Th ppm	Th/U	#f206Pb %	207Pb/206Pb	238U/206Pb	Age, Ma
1.1	512	345	0.674 ± 0.005	3.41	0.08480 ± 0.00092	12.49 ± 0.20	480.3 ± 7.5
2.1	704	185	0.263 ± 0.002	0.10	0.05852 ± 0.00076	11.94 ± 0.17	518.0 ± 7.2 •
3.1	952	417	0.439 ± 0.002	0.23	0.05813 ± 0.00067	13.40 ± 0.18	462.8 ± 5.9
4.1	864	286	0.331 ± 0.001	0.80	0.06320 ± 0.00069	12.96 ± 0.17	475.5 ± 5.9
5.1	317	261	0.823 ± 0.008	0.61	0.06130 ± 0.00156	13.37 ± 0.25	462.3 ± 8.5
6.1	1098	328	0.299 ± 0.002	0.14	0.05779 ± 0.00042	13.01 ± 0.19	476.9 ± 6.7
7.1	1033	295	0.286 ± 0.001	0.20	0.05836 ± 0.00050	12.90 ± 0.17	480.3 ± 6.1
8.1	927	300	0.323 ± 0.001	1.56	0.06907 ± 0.00074	13.28 ± 0.18	460.8 ± 6.0
9.1	1532	508	0.332 ± 0.002	0.26	0.05832 ± 0.00079	13.50 ± 0.19	459.5 ± 6.4
10.1	470	180	0.383 ± 0.003	0.16	0.05959 ± 0.00085	11.43 ± 0.18	539.8 ± 8.2 •
11.1	707	467	0.660 ± 0.008	0.33	0.05924 ± 0.00064	13.11 ± 0.22	472.5 ± 7.7
12.1	1452	365	0.252 ± 0.001	0.28	0.05882 ± 0.00042	13.12 ± 0.18	472.2 ± 6.1
13.1	292	144	0.491 ± 0.003	0.27	0.06139 ± 0.00106	10.73 ± 0.15	572.7 ± 7.7 •
14.1	418	222	0.531 ± 0.005	0.44	0.05994 ± 0.00106	13.29 ± 0.24	465.7 ± 8.2
15.1	118	54	0.459 ± 0.007	1.27	0.06694 ± 0.00238	12.99 ± 0.37	472.4 ± 13.1
16.1	1244	242	0.194 ± 0.001	0.08	0.05774 ± 0.00080	12.51 ± 0.17	495.5 ± 6.4 •
17.1	534	623	1.167 ± 0.008	0.03	0.05725 ± 0.00120	12.59 ± 0.20	492.7 ± 7.6 •
18.1	315	232	0.736 ± 0.005	0.51	0.06067 ± 0.00086	13.17 ± 0.22	469.6 ± 7.5
19.1	747	204	0.273 ± 0.003	0.49	0.05935 ± 0.00104	14.51 ± 0.32	427.5 ± 9.2 •
20.1	1403	265	0.189 ± 0.002	0.52	0.06061 ± 0.00041	13.28 ± 0.26	465.7 ± 8.9

$^{206}Pb/^{238}U$ AGE
Weighted Mean: (n=14/20; MSWD = 0.99) 469.8 ± 1.9 (1σ)
Error in standard: 0.35% (1σ)
Final age: 469.8 ± 5.0 Ma (2σ)

Notes:
Uncorrected for common Pb.
All errors in table are 1σ.
#f206Pb% is percentage common ^{206}Pb in the total measured ^{206}Pb.
Common Pb: $^{204}Pb/^{206}Pb$ correction.
Age: $^{206}Pb/^{238}U$.
Outliers •

Standards
The weighted mean of all 16 standards has an MSWD of 1.39. The most deviant point is -2.1σ and was not rejected. The error of the weighted mean for 16 standards is 3.9 Ma which is 0.35%.

AC060

Label	U ppm	Th ppm	Th/U	#f206Pb %	$^{207}Pb/^{206}Pb$	$^{238}U/^{206}Pb$	Age, Ma
1.1	922	455	0.493 ± 0.003	0.02	0.05740 ± 0.00046	12.39 ± 0.15	500.3 ± 5.9 £
2.1	405	157	0.387 ± 0.007	0.26	0.05921 ± 0.00087	12.56 ± 0.25	492.4 ± 9.3 £
3.1	569	208	0.366 ± 0.008	0.33	0.05932 ± 0.00064	13.05 ± 0.46	474.4 ± 16.1 ¢
4.1	280	63	0.225 ± 0.002	0.31	0.06612 ± 0.00105	8.36 ± 0.16	726.0 ± 13.1 •
5.1	334	137	0.409 ± 0.005	1.31	0.06781 ± 0.00148	12.44 ± 0.30	492.3 ± 11.6 £
6.1	168	116	0.691 ± 0.010	1.67	0.07039 ± 0.00223	12.75 ± 0.27	478.9 ± 9.8 ¢
7.1	820	538	0.656 ± 0.006	0.47	0.06084 ± 0.00066	12.61 ± 0.18	489.6 ± 6.6 £
8.2	809	369	0.456 ± 0.004	1.45	0.06802 ± 0.00068	13.46 ± 0.17	455.6 ± 5.4 ¢
9.1	336	136	0.405 ± 0.005	1.83	0.07241 ± 0.00138	12.09 ± 0.34	503.1 ± 13.7 £
10.1	576	216	0.375 ± 0.007	4.36	0.09256 ± 0.00150	12.50 ± 0.35	475.4 ± 12.8 ¢
11.1	593	298	0.504 ± 0.007	27.77	0.29061 ± 0.01154	8.89 ± 0.22	503.5 ± 15.1 £
12.1	544	270	0.496 ± 0.003	0.50	0.06050 ± 0.00112	13.18 ± 0.18	469.3 ± 6.3 ¢
13.1	705	334	0.473 ± 0.002	0.93	0.06430 ± 0.00058	12.87 ± 0.15	477.9 ± 5.5 ¢
14.1	520	282	0.542 ± 0.003	0.98	0.06449 ± 0.00132	13.10 ± 0.19	469.7 ± 6.6 ¢
15.1	96	45	0.463 ± 0.005	2.58	0.08870 ± 0.00197	7.13 ± 0.15	825.3 ± 15.9 •
16.1	279	163	0.584 ± 0.004	1.23	0.06655 ± 0.00149	13.03 ± 0.19	470.9 ± 6.6 ¢
17.1	772	264	0.341 ± 0.003	0.57	0.06044 ± 0.00053	13.99 ± 0.28	442.7 ± 8.5 •
18.1	209	132	0.634 ± 0.004	1.15	0.06577 ± 0.00156	13.23 ± 0.22	464.5 ± 7.6 ¢
19.1	640	405	0.632 ± 0.003	0.78	0.06264 ± 0.00106	13.36 ± 0.20	461.9 ± 6.8 ¢

£ $^{206}Pb/^{238}U$ AGE
Weighted Mean: (n=6/19; MSWD = 0.44) 495.8 ± 3.5 (1σ)
Error in standard: 0.45% (1σ)
Final age: 495.8 ± 8.3 Ma (2σ)

¢ $^{206}Pb/^{238}U$ AGE
Weighted Mean: (n=10/19; MSWD = 1.27) 468.2 ± 2.3 (1σ)
Error in standard: 0.45% (1σ)
Final age: 468.2 ± 6.2 Ma (2σ)

Notes:
Uncorrected for common Pb.
All errors in table are 1σ.
#f206Pb% is percentage common ^{206}Pb in the total measured ^{206}Pb.
Common Pb: $^{204}Pb/^{206}Pb$ correction.
Age: $^{206}Pb/^{238}U$.
£,¢ two components
Outliers •

Standards

The weighted mean of all 12 standards has an MSWD of 2.49. The most deviant point is analysis 1.8 which is 3.2σ below the mean. Rejecting this point gives an MSWD of 1.54. The most deviant point after rejection is -2.0σ and was not rejected. The error of the weighted mean for 11 standards is 4.9 Ma which is 0.45%.

Uncorrected for common Pb

JS033

Label	U ppm	Th ppm	Th/U	#f206Pb %	$^{207}Pb/^{206}Pb$	$^{238}U/^{206}Pb$	Age, Ma
1.1	2322	635	0.273 ± 0.002	0.14	0.05629 ± 0.00026	14.84 ± 0.23	420.0 ± 6.3
2.1	1306	514	0.394 ± 0.004	0.06	0.05566 ± 0.00056	14.85 ± 0.34	420.0 ± 9.4
3.1	1156	1560	1.349 ± 0.010	0.11	0.05556 ± 0.00063	15.66 ± 0.35	398.6 ± 8.7
4.1	2996	439	0.146 ± 0.005	3.95	0.07987 ± 0.00151	49.94 ± 2.89	122.8 ± 7.1 •
5.1	1022	460	0.450 ± 0.010	3.44	0.07640 ± 0.00171	42.24 ± 2.03	145.7 ± 6.9 •
6.1	641	1364	2.126 ± 0.018	0.30	0.05724 ± 0.00086	15.37 ± 0.28	405.2 ± 7.1
7.1	1004	1220	1.215 ± 0.024	1.07	0.05990 ± 0.00097	24.53 ± 0.66	254.9 ± 6.8 •
8.1	361	386	1.070 ± 0.019	0.52	0.05857 ± 0.00181	16.23 ± 0.62	383.5 ± 14.3
9.1	839	1722	2.052 ± 0.022	0.46	0.05849 ± 0.00068	15.51 ± 0.29	401.0 ± 7.2
10.1	919	800	0.871 ± 0.008	0.20	0.05657 ± 0.00047	15.29 ± 0.34	407.6 ± 8.7
11.1	2034	658	0.324 ± 0.002	1.57	0.06728 ± 0.00070	15.73 ± 0.22	391.2 ± 5.4
12.1	286	412	1.444 ± 0.018	0.66	0.06058 ± 0.00145	14.79 ± 0.29	419.0 ± 8.0
13.1	679	873	1.286 ± 0.016	0.39	0.05789 ± 0.00057	15.59 ± 0.31	399.3 ± 7.6
14.1	703	890	1.266 ± 0.015	0.38	0.05756 ± 0.00100	16.01 ± 0.46	389.1 ± 10.9
15.1	128	78	0.612 ± 0.009	1.23	0.06443 ± 0.00196	15.98 ± 0.42	386.5 ± 9.9
16.1	429	683	1.592 ± 0.017	0.67	0.06030 ± 0.00081	15.29 ± 0.33	405.7 ± 8.5
17.1	3257	1055	0.324 ± 0.004	0.61	0.05962 ± 0.00071	15.60 ± 0.29	398.1 ± 7.3
18.1	484	925	1.910 ± 0.024	0.38	0.05790 ± 0.00103	15.36 ± 0.35	405.0 ± 9.0
19.1	493	723	1.465 ± 0.016	0.66	0.06026 ± 0.00125	15.29 ± 0.37	405.7 ± 9.5

$^{206}Pb/^{238}U$ AGE
Weighted Mean: (n=16/19; MSWD = 1.81) 403.1 ± 2.0 (1σ)
Error in standard: 0.47% (1σ)
Final age: 403.1 ± 5.5 Ma (2σ)

Notes:
Corrected for common Pb.
All errors in table are 1σ.
#f206Pb% is percentage common ^{206}Pb in the total measured ^{206}Pb.
Common Pb: $^{204}Pb/^{206}Pb$ correction.
Age: $^{206}Pb/^{238}U$.
Outliers •

Standards

The weighted mean of all 15 standards has an MSWD of 0.99 and all data lies within 2σ of the mean. The error of the weighted mean for 15 standards is 5.2 Ma which is 0.47%.

AC054

Label	U ppm	Th ppm	Th/U	#f206Pb %	207Pb/206Pb	238U/206Pb	Age, Ma
1.1	278	83	0.299 ± 0.003	1.91	0.07023 ± 0.00123	15.42 ± 0.39	397.5 ± 9.7
2.1	682	359	0.526 ± 0.006	0.68	0.05990 ± 0.00083	16.15 ± 0.30	384.7 ± 6.9
3.1	1046	416	0.397 ± 0.011	0.36	0.05736 ± 0.00086	16.05 ± 0.57	388.3 ± 13.4
4.1	197	168	0.852 ± 0.006	2.30	0.07314 ± 0.00145	15.79 ± 0.37	386.9 ± 8.9
5.1	1117	356	0.319 ± 0.002	0.57	0.05950 ± 0.00049	15.36 ± 0.20	404.2 ± 5.1
6.1	796	355	0.447 ± 0.002	0.57	0.05918 ± 0.00051	15.81 ± 0.25	393.2 ± 6.1
7.1	1094	568	0.519 ± 0.002	0.60	0.05941 ± 0.00041	15.90 ± 0.25	390.9 ± 5.9
8.1	494	349	0.707 ± 0.004	1.18	0.06431 ± 0.00105	15.45 ± 0.26	399.7 ± 6.4
9.1	262	151	0.576 ± 0.005	2.02	0.07119 ± 0.00147	15.31 ± 0.39	399.8 ± 9.8
10.1	753	250	0.332 ± 0.003	0.26	0.05688 ± 0.00063	15.53 ± 0.32	401.2 ± 8.1
11.1	1407	625	0.444 ± 0.006	1.92	0.06937 ± 0.00080	17.05 ± 0.52	360.6 ± 10.8 •
12.1	139	194	1.396 ± 0.009	2.50	0.07531 ± 0.00146	15.02 ± 0.28	405.4 ± 7.3
14.1	1384	630	0.455 ± 0.002	2.07	0.07036 ± 0.00218	17.52 ± 0.40	350.6 ± 7.8 •
15.1	1146	353	0.308 ± 0.003	3.92	0.08310 ± 0.00194	23.17 ± 0.55	262.0 ± 6.1 •
16.1	162	213	1.311 ± 0.014	2.25	0.07262 ± 0.00118	15.97 ± 0.29	383.0 ± 6.7
17.1	946	326	0.345 ± 0.002	0.25	0.05626 ± 0.00092	16.37 ± 0.25	381.4 ± 5.7
18.1	1193	381	0.319 ± 0.002	-0.07	0.05364 ± 0.00073	16.46 ± 0.32	380.4 ± 7.2

$^{206}Pb/^{238}U$ AGE
Weighted Mean: (n=12/13; MSWD = 0.99) 392.7 ± 1.9 (1σ)
Error in standard: 0.37% (1σ)
Final age: 392.7 ± 4.8 Ma (2σ)

Notes:
Corrected for common Pb.
All errors in table are 1σ.
#f206Pb% is percentage common ^{206}Pb in the total measured ^{206}Pb.
Common Pb: $^{204}Pb/^{206}Pb$ correction.
Age: $^{206}Pb/^{238}U$.

Standards (with PL063)
The weighted mean of all 15 standards has an MSWD of 1.85 (c.f. 1.57 for F distribution with 20 degrees of freedom). Analysis 9.1 is 3.0σ below the mean and rejecting this point gives a MSWD of 1.23. The most deviant point after rejection is -2.5σ and was not rejected.The error of the weighted mean for 14 standards is 4.1 Ma which is 0.37%.

Uncorrected for common Pb

PS167

Label	U ppm	Th ppm	Th/U	#f206Pb %	$^{207}Pb/^{206}Pb$	$^{238}U/^{206}Pb$	Age, Ma
1.1	294	230	0.783 ± 0.008	0.79	0.06101 ± 0.00098	15.70 ± 0.32	394.9 ± 7.9
2.1	337	124	0.369 ± 0.004	0.75	0.06052 ± 0.00169	16.03 ± 0.28	387.4 ± 6.6
3.1	610	337	0.552 ± 0.007	0.55	0.05878 ± 0.00095	16.27 ± 0.33	382.5 ± 7.6
4.1	354	276	0.780 ± 0.015	0.80	0.06058 ± 0.00092	16.56 ± 0.56	375.0 ± 12.4
5.1	425	26	0.061 ± 0.001	0.51	0.05836 ± 0.00112	16.35 ± 0.37	380.9 ± 8.5
6.1	410	25	0.061 ± 0.001	0.75	0.06058 ± 0.00104	15.97 ± 0.26	388.6 ± 6.2
7.1	570	477	0.837 ± 0.010	0.31	0.05668 ± 0.00084	16.59 ± 0.40	376.2 ± 8.8
8.1	158	141	0.888 ± 0.014	5.85	0.10170 ± 0.00348	15.97 ± 0.51	369.3 ± 11.6
9.1	575	115	0.200 ± 0.006	0.50	0.05810 ± 0.00109	16.70 ± 0.38	373.0 ± 8.2
10.1	315	83	0.265 ± 0.005	0.64	0.05936 ± 0.00154	16.46 ± 0.47	377.8 ± 10.6
11.1	204	119	0.586 ± 0.007	1.24	0.06420 ± 0.00150	16.48 ± 0.37	375.2 ± 8.3
12.1	356	128	0.359 ± 0.005	0.63	0.05937 ± 0.00136	16.37 ± 0.27	379.8 ± 6.1
13.1	222	124	0.559 ± 0.009	0.57	0.05914 ± 0.00329	15.86 ± 0.44	392.0 ± 10.8

$^{206}Pb/^{238}U$ AGE
Weighted Mean: (n=13/13; MSWD = 0.79) 382.2 ± 2.3 (1σ)
Error in standard: 0.47% (1σ)
Final age: 382.2 ± 5.8 Ma (2σ)

Notes:
Corrected for common Pb.
All errors in table are 1σ.
#f206Pb% is percentage common ^{206}Pb in the total measured ^{206}Pb.
Common Pb: $^{204}Pb/^{206}Pb$ correction.
Age: $^{206}Pb/^{238}U$.
Outliers •

Standards
The weighted mean of all 15 standards has an MSWD of 0.99 and all data lies within 2σ of the mean. The error of the weighted mean for 15 standards is 5.2 Ma which is 0.47%.

REFERENCES CITED

Aceñolaza, F. G., and Toselli, A. J., 1976, Consideraciones estratigráficas y tectónicas sobre el Paleozoico inferior del Noroeste Argentino: Memoria, II Congreso Latinoamericano de Geología v. 2, p. 755–764.

Aceñolaza, F. G., and Toselli, A. J., 1981, Geología de Noroeste Argentino. Publicación Especial Fac. Ci. Nat. U.N.T., Tucumán, v. 1287, p. 212.

Astini, R. A., 1996, Las fases diastróficas del Paleozoico medio en La Precordillera del oeste Argentino—evidencias estratigráficas: XIII Congreso Geológico Argentino y III Congreso de Exploración de Hidrocarburos, Actas V, p. 509–526.

Astini, R. A., Ramos, V. A., Benedetto, J. L., Vaccari, N. E., and Cañas, F. L., 1996, La Precordillera: Un terreno exótico a Gondwana: XIII Congreso Geológico Argentino y III Congreso de Exploración de Hidrocarburos, Actas V, p. 293–324.

Baldo, E., and Casquet, C., 1996, Garnet zoning in migmatites, and regional metamorphism, in the Sierra Chica de Córdoba (Sierras Pampeanas, Argentina): XIII Congreso Geológico Argentino y III Congreso de Exploración de Hidrocarburos, Actas V, p. 507.

Bergstrom, S. M., Huff, W. D., Kolata, D. R., Krekeler, M. P. S., Cingolani, C., and Astini, R. A., 1996, Lower and Middle Ordovician K-Bentonites in the Precordillera of Argentina: a progress report: XIII Congreso Geológico Argentino y III Congreso de Exploración de Hidrocarburos, Actas V, p. 481–490.

Cumming, G. L., and Richards, J. R., 1975, Ore lead isotope ratios in a continuously changing earth: Earth and Planetary Science Letters, v. 28, p. 155–171.

Dahlquist, J. A., and Baldo, E. G. A., 1996, Metamorfismo y deformación famatinianos en la Sierra de Chepes, La Rioja, Argentina: XIII Congreso Geológico Argentino y III Congreso de Explorarión de Hidrocarburos, Actas V, p. 393–409.

Dalla Salda, L., 1987, Basement tectonics of the southern Pampean ranges, Argentina. Tectonics, v. 6, p. 249–260

Dalla Salda, L. H., Cingolani, C., and Varela, R., 1992, Early Paleozoic orogenic belt of the Andes in southwestern South America: result of Laurentia-Gondwana collision: Geology, v. 20, p. 617–620.

Dalla Salda, L. H., Cingolani, C., Varela, R., and Lopez de Luchi, M., 1995, The Famatinian orogenic belt in South-western South America. Granites and metamorphism: an Appalachian similitude? IX Congreso Latinoamericano de Geología, Caracas, Resúmenes.

Dalziel, I. W. D., 1991, Pacific margins of Laurentia and East Antartica–Australia as a conjugate rift pair: evidence and implications for an Eocambrian supercontinent. Geology, v. 19, p. 598–601.

Dalziel, I. W. D., Dalla Salda, L. H., and Gahagan, L. M., 1994, Paleozoic Laurentia-Gondwana interaction and the origin of the Appalachian-Andean mountain system: Geological Society of America Bulletin, v. 106, p. 243–252.

Dalziel, I. W. D., Dalla Salda, L. H., Cingolani, C., and Palmer, P., 1996, The Argentine Precordillera: a Laurentian terrane? Penrose Conference Report: GSA Today, February 1996, p. 16–18.

Demange, M., Baldo, E. G., and Martino, R. D., 1993, Structural evolution of the Sierras de Córdoba (Argentina): II I.S.A.G., Oxford, United Kingdom, v. 21, p. 513–516.

Escayola, M. P., Rame, G. A., and Kraemer, P. E., 1996, Caracterización y significado geotectónico de las fajas ultramáficas de las Sierras Pampeans de Córdoba: XIII Congreso Geológico Argentino y III Congreso de Exploración de Hidrocarburos, Actas III, p. 421–438.

Halpern, N. M., Linares, E., and Latorre, C. O., 1970, Estudio preliminar por el método rubidio-estroncio de rocas metamórficas y graníticas de la provincia de San Luis: Revista de la Asociación Geológica Argentina, v. 25, p. 293–302.

Keller, M., and Dickerson, P. W., 1996, The missing continent of Llanoria—was it the Argentine Precordillera?: XIII Congreso Geológico Argentino y III Congreso de Exploración de Hidrocarburos, Actas V, p. 355–367.

Kraemer, P., Escayola, M. P., and Martino, R. D., 1995, Hipótesis sobre la evolucíon tectónica neoproterozoica de las Sierras Pampeanas de Córdoba (30°40´–32°40´), Argentina: Revista de la Asociación Geológica Argentina, v. 50, p. 47–59.

Linares, E., and Gonzales, R. R. 1990, Catálogo de edades radimétricas de la República Argentina 1957-1987: Asociación Geológica Argentina Publicaciones Especiales, Ser. B, v. 19, 628 p.

Lira, R., Millone, H. A., Kirschbaum, A. M., and Moreno, R. S., 1996, Granitoides calcoalcalinos de magmático en la Sierra Norte de Córdoba: XIII Congreso Geológico Argentino y III Congreso de Exploración de Hidrocarburos, Actas III, p. 497.

Llambías, E. J., Cingolani, C., Varela, R., Prozzi, C., Ortíz Suárez, A., Toselli, A., and Saavedra, J., 1991, Leucogranodioritas sin-cinemáticas ordovícicas en la Sierra de San Luis: VI Congreso Geológico Chileno Resúmenes Expandidos, p. 187–191.

Lork, A., Miller, H., and Kramm, U., 1989, U-Pb zircon and monazite ages of the La Angostura Granite and the orogenic history of the northwest Argentina basement: Journal of South America Earth Sciences, v. 2, p. 147–153.

Lyons, P., Skirrow, R. G., and Stuart-Smith, P. G., 1997, Informe geológico y metalogénico de las Sierras Septentrionales de Córdoba (provincia de Córdoba), 1:250.000: Buenos Aires, Instituto de Geología y Recursos Minerales, SEGEMAR, Anales 27.

Martino, R. D., Simpson, C., and Law, R. D., 1994, Ductile thrusting in Pampean ranges: its relationships with the Ocloyic deformation and tectonic significance: IGCP, Novia Scotia Projects 319/376, Abstracts.

Martino, R., Kraemer, P., Escayola, M., Giambastiani, M., y Arnosio, M., 1995, Transecta de Las Sierras Pampeanas de Córdoba a los 32° S: Revista de la Asociación Geológica Argentina, v. 50, p. 60–77.

Muir, R. J., Ireland, T. R., Weaver, S. D., and Bradshaw, J. D., 1996, Ion microprobe dating of Paleozoic granitoids: Devonian magmatism in New Zealand and correlations with Australia and Antarctica: Chemical Geology (Isotope Geoscience), v. 127, p. 191–210.

Pankhurst, R., Rapela, C. W., Saavedra, J., Baldo, E., Dahlquist, J., and Pascua, I., 1996, Sierras de Los Llanos, Malanzán and Chepes: Ordovician I- and S-type granitic magmatism in the Famatinian Orogen: XIII Congreso Geológico Argentino y II Congreso de Exploración de Hidrocarburos, Actas V, p. 415.

Perez, M. B., Rapela, C. W., and Baldo, E. G., 1996, Geología de los granitoides del sector septentrional de la Sierra Chica de Córdoba: XIII Congreso Geológico Argentino y II Congreso de Exploración de Hidrocarburos, Actas V, p. 493–505.

Pieters, P., Skirrow, R. G., and Lyons, P., 1997, Informe geológico y metalogénico de las Sierras de Chepes, Las Minas y Los Llanos (provincia de La Rioja), 1:250 000: Buenos Aires, Instituto de Geología y Recursos Minerales, SEGEMAR, Anales 26.

Ramos, V., 1988, Late Proterozoic–Early Paleozoic of South America—a collisional history: Episodes, v. 11, p. 168–174.

Ramos, V. A., Jordan, T. E., Allmendinger, R. W., Mpodozis, C., Kay, S., Cortes, J. M., and Palma, M. A., 1986, Paleozoic terranes of the Central Argentine–Chilean Andes: Tectonics, v. 5, p. 855–880.

Ramos, V. A., Vujovich, G. I., and Dallmeyer, R. D., 1996, Los klippes y ventanas tectónicas preándicas de La Sierra de Pie de Palo (San Juan): edad e implicaciones tectónicas: XIII Congreso Geológico Argentino y III Congreso de Exploración de Hidrocarburos, Actas V, p. 377–391.

Rapela, C. W., Pankhurst, R. J., Baldo, E., and Saavedra, J., 1995, Cordierites in S-type granites: restites following low pressure, high degree partial melting of metapelites, *in* The origin of granites and related rocks, 3rd Hutton Symposium: U.S. Geological Survey Circular, 1129, p. 120–121.

Rapela, C. W., Heaman, L. M., and Nutti, J. C. M., 1982, Rb-Sr geochronology of granitoid rocks from the Pampean Ranges, Argentina: Journal of Geology, v. 90, p. 574–582.

Rapela, C. W., Coira, A., Toselli, A., and Saavedra, J., 1992, El magmatismo del Paleozoico Inferior en el Sudoeste de Gondwana, *in* Kay, S., and C. W. Rapela, eds., Plutonism from Antarctica to Alaska: Geological Society of America Special Paper 241, p. 67–76.

Rapela, C. W., and Pankhurst, R. J., 1996, The Cambrian plutonism of the Sierras de Córdoba: pre-Famatinian subduction? and crustal melting: XIII Congreso Geológico Argentino y III Congreso de Exploración de Hidrocarburos, Actas V, p. 491.

Rapela, C. W., Saavedra, J., Toselli, A., and Pellitero, E., 1996. Eventos magmáticos fuertemente peraluminosos en las Sierras Pampeanas: XIII Congreso Geológico Argentino y III Congreso de Exploración de Hidrocarburos, Actas V, p. 337–353.

Rapela, C. W., Pankhurst, R. J., Casquet, C., Baldo, E., Saavedra, J., Galindo, C., and Fanning, C. M., 1998, The Pampean Orogeny of the southern Proto-Andes: Cambrian continental collision in the Sierras de Córdoba, *in* Pankhurst, R. J., and Rapela, C. W., eds., The proto-Andean margin of Gondwana: Geological Society, London, Special Publication, v. 142, p. 181–218.

Sambridge, M. S., and Compston, W., 1994, Mixture modeling of multi-component datasets with application to ion-probe zircon ages: Earth and Planetary Science Letters, v. 128, p. 373–390.

Sato, A. M., Ortíz Suárez, A., Llambías, E. J., Cavarozzi, C. E., Sanchez, V., Varela, R., and Prozzi, C., 1996, Los plutones pre-Oclóyicos del sur de la Sierra de San Luis: Arco magmático al início del ciclo Famatiniano: XIII Congreso Geológico Argentino y III Congreso de Exploración de Hidrocarburos, Actas V, p. 259–272.

Secretaría de Minería, 1995, Mapa Geológico de la Provincia de Córdoba, República Argentina:, Cordoba, Argentina, Dirección Nacional del Servício Geológico, Secretaría de Minería, Escala 1:500,000.

Sims, J. P., Skirrow, R. G., Stuart-Smith, and P. G., Lyons, P., 1997, Informe geológico y metalogénico de las Sierras de San Luis y Comechingones (provincias de San Luis y Córdoba), 1:250 000: Buenos Aires, Instituto de Geología y Recursos Minerales, SEGEMAR, Anales 28.

Sims, J. P., Ireland, T. R., Camacho, A., Lyons, P., Pieters, P. E., Skirrow, R. G., Stuart-Smith, P. G., and Miró, R., 1998, U-Pb, Th-Pb and Ar-Ar geochronology from the southern Sierras Pampeanas, Argentina: implications for the Paleozoic tectonic evolution of the western Gondwana margin, *in* Pankhurst, R. J., and Rapela, C. W., eds., The proto-Andean margin of Gondwana: Geological Society, London, Special Publications, v. 142, p. 259–281.

Toselli, A. J., Dalla Salda, L., and Caminos, R., 1992, Evolución metamórfica del Paleozoico Inferior de Argentina, *in* Gutiérrez Marco, J. G., Saavedra, J., and Rábano, I., eds., Paleozoico Inferior de Ibero-América: Universidad de Extremadura, p. 279–309.

Toselli, A. J., Durand, F. R., Rossi de Toselli, J. N., and Saavedra, J., 1996, Esquema de evolución geotectónica y magmática eopaleozoica del Sistema de Famatina y sectores de Sierras Pampeanas: XIII Congreso Geológico Argentino y III Congreso de Exploración de Hidrocarburos, Actas V, p. 443–462.

Van der Voo, R., 1993, Paleomagnetism of the Atlantic, Tethys and Iapetus oceans: London: Cambridge University Press, 411 p.

Williams, I. S., 1992, Some observations on the use of zircon U-Pb geochronology in the study of granitic rocks: Transaction of the Royal Society of Edinburgh Earth Sciences, v. 83: p. 447–458.

Willner, 1990, División tectonometamórfica del basamento del noreste argentino, *in* Aceñolaza, F., Miller, H., and Toselli, A., eds., El Ciclo Pampeano en el Noreste Argentino: Serie Correlación Geológica, v. 4, p. 113–159.

Willner, A. P., and Miller, H., 1986, Structural division and evolution of the lower Paleozoic basement in the NW Argentine Andes: Zentralblat fur Geologie und Paläontologie, v. 1, p. 1245–1255.

Zeller, R. A., 1965, Stratigraphy of the big Hatchet mountains area, New Mexico: New Mexico Bureau of Mines Research Memoir 16, 128 p.

Zendejas, M. S. 1973, Exploración de Cobre Diseminado en el Proyecto Catalinas, del Distrito Minero de Cananea, Sonora [Tesis Profesional]: México City, Instituto Politécnico Nacional, Escuela Superior de Ingeniería y Arquitectura, 47 p.

Manuscript Accepted by the Society September 4, 1998

Geological Society of America
Special Paper 336
1999

Closing the ocean between the Precordillera terrane and Chilenia: Early Devonian ophiolite emplacement and deformation in the southwest Precordillera

J. Steven Davis* and Sarah M. Roeske
Department of Geology, University of California, One Shields Avenue, Davis, CA 95616
William C. McClelland
Department of Geology and Geological Engineering, University of Idaho, Moscow, Idaho 83844
Lawrence W. Snee
U.S. Geological Survey, Box 25046, MS 913, Federal Center, Denver, Colorado 80225

ABSTRACT

New geologic mapping, structural, and geochronologic analyses of the pre-Carboniferous ultramafic, mafic, and metasedimentary rocks exposed on the southwest margin of the Precordillera terrane, western Argentina, show that the ultramafic and mafic rocks belong to at least four distinct tectonic units, not one ophiolite pseudostratigraphy, as previously thought. One unit comprises gabbro, microgabbro, diabase, and minor plagiogranite that are interpreted as the mafic crustal section of an ophiolite pseudostratigraphy. Another unit consists of serpentinized peridotite, ultramafic cumulate, layered gabbro, and quartzofeldspathic gneiss that experienced granulite facies metamorphism in a deep continental crust environment. Other mafic rocks include highly altered basaltic flows interlayered with low-grade clastic metasedimentary rocks and diabase and microgabbro sills that intrude the metasedimentary rocks.

The pre-Carboniferous rock units were juxtaposed along synmetamorphic, ductile, top-to-the-east shear zones in the Early to Middle Devonian. The intense deformational fabrics formed during the juxtaposition event were overprinted by strong Permian folding and top-to-the-west brittle thrust faulting, Late Permian to middle Mesozoic extension, and Late Tertiary west-vergent folding and top-to-the-west brittle thrust faulting. A speculative tectonic scenario for the juxtaposition of the pre-Carboniferous in the Early to Middle Devonian places the deep continental crust at the base of the upper plate (eastern Chilenia) of a west-dipping subduction zone, the mafic ophiolite crustal section in an ocean basin between the western Precordillera terrane and eastern Chilenia, and the mafic sills and flows on the extended western margin of the Precordillera terrane. Early to Middle Devonian closure of the ocean basin resulted in juxtaposition of the pre-Carboniferous rock units.

*Present Address: (Davis) Exxon Mobil Upstream Research Co., P.O. Box 2189, Houston, TX 77252.

Davis, J. S., Roeske, S. M., McClelland, W. C., and Snee, L. W., 1999, Closing the ocean between the Precordillera terrane and Chilenia: Early Devonian ophiolite emplacement and deformation in the southwest Precordillera, *in* Ramos, V. A., and Keppie, J. D., eds., Laurentia-Gondwana Connections before Pangea: Boulder, Colorado, Geological Society of America Special Paper 336.

INTRODUCTION

Ramos et al. (1984, 1986) first noted that the Precordillera mountain range of western Argentina (Fig. 1) could be an exotic tectonostratigraphic terrane (sensu Howell et al., 1985). They pointed out that the Precordillera is fault bounded and has an early Paleozoic geology distinct from that of surrounding regions. Recent tectonic models for the Precordillera terrane agree that it is derived from eastern Laurentia (Dalla Salda et al., 1992a,b; Dalziel et al., 1994, 1996; Astini et al., 1995, 1996; Dalziel, 1997) and cite the strong Laurentian affinities of Cambrian to Lower Ordovician trilobite, brachiopod, and sponge faunas of the central and eastern Precordillera (Benedetto, 1993; Astini et al., 1995; Benedetto et al., 1995). In addition, basement xenoliths from Miocene volcanics on the Precordillera have Grenville ages (1.1 ± 0.1 Ga) and whole-rock Pb isotopic ratios with depleted signatures, a characteristic of the North American Grenville basement not seen elsewhere in South America (Abruzzi et al., 1993; Kay et al., 1996).

Two types of models have been proposed to explain the transfer of the Precordillera from Laurentia to Gondwana: rift-drift-accrete models and continental collisional models. Both types of model are based on the paleogeographic reconstruction of Dalziel (1991) in which the early Paleozoic Iapetus ocean was a narrow ocean between eastern Laurentia and western Gondwana (western South America). In both types of model the Precordillera is thought to have accreted to western Gondwana when the eastern margin of the Precordillera collided with an east-dipping subduction zone along western Gondwana in the Middle to Late Ordovician (Ramos et al., 1984, 1986, 1993; Dalla Salda et al., 1992a,b; Dalziel et al., 1994; Astini et al., 1995; Thomas and Astini, 1996; Dalziel, 1997). According to rift-drift-accrete transfer models (Ramos et al., 1984, 1986; Astini et al., 1995, 1996; Thomas and Astini, 1996), the Precordillera terrane rifted from Laurentia in the Early Cambrian (Astini et al., 1995, Thomas and Astini, 1996) and drifted across the Iapetus ocean to western Gondwana. In the collisional models (Dalla Salda et al., 1992a,b; Dalziel et al., 1994), the Iapetus ocean closed, resulting in Middle Ordovician collision between western Gondwana and eastern Laurentia (or a promontory of Laurentia) (Dalziel, 1997). In the collisional models, the Precordillera terrane was left attached to Gondwana when the two supercontinents rifted apart in the Late Ordovician.

A key component of these tectonic models is a belt of deformed mafic and ultramafic rocks that crop out along the western margin of the Precordillera range (Fig. 1). Haller and Ramos (1984) and Ramos et al. (1984) lump all of the mafic and ultramafic rocks in the western Precordillera into one ophiolite unit called the Famatinian ophiolite, assign it an Ordovician age, and relate it to subduction of an oceanic spreading ridge. All existing models for the early Paleozoic tectonics of the western Precordillera (e.g., Ramos et al., 1984, 1986; Dalla Salda et al., 1992a,b; Dalziel et al., 1994; Astini et al., 1995; Dalziel, 1997) incorporate this concept of a Late Ordovician ophiolite. The ophiolite may represent the ocean basin that opened west of the

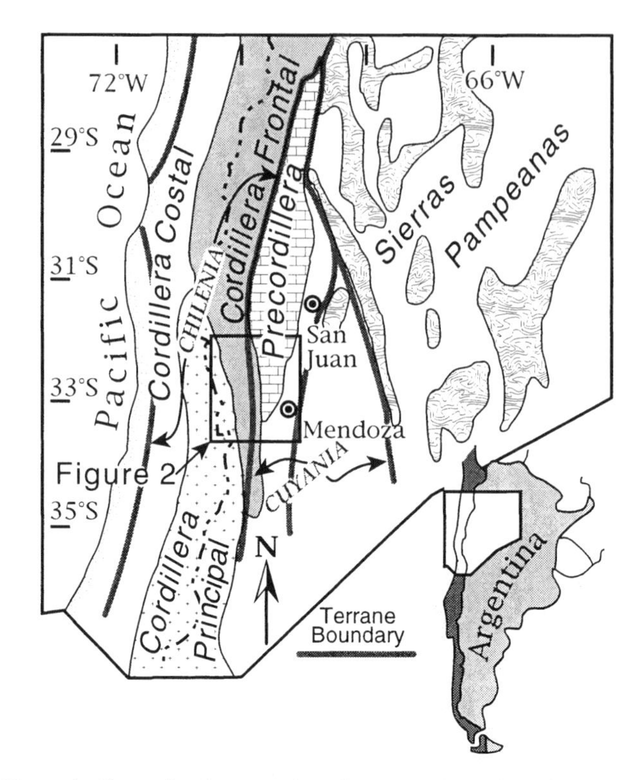

Figure 1. Generalized geography of western Argentina after Ramos (1988). Terrane boundaries after Astini et al. (1996).

Precordillera as it rifted and drifted away from Laurentia (Astini et al., 1995, 1996; Thomas and Astini, 1996) and that closed in the Middle Devonian when another exotic terrane, Chilenia, collided with the Precordillera (Ramos et al., 1984, 1986). This ocean basin was consumed either by east-dipping subduction under the western margin of the Precordillera (Ramos et al., 1984, 1986), or by west-dipping subduction under eastern Chilenia (Astini et al., 1995). In the collisional models, the belt of mafic rocks represents either a small interior rift basin that formed as Laurentia rifted away from Gondwana in the late Ordovician, leaving the Occidentalia terrane (composed of Chilenia and the Precordillera terranes) behind (Dalla Salda et al., 1992a) or the ocean that opened between Laurentia and Gondwana as they rifted apart in the Late Ordovician (Dalziel et al., 1994). An alternate model was proposed by Dalziel (1997), who suggested that a promontory of Laurentia collided with western Gondwana in the Ordovician. In this model the Precordillera terrane occupied the tip of the promontory and the ultramafic and mafic rocks formed in a failed oceanic rift west of the Precordillera terrane (cf. to the Malvinas Plateau) (Dalziel, 1997). In all of these models, the mafic and ultramafic rocks were emplaced in an east-dipping subduction zone following Ramos et al. (1984, 1986).

The southwest Precordillera contains the best exposed and most complete sections of mafic and ultramafic rocks in the western Precordillera. Despite the excellent exposures, little work has been done on these rocks. As a result, the relationships of these mafic and ultramafic units to each other, as well as to other mafic rocks on the western margin of the Precordillera, have remained entirely specu-

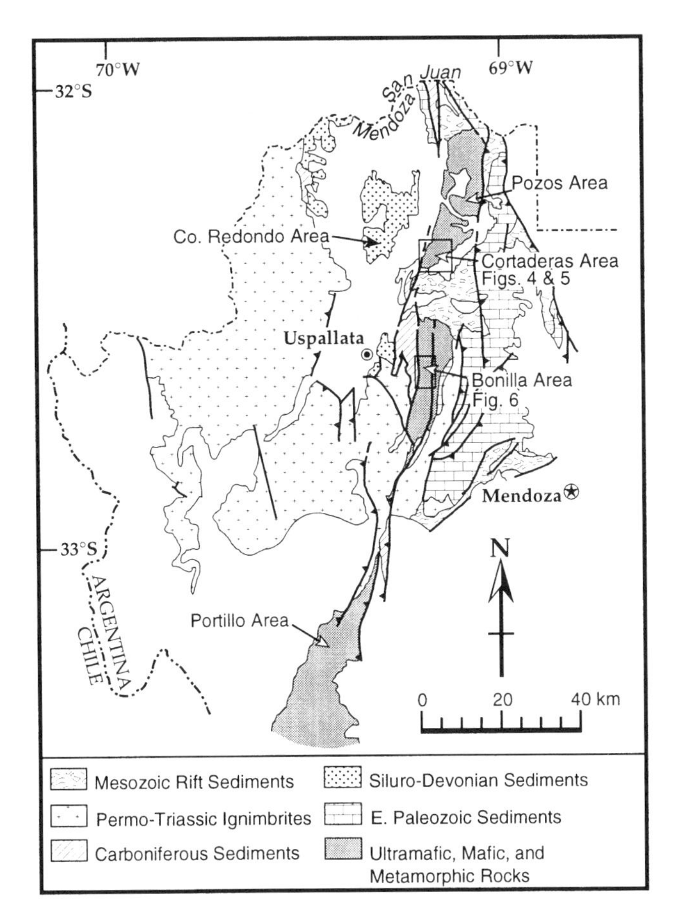

Figure 2. Generalized geology and locations of areas discussed in the southern Precordillera and southern Cordillera Frontal. Note location of boundary between the provinces of Mendoza and San Juan. Geology after Caminos et al. (1993)

lative. Through detailed geologic mapping, structural analysis, and geochronology, this chapter addresses the structural and genetic relationships between the mafic, ultramafic, metasedimentary and metavolcanic rocks exposed in the Cortaderas and Bonilla areas of the southwestern Precordillera (Fig. 2), as well as the deformational and tectonic histories of the southwest Precordillera. We show that there are four independent rock units that contain mafic to ultramafic rocks. These include an ultramafic and layered gabbro complex that has experienced granulite facies metamorphism, an upper crustal mafic ophiolite section, a metasediment/metavolcanic complex that contains mafic flows, and a carbonate-metasiltstone unit that, along with the metasediment/metavolcanic complex, contains fine-grained mafic sills. These units were tectonically juxtaposed in the Early to Middle Devonian along ductile, top-to-the-east shear zones that we suggest were active in a west-dipping subduction zone under eastern Chilenia.

GENERAL STRATIGRAPHY AND STRUCTURE OF THE PRECORDILLERA

An outstanding feature of the Precordillera are the thick, Cambrian to Lower Ordovician shelf carbonate sequences of the central and eastern Precordillera. These shelf sequences host the Laurentian trilobite and brachiopod faunas that define the Precordillera as an allochthonous terrane. In contrast, the upper Lower Ordovician to Carboniferous in the central and eastern Precordillera is dominated by siliciclastic sedimentary rocks. In the western Precordillera, lower Paleozoic sedimentary rocks are restricted to the upper Lower to Upper Ordovician, and are interpreted as outer shelf and slope facies siliciclastics (Keller, 1995). Silurian to Devonian sedimentary rocks have been documented only from a few localities in the western Precordillera (Astini, 1996). The Upper Ordovician slope facies rocks include turbidite units that contain both mafic volcanic flows and mafic and ultramafic sills (Haller and Ramos, 1984). Carboniferous glaciogenic to marine and continental sedimentary rocks overlie the lower Paleozoic section with a pronounced angular unconformity in the western and eastern Precordillera. These in turn are unconformably overlain by a variety of Permian to Tertiary marine and continental sedimentary rocks as well as by the voluminous Permo-Triassic Choiyoi volcanics, which are found only in the western Precordillera and farther west (Figs. 1 and 2). Other than the Choiyoi volcanics, igneous rocks are scarce in the Precordillera, being limited to a few small Carboniferous plutons and scattered centers of Late Tertiary volcanism and shallow-level intrusions.

The Precordillera records a long history of deformation, starting with pre-Carboniferous compressional deformation demonstrated by the angular unconformity discussed above. Carboniferous and younger deformations include a compressional event in the late Paleozoic, extension in the early to middle Mesozoic, and a major Late Tertiary to Recent compressional episode (see reviews by Ramos, 1988; Mpodozis and Ramos, 1989). The modern Precordillera range is the topographic expression of an Andean-age (Late Tertiary to Recent) east-vergent fold and thrust belt. In addition to the well-documented Late Tertiary east-vergent structures, west-directed back thrusts of similar age have been found in the central and western Precordillera (Fig. 2) (von Gosen, 1992).

ROCK UNITS OF THE SOUTHWEST PRECORDILLERA

Previous work

A wide variety of nomenclature schemes have been proposed for the pre-Carboniferous ultramafic, mafic, and metasedimentary rocks exposed in the southwest Precordillera (e.g., Keidel, 1939; de Romer, 1964; Harrington, 1971; Cucchi, 1972; Varela, 1973). Figure 3 provides a comparison of the previous rock unit nomenclature in the Cortaderas and Bonilla areas with the rock units defined by this study, based on our detailed mapping in these two areas. The variety of nomenclatural schemes reflects the lack of control on the depositional ages of the metasedimentary units, as well as the intense deformation the areas have experienced. In the Cortaderas area, the sedimentary ages of the rocks in the two facies of the Villavicencio Group proposed by Harrington (1971) (Fig. 3) were estimated to be Devonian based on a tie to the Devonian Villavicencio Formation. North of the Cortaderas and Pozos

A

Cortaderas Area

Harrington (1971)	Cucchi (1972)	This Study
Villavicencio Grp. 'Normal Facies'	Villavicencio Fm.	Metasediment/Metavolcanic Complex
Villavicencio Grp. Alojamiento Facies	Alojamiento Fm.	??
Villavicencio Grp. Cortaderas Facies	Upper Member, Cortaderas Fm.	Carbonate Metasiltstone
	Lower Member, Cortaderas Fm.	Metasediment/Metavolcanic Complex

B

Bonilla Area

Keidel (1939)	de Roemer (1968)	von Gosen (1995)	This Study
Manto de Jagüelito	Upper Unit	Bonilla Group	Metasediment/Metavolcanic Complex
Manto de Buitre			??
Bonilla Unit	Lower Unit		Metasediment/Metavolcanic Complex
			Carbonate Metasiltstone
Farallones Unit			Metasediment/Metavolcanic Complex

Figure 3. Map unit correlations between metasedimentary units discussed by previous authors and this study.

areas (Fig. 2), Cuerda et al. (1987) found poorly preserved Early to Late Ordovician graptolites in rocks they correlate to the lower Cortaderas Formation of Cucchi (1972) (Fig. 3). Cuerda et al. (1987) and Kury (1993) now regard the Devonian Villavicencio Formation (sensu strictu) as unrelated to the Cortaderas and Alojamiento Facies of Harrington (1971). In the Bonilla area (Fig. 2), Keidel (1939), and later authors, assigned the low-grade metasedimentary rocks a Late Precambrian to lower Paleozoic age, but there are neither strong regional correlations nor any fossils on which to base that age estimate.

Haller and Ramos (1984) considered all of the mafic and ultramafic rocks in the Cortaderas and Bonilla areas to be related and were the first to suggest that they belonged to an ophiolite pseudostratigraphic sequence. Trace element geochemical analyses of some of these mafic rocks (Haller and Ramos, 1984; Kay et al., 1984) plot in the Enriched–Mid-Ocean Ridge Basalt (E-MORB)/within-plate basalt field on the Th/2-Ta-Hf discrimination diagram of Wood et al. (1979), and confirm their oceanic character. Based on the geochemistry alone, Haller and Ramos (1984) correlated the mafic rocks of the southwest Precordillera to mafic flows and sills in Late Ordovician sedimentary rocks on trend over 100 km to the north. They defined this suite of mafic and ultramafic rocks along the western Precordillera as the Late Ordovician Famatinian ophiolite. Detailed lithologic mapping in the Cortaderas area by Dias and de Tonel (1987) allowed them to delineate an almost complete pseudostratigraphic sequence of ophiolitic rocks, from basal serpentinized ultramafic mantle rocks up through diabasic rocks. However, they did not discuss the ordering in which the ophiolitic rocks were found, nor their structural disposition with respect to each other or the underlying metasediments. No published studies have documented the crystallization ages of the mafic and ultramafic rocks.

This study

The results of our detailed mapping allow us to group the pre-Carboniferous rocks in the Cortaderas and Bonilla areas into four basic units: an ultramafic and layered gabbro complex, a gabbro and diabase complex, a carbonate-metasiltstone unit, and a metasediment/metavolcanic complex (Figs. 4–6). Each of these units is structurally separated by faults or shear zones from the others, and there is no indication that they were ever in a primary igneous or depositional relationship with one another. Mafic sills intrude the two metasedimentary units but do not cross-cut any of the contacts between the units. Because all mapped contacts are tectonic, and because of internal deformation, the unit thicknesses are unknown. Despite pervasive deformation, the units do display a consistent structural sequence and are not juxtaposed in a chaotic (mélange-like) manner. With a few exceptions in the Cortaderas area (Fig. 4) the mafic and

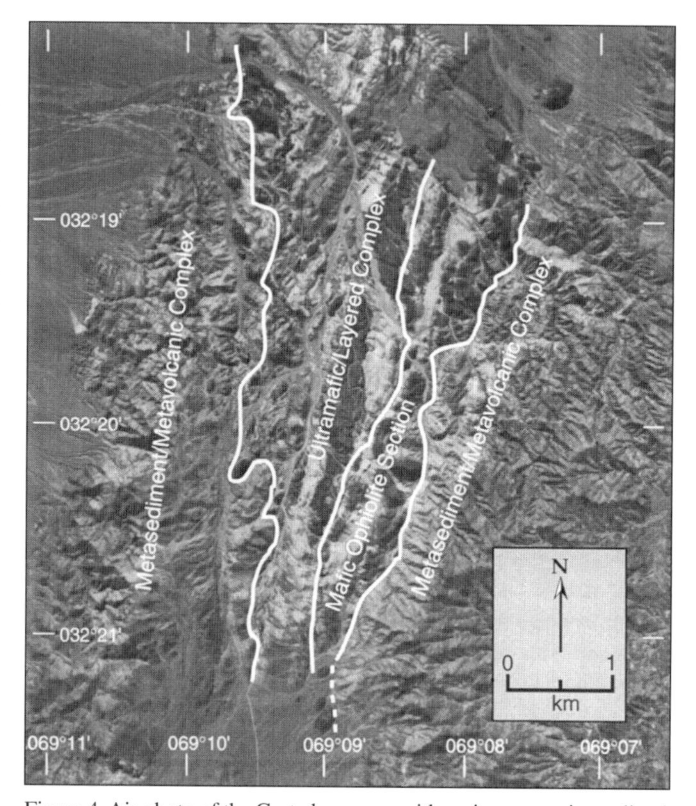

Figure 4. Air photo of the Cortaderas area with major map units outlined. Note that the carbonate metasiltstone unit, which is light gray on the airphoto, is included within the ultramafic/layered complex and mafic ophiolite section, and is not outlined. The ultramafic/layered complex, mafic ophiolite section, and carbonate-metasiltstone unit comprise a central belt flanked by eastern and western belts of the metasediment/metavolcanic complex. See Figure 2 for location. Scale is 1:41,667.

ultramafic rocks are the highest structural units and are everywhere underlain by the carbonate-metasiltstone unit and metasediment/metavolcanic complex.

Our initial field studies in the Cortaderas area indicated that the ultramafic and layered gabbro complex, in combination with the gabbro and diabase complex, appeared to form an in-order, near-complete ophiolite pseudostratigraphy (Davis et al., 1994, 1995a,b). More recent field observations indicate that the apparent ophiolite pseudostratigraphy likely consists of two distinct igneous units that are tectonically juxtaposed. One section consists of serpentinized peridotites, ultramafic cumulates, and layered gabbros, which we refer to as the ultramafic/layered complex. The ultramafic/layered complex has experienced granulite facies metamorphism and locally intrudes quartzofeldspathic gneiss. We now interpret this complex as deep continental crust. The other section, referred to as the mafic ophiolite section, consists of bodies of coarse-grained gabbros to microgabbros and diabase. The mafic ophiolite section has attributes of the upper crustal section of an ophiolite. We have found no evidence of a link between the ultramafic/layered complex and the mafic ophiolite section; they are tectonically juxtaposed.

In both the Cortaderas and Bonilla areas, the map units crop out with similar lateral distributions. Essentially, the ultramafic/layered complex, mafic ophiolite section, and underlying carbonate-metasiltstone unit form a central belt that is bounded on the eastern and western sides by belts of the underlying metasediment/metavolcanic complex (Figs. 4 and 5). These belts trend northerly, ranging in width from hundreds of meters to kilometers. We report preliminary petrographic and field descriptions in this chapter. The unit descriptions are presented in order from structurally lowest to structurally highest.

Metasediment/metavolcanic complex. Thin- to medium-bedded metasandstones and metasiltstones with interlayered intermediate to mafic metavolcanic flows and tuffs are the most widely exposed rock types in the Cortaderas and Bonilla areas. The gray-green to orange metasiltstones have a distinct phyllitic sheen, which is most strongly developed in proximity to the ultramafic/layered complex. The dominant rock type is fine-grained metasandstone, although portions of this unit contain significant amounts of metasiltstones. The sedimentary protoliths of this unit were medium- to fine-grained sandstones, siltstones, and rare chert. Compositions of the metaclastic rocks range from quartz-rich to graywacke, but the predominant rock types are intermediate between these. Monomineralic detrital grains commonly include quartz, plagioclase, alkali feldspar, and white mica, along with minor to trace amounts of chlorite, apatite, and zircon. Lithic particles primarily consist of fine-grained siliciclastic sediments such as mudstone and siltstone, but include minor amounts of phyllite. The cherts are recrystallized and very fine grained. Carbonate minerals are found only as vein fillings and secondary cements in this unit. Because the sedimentary structures are strongly overprinted by tectonic deformation, we were not able to determine original depositional topping directions. However, despite the deformational overprint these rocks partially retain their clastic texture in hand sample and thin section.

Intermediate to mafic tuffs and flows, most abundant near the structural top of the section, are interlayered with the metasedimentary rocks. The flows have thicknesses in the meter range. Although the tuffs and flows experienced very low-grade metamorphism, they locally contain relict phenocrysts of clinopyroxene. Relict igneous textures include feldspar phenocrysts, vesicles (filled with calcite and pumpellyite), and crude pilotaxitic and porphyritic-aphanitic textures in the mafic flows.

As the metasediment/metavolcanic complex has yielded no fossils, its depositional age is unknown. The metasediment/metavolcanic complex correlates with the Lower Cortaderas Facies of Cucchi (1972) (Fig. 3). Cuerda et al. (1987) assigned the Lower Cortaderas Facies to an age of Middle to Late Ordovician, based on a correlation to similar appearing metasedimentary rocks north of the Pozos area (Fig. 2).

Carbonate-metasiltstone unit. The carbonate-metasiltstone unit crops out extensively in the Cortaderas area, but less so in the Bonilla area. The unit consists of light colored, slaty, and phyllitic to finely schistose quartz-rich metaclastic rocks with scattered thin (<1 m thick) micritic to recrystallized fine-grained dolomitic limestone layers. A strong layer-parallel foliation has almost completely overprinted the original sedimentary structures and layering. Where preserved, sedimentary layering ranges in thickness from a few centimeters to tens of centimeters-scale, and is characterized by thin layers of very fine-grained metasandstone in a dominantly slate to phyllite matrix. The metaclastic rocks consist almost entirely of very fine-grained recrystallized and clastic quartz and neocrystallized fine-grained white mica. The predominant protolith for this unit was quartz-rich siltstone. Relict detrital white mica, sand-sized quartz grains, and rare relict plagioclase grains are variably present. These metasediments in some places contain recrystallized calcite that occurs as a cement or matrix, suggesting a marl or calcareous argillite protolith.

Three poorly preserved microfossils from dolomitic limestone layers were identified by workers at the Branch of Paleontology and Stratigraphy, U.S. Geological Survey, Reston. A sample from the Cortaderas area yielded a protoconodont with an age range of uppermost Precambrian or lowermost Cambrian to Lower Ordovician (J. Repetski, personal communication, 1995). Another Cortaderas sample yielded a zooecial lining of a phosphatized bryozoan that could be as old as Lower Ordovician, but is more likely Middle Ordovician to Devonian (J. Repetski, personal communication, 1994). A Bonilla area sample yielded a conodont of probable Ordovician, possibly Early Ordovician, age (J. Repetski, personal communication, 1994). The age ranges for all three microfossils overlap in the Early Ordovician, which strongly suggests that is the depositional age of the sediments. However, both protoconodonts and phosphatic bryozoans are relatively rare in Early Ordovician rocks, thus a Middle to Upper Ordovician age may be more appropriate if it is assumed that the protoconodont represents a reworked fossil.

Mafic sills. Fine-grained mafic sills occur in the metasedimentary and metavolcanic unit in the Cortaderas area and in the carbonate-metasiltstone unit in the Pozos area, immediately north

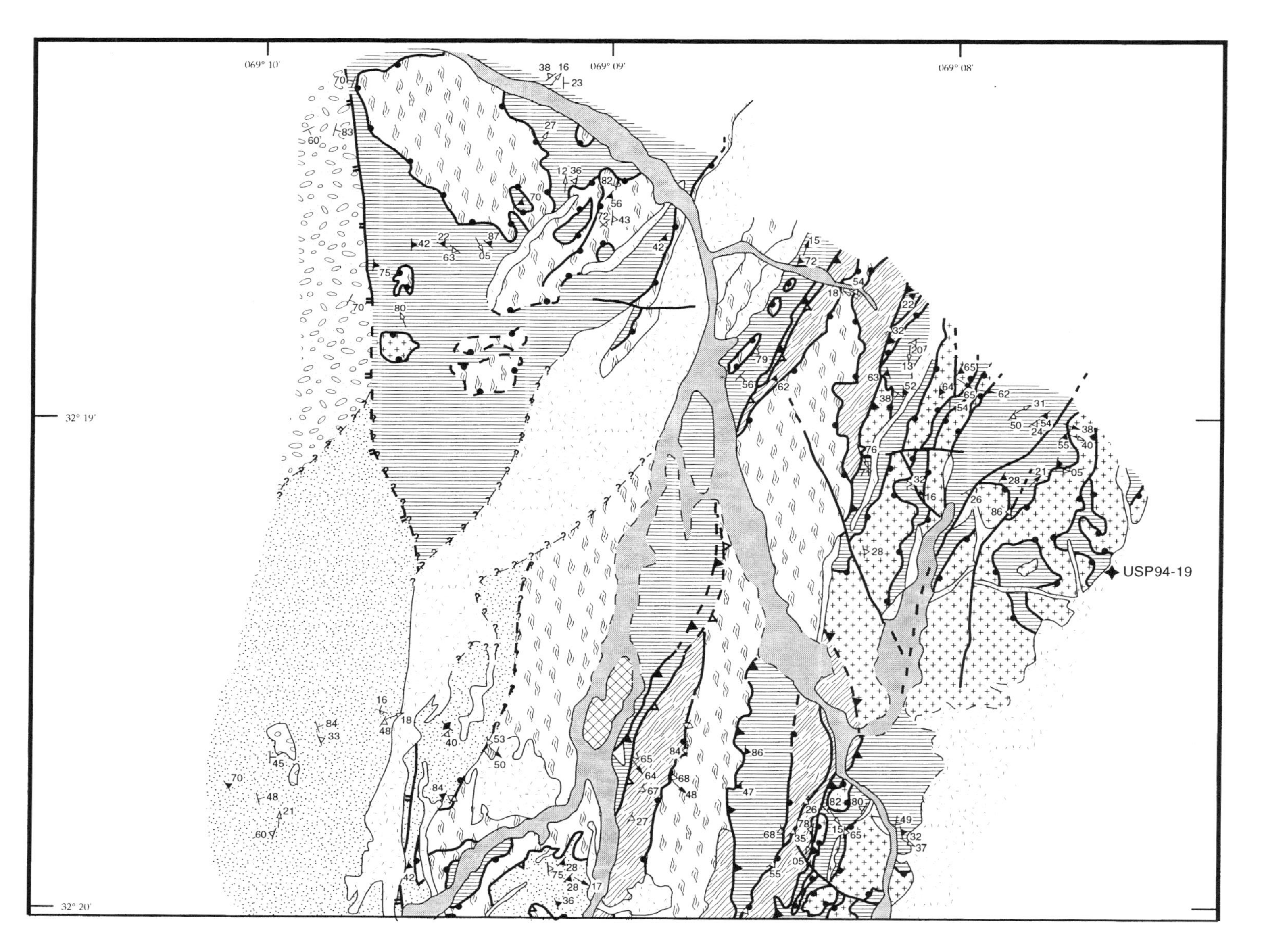

-Figure 5 (on this and facing page). Simplified geologic map of the Cortaderas area. Original mapping was done on aerial photos at a scale of 1:28,100. See Figure 2 for location.

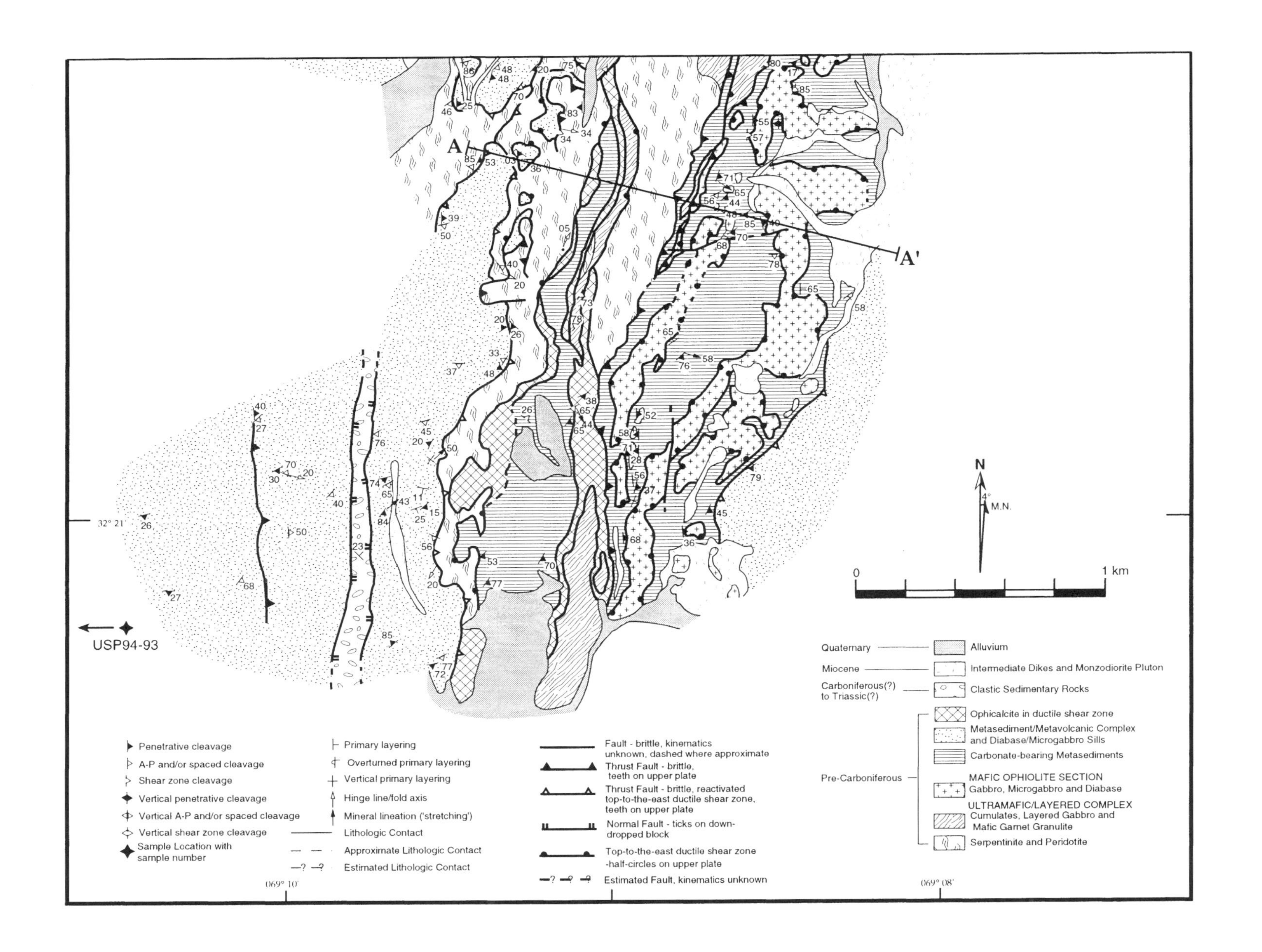

A
A'
N
M.N.
4°
0
1 km
32° 21'
069° 10'
069° 08'
USP94-93
Quaternary
Alluvium
Miocene
Intermediate Dikes and Monzodiorite Pluton
Carboniferous(?) to Triassic(?)
Clastic Sedimentary Rocks
Pre-Carboniferous
Ophicalcite in ductile shear zone
Metasediment/Metavolcanic Complex and Diabase/Microgabbro Sills
Carbonate-bearing Metasediments
MAFIC OPHIOLITE SECTION
Gabbro, Microgabbro and Diabase
ULTRAMAFIC/LAYERED COMPLEX
Cumulates, Layered Gabbro and Mafic Garnet Granulite
Serpentinite and Peridotite
Penetrative cleavage
A-P and/or spaced cleavage
Shear zone cleavage
Vertical penetrative cleavage
Vertical A-P and/or spaced cleavage
Vertical shear zone cleavage
Sample Location with sample number
Primary layering
Overturned primary layering
Vertical primary layering
Hinge line/fold axis
Mineral lineation ('stretching')
Lithologic Contact
Approximate Lithologic Contact
Estimated Lithologic Contact
Fault - brittle, kinematics unknown, dashed where approximate
Thrust Fault - brittle, teeth on upper plate
Thrust Fault - brittle, reactivated top-to-the-east ductile shear zone, teeth on upper plate
Normal Fault - ticks on down-dropped block
Top-to-the-east ductile shear zone -half-circles on upper plate
Estimated Fault, kinematics unknown

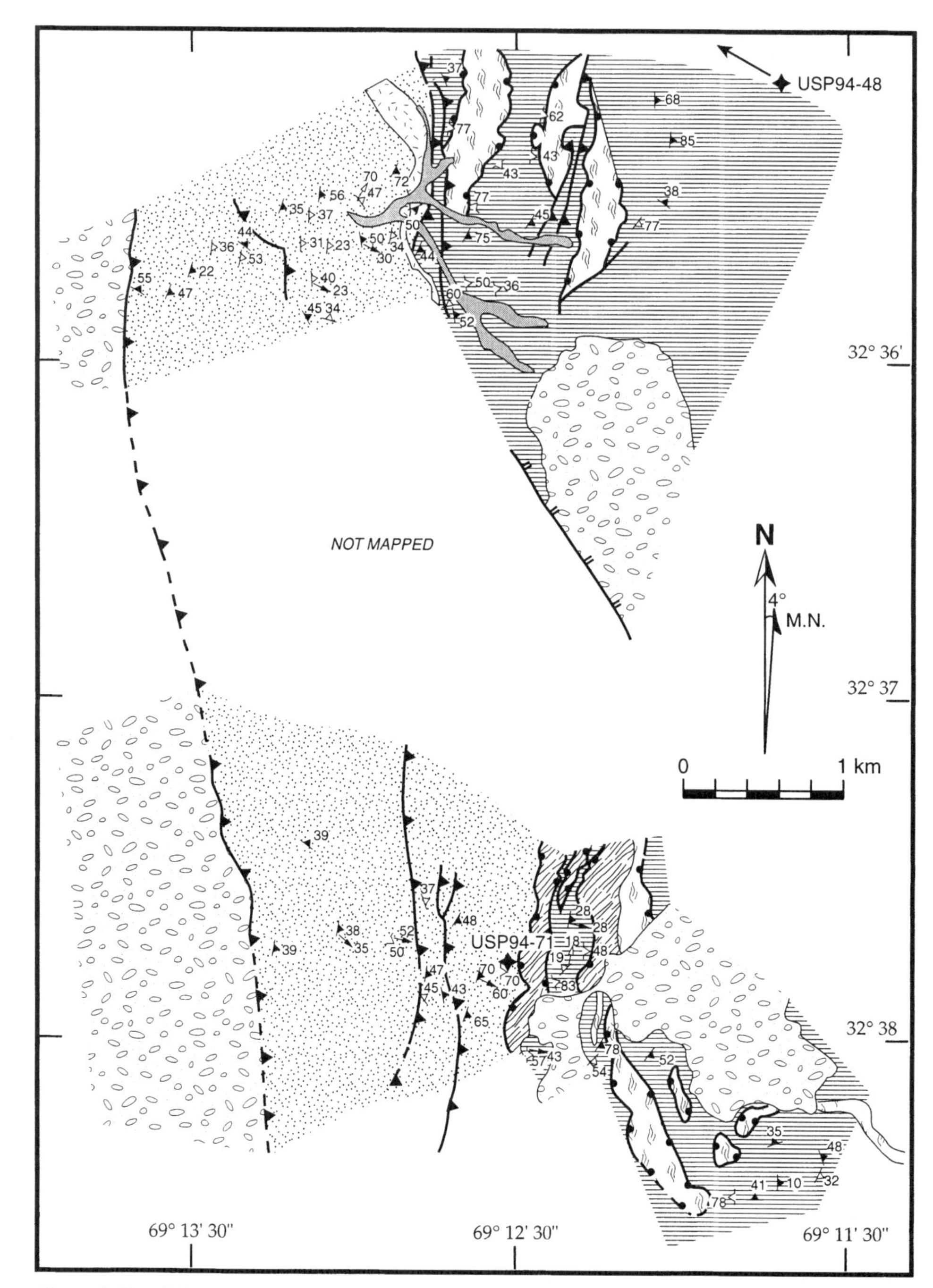

Figure 6. Simplified geologic map of the Bonilla area. Refer to Figure 4 for map unit patterns. Original mapping was done on aerial photos at a scale of 1:42,400. See Figure 2 for location.

of the Cortaderas. The sills in the Cortaderas area are concordant to the lithologic layering and range in thickness from one to tens of meters. They are most common in the western belt of the metasedimentary and metavolcanic unit, increasing in abundance to the west and down section. Sills are very rare in the eastern belt of the metasedimentary and metavolcanic unit.

Metasedimentary rocks in immediate contact with the sills locally exhibit the effects of contact metamorphism, particularly in the carbonate-metasiltstone unit in the Pozos region. Metasiltstones within 2–4 m of some of the mafic sills have relict bedding subparallel to the sill margin and the siltstones are hard and massive. Metasediments from within a meter of one the sills preserve radiating sprays of biotite. In other samples neocrystallized white mica cross-cuts the contact metamorphic textures, indicating that the contact metamorphism occurred prior to the regional low-grade metamorphism and deformation. In thin section the sills display moderately well-preserved clinopyroxene, moderately to poorly preserved plagioclase, and poorly preserved olivine and orthopyroxene. Relict igneous features and textures include locally preserved chilled margins, columnar jointing, and intergranular to subgranular textures.

Ultramafic/layered complex. The ultramafic/layered complex crops out extensively in the central Cortaderas (Fig. 5) where it occurs in two structural sheets. In general, the ultramafic rocks are highly serpentinized and few relict bodies of ultramafic rocks remain. From west to east, each structural sheet of the ultramafic/layered complex is composed of serpentinized ultramafic rocks, ultramafic cumulates, and layered gabbros. Despite the extensive serpentinization, we are able to distinguish a variety of ultramafic rocks by identifying relict phases (predominantly relict clinopyroxene, chromite, and spinel, along with minor amounts of relict olivine and orthopyroxene) and serpentine replacement textures in thin section as well as in the field. Rock types identified include wehrlite, harzburgite, lherzolite websterite, and small amounts of dunite in veins cross-cutting wehrlite. In rare instances, relict high-temperature mantle tectonite textures are visible in the ultramafics, but the extensive serpentinization has obliterated such textures almost completely.

Moderately to very well-preserved bodies of coarsely crystalline (0.5–1 cm) ultramafic cumulates crop out along the eastern edge of the serpentinite exposures. Field relationships show that they form the base of the layered gabbro section, and grade downward into the serpentinized ultramafic ophiolite section. This basal cumulate section ranges from 30 to 100 m thick, but it is everywhere bounded by at least one tectonic contact. Most of the ultramafic cumulate rocks have equigranular adcumulate textures, and are composed of subhedral grains of undeformed clinopyroxene and hornblende. In some samples the hornblende is pargasite and is a primary igneous phase. In other samples hornblende occurs as a high-temperature isomorphic replacement of clinopyroxene. A small proportion of the cumulates are composed of equant, subhedral clinopyroxene, minor subhedral to anhedral orthopyroxene, and rare olivine. The olivine and orthopyroxene are almost completely serpentinized. Thin dikes and sills(?) of altered clinopyroxenite pegmatites (crystal size as large as 7 cm) cross-cut the ultramafic cumulates and layered gabbros.

The layered gabbros constitute the uppermost section of the ultramafic/layered complex. Their lower contact with the ultramafic cumulate section is intrusive, but the upper contact everywhere is tectonic. Layered gabbro thicknesses range from a few to 150 m. These rocks consist of alternating, plagioclase and clinopyroxene-rich layers 1–10 cm thick. The original igneous mineralogy of these layers is commonly replaced by very low-grade metamorphic minerals (albite, clinozoisite, chlorite). Rare relict olivine is present in clinopyroxene-rich layers. Most of the layers are modally graded but a few are isomodal (see Nicolas, 1989). The clinopyroxene commonly is elongate lozenge-shaped with wispy tails, implying high temperature solid-state deformation. Numerous thin (centimeter-scale), fine- to medium-grained clinopyroxenite dikes and sills intrude the layered gabbro. Preliminary major and trace element geochemical data (S. M. Roeske, unpublished data; S. M. Kay, unpublished data) are interpreted by Davis (1997) to indicate that the layered gabbro is of calc-alkaline nature and is not related to the ultramafic rocks.

High-grade metamorphism of the layered gabbro complex has changed the mineral assemblage to a mafic garnet granulite, most noticeably in the more feldspathic layers. The granulite consists of medium-sized garnet (2–5 mm) and fine clinopyroxene; both are variably preserved and often partially or completely replaced by chlorite. The remaining mineralogy is very fine-grained albite, clinozoisite, and pumpellyite, which appear to have replaced plagioclase. Microprobe data of the relict garnet and clinopyroxene show that the garnets are 33–45% pyrope and that the clinopyroxenes have high Na and Al^{vi} compared to igneous clinopyroxene (Davis et al., 1995b). The mineral chemistry suggests metamorphic T and P conditions of approximately 850°–1000°C and a minimum of 9 kb (S. M. Roeske, unpublished data). Field observation of a thin (<5 cm), undeformed mafic garnet granulite sill within the basal ultramafic cumulate section suggests not only that the layered gabbro postdates the ultramafic rocks, but that the entire cumulate section was metamorphosed under high P-T conditions, probably under postsolidus static recrystallization.

Field evidence a few kilometers north of the Cortaderas region provides an explanation for the unusual P-T setting for the layered gabbro complex. In the region between the Pozos and Cortaderas areas (Fig. 2), the layered gabbro intrudes quartzofeldspathic gneiss, in addition to intruding and grading into ultramafic rocks. The quartzofeldspathic gneiss ranges in mineralogy from almost pure quartzite to quartz + plagioclase + garnet gneiss. Relict plagioclase and garnet compositions indicate amphibolite or higher grade metamorphic conditions occurred. The compositional banding in the layered gabbro is subparallel to obliquely cross-cutting the metamorphic layering in the quartzofeldspathic gneiss, indicating that no high-T penetrative deformation of the layered gabbro occurred after it intruded the quartzofeldspathic gneiss. This field evidence supports our interpretation that the high-grade metamorphism in the layered gabbro occurred during postsolidus static recrystallization and the P-T conditions of the mafic garnet granulite reflect the depth at which the layered gabbro intruded.

Mafic ophiolite section. The mafic ophiolite section crops out east of the ultramafic/layered complex (Figs. 4 and 6) in north-striking elongate bodies that range from a few meters to more than a kilometer in length and few to hundreds of meters in thickness. The mafic rocks are dominantly microgabbro and fine-grained hypabyssal diabase with subordinate coarse-grained gabbro (crystals as large as 0.5–1 cm). The mafic ophiolite section has experienced very low-grade to low-grade metamorphism, such that relict clinopyroxene and minor relict plagioclase are the only remaining igneous phases. Magma flow fabrics, defined by relict textures of aligned igneous minerals, are developed in some areas. However, overall the fabrics are quite homogenous and ophitic to subophitic. Rare, small irregular dikes or sills of plagiogranite are found within the gabbro, but are too small to map at this scale. Contacts between the coarse gabbro, microgabbro, and diabase are gradational over distances of less than a meter but in some instances are sharp. In the diabase and microgabbro, columnar joints are locally present as are poorly preserved chilled margins on half dikes in a few locations. Basaltic pillow lavas are uncommon but are exposed in the northeast corner of the Cortaderas area where they form screens between dikes at the top of what we interpret as a sheeted dike complex. Preliminary trace and rare earth element (REE) analyses of four samples from the mafic ophiolite section indicate an E-MORB or within-plate origin (Davis et al., 1995a,b).

The rocks mapped as mafic ophiolite section in the Cortaderas area (Fig. 4) probably include some mafic sills intruded into the carbonate-metasiltstone unit. Because all of the mafic/metasedimentary contacts are ductilely or brittly deformed in the Cortaderas area, determining how many of the original contacts are tectonic or intrusive is difficult. The mafic ophiolite section, sensu stricto, contains features typical of the upper crustal portion of an ophiolite pseudostratigraphy, such as massive coarse gabbro and microgabbro, small plagiogranite intrusions, diabase, and diabase half-dikes with pillow screens. However, homogeneous diabase bodies that are sills are texturally indistinguishable from the diabase bodies that are part of the mafic ophiolite section. Thus, only where contact metamorphic textures or chilled margins are preserved can we be confident that they are sills and not part of the mafic ophiolite section. Almost none of the diabase bodies in the Cortaderas area meet this criteria. The only solid evidence that some of the mafic bodies in the Cortaderas area might be intrusive into the metasediments is one location where a small microgabbro body contains a block of the carbonate-metasiltstone within it as a xenolith. The metasediment block shows primary layering and is not in contact with the other metasediments.

Other rock units. Most of the other rock units are demonstrably younger than those discussed above based on cross-cutting or depositional relationships. Rusty orange ophicalcite is extensively developed along the contacts between the carbonate-metasiltstone unit and the ultramafic/layered complex, as well as between the serpentine and mafic garnet granulite (Fig. 4). The ophicalcite consists of pervasively deformed serpentinite and mafic garnet granulite with a variably developed calcitic to dolomitic matrix and extensive dolomite and calcite veining.

In the Cortaderas area, younger sedimentary rocks consist primarily of coarse clastics varying from quartzites to conglomerates of the Carboniferous Jagüel Formation (Harrington, 1971) in the northwestern part of the area, and rift-related conglomerates and volcaniclastics of the Upper Triassic Cacheuta Group (Harrington, 1971) in the southwestern part (Fig. 4). In the Bonilla area, the sedimentary rocks consist of glacial, fluvial and marine conglomerates, sandstones and shales of the Carboniferous Santa Elena Formation (Keidel, 1939) (Fig. 5).

Younger igneous rocks in the Cortaderas area consist of Lower to Middle Miocene hornblende andesite plutons and dikes that crop out throughout the area, and a Lower Miocene monzodiorite pluton that crops out in the eastern part of the area (Fig. 4). In general, the andesites are quite fresh, but the monzodiorite is variably altered. The Bonilla area has similar Miocene andesite intrusions, as well as tuffs, in addition to a very thick sections of the Permo-Triassic ignimbrites and flows of the Choiyoi volcanics (Fig. 2).

Low-grade metamorphism

In contrast to the Carboniferous and younger rock units, all of the pre-Carboniferous rock units discussed above manifest the effects of a single low-grade metamorphism. In the Cortaderas, area low-grade regional metamorphism has partially replaced the igneous and high-temperature assemblages in the mafic rocks, layered gabbros, and mafic garnet granulites, with albite + pumpellyite + clinozoisite + chlorite + titanite + rutile ± white mica ± quartz. Although similar low-grade metamorphism has affected the mafic sills in the metasediment/metavolcanic complex, primary igneous phases and textures are better preserved. In the Bonilla area, relict igneous phases are extremely rare and chlorite + actinolite + calcite + albite + quartz + titanite is the ubiquitous assemblage in the mafic rocks of the ultramafic/layered complex and mafic ophiolite section, as well as the mafic sills in the metasediment/metavolcanic complex. In both the Cortaderas and Bonilla areas, the textural grade (and metamorphic grade?) in the metasediment/metavolcanic complex in the eastern and western belts increases toward the structural contact with the central belt. There is also a slight increase in textural grade from east to west in the carbonate-metasiltstone unit, from slaty to phyllitic. The low-grade metamorphism appears to lie in the prehnite-pumpellyite facies to lowest greenschist facies in the Cortaderas area, and in the lower to middle greenschist facies in the Bonilla area. The CAI of the conodont from the Bonilla area is 5.5, indicating that the rocks were heated to 300°–350°C, a finding consistent with the estimated metamorphic grade.

DEFORMATION

This study documents three temporally distinct deformations, which are discussed below from oldest to youngest. These deformations include a pre-Carboniferous, synmetamorphic top-to-the-east deformation, a Mesozoic brittle extensional deformation, and a Late Tertiary brittle, top-to-the-west deformation. In addition, a fourth deformation was documented by von Gosen (1995), who concluded that the unmetamorphosed Carboniferous rocks, as well as the unconformably underlying ultramafic, mafic, metasedimenary and metavolcanic rocks in the Bonilla area were affected by intense Permian compressional deformation. Our reconnaissance of the Carboniferous section in the Bonilla area and the detailed work of von Gosen (1995) show that the strongly deformed Carboniferous rocks are unconformably overlain by gently folded Permo-Triassic Choiyoi volcanics. A consequence of the Permian folding and top-to-the-west thrust faulting (e.g., the Bonilla Fault of von Gosen, 1995) is that the pre-Carboniferous structures in the Bonilla area are strongly overprinted such that they now have a fairly uniform easterly dip. Because

Permian deformation was regionally extensive (Mpodozis and Ramos, 1989), we infer that this deformation also affected the rock units in the Cortaderas area, although the absence of Carboniferous rocks does not permit us to directly document the Permian deformation there. The general structural trend for each of the four deformation events is north to north-northeast, which renders difficult the distinction of one deformation event from another based on orientation data alone. It is worth noting that, in general way, Keidel (1939) recognized all of these deformation events, although without much kinematic or temporal control.

Ductile top-to-the-east deformation

The most important result of our field and thin-section structural analysis is the recognition of previously unrecognized top-to-the-east synmetamorphic ductile shear zones at the contacts between the ultramafic/layered complex and mafic ophiolite section and the two metasedimentary units, as well as between and within the metasedimentary units themselves. In the Cortaderas area, the ductile shear zones are best developed within the carbonate-bearing metasediment unit. The ductile shear zones are tens of centimeters to meters thick, strike northwest to northeast, and dip both to the east and west, although southeast dips predominate (Fig. 7D). The variable orientations reflect the intense folding of these shear zones. In the Bonilla area, the shear zones are more widely developed, thicker (meters to tens of meters), and include a thick (ca. 100 m) ductile shear zone between the metasediment/metavolcanic complex and the overlying ultramafic/layered complex and carbonate-metasiltstone unit. There, the ductile shear zones strike more uniformly north to northeast and dip to the east and southeast (Fig. 8D). In general, the shear zones are subparallel to parallel to the foliations described below.

Quartz in the ductile shear zones in the Cortaderas and Bonilla areas shows crystal plastic deformation textures including undulose extinction, subgrains, irregular embayed grain boundaries, core-mantle textures, quartz ribbons, and grain-shape and lattice preferred orientations. Calcite exhibits strong grain-shape preferred orientation and irregular deformation twins, and albite shows variably developed deformation twins. The quartz textures imply deformation in the Regime 1 and Regime 2 of Hirth and Tullis (1992), which is consistent with the prehnite-pumpellyite to lower greenschist metamorphic grade of the mafic and metasedimentary rocks. Shear zone fabrics include fine-grained S-C mylonite fabrics, where the S foliations are defined by either very fine-grained aligned neocrystallized white mica or strong quartz grain-shape preferred orientation. The C foliations are defined by

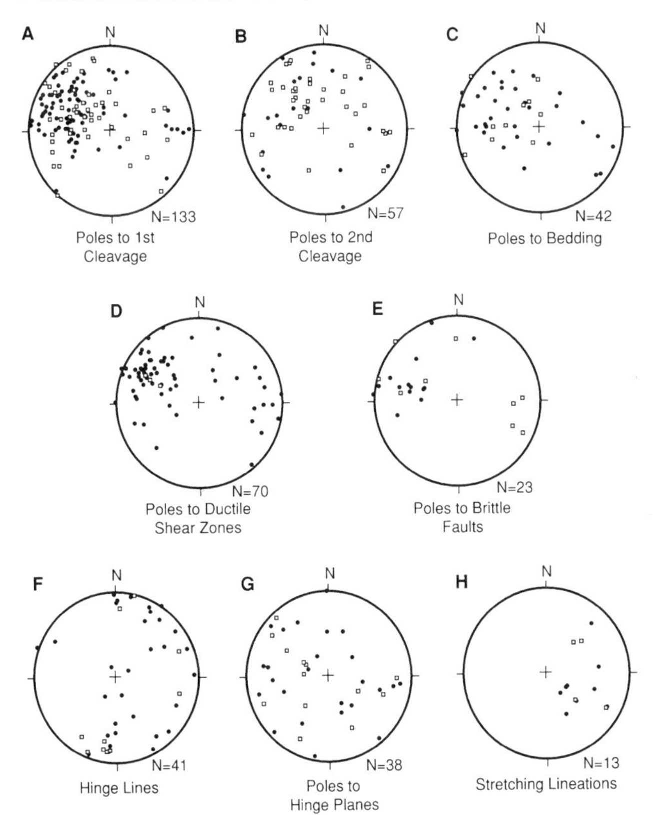

Figure 7. Lower hemisphere equal area projections showing orientations of various structural elements in the Cortaderas area.

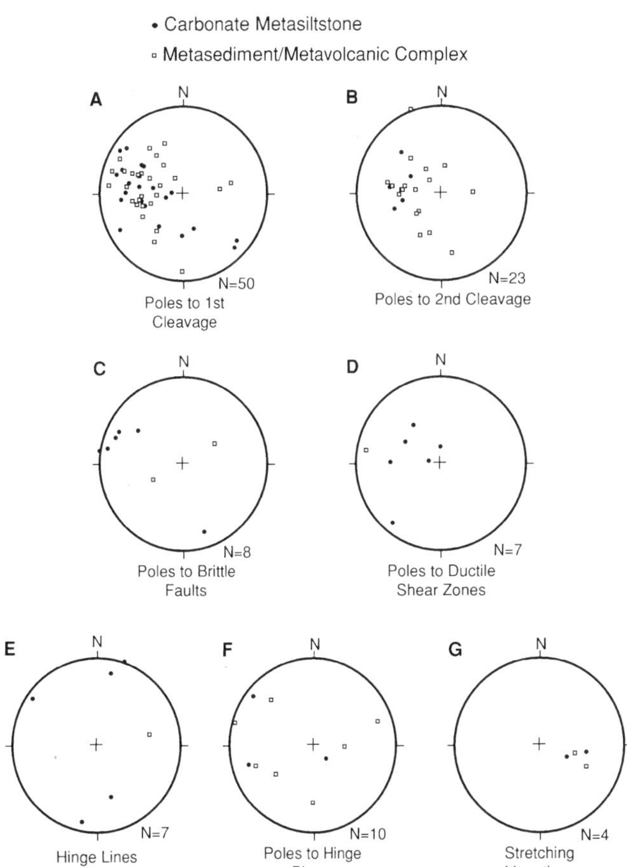

Figure 8. Lower hemisphere equal area projections showing orientations of various structural elements in the Bonilla area.

shear bands containing strongly aligned white mica and very fine-grained dynamically recrystallized quartz. The S foliations lie at angles of 25°–45° to the C foliation, but in places they are nearly parallel, indicating large strains across some of the shear zones.

The mafic rocks in the Cortaderas area commonly exhibit a quartz- and calcite-cemented brecciated texture (millimeter- to centimeter-scale cataclastic products) within a meter of the ductile shear zones. Quartz and calcite veins are deflected into the ductile shear zones and highly attenuated. In the Bonilla area, the higher grade of metamorphism promoted ductile deformation in some of the mafic bodies. In these, shearing deformation was dominantly by slip along closely spaced chlorite and white mica-rich shear bands. In the mafic rocks of the Bonilla area, relict plagioclase and pyroxene crystals, as well as albite, commonly exhibit fibrous chlorite and albite-filled fractures and microfaults.

Field and thin-section kinematic studies of the ductile shear zones between and within the various pre-Carboniferous map units in the Cortaderas and Bonilla areas provide overwhelming evidence for top-to-the-east shear sense of the ductile shear zones. Field kinematic indicators include small-scale (<1 m) thrusts, foliations and veins dragged into the shear zones, and pyrite porphyroclasts with asymmetric quartz-filled strain fringes. Lineations associated with the shear zones are dominantly parallel to dip, and are defined by pyrite with quartz strain fringes, stretched single quartz grains, and elongate polycrystalline quartz or white mica aggregates. Thin section-scale kinematic indicators include S-C fabrics, quartz porphyroclasts with asymmetric recrystallized tails (σ porphyroclasts of Simpson and Schmid, 1983), pyrite with asymmetric strain fringes, and fractured and rotated albite grains with antithetic shear sense on the cross-cutting fractures. Of 19 thin sections for which we were able to unambiguously determine shear sense, 17 show top-to-the-east shear sense and 2 top-to-the-west (Fig. 9). When combined with the shear sense determinations made in the field, 22 of 25 determinations from the ductile shear zones in the carbonate-metasiltstone unit show top-to-the-east shear sense, and 12 of 12 determinations for ductile shear zones in the metasediment/metavolcanic complex show top-to-the-east shear sense. The shear sense determinations include several from the ductile shear zone between the metasediment/metavolcanic complex and overlying ultramafic/layered complex in both the Cortaderas and Bonilla areas, as well as many from shear zones between the ultramafic/layered complex and mafic ophiolite section and the underlying carbonate-metasiltstone unit in both areas. To ensure that the kinematic indicators were not the product of interlayer slip during folding of the various units, we determined the shear sense for the folded shear zones on both limbs of a folded shear zone in several instances. In all of these cases, the shear sense was uniformly top-to-the-east. If interlayer slip had been responsible for the development of the shear zones, kinematic indicators would yield opposite movement senses on opposite fold limbs.

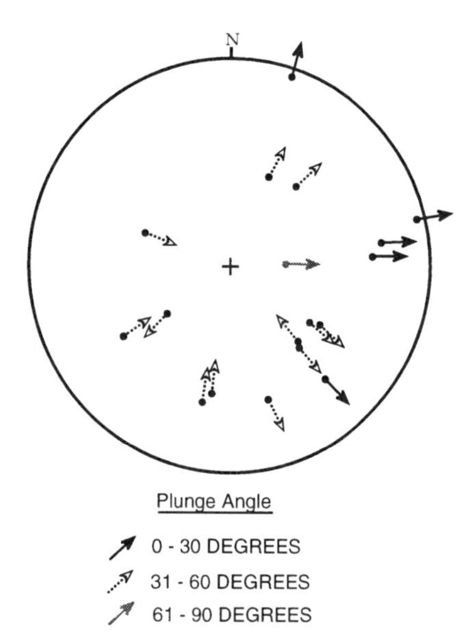

Figure 9. Lower hemisphere equal area projection showing upper plate kinematics of synmetamorphic ductile shear zones in the Cortaderas area. Dots show trend and plunge of the movement lineation; arrows show direction of movement of the upper plate parallel to that lineation.

The earliest foliation is a penetrative layer-parallel foliation that defines the hinge planes of isoclinal to tight folds. It is well developed in both of the metasedimentary units. In the Cortaderas and Bonilla areas, the foliation strikes predominantly northeast-southwest to north-south and dips moderately to steeply to the southeast and east (Figs. 7A and 8A). Scatter in the orientations and the great circle distribution of poles to foliation, suggests later folding of the penetrative foliation about a northeast-plunging fold axis. In the phyllitic to finely schistose rocks, the foliation is developed as a continuous foliation defined by aligned fine-grained, neocrystallized white mica and a moderately developed quartz grain-shape preferred orientation. Where the metamorphic grade is lower in the western parts of the metasediment/metavolcanic complex, this early foliation occurs as a subparallel to anastamosing disjunctive foliation defined by pressure solution surfaces containing insoluble residues (primarily oxides) and partially aligned tabular detrital particles.

The penetrative layer parallel foliation is cross-cut by a second foliation represented by either a crenulation foliation or a disjunctive foliation in the carbonate-metasiltstone unit and the metasediment/metavolcanic complex. Field observations in both the Cortaderas and Bonilla areas demonstrate that the second foliation approximates the hinge plane of a second generation of tight to isoclinal folds that deformed the first foliation. The cross-cutting foliation strikes between north and east, and dips both southeast and northwest (Figs. 7 and 8). In both the carbonate-metasiltstone unit and the metasedimentary/metavolcanic unit, this anastamosing to subparallel disjunctive foliation is defined by pressure solution surfaces containing insoluble residues, including oriented white mica. In the Cortaderas and Bonilla areas, the crenulation foliation involves the growth of white mica within the cleavage domains that separate the microlithons. The microlithons preserve the first foliation that was folded by the crenulations. We propose that the two foliations represent a composite foliation (Tobisch and Paterson, 1988) that formed during one progressive synmetamorphic deformation event. We base this interpretation on the following: (1) there was only one low-grade regional metamorphic event in the Cortaderas and Bonilla areas, and both the first and second foliations involved the growth of neocrystalline white mica and crystal-plastic deformation of quartz; (2) the foliations have roughly similar orientations; and (3) the foliations formed as hinge plane foliations to at least two generations of tight to isoclinal folds that have similar hinge orientations (Figs. 7 and 8).

Tight to isoclinal folds occur on a variety of scales. Fold amplitudes and wavelengths range from tens of centimeters to hundreds of meters. Fold hinges for both generations of folds plunge gently to moderately steeply to both the north-northeast and the south-southwest (Figs. 7F and 8E), but there is scatter in the data. Although description of fold vergence and asymmetry are not relevant for folds this tight, the general east dip of the hinge plane foliations has led previous workers in the Bonilla and Cortaderas areas, and elsewhere in the western Precordillera, to propose that they have a west-vergent asymmetry (e.g., de Romer, 1964; Harrington, 1971; Cucchi, 1972; Ramos et al, 1984; von Gosen, 1992, 1995). However, our hinge plane orientation data do not support a uniform west vergence for these folds, and von Gosen (1995) has documented both eastward and westward vergences for the second generation of tight to isoclinal folds. In some locations in the Bonilla area and elsewhere in the western Precordillera, restoration of the Carboniferous unconformity to horizontal results in rotation of the easterly dip of the pre-Carboniferous, synmetamorphic foliations (i.e., hinge planes) to a westerly dip that is compatible with eastward fold vergence. In most locations the ductile shear zone fabrics are parallel to the continuous foliation developed during the first phase of tight to isoclinal folding, but in rare instances, the shear zones cross-cut the continuous foliation, which suggests that the shear zones were active contemporaneously with, or slightly postdated, the first isoclinal folding.

Brittle extensional deformation

The second deformation that we were able to delineate is extensional in nature and is manifested by several normal faults that clearly cross-cut the structures developed during the first deformation event. The normal faults strike north to northwest, and dip steeply east or west. Thickness of the normal fault zones range from tens of centimeters to meters. In the Cortaderas area, a major north-striking normal fault crosses the western part of the map area (Fig. 4). In the southern part of the map area, this fault is expressed as a narrow graben ca. 100 m wide. In both the Cortaderas and Bonilla areas, these faults are brittle structures, and typical fault rocks include unconsolidated to consolidated gouge, breccias, and large amounts of disrupted quartz veins. The breccia clasts range in size from centimeters to meters and comprise pieces of ultramafic and mafic rocks, phyllite, mafic sills, and Carboniferous to Upper Triassic sedimentary rocks. The numerous disrupted quartz veins suggest repeated vein formation and deformation during faulting. Fault kinematic indicators include younger over older relationships and drag folds.

Brittle top-to-the-west deformation

The third deformation event comprises top-to-the-west thrust faults and related folds that affect all rock units in the Cortaderas and Bonilla areas. As with the previous deformation events, this event also resulted in north to northeast striking faults. The thrust faults generally dip moderately to the southeast, but several were identified that had southwest dips (Figs. 7E and 8C). Fault zone thicknesses range from a few centimeters to a few meters. Deformation in these thrusts is brittle, and fault rocks almost uniformly consist of loose, powdery gouge with clasts and blocks of footwall and hangingwall rocks, and abundant coarsely crystalline quartz veins. The kinematic indicators of the top-to-the-west faults include ramp-flat geometries, drag folds, and older over younger relationships. Ramp-flat geometries at scales ranging from meters to hundreds of meters were observed in both the Cortaderas and Bonilla regions, and consistently indicated top-to-the-west movement. For example, in the western portion of the metasediment/metavolcanic complex, just west of the map area in the Cortaderas area, the Upper Triassic Cacheuta Group and unconformably underlying metasediment/metavolcanic complex display at least two repetitions. In at least one instance, a clear ramp and flat geometry is discernable in one of these, and it indicates top-to-the-west thrusting. In the Bonilla area, repetitions of the lower Paleozoic structures occur across several thrusts ramps whose geometries yield top-to-the-west shear sense. In some instances, the thrusts place the ultramafic, mafic and metasedimentary rocks over Carboniferous sedimentary rocks.

In both the Cortaderas and Bonilla areas, post-Carboniferous sedimentary and igneous rocks describe open to gentle, kilometer-scale folds. As these rock units unconformably overlie the ultramafic, mafic, metasedimentary and metavolcanic rocks, we infer that all of these units have been affected by this gentle folding event.

GEOCHRONOLOGY

We undertook $^{40}Ar/^{39}Ar$ and U/Pb geochronology in order to obtain information on the chronology of deformational and metamorphic events in the Cortaderas and Bonillla areas. $^{40}Ar/^{39}Ar$ analyses of two samples of metasedimentary rocks were performed in order to define the approximate age of the synmetamorphic, east-vergent deformation. One of these $^{40}Ar/^{39}Ar$ analyses was performed on a sample of metasedimentary rock from the Portillo area (Fig. 2) in order to ascertain whether the metamorphism that affected the ultramafic, mafic, and metasedimentary rocks in the southwest Precordillera was regionally extensive. Igneous rocks that show cross-cutting relationships with brittle, west-vergent structures were analyzed by

TABLE 1. ^{45}Ar/^{39}Ar DATA FOR SAMPLES FROM CORTADERAS, BONILLA, AND PORTILLO AREAS

USP94-71, White Mica; j = 0.009390 ± 0.1%; 17.3 mg

Temp. (°C)	Radiogenic ^{40}Ar•†	K-derived ^{39}Ar*†	$^{40}Ar_R/^{39}Ar_K$§	Radiogenic Yield (%)	^{39}Ar Total (%)	Apparent Age ± 1σ Error (Ma)**
600	0.01755	0.00177	9.935	31.2	0.1	160.91 ± 19.81
700	0.24823	0.01341	18.504	87.8	1.1	289.09 ± 3.33
750	0.28331	0.01290	21.961	94.9	1.1	338.23 ± 1.48
800	0.51273	0.02250	22.786	82.4	1.9	349.78 ± 1.02
850	0.74889	0.03067	24.414	97.4	2.5	372.35 ± 1.97
900	1.43745	0.05841	24.608	98.8	4.8	375.02 ± 1.11
950	3.29889	0.13181	25.028	99.2	10.8	380.80 ± 0.54
1000	8.20035	0.32513	25.222	99.4	26.8	383.45 ± 0.54
1050	11.91288	0.47076	25.306	99.5	38.7	384.60 ± 0.54
1100	2.59635	0.10271	25.279	99.0	8.5	384.24 ± 0.54
1150	0.54929	0.02203	24.933	96.8	1.8	379.49 ± 0.71
1200	0.30495	0.01230	24.797	95.9	1.0	377.62 ± 3.88
1350	0.27223	0.01085	25.099	95.7	0.9	381.77 ± 3.92
Total Gas			25.002			380.43 ± 0.76

Plateau Age 384.5 ± 0.5
Isochron Age 385.3 ± 0.4
(^{40}Ar/^{39}Ar)$_i$ 188 ± 4

CM94-18, White Mica; j = 0.009382 ± 0.1%; 23.2 mg.

Temp. (°C)	Radiogenic ^{40}Ar•†	K-derived ^{39}Ar*†	$^{40}Ar_R/^{39}Ar_K$§	Radiogenic Yield (%)	^{39}Ar Total (%)	Apparent Age ± 1σ Error (Ma)**
600	0.03841	0.00669	5.745	61.8	0.4	94.70 ± 6.1
700	0.12759	0.01290	9.894	90.1	0.8	160.13 ± 3.56
750	0.29600	0.01335	15.877	94.2	0.9	250.49 ± 3.96
800	0.41928	0.02192	19.131	76.5	1.4	297.80 ± 2.83
850	0.69028	0.03068	22.496	97.2	2.0	345.46 ± 1.67
900	2.36478	0.09910	23.863	98.2	6.3	364.46 ± 0.83
950	6.52309	0.26366	24.740	99.3	16.9	376.55 ± 0.53
1000	7.73661	0.31216	24.784	99.7	20.0	377.15 ± 0.53
1050	5.47209	0.22014	24.858	99.7	14.1	378.16 ± 0.53
1100	9.10905	0.24624	24.809	99.4	15.8	377.5 ± 0.53
1150	4.17497	0.16903	24.700	99.5	10.8	37.99 ± 0.53
1200	1.93848	0.07833	24.747	99.5	6.0	376.65 ± 0.57
1350	2.19561	0.08845	24.822	99.2	5.7	377.68 ± 0.57
Total Gas			24.319			370.76 ± 0.77

Plateau Age 377.6 ± 0.5
Isochron Age 377.6 ± 0.4
(^{40}Ar/^{39}Ar)$_i$ 264 ± 1.45

USP94-48, Biotite; j = 0.009397 ± 0.1%; 32.1 mg.

Temp. (°C)	Radiogenic ^{40}Ar•†	K-derived ^{39}Ar*†	$^{40}Ar_R/^{39}Ar_K$§	Radiogenic Yield (%)	^{39}Ar Total (%)	Apparent Age ± 1σ Error (Ma)**
600	0.07558	0.09345	0.809	31.4	1.6	13.66 ± 0.10
700	0.08525	0.10317	0.836	51.0	1.8	13.95 ± 0.36
750	0.05010	0.06788	0.738	39.3	1.2	12.47 ± 0.25
800	0.08920	0.10950	0.739	15.5	1.9	12.48 ± 0.47
850	0.21180	0.30426	0.696	50.9	5.2	11.76 ± 0.28
900	0.19975	0.28763	0.694	69.6	4.9	11.73 ± 0.40
950	0.39359	0.59310	0.664	77.4	10.2	11.21 ± 0.02
1000	0.40795	0.62269	0.658	80.2	10.7	11.12 ± 0.07
1050	0.40916	0.62023	0.660	78.4	10.6	11.15 ± 0.06
1100	0.40090	0.60815	0.659	79.3	10.4	11.14 ± 0.07
1150	0.74703	1.12731	0.663	84.9	19.3	11.20 ± 0.02
1350	0.85933	1.29372	0.664	86.8	22.2	11.23 ± 0.07
Total Gas			0.673			11.37 ± 0.09

Plateau Age 11.2 ± 0.1
Isochron Age 11.1 ± 0.01
(^{40}Ar/^{39}Ar)$_i$ 309 ± 0.03

USP94-93, Hornblende; j = 0.009380 ± 0.1%

Temp. (°C)	Radiogenic ^{40}Ar•†	K-derived ^{39}Ar*†	$^{40}Ar_R/^{39}Ar_K$§	Radiogenic Yield (%)	^{39}Ar Total (%)	Apparent Age ± 1σ Error (Ma)**
800	0.00785	0.00400	1.961	5.2	0.2	32.89 ± 17.49
900	0.00396	0.00760	0.521	2.3	0.4	8.79 ± 2.29
950	0.01323	0.00887	1.492	7.8	0.5	25.07 ± 10.67
1000	0.00131	0.00236	0.554	6.4	0.1	9.34 ± 8.18
1025	0.00118	0.00200	0.591	9.7	0.1	9.98 ± 14.65
1050	0.00110	0.00313	0.351	8.2	0.2	5.93 ± 10.0
1075	0.00326	0.00683	0.477	17.6	0.4	8.06 ± 0.77
1100	0.05334	0.04838	1.102	75.0	2.8	18.56 ± 0.65
1125	0.15273	0.14580	1.048	82.5	8.5	17.64 ± 0.45
1150	0.23354	0.22380	1.044	86.1	13.0	17.57 ± 0.38
1175	0.30046	0.28858	1.041	88.0	16.8	17.53 ± 0.18
1200	0.65773	0.62510	1.052	91.9	36.4	17.72 ± 0.05
1250	0.22399	0.20887	1.072	83.1	12.2	18.06 ± 0.19
1300	0.11753	0.10823	1.086	73.5	6.3	18.28 ± 0.09
1400	0.04964	0.03347	1.483	38.4	1.9	24.92 ± 1.43
Total Gas			1.060			17.86 ± 0.4

Plateau Age 17.6 ± 0.3
Isochron Age 17.3 ± 0.02
(^{40}Ar/^{39}Ar)$_i$ 347 ± 0.05

*See Figure 3. Samples of phyllite, schist, intermediate tuff, and an intermediate sill were collected for white mica, hornblende, and biotite. All samples were crushed, ground, and sieved, and the 100- to 200-mesh-size sieve fractions were passed through heavy liquids, and a magnetic separator to obtain concentrates. The concentrates were hand picked to 100% purity. The four mineral separates were cleaned in reagent-grade ethanol, acetone, and de-ionized water in an ultrasonic bath, air dried, wrapped in aluminum packages, and sealed in silica vials along with monitor minerals prior to irradiation. Samples were irradiated in the TRIGA reactor at the U.S. Geological Survey in Denver, Colorado, for 39.87 hours at 1 megawatt.

†A Mass Analyzer Products 215 Rare Gas mass spectrometer with a Faraday cup was used to measure argon-isotope abundances. Abundances of 'radiogenic ^{40}Ar and K-derived ^{39}Ar are reported in volts. Conversion to moles can be made using 9.74×10^{-13} moles argon per volt of signal. Detection limit at the time of this experiment was 2×10^{-17} moles argon. Analytic data for radiogenic ^{40}Ar and K-derived ^{39}Ar are calculated to 5 decimal places; $^{40}Ar_R/^{39}Ar_K$ is calculated to three decimal places. Radiogenic ^{40}Ar and K-derived ^{39}Ar are rounded to significant figures using calculated analytic precisions. Apparent ages and associated errors were calculated from unrounded analytical data and then each rounded using associated errors. All analyses were done in the Argon Laboratory, U.S. Geological Survey, Denver, Colorado. Decay constants are those of Steiger and Jäger, 1977. The irradiation monitor, hornblende MMhb-I with percent K = 0.555, $^{40}Ar_R = 1.624 \times 10^{-9}$ mol/g, and K/Ar age = 520.4 Ma (Samson and Alexander, 1987) was used to calculate J values for this experiment.

§$^{40}Ar_R/^{39}Ar_K$ has been corrected for all interfering isotopes including atmospheric argon. Mass discrimination in the mass spectrometer was determined by measuring the ^{40}Ar/^{36}Ar ratio of atmospheric argon; our measured value is 298.9 during the period of this experiment, the accepted atmospheric ^{40}Ar/^{36}Ar ratio is 295.5. Abundances of interfering isotopes of argon from K and Ca were calculated from reactor production ratios determined by irradiating and analyzing pure CaF_2 and K_2SO_4 simultaneously with these samples. The measured production

TABLE 1. $^{45}Ar/^{39}Ar$ DATA FOR SAMPLES FROM CORTADERAS, BONILLA, AND PORTILLO AREAS (continued)

ratios are $(^{40}Ar/^{39}Ar)_K = 7.92 \times 10^{-3}$, $(^{38}Ar/^{39}Ar)_K = 1.309 \times 10^{-2}$, $(^{37}Ar/^{39}Ar)_K = 1.75 \times 10^{-4}$, $(^{36}Ar/^{37}Ar)_{Ca} = 2.75 \times 10$-4, $(^{39}Ar/^{37}Ar)_{Ca} = 6.85 \times 10^{-4}$, and $(^{39}Ar/^{37}Ar)_{Ca} = 4.4\ 10^{-5}$. Corrections were also made for additional interfering isotopes of argon produced from irradiation of chlorine using the method of Roddick, 1983. The reproducibility of split gas fractions from each monitor (0.05–0.25%, 1) was used to calculate imprecisions in J. J values for each sample were interpolated from adjacent monitors and have similar uncertainties to the monitors. Uncertainties in calculation for the date of individual age steps in a spectrum were calculated using modified equations of Dalrymple et al., 1981.
**1σ error reported for raw age spectrum data. Error for plateau and isochron dates is 2s.

both $^{40}Ar/^{39}Ar$ and U/Pb methods. These analyses provide chronologic constraints on the age of the most recent west-vergent deformation.

$^{40}Ar/^{39}Ar$ geochronology

We applied stepwise $^{40}Ar/^{39}Ar$ analyses to samples of phyllite and subvolcanic to volcanic rocks of intermediate composition from the Cortaderas and Bonilla areas, as well as to a sample of schist from the Portillo area. Mineral separation, lab analytic procedures and data reduction information are given in footnotes to Table 1. Age spectra and isotope correlation diagrams ($^{40}Ar/^{36}Ar$ vs. $^{39}Ar/^{36}Ar$) were generated from the data for each sample (see Table 1) and are presented in Figure 10. Ages for the age spectra plateau and the isochron ages from the isotope correlation diagrams have errors of 2σ.

Sample USP94-71, from the Bonilla area, is white mica from a quartz–white mica–albite phyllite to fine-grained schist from the ductile shear zone separating the ultramafic/layered complex and carbonate-metasiltstone unit from the structurally underlying metasediment/metavolcanic complex. Thin section examination shows that all the white mica is neocrystallized and that it defines the foliation. The age spectrum (Fig. 10A) that resulted from the $^{40}Ar/^{39}Ar$ analysis indicates some argon loss from the low- and high-temperature fractions, and a near-plateau from heating steps 8–10. Steps 9 and 10 yield a near-plateau cooling age of 384.5 ± 0.5 Ma. An isotope correlation diagram (Fig. 10A) constructed using a York regression analysis (York, 1969) of steps 8–13 of this analysis (steps 1–7 excluded because of obvious Ar loss) yields an isochron age of 385.3 ± 0.4 Ma , identical within error to the near-plateau cooling age of 384.5 ± 0.5.

Sample CM94-18 is white mica from a quartz-muscovite schist from the north-central Portillo area. Our reconnaissance work suggests that this area underwent a regional lower amphibolite facies metamorphism. Although its relationship to regional deformation is not clear, the age was obtained, in part, to determine if the numerous intrusions in the area had regionally affected the Ar systematics, as well as to compare the age of this metamorphic event to that in the Bonilla. Thin-section examination of this sample showed that the white mica forms the foliation, and that it is a neocrystallized phase. The age spectrum (Fig. 10D) for CM94-18 shows Ar loss from the low temperature heating steps and then a near-plateau over the remaining steps. Heating steps 8 through 10 define a plateau cooling age at 377.6 ± 0.5 Ma. An isotope correlation plot (Fig. 10D) of heating steps 7–13 (steps 1–6 excluded because of obvious Ar loss) yield an isochron age of 377.6 ± 0.4 Ma, identical within error to the plateau cooling age of 377.6 ± 0.5 Ma.

Sample USP94-48 is biotite from a tuff of intermediate composition from the Bonilla area. The tuff is structurally overlain by the carbonate-metasiltstone unit along a top-to-the-west thrust in the core of a kilometer-scale syncline that has west-vergent asymmetry. The gas release spectrum (Fig. 10B) shows that the low temperature steps contain excess Ar, but the sample yields a plateau cooling age of 11.2 ± 0.1 Ma. The isotope correlation plot for steps 3 to 12 (steps 1 and 2 excluded because of obvious excess argon content) produces an isochron age of 11.1 ± 0.01 Ma, which overlaps the plateau cooling age within error.

Sample USP94-93 is hornblende from a sill of intermediate composition with hornblende phenocrysts. The sill crops out in the metasediment/metavolcanic complex in the western Cortaderas area, and is located in the footwall of a top-to-the-west thrust. The simple K/Ca vs. gas relationship and the age spectrum indicate that the hornblende consisted of one phase that released gas uniformly over the middle- to high-temperature heating steps. The age spectrum (Fig. 10C) shows a well-defined plateau for the intermediate- to high-temperature heating steps that gives a cooling age of 17.6 ± 0.3 Ma. The isotope correlation diagram yields an isochron age of 17.3 ± 0.02, which overlaps with the plateau cooling age.

U/Pb geochronology

We analyzed zircons separated from a sample (USP94-19) of the small monzodiorite pluton in the eastern Cortaderas area (Fig. 4) in order to provide age constraints on the deformation chronology of the southwest Precordillera. Zircon separation and analytic techniques are summarized in McClelland and Mattinson (1996). Four multi-grain fractions consisting of clear, euhedral zircons yielded nearly concordant to strongly discordant results that define a nonlinear array (Table 2; Fig. 11). The nearly concordant analysis (fraction b in, Table 2) was obtained from clear inclusion-free zircons that formed a minor component of the sample population. Fractions that included zircon grains with minor inclusions and possible xenocrystic compo-

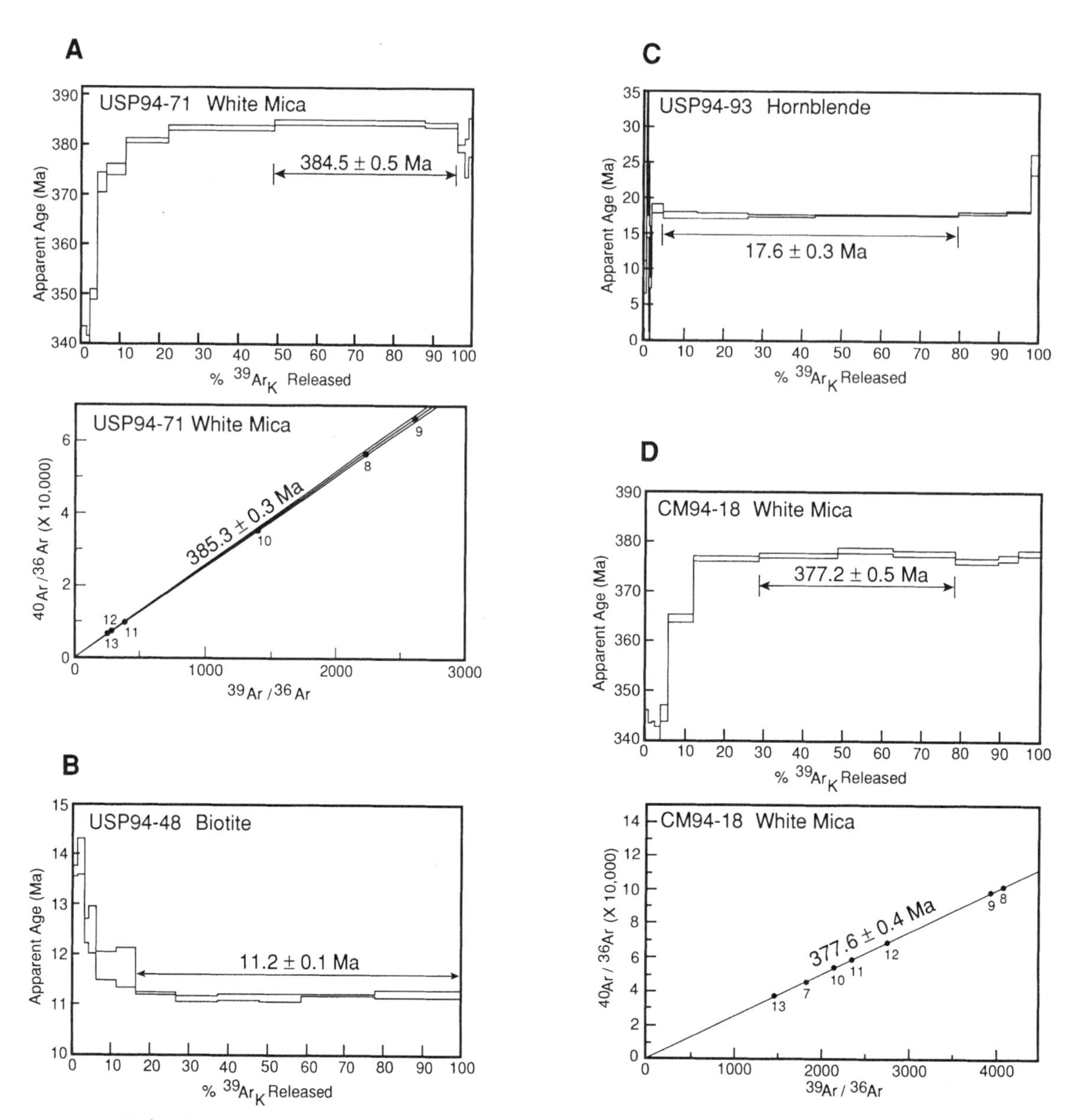

Figure 10. Gas release spectra and isotope correlation diagrams for $^{40}Ar/^{39}Ar$ analyses. Arrows adjacent to ages indicate temperature steps used to determine age. An age plateau is defined as consisting of contiguous heating steps with apparent ages that are identical within error, and that contain a total of at least 50% of the gas released (Snee et al., 1988). A, Sample USP94-71, white mica separated from synmetamorphic east-vergent ductile shear zone in the Bonilla area. B, Sample USP94-48, biotite separated from andesite tuff in the footwall of a brittle west-vergent thrust, which places the carbonate-bearing metasediments over it in the Bonilla area. C, Sample USP94-93, hornblende from an andesite sill in the footwall of brittle west-vergent thrust, which places Mesozoic sedimentary rocks over it in the western Cortaderas area. D, CM94-18, white mica separated from schist in the Portillo area.

nents yielded discordant analyses. A crystallization age of 18 ± 2 Ma is interpreted for the monzodiorite pluton on the basis of the nearly concordant analysis and the interpretation that discordance is related to inheritance. Upper intercept ages, using the inferred crystallization age of 18 ± 2 Ma and individual strongly discordant analyses, range from 925–1,125 Ma. We interpret the spread of upper intercept ages to reflect incorporation of Grenvillian-age xenocrystic components.

Other geochronologic studies

K-Ar whole-rock ages reported from early geochronologic studies of the metasedimentary rocks in the Bonilla area (Cucchi, 1971; Caminos et al., 1979) range from 88 to 630 Ma, with errors as large as 330 Ma. The widespread range of ages probably reflects the presence of detrital white mica, as well as argon loss in the vicinity of younger intrusive bodies. As the studies did not

TABLE 2. U-Pb ISOTOPIC DATA AND APPARENT AGES FOR USP94-19

Fraction-Size[†]		Wt.	Concentrations[§]		Isotopic Composition**			Apparent Ages[‡]			Th-corrected Ages[§§]	
	(m m)	(mg)	U	Pb*	$^{206}Pb/^{204}Pb$	$^{206}Pb/^{207}Pb$	$^{206}Pb/^{208}Pb$	$^{206}Pb^*/^{238}U$ (Ma)	$^{207}Pb^*/^{235}U$ (Ma)	$^{207}Pb^*/^{206}Pb^*$ (Ma)	$^{206}Pb^*/^{238}U$ (Ma)	$^{207}Pb^*/^{206}Pb^*$ (Ma)
a	45–63	0.1	284	2.7	1422 ± 2	13.328	6.481	58.9	80.2 ± 0.2	773	58.9 ± 0.1	770 ± 4
b	63–100	0.1	519	1.8	692 ± 1	14.703	2.547	18.3	18.5 ± 0.1	40	18.4 ± 0.1	29 ± 8
c	63–100A	0.3	218	2.0	10.16 ± 3	13.201	2.786	49.9	64.9 ± 0.2	658	50.0 ± 0.1	655 ± 4
d	100–350A	0.5	30	0.6	2669 ± 12	12.787	6.556	120.3	176.0 ± 0.4	1011	120.4 ± 0.2	1010 ± 2

[†]a, b, etc. designate conventional fractions. Conventional fractions were washed in warm 3N HNO_3 and 3N HCl for 15 minutes each, spiked with ^{205}Pb=^{235}U tracer, and dissolved in a 50% Hf>>14N HNO_3 solution within 3 ml Savillex capsules placed in 45 ml TFE Teflon-lined Parr acid digestion bomb which is similar to procedure outlined by Parrish et. al., 1987. Following evaporation and dissolution in HCl, Pb and U for all fractions were separated following techniques modified from Krogh, 1973. Pb and U were combined and loaded with H^3PO_4 and silica gel onto single degassed Re filaments. Isotopic composition of Pb and U were determined through static collection on a Finnigan-MAT261 multi-collector mass spectrometer utilizing a secondary multiplier (SEM) for collection of the ^{204}Pb beam. SEM gain was determined by simultaneous collection of the ^{205}Pb beam in the SEM and ^{206}Pb in the adjacent Faraday cup before and after each sample and comparison of the average ratio thus measured with the comparable Faraday ratio determined during sample analysis. Zircon fractions are nonmagnetic on Frantz magnetic separator at 1.6 amps. 15° forward slope and side slope of 5°.

[§]Pb* is radiogenic Pb expressed as ppm. U is expressed as ppm.

**Reported ratios corrected for fractionation (0.125 ± 0.038%/AMU) and spike Pb. Ratios used in age calculation were adjusted for 6 to 10 pg of blank Pb with isotopic composition of $^{206}Pb/^{204}Pb$ = 18.6, $^{207}Pb/^{204}Pb$ = 15.5, and $^{208}Pb/^{204}Pb$ = 38.4, 2 pg of blank U, 0.25 ± 0.049%/AMU fractionation for UO_2, and initial common Pb with isotopic composition approximated from Stacey and Kramers, 1975, with an assigned uncertainty of 0.1 to initial $^{207}Pb/^{204}Pb$ ratio.

[‡]Uncertainties reported as 2σ, error assignment for individual analyses follows Mattinson, 1987, and is consistent with Ludwig, 1991. An uncertainty of 0.2% is assigned to the $^{206}Pb/^{238}U$ ratio based on our estimated reproductibility unless this value is exceeded by analytic uncertainties. Calculated uncertainty in the $^{207}Pb/^{204}Pb$ ratio, and composition and amount of blank. For appropriate samples, linear regression of discordant data utilized Ludwig, 1992. Decay constants used ^{238}U = 1.5513 E-1-, ^{235}U = 9.8485 E-10, $^{238}U/^{235}U$ = 137.88.

[§§]A 75% ± 25% efficiency in ^{230}Th exclusion during zircon crystallization is assumed and $^{207}Pb/^{206}Pb$ and $^{206}Pb/^{238}U$ ratios have been adjusted accordingly. Age assignments presented are derived from the Th-corrected ratios.

relate ages to specific microtextures or structures, it is difficult to evaluate the significance of the dates.

Buggisch et al. (1994) performed K-Ar dating of 2–6μ illite/muscovite separated from the metasedimentary rocks in the Bonilla area. Their microstructural observations indicate that the white mica lies in the foliation and thus may be interpreted as a neocrystallized phase that represents the syndeformational metamorphism. Buggisch et al. (1994) reported three K-Ar ages from the Bonilla Formation (the carbonate-metasiltstone unit and metasediment/metavolcanic complex) that range from 353.1 to 399.6 Ma, and three ages from samples of the "early Paleozoic east of Uspallata" that range from 384.5 to 437.4 Ma. In their discussion of the K-Ar data from the early Paleozoic east of Uspallata, Buggisch et al. (1994) pointed out that there is significant detrital white mica in the samples, and that the ages derived from them may reflect some contamination by detrital phases.

CHRONOLOGY OF DEFORMATION

We combine our structural and geochronologic data to outline the temporal framework for the structural evolution of the southwest Precordillera. Our $^{40}Ar/^{39}Ar$ data, in conjunction with the K-Ar dates of Buggisch et al. (1994), strongly suggest an Early to Middle Devonian age for the synmetamorphic ductile deformation that juxtaposed all the pre-Carboniferous rock units. Furthermore, the Middle Devonian age from the Portillo area suggests that synmetamorphic deformation was regionally extensive. We interpret the 384.5 ± 0.5 Ma date from the Bonilla area as a crystallization age for this sample since closure to argon loss in white mica occurs at approximately 350°, which we estimate was the temperature achieved during the syndeformational metamorphism. This age falls in the middle of the 353.1–399.6 Ma range of K-Ar ages presented by Buggisch et al. (1994). Because of the difficulty in determining the presence or absence of excess argon or argon loss in K-Ar analyses, we interpret the 384.5 Ma date as the best estimate of the age the first deformation.

The timing of the first episode of top-to-the-west thrusting and associated folding is well documented by von Gosen (1995), who showed that the intense folding and top-to-the-west thrusting affects the Carboniferous units, but not the unconformably overlying Permo-Triassic Choiyoi Group volcanic rocks. This observation is consistent with our observations of the gently folded Triassic Cacheuta Group rocks immediately south of the Cortaderas area, which unconformably overlie the strongly deformed pre-Carboniferous rock units. Therefore, the intense deformation of the Carboniferous rocks must represent the Permian deformation associated with the San Rafaelic orogenic phase (see review in Mpodozis and Ramos, 1989).

The timing of normal faulting is best established in the Cortaderas area, where the major normal fault in the western part

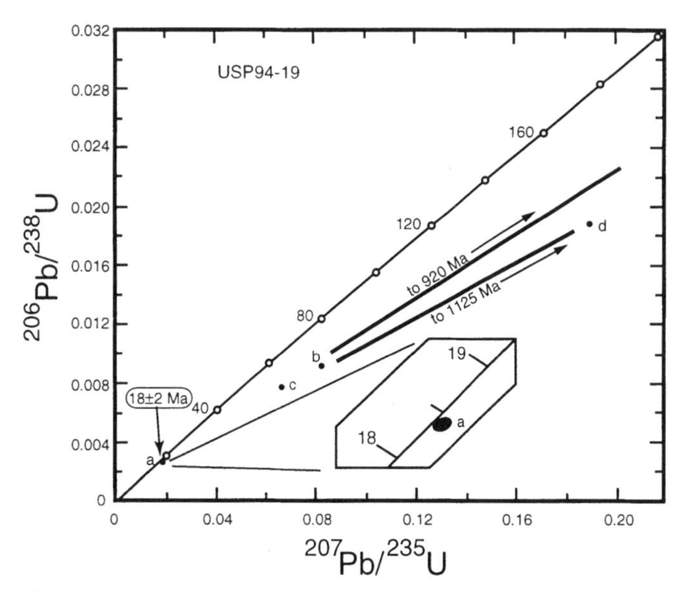

Figure 11. U/Pb concordia plot of zircon data from USP94-19. All four fractions are discordant, defining a nonlinear array with a lower intercept of 18 ± 2 Ma defined by near-concordant zircon size fraction b, and upper intercepts of 925–1125 Ma.

of the field area preserves part of the Upper Triassic Cacheuta Group in the narrow graben discussed above. In its central reaches the fault is cross-cut by a hornblende-phyric andesite stock that appears very similar to the hornblende-phyric intermediate sill for which we obtained a $^{40}Ar/^{39}Ar$ plateau cooling age of 17.6 ± 0.3 Ma. Thus, normal faulting in the Cortaderas area is constrained to be post-Triassic and pre–Middle Miocene. In the Bonilla area, our field observations record that normal faults clearly deform the Carboniferous Santa Elena Formation. von Gosen (1995), who mapped a large area in the Bonilla region, constrained the normal faulting to Late Permian to Triassic based on cross-cutting relationships with the Permo-Triassic Choiyoi Group volcanics in the Bonilla area. Normal faulting in the Bonilla area probably is coeval with that in the Cortaderas, and it may be that the somewhat later normal faulting constraints documented in the Cortaderas relate to the presence of outcrops of younger sedimentary formations in the Cortaderas that record continuing normal fault deformation. Early to middle Mesozoic extensional deformation is widely documented in western Argentina (see reviews by Mpodozis and Ramos, 1989, and Ramos and Kay, 1991), and the normal faults in the Cortaderas and Bonilla areas probably belong to that extension episode.

The most recent episode of deformation in the Cortaderas and Bonilla areas is Middle Miocene and younger top-to-the-west thrusting and west-vergent folding. In the Bonilla area, the 11.2 ± 0.1 Ma biotite-phyric andesite tuff (sample USP94-48) is overlain by the carbonate-metasiltstone unit along a brittle top-to-the-west thrust. Thus, at least some of the west-vergent deformation in the Bonilla area is Middle Miocene or younger. The Cortaderas area also has evidence of Middle Miocene and younger west-vergent deformation. Dikes from the 18 ± 2 Ma monzodiorite pluton (sample USP94-19) are cross-cut by a top-to-the-west thrust in the eastern part of the Cortaderas area and a 17.6 ± 0.3 Ma andesite sill (sample USP94-93) is cut by a thrust that repeats the Cacheuta Group sediments and the metasediment/metavolcanic complex just west of the map area.

DISCUSSION

Tectonic settings of the ultramafic and mafic rocks of the southwest Precordillera

The possible tectonic settings responsible for generation of the ultramafic/layered complex and mafic ophiolitic section provide critical constraints for discussion of the tectonic history of the western margin of the Precordillera. The settings must account for both the igneous and metamorphic characteristics of the ultramafic and mafic rocks. The rocks of the ultramafic/layered complex have similarities to the basal sections of many ophiolites, but the high-pressure granulite facies metamorphism and intrusive relationship with quartzofeldspathic gneiss requires formation in either an anomalous oceanic setting or a deep continental crust environment. The presence of garnet granulite metamorphic rocks in the layered complex implies that the layered complex was heated to temperatures in the range of 850–1000°C at a depth of about 30 km (9-11 kb). Thirty kilometers of tectonic burial in a subduction zone at subduction zone thermal gradients would not achieve the temperatures required. Burial by tectonic stacking in the hinterland during an orogenic event could yield the required P-T conditions and is possible, but the well-preserved igneous layering and textures do not support a major high-temperature deformation event.

Formation of the ultramafic and cumulate sections in an anomalous oceanic setting or deep continental environment could include formation at the base of a thick oceanic plateau, formation at the base of transitional oceanic crust during continental rifting, or formation at the base of volcanic arc in a subduction zone setting. Previously cited data pertinent to the problem of deciphering the setting of formation of the ultramafic/layered complex includes field observations that show that the ultramafic cumulates (and hence the associated peridotite [serpentinized]) intrude the quartzofeldspathic gneiss, that the layered gabbro intrudes both the ultramafic rocks and the quartzofeldspathic gneiss, and that preliminary geochemical data suggest that the layered gabbro is of calc-alkaline nature and not genetically related to the ultramafic rocks. Although granulite facies metamorphic conditions could be achieved at the base of a thick oceanic plateau, the calc-alkaline nature of the layered gabbro and the association of the ultramafic/layered complex with quartzofeldspathic gneiss better support a continental marginal environment. The geochemistry and metamorphic grade of the layered gabbro (mafic garnet granulite), as well as the presence of the pyrope-rich garnet and the association of the layered gabbro to quartzofeldspathic gneiss are comparable to the Jilal complex that formed at the base of the Kohistan arc (Jan and Jabeen,

1991). The Jilal complex is interpreted to represent early subduction-related layered gabbro that was intruded into deep continental crust before construction of the volcanic arc edifice. Therefore, we suggest an early history in which ultramafic rocks were under-or intraplated to deep continental crust (perhaps during a rifting event), and that the layered gabbro was intruded as an early subduction-related igneous product during subduction under the continental crust represented by the ultramafic rocks and quartzofeldspathic gneiss.

The mafic ophiolite section contains features expected from an ophiolite generated at an oceanic spreading ridge. First, the presence of a half-dike complex with pillow screens is strongly indicative of the transition between the sheeted dike complex and overlying pillow lava volcanic section observed in many ophiolites (Moores, 1982). Second, small plagiogranite bodies are found within the microgabbros, another feature common to ophiolites formed within an oceanic spreading ridge environment. Last, the results of preliminary trace and rare earth element geochemistry (Davis et al., 1995a,b) show that the diabase and microgabbros have an E-MORB signature, typical of some spreading ridges.

Fine-grained mafic sills within the metasedimentary units could form in a variety of settings. One setting would be a sedimented oceanic spreading ridge in which the upper crustal ophiolite section is overlain by fine-grained, terrigenous, hemipelagic siliciclastic sediments (the carbonate-metasiltstone unit) that contain diabase sills intruded into them. The sedimented ridge model may explain the presence of sills in (meta)sedimentary rocks instead of a well-developed volcanic section above the mafic ophiolite section since sedimented ridges typically lack a volcanic sequence but do contain mafic sills (e.g., the Escanaba trough on the Gorda Ridge, Zierenberg et al., 1993). A second possibility is within-plate magmatism intruding through an ophiolitic section and into hemipelagic sediments. A third scenario is that the hemipelagic sediments were deposited on transitional or continental crust and were invaded by mafic sills related to an extensional event. At the current level of data, there is no way to distinguish between these possibilities. An alternate possibility is that the sills could represent intrusions into an accretionary complex during subduction of a spreading ridge as suggested by Haller and Ramos (1984). Ridge subduction typically results in a distinct thermal event registered by high-temperature/low-pressure metamorphism of the accretionary complex and/or anomalous near-trench plutonism with magmatic products ranging from mafic to silicic (see review in Sisson et al., 1994). The carbonate-bearing metasedimentary rocks do not resemble an accretionary complex, and no intense thermal event that might be associated with ridge subduction is recorded in the rocks of the western Precordillera; thus, ridge subduction is an unlikely mechanism for generation of the mafic sills in the carbonate-bearing metasedimentary rocks.

The Early to Middle Devonian deformation in the southwestern Precordillera resulted in juxtaposition of ultramafic and mafic rocks that formed in a wide variety of tectonic environments. In contrast to the Cortaderas and Bonilla regions, most of the mafic rocks reported from the western Precordillera crop out as flows or sills within Ordovician basin sedimentary deposits. All tectonic models for the early Paleozoic of the western margin of the Precordillera are based on the interpretations of Haller and Ramos (1984) and Ramos et al. (1984), who combined the ultramafic rocks with the mafic rocks consisting primarily of the mafic flows and sills in sedimentary rocks, to form an ophiolite. However, most of the mafic flows and sills they considered may be equivalent to the mafic flows and sills in the metasediment/metavolcanic complex in the Cortaderas and Bonilla areas. As we have shown, the mafic flows and sills in the metasediment/metavolcanic complex are structurally separated

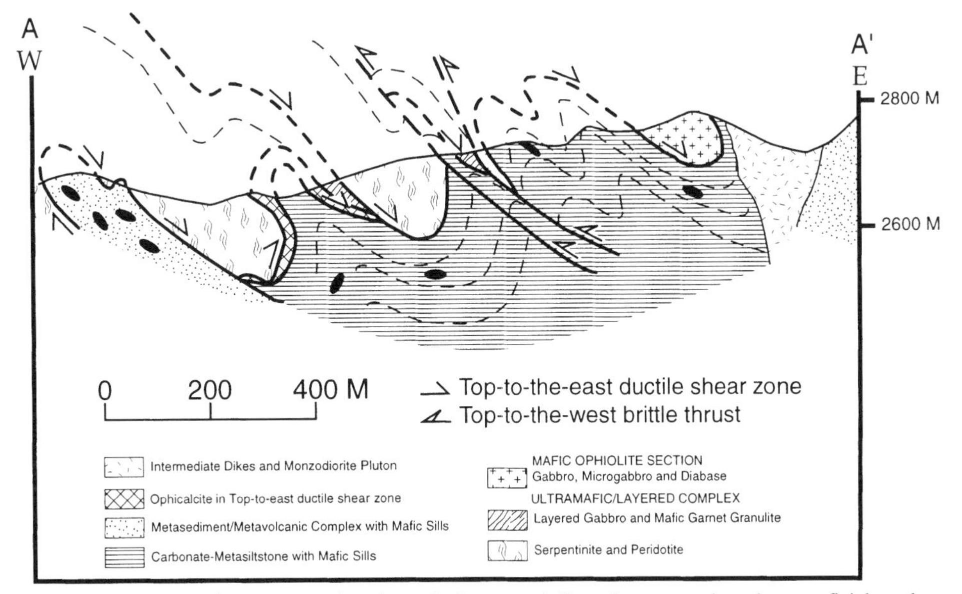

Figure 12. Interpretive cross-section through the central Cortaderas area, based on surficial geology. Important relationships discussed in the text.

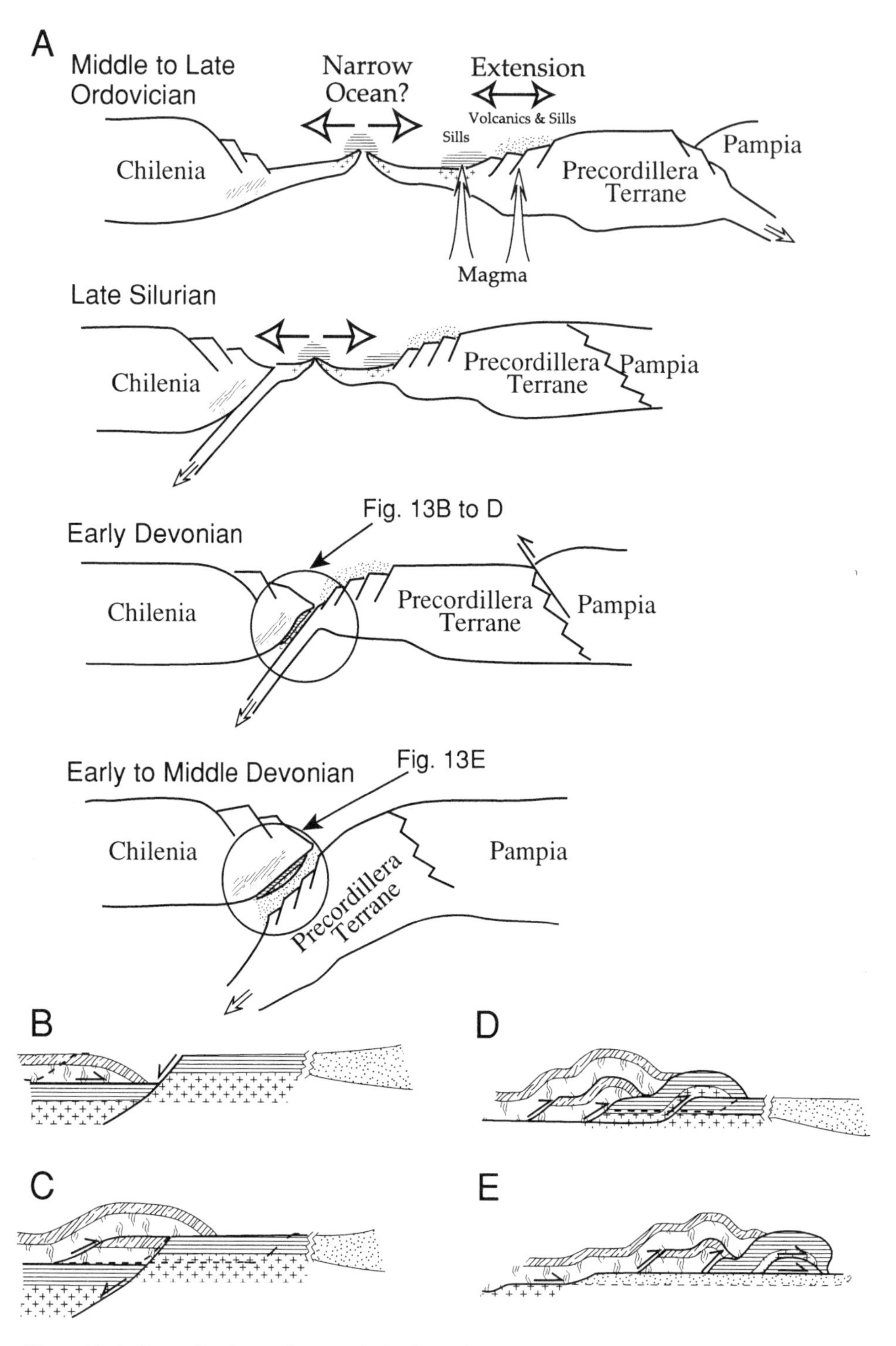

Figure 13. A, Generalized tectonic scenario for the early Paleozoic tectonic development of the western Precordillera. Note references in A that show the proposed tectonic setting and timing for parts B–E of this figure. Parts B–E depict the general stacking sequence required to reproduce the structural relationships seen in the cross section (Fig. 12), and on the Cortaderas area map. Assumptions discussed in text. Unit patterns same as in Figure 12.

from and not related to the ultramafic/layered complex or the mafic ophiolite section in the Cortaderas and Bonilla areas. Thus, the existing tectonic models for the western Precordillera are based on a definition of an early Paleozoic ophiolite that includes miscorrelations and misinterpretations of the mafic and ultramafic rocks along the western margin of the Precordillera. We suggest that the idea of an early Paleozoic ophiolite be abandoned. The only tectonic unit that could be assigned to an ophiolite is the mafic opholite section.

Devonian deformation history and tectonic scenarios for the early Paleozoic

In this section, we present an interpretive cross section through the central Cortaderas area (Fig. 12) and a working model for the Ordovician through Middle Devonian tectonics of the western Precordillera (Fig. 13). The tectonic model is predicated on a structural stacking sequence that is inferred from the cross section, as well as map relationships. The key relationships with regard to the tectonic model presented below are repetition of the ultramafic/layered complex across a top-to-the-east ductile shear zone, contact of the ultramafic/layered complex and mafic ophiolite section with the carbonate-metasiltstone unit across top-to-the-east ductile shear zones, and the emplacement of the ultramafic/layered complex, mafic ophiolite section, and carbonate-metasiltstone unit over the metasediment/metavolcanic complex along top-to-the-east ductile shear zones. As discussed earlier, we infer that all of the top-to-the-west deformation, and the associated easterly dip, shown on the cross section resulted from the Permian or Miocene deformations.

We present one tectonic model for the juxtaposition of the pre-Carboniferous units (Fig. 13) that, for simplicity, uses a foreland propagating stacking sequence where possible. The tectonic scenario incorporates the following assumptions: (1) Chilenia exists, and was west of the Precordillera terrane in the early Paleozoic; (2) the ultramafic/layered complex was emplaced within continental margin crust of eastern Chilenia; (3) the mafic ophiolite section and carbonate-metasiltstone unit with mafic sills may have been originally contiguous (i.e., a sedimented ridge), but most likely represent the upper crust of an ophiolite overlain by younger within-plate hemipelagic sediments intruded by mafic sills; (4) the metasediment/metavolcanic complex, which has a recycled orogen or continental protolith composition (cf., Kury, 1993; Loske, 1993), was deposited on the western margin of the Precordillera terrane inboard of the carbonate-metasiltstone unit during Ordovician margin–normal extension (von Gosen, 1992; Keller et al., 1993; Keller, 1995); (5) juxtaposition of the pre-Carboniferous units occurred after the ultramafic/layered complex had cooled significantly; and (6) the top-to-the-east ductile shear zones formed in a west-dipping subduction zone, this being the simplest setting in which to generate these structures during stacking of the various units.

Figure 13A outlines the tectonic model for the Ordovician through Devonian evolution of the western margin of the Precordillera. After initial juxtaposition of the ultramafic/layered complex with the carbonate-metasiltstone unit (Fig. 13B), an Early to Middle Devonian top-to-the-east foreland propagating stacking sequence for the pre-Carboniferous units requires that the ultramafic/layered complex was repeated first (Fig. 13C), then followed by repetition of the upper mafic ophiolite section and overlying sediments (Fig. 13D). Finally, that stack is emplaced over the metasediment/metavolcanic complex (Fig. 13E). If this sequence of Early to Middle Devonian top-to-the-east juxtapositions occurred in a subduction zone environment, then the simplest model is that the subduction zone was west-dipping under Chilenia. Note that the exact placement of the hemipelagic sediments and underlying mafic ophiolite section is not important, as long as they formed east of the ultramafic/layered complex.

SUMMARY AND CONCLUSION

In contrast with earlier studies that combined all the mafic and ultramafic rocks of the western Precordillera into one ophiolite unit (e.g., Haller and Ramos, 1984; Ramos et al., 1984), this study shows that the mafic and ultramafic rock units in the southwest Precordillera occur as four distinct units. One of the units is an ultramafic/layered complex associated with deep continental quartzofeldspathic gneiss and subjected to an early granulite facies metamorphism. Another unit, the mafic ophiolite section, consists of gabbro, plagiogranite, and portions of the upper sheeted diabase dike complex and lower pillow basalt section, all affected by a low-grade metamorphic event. The other two mafic rock units occur as highly altered basalt flows within the metasediment/metavolcanic complex, and as diabase to microgabbro sills housed within both the metasediment/metavolcanic complex and the carbonate-metasiltstone unit.

We assign different tectonic settings to the formation of each of the ultramafic and mafic units. The ultramafic/layered complex and associated quartzofeldspathic gneiss may represent deep continental crust at the eastern margin of Chilenia. This deep continental crust first experienced under or intraplating of ultramafic rocks (possibly in an extensional setting) followed by intrusion of early subduction-related magmatism during west-dipping subduction under eastern Chilenia. The mafic ophiolite section formed at a mid-ocean spreading ridge in the ocean between the western Precordillera terrane and eastern Chilenia. The metasediment/metavolcanic complex may have formed in an Ordovician extensional environment along the western margin of the Precordillera terrane, and the carbonate-metasiltstone unit with mafic sills might represent hemipelagic sediments deposited in either a sedimented ridge setting or on oceanic crust and later invaded by mafic sills.

The four pre-Carboniferous rock units experienced low-grade regional metamorphism and were juxtaposed along previously unrecognized ductile, top-to-the-east ductile shear zones in the Early to Middle Devonian. That event was followed by Permian folding and top-to-the-west brittle thrust faulting (documented in detail by von Gosen, 1995), Mesozoic brittle normal faulting, and Miocene and younger top-to-the-west brittle thrusting and folding. Existing tectonic models (e.g., Ramos et al., 1984, 1986; Dalla Salda et al., 1992a,b; Dalziel et al., 1994; Astini et al., 1996) assume Late Ordovician and Late Silurian or Devonian top-to-the-west deformations, but this study shows that the top-to-the-west deformation is younger, and the pre-Carboniferous deformation is top-to-the-east. Furthermore, there is no evidence for an Ordovician compressional event; recent studies (von Gosen, 1992, Keller et al., 1993, Keller, 1995) suggest that the western Precordillera experienced Middle to Late Ordovician extension.

Most models for the tectonics of the western margin of the Precordillera propose Ordovician to Devonian east-dipping subduction under the western margin of the Precordillera terrane (e.g., Ramos et al., 1984, 1986; Dalla Salda et al., 1992a,b; Dalziel et al., 1994). These models assumed top-to-the-west transport based on the apparent fold vergence of structures thought to be of Late Ordovician to Silurian age and Devonian age. In contrast, Astini et al. (1995) relied on the absence of volcanic arc rocks in the Precordillera as evidence for west-dipping subduction under Chilenia, but volcanic arc rocks of appropriate age are not known from Chilenia either. The simplest tectonic scenario for the top-to-the-east shear sense of ophiolite emplacement and structural stacking documented by this study is for the emplacement to be related to west-dipping subduction; thus we corroborate the assertion by Astini et al. (1995).

Several issues remain to be solved before a detailed tectonic model for the western margin of the Precordillera may be proposed with confidence. Primary among these issues are (1) the crystallization ages of all mafic and ultramafic rock units along the western margin of the Precordillera terrane; (2) the age of high-grade metamorphism of the ultramafic/layered complex; (3) the settings of formation of the mafic and ultramafic rocks; (4) the depositional settings and ages of the carbonate-bearing metasedimentary unit, the metasediment/metavolcanic complex and the "early Paleozoic east of Uspallata" (Buggisch et al., 1994); (5) the original geographic distribution of the four pre-Carboniferous lithologic units; (6) the lack of an identified volcanic arc of Silurian to Devonian age west or east of the western Precordillera; and (7) the character and size of Chilenia. Without resolution of these issues, the significance of the western margin of the Precordillera with respect to global terrane reconstructions and tectonic scenarios remains unknown.

ACKNOWLEDGMENTS

This work represents part of J. S. D.'s doctoral dissertation. J. S. D thanks Victor Ramos, Suzanne Kay, Miguel Haller, Roberto Caminos, and Daniel Gregori, who helped get this project off the ground and provided vital background information on the geology of the southwestern Precordillera. Conversations with numerous other geologists have provided important insights into the geology of the Precordillera. Special thanks are extended to Patricio Figueredo, who gave invaluable assistance with logistics and companionship in the field. Ross Yeoman is thanked for patiently teaching J. S. D. how to use the Argon Isotope Lab at the U.S. Geological Survey in Denver, and Rick Allmendinger is thanked for the use of his STEREONET program. Eldridge Moores and Robert Twiss provided many useful comments on, and discussions of, earlier versions of this manuscript. Funding for this project was provided by grants from the Geological Society of America, Sigma Xi, and the University of California, Davis, Department of Geology (to J. S. D), and also by National Science Foundation Grants EAR92-19385 (to E. M. Moores and S. M. Roeske) and EAR96-14826 (S. M. Roeske and E. M. Moores). This is a contribution to International Geological Correlations Program Project 376.

REFERENCES CITED

Abruzzi, J., Kay, S. M., and Bickford, M. E., 1993, Implications for the nature of the Precordilleran basement from Precambrian xenoliths in Miocene volcanic rocks, San Juan Province, Argentina, Mendoza, XII Congreso Geológico Argentino y II Congreso de Exploración de Hidrocarburos, tomo III, p. 331–339.

Astini, R. A., 1996, Las fases diastroficas del Paleozoico medio en la Precordillera del oeste Argentino-evidencias estratigráficas, XIII Congreso Geológico Argentino y III Congreso de Exploración de Hidrocarburos, tomo V, p. 509–526.

Astini, R. A., Benedetto, J. L., and Vaccari, N. E., 1995, The early Paleozoic evolution of the Argentine Precordillera as a Laurentian rifted, drifted, and collided terrane: a geodynamic model: Geological Society of America Bulletin, v. 107, p. 253–273.

Astini, R. A., Ramos, V. A., Benedetto, J. L., Vaccari, N. E., and Cañas, F. L., 1996, La Precordillera: Un terreno exótico a Gondwana, XIII Congreso Geológico Argentino y III Congreso de Exploración de Hidrocarburos, tomo V, p. 293–324.

Benedetto, J. L., 1993, La hipótesis de la aloctonia de la Precordillera Argentina: Un test estratigrafico y biogeográfico, XIII Congreso Geológico Argentino y III Congreso de Exploración de Hidrocarburos, tomo V, p. 375–384.

Benedetto, J. L., Vaccari, N. E., Carrera, M. G., and Sanchez, T. M., 1995, The evolution of faunal provicialism in the Argentine Precordillera during the Ordovician: new evidence and paleontological implications, *in* Cooper, J. D., Droser, M. L., and Finney, S. C., eds., Ordovician odyssey, short papers for the 7th International Symposium on the Ordovician System: Las Vegas, Nevada, SEPM, Pacific Section, p. 181–184.

Buggisch, W., Gosen, W. von, Henjes-Kunst, F., and Krumm, S., 1994, The age of early Paleozoic deformation and metamorphism in the Argentine Precordillera—evidence from K-Ar data: Zentralblat Geolische und Paläontologie, Teil I, p. 275–286.

Caminos, R., Cordani, U. G., and Linares, E., 1979, Geología y geocronología de las rocas metamórficas y eruptivas de la Precordillera y Cordillera Frontal de Mendoza, Republica Argentina: Actas Segundo Congresso Geológico Chileno, v. 2, p. F43–60.

Caminos, R., Nullo, F. E., Panza, J. L., and Ramos, V. A., 1993, Mapa Geológico de la Provincia de Mendoza: Secretaria de Mineria, Direccion Nacional del Servicio Geologico, Buenos Aires, Argentina.

Cucchi, R. J., 1971, Edades radimétricas y correlacion de metamorfitas de la Precordillera San Juan-Mendoza, República Argentina: Revista de la Asociación Geológica Argentina, v. XXVI, p. 503–515.

Cucchi, R. J., 1972, Geología y estructura de la Sierra de Cortaderas, San Juan–Mendoza, República Argentina: Revista de la Asociación Geológica Argentina, v. XXVII, no. 2, p. 229–248.

Cuerda, A., Cingolani, C., Varela, R., and Schauer, O., 1987, Graptolitos Ordovícicos del "Grupo Villavicencio," flanco sudoriental de la Sierra del Tontal en el area de Santa Clara, Precordillera de San Juan-Mendoza, República Argentina: IV Congreso Latinoamericano de Paleontología, Bolivia, tomo I, p. 111–118.

Dalla Salda, L. H., Cingolani, C. A., and Varela, R., 1992a, Early Paleozoic orogenic belt of the Andes in southwestern South America: result of Laurentia-Gondwana collision?: Geology, v. 20, p. 616–620.

Dalla Salda, L. H., Dalziel, I. W. D., Cingolani, C. A., and Varela, R., 1992b, Did the TaconicAppalachians continue into South America?: Geology, v. 20, p. 1059–1062.

Dalrymple, G. B., Alexander, E. C., Jr., Lanphere, M. A., and Kraker, G. P., 1981, Irradiation of samples for $^{40}Ar/^{39}Ar$ dating using the Geological Survey TRIGA reactor: U.S. Geological Survey Professional Paper 1176.

Dalziel, I. W. D., 1991, Pacific margins of Laurentia and East Antarctica-Australia as a conjugate rift pair: evidence and implications for an Eocambrian supercontinent; Geology: v. 19, p. 598–601.

Dalziel, I. W. D., 1997, Neoproterozoic-Paleozoic geography and tectonics: review, hypothesis, environmental speculation: Geological Society of America Bulletin, v. 109, p. 16–42.

Dalziel, I. W. D., Dalla Salda, L. H., and Gahagan, L. M., 1994, Paleozoic Laurentia-Gondwana interaction and the origin of the Appalachian-Andean mountain system: Geological Society of America Bulletin, v. 106, p. 243–252.

Dalziel, I. W. D., Dalla Salda, L. H., Cingolani, C. A., and Palmer, P., 1996, The Argentine Precordillera: a Laurentian terrane? Penrose Conference Report: GSA Today, v. 6, p. 16–18.

Davis, J. S., 1997, The tectonics of ancient plate margins of the western Americas [Ph.D. thesis]: Davis, University of California, 199 p.

Davis, J. S., Roeske, S. M., and Moores, E. M., 1994, A nearly complete pre-Carboniferous ophiolite in the S.W. Precordillera, Argentina requires new tectonic models: Geological Society of America Abstracts with Programs, v. 26, no. 7, p. A503.

Davis, J. S., Moores, E. M., Kay, S. M., and McClelland, W. C., 1995a, New information sheds light on the nature and emplacement history of the Precordillera Ophiolite, western Argentina: Geological Society of America Abstracts with Programs, v. 27, no. 6, p. 124.

Davis, J. S., Moores, E. M., Roeske, S. M., Kay, S. M., McLelland, W. C., and Snee, L. W., 1995b, New data from the western margin of the Precordillera terrane, Argentina, constrain scenarios for the middle Paleozoic tectonics of western South America: Laurentian-Gondwanan connections before Pangea Field Conference, Jujuy, Argentina, Program with Abstracts, p. 15–16.

de Romer, H. S., 1964, Sobre la geología de la zona de "El Choique" entre el Cordón de Farellones y el Cordón de Bonilla, Quebrada de Santa Elena, Uspallata, Mendoza: Revista de la Asociación Geológica Argentina, tomo XIX, p. 9–18.

Dias, H. D., and Tonel, M. Zanoni de, 1987, La filiación ofiolítica de las rocas ultramáficas de la Sierra de Cortaderas (Depto. Las Heras, Mendoza) y su significación metalogenética en la fijación de pautas de prospección: X° Congreso Geológico Argentino, tomo I, p. 61–64.

Haller, M., and Ramos, V. A., 1984, Las ofiolitas Famatinianas (Eopaleozoico) de las provincias de San Juan y Mendoza: Noveno Congreso Geológico Argentino, S.C. Bariloche, tomo II, p. 66–83.

Harrington, H. J., 1971, Descripción Geológica de la Hoja 22c, "Ramblon," Provincias de Mendoza y San Juan: Dirección Nacional de Geología y Minería, Boletín No. 114, Buenos Aires.

Hirth, G., and Tullis, J., 1992, Dislocation creep regimes in quartz aggregates: Journal of Structural Geology, v. 14, p. 145–159.

Howell, D., Jones, D. L., and Schermer, E. R., 1985, Tectono-stratigraphic terranes of the Circum-Pacific region, *in* Howell, D. G., ed., Tectonostratigraphic terranes of the Circum-Pacific region, Circum-Pacific Council for Energy and Mineral Resources, Earth Science Series, no. 1, p. 3–30.

Jan, M. Q., and Jabeen, N., 1991, A review of mafic-ultramafic plutonic complexes in the Indus suture zone of Pakistan, *in* Sharma, K. K., ed., Geology and geodynamic evolution of the Himalayan collision zone: Physics and Chemistry of the Earth, v. 17, p. II, p. 93–113.

Kay, S. M., Ramos, V. A., and Kay, R., 1984, Elementos mayoritarios y trazas de las vulcanitas Ordovicicas de la Precordillera Occidental: Basaltos de rift oceánico temprano(?) próximos al margen continental: Noveno Congreso Geológico Argentino, S.C. Bariloche, tomo II, p. 48–65.

Kay, S. M., Orrell, S., and Abbruzzi, J. M., 1996, Zircon and whole rock Nd-Pb isotopic evidence for a Grenville age and a Laurentian origin for the basement of the Precordilleran terrane in Argentina: Journal of Geology, v. 104, p. 637–648.

Keidel, J., 1939, Las estructuras de corrimientos Paleozoicos: Revista de la Sociedad Argentina de Ciencias Naturales, v. 14, 96 p.

Keller, M., 1995, Continental slope deposits in the Argentine Precordillera: sediments and geotectonic significance, *in* Cooper, J. D., Droser, M. L., and Finney, S. C., eds., Ordovician odyssey, short papers for the 7th International Symposium on the Ordovician System: Las Vegas, Nevada, SEPM, Pacific Section, p. 211–215.

Keller, M., Eberlein, S., and Lehnert, O., 1993, Sedimentology of Middle Ordovician carbonates in the Argentine Precordillera: evidence of regional relative sea-level changes: Geologische Rundshau, v. 82, p. 362–377.

Krogh, T. E., 1973, A low-contamination method for hydrothermal decomposition of zircon and extraction of U and Pb for isotopic age determinations: Geochimica et Cosmochimica Acta, v. 37, p. 485–494.

Kury, W., 1993, Características composicionales de al Formación Villavicencio, Devónico, Precordillera de Mendoza: XII Congreso Geológico Argentino y III Congreso de Exploración de Hidrocarburos, tomo III, p. 321–328.

Loske, W., 1993, La Precordillera del oeste Argentino: Una cuenca de back-arc en el Paleozoico: XII Congreso Geológico Argentino y III Congreso de Exploración de Hidrocarburos, tomo III, p. 5–13.

Ludwig, K. R., 1991, Isoplot; a plotting and regression program for radiogenic-isotope data; Version 2.53: U.S. Geological Survey Open-File Report, OF91-0445.

Ludwig, K. R., 1992, User's manual for Analyst Version 2.00; an IBM-PC computer program for control of thermal-ionization, single collector mass-spectrometer: U.S. Geological Survey Open-File Report, OF92–0543.

Mattinson, J. M., 1987, U-Pb ages of zircons: a basic examination of error propagation: Chemical Geology, v. 66, p. 151–162.

McClelland, W. C., and Mattinson, J. M., 1996, Resolving high precision ages from Tertiary plutons with complex zircon systematics: Geochimica et Cosmochimica Acta, v. 60, p. 3955–3965.

Moores, E. M., 1982, Origin and emplacement of ophiolites: Reviews of Geophysics and Space Physics, v. 2, p. 735–760.

Mpodozis, C., and Ramos, V. A., 1989, The Andes of Chile and Argentina, in Erickson, G. E., Cañas Pinochet, M. T., and Reinemud, J. A., eds., Geology of the Andes and its relation to hydrocarbon and mineral resources, Circum-Pacific Council for Energy and Mineral Resources, Earth Sciences Series, v. 11, p. 59–90.

Nicolas, A., 1989, Structures of ophiolites and dynamics of oceanic lithosphere: Dordrecht, Netherlands, Kluwer, Petrology and Structural Geology v, 4, 367 p.

Parrish, R. R., Roddick, J. C., Loveridge, W. D., and Sullivan, R. W., 1987, Uranium-lead analytical techniques at the geochronology laboratory, Geological Survey of Canada, *in* Radiogenic age and isotope studies, Report 1: Geological Survey of Canada Professional Paper 87-2, p. 3–7.

Ramos, V. A., 1988, The tectonics of the central Andes; 30 to 33° S. Latitude: Geological Society of America Special Paper 218, p. 31–54.

Ramos, V. A., and Kay, S. M., 1991, Triassic rifting and associated basalts in the Cuyo basin, central Argentina, *in* Harmon, R. S., and Rapela, C. W., eds., Andean magmatism and its tectonic setting: Geological Society of America Special Paper 265, p. 79–91.

Ramos, V. A., Jordan, T. E., Allmendinger, R., and Kay, S. M, 1984, Chilenia: Un terreno alóctono en la evolución de los Andes centrales: Noveno Congreso Geológico Argentino, S.C. Bariloche, tomo 2, p. 84–106.

Ramos, V. A., Jordan, T. E., Allmendinger, R. W., Mpodozis, C., Kay, S. M., Cortes, M., and Palma, M., 1986, Paleozoic terranes of the central Argen-

tine-Chilean Andes: Tectonics, v. 5, no. 6, p. 855–880.
Ramos, V. A., Vujovich, G., Kay, S. M., and McDonough, M., 1993, La orogénisis de Greville en las Sierras Pampeansas Occidentales: La Sierra de Pie de Palo y su integración al supercontinente Proterozoico: XII Congreso Geológico Argentino y III Congreso de Exploración de Hidrocarburos, tomo III, p. 343–357.
Roddick, J. C., 1983, High precision intercalibration of Ar-Ar standards: Geochimica et Cosmochimica acta, v. 47, p. 887–898.
Samson, S. D., and Alexander, E. C., 1987, Calibration of the interlaboratory $^{40}Ar/^{39}Ar$ dating standard MMhb-1: Chemical Geology, v. 66, p. 27–34.
Simpson, C., and Schmid, S., 1983, An evaluation of criteria to deduce the sense of movement in sheared rocks: Geological Society of America Bulletin, v. 94, p. 1281–1288.
Sisson, V. B., Pavlis, T. L., and Prior, D. J., 1994, Effects of triple junction interactions at convergent plate margins, Penrose Conference Report: GSA Today, v. 4, p. 248–249.
Snee, L. W., Sutter, J. F., and Kelly, W. C., 1988, Thermochronology of economic mineral deposits: dating the stages of mineralization at Panasqueira, Portugal, by high-precision $^{40}Ar/^{39}Ar$ age spectrum techniques on muscovite: Economic Geology, v. 83, p. 335–354.
Stacey, J. S., and Kramers, J. D., 1975, Approximation of terrestrial lead isotope evolution by a two-stage model: Earth and Planetary Science Letters, v. 26, p. 207–211.
Steiger, R. H., and Jäger, E., 1977, Subcommission on geochronolgy, convention on the use of decay constants in geo- and cosmochronology: Earth and Planetary Science Letters, v. 36, p. 359–362.
Thomas, W. A., and Astini, R. A., 1996, The Argentine Precordillera: a traveler from the Ouachita embayment of North American Laurentia: Science, v. 273, p. 752–757.
Tobisch, O. T., and Paterson, S. R., 1988, Analysis and interpretation of composite foliations in areas of progressive deformation: Journal of Structural Geology, v. 10, no. 7, p. 745–754.
Varela, R., 1973, Estudio Geotectónico del Extremo Sudoeste de la Precordillera de Mendoza, Republica Argentina: Revista de la Asociación Geológica Argentina, v. 28, no. 3, p. 241–267.
von Gosen, W., 1992, Structural evolution of the Argentine Precordillera: Journal of Structural Geology, v. 14, no. 6 p. 643–667.
von Gosen, W., 1995, Polyphase structural evolution of the southwestern Argentine Precordillera: Journal of South American Earth Sciences, v. 8, p. 377–404.
Wood, D. A., Joron, J.-L., and Treuil, M., 1979, A reappraisal of the use of trace elements to classify and discriminate between magma seris erupted in different tectonic settings: Earth and Planetary Science Letters, v. 45, p. 326–336.
York, D., 1969, Least squares fitting of a straight line with correlated errors: Earth and Planetary Science Letters, v. 5, p. 320–324.
Zierenberg, R. A., Koski, R. A., Morton, J. L., and Bouse, R. M., 1993, Genesis of massive sulfide deposits on a sediment-covered spreading center, Escanaba Trough, southern Gorda Ridge: Economic Geology and the Bulletin of the Society of Economic Geologists, v. 88, p. 2065–2094.

Manuscript Accepted by the Society September 4, 1998

Printed in USA

Geological Society of America
Special Paper 336
1999

Puncoviscana Formation of northwestern and central Argentina: Passive margin or foreland basin deposit?

J. Duncan Keppie
Instituto de Geología, Universidad Nacional Autónoma de México, Ciudad Universitaria, Delegación Coyoacan, 04510 México D.F., México
Heinrich Bahlburg
Geologisch-Paläontologisches Institut, Westfälische Wilhelms-Universität, Conenstrasse 24, 48149 Münster, Germany

ABSTRACT

Turbidites of the Late Neoproterozoic–Early Cambrian Puncoviscana Formation are currently inferred to represent the passive margin deposits built on the western edge of the South American Pampia craton that formed the conjugate margin for the eastern Laurentian passive margin. However, recent data in the Pampean orogen indicates that it is contemporaneous with deposition of the Puncoviscana Formation. This fact, together with the upward increasing rate of sedimentation and the orogenic provenance of the Puncoviscana Formation, suggests that it was deposited in a foreland basin. This is supported by the Middle Cambrian polyphase deformation of the Puncoviscana Formation as the deformation front advanced into the foreland basin, and with the post-tectonic intrusion of peraluminous and calc-alkaline granites that may be the product of melting overthickened crust. Such a reinterpretation of the Puncoviscana Formation suggests that, if a conjugate passive margin existed at all on the western edge of Late Proterozoic–Early Cambrian South America, it was not located on the western edge of the Pampia craton but farther west beneath the Andean overthrusts, possibly along the western margin of the Arequipa-Antofalla Terrane. Alternatively, it may bring into question the validity of the terminal Neoproterozoic reconstructions that place eastern Laurentia adjacent to western South America.

INTRODUCTION

The Puncoviscana Formation is generally considered to represent a passive margin deposit of latest Precambrian–Early Cambrian age on the western margin of the Pampia Craton (Figs. 1 and 2) (e.g., Jezek et al., 1985). This passive margin setting was taken up in most Late Proterozoic global reconstructions that place western South America adjacent to eastern Laurentia, which then rifted apart to form the southern Iapetus Ocean (Fig. 1) (e.g., Dalziel, 1997). However, the presence of the Late Proterozoic–Early Cambrian Pampean Orogen through this part of western South America raises questions about this hypothesis (Kraemer et al., 1995). This problem was generally ascribed to poor age control on the age of the Pampean Orogeny, but recent geochronologic data (Rapela and Pankhurst, 1996) indicate that orogenic activity does indeed extend from the Late Proterozoic into the Cambrian and requires a reexamination of the basis for Late Precambrian global reconstructions in this region. In this context, the depositional environment of the Puncoviscana Formation is critical, and forms the topic of this chapter.

GEOLOGIC SETTING

Southern South America is cored by an Archean–Middle Proterozoic craton bordered on its western side by two Late Proterozoic terranes: the Córdoba and Pampian arc terranes, which were accreted during latest Proterozoic–Early Cambrian times (Fig. 2) (Astini et al., 1996; Rapela and Pankhurst, 1996).

Keppie, J. D., and Bahlburg, H., 1999, Puncoviscana Formation of northwestern and central Argentina: Passive margin or foreland basin deposit? *in* Ramos, V. A., and Keppie, J. D., eds., Laurentia-Gondwana Connections before Pangea: Boulder, Colorado, Geological Society of America Special Paper 336.

Figure 1. Terminal Neoproterozoic (550 Ma) reconstruction showing the location of the Pampean orogen and the Puncoviscana Formation modeled as a passive margin deposit (modified after Dalziel, 1997). RP = Río de la Plata craton; PV = Puncoviscana Formation.

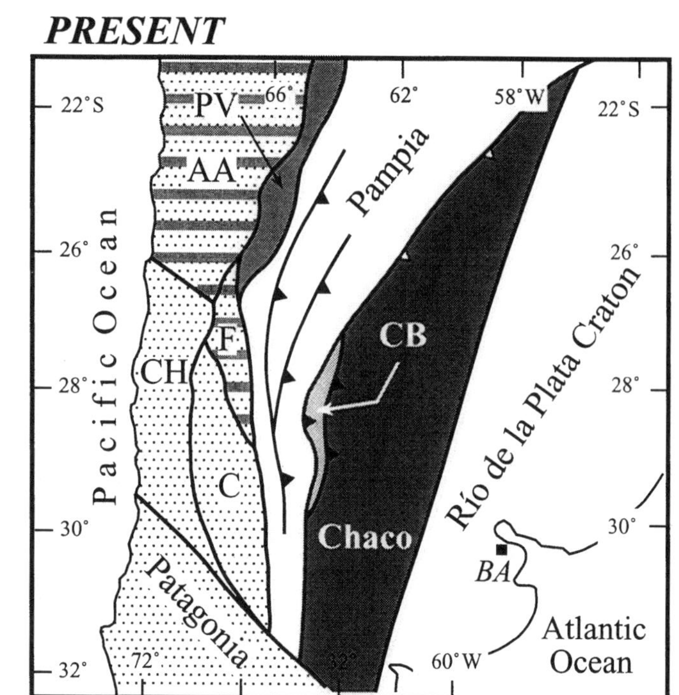

Figure 2. Terrane map of southern South America showing the possible extent of the latest Neoproterozoic–Early Cambrian Pampean Orogen, polarity of overthrusting, and the Puncoviscana Formation modeled as a foreland basin deposit (modified from Astini et al., 1996, and Bahlburg and Hervé, 1997). AA = Arequipa-Antofalla Terrane; *BA* = Buenos Aires; C = Cuyania Terrane; CB = Córdoba Terrane; CH = Chilenia; F = Famatina Terrane; PV = Puncoviscana belt.

These are bounded farther west by several terranes accreted during the Ordovician: (1) the Famatina Terrane, an Ordovician arc built on the Pampian Terrane (Breitkreuz et al., 1989; Mannheim, 1993; Rapela et al., 1996; Pankhurst et al., 1998); (2) the Cuyania Terrane, a fragment of southern Laurentia (Keppie, 1991; Astini et al., 1995; Keppie et al., 1996); and (3) the Arequipa-Antofalla Terrane, a microcontinental fragment of South America, which includes the Ordovician Puna retroarc-foreland basin (Bahlburg, 1991; Forsythe et al., 1993; Astini et al., 1996; Conti et al., 1996; Lucassen et al., 1996; Bahlburg and Herve, 1997; Bahlburg, 1998). The Famatina and Arequipa-Antofalla Terranes are generally inferred to represent part of the Pampia Terrane in the latest Proterozoic–Early Cambrian, which were rifted off Pampia during the Cambrian, only to be accreted again in the Ordovician (Forsythe et al., 1993; Conti et al., 1996). Geochemical data indicate that the Ordovician Puna mafic and ultramafic rocks represent arc and backarc tectonic settings (Bahlburg et al., 1997). In the Devonian, Chilenia was accreted to the western margin of Cuyania (Astini et al., 1996).

The Puncoviscana Formation outcrops near the northwestern margin of the Pampia Craton (Figs. 2 and 3). It consists mainly of sand- and gravel-dominated channel-fill turbidites and debris flow deposits, increasingly mud-rich turbidite deposits of middle and outer fan facies, and red pelites of hemipelagic origin reaching thicknesses of 1000 m (Jezek et al., 1985; Jezek, 1986, 1990). Willner et al. (1987) showed that the Puncoviscana Formation youngs northward along the strike of the basin. Jezek (1986, 1990) has the Puncoviscana Formation subdivided into three stages: (1) the first stage is characterized by sand-rich and partly channellized turbidites of middle fan channel and lobe facies; followed by (2) deposits of marginal middle fan to outer fan regions alternating with hemipelagic red pelites; overlain by (3) spatially more restricted deposits showing evidence of slumping, and including slide masses, debris flow deposits, and widespread water escape structures indicating an increase of sedimentation rates and tectonic segmentation of the basin in this third interval. The increasing rate of sedimentation may be the result of increasing relief caused by tectonic loading and thrust advance. In peripheral foreland basins, deposition of thick turbidite units marks a stage of sediment underfilling as the creation of accommodation space through thrust advance outpaces sedimentation rates. Once this relationship is reversed, basins tend to be overfilled (Sinclair and Allen, 1992). In such a context, the third stage of Puncoviscana deposition may mark an increased rate of collision-related thrust advance. In places, the turbidites are associated with generally fault-bounded carbonate deposits up to 280 m thick (Salfity et al., 1975). They may either

CAMBRIAN

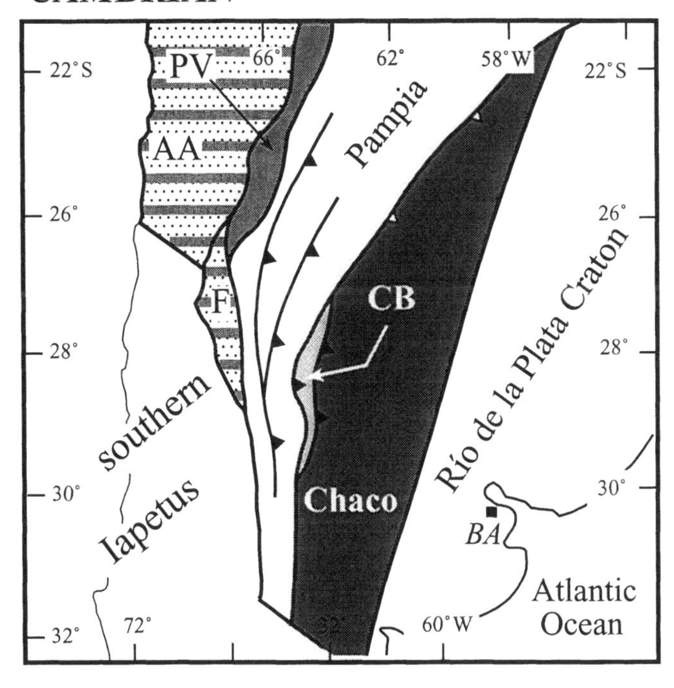

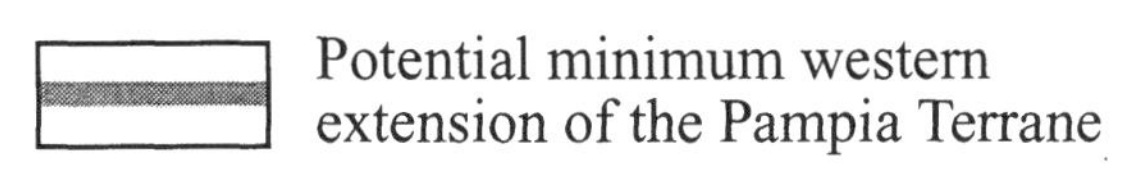

Figure 3. Terrane and continental margin configuration of southwestern South America in the Early Cambrian after the collision of the Chaco and Córdoba Terranes with the Pampia Terrane leading to the formation of the Puncoviscana foreland basin and foldbelt. For abbreviations, see Figure 2.

represent olistostromal blocks or autochthonous deposits on submarine swells (Jezek, 1986).

Current directions are essentially unipolar and directed toward the northwest and west. QFL and QmFLt compositions of turbidite sandstones generally fall in the field of "recycled orogen provenance" of Dickinson and Suczek (1979; Jezek, 1990). The QpLvLs compositions plot in the field of "collision orogen sources" and display an almost complete absence of volcanic rock fragments in the Puncoviscana rocks. The Puncoviscana Formation has been interpreted as a prograding submarine fan deposit by Omarini (1983).

The Puncoviscana Formation contains a diverse ichnofauna, including medusoid impressions and *Oldhamia*, which indicate a Vendian–Early Cambrian (Tommotian) age of the respective strata (Aceñolaza and Durand, 1986). Minor intercalations of trachytes and partly alkaline basalts and hyaloclastic rocks also occur (Toselli and Aceñolaza, 1984; Omarini and Alonso, 1987). Clear, euhedral, long prismatic and idiomorphic, detrital zircons, which have not suffered significant transport from a rhyolitic source, occur in the third-stage deposits and have lower intercept ages of between 530 and 560 Ma (Lork et al., 1990). These and several populations of rounded zircons have upper intercepts between ca. 1700 and 1800 Ma (Lork et al., 1990).

The Puncoviscana Formation was complexly deformed during the Pampean Orogeny (Early Cambrian) before being unconformably overlain by quartz arenites and shales of the Late Cambrian Mesón Group (Turner, 1960). The Puncoviscana Formation was also intruded by several calc-alkaline and peraluminous granites, including the Santa Rosa de Tastil and Cañaní plutons. Zircon geochronology of the former resulted in a U-Pb age of 536 ± 7 Ma, whereas the latter yielded ages of 519 and 534 Ma (Bachmann et al., 1987). The smaller Tipayoc stock, in turn, gave a K-Ar muscovite age of 550 ± 26 Ma (Omarini et al., 1996), which is essentially the same as the U-Pb ages within errors. These data indicate that deposition of the third stage of the Puncoviscana Formation occurred between 560 and 536 Ma.

East and southeast of the Puncoviscana outcrops lie the Pampia and Córdoba Terranes (Figs. 2 and 3), which are made up of gneisses and schists intruded by voluminous 560–520-Ma plutons that have arc-collisional characteristics in the Córdoba Terrane (Ramos et al., 1993; Rapela and Pankhurst, 1996; Llambias et al., 1996; Sato et al., 1996; Pankhurst et al., 1997). The Córdoba Terrane records a complex tectonic and magmatic history (Pankhurst et al., 1997): (1) mafic magmatism; (2) 530-Ma Cambrian granulite-grade metamorphism at ±6 kb and 800°C; and (3) widespread anatexis resulting in intrusion of cordierite-rich and S-type granites at 522 ± 2 Ma. These events are interpreted in terms of subduction of the Pampia Terrane beneath the Córdoba and Chaco Terranes followed by collision (Fig. 4) (Ramos et al., 1993; Astini et al., 1996; Pankhurst et al., 1997, 1998). This produced an orogenic belt about 2,000 by 300 km that underwent ca. 23 km of exhumation to expose the granulite facies rocks. Peripheral effects of this collision may be represented by the mid-Cambrian deformation of the Puncoviscana Formation and followed by anatexis, producing peraluminous and calc-alkaline granites that intrude the Puncoviscana Formation.

TECTONIC INTERPRETATION

Almost all previous interpretations of the Puncoviscana Formation have referred it to a passive margin sequence. Only Kraemer et al. (1995), based on new structural data in the southernmost Pampia Terrane, suggested that it might represent a foreland basin deposit. The improved geochronologic data base supports the interpretation that deposition of the Puncoviscana Formation was synchronous with the formation of a magmatic arc and a collisional orogen in the Sierras Pampeanas centered on the Córdoba Terrane. In such a context, the Puncoviscana Formation may have been deposited in a periarc basin (forearc or backarc, and peripheral foreland or retroarc basin, depending on the arc polarity). Assuming the inference by Astini et al. (1996) that the Pampia Terrane was subducted beneath the Córdoba and Chaco Terranes, this would place the Puncoviscana in a peripheral foreland basin setting (Fig. 4). Pankhurst et al. (1997) suggest that this was followed by collision culminating with

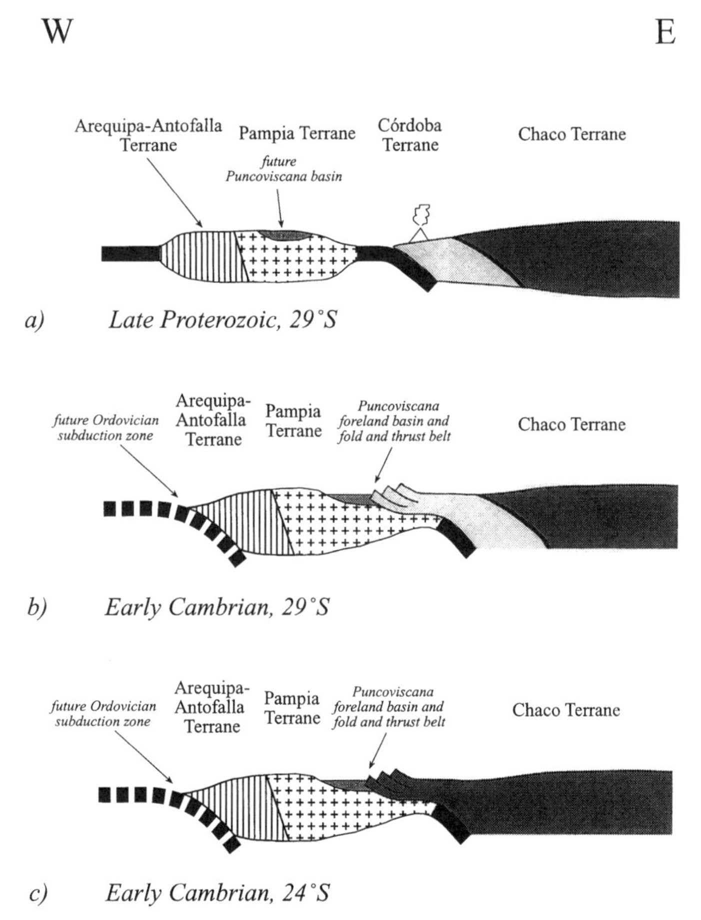

Figure 4. Late Neoproterozoic–Early Cambrian tectonic model for the Puncoviscana Formation, and the Pampia, Córdoba and Chaco Terranes at 29°S and 24°S.

granulite facies metamorphism at 530 Ma and anatexis at 522 Ma. The time difference between collision, peak metamorphism, and anatexis may be explained by depression of the isotherms followed by gradual reequilibration. The vast volume of sediment produced by latest Proterozoic–Early Cambrian exhumation could have been deposited in the Puncoviscana foreland basin, an interpretation consistent with the upward increasing rate of sedimentation inferred by Jezek (1986). The advance of the deformation front westward could explain the mid-Cambrian deformation of the Puncoviscana belt.

The interpretation of the Puncoviscana Formation as a foreland basin deposit has implications for the palinspastic reconstructions that place eastern Laurentia adjacent to western South America in the Late Proterozoic, that is, Pannotia (e.g., Dalziel, 1997). Proof of such reconstructions commonly suggest that the conjugate passive margin to the eastern Laurentian miogeocline is the Puncoviscana Formation, an invalid proof if the Puncoviscana Formation was deposited in a foreland basin. Nevertheless, such a reconstruction may still be valid if the South American passive margin lies farther to the west, such as on the western margin of the Arequipa-Antofalla Terrane (Figs. 1 and 3). Paleomagnetic data suggests that during the Cambrian and Early Ordovician, the Arequipa-Antofalla Terrane was a peninsula of the Amazon craton with which it was telescoped during the mid-Ordovician (Forsythe et al., 1993). On the other hand, recent dating in Antarctica indicates that the assumed Grenvillian–older basement piercing point is also invalid (Dalziel, 1997). This suggests that alternate Late Proterozoic reconstructions, such as that proposed by Keppie et al. (1996), which places eastern Laurentia adjacent to North Africa, might be considered more carefully.

ACKNOWLEDGMENTS

We thank A. Lork of Hamburg and W. Loske of Münster for the critical discussion of radiometric data, our ideas, and the manuscript itself, and Pablo Kraemer and an anonymous reviewer for their constructive reviews of the manuscript. This work is a contribution to International Geological Correlation Program Projects 345 and 376.

REFERENCES CITED

Aceñolaza, G. F., and Durand, F. R., 1986, The biota of the Upper Precambrian–Lower Cambrian from the northwest of Argentina: Geological Magazine, v. 123, p. 367–375.

Astini, R. A., Benedetto, J. L., and Vaccari, N. E., 1995, The Early Paleozoic evolution of the Argentine Precordillera as a Laurentian rifted, drifted and collided terrane: Geological Society of America Bulletin, v. 107, p. 253–273.

Astini, R. A., Ramos, V. A., Benedetto, J. L., Vaccari, N. E., and Cañas, F. L., 1996, La Precordillera: un terreno exótico a Gondwana: XIII Congreso Geológico Argentino Actas, v. 5, p. 293–324.

Bachmann, G., Grauert, B., Kramm, U., Lork, A., and Miller, H., 1987, El magmatismo del Cámbrico Medio/Cámbrico Superior en el basamento del noroeste Argentino: Investigaciónes isotópicas y geochronológicas sobre los granitoides de los complejos intrusivos de Santa Rosa de Tastil y Cañani: X Congreso Geológico Argentino Actas, v. 4, p. 125–127.

Bahlburg, H., 1991, The Ordovician back-arc to foreland successor basin in the Argentinian-Chilean Puna: tectonosedimentary trends and sea-level changes, *in* Macdonald, D. I. W., ed., Sedimentation, tectonics, and eustasy: International Association of Sedimentologists Special Publications, v. 12, p. 465–484.

Bahlburg, H., 1998, The geochemistry and provenance of Ordovician turbidites in the Argentinian Puna, *in* Pankhurst, R. J., and Rapela, C. W., eds., The Proto-Andean margin of Gondwana: Geological Society of London Special Publication 142, p. 127–142.

Bahlburg, H., and Herve, F., 1997, Geodynamic evolution and tectonostratigraphy of northwestern Argentina and northern Chile: Geological Society of America Bulletin, v. 109, p. 869–884

Bahlburg, H., Kay, S. M., and Zimmerman, U., 1997, New geochemical and sedimentological data on the evolution of the Early Paleozoic Gondwana margin in the southern central Andes: Geological Society of America Annual Meeting Abstracts with Programs, v. 29, no. 6, p. A378.

Breitkreuz, C., Bahlburg, H., Delakowitz, B., and Pichowiak, S., 1989, Volcanic events in the Paleozoic central Andes: Journal of South American Earth Sciences, v. 2, p. 171–189.

Conti, C. M., Rapalini, A. E., Coira, B., and Koukharsky, M., 1996, Paleomagnetic evidence of an early Paleozoic rotated terrane in northwest Argentina: a clue for Gondwana-Laurentia interaction?: Geology, v. 24, p. 953–956.

Dalziel, I. W. D., 1997, Neoproterozoic-Paleozoic geography and tectonics: review, hypothesis, environmental speculation: Geological Society of America Bulletin, v. 109, p. 16–42.

Dickinson, W. R., and Suczek, C. A., 1979, Plate tectonics and sandstone composition: American Association of Petroleum Geologists Bulletin, v. 63, p. 2164–2182.

Forsythe, R. D., Davidson, J., Mpodois, C., and Jesinkey, C., 1993, Lower Paleozoic relative motion of the Arequipa block and Gondwana: paleomagnetic evidence from Sierra de Almeida of northern Chile: Tectonics, v. 12, p. 219–236.

Jezek, P., 1986, Petrographie und Fazies der Puncoviscana Formation, einer turbiditischen Folge im Jungpräkambrium und Unterkambrium Nordwest-Argentiniens: Inaugural-Dissertation, Fachbereich Geowissenschaften: Muenster, Germany, Westfaelische Wilhelms-Universitaet, 144 p.

Jezek, P., 1990, Análisis sedimentológico de la Formación Puncoviscana entre Tucumán y Salta, *in* Aceñolaza, F., Miller, H., and Toselli, A., eds., El Ciclo Pampeano en el noroeste Argentino: Serie Correlación Geológica, v. 4, p. 9–36.

Jezek, P., Willner, A. P., Aceñolaza, F., and Miller, H., 1985, The Puncoviscana trough—a large basin of late Precambrian to Early Cambrian age on the Pacific edge of the Brasilian Shield: Geologische Rundschau, v. 74, p. 573–584.

Keppie, J. D., 1991, Avalon, an exotic Appalachian-Caledonide terrane of western South American provenance, *in* 5th Circum-Pacific Terrane Conference: Santiago, Chile, Resúmenes Expandidos, Comunicaciones Universidad de Chile, v. 42, p. 109–111.

Keppie, J. D., Dostal, J., Murphy, J. B., and Nance, R. D., 1996, Terrane transfer between eastern Laurentia and western Gondwana in the early Paleozoic: constraints on global reconstructions, *in* Nance, R. D., and Thompson, M. D., eds., Avalonian and related peri-Gondwanan terranes of the circum–North Atlantic: Geological Society of America Special Paper, v. 304, p. 369–380.

Kraemer, P. E., Escayola, M. P., and Martino, R. D., 1995, Hipótesis sobre la evolución neoproterozoica de las Sierras Pampeanas de Córdoba (30°40′–32°40′S), Argentina: Revista de la Asociación Geológica Argentina, v. 50, p. 47–59.

Llambías, E. J., Sato, A. M., Prozzi, C., and Sánchez, V., 1996, Los pendentes en el Plutón Gasarillo: evidencia de un metamorfismo pre-Famatiniano en las Sierras de San Luis: XIII Congreso Geológico Argentino Actas, v. 5, p. 369–376.

Lork, A., Miller, H., Kramm, U., and Grauert, B., 1990, Sistemática U-Pb de circones detríticos de la Fm. Puncoviscana y su significado para la edad máxima de sedimentación en la Sierra de Cachi (Prov. de Salta, Argentina), *in* Aceñolaza, F., Miller, H., and Toselli, A., eds., El Ciclo Pampeano en el noroeste Argentino: Serie Correlación Geológica, v. 4, p. 199–208.

Lucassen, F., Wilke, H. G., Viramonte, J., Beccio, R., Franz, G., Laber, A., Wemmer, K., and Vroon, P., 1996, The Paleozoic basement of the central Andes (18°–26°S): a metamorphic view: Paris, 3rd International Symposium on Andean Geodynamics, collection Colloques et Séminaires, p. 779–782.

Mannheim, R., 1993, Genese der vulkanite und subvulkanite des altpaläozoischen Famatina-syestems, NW-Argentinien, und seine geodynamische entwicklung: Münchner Geologische Hefte, v. 9, p. 130.

Omarini, R. H., 1983, Caracterización litológia, diferenciación y génesis de la Formación Puncoviscana entre el Valle de Lerma y la Faja Eruptiva de la Puna: [Ph.D. thesis]: Salta, Argentina, Universidad Nacional de Salta, 202 p.

Omarini, R. H., and Alonso, R. N., 1987, Lavas en la Formación Puncoviscana, Río Blanco, Salta, Argentina: X Congreso Geológico Argentino Actas, v. 4, p. 292–295.

Omarini, R. H., Torres, J. G., and Moya, L. S., 1996, El stock Tipayoc: génesis magmática en un arco de Islas del Cámbrico Inferior en el Noroeste de Argentina: XIII Congreso Geológico Argentino Actas, v. 5, p. 545–560.

Pankhurst, R. J., Rapela, C. W., and Saavedra, J., 1997, The Sierras Pampeanas of NW Argentina—growth of the Pre-Andean margin of Gondwana, European Union of Geosciences, EUG Abstract Supplement 1: Terra Nova, v. 9, p. 162.

Pankhurst, R. J., Rapela, C. W., Saavedra, J., Baldo, E., Dahlquist, J., Pascula, I., and Fanning, C. M., 1998, The Famatinian magmatic arc in the central Sierras Pampeanas, *in* Pankhurst, R. J., and Rapela, C. W., eds., The Proto-Andean margin of Gondwana: Geological Society of London Special Publication 142, p. 343–367.

Ramos, V. A., Vujovich, G., Mahlburg Kay, S., and McDonough, M., 1993, La orogénesis de Grenville en las Sierras Pampeanas occidentales: la Sierra Pie de Palo y su integración al supercontinente Proterozoico: XII Congreso Geológico Argentino y II Congreso de Exploración de Hidrocarburos Actas, no. III, p. 343–358.

Rapela, C. W., and Pankhurst, R. J., 1996, The Cambrian plutonism of the Sierra de Córdoba: pre-Famatinian subduction? and crustal melting: XIII Congreso Geológico Argentino Actas, v. 5, p. 491–492.

Rapela, C. W., Saavedra, J., Toselli, and Pellitero, E., 1996, Eventos magmáticos fuertemente peraluminosos en las Sierras Pampeanas: XIII Congreso Geológico Argentino Actas, v. 5, p. 337–354.

Salfity, J. A., Omarini, R. H., Baldis, B., and Gutierrez, W. J., 1975, Consideraciones sobre la evolución geológica del Precámbrico y Paleozoico del norte Argentino: II Congreso Iberoamericano de Geología Económica, v. 4, p. 341–361.

Sato, A. M., Tickyi, H., and Llambias, E. J., 1996, Geología de los granitoides aflorantes en el Sur de la Provincia de la Pampa, Argentina: XIII Congreso Geológico Argentino Actas, v. 5, p. 429–440.

Sinclair, H. D., and Allen, P. A., 1992, Vertical versus horizontal motions in the Alpine orogenic wedge: stratigraphic response in the foreland basin: Basin Research, v. 4, p. 215–232.

Toselli, A. J., and Aceñolaza, F. G., 1984, Presencia de eruptivas basálticas en afloramientos de la Formación Puncoviscana, en Coraya, Departamento Humahuaca, Jujuy: Revista de la Asociación Geológica Argentina, v. 39, p. 158–159.

Turner, J. C. M., 1960, Estratigrafía de la Sierra de Santa Victoria y adyaciencias: Boletín de la Academia Nacional de Ciencias de Córdoba, v. 41, p. 163–196.

Willner, A., Lottner, U., and Miller, H., 1987, Early Paleozoic structural development in the NW Argentine basement of the Andes and its implications for geodynamic reconstructions, in McKenzie, G. D., ed., Gondwana six: Structure, tectonics and geophysics: American Geophysical Union Monograph, v. 40, p. 229–239.

MANUSCRIPT ACCEPTED BY THE SOCIETY SEPTEMBER 4, 1998

Printed in U.S.A.

Geological Society of America
Special Paper 336
1999

Magmatic sources and tectonic setting of Gondwana margin Ordovician magmas, northern Puna of Argentina and Chile

Beatriz Coira, Belén Peréz, Patrocinio Flores
CONICET and Universidad Nacional de Jujuy, Casilla de Correo 258, (4600) S.S. de Jujuy, Argentina
Suzanne Mahlburg Kay, Bryn Woll, Maura Hanning
Department of Geological Sciences and Institute for the Study of the Continents, Snee Hall, Cornell University, Ithaca, New York 14853

ABSTRACT

Understanding the sources of Ordovician magmatic rocks in the broad western Faja Eruptiva Occidental and eastern Faja Eruptiva Oriental magmatic belts in the northern Puna of Argentina and Chile is important to Ordovician geodynamic models for western Gondwana evolution and Gondwana-Laurentian terrane interactions. A critical evaluation of existing chemical and age data, along with new major trace element data, and field observations leads to a working model in which magmatism in the western belt progressively occurred in an active to waning arc to collisional regime, whereas that in the eastern belt occurred in an oblique fault regime transitional to a subduction zone to the south. The model is hampered by data gaps in all areas, and particularly by lack of ages in western and southern regions. The best understood area is the northern Faja Eruptiva Oriental where dacitic and mafic units with published ages of 476–467 Ma occur in linear fault-controlled trends. Their chemistry is consistent with dacitic magma sources being dominated by sedimentary-type protoliths melted in association with emplacement of mantle-derived alkaline mafic magmas. These dacitic units represent lava-dome complexes emplaced contemporaneously with outer shelf and slope basin sedimentation. Farther south where deeper crustal levels are exposed, volcanic/subvolcanic units grade into poorly studied plutonic facies. Trondhjemitic bodies are reported to the east. Contemporaneous units in the Faja Eruptiva Occidental are plutons and mafic to rhyolitic lavas in sequences of volcaniclastic sediments. Chemical signatures in the poorly dated Cordón de Lila lavas and Choschas dioritic to leucogranodioritic pluton in Chile are consistent with emplacement in a magmatic arc on thinned continental or oceanic crust. Mafic volcanic units in volcanic-sedimentary sequences just to the east could represent arc volcanism behind the front. An Arenig-Llanvirn change to bimodal and predominantly silicic magmatism, the emplacement of shoshonitic and mafic andesitic flows with weak arc signatures followed by alkaline dikes in the eastern Faja Eruptiva Occidental, and the formation of the Faja Eruptiva Oriental dacitic-mafic sequences could signal a change to a very oblique subduction regime. This change would be roughly contemporaneous with the arrival of the Laurentia-derived Precordillera terrane to the south. A post–Arenig-Llanvirn compressional regime seems required to explain Ocloyic deformation in the east and late plutons to the west. The nature of the eastern boundary of a peri-Gondwana Arequipa terrane and the existence of a peri-Gondwana Famatina-Puna Oriental terrane remain unclear.

Coira, B. L., Kay, S. Mahlburg, Peréz, B., Woll, B., Hanning, M., and Flores, P., 1999, Magmatic sources and tectonic setting of Gondwana margin Ordovician magmas, northern Puna of Argentina and Chile, *in* Ramos, V. A., and Keppie, J. D., eds., Laurentia-Gondwana Connections before Pangea: Boulder, Colorado, Geological Society of America Special Paper 336.

INTRODUCTION

The possibility that parts of northwest Argentina and northern Chile were allocthonous or peri-Gondwana terranes accreted to the southwestern Gondwana margin during the early Paleozoic has been considered by numerous authors (e.g., Coira et al., 1982; Allmendinger et al., 1982, 1983; Ramos et al., 1986; Dalziel and Forsythe, 1985; Ramos, 1986, 1988; Bahlburg, 1990; Bahlburg and Breitkreuz, 1991; Forsythe et al., 1993; Conti et al., 1996; Bahlburg and Hervé, 1997). Although discussion of possible terranes in this region has been widespread, there has been little agreement as to whether they all exist and where their boundaries or what their times of accretion are. Recent models based largely on paleomagnetic results argue for peri-Gondwana terranes (see Fig. 1). In the model of Forsythe et al. (1993), the Arequipa terrane rotated southwestward (present-day direction) away from Gondwana in the late Precambrian to be resutured in the Silurian. In the model of Conti et al. (1996), a Puna Oriental-Famatina rotated terrane now inboard to the Arequipa-Antofalla terrane is accreted in mid-Ordovician time. A Cambrian-Ordovician subduction zone extending north from modern Patagonia along the Gondwana margin terminates east of the Puna Oriental-Famatina terrane (e.g., Ramos, 1988). Interest in the origin and position of possible terranes has been fueled by suggestions that parts of western Argentina and Chile are derived from Laurentia (e.g., Dalla Salda et al., 1992; Penrose Conference 1996). The derivation of the Precordillera terrane to the south (Figs. 1 and 2) from Laurentia is well accepted (e.g., Astini et al., 1995). A driving force for these models is the early Paleozoic plate reconstructions of Dalziel (e.g., 1991, 1997) that show Laurentia and Gondwana separated by the western Iapetus ocean (see Fig. 1).

Important to deciphering possible terrane interactions and boundaries in northwest Argentina and Chile is an understanding of the significance of the magmatic rocks in the two generalized Ordovician magmatic belts exposed in the Puna plateau region (see Fig. 2). These belts, known as the Faja Eruptiva Occidental in the west (Palma et al., 1986) and the Faja Eruptiva Oriental in the east (e.g., Méndez et al., 1973), have been defined and interpreted in various ways. As shown in the inset in Figure 2, the western belt is mostly in the Arequipa-Antofalla terrane of Ramos (1988) and Forsythe et al. (1993), whereas the eastern belt is in the Puna Oriental-Famatina block of Conti et al. (1996). The origins of these belts place limits on possible terrane scenarios. In the models of Dalziel and Forsythe (1985), Hervé et al. (1987), and Forsythe et al. (1993), these magmatic belts are shown as arcs above western west-dipping and eastern east-dipping subduction zones, respectively. In the models of Ramos (1988) and Rapela et al. (1992), both subduction zones dip east. In other models, the eastern belt is not associated with an arc. For Davidson and Mpodozis in Coira et al. (1982) and Aceñolaza and Toselli (1984), the eastern belt formed in a continental extensional regime between the Arequipa terrane to the west and the Brazilian craton to the east. Hanning (1987) used geochemical data to support an extensional model for the northern part of this belt. Bahlburg (1990, 1993) prescribed an extensional backarc setting for Ordovician sedimentary rocks in this region but, along with Damm et al. (1990), argued that the eastern magmatic belt represents a Late Ordovician–Silurian syntectonic intrusive sequence associated with terrane collision. Bahlburg and Hervé (1997) interpreted the northern part of the eastern belt as an extensional fault zone, and the southern part as an east-dipping arc associated with subduction. The connection is unclear.

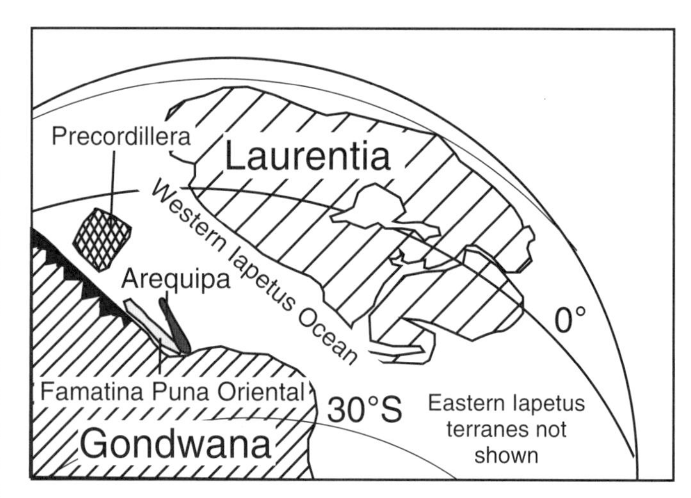

Figure 1. Map showing early Ordovician (Arenig) configuration of Laurentia and Gondwana relative to the Iapetus ocean (after Dalziel, 1997), the arriving Laurentian-derived Precordillera terrane (e.g., Astini et al., 1995; Dalziel, 1997), and previously proposed peri-Gondwana terranes, which include the rotated Arequipa terrane, as suggested by Forsythe et al. (1993), and the Famatina Puna Oriental terrane, as suggested by Conti et al. (1996). Saw-toothed pattern shows east-dipping subduction zone (e.g., Ramos, 1988).

This chapter critically reviews the field, geochronologic, petrologic, and geochemical characteristics of the Faja Eruptiva Oriental and Occidental and related magmatic rocks of northern Argentina and Chile to place limits on early Paleozoic tectonic models and terrane configurations. Sixty new major and trace element geochemical analyses from the eastern and western magmatic belts are integrated with published data to place constraints on magmatic sources and tectonic setting of the magmas. A new synthesis of field descriptions of the style of emplacement of the controversial northern Faja Eruptiva Oriental magmatic units establishes them as synsedimentary volcanic/subvolcanic rocks. A tentative three-stage tectonic scenario is presented that takes into account the available data. The tectonic details of peri-Gondwana terranes in the region remain unclear. A prevailing theme is the inadequacy of the existing data base for testing tectonic models.

ORDOVICIAN UNITS OF NORTHWESTERN ARGENTINA AND ADJACENT CHILE

The late Precambrian and early Paleozoic magmatic and sedimentary sequences in northwestern Argentina and adjacent regions in Chile discussed here are shown in Figure 2. They crop out in the Western Cordillera, Puna plateau, Eastern Cordillera, and Sierras Pampeanas geologic provinces of the modern Andes. The Argen-

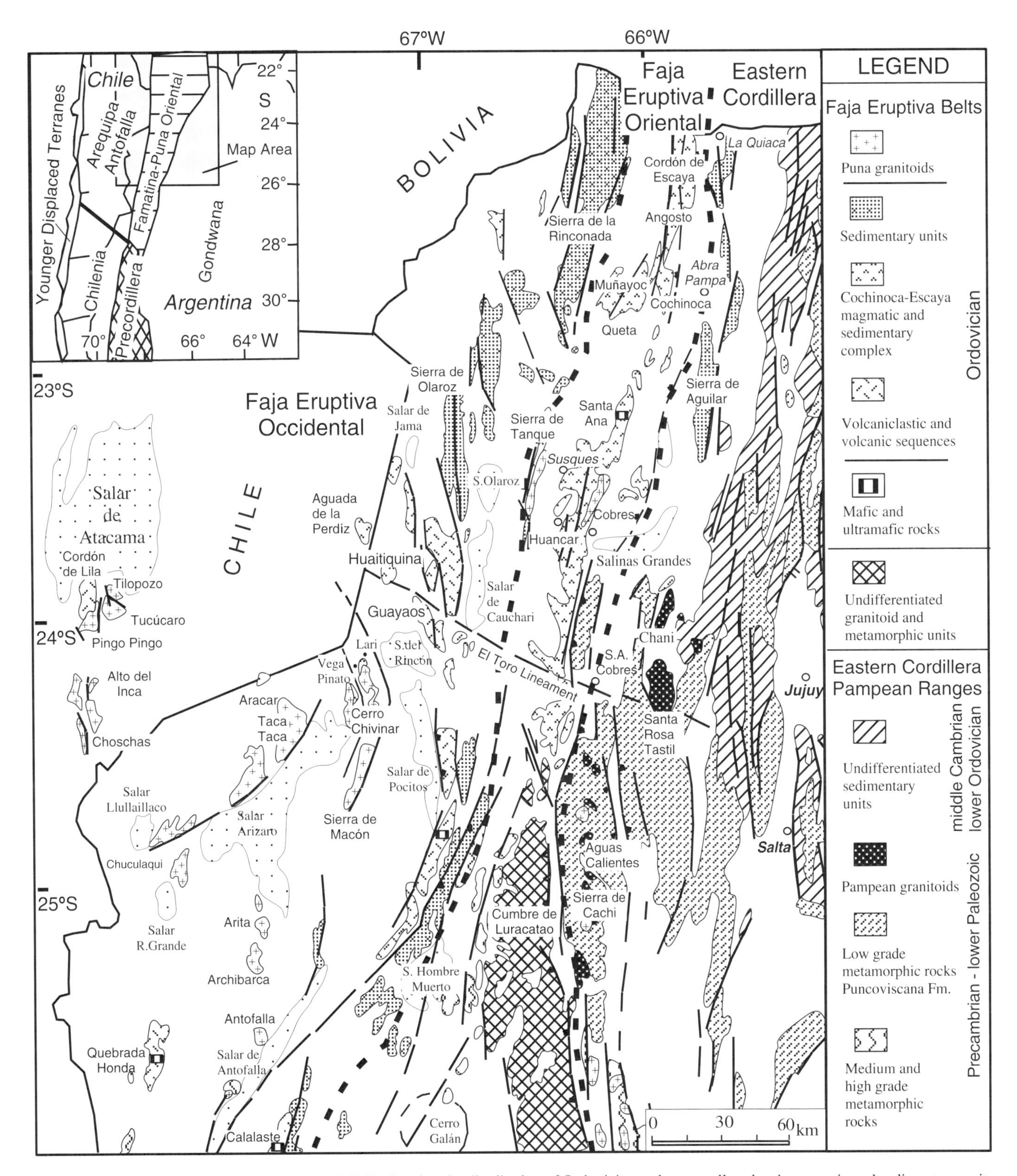

Figure 2. Map of part of northern Argentina and Chile showing the distribution of Ordovician and temporally related magmatic and sedimentary units in the region of the modern central Andean Puna plateau. Principal faults, salars (salt basins, shown by spaced dots), and towns (open circles, names in italics) are shown for reference. Generalized boundaries between the Faja Eruptiva Occidental, Faja Eruptiva Oriental, and Eastern Cordillera are shown as heavy dashed lines. Labeled localities are discussed in text. Figure is modified from Rapela et al. (1992). Inset shows the proposed Paleozoic Chilenia, Arequipa-Antofalla, Precordillera (Cuyania), and Famatina-Puna Oriental terranes in the Puna region after Ramos (1988), Bahlburg and Hervé (1997), and Conti et al. (1996).

tine Puna with the Bolivian Altiplano forms one of the largest plateaus on earth with an average elevation of 3,700 m above sea level. The plateau has been uplifted since the Early Miocene in association with Neogene crustal shortening over a steepening subduction zone (e.g., Kay et al. 1999). The Puna is bordered to the west by the Andean Central Volcanic Zone, to the south and southeast by the block-faulted Sierras Pampeanas, and to the east by the Eastern Cordillera and SubAndean fold-thrust belts. The exposed basement is generally upper Precambrian to lower Paleozoic units with much consisting of Ordovician Tremadoc to Arenig-Llanvirn magmatic and sedimentary sequences. In some regions, exposures are extremely limited as these units are covered by Tertiary volcanic rocks and salt basins (salars).

The discussion here focuses on the lower Paleozoic magmatic units exposed from just south of the El Toro lineament to the Bolivian border. They are considered relative to the three broad belts shown in Figure 2: the generalized Faja Eruptiva Occidental in the west (Palma et al., 1986), the generalized Faja Eruptiva Oriental (Méndez et al., 1973), and the Eastern Cordillera. Individual units are shown in Figure 2 and summarized in Figure 3. Critical to understanding these units is their association with surrounding sedimentary sequences and the Lower Paleozoic units of the southern Puna whose key aspects are summarized below.

A review of the sedimentary sequences in northern Puna region is given by Bahlburg (1990); those important to the discussion are shown in Figure 3. Beginning in the Faja Eruptiva Occidental, key sequences in Chile occur in the Cordón de Lila complex (see Niemeyer 1989), which consists of interlayered pre–Siluro-Devonian sedimentary and volcanic rocks. Farther east, the base of the Ordovician sequence is defined by the Las Vicuñas Formation (Moya et al., 1993), which crops out southwest of the Salar del Rincón (Fig. 2), and consists of shallow marine shelf sequences containing lower Tremadocian fossils. Correlative quartzitic sequences in the Tolar Chico Formation (Zapettini et al., 1994) and the overlying clastic units of the Tolillar Formation in the Salar del Rincón region are interlayered with mafic and ultramafic rocks (Ojo de Colorado complex) whose minimum age is poorly constrained by whole-rock K-Ar ages of 470 ± 20 and 490 ± 20 Ma in a cross-cutting dioritic stock (Blasco et al., 1996). Overlying volcaniclastic sequences with middle Arenig to Llanvirn faunas are assigned to the Aguada de la Perdiz (García et al., 1962) and equivalent formations. These units underlie the Arenig to lower Caradoc? sediments of the Puna Turbidite complex of Bahlburg (1990), which incorporates the Coquena (Schwab, 1973), Falda Ciénaga (Aceñolaza et al., 1975) and Lina (Ramos, 1972) Formations. To the east in the Faja Eruptiva Oriental, sedimentary units correlating with the Aguada de la Perdiz and Coquena Formations occur in the Cochinoca-Escaya complex of Coira et al. (1999). Still farther east, Ordovician sequences in the Eastern Cordillera are in the Tremadoc to lower Llanvirn Santa Victoria Group of Turner (1960), which consists of distal platform sandstone and shale facies (Fig. 2). They overlie Middle to Late Cambrian shallow marine clastic shelf sediments of the Mesón Group, and latest Precambrian to Cambrian (Vendian to Tommotian) turbiditic sequences of the Puncoviscana Formation. Puncoviscana sequences were deformed in the Pampean orogenic event. The easternmost deformation associated with the late Ordovician-Silurian Ocloyic event (Fig. 3) coincides with the eastern limit of the Faja Eruptiva Oriental.

Important among southern Puna magmatic units are mafic/ultramafic sequences from the Sierra de Calalaste and south of the Salar de Pocitos (Fig. 2) that occur in metamorphosed turbiditic sequences (Blasco et al., 1996; Bahlburg et al., 1997). These rocks are of oceanic origin and have been used as terrane sutures (e.g., Ramos, 1988; Conti et al., 1996). Recent work has shown that the magmatic rocks have backarc-like geochemical characteristics with the strongest arc-like tendencies in the Salar de Pocitos units (Bahlburg et al., 1997). In models such as the one of Forsythe et al. (1983), these units represent oceanic crust associated with greater extension in the south as the Arequipa terrane rotated away from Gondwana. They are assumed to be Ordovician in age (Allmendinger et al., 1982), although Mon and Hongn (1991) prefer a pre-Ordovician age based on structural style. The issue is complicated as sedimentary contacts are structural in nature. Poorly known gabbros and spilitized mafic lavas in turbiditic sediments also occur in the Quebrada Honda west of Calalaste (e.g., Kay et al., 1984) (see Fig. 2).

ORDOVICIAN AND RELATED MAGMATIC UNITS OF THE NORTHERN PUNA

The northern Puna region magmatic units that are the focus here are described from west to east (see Fig. 2). Those discussed in the Faja Eruptiva Occidental include plutons exposed from south of the Salar de Atacama to west of the Salar de Antofalla, and volcanic units interlayered in sedimentary sequences from the Guayaos region to Bolivia. Those in the Faja Eruptiva Oriental include magmatic rocks in the Cochinoca-Escaya complex and related plutons. Those in the Eastern Cordillera include late Precambrian to Ordovician plutons. Chemical analyses in Tables 1–8 are plotted with published data in Figures 4–6.

The discussion uses chemical signatures as guides to magmatic sources that provide clues to tectonic setting. Generic interpretations are based on entire analyses, not discrimination diagrams. Plots of Na_2O vs. K_2O (Fig. 4) showing fields for I (igneous)-, S (sedimentary)-, and A (anorogenic or collisional)-type granites are used for classification purposes. Some key characteristics guiding interpretations are: (1) high molar $Al_2O_3/(K_2O + Na_2O + CaO)$ ratios (>1.1) which imply Al-rich sources typically associated with sediments; (2) high La (light rare earth element [LREE]) or Ba or Th to Ta/Nb/Ti (high field strength element) ratios that can imply subduction environments (arc magmas generally have La/Ta >25, Ba/La >18); and (3) REE characteristics that are guides to source region enrichment, melting percentages, and fractionation. The Hf-Th-Ta (Wood et al., 1979) in Figure 6 provides a summary of implied sources for mafic units.

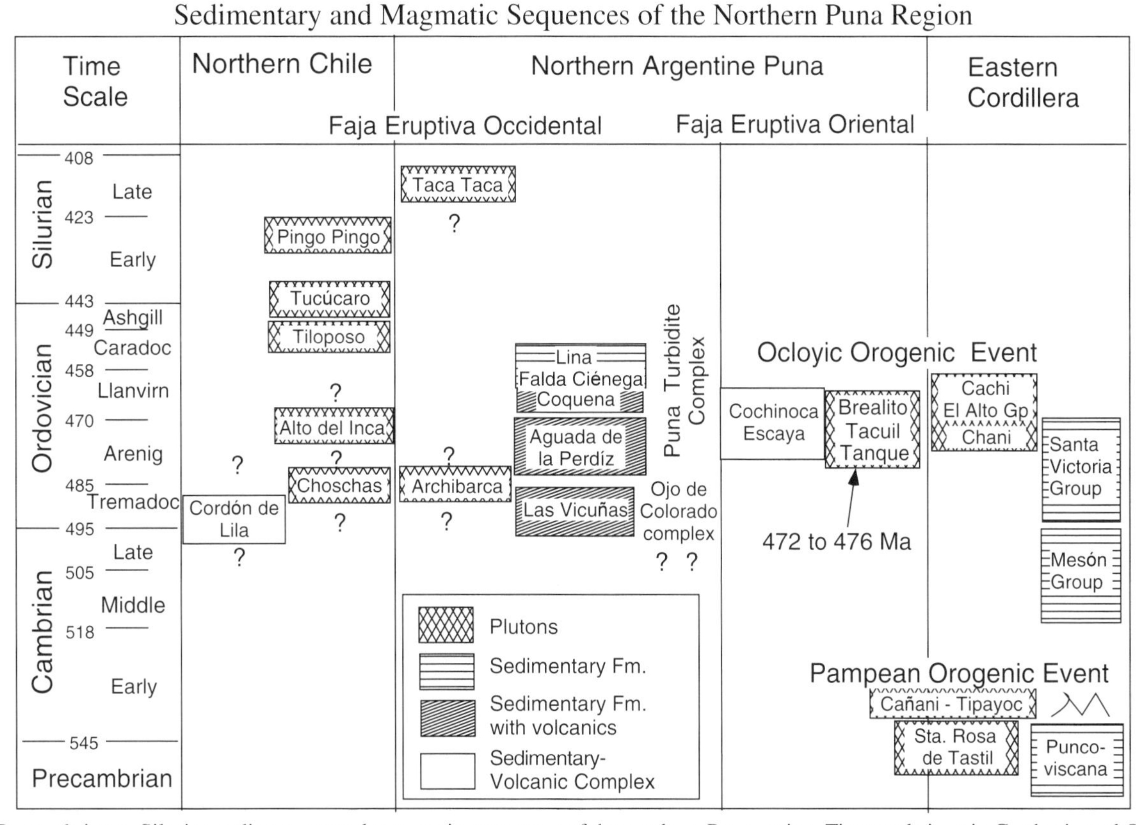

Figure 3. Precambrian to Silurian sedimentary and magmatic sequences of the northern Puna region. Time scale is as in Gradstein and Ogg (1996).

Faja Eruptiva Occidental magmatic units

Cordón de Lila complex. Starting in the west, basaltic andesitic and rhyodacitic units interlayered with sediments in the Cordón de Lila complex south of the Salar de Atacama are argued by Niemeyer (1989) to be submarine flows extruded in water depths of less than 500 m. Radiometric age constraints are restricted to a poorly constrained Nd-Sm model age of 448 ± 145 Ma ($\{^{143}Nd/^{144}Nd\}_i$ = 0.512319; ε_{Nd} = +5) in Damm et al. (1990). Niemeyer et al. (1989) have suggested that Ordovician fossils in fragments in overlying Paleozoic sediments could be from Cordón de Lila complex sediments, which would indicate an Ordovician age. The presence of Cambrian/late Precambrian basement under the western Faja Eruptiva Occidental is consistent with $^{207}Pb/^{206}Pb$ ages in individual zircon fractions from the Cordón de Lila diorite and the Choschas, Tucúcaro, and Pingo Pingo plutons (Damm et al., 1990).

A minimum Devonian to lower Carboniferous age for the Cordón de Lila complex is indicated by overlying sedimentary sequences (Mpodozis et al., 1983). Evidence for an older minimum age comes from concordant Silurian K-Ar ages of 425 ± 11 Ma on biotite and 429 ± 12 Ma on hornblende in the cross-cutting Pingo Pingo pluton reported by Mpodozis et al. (1983). A Rb-Sr errorchron of 288 ± 15 Ma (initial $^{87}Sr/^{86}Sr$ = 0.7118) in Mpodozis et al. (1983) and a lower intercept Devonian U-Pb zircon age of 338+14/–18 interpreted as an intrusion age for the Pingo Pingo pluton by Damm et al. (1990) appear to reflect other causes. Further evidence for a minimum age comes from a lower intercept U-Pb zircon age of 434 ± 2 Ma reported by Damm et al. (1990) for a diorite cutting the Cordón de Lila complex. This age is a revision of a 491+80/–59 Ma age based on individual zircon fractions yielding $^{207}Pb/^{206}Pb$ ages of 562, 483, and 458 Ma in Damm et al. (1986).

Chemical analyses for the Cordón de Lila volcanic sequence are given by Niemeyer (1989), Breitkreuz et al. (1989), and Damm et al. (1986, 1990, 1991). The lavas range from mafic to silicic in composition (42–68% SiO_2) and have arc-like characteristics in that they plot in the low-K arc field in Figure 4A, as I-type metaluminous rocks in Figure 4B, and in the arc field on the Hf-Th-Ta diagram in Figure 6. As argued by Niemeyer (1989), based on low TiO_2 concentrations (<1.0%), the chemistry of the Cordón de Lila units fits that of a low-K arc sequence erupted on oceanic or thinned continental crust. As expected in this setting, extended trace element patterns show little light REE enrichment and flat heavy REE, leading to low La/Yb ratios (<4) (Fig. 5A), arc-like Th and U enrichments, and arc-like Nb and Ta depletions (La/Ta ~25–36, assuming Nb/Ta = 17) (Fig. 5B). Amphibole phenocrysts in andesitic and dacitic units and An-rich phenocrysts in mafic rocks (Damm et al., 1991) are consistent with an arc origin.

Damm et al. (1991) suggested that Cordón de Lila lavas with ~42% SiO_2 and 29% MgO at the base of the sequence are komatiitic flows related to crustal extension. However, their chemical characteristic are atypical of komatiites as chondrite-normalized analyses show light REE–depleted patterns, high

TABLE 1. ANALYSES OF FAJA ERUPTIVA OCCIDENTAL PLUTONIC UNITS

	Choschas Pluton						Alto de Inca Pluton			Tucúcaro Pluton	Pingo Pingo Pluton
Sample	SO814	SO1121	SO1123	SO631	SO633	SO634	SO44	SO51	SO46	SO27	SO37
SiO_2	53.45	58.79	63.02	65.13	66.89	75.01	68.19	70.73	77.79	78.18	65.47
TiO_2	1.42	0.67	0.50	0.50	0.40	0.10	0.35	0.38	0.07	0.07	0.87
Al_2O_3	15.62	16.66	16.09	14.70	14.60	13.45	14.24	14.07	12.37	11.67	15.62
FeO	8.56	7.40	5.63	4.80	4.35	0.90	4.07	3.50	0.68	0.79	5.93
MnO	0.15	0.15	0.12	0.10	0.09	0.01	0.09	0.08	0.04	0.03	0.09
MgO	5.21	3.38	2.42	2.60	2.09	0.30	1.49	0.90	0.03	0.10	2.04
CaO	7.65	6.35	5.33	4.28	3.82	1.18	3.47	3.16	0.29	0.49	1.93
Na_2O	3.15	2.04	2.56	2.29	2.39	3.57	3.15	3.13	3.73	3.21	2.02
K_2O	1.87	2.05	1.71	3.34	3.06	4.52	2.68	2.70	4.31	4.57	3.52
P_2O_5	0.24	0.13	0.14	0.10	0.09	0.20	0.08	0.09	0.02	0.02	0.16
Volatiles	2.16	1.30	2.23	1.83	2.09	0.85	1.52	1.02	0.35	0.42	1.78
Total	99.48	98.92	99.75	99.67	99.87	100.09	99.33	99.76	99.68	99.55	99.43
La	21.7	20.8	14.6	17.4	29.8	15.2	19.8	52.4	17.5	17.1	54.3
Ce	46.1	48.6	32.9	37.8	58.5	33.8	38.4	75.7	46.5	34.2	112.7
Nd	24.1	24.8	20.2	20.8	23.3	16.8	16.7	28.5	26.7	15.4	57.7
Sm	5.44	4.85	3.93	4.30	4.07	3.47	3.16	5.43	7.66	2.69	10.8
Eu	1.50	0.917	0.796	0.772	0.706	0.406	0.606	0.763	0.147	0.428	1.90
Tb	0.929	0.731	0.604	0.656	0.586	0.492	0.438	0.538	1.18	0.307	1.60
Yb	3.29	2.53	2.57	2.52	2.12	2.36	2.12	2.68	5.67	1.59	5.62
Lu	0.462	0.378	0.363	0.375	0.303	0.329	0.311	0.369	0.753	0.221	0.753
Sr	290	160	235	223	164	66	143	133	24	60	206
Ba	419	321	497	430	395	251	410	785	113	625	822
Cs	1.4	3.6	1.4	3.8	5.2	4.7	2.4	1.7	2.8	3.1	3.5
U	1.1	1.0	0.9	1.4	1.3	3.0	1.4	1.6	3.4	1.5	2.6
Th	4.7	6.0	5.1	8.2	13.3	21.4	8.8	13.1	17.7	7.3	14.7
Hf	4.6	3.7	3.7	4.3	3.5	3.0	3.4	4.0	2.8	2.3	9.3
Ta	0.65	0.71	0.67	1.1	0.92	3.6	0.69	0.68	1.9	0.44	1.1
Sc	29.3	26.4	22.8	18.3	17.2	3.4	14.3	9.6	6.1	2.6	13.8
Cr	103	28	21	94	55	6	8	4	4	2	60
Ni	83	14	14	25	17	4	6	3	3	3	31
Co	80	97	136	154	148	169	181	114	208	249	178
Ratios											
FeO/MgO	1.64	2.19	2.33	1.85	2.08	3.00	2.73	3.89	22.67	7.90	2.91
K_2O/Na_2O	0.59	1.00	0.67	1.46	1.28	1.27	0.85	0.86	1.16	1.42	1.74
Ba/La	19.3	15.5	34.1	24.8	13.3	16.5	20.7	15.0	6.4	36.6	15.1
La/Sm	4.0	4.3	3.7	4.0	7.3	4.4	6.3	9.7	2.3	6.3	5.0
La/Yb	6.6	8.2	5.7	6.9	14.0	6.4	9.4	19.5	3.1	10.8	9.7
Eu/Eu*	0.84	0.60	0.64	0.57	0.56	0.38	0.62	0.51	0.06	0.55	0.06
Sm/Yb	1.7	1.9	1.5	1.7	1.9	1.5	1.5	2.0	1.4	1.7	1.9
Ba/Ta	646	456	741	394	429	69	592	1158	60	1413	728
La/Ta	33	29	22	16	32	4	29	77	9	39	48
Th/U	4.4	5.9	5.5	5.9	10.4	7.2	6.4	8.4	5.2	5.0	5.5
Molar $Al_2O_3/(CaO+K_2O+N_2O)$	0.74	0.97	1.02	0.97	1.03	1.04	0.99	1.02	1.09	1.05	1.09

normalized Th (and U) concentrations, and slight positive Eu anomalies. The FO_{75} olivine reported by Damm et al. (1991) is not in accord with komatiitic magmas. These units seem best interpreted as mafic arc lavas with excess olivine and pyroxene.

Sierra de Almeida and Cordón de Lila plutonic units. Faja Eruptiva Occidental plutons of probable Ordovician age occur to the south in the Sierra de Almeida (Fig. 2) (Davidson et al. 1981; Mpodozis et al. 1983; Damm et al. 1986, 1990, 1991). The largest and most diverse of these plutons is the Choschas whose units range from hornblende-biotite diorite to leucogranodiorite (Davidson et al., 1981). The age of this pluton is poorly constrained by imprecise Pb alpha dates of 487 ± 50 and 467 ± 50 Ma in Mpodozis et al. (1983), and a $^{207}Pb/^{206}Pb$ model age of 502 Ma on a single zircon fraction from a granodiorite in Damm et al. (1990). A question is whether this granodiorite is really from the same pluton as the averaged chemical analyses presented by Damm et al. (1990) has a very steep REE pattern (La/Yb = 52), making it chemically distinct from the six analyzed samples presented in Table 1.

TABLE 2. ANALYSES OF FAJA ERUPTIVA OCCIDENTAL AND EASTERN CORDILLERA PLUTONIC UNITS

	Taca Taca Pluton								Sta. Rosa de Tastil Pluton
	Diorite	Monzogranite			Cori Syenogranite				
Sample	Taca 4	Taca 3	Taca 2	Taca 1	SO1208	SO1209	SO1211	SO1210	TAS
SiO_2	58.99	71.58	73.35	73.57	75.50	75.69	75.65	76.33	66.36
TiO_2	1.08	0.35	0.30	0.28	0.20	0.05	0.17	0.17	0.83
Al_2O_3	16.13	13.64	13.09	13.24	12.30	12.45	12.30	11.69	15.53
FeO	7.14	2.60	2.18	2.12	2.11	0.35	1.70	1.68	4.55
MnO	0.25	0.06	0.04	0.04	0.05	0.03	0.05	0.13	0.09
MgO	2.82	0.86	0.65	0.68	0.08	0.02	0.08	0.07	1.81
CaO	5.22	2.34	1.83	1.78	0.59	1.96	0.56	0.48	2.70
Na_2O	3.29	2.56	2.48	2.45	3.25	3.56	3.54	3.63	2.98
K_2O	2.76	4.27	4.81	4.68	5.06	5.30	5.09	4.92	3.54
P_2O_5	0.21	0.08	0.06	0.06	0.03	0.02	0.03	0.03	0.23
Volatiles	1.49	1.13	0.90	0.89	0.75	0.40	0.48	0.72	1.33
Total	99.38	99.47	99.69	99.79	99.92	99.83	99.75	99.85	99.95
La	26.8	34.8	34.0	33.2	50.5	14.7	60.9	56.5	37.2
Ce	56.1	65.7	63.2	64.4	73.8	35.5	75.3	83.5	74.9
Nd	30.6	26.4	24.4	23.2	37.5	6.77	57.6	52.7	37.0
Sm	6.60	4.47	3.89	3.71	9.38	7.74	13.18	12.90	7.49
Eu	1.43	0.783	0.733	0.693	0.562	0.227	0.699	0.700	1.36
Tb	0.993	0.525	0.429	0.425	1.76	2.21	2.15	2.03	1.01
Yb	3.51	2.50	1.99	1.97	10.1	18.9	10.3	9.57	3.09
Lu	0.477	0.360	0.293	0.280	1.39	2.61	1.36	1.26	0.427
Sr	233	133	122	119	68	25	70	39	167
Ba	328	529	450	429	475	155	550	379	423
Cs	7.1	7.0	9.4	8.7	6.7	2.3	2.5	2.0	11.9
U	2.7	4.2	3.0	3.5	5.5	9.2	7.9	4.8	1.6
Th	9.9	19.8	18.4	19.9	31.2	82.7	33.7	32.8	12.1
Hf	4.4	4.1	3.1	3.3	10.5	13.7	10.0	10.1	6.2
Ta	0.88	1.6	1.3	1.2	4.4	16.4	3.8	3.9	1.5
Sc	24.9	7.3	6.0	5.6	3.9	0.4	2.5	2.5	12.8
Cr	22	7	8	20	7	3	13	3	37
Ni	18	7	7	8	3	1	7	3	25
Co	113	200	177	165	178	142	218	210	107
Ratios									
FeO/MgO	2.53	3.02	3.35	3.12	26.38	17.50	21.25	24.00	2.51
K_2O/Na_2O	0.84	1.67	1.94	1.91	1.56	1.49	1.40	1.36	1.19
Ba/La	12.2	15.2	13.2	12.9	9.4	10.5	9.0	6.7	11.4
La/Sm	4.1	7.8	8.7	9.0	5.4	1.9	4.6	4.4	5.0
La/Yb	7.6	13.9	17.1	16.9	5.0	0.8	5.9	5.9	12.1
Eu/Eu*	0.69	0.60	0.66	0.65	0.18	0.08	0.16	0.17	0.60
Sm/Yb	1.9	1.8	2.0	1.9	0.9	0.4	1.3	1.3	2.4
Ba/Ta	375	333	359	348	109	9	146	96	286
La/Ta	31	22	27	27	12	1	16	14	25
Th/U	3.7	4.7	6.2	5.6	5.7	9.0	4.3	6.9	7.5
Molar $Al_2O_3/(CaO+K_2O+N_2O)$	0.90	1.04	1.04	1.07	1.03	0.82	0.98	0.96	1.14

Major elements for Choschas units in Table 1 indicate a metaluminous to marginally peraluminous (molar $Al_2O_3/(K_2O + Na_2O + CaO) = 0.74 - 1.04$) and medium- to high-K character (Fig. 4C,D). REE patterns (Fig. 6A,B) of the dioritic to leucogranodioritic units (53–67% SiO_2) confirm a generally I-type arc-like signature with normalized light REEs at ~35–60 times chondrites. REE patterns are relatively flat (La/Yb = 6–7) with the steeper pattern of leucogranodiorite SO633 (La/Yb = 14) being attributable to hornblende fraction (Fig. 5A). An arc or backarc-like source is consistent with La/Ta ratios mostly from 22–32. As a whole, the gross features of these units are like those of calc-alkaline plutons in the Aleutian oceanic island arc in Alaska (Kay et al., 1990) consistent with the Choschas pluton being intruded into thin crust. A difference is that Choschas units have lower Sr and Na_2O concentrations at 59–65% SiO_2 than Aleutian plutons (2.1–2.6% Na_2O compared to >3.5% Na_2O), which can be attributed to contamination in a thinned continental crust.

Other plutons in the region are generally more silicic. One intruding the Choschas pluton is the medium- to high-K (Fig. 4C) Alto del Inca pluton (Davidson et al., 1981) for which a Rb-Sr whole-rock age of 468 ± 100 Ma (initial $^{87}Sr/^{86}Sr$ = 0.7045) is reported by Halpern et al. (1978), and recalculated to 478 ± 44 Ma by Mpodozis et al. (1983). Analyses in Table 1 show that leucogranodioritic units (68–71% SiO_2) are of transitional metaluminous character (Al_2O_3/(K_2O + Na_2O + CaO) ± 0.99–1.02), and have generally I-type Na_2O/K_2O ratios (Fig. 4D). Samples plotted in Figure 7C represent distinct units—SO51 is more Choschas-like (La/Ta = 29; Ba/La = 21; Ce = 29 ppm), whereas SO44 has distinctly lower Ba/La (15) and higher La/Ta (77) ratios, and higher LREE concentrations (75 ppm Ce). Granite SO46 (Fig. 7C) has the characteristics of a highly differentiated magma

TABLE 3. ANALYSES OF FAJA ERUPTIVA OCCIDENTAL VOLCANIC UNITS

	Rio Huaitiquina*							Sierra de Guayaos		
Sample	Dike J72	Dike J71	Unit A J43	Unit B J31	Unit C J10	Unit C J76	Unit C J11	Silicic Tuffs G121	G130	Shoshonite G125
SiO_2	45.79	46.31	63.06	74.08	58.98	58.88	68.53	77.87	75.49	53.56
TiO_2	3.70	3.39	0.82	0.37	1.28	1.29	0.49	0.28	0.21	1.14
Al_2O_3	12.19	10.71	17.20	14.27	17.48	17.59	14.52	12.82	14.65	18.31
FeO	12.10	13.31	4.63	2.89	6.33	6.17	5.57	1.93	1.60	8.27
MnO	0.21	0.21	0.21	0.00	0.13	0.14	0.18	0.00	0.04	0.16
MgO	10.49	11.20	2.78	0.74	4.37	4.01	2.28	0.40	0.67	4.77
CaO	10.33	11.58	6.87	0.93	5.53	4.21	2.33	0.64	1.09	8.80
Na_2O	2.81	1.97	4.03	4.92	4.35	5.12	4.76	5.42	3.63	1.99
K_2O	1.43	1.30	0.41	1.81	1.53	2.29	1.35	0.64	2.63	3.01
P_2O_3										0.30
Anhydrous analyses										
Total	99.05	99.98	100.00	97.12	99.98	93.53	100.01	100.00	100.01	100.01
La	55.6	45	29.4	28.4	29.9	29.6	20.7	21.8	51.6	51.8
Ce	111.8	90.6	64.1	55.3	63.5	62.3	37.3	47.3	114.1	105.3
Nd	55.5	41.1	25.5	25.3	26.8	25.1	16.1	19.7	45	43.6
Sm	10.9	8.90	6.10	6.00	5.90	5.80	4.20	4.10	8.50	9.30
Eu	3.17	2.71	1.38	0.970	1.44	1.48	0.830	0.730	1.63	2.00
Tb	1.26	1.18	0.769	0.775	0.780	0.795	0.583	0.646	1.27	0.971
Yb	2.16	2.07	2.47	3.41	2.9	2.85	2.9	2.43	3.82	2.34
La	0.272	0.229	0.353	0.444	0.407	0.404	0.38	0.332	0.554	0.303
Sr	743							*205*	*233*	*450*
Ba	545	421	343	352	428	500	452	235	1367	1239
Rb								*32*	*32*	*80*
Cs	3.4	8.1	11.1	2.5	4.7	0.6	1.3	1.4	12.1	6.3
U	1.9	1.5	3.7	2.4	3.5	3.4	2.6	2.1	2.9	3.4
Tb	6.9	5.7	14.9	9.4	12.1	12	8.2	8.7	17.5	15.8
Hf	7.4	6.4	2.6	3.9	4.9	4.9	4.1	3.4	5.5	3.5
Ta	5.3	4	0.3	0.7	1.4	1.4	0.4	0.8	1.7	0.9
Sc	19.7	26	13.6	10.5	18.9	18.8	15.6	9.1	9.5	29.2
Cr	202	463	18	9	58	60	23	18	6	78
Ni	164	283		1	27	21	9	5	4	36
Co	51	71	13	2	21	20	14	7	3	33
'Ratios										
FeO/MgO	1.15	1.19	1.67	3.91	1.45	1.54	2.44	4.83	2.39	1.73
K_2O/Na_2O	0.10	0.09	0.03	0.12	0.10	0.15	0.09	0.04	0.18	0.20
Ba/La	9.8	9.4	11.7	12.4	14.3	16.9	21.8	10.8	26.5	23.9
La/Th	8.1	7.9	2.0	3.0	2.5	2.5	2.5	2.5	2.9	3.3
La/Sm	5.1	5.1	4.8	4.7	5.1	5.1	4.9	5.3	6.1	5.6
La/Yb	25.7	21.7	11.9	8.3	10.3	10.4	7.1	9.0	13.5	22.1
Eu/Eu*	1.01	1.01	0.76	0.54	0.81	0.83	0.64	0.55	0.61	0.77
Sm/Yb	5.0	4.3	2.5	1.8	2.0	2.0	1.4	1.7	2.2	4.0
Ba/Ta	102	105	1143	503	306	357	1130	294	804	1377
La/Ta	10	11	98	41	21	21	52	27	30	58
Th/U	3.7	3.8	4.0	3.9	3.5	3.5	3.2	4.1	6.0	4.6
Molar Al_2O_3/(CaO+K_2O+N_2O)	0.49	0.42	0.88	1.22	0.93	0.95	1.07	1.19	1.36	0.81

*See section in Figure 8.

TABLE 4. ANALYSES OF FAJA ERUPTIVA OCCIDENTAL VOLCANIC UNITS

	Aracar		
Sample	SO746	SO741	SO739
SiO_2	56.71	57.50	58.21
TiO_2	1.08	1.07	1.07
Al_2O_3	18.46	18.13	18.36
FeO	6.16	5.95	5.36
MnO	0.19	0.21	0.10
MgO	2.24	2.16	1.87
CaO	6.17	6.18	4.99
Na_2O	3.44	2.98	3.77
K_2O	3.40	3.83	3.86
P_2O_3	0.29	0.31	0.32
Volatiles	1.60	1.43	1.53
Total	99.74	99.75	99.44
La	43.2	45.0	47.2
Ce	99.4	95.4	99.9
Nd	45.7	41.4	39.2
Sm	7.63	7.93	7.88
Eu	1.51	1.44	1.53
Tb	0.989	0.976	1.04
Yb	2.98	3.01	3.15
Lu	0.437	0.433	0.429
Sr	456	436	395
Ba	675	677	790
Rb			
Cs	8.5	7.1	22.6
U	2.0	1.9	2.1
Th	15.2	14.9	16.4
Hf	6.9	6.6	7.2
Ta	1.0	0.9	1.1
Sc	17	16.6	13
Cr	34	26	21
Ni	32	28	26
Co	67	76	90
Ratios			
FeO/MgO	2.75	2.75	2.87
K_2O/Na_2O	0.23	0.26	0.26
Ba/La	15.6	15.1	16.7
La/Th	2.8	3.0	2.9
La/Sm	5.7	5.7	6.0
La/Yb	14.5	15.0	15.0
Du/Eu*	0.66	0.62	0.64
Sm/Yb	2.6	2.6	2.5
Ba/Ta	704	726	738
La/Ta	45	48	44
Th/U	7.6	7.9	7.8
Molar $Al_2O_3/(CaO+K_2O+N_2O)$	0.90	0.89	0.94

with a very flat REE pattern (La/Yb = 3.1), an extreme negative Eu anomaly (0.06), and very low Sr (24 ppm) and high Th (17.7 ppm) concentrations.

Additional plutons in the region are the Silurian Pingo Pingo pluton whose age is discussed above and the monzogranitic Tilopozo and Tucúcaro plutons for which Mpodozis et al. (1983) have reported Ordovician Rb-/Sr whole-rock ages of 452 ± 4 and of 441 ± 8 Ma, respectively. A lower intercept U-Pb zircon age of 450 ± 11–12 Ma for the Tucúcaro pluton in Damm et al. (1990) agrees within error. Initial $^{87}Sr/^{86}Sr$ ratios of 0.7102 for the Tucúcaro pluton and 0.7113 for the Tilopozo muscovite-bearing monzogranite pluton (Mpodozis et al., 1983) are consistent with old crustal components in the magma source region.

Chemical characteristics of these younger plutonic units are consistent with melts formed in a collisional setting. An analysis of a coarse-grained leucocratic Tucúcaro monzogranite SO27 (78% SiO_2) with sparse biotite is given in Table 1. This sample is marginally peraluminous ($Al_2O_3/(K_2O + Na_2O + CaO) = 1.05$), has a relatively flat REE pattern (La/Yb = 11) with Ce (LREE) at ~35 X chondrites, and an arc-like La/Ta (39) ratio (see Figs. 4, 5, and 7D). An averaged analysis of the Tucúcaro pluton (75% SiO_2) from Damm et al. (1990) is marginally peraluminous (1.02) and has a high average light REE concentration (e.g., Ce = 76 ppm) that is typical of so-called A-type (collisional) granites. An analysis of a Pingo Pingo granodiorite from Table 1 plotted in Figure 7D also has A-type characteristics.

Western Northern Puna plutonic units. Faja Eruptiva Occidental plutons to the east in Argentina (Fig. 2) emplaced in Ordovician or older sedimentary sequences and metamorphic basement include those in the Taca-Taca, Arita, Cerro Chivinar, Sierra de Macon, Antofalla, and Archibarca regions (e.g., Palma et al., 1986; Damm et al., 1990; Koukharsky and Lanes, 1994). Age constraints are sparse, and it is uncertain if any of these plutons is Ordovician in age. A biotite K/Ar age for the Taca Taca pluton suggests a Late Silurian age (M. Koukharsky, personal communication, 1998). The Arita pluton across a major fault to the south has some chemical tendencies (analyses in Damm et al., 1990) like the Taca Taca units. The meaning of a biotite K-Ar age of 180 ± 10 Ma from the Arita pluton in Méndez et al. (1979) is uncertain. Continuing eastward, the Macón pluton is thought to be of late Precambrian age based on biotite K-Ar dates of 630 ± 20 and 670 ± 20 Ma reported in Méndez et al. (1973). Average analyses in Damm et al. (1990) indicate that Macón biotite-hornblende–bearing granodioritic and monzogranitic units are strongly metaluminous and have arc-like chemical tendencies (e.g., Ba/La = 27, La/Ta = 47). Farther south, Palma et al. (1986) reported a K-Ar age of 485 ± 15 Ma on altered biotite with 4% K_2O from the Archibarca pluton. At best, this is a minimum age.

The best studied of these is the high-K Taca Taca pluton, which is composed of monzogranitic and syenogranitic units with dioritic inclusions (Koukharsky and Lanes, 1994). New analyses in Table 2 are plotted in Figure 8A,B. The diorite (59% SiO_2) has an I-type calc-alkaline signature with molar ($Al_2O_3/(K_2O + Na_2O + CaO) = 0.90$, an arc-like La/Ta ratio (31), and a REE pattern like that of Faja Eruptiva Occidental plutonic units to the west (La/Yb = 8) (Figs. 5A and 8B). In contrast, the widespread monzogranitic unit (71–74% SiO_2) has 4.3–4.7% K_2O (Fig. 4), is marginally peraluminous ($Al_2O_3/(K_2O + Na_2O + CaO) = 1.04$–1.07) and plots in the S-type field on a Na_2O–K_2O plot (Fig. 4D). La/Ta (22–27) and Ba/La ratios (12–15) are like those of centers behind a main arc front. Unlike those of plutons to the west, REE

TABLE 5. ANALYSES OF SILICIC UNITS IN THE FAJA ERUPTIVA ORIENTAL

	Cordon de Escaya			Q. Casa	Questa	Muñayoc			Cochinoca	
	Breccia	Breccia	Lava	Colorado	Lava	Breccia	Lava	Lava	Dome	Lava
Sample	Esca52	Esca34	Esca73	CAS1	QUE23	MU45	MU18	MU36	P19	CO_2
SiO_2	68.64	70.99	68.37	68.07	70.00	68.35	69.20	69.87	72.27	67.55
TiO_2	0.62	0.64	0.78	0.76	0.66	0.68	0.61	0.62	0.70	0.78
Al_2O_3	14.79	14.23	14.93	14.66	14.95	14.70	14.97	15.41	13.61	15.41
FeO	4.96	4.49	4.46	4.99	4.13	4.70	3.82	3.66	4.68	5.41
MnO	0.08	0.08	0.11	0.06	0.05	0.08	0.16	0.05	0.00	0.07
MgO	3.07	2.45	1.73	2.24	1.24	2.09	1.36	1.45	1.81	2.41
CaO	2.31	1.00	1.63	1.76	0.18	0.97	1.43	0.32	1.98	1.29
Na_2O	2.04	3.98	3.20	2.38	2.78	2.39	2.94	3.57	2.51	2.28
K_2O	2.61	1.07	4.06	4.10	5.43	5.46	5.12	4.38	2.44	4.03
P_2O_5	0.21	0.31	0.24	0.22	0.26	0.22	0.23	0.23		0.24
Anhydrous analyses										
Total	99.32	99.24	99.50	99.24	99.67	99.62	99.84	99.56	100.00	99.46
La	29.9	43.5	35.5	27.1	26.3	38.4	41.4	35.8	39.6	38.3
Ce	64.8	80.3	75.8	60.3	61.6	87.3	87.8	81.7	88.8	83.9
Nd	28.6	38.5	36.6	24.6	28.6	35.6	34.3	32.2	38.4	35.4
Sm	6.50	7.64	8.00	7.09	6.47	8.53	7.94	7.39	8.00	8.23
Eu	0.99	1.16	1.10	1.10	0.99	1.20	1.27	1.05	1.26	1.19
Tb	0.985	1.11	1.18	1.10	1.07	1.32	1.21	1.05	1.21	1.19
Yb	3.26	3.19	3.82	3.45	3.58	4.28	3.91	3.74	3.5	3.65
Lu	0.444	0.457	0.498	0.499	0.578	0.571	0.531	0.503	0.464	0.541
Sr	122	131	120	126	131	97	130	111		130
Ba	390	224	407	342	559	403	556	406	407	438
Rb										
Cs	7.3	2.9	3.2	5.6	7.1	5.3	6.5	6.2	8.3	4.7
U	3.4	3.2	3.5	4.0	2.8	4.4	4.3	4.3	3.6	4.3
Th	12.9	12.0	13.7	15.6	14.6	15.4	15.6	16.0	16.1	15.3
Hf	5.2	5.0	5.6	6.4	6.2	6.1	5.8	6.1	6.2	6.4
Ta	1.2	1.1	1.2	1.5	1.4	1.4	1.4	1.4	1.5	1.5
Sc	10.9	9.9	11.1	12.5	11.3	12.7	12.0	12.1	12.9	13.3
Cr	44	41	47	88	43	49	41	46	64	53
Ni	21	24	23	16	24	26	24	25	26	32
Co	12	12	12	14	12	13	10	11	12	14
Ratios										
FeO/MgO	1.61	1.83	2.58	2.23	3.34	2.25	2.82	2.52	2.59	2.25
K_2O/Na_2O	0.17	0.07	0.27	0.27	0.36	0.36	0.34	0.29	0.16	0.27
Ba/La	13.1	5.2	11.5	12.6	21.3	10.5	13.4	11.3	10.3	11.4
La/Th	2.3	3.6	2.6	1.7	1.8	2.5	2.6	2.2	2.5	2.5
La/Sm	4.6	5.7	4.4	3.8	4.1	4.5	5.2	4.9	5.0	4.7
La/Yb	9.2	13.7	9.3	7.9	7.3	9.0	10.6	9.6	11.3	10.5
Eu/Eu*	0.48	0.49	0.44	0.49	0.47	0.44	0.5	0.46	0.5	0.46
Sm/Yb	2.0	2.4	2.1	2.1	1.8	2.0	2.0	2.0	2.3	2.3
Ba/Ta	325	206	336	229	398	279	405	288	271	289
La/Ta	25	40	29	18	19	27	30	25	26	25
Th/U	3.7	3.7	3.9	3.9	5.3	3.5	3.6	3.7	4.5	3.6
Molar $Al_2O_3/(CaO+K_2O+N_2O)$	1.43	1.49	1.18	1.27	1.39	1.27	1.15	1.38	1.31	1.47

patterns are characterized by moderate Eu anomalies (0.6–0.66), relatively flat, heavy REEs (Sm/Yb ~2.0), and extreme, light REE enrichment (La/Sm = 8–9) resulting in high La/Yb ratios (14–17) (Fig. 5A). A question is whether the diorite is older (Middle Ordovician or older?) than the monzogranite.

Samples from the Cori unit in the Taca Taca region just across the border in Chile are highly fractionated, metaluminous biotite granites (76% SiO_2, molar $Al_2O_3/(K_2O + Na_2O + CaO)$ = 0.82–1.03) that plot in the A-type field in Figure 4D. High, light REE concentrations (e.g., Ce = 74-84 ppm) are in accord with an A-type classification. La/Ta (12–16) and Ba/La ratios (6–10) are low. Flat REE patterns (La/Yb < 6) (Figs. 5A and 8B), large negative Eu anomalies, and low Sr concentrations are consistent with highly evolved magmas, as argued by Koukharsky and Lanes (1994).

Western Puna volcanic units. In the central and eastern Faja Eruptiva Occidental, mafic to acidic volcanic sequences occur intercalated in volcaniclastic turbiditic sequences (Figs. 2 and 9). Well-described sequences are those in the Aguada de la Perdiz (Breitkreuz et al., 1989), Huaitiquina (Coira and Barber, 1989; Bahlburg, 1990), Salina de Jama (Coira and Nullo, 1989), and Guayaos (Coira et al., 1987; Koukharsky et al., 1989) localities. Graptolite fossil assemblages in the turbiditic sequences in these regions have Arenig to Llanvirn ages. Tremadoc fossil assemblages are reported at Vega Pinato and Lari, southwest of Huaitiquina by Koukharsky et al. (1996). Another volcanic sequence east of Cerro Aracar just across the Chilean border is cut by units of the Taca Taca pluton. Chemical analyses of the Huaitiquina, Guayaos, and Aracar sections are given in Tables 3 and 4 and are plotted in Figure 10.

Basaltic and andesitic units in the Huaitiquina section can be put in three groups that show a temporal progression from oceanic arc or backarc-like to alkaline within-plate–type chemistry. The first group (e.g., J43 in Table 3; Koukharsky et al., 1988; Bahlburg, 1990) near the base of the section in Huaitiquina unit A (Fig. 9) consists of basaltic andesitic to andesitic (~53–63% SiO_2) pillow breccias and flows with substantial alteration. These samples have medium-K calc-alkaline arc chemical characteristics (Fig. 4A), including high Al_2O_3 (16–20%) and relatively low TiO_2 (<1.3%) concentrations, low FeO/MgO ratios <2, high La/Ta (=38–55) (Fig. 5B) and low Th/La ratios, and Hf-Ta-Th relations (Fig. 6). Low Ba/La ratios (~12) in some could indicate they did not erupt in the frontalmost arc, but could also reflect alteration effects. The second group in Huaitiquina unit C (Fig. 9) includes mafic andesitic flows (59% SiO_2) (J10 and J76 in Table 3). These samples have weaker arc signatures (Figs. 6, and 10A,B) as shown by their higher TiO_2 and K_2O (Fig. 4A) concentrations, lower La/Ta (=21) (Fig. 5B) and higher La/Th (=2.5) ratios, and Hf-Th-Ta relations (Fig. 6). Their REE patterns are steeper (La/Yb = 10–12) (Fig. 10A). The third group includes the mafic dikes (~46% SiO_2, HD, J72, J71 in Table 3; Fig. 10A) that intrude unit C. As discussed in the next section, these magmas have no arc-like characteristics (e.g., Fig. 6) and are best explained as small percentage melts of a garnet-bearing intraplate alkaline mantle source.

Dacitic and rhyolitic samples in Huaitiquina units B and C (Fig. 9; analyses in Table 3, Koukharsky et al., 1988; Bahlburg, 1990) have characteristics of silicic units generated in arc environments. Arc-like characteristics include concentrations of TiO_2 < 0.5%, Na_2O > 4.0%, and K_2O < 2% (Figs. 4A,B), and high La/Ta ratios (40–90) (Figs. 5B and 10B). Parental magmas are best interpreted as extreme differentiates of arc magmas or as melts of arc-like crust. The stronger arc-like signature in unit C dacite J11 (68% SiO_2) than in unit C (group 2) mafic lavas supports the dacitic magmas having an important component from an arc-like crustal basement. Comparing dacite J11 in Figure 10B to Faja Eruptiva Oriental dacites in Figure 11 shows that J11 has a flatter REE pattern, lower incompatible element abundances—particularly REE, Th, and U, and higher Ba/La (22) and La/Ta (52) ratios. This is interpreted as signifying that Huaitiquina silicic units are more closely linked to an arc setting than are Cochinoca-Escaya complex dacitic units.

TABLE 6. ANALYSES OF SILICIC UNITS IN THE WESTERN FAJA ERUPTIVA ORIENTAL

	Huancar		Angosto		Cobres
	Breccia	Lava	Lava	Lava	Pluton
Sample	HUA19	HUA22	AN16	AN9	COB15
SiO_2	73.03	68.99	69.35	64.38	68.23
TiO_2	0.57	0.83	0.71	0.73	0.72
Al_2O_3	13.35	14.79	14.38	18.83	15.48
FeO	4.55	4.23	4.62	5.22	3.98
MnO	0.07	0.07	0.12	0.03	0.07
MgO	3.47	1.77	1.45	3.24	1.26
CaO	0.42	0.14	1.27	0.19	1.77
Na_2O	1.99	2.69	2.49	4.37	3.21
K_2O	1.72	4.69	4.85	1.82	4.72
P_2O_5	0.27	0.25	0.23	0.21	0.26
Anhydrous analyses					
Total	99.44	99.46	99.47	99.02	99.69
La	16.9	34.9	38.5	39.3	41.4
Ce	41.5	74.4	85.9	86.0	89.6
Nd	14.6	32.5	34.7	38.7	36.8
Sm	4.57	7.45	8.43	8.08	8.59
Ba	0.70	1.16	1.23	1.22	1.19
Tb	0.854	1.05	1.21	1.19	1.27
Yb	3.78	4.21	4.15	3.78	4.05
La	0.513	0.559	0.555	0.524	0.637
Sr	146	113	99	226	122
Ba	570	666	704	563	430
Rb					
Cs	3.3	4.4	7.0	4.8	16.9
U	3.2	4.2	4.3	4.0	1.9
Th	11.2	14.1	15.1	15.4	17.0
Hf	4.4	5.8	6.6	6.4	6.6
Ta	1.0	1.5	1.5	1.5	1.9
Sc	9.8	11.8	11.8	12.6	10.0
Cr	39	43	52	53	25
Ni	21	21	24	40	15
Co	10	11	12	13	9
Ratios					
FeO/MgO	1.31	2.39	3.18	1.61	3.16
K_2O/Na_2O	0.11	0.31	0.32	0.12	0.31
Ba/La	33.8	19.1	18.3	14.3	10.4
La/Th	1.5	2.5	2.6	2.6	2.4
La/Sm	3.7	4.7	4.6	4.9	4.8
La/Yb	4.5	8.3	9.3	10.4	10.2
Eu/Eu*	0.45	0.50	0.47	0.48	0.44
Sm/Yb	1.2	1.8	2.0	2.1	2.1
Ba/Ta	555	457	472	380	225
La/Ta	16	24	26	27	22
Th/U	3.5	3.4	3.5	3.8	9.2
Molar $Al_2O_3/(CaO+K_2O+N_2O)$	2.26	1.28	1.23	1.98	1.14

Analyses of altered igneous units from the Aguada de la Perdiz volcanic-sedimentary section just to the northwest in Chile are presented by Breitkreuz et al. (1989). A basaltic andesite lava (AP20 » 54% SiO_2) in that group has arc-like incompatible, immobile trace element ratios (La/Ta = 32 {Ta approximated as Nb/17}, Ba/La = 58, La/Yb = 8). Other samples have SiO_2 contents >77% and are either very altered or reworked volcaniclastic units. Some are described as clasts from submarine siliceous volcaniclastic deposits that formed in a sedimentary apron environment in a backarc region.

Silicic volcanic tuffs also occur in the Sierra de Guayaos section to the south (see Fig. 2). Chemical analyses are presented in

TABLE 7. ANALYSES OF MAFIC UNITS IN THE FAJA ERUPTIVA ORIENTAL

	Q. Casa Colorada	Queta	Sierra de Tanque		Cordón de Escaya			Huancar	Cochinoca	Cobres
	Lava	Lava	Lava	Lava	Lava	Lava	Lava	Dike	Lava	Diabase
Sample	CO6	QUE3	BET 34	BET 35	Esca23	Esca95	Esca14	PCH1	PCO425	COB10
SiO_2	49.24	49.93	49.29	50.16	46.20	52.90	50.30	41.90	44.09	46.73
TiO_2	1.16	2.04	1.18	1.26	1.10	1.79	2.91	3.10	3.93	2.55
Al_2O_3	15.28	16.32	16.09	15.87	19.35	15.48	13.69	14.21	17.50	15.01
FeO	8.34	9.83	9.32	9.05	11.04	8.31	13.25	12.02	11.50	10.61
MnO	0.14	0.16	0.19	0.20	0.16	0.16	0.22	0.22	0.15	0.18
MgO	9.93	7.52	9.28	8.71	8.47	6.21	5.78	6.64	6.81	9.35
CaO	10.60	9.09	9.68	9.48	4.00	4.14	6.96	17.47	12.41	9.15
Na_2O	3.30	3.40	1.88	1.81	3.72	5.65	3.10	2.95	2.98	3.00
K_2O	0.14	0.06	0.56	1.13	0.08	0.32	0.56	1.50	0.62	1.48
P_2O_5	0.22	0.30	0.15	0.14	0.19	0.27	0.36			0.58
Volatiles			2.65	2.44	5.16	3.55	2.26			1.86
Total	98.35	98.66	100.27	100.25	99.47	98.78	99.39	100.01	99.99	98.65
La	17.0	25.8			7	24.9	18	201.1	38.9	42.9
Ce	35.7	52.3			5	46.2	38	352.3	74.8	75.8
Nd	18.8	23.4				23.0		122.6	33.8	30.6
Sm	3.87	5.43				5.40		19.90	6.70	6.73
Eu	1.15	1.77				1.60		5.75	2.57	1.91
Tb	0.621	0.786				0.90		2.08	0.928	0.849
Yb	1.78	2.01				2.40		3.66	2.09	1.91
La	0.235	0.293				0.300		0.455	0.271	0.212
Sr	463	530	163	166	141	455	194			734
Ba	184	77	128	232	23	845	224	2801	248	1609
Rb			31	52	12		27			
Cs	0.4	0.2				0.8		227.2	10.3	1.3
U	0.5	0.9				0.80		5.1	0.9	1.4
Tb	2.2	3.1				3.2		22.2	3.9	5.5
Hf	2.5	4.0				3.8		10	4.1	4.3
Ta	1.5	2.5				2.2		11.3	3.3	5.1
Sc	37.0	31.0				37.4		18	31.1	25.0
Cr	570	109	525	465	396	213	41	89	243	418
Ni	142	27	115	94	104	90	6		83	98
Co	44	43	46	42	89	34	70	39	52	52
Ratios										
FeO/MgO	0.84	1.31	1.00	1.04	1.30	1.34	2.29	1.81	1.69	1.13
K_2O/Na_2O	0.01	0.00	0.04	0.08	0.01	0.02	0.04	0.10	0.04	0.10
Ba/La	10.8	3.0			3.3	33.9	12.4	13.9	6.4	37.5
La/Th	7.8	8.2				7.8		9.1	10.0	7.8
La/Sm	4.4	4.7				4.6		10.1	5.8	6.4
La/Yb	9.5	12.8				10.4		54.9	18.6	22.4
Eu/Eu*	0.93	1.05				0.91		1.03	1.25	0.95
Sm/Yb	2.2	2.7				2.3		5.4	3.2	3.5
Ba/Ta	121	30				384		248	75	313
La/Ta	11	10				11		18	12	8
Th/U	4.3	3.3				4.0		4.4	4.3	3.8
Molar $Al_2O_3/(CaO+K_2O+N_2O)$	0.61	0.74	0.77	0.76	1.44	0.90	0.75	0.37	0.62	0.65

Table 3 and in Koukharsky et al. (1989). Those with >79% SiO_2 and low Al_2O_3 (<11%) and Na_2O (to 0.2%) concentrations do not represent magmatic liquids. The two analyses in Table 3 are the most melt-like, but the high Na_2O (5.4%) concentration in G121 and the high Th/U ratio (6) in G130 are unlikely to be primary. Taking a wider view, high Na_2O/K_2O ratios (Fig. 4B), arc-like La/Ta (27–35) (Fig. 5B) and La/Th (2.5–2.9) ratios, and moderate REE slopes (La/Yb = 9–14) (Figs. 5A and 10C) are consistent with these tuffs forming in an arc to backarc region on thin continental or oceanic crust.

More diagnostic of the tectonic setting of the Guayaos section is the composition of mafic (~54% SiO_2) lavas G125 (Table 3) and

TABLE 8. ANALYSES OF MAFIC UNITS IN THE FAJA ERUPTIVA ORIENTAL

	Sierra de Tanque				Las Burras	Santa Ana
	Gabbro	Gabbro	Gabbro Dikes		Gabbro Sill	Gabbro Sill
Sample	BET 56	BET 78	AC-1H	AC-11	RBU	SA4
SiO_2		50.24	46.26	53.65	50.50	51.09
TiO_2		2.87	2.23	1.70	2.69	2.12
Al_2O_3		15.65	14.05	13.97	14.87	18.74
FeO	9.09	10.61	11.32	9.43	10.20	9.22
MnO		0.16	0.27	0.21	0.17	0.30
MgO		6.39	6.24	6.69	5.42	5.34
CaO		6.13	9.41	8.68	9.77	8.67
Na_2O	2.87	3.60	2.09	2.60	2.75	3.12
K_2O		2.35	0.36	0.32	0.63	1.39
P_2O_5		0.62	0.35	0.28	0.36	
Volatiles		1.90	5.95	0.97	1.19	
Total		100.52	98.53	98.50	98.55	99.99
La	46.1		18.7	15.3		22.9
Ce	89.7		39.0	31.0		51.9
Nd	45.9		20.0	18.0		24.5
Sm	8.04		5.30	4.50		5.70
Eu	1.8		1.90	1.50		1.8
Tb	1.06		1.10	1.00		0.958
Yb	2.29		3.20	2.70		2.61
La	0.301		0.400	0.400		0.359
Sr	707	489			247	
Ba	477	764			253	268
Rb		68	21	6	35	
Cs	6.9		2.8	2.2		3
U	2.0		0.5	0.5		1.1
Th	5.0		1.6	2.9		4
Hg	5.1		3.7	3.9		4.9
Ta	4.6	4	2.0	3.0		1.8
Sc	27.5		33	28.0		25.3
Cr	183	141	140	250	53	164
Ni	92	48	75	124	3	46
Co	2	38	45	45	47	36
Ratios						
FeO/MgO		1.66	1.81	1.41	1.88	1.73
K_2O/Na_2O		0.16	0.02	0.02	0.04	0.09
Ba/La	10.3					11.7
La/Th	9.2		11.7	5.3		5.7
La/Sm	5.7		3.5	3.4		4.0
La/Yb	20.1		5.8	5.7		8.8
Eu/Eu*	1.03					0.96
Sm/Yb	3.5		1.7	1.7		2.2
Ba/Ta	105					149
La/Ta	10	0	9	5		13
Th/U	2.6		3.2	5.8		3.6
Molar $Al_2O_3/(CaO+K_2O+N_2O)$		0.81	0.67	0.68	0.65	0.84

125B (in Koukharsky et al., 1989) whose high K_2O (~3%) concentrations put them in the shoshonitic field in Figure 4A. High light REE (Ce ~105 ppm), Ba (1239 ppm), Th (15.8 ppm), and Rb (80 ppm) concentrations and very LREE-enriched patterns (La/Sm = 5.6; La/Yb = 22) are in accord with a shoshonitic classification (Fig. 10C). An arc-like signature as is common in shoshonites is indicated by high Ba/La (24), La/Ta (58) (Fig. 5B), and La/Th (3.3) ratios, and high Al_2O_3 (18.3%) and low TiO_2 (1.14%) concentrations. A low Sr concentration (450 ppm) and a negative Eu anomaly reflect plagioclase fractionation (Fig. 10D).

A suggestion that the shoshonitic character of the Guayaos lavas is more than a local feature comes from similarities to lavas (57–58% SiO_2) that crop out near Aracar (Table 4) along the Chilean border in the Taca Taca region (Fig. 2). Although their ages are not well established, shoshonitic-like K_2O (3.4–3.9%) (Fig. 4A), Ba (675–790 ppm), and Th (15–16 ppm) concentrations, light REE–enriched (La/Sm = 5.7–6.0; La/Yb = 14–15) patterns, and arc-like La/Ta (44–48) (Fig. 5B) and La/Th (2.8–3.0) ratios (see Fig. 10D) suggest an affinity with the Guayaos lavas. Higher FeO/MgO ratios (2.8) and larger negative Eu anomalies (Eu/Eu* = 0.62–0.66) indicate the Aracar region lavas are more evolved. Like the Guayaos lavas, they plot in the calc-alkaline arc field in the Th-Ta-Hf diagram in Figure 6.

Faja Eruptiva Oriental magmatic units

Magmatic units in the northern part of the Faja Eruptiva Oriental are found from near 17°S latitude in Bolivia to just south of San Antonio de los Cobres near 24°S latitude (see Fig. 2). In northern Argentina, this belt is defined by the magmatic-sedimentary Cochinoca-Escaya complex, which is typified by dacitic and mafic magmatic units interspersed with thick sand to shale sequences containing Arenig to Llanvirn graptolite faunas (Aceñolaza and Toselli, 1984; Bahlburg et al., 1990; Gutierrez Marco et al., 1996). These units are cut by mafic alkaline dikes and sills (Coira, 1975, 1979; Koukharsky and Mirré, 1974; Coira and Koukharsky, 1994) that are chemically similar to the Huaitiquina dikes. The ages of the dikes and sills are uncertain, but it is reasonable from their chemistry (see below) that they represent the last stage of the magmatic episode.

Near 23.5°S latitude, the Cochinoca-Escaya complex is cut by the north-south–elongated Sierra de Tanque and Cobres plutons (Fig. 2). The epizonal Sierra de Tanque pluton is composed primarily of granodiorite and granite containing biotite, muscovite, zircon, and rare titanite and apatite. Alkaline gabbroic stocks and dikes containing kaersutite, pyroxene, apatite, and ilmenite occur in the northern part. Hybrid and mechanical mixtures of silicic and mafic phases near contact zones indicate mingling and mixing, and thus similar ages for gabbroic and granitic units. The coarse-grained Cobres granodiorite is typified by euhedral orthoclase phenocrysts in an aggregate of oligoclase, quartz biotite, and minor muscovite. This pluton is surrounded by a contact metamorphic aureole and is deformed as shown by cataclastic textures and bent feldspar twins. Both plutons are cut by late-stage lamprophyric dikes.

Farther south, the northern Faja Eruptiva Oriental magmatic units merge into extensive plutonic units south of the El Toro lineament (Fig. 2). As exposure levels deepen, Ordovician plutonic rocks become difficult to differentiate from polydeformed

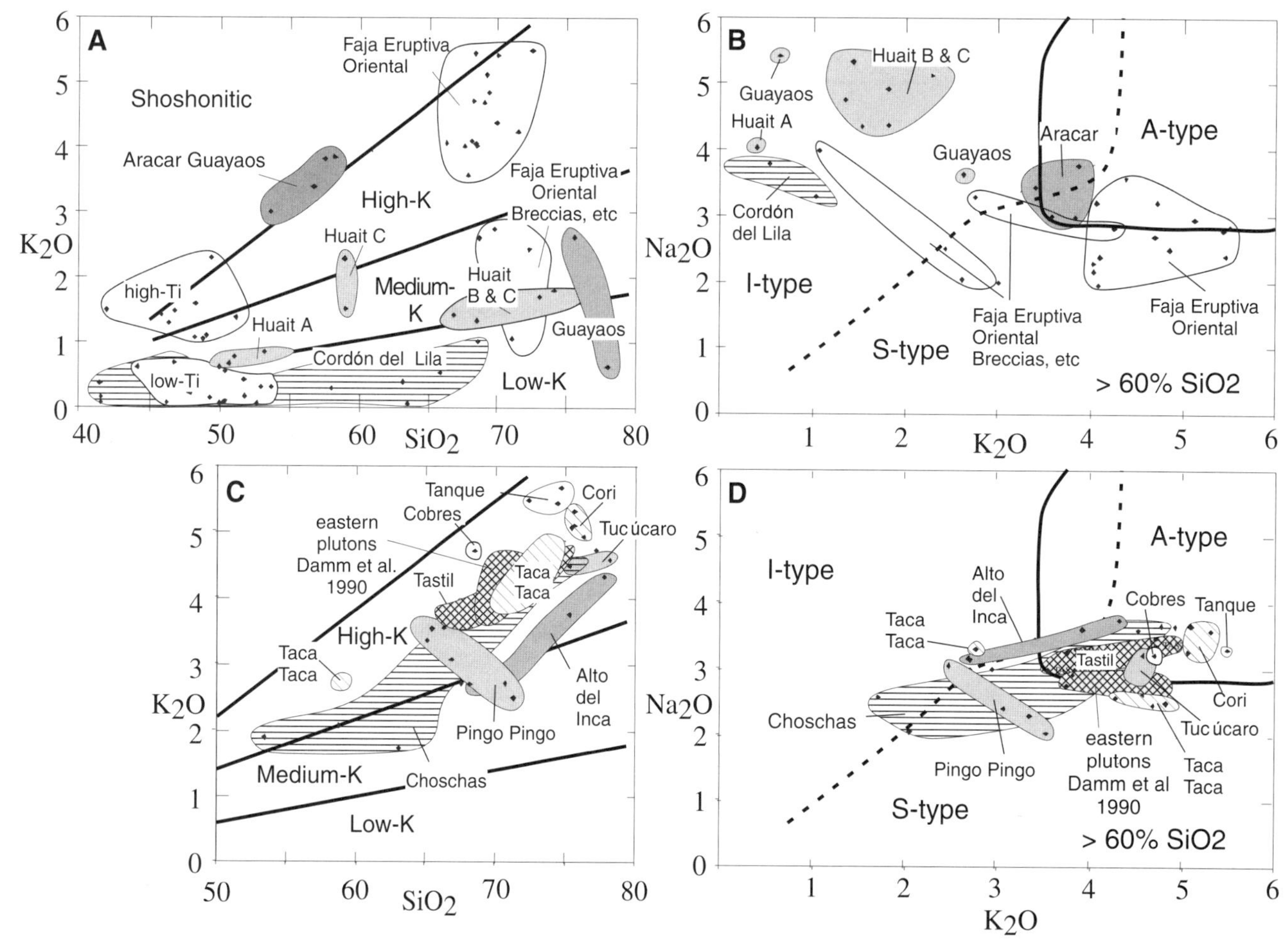

Figure 4. Plots of K_2O vs. SiO_2 and Na_2O vs. K_2O (all in wt%) for Ordovician volcanic (A and B) and plutonic samples (C and D) from the Faja Eruptiva Oriental, Faja Eruptiva Oriental, and Eastern Cordillera. Fields on K_2O vs. SiO_2 plot are those commonly used to classify low-, medium-, and high-K arc rocks. Classification fields on Na_2O–K_2O plots for I (igneous), S (sedimentary), and A (alkaline or anorogenic) -type granites from White and Chappell (1983). Only samples with $SiO_2 > 60\%$ are plotted. Fields are for classification, not generic purposes. Data from Tables 1 to 8, and from Bahlburg (1990) and Damm et al. (1986, 1990, 1991). See text for discussion. Huait refers to Huaitiquina samples.

upper Precambrian to lower Paleozoic igneous-metamorphic complexes. A major unknown in the regional tectonic picture is the character and age of this extensive belt of granitoids, which extends southward through the Cumbres de Luracatao to the Sierra de Chango Real near 27°S latitude.

Silicic magmatic units. The relation between silicic magmatic and sedimentary units in the northern Faja Eruptiva Oriental has been the cause of much debate. One side has maintained that the Cochinoca-Escaya complex silicic magmatic units are Late Ordovician to Silurian porphyries and shallow level plutons intruded into Ordovician clastic sequences (e.g., Turner, 1964; Schwab, 1973; Méndez et al., 1973; Omarini et al., 1979; Bahlburg, 1990, 1993). Initially, the opposing view was that the silicic units represented ignimbritic and submarine pyroclastic flows and lavas erupted at the time of sediment deposition (e.g., Coira, 1973, 1975; Koukharsky and Mirré, 1974; Coira et al., 1982). More recently, Coira and Koukharsky (1991, 1994) and Coira (1996) have argued these units represent synsedimentary lava dome complexes. In the syndepositional model, the silicic and mafic units form a bimodal magmatic province.

A problem in interpreting the silicic units is that they occur in the zone of most intense deformation at the eastern edge of the Ocloyic deformational front (Mon and Hongn, 1991). The resulting structural overprint makes deciphering magmatic and sedimentary relations difficult as the competency contrast among these units has resulted in shear zones developing along contact zones. In places, the rocks are protomylonites. Hydrothermal alteration associated with formation of massive sulfide deposits (Zn, Cu, Pb, Au-Ag) has further obscured original textures and relationships.

Radiometric ages (Fig. 3) are important in resolving the question of the origin of the silicic units. The best ages available are U-Pb monazite ages from Lork and Bahlburg (1993): 467 ± 7 Ma for a Cochinoca-Escaya complex dacitic unit in the Cochinoca region; 476 ± 2 Ma for the Sierra de Tanque pluton; and 472 ± 3 and 472 ± 1 Ma, respectively, for the Tacuil and

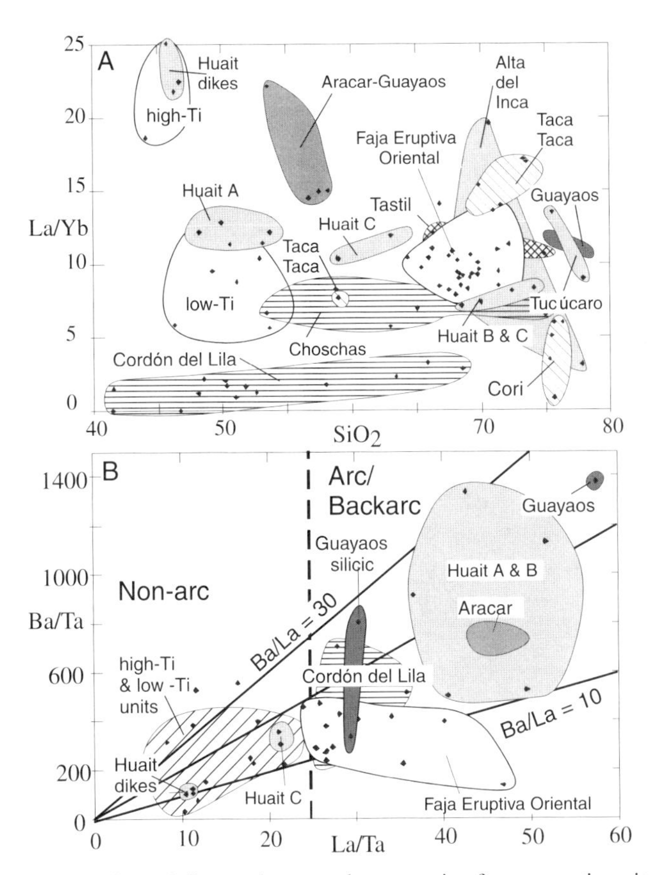

Figure 5. Plots of diagnostic trace element ratios for magmatic units from the Faja Eruptiva Oriental and Occidental. A, La/Yb ratio plotted as a measure of slope of REE pattern vs. weight percent SiO_2 for volcanic and plutonic units. Lower La/Yb in low-Ti compared to high-Ti mafic units is interpreted as reflecting higher melting percentages of a similar garnet-bearing ultramafic mantle source to that producing the high-Ti group. Other differences in La/Yb ratios largely reflect variations in light REE enrichment of mantle arc and crustal sources. B, Ba/Ta vs. La/Ta ratios for volcanic samples. Plot highlights high field strength element depletion (high La/Ta) and alkali enrichment (high Ba/Ta and Ba/La) common in magmatic rocks with arc-like affinities. As Ta was contaminated by grinding of Cordón de Lila samples in Damm et al. (1986, 1990, 1991), Ta is calculated as Nb/17 (see Sun and McDonough 1989). Data sources as in Figure 4.

Brealito plutons south of San Antonio de los Cobres. These ages are in accord with an Rb-Sr whole-rock isochron composite age of 471 ± 12 Ma cited by Omarini et al. (1984) for dacitic units near the El Toro lineament, and upper Arenig to Llanvirn graptolite ages in the sediments. Similarities in chemistry (see below) are consistent with a common age for the granitoids and the disputed silicic units. Although the age data support a syndepositional origin for the silicic magmas, Bahlburg and Hervé (1997) have maintained there could still be a problem as Bahlburg et al. (1990) have argued that a Cochinoca region dacitic unit intrudes folded graptolite-bearing Late Ordovician sediments.

Because of the importance of the Cochinoca-Escaya complex to tectonic models, evidence for the nature of the emplacement of the silicic magmatic units and their chemistry is presented below. Stratigraphic sequences in the Huancar, Muñayoc, Queta, and Cordón de Escaya regions are presented in Figure 9. Chemical analyses for samples from the Escaya, Angosto, Cochinoca, Muñayoc, and Queta areas in the north, and the Huancar, Sierra de Cobres, Sierra de Tanque, and Santa Ana areas in the south (Fig. 2), are given in Tables 5 and 6 and plotted in Figure 11. Other Queta region analyses can be found in Bahlburg (1990).

The evidence indicates that the silicic units are best interpreted as porphyritic lava flows and hyaloclastites, sills, dikes, and cryptodomes that constitute near-vent nonexplosive submarine units. Field and microscopic studies show that their textures and structures can be attributed to thermal stress fragmentation at lava-water interfaces and interaction between magma and wet sediments. Submarine emplacement is consistent with quenching processes producing in situ hyaloclastites and volcanic breccias with jigsaw block textures and peperites. Volcaniclastic sediments composed of reworked hyaloclastites occur on top of and laterally outward from flows and cryptodomes.

In detail, individual flows are 20–150 m thick and occur in complexes up to 500 m thick. They are composed of porphyritic dacite with 30–35% phenocrysts of alkali feldspar (Ab70–100), oligoclase (An15–20), and biotite in a gray-green perlitic, devitrified, glassy to microgranular groundmass that contains aggregates

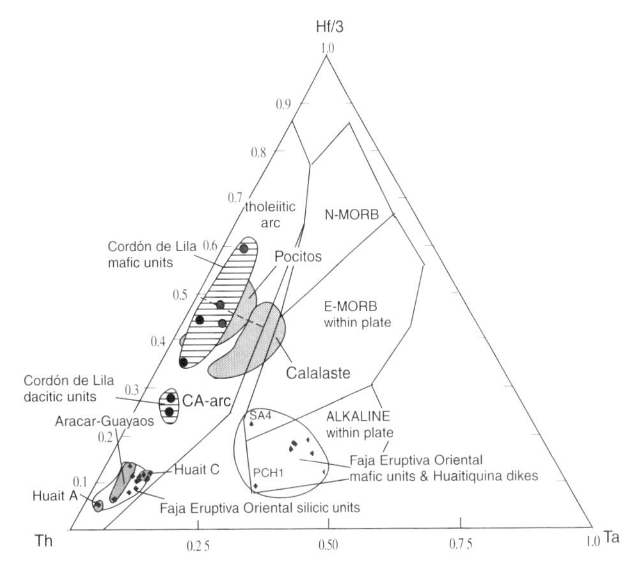

Figure 6. Ternary Hf/3-Th-Ta plot for mafic units from the Faja Eruptiva Oriental (data from Tables 7 and 8) and the Faja Eruptiva Occidental (Huaitiquina and Guayaos data from Table 3, and Cordón de Lila (data from Damm et al., 1986, 1991), Shown for comparison are data fields for mafic/ultramafic units from the central and southern Puna Pocitos and Calalaste regions (from Bahlburg et al., 1997, unpublished data). N-MORB, within-plate, and arc fields from Wood et al. (1979) are shown for comparison. Also plotted for reference are dacitic/silicic samples from the Faja Eruptiva Oriental (data from Tables 5 and 6) and Cordón de Lila areas (data from Damm et al., 1986, 1991), whose location on the diagram should not be correlated with mafic rock fields. As in Figure 5, Ta concentrations for samples analyzed by Damm et al. are calculated as Nb/17.

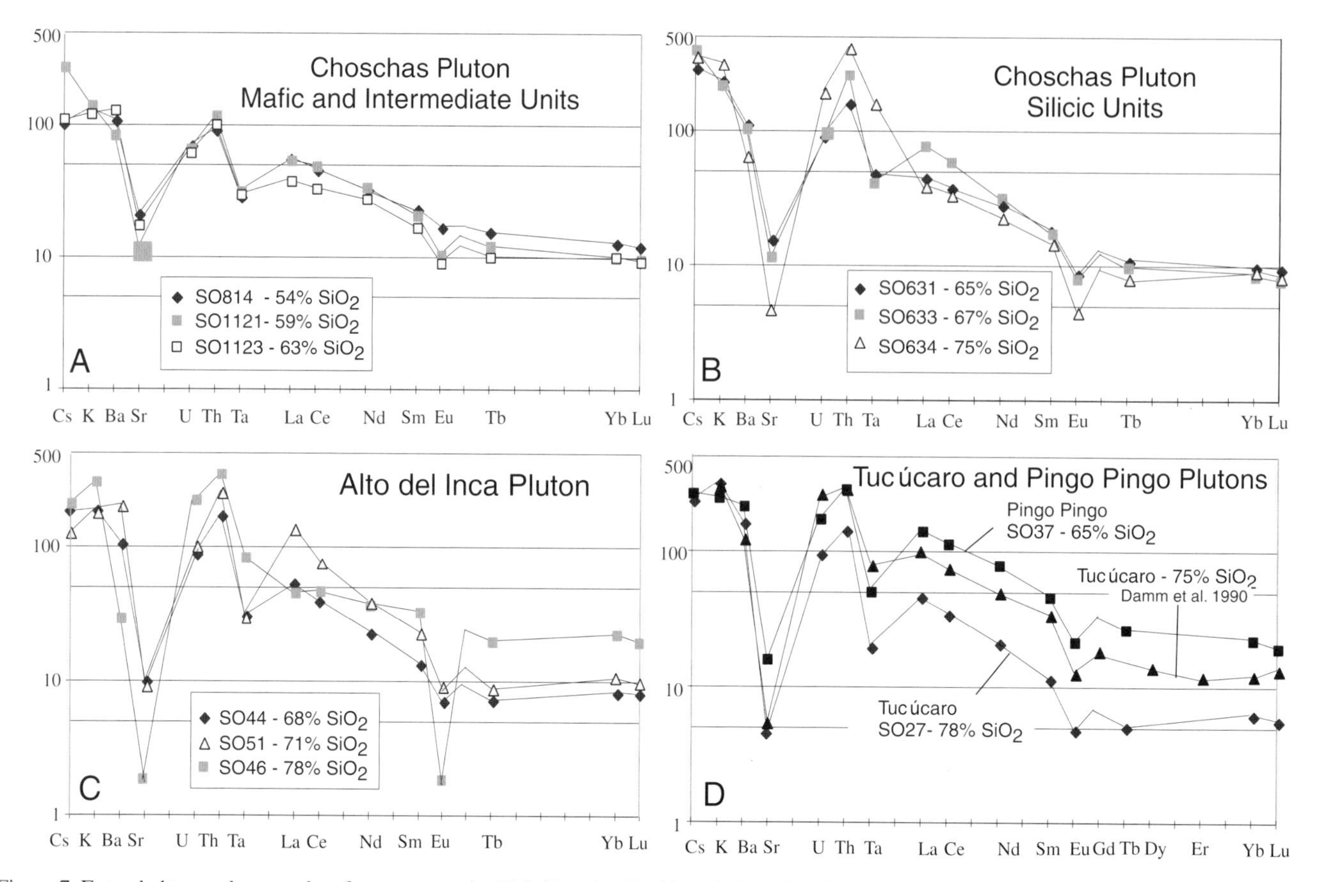

Figure 7. Extended trace element plots for representative Faja Eruptiva Occidental plutonic units from the Sierra de Almeida region in Chile. Normalization factors are Cs (0.013), Ba (3.77), Sr (14), U (0.015), Th (0.05), Ta (0.022), La (0.378), Ce (0.976), Nd (0.716), Sm (0.23), Eu (0.0866), Tb (0.0589), Yb (0.249), and Lu (0. 0383). Data from Table 1 except for Tucúcaro composite analyses, which are from Damm et al. (1990). See Figure 3 for locations and text for discussion.

of quartz, alkali feldspar, chlorite, and sericite. The presence of alkali feldspar phenocrysts up to 5 cm across suggests that magmas were stored in crustal magma chambers before erupting. Perlitic textures are attributed to hydration and chilling effects.

On a macroscopic level, the tops and margins of the lava flows are characterized by in situ jigsaw-type autobreccias composed of blocks ranging from 0.5 to 5 m in diameter, and hyaloclastite with angular to subangular fragments up to 0.5 m, dispersed in a hyaloclastitic matrix. The hyaloclastite is attributed to shattering of volcanic glass. It consists of angular fragments of porphyric dacite or glass altered to chlorite and sericite, and broken feldspar, quartz, and biotite crystals. The percentage of hyaloclastite increases as flow margins are approached. In extreme cases, sections of flows up to 30 m thick are composed of hyaloclastite with individual zones containing angular to subangular fragments up to 40 cm across (see Huancar profile in Fig. 9). Where primary textures are preserved, the base of the flows contain distorted sedimentary fragments, and primary structures in the underlying sediments are disturbed or completely destroyed.

Also present in the sedimentary sequences are 10–30-m-thick tabular bodies that commonly display columnar jointing. They are interpreted as sills intruded into wet sediments. Structures indicating interaction between wet sediment and magmas are present on their top and bottom surfaces. Contacts with sediments are characterized by peperites composed of quenched dacitic fragments in an indurated sedimentary matrix lacking primary sedimentary structures (Huancar profile in Fig. 9). The peperites have been silicified by circulation of hot pore fluids through expansion cracks.

Other units are interpreted as cryptodomes, as shown in the Muñayoc profile in Figure 9. The top 10–20 m consist of jigsaw autobreccia containing fragments 0.1–3 m across. Upper contacts are characterized by megapeperites containing subangular to subrounded dacitic blocks up to 0.2–1.2 m across that are enclosed in a fine-grained silicified matrix (15–20% by volume). Facies and structures like these form as cryptodomes ascend through unconsolidated sediments and breach the sediment-water interface, producing phreatomagmatic explosions. The cryptodomes are cut by numerous opaline silica veins.

Sequences of reworked hyaloclastites separated by erosional hiatuses occur above the cryptodomes as shown in the Muñayoc profile (Fig. 9). In detail, individual sequences up to 10 m thick

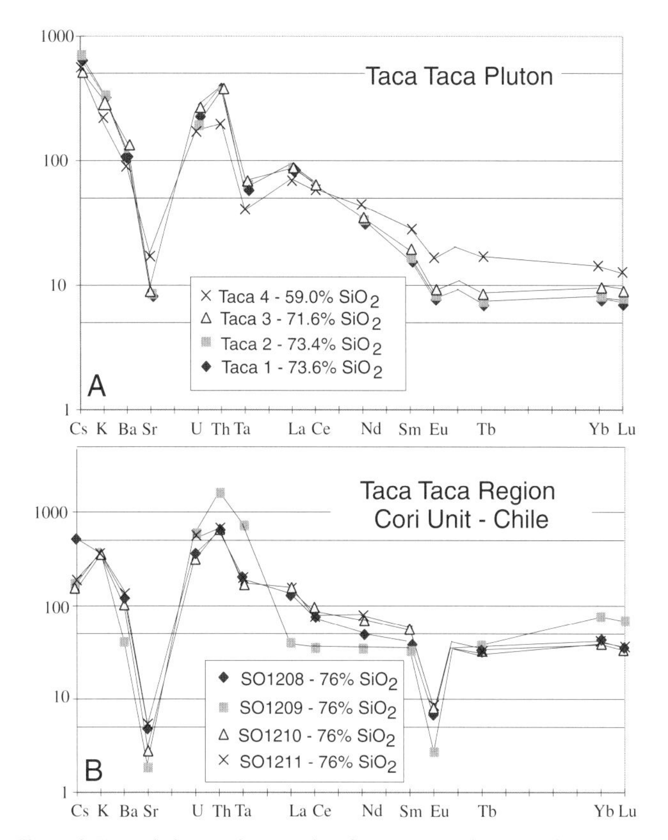

Figure 8. Extended trace element plots for representative plutonic units from the Taca Taca region in the Faja Eruptiva Occidental. A, Taca Taca diorite and monzogranite. B, Cori unit syenogranite. Normalization as in Figure 7. Data from Table 2. See Figure 3 for locations and text for discussion.

are characterized by cross-strata with amplitudes up to 8 m and internal graded bedding. Finer units, typically 5–40 cm thick, are composed of siliceous glass fragments. Sandy graded units, typically 10–50 cm thick, consist of fragments of quartz, sodic oligoclase, alkali feldspar, and biotite, along with devitrified glass replaced by chlorite, quartz, and alkali feldspar. The presence of discrete reworked hyaloclastitic sequences signifies periods of emergence and erosion of the tops of the domes. These sequences contrast with volcaniclastic-poor sedimentary units interpreted as having formed during periods of rapid continental detrital sediment accumulation.

Regardless of eruptive style, the silicic magmas are mostly of dacitic composition with 68–71% SiO_2, 2.2–3.2% Na_2O, and >4% K_2O (Fig. 4; Tables 5 and 6). On granitoid classification diagrams, they generally plot in the S-type field (Fig. 4B). Most are peraluminous with high molar (Al_2O_3/(K_2O + Na_2O + CaO) = 1.1–1.4 consistent with melts derived predominantly from sedimentary sources. Samples with the lowest Na_2O and K_2O concentrations are breccias (Fig. 4A,B). Comparison of analyses of these breccias with a flow from Muñayoc in Table 5 shows that alteration-resistant trace elements are similar, consistent with low K and Na concentrations being due to secondary alteration. Chemical signatures of the Sierras de Cobres (Table 6) and Tanque plutons are generally like those of the volcanic rocks (see Fig. 4).

Trace element characteristics of these units (Figs. 5A and 11) are also consistent with an important sedimentary crustal component in the magmatic source. High Cr (40–52 ppm, 1 >80) and Ni (20–30 ppm), low Sr, and high normalized Cs, U, and Th (200–300 X chondrites) concentrations, and La/Th ratios from 1.5 to 2.6 provide the most convincing evidence. Other evidence comes from extended element plots showing moderately enriched light REE patterns and La/Yb ratios ranging from 8 to 10.5 (Fig. 5A) like those of typical Proterozoic shales. Large negative Eu anomalies indicate melting of a source whose previous history involved removal of substantial amounts of feldspar as expected for sediments derived from magmatic rocks. La/Ta ratios, which are mostly between 22 and 30 (range is 18–40) and Ba/La ratios from 10 to 13 (Fig. 5B) show the weak arc-type signature typical of many sediments.

The few published analyses of the extensive plutonic outcrops south of the El Toro lineament include those for the strongly sheared and mylonitized Quebrada Tajamár and Salar de Diablillos monzogranite and granodiorite plutons and undeformed porphyritic, two mica granites from Ochaquí and Salar de Diablillo in Damm et al. (1990). These analyses are like those of the Cochinoca-Escaya complex dacitic units (e.g., see eastern plutons in Fig. 4), indicating a generally similar source for all northeastern Faja Eruptiva Oriental silicic units. A sedimentary crustal component is supported by $\delta^{18}O/^{16}O$ ratios of 10–12.4‰ in units studied by Damm et al. (1990), and initial $^{87}Sr/^{86}Sr$ ratio near 0.710 inferred from the Rb-Sr whole-rock isochron composite in Omarini et al. (1984).

Mafic magmatic units. Mafic units in the Cochinoca-Escaya complex consist of massive and pillowed flows, sills, and dikes. The most abundant units are 6–25-m-thick massive flows with microporphyritic to finely granular to ophitic textures. Coarser grained zones have variolitic textures with albite crystals in a groundmass of chlorite (pennite), epidote (pistacite), Ti- magnetite, and interstitial quartz. Finer grained zones contain calcite and chlorite amygdules. The pillowed flows range from 20 to 30 m thick. Individual pillows have the following characteristics: aphyric chilled margins with suboval quartz and chlorite-filled amygdules; intermediate zones with aligned amygdules, mafic phenocrysts replaced by chlorite, quartz and pistacite, and plagioclase (now An_{8-10}); and cores with amygdules and veins of quartz, calcite, siderite, chlorite, and pyrite. These pillows were likely emplaced at water depths of <500 m consistent with associated sediments whose characteristics indicate deposition in a medial to distal platform environment below wave base. The predominance of massive flows suggests either a near-vent location or high rates of extrusion.

Also present are fine-grained mafic and porphyritic gabbroic dikes and layered gabbroic sills cutting sandy and pelitic sequences. They are interpreted as being emplaced in the final magmatic stages of the Cochinoca-Escaya complex as boudin and other structures that would be expected from their ductility contrast with the surrounding sedimentary units indicate emplacement before the Ocloyic deformational event. In detail, the fine-grained dikes and sills are 1–3 m thick and are composed of aphyric to scarcely porphyritic alkaline basalt and basanite. Mafic minerals include augite, kaersutite, and in some cases, biotite. These units are well exposed in the Huancar and Sierra de Tanque sections, as well at Huaitiquina in the Faja Eruptiva Occidental (Fig. 9). Also present are 1–3-m-thick, weakly discordant porphyritic gabbroic dikes that contain scattered hornblende phenocrysts in a matrix of plagioclase, hornblende, tremolite, ilmenite, and leucoxene. Layered gabbroic sills 5–500-m-thick at the Santa Ana locality (Fig. 2) are characterized by coarse bands of serpentinized olivine and augite alternating with bands of plagioclase (An_{36-40}) that are partially replaced by sericite and epidote, and mafic minerals replaced by chlorite.

Chemical analyses of these mafic units in Tables 7 and 8 and in Figure 12 all show subalkaline to alkaline characteristics, and non arclike La/Ta ratios (most 8–13). All fall in the within-plate mafic fields on the Hf-Ta-Th plot in Figure 6. They can be roughly divided into low-Ti and high-Ti groups. The low-Ti group consists of spilitized lavas from the Casa Colorada near Cerro Queta (CO6), Cerro Queta (QUE3), Cordón de Escaya (ESCA 23, 95) and Sierra de Tanque (Bet34, Bet 35) localities in Figure 2. This group generally has ~48–53% SiO_2 (anhydrous), 1.2–2% TiO_2, and <0.6 K_2O (Fig. 4), and falls in the low-K to normal subalkaline field. As shown in Figures 5, 6 and 12A, trace element characteristics include enriched light REEs,

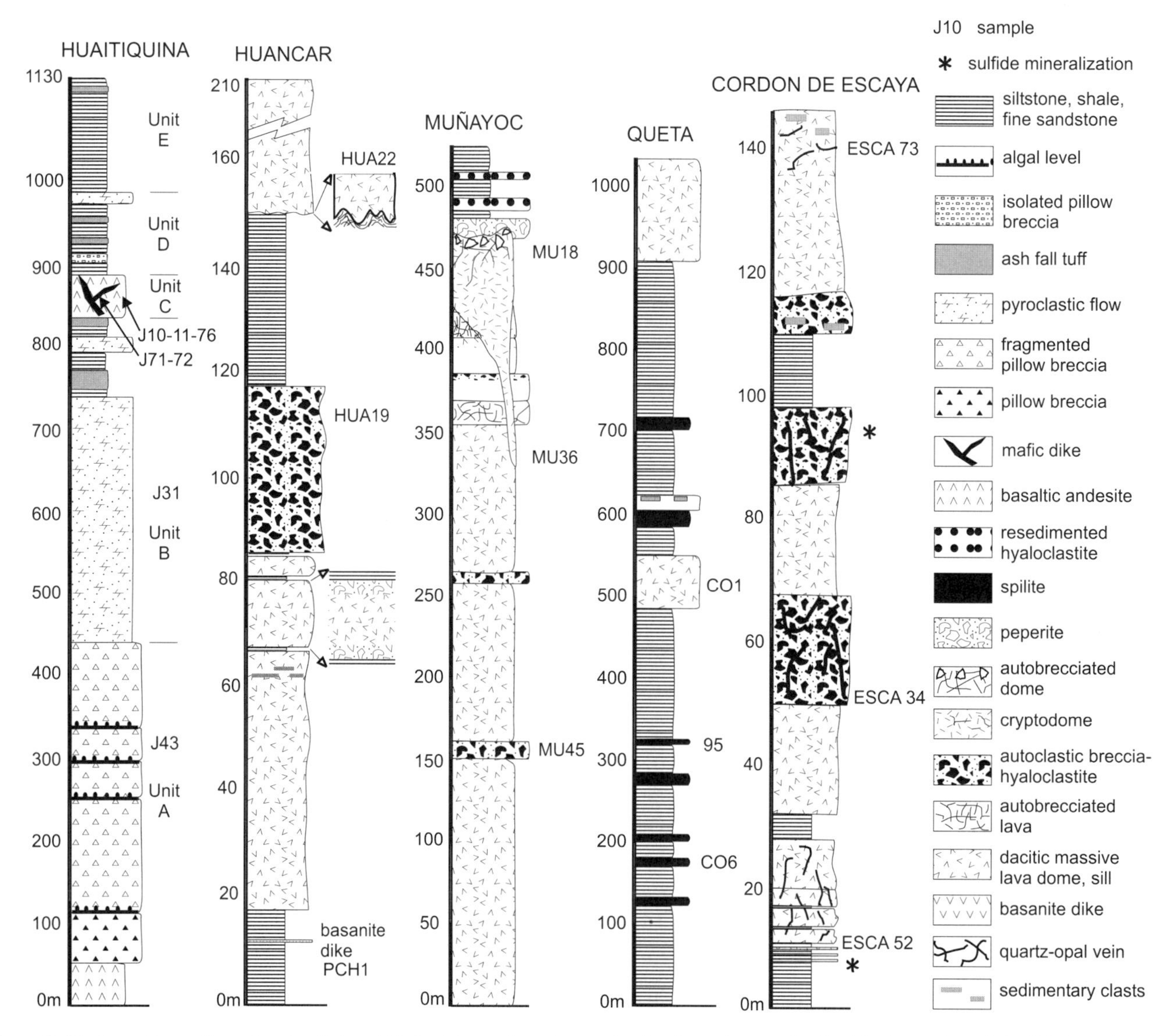

Figure 9. Stratigraphic sections showing facies of Ordovician sedimentary and volcanic sections in the Huaitiquina region of the Faja Eruptiva Occidental and representative sections of the Cochinoca-Escaya complex of the Faja Eruptiva Oriental. Numbers on left side of columns are thicknesses in meters. Sample localities for samples in Tables 3 and 5–7 shown on the right of the sections. Location of sections shown in Figure 3. Discussion in text.

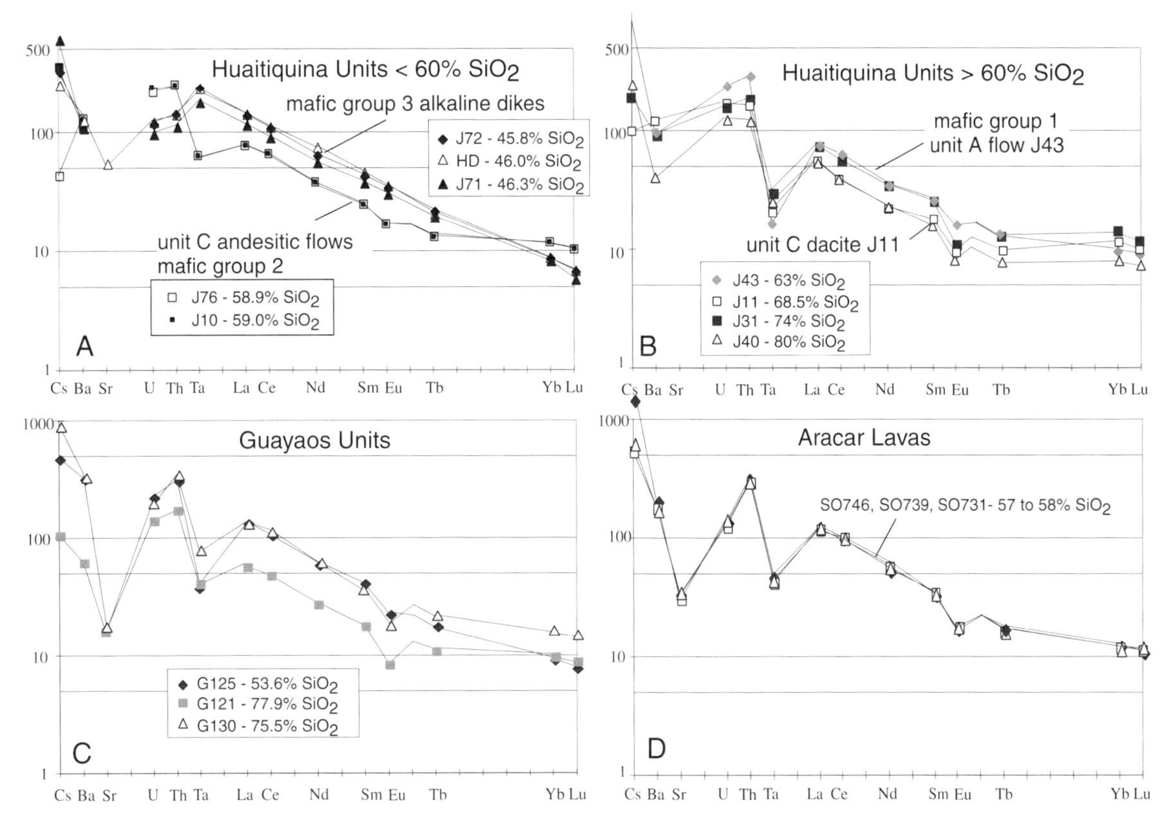

Figure 10. Extended trace element plots for representative volcanic units from: Huaitiquina (A and B), Guayaos (C), and Aracar regions (D) in the Faja Eruptiva Occidental. Normalization as in Figure 7. Data from Tables 3 and 4. See Figure 3 for locations and text for discussion.

La/Yb ratios from 10 to 12, small Eu and no Sr anomalies, and La/Th ratios near 8. The high-Ti group includes Ti- and K-rich alkaline mafic dikes in the Huancar (PCH1) and Sierra de Cobres (COB10) regions, lavas at Cochinoca (PCO425) and Escaya (ESCA 14), the gabbros and sills in the Sierra de Tanque (Bet78, 56), and the dikes cutting the Huaitiquina section in the Faja Eruptiva Occidental. These units are characterized by lower SiO_2 (42–50%), higher TiO_2 (2.8–4%) and generally higher K_2O (0.36–2.3% in Fig. 4A) concentrations. They have steeper REE patterns (La/Yb = 19–55) (Fig. 5A) and higher incompatible element concentrations (Fig. 12B) than the low-Ti units. Similarities in incompatible and immobile element ratios between the two groups are consistent with a generally similar garnet-bearing mantle source for both groups with the low-Ti magmas representing higher percentages of melting. The higher La/Ta ratio (19) of the high-Ti Huancar dike could reflect its proximity to an arc region to the west.

Gabbroic dikes from the Sierra de Tanque (ACI, ACH) and Rio de Las Burras (RBU) near Sierra de Cobres, and gabbros from Santa Ana (SA4 in Fig.12A) all have near 50% SiO_2, 1.7–2.7 TiO_2, and <1.5% K_2O (Table 8). Their chemical characteristics are transitional between those of the low-Ti spilites and the high-Ti alkaline intrusives, consistent with derivation from a generally similar source.

Eastern Cordillera plutonic belt

Farther east, a number of plutons cut the late Precambrian Puncoviscana Formation, Cambrian Mesón Group, and Ordovician Santa Victoria Formation in the eastern Puna and Eastern Cordillera (see Fig. 3). Important to the discussion here is the Chañi pluton (see Fig. 2), which is considered to be Ordovician in age based on K-Ar biotite dates of 477 ± 20 and 463 ± 6 Ma in Méndez et al. (1979). No chemical data are published.

Some of the plutons in this region are unconformably overlain by Cambrian and Ordovician strata. From north to south, these are: (1) Cañani, which is assigned an age between 534 and 519 Ma, based on U-Pb zircon ages in Bachmann et al. (1987); (2) the muscovite-bearing granodioritic to monzogranitic Tipayoc stock, which is assigned an age of 550 ± 26 Ma, based on a K-Ar muscovite age in Omarini et al. (1996); and (3) the granodioritic to monzogranitic Santa Rosa de Tastil pluton, which is considered to be 536 ± 7 Ma, based on a U-Pb zircon age in Bachmann et al. (1987). The Puncoviscana Formation adjacent to the Santa Rosa de Tastil pluton yields a whole-rock K-Ar metamorphic age of 535 ± 6 Ma (Adams et al., 1990). Chemical analyses in Damm et al. (1990) and Table 2 for the Santa Rosa de Tastil granodiorite (66% SiO_2) indicate a peraluminous (molar $Al_2O_3/(K_2O + Na_2O + CaO)$ = 1.14, 3% Na_2O, 3.6% K_2O) character (Fig. 4). An

Rb-Sr mineral isochron of 514 ± 4 Ma yields an initial $^{87}Sr/^{86}Sr$ ratio of ~0.710 (Bachmann et al., 1987).

Farther south, other plutons cut the Puncoviscana Formations. This group includes those in the Cachi Range (Fig. 2), which Galliski et al. (1990) described as high-Al metaluminous trondhjemites (69–73% SiO_2, >4.7% Na_2O, <1.35 K_2O; >550 ppm Sr) and interpreted as partial melts of mafic rocks, probably amphibolite, in a continental margin arc setting. A provisional age of 564 ± 25 or 515 ± 20 Ma based on K-Ar ages of stocks and pegmatites associated with trondhjemitic bodies was discussed by Galliski et al. However, Lork et al. (1991) has argued that U-Pb systematics for zircons from the tops and margins of these bodies reflect incorporation of old crustal zircons into magmas with mantle-like initial $^{87}Sr/^{86}Sr$ (near 0.7032–0.7034). They argued that a lower intercept U-Pb zircon age of 488 +14/−16 Ma for the El Vallecito trondhjemite and U-Pb monazite ages of 466 ± 1–468 ± 1 Ma for the El Alto and of 478–481 ± 1 Ma for the Aguas Calientes trondhjemite indicate Ordovician crystallization ages. Care is needed as recent studies show that monazite is more sensitive to recrystallization than once thought (e.g., Miller et al., 1997).

Well-studied magmatic rocks in the Sierra de Fiambalá south of 27°S discussed by Grissom (1991) and De Bari (1990) could be important in the regional picture. In this area, greenschist to amphibolite facies metamorphism lasting from 550 to 540 Ma is associated with mildly peraluminous orthogneisses (e.g., La Puntilla) that are chemically similar to the Santa Rosa de Tastil pluton. A second event between 515 and 470 Ma resulted in amphibolite to granulite facies metamorphism and deformation. This metamorphism is inferred by Grissom (1991) to be related to long-lasting thermal effects from the arc magmatism, which produced the 514-Ma Fiambalá gabbro norite (De Bari, 1990). The final stages overlap the emplacement of the Chañi pluton and Faja Eruptiva Oriental magmatic units.

DISCUSSION: ORDOVICIAN MAGMATIC AND TECTONIC HISTORY OF THE FAJA ERUPTIVA BELTS

The distribution, age, chemistry, and inferred setting of the Faja Eruptiva Occidental and Oriental, and Eastern Cordillera magmatic units provide constraints on the tectonic history of the

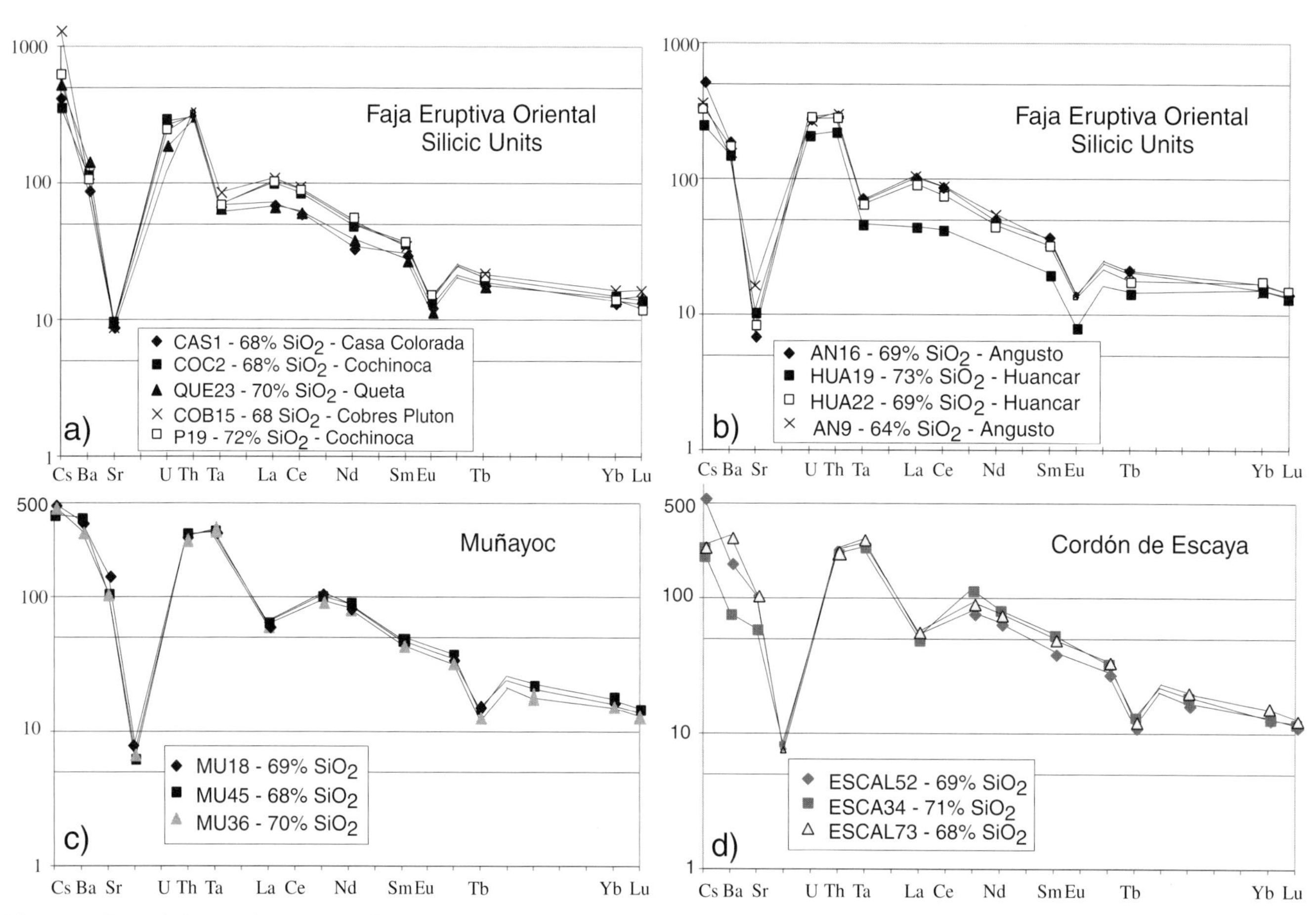

Figure 11. Extended trace element plots for representative silicic units from the Faja Eruptiva Oriental sequences at: Cochinoca, Cerro Queta, and Quebrada Casa Colorado near Cerro Queta, and the Cobres Pluton (A); Abra del Angosto and Cerro Huancar (B); Muñayoc (C); and Cordón de Escaya (D). Localities shown in figures, profiles in Figure 9. Data from Tables 5 and 6. Normalization as in Figure 7.

northern Puna. These data are discussed below and used in shaping the three sequential map views in Figure 13 that refine and modify the Ordovician tectonic model presented by Bahlburg and Hervé (1997).

Faja Eruptiva Occidental

The Ordovician magmatic and tectonic events in the Faja Eruptiva Occidental are difficult to interpret due to poor age control and the fact that the location of the Cordón de Lila and Sierra de Almeida magmatic units relative to units farther east is uncertain. Two major immediate problems are a lack of age data and the fact that the blank region west of the Salar de Atacama in Figure 2 is almost entirely covered by Neogene volcanic rocks.

A first step is to unravel the tectonic setting and temporal sequence of the Cordón de Lila and the Sierra de Almeida units and to try to correlate them with units farther east. A reasonable proposition given existing ages and chemistry is that the Cordón de Lila complex contains the oldest early Paleozoic magmatic unit in the region. This view is supported by the low-K arc/backarc-like chemistry of the magmatic rocks that fit with extrusion on very thin continental or oceanic crust. This Cordón de Lila sequence differs chemically from the medium- to high-K field transitional metaluminous to peraluminous Choschas plutonic units. However, attention has to be paid to the fact that volcanic and plutonic units are being compared. In particular, Kay et al. (1990) showed that Aleutian plutonic units have higher K_2O (and other incompatible element) concentrations than volcanic rocks at the same SiO_2, and argued the plutons evolved in a closed system at the end of a magmatic cycle. The similarity of the gross features of Choschas and Aleutian plutonic units could be consistent with the Choschas pluton representing the end of the magmatic cycle associated with the Cordón de Lila complex. Marginal S-like chemical characteristics of the Choschas pluton can be attributed to crustal contamination. Similarities of arc-like La/Ta ratios in the Cordón de Lila complex (mostly 20–36) and in the Choschas plutonic units (22–32) are consistent with this interpretation.

Going a step further, the Choschas pluton could be more or less contemporaneous with the basal arc/backarc Huaitiquina unit A volcanic sequence. If so, the Choschas pluton could be late Tremadoc/Arenig and the Cordón de Lila complex slightly older. The shoshonitic units at Aracar and Guayaos fit a late postarc/backarc setting. Support for such a model comes from the basal Huaitiquina sequence and the Choschas pluton both being succeeded by predominantly siliceous magmatic sequences. The silicic unit postdating the Choschas pluton is the cross-cutting Alto del Inca pluton. At face value, its reported radiometric ages overlap the Arenig-Llanvirn graptolite ages of the western belt Huaitiquina, Aguada de la Perdiz, and Guayaos silicic volcanic/volcaniclastic sequences. The Alto de Inca and associated plutons could be related to the centers from which these volcaniclastic units were derived.

Finally, the Late Ordovician Caradoc to Silurian ages of the Tucúcaro, Tilopozo, Pingo Pingo, and Taca Taca plutons are consistent with these plutons representing younger magmas associated with compressive deformation in the northern Puna. Following Mpodozis et al. (1983) and Forsythe et al. (1993), these units would be associated with closure of a western marine basin. These plutons all lie above the so-called "gravity terrane" of Goetze et al. (1994), which denotes a regional gravity high running from the southern Puna through the northern half of the Salar de Antofalla northwestward through the Salar de Atacama. The gravity terrane has been correlated with the Faja Eruptiva Occidental plutonic belt of Palma et al. (1986) and interpreted as a Paleozoic feature by Goetze et al. (1994). It could represent a suture along which the Arequipa terrane was joined to Gondwana. In detail, the vergence and number of arcs that were present in this area are unknown.

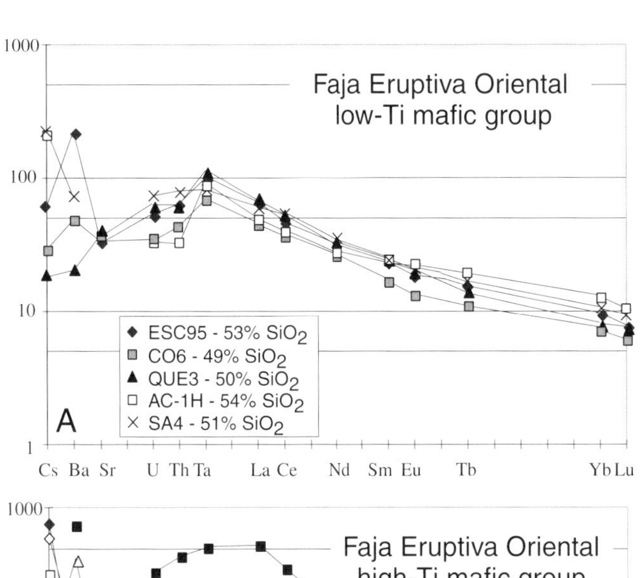

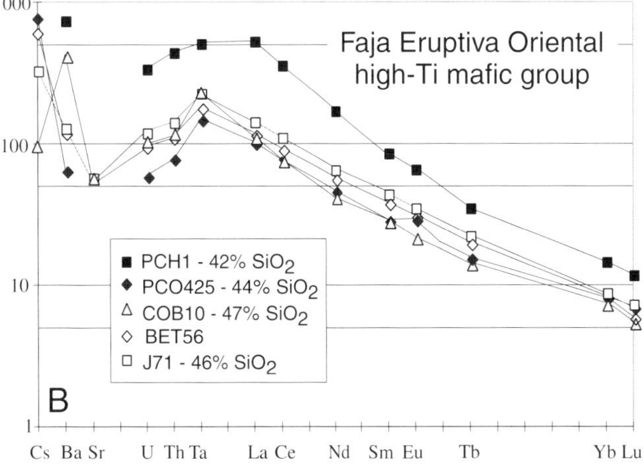

Figure 12. Extended trace element plots for representative mafic units from the Faja Eruptiva Oriental. Samples are divided into a low-Ti and high-Ti group as discussed in text. Late-stage Huaitiquina dike J71 in the Faja Eruptiva Occidental shown for reference. Note the relative constancy of the relationships between Th, Ta, and La, which suggests that the magmas are derived from generally similar sources. Normalization as in Figure 7. Data from Tables 7 and 8. See Figure 2 for locations.

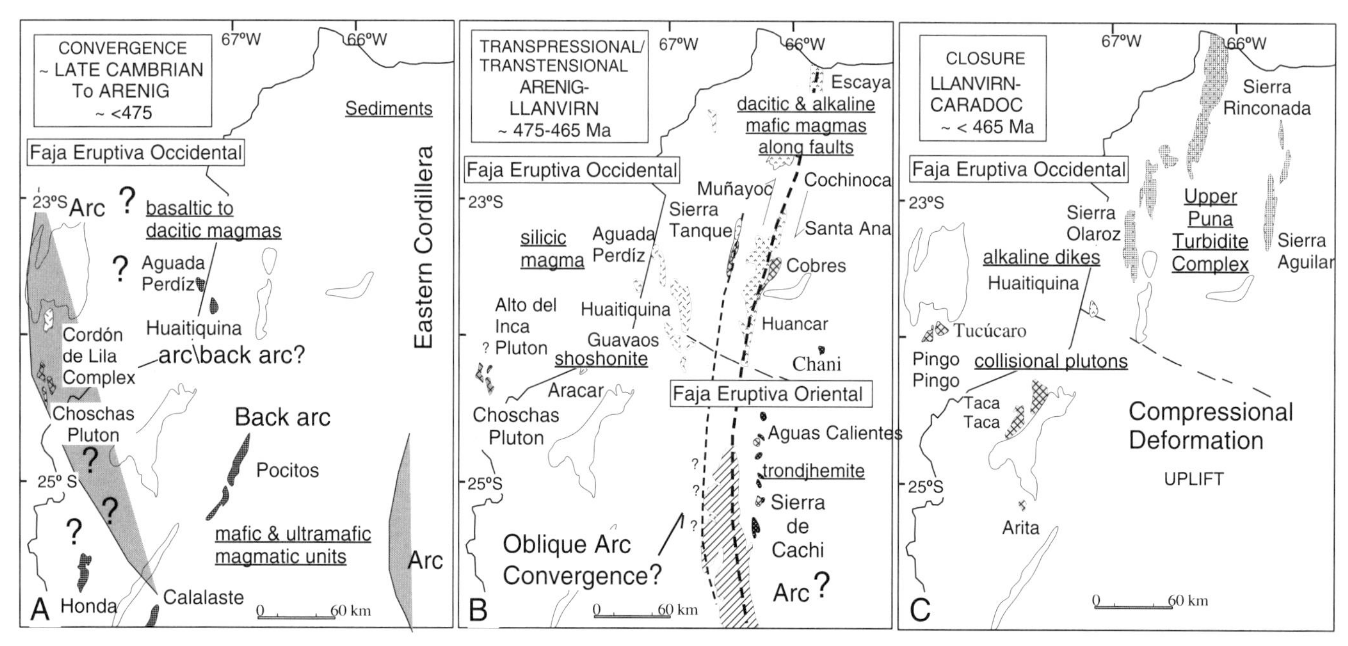

Figure 13. Map views discussed in text: Late Cambrian to Arenig (A), Arenig-Llanvirn (B), and Llanvirn-Caradoc (C) times for the northern Puna. Modern salars are outlined as in Figure 2 for reference.

Faja Eruptiva Oriental and Eastern Cordillera

To the east, the best understood Faja Eruptiva Oriental magmatic units are the Cochinoca-Escaya complex mafic and dacitic volcanic/subvolcanic units and related plutons (Figs. 2 and 13). As discussed, the dacitic units can be interpreted as volcanic and shallow level intrusive bodies emplaced contemporaneously with sediment deposition. Their chemical characteristics are consistent with an origin that included melting of older continental crust containing an important sedimentary component. A sedimentary component is supported by weak arc-like chemical characteristics that are typical of continental crust or sediments derived from continental crust (e.g., Taylor and McClennan, 1985). The general chemical homogeneity of the dacitic magmas in an elongated belt over 200 km long running from the Cordón de Escaya to the Sierra de Cobres supports a similar magmatic source region along the entire belt. Radiometric and fossil age constraints indicate these magmas formed over a relatively restricted time span near 476–467 Ma in Arenig to Llanvirn time.

Associated mafic spilitic pillow and massive lava flows and cross-cutting dikes and sills have alkaline intraplate-like characteristics with no evidence of an arc component. The mafic flows can be interpreted as higher percentage melts of a similar garnet-bearing mantle source to that which produced the more strongly alkaline dikes and sills. Association of the pillowed and massive flows with the dacitic and sedimentary units is consistent with the heat source that caused the melting of the crustal components in the dacitic magmas being associated with the processes that produced the mafic magmas.

The chemistry of the dacitic and mafic alkaline rocks and their linear distribution suggest a generally extensional regime. The magmatic heat source could be related to adiabatic melting of mantle along a zone of crustal extension. Pronounced mylonitic textures in metasedimentary basement xenoliths in Tertiary volcanic rocks (P. Caffe, personal communication, 1997) and in outcrop at the Sierra de Tanque could be residual from crustal melting associated with shearing at this time. The question is the type of fault: strike-slip, transtensional-transpressive, or extensional. Bahlburg (1990) recognized that the dacitic units in the Cordón de Escaya are foliated along subvertical north-south–striking left-lateral shear zones. A case for melt zones associated with alkaline magmas along a major transpressive-transpressive strike-slip fault in British Columbia has been presented by Hollister and Andrinocos (1997).

A relevant question is what happens to this magmatic belt south of the El Toro lineament. The long prevailing theory from Méndez et al. (1973) is that the belt continues southward through the Cumbres de Luracatao (Fig. 2) to the Sierra de Chango Real near 27°S with plutons appearing as erosion levels increase. We accept this as a possible interpretation and suggest that Ordovician plutons in this belt could be related to a southward continuation of a northern Puna fault zone (Fig. 13B).

Any regional model for the Faja Eruptiva Oriental must address the question of arc magmatism in the Eastern Cordillera. In this region, Ordovician magmatism appears to be restricted to south of 24°S with the northernmost pluton being Chañi, just north of the El Toro lineament (Fig. 13B). Biotite K-Ar ages allow the Chañi pluton to be between 497 and 457 Ma, most probably near 470 Ma. Farther south, the Cachi trondhjemite group (Fig. 13B), which Galliski et al. (1990) interpreted as a continental margin arc suite, and for which Lork et al. (1991) argued Arenig-Llanvirn ages (481–466 Ma) requires explanation.

In the model of Bahlburg and Hervé (1997), the Ordovician tectonic setting of the central and southern Puna is modeled by subduction zones beneath both the Faja Eruptiva Occidental and Oriental. Evidence for the eastern arc comes from the Cachi trondhjemite suite, and apparently from the assumption that the Cumbres de Luracatao granitoids, whose chemistry and age are largely unknown, formed in an arc setting. A problem is that the Cachi plutons are not the typical granodiorite suite of a simple subduction-related continental margin arc. The issue of the existence and nature of an Arenig-Llanvirn arc in the eastern Faja Eruptiva Oriental remains an outstanding problem.

A model that could combine a subduction zone to the south with a left-lateral strike-slip fault to the north is an oblique left-lateral transpressional subduction zone that extends northward into an oblique strike-slip fault (see Fig. 13B). This is more or less the case from east to west in the modern Aleutian arc (e.g., Kay and Kay, 1994). In the southern Puna, the obliquity could have been such that a subducting slab passing under the Cumbres de Luracatao belt produced a limited amount of subduction-related magmatism to the east. Decreasing obliquity to the south could account for an increase in arc-related magmatism in that direction (e.g., Rapela et al., 1992). A nearly margin parallel strike-slip fault to the north would explain the lack of arc-related magmatism in that region. Pull-apart like basins associated with shearing could account for the alkaline volcanism and related crustal melting.

Temporal and geologic constraints are not sufficient to rule out a multitude of other models. One possibility is that the whole region was subject to backarc extension, with the greatest extension in the south. In this case, northern Puna magmas could be the northern manifestation of the backarc extensional regime that produced Calalaste and Pocitos mafic/ultramafic sequences in the southern Puna. Such a model requires oblique extension. This model is not without problems, one of which is why mafic lavas in a backarc setting to the north have no slab signature. A way around this is for these magmas to be entirely related to continental extension. Larger amounts of extension in the south could have opened an oceanic basin that never developed to the north, as would be favored by the rotating Arequipa terrane model of Forsythe et al. (1993). This model requires discounting the paleomagnetic data of Conti et al. (1996) that indicates a peri-Gondwana Famatina-Puna Oriental terrane, and developing a complicated scenario involving arc and backarc regimes in the south. A problem with the interpretation of Conti et al. (1996) is that there is no evidence for a suture between the Faja Eruptiva Oriental and Eastern Cordillera in the northern Puna.

Another model is to invoke ridge subduction to explain magmatism in the eastern belt. Such a model could explain a linear zone of high-temperature magmas and melting of sediments in the oceanic trench. In this case, subduction would die out to the north as the subducting crust became very young and hot. An advantage of this model is that it is consistent with the occurrence of Ordovician trondhjemitic magmatism in the Sierra de Cachi as trondhjemites in other regions have been associated with subduction of very young ocean crust (Drummond and Defant, 1990). Additional data, particularly temporal constraints, are required before any of these scenarios or others can be adequately proposed.

ORDOVICIAN TECTONIC MODEL FOR THE NORTHERN PUNA

Putting together all of the information leads to a revised tectonic model for the Ordovician in the northern Puna. The general model proposed is the simplest model consistent with the existing data; it is discussed below in the three time frames shown in Figure 12:

1. Late Cambrian to Early Arenig (Fig. 13A). Active subduction in the western Faja Eruptiva Occidental would be represented by the Cordón de Lila volcanic units in the arc. The position of the Cordón de Lila arc, which could have been off to the west, as suggested by Forsythe et al. (1993), is uncertain. The end of the Cordón de Lila arc magmatic cycle could be signaled by emplacement of the Choschas pluton. At the time the Cordón de Lila complex and Choschas plutons were forming in the arc, the basal mafic volcanic units at Huaitiquina could be erupting in a zone of broad arc/backarc extension. These Huaitiquina lavas could be synchronous with the Salar de Calalaste and Pocitos mafic-ultramafic volcanics, which erupted in a more extended arc/backarc regime to the south. A major unknown is the tectonic setting of the region between the Cordón de Lila and the Aguada de la Perdiz region and the vergence of an arc associated with the Cordón del Lila complex and the Choschas pluton. More than one arc in this region cannot be ruled out.

2. Arenig-Llanvirn (Fig. 13B). The picture for the Arenig to Llanvirn setting suggests a significant tectonic change with the most profound effect taking place between about 476 and 467 Ma. Among the differences from the Early Arenig is a switch from the broad range of plutonic units in the Choschas pluton to the silicic units of the Alto de Inca pluton. Such a magmatic switch in a waning subduction zone could explain silicic detritus being shed into the Huaitiquina, Guayaos, and Aguada de la Perdiz regions to the east. A waning arc regime would be consistent with weakening of the arc magmatic signature in the Huaitiquina unit C andesitic lavas. One mechanism to shut off arc volcanism is to have a change to a more oblique convergence or even strike-slip regime along the margin. In such a setting, faults could form in several parallel zones along which crustal melting could occur. Alkaline magmatism produced in association with mantle decompression could form along pull-apart regions in the former backarc. Such a regime could explain fault zones associated with dacitic and basaltic magmas in the Cochinoca-Escaya complex of the Faja Eruptiva Oriental. As such, the map in Figure 13B shows an oblique fault cutting attenuated continental crust in the northern Puna and extending into extremely thin continental or oceanic crust to the south. Such a configuration could allow for a southern Puna arc, east of the fault. The cause of such a change in plate motion most likely reflects some type of plate reorganization in the western Iapetus ocean. The timing roughly corresponds with the arrival of the Precordillera terrane to the south (e.g., Astini et al., 1995; Dalziel, 1997). In this model, initial formation of the Puna Turbidite complex is controlled by transpressional basins, as well as loading.

3. Late and post–Llanvirn-Caradoc (Fig. 13C). This could be the time of compressional deformation, and the end of Ordovician magmatism. Magmatic activity includes the late-stage alkaline dikes in the western Huaitiquina and Cochinoca-Escaya complex, which are produced by low percentages of melting in the cooling mantle. A collision of the Arequipa terrane containing the Cordón de Lila in the Late Ordovician–Silurian (e.g., Ramos, 1988; Forsythe et al., 1993) could produce the magmatism exemplified in the Tucúcaro. Pingo Pingo, and Taca Taca plutons in the west. The result to the east would be deformation associated with the Ocloyic orogeny, and continued basin evolution associated with deposition of the Upper Puna Turbidite complex (Bahlburg, 1990, 1993; Bahlburg and Furlong, 1996).

CONCLUSIONS

The presentation above outlines what is known about the timing, distribution, and chemistry of Early to Middle Ordovician magmatism in the northern Puna region. These data are combined with geologic information to arrive at the generalized three-stage tectonic model shown in Figure 13. The model consists of an early arc stage, with arc, behind the arc, and backarc volcanism; an intermediate stage, characterized by an oblique convergent regime in the south and a strike-slip regime in the north; and a late compressional stage, involving basin closure, deformation, cessation of volcanism in the northeast, and late-stage collisional-type plutonism in the west. The model is provisional as age control is lacking, particularly in the Cordón de Lila and plutonic complexes of the Faja Eruptiva Occidental in the west, and in the plutonic rocks of the southern Faja Eruptiva Oriental. What is well constrained is the Arenig-Llanvirn age of the dacitic and mafic magmatic rocks of the Cochinoca-Escaya complex that are shown here to have been emplaced in a submarine environment essentially contemporaneously with their enclosing sediments. The chemistry of these dacitic units is consistent with their being derived largely from melting of continental crustal sources with heat provided by the processes that produced mafic alkaline magmatism along a major Arenig-Llanvirn fault zone. The correlations with units in the Cordón de Lila and plutonic complexes of the Faja Eruptiva Occidental are tentative and are largely based on magmatic style, chemistry, and regional geology.

Extending this tectonic model to the southern Puna is hampered by a lack of information. For example, the age and chemistry of the magmatic rocks of the vast Cumbres de Luracatao and their relation to the trondhjemitic suite in the Eastern Cordillera are virtually unknown. Farther west, the chemistry and age of the important mafic complex at Quebrada Honda is almost unknown. The ages of the ultramafic/mafic suites and plutons in the southern Puna are not well constrained. Tectonic models defining peri-Gondwana terranes in the western Iapetus ocean along the early Paleozoic border of Gondwana will remain problematic until better information is available. Models need to be tested with new data, not by continued reinterpretation of the existing inadequate data base.

ACKNOWLEDGMENTS

We thank H. Bahlburg, M. Koukharsky, V. Ramos, and C. Mpodozis for discussion of Ordovician tectonic problems. John Davidson (Chile) is thanked for providing samples of the Sierra de Almeida and Aracar and Taca Taca regions. Reviews by Calvin Miller and C. Mpodozis are gratefully acknowledged. This project was supported by Argentinian Grant CONICET PICT0081Prestamo BID802 OC/AR, and by U.S. National Science Foundation Grant EAR92-05042 and EAR9508023. This work is a contribution to International Geological Correlation Program Project 376.

APPENDIX ON ANALYTIC METHODS IN TABLES 1–8

Whole-rock samples were pulverized in an aluminum oxide ceramic shatterbox. Major element analyses of the Faja Eruptiva Oriental samples (except PCO425, SA4, PCH1, and P19), trace element analyses for which REE are not available, and Rb analyses were done by x-ray fluorescence (XRF) at Jujuy University, Argentina. Analyses are based on comparison with U.S. Geological Survey and Japanese Geological Survey standards. Analyses not summing to near 100% reflect volatile components. Major element analyses in Table 1 were done at the Chilean Geologic Survey (SERNAGEOMIN) in Santiago, Chile, by a combination of XRF, wet chemical, and atomic absorption methods. All other analyses were done at Cornell University, Ithaca, New York. Na and Fe analyses of all samples with REE analyses are duplicated by instrument neutron activation (INAA) methods. Major elements were determined on glass prepared from powder fused in a molybdenum strip furnace or with Li metaborate flux in carbon crucibles. Analyses were carried out on a JEOL-733 Superprobe operating in wavelength dispersive spectrometric mode at 15 kV accelerating voltage and 15 nA beam current and are based on four to six 30-mm spot analyses. Standards included natural minerals and basaltic glasses. Typical 2s-precision and accuracy is 1–5% for major elements (>1 wt%) and ~5–10% from for minor elements (<0.5 wt%) based on replicate analysis. Data were reduced using a Bence-Albee matrix correction. REE and other trace elements were analyzed by INAA at Ward Laboratory, Cornell University, Ithaca, New York. Sample powders (0.5 g) were packed in ultrapure Suprasil quartz tubing with three internal standards and irradiated in a TRIGA reactor at a power level of ~400 kW for 3–4 h. Samples and standards were counted for 4–10 h on an Ortec Intrinsic Ge detector at 6 and ~40 days after irradiation. Precision and accuracy (2σ) based on replicate standard analysis is 2–5% for all elements except U, Sr, and Nd, which are ~8%. Further information on INAA analyses is in Kay et al. (1987).

REFERENCES CITED

Aceñolaza, F., and Toselli, A., 1984, Lower Ordovician volcanism in northwest Argentina, *in* Bruton, D. L., ed., Aspects of the Ordovician system: Paleontological Contributions of the University of Oslo 295, p. 203–209.

Aceñolaza, F. G., Toselli, A., and Durand, F. R., 1975, Estratigrafía y paleontología de la región de Hombre Muerto, provincia de Catamarca, Argentina: I Congreso Argentino de Paleontología y Bioestratigrafía Actas, no. 1, p. 109–123.

Adams, C., Miller, H., and Toselli, A. J., 1990, Nuevas edades de metamorfismo por el método K-Ar de la Formación Puncoviscana y equivalentes, NW de Argentina, *in* Aceñolaza, F. G., Miller, H., and Toselli, A. J., eds., El ciclo Pampeano en el Noroeste Argentino: Correlación Geológica, no. 4, p. 209–219.

Allmendinger, R. W., Jordan, T., Palma, M., and Ramos, V., 1982, Perfil estructural de la Puna catamarqueña (25°–27°S), Argentina: V Congreso Latinoamericano de Geología Actas, v. 1, p. 499–518.

Allmendinger, R. W., Ramos V. A., Jordan, T. E., Palma, M., and Isacks, B. L., 1983, Paleogeography and Andean structural geometry, northwest Argentina: Tectonics, v. 2, p. 1–16.

Astini, R. A., Benedetto, J. L. and Vaccari, N. E., 1995, The early Paleozoic evolution of the Argentine Precordillera as a Laurentian rifted, drifted and collided terrane: a geodynamic model. Geological Society of America Bulletin, v. 107, p. 235–273.

Bachmann, G., Grauert, B., Kramm, U., Lork, A., and Miller, H., 1987, El magmatismo del Cámbrico medio/Cámbrico superior en el basamento del noroeste argentino: Investigaciones isotópicas y geocronológicas sobre los granitoides de los complejos intrusivos de Santa Rosa de Tastil y Cañani: X Congreso Geológico Argentino Actas, no. 4, p. 125–127.

Bahlburg, H., 1990, The Ordovician Puna basin in the Puna of NW Argentina and N Chile: Geodynamic evolution from back-arc to foreland basin: Geotektonische Forschungen, v. 75, p. 1–107.

Bahlburg, H., 1993, Hypothetical southeast Pacific continent revisited: new evidence from middle Paleozoic basins of northern Chile: Geology, v. 21, p. 909–912.

Bahlburg, H., and Breitkreuz, C., 1991, The evolution of marginal basins in the southern central Andes of Argentina and Chile during the Paleozoic: Journal of South American Earth Sciences, v. 4, p. 171–188.

Bahlburg, H., and Furlong, K. P., 1996, Lithospheric modeling of the Ordovician foreland basin in the Puna of northwestern Argentina: on the influence of arc loading on foreland basin formation: Tectonophysics, v. 259, p. 245–258.

Bahlburg, H., and Hervé, F., 1997, Geodynamic evolution and tectonostratigraphic terranes of northwestern Argentina and northern Chile: Geological Society of America Bulletin, v. 109, p. 869–884.

Bahlburg, H., Breitkreuz, C., Maletz, J., Moya, M. C., and Salfity, J. A., 1990, The Ordovician sedimentary rocks in the northern Puna of Argentina and Chile: new stratigraphical data based on graptolites: Newsletters on Stratigraphy, v. 23, p. 69–89.

Bahlburg, H., Kay, S. M., and Zimmerman, U., 1997, New geochemical and sedimentological data on the evolution of the early Paleozoic margin in the southern central Andes: Geological Society of America Abstracts with Programs, v. 28, no. 7, p. A380.

Blasco, G., Villar, L., and Zappettini, E. O., 1996, El complejo ofiolítico desmembrado de la Puna argentina, provincias de Jujuy, Salta y Catamarca: XIII Congreso Geológico Argentino y III Congreso de Exploración de Hidrocarburos Actas, no. 3, p. 653–667.

Breitkreuz, C., Bahlburg, H., Delakowitz, B., and Pichowiak, S., 1989, Volcanic events in the Paleozoic central Andes: Journal of South American Earth Sciences, v. 2, p. 171–189.

Coira, B., 1973, Resultados preliminares sobre la petrología del ciclo eruptivo ordovícico concomitante con la sedimentación de la Formación Acoite, en la zona de Abra Pampa, Provincia de Jujuy: Revista de la Asociación Geológica Argentina, no. 28, p. 85–88.

Coira, B., 1975, Ciclo efusivo ordovícico registrado en la Formación Acoite, Abra Pampa, Provincia de Jujuy, Argentina: II Congreso Iberoamericano de Geología Económica Actas, no. 1, p. 37–56.

Coira, B., 1979, Descripción Geológica de la Hoja 3C Abra Pampa, provincia de Jujuy: Servicio Geológico Nacional Boletín, v. 170, p. 1–90.

Coira, B., 1996, Volcanismo submarino silíceo ordovícico en la Puna nororiental (22°–24°S, 65°45′– 66°45′O), Argentina: XII Congreso Geológico de Bolivia Memorias, no. 3, p. 1003–1009.

Coira, B., and Barber, E., 1989, Volcanismo submarino ordovícico (Arenigiano-Llanvirniano) del río Huaitiquina, Provincia de Salta: Revista de la Asociación Geológica Argentina, v. 44, p. 68–77.

Coira, B., and Koukharsky, M., 1991, Lavas en almohadillas ordovícicas en el Cordón de Escaya, Puna septentrional, Argentina: 6° Congreso Geológico Chileno Actas, no. 1/A-5, p. 674–678.

Coira, B., and Koukharsky, M., 1994, Complejos submarinos dómicos-lávicos silíceos de edad ordovícica en el sector oriental de Puna Jujeña (22°, 23°45′S). Sus implicancias: 7° Congreso Geológico Chileno Actas, no. 2, p. 1000–1004.

Coira, B., and Nullo, F., 1989, Facies piroclásticas del volcanismo ordovícico (Arenigiano-Llanvirniano) Salina de Jama, Jujuy: Revista de la Asociación Geológica Argentina, v. 44, p. 89–95.

Coira, B., Davidson, J., Mpodozis, C., and Ramos, V., 1982, Tectonic and magmatic evolution of the Andes northern Argentina and Chile: Earth Science Reviews, v. 18, p. 303–332.

Coira, B., Koukharsky, M., and Pérez, A. J., 1987, Rocas volcaniclásticas ordovícicas de la Sierra de Guayaos, Provincia de Salta, Argentina: X Congreso Geológico Argentino Actas, no. 4, p. 312–315.

Conti, C. M., Rapalini, A. E., Coira, B., and Koukharsky, M., 1996, Paleomagnetic evidence of an early Paleozoic rotated terrane in northwest Argentina: a clue for Gondwana-Laurentia interaction?: Geology, v. 24, p. 953–956.

Coira, B., Caffe, P. Ramirez, A., Chayle, W., Diaz, A., Rosas, S., Pérez, A., Pérez, B., Orosco, O., and Martinez, 1999, Descripción de la Hoja Geológica 2366-I Mina Pirquitas: Servicio Geológico y Minero Argentino (in press).

Dalla Salda, L., Cingolani, C., and Varela, R., 1992, Early Paleozoic orogenic belt of the Andes in southwestern South America: result of Laurentia-Gondwana collision? Geology, v. 20, p. 617–620.

Dalziel, I. W. D., 1991, On the organization of American plates in the Neoproterozoic and the breakout of Laurentia: GSA Today, v. 2, no. 11, p. 237–241.

Dalziel, I. W. D., 1997, Neoproterozoic-Paleozoic geography and tectonics: review, hypothesis, environmental speculation: Geological Society of America Bulletin, v. 109, p. 16–42.

Dalziel, I. W. D., and Forsythe, R. D., 1985, Andean evolution and the terrane concept, *in* Howell, D. G., ed., Tectonostratigraphic terranes of the Circum-Pacific Region: Circum-Pacific-Council for Energy and Mineral Resources Earth Science Series, v. 1, p. 565–581.

Damm, K. W., Pichowiak, S., and Todt, W., 1986, Geochimie, Petrologie und Geochronologie der plutonite und des metamorphen grundgebirges in Nordchile: Berliner Growiss Abh. (A), v. 66, p. 73–146.

Damm, K. W., Pichowiak, S., Harmon, R. S., Todt, W., Kelley, S., Omarini, R., and Niemeyer, H., 1990, Pre-Mesozoic evolution of the central Andes; the basement revisited, *in* Kay, S. Mahlburg, and Rapela, C. W., eds., Plutonism from Antarctica to Alaska: Geological Society of America Special Paper 241, p. 101–126.

Damm, K. W., Pichowiak, S., Breitkreuz, C., Harmon, R. S., Todt, W., and Buchelt, M., 1991, The Cordón de Lila complex, central Andes, northern Chile; an Ordovician continental volcanic province, *in* Harmon, R. S., and Rapela C. W., eds., Andean magmatism and its tectonic setting: Geological Society of America Special Paper 265, p. 179–188.

Davidson M. J., Mpodozis, C., and Rivano, S., 1981, El Paleozoico de Sierra de Almeida, al oeste de Monturaqui, Alta Cordillera de Antofagasta, Chile: Revista Geológica de Chile, no. 12, p. 3–23.

De Bari, S. M., 1990, Comparative field and petrogenetic study of arc magmatism in the lower crust; exposed examples from a continental and an intraoceanic setting [Ph.D. thesis]: Palo Alto, California, Stanford University, 186 p.

Drummond, M. S., and Defant, M. J., 1990, A model for trondhjemite-tonalite-dacite genesis and crustal growth via slab melting: Archean to modern comparisons: Journal of Geophysical Research, v. 95, p. 21503–21521.

Forsythe, R. D., Davidson, J., and Mpodozis, C., 1993, Lower Paleozoic relative motion of the Arequipa block and Gondwana: paleomagnetic evidence from Sierra de Almeida of northern Chile: Tectonics, v. 12, p. 219–236.

Galliski, M. A., Toselli, A. J., and Saavedra, J., 1990, Petrology and geochemistry of the Cachi high-alumina trondhjemites, northwestern Argentina, *in* Kay, S. M., and Rapela, C. W., eds., Plutonism from Antarctica to Alaska: Geological Society of America Special Paper 241, p. 91–100.

García, A. F., Perez D'Angelo, E., and Ceballos, S. E., 1962, El Ordovícico de Aguada de la Perdiz, Puna de Atacama, provincia de Antofagasta: Revista Mineralógica, v. 77, p. 52–61.

Goetze, H.-J., Lahmeyer, B., Schmidt, S., and Strunk, S., 1994, The lithospheric structure of the central Andes (20°–26°S) as inferred from interpretation of regional gravity data, *in* Reuter, K.-J., Scheuber, E., and Wigger, P. J., eds., Tectonics of the southern central Andes: Berlin, Springer-Verlag, p. 7–22.

Gradstein, F. M., and Ogg, J., 1996, A Phanerozoic time scale: Episodes, v. 19, p. 23–25.

Grissom, G. C., 1991, Empirical constraints on thermal processes in the deep crust of magmatic arcs: Sierra de Fiambalá, northwestern Argentina [Ph.D. thesis]: Palo Alto, California, Stanford University, 262 p.

Gutierrez Marco, J. C., Aceñolaza, G. F., and. Esteban, S. B., 1996, Revisión de algunas localidades con graptolitos ordovícicos en la Puna salto-jujeña (noroeste de Argentina): 12° Congreso Geológico de Bolivia Memorias, no. 2, p. 725–731.

Halpern, M., 1978, Geological significance of Rb-Sr isotopic data of northern Chile. Crystalline rocks of the Andean orogen between 23° and 27°S: Geological Society of America Bulletin, v. 89, p. 522–532.

Hanning, M., 1987, The geochemistry of Ordovician volcanics from the Faja Eruptiva, Argentina: implications for the tectonics setting [B.A. thesis]: Ithaca, New York, Cornell University, 44 p.

Hervé, F., Godoy, E., Parada, M., Ramos, V. A., Rapela, C., Mpodozis, C., and Davidson, J., 1987, A general view of the Chilean-Argentine Andes, with emphasis on their early history, *in* Monger, J., and Francheteau, J. eds., Circumpacific orogenic belts and evolution of the Pacific Ocean basin: Geodynamics Series, v. 18, p. 97–113.

Hollister, L., and Andrinocos, C., 1997, A candidate for the Baja British Columbia fault system in the coast plutonic complex: GSA Today, v. 7, p. 1–7.

Kay, S. M and Kay, R. W., 1994, Aleutian magmatism in space and time, *in* Plafker. G., and Berg, H. C., eds., The Geology of Alaska: Boulder, Colorado, Geological Society of America, Geology of North America, v. G-1, p. 687–722.

Kay, S. M., Ramos, V. A., and Kay, R. W., 1984, Elementos mayoritarios y trazas de las vulcanitas ordovícicas de la Precordillera Occidental: basaltos del rift oceanico temprano (?) proximos al margen continental: Actas del IX Congreso Geológico Argentino, vol. II, p. 48-65.

Kay, S. M., Maksaev, V., Mpodozis, C., Moscoso, R., and Nasi, C., 1987, Probing the evolving Andean lithosphere: middle to late Tertiary magmatic rocks in Chile over the modern zone of subhorizontal subduction (29°–31.5°S): Journal of Geophysical Research, v. 92, p. 6173–6189.

Kay, S. M., Kay, R. W., Citron, G. P., and Perfit, M., 1990, Calc-alkaline plutonism in the intra-oceanic Aleutian Arc, Alaska, *in* Kay, S. M., and Rapela, C. W., eds., Plutonism from Antarctica to Alaska: Geological Society of America Special Paper 241, p. 233–255.

Kay, S. Mahlburg, Mpodozis, C., and Coira, B., 1999, Magmatism, tectonism, and mineral deposits of the central Andes (22°–33°S latitude), *in* Skinner, B., ed., Geology and ore deposits of the central Andes: Society of Economic Geology Special Publication 7, p. 27–59.

Koukharsky, M., and Mirré, J. C., 1974, Nuevas evidencias de vulcanismo ordovícico en la Puna: Revista de la Asociación Geológica Argentina, no. 29, p. 128–134.

Koukharsky, M., and Lanes, S., 1994, Las rocas graníticas paleozoicas de la sierra de Taca Taca, Puna Salteña (24°10′–24°30′S), Argentina: 7° Congreso Geológico Chileno Actas, v. 2, p. 1071–1075.

Koukharsky, M., Coira, B., Barber, E., and Hanning, M., 1988, Geoquímica de las volcanitas ordovícicas de la Puna (Argentina) y sus implicancias tectónicas: 5° Congreso Geológico Chileno Actas, v. 3, p. 137–151.

Koukharsky, M., Coira, B., and Morello, O., 1989, Volcanismo ordovícico de la Sierra de Guayaos, Puna salteña, características petrológicas e implicancias tectónicas: Revista de la Asociación Geológica Argentina, no. 44, p. 207–216.

Koukharsky, M., Torres Clara, R., Etcheverria, M., Vaccari, N. E., and Waisfeld, B. G., 1996, Episodios volcánicos del Tremadociano y del Arenigiano en Vega Pinato, Puna salteña, Argentina: XIII Congreso Geológico Argentino y III Congreso de Exploración de Hidrocarburos Actas, no. 5, p. 535–542.

Lork, A., and Bahlburg, H., 1993, Precise U-Pb ages of monazites from the Faja Eruptiva de la Puna Oriental and the Cordillera Oriental, NW Argentina: XII Congreso Geológico Argentino y II Congreso de Exploración de Hidrocarburos Actas, no. 4, p. 1–6.

Lork, A., Grauert, B., Kramm, U., and Miller, H., 1991, U-Pb investigations of monazite and polyphase zircon: implications for age and petrogenesis of trondhjemites of the southern Cordillera Oriental, NW Argentina: 6° Congreso Geológico Chileno Actas (Vina del Mar), v. 1, p. 398–402.

Méndez, V., Navarini, A., Plaza, D., and Viera, O., 1973, Faja Eruptiva de la Puna Oriental: V Congreso Geológico Argentino Actas, no. 4, p. 147–158.

Méndez, V., Turner, J. C., Navarini, A., Amengual, R., and Viera, O., 1979, Geología de la Región Noroeste, provincias de Salta y Jujuy, República Argentina: Dirección General de Fabricaciones Militares, 118 p.

Miller, C. F., D'Andrea, J. L., Ayers, J. C., Coath, C. D., and Harrison, T. M., 1997, BSE imaging and ion probe geochronology of zircon and monazite from plutons of the Eldorado and Newberry Mountains: age, inheritance, and subsolidus modification: Eos (Transactions, American Geophysical Union), v. 78, p. F783.

Mon, R., and Hogn, F., 1991, The structure of the Precambrian and lower Paleozoic basement of the central Andes between 22° and 32°S Lat: Geologische Rundschau, v. 80, p. 745–758.

Moya, M. C., Malanca, S., Hongn, F. D., and Bahlburg, H., 1993, El Tremadoc temprano en la Puna occidental argentina: XII Congreso Geológico Argentino y II Congreso de Exploración de Hidrocarburos Actas, no. 2, p. 20–30.

Mpodozis, C., Hervé, F., Davidson, J., and Rivano, S., 1983, Los granitoides de Cerro Lila, manifestaciones de un episodio intrusivo y termal del Paleozoico inferior en los Andes del Norte de Chile: Revista Geológica de Chile, v. 18, p. 3–14.

Niemeyer, H., 1989, El complejo ígneo-sedimentario del Cordón de Lila, región de Antofagasta. Significado tectónico: Revista Geológica de Chile, v. 16, p. 163–181.

Omarini, R., Cordani, U., Viramonte, G., Salfity, J., and Kawashita, K., 1979, Estudio isotópico Rb-Sr de la Faja Eruptiva de la Puna a los 22°35′ L.S., Argentina: 2° Congreso Geológico Chileno Actas, v. 2, p. 257–269.

Omarini, R. H., Viramonte, J. G., Cordani, U., Salfity, J. A., and Kawashita, K., 1984, Estudio geocronológico Rb-Sr de la Faja Eruptiva de la Puna en el sector de San Antonio de los Cobres, provincia de Salta: V Congreso Geológico Argentino Actas, no 3, p. 146–158.

Omarini, R. H., Torres, J. G., and Moya, L. A., 1996, El stock Tipayoc: génesis magmática en un arco de islas del Cámbrico inferior en el noroeste de Argentina: XIII Congreso Geológico Argentino y III Congreso de Exploración de Hidrocarburos Actas, no. 5, p. 545–560.

Palma, M. A., Parica, P., and Ramos, V., 1986, El granito Archibarca: su edad y significado tectónico, provincia de Catamarca: Revista de la Asociación Geológica Argentina, v. 41, p. 414–419.

Penrose Conference, 1996, The Argentine Precordillera: a Laurentian terrane? GSA Today, v. 6, p. 16–18.

Ramos, V. A., 1972, El Ordovícico fosilífero de la Sierra de Lina, Departamento Susques, Provincia de Jujuy: Revista de la Asociación Geológica Argentina, v. 27, p. 84–94.

Ramos, V. A., 1986, El diastrofismo oclóyico: un ejemplo de tectónica de colisión durante el eopaleozoico en el Noroeste Argentino: Revista del Instituto de Geología y Minería, v. 6, p. 13–28.

Ramos, V., 1988, Late Proterozoic–Early Paleozoic of South America: a collisional history: Episodes, v. 11, p. 168–174.

Ramos, V. A., Jordan, T. E., Allmendinger, R. W., Mpodozis, C., Kay, S. M.,. Cortés, J. M., and Palma, M. A., 1986, Paleozoic terranes of the central Argentine-Chilean Andes: Tectonics, v. 5, p. 855–880.

Rapela, C. W., Coira, B., Toselli, A., and Saavedra, J., 1992, The lower Paleozoic magmatism of southwestern Gondwana and the evolution of the Famatinian orogen: International Geology Review, v. 34, p. 1081–1142.

Schwab, K., 1973, Die Stratigraphie in der Umgebung des Salar de Cauchari (NW Argentinien). Ein Beitrag zur erdgeschichtlichen Entwicklung der Puna: Geotektonische Forschungen, v. 43, p. 1–168.

Sun, S. S., and McDonough, W. F., 1989, Chemical and isotopic systematics of oceanic basalts: implications for mantle composition and processes, *in* Saunders, A. D., and Norry, M. J., eds., Magmatism in ocean basins: Geological Society of London Special Publication 42, p. 313–345.

Taylor, S. R., and McLennan, S. M., 1985, The continental crust: its composition and evolution: Boston, Blackwell Scientific Publications, 312 p.

Turner, J. C., 1960, Estratigrafía de la Sierra de Santa Victoria y adyacencias: Boletín de la Academia Nacional de Ciencias de Córdoba, no. 41, p. 1464–1467.

Turner, J. C., 1964, Descripción de la Hoja Geológica 2b, La Quiaca (Provincia de Jujuy): Instituto Nacional de Geología y Minería Boletín, no. 103, 109 p.

White, A. J. R., and Chappell, B. W., 1983, Granitoid types and their distribution in the Lachlan Fold Belt, southeastern Australia: Geological Society of America Memoir 159, p. 21–34.

Wood, D. A., Joron, J. L., and Trevil, M., 1979, A reappraisal of the use of trace elements to classify and discriminate between magma series erupted in different tectonic settings: Earth and Planetary Science Letters, v. 45, p. 326–336.

Zapettini, E. O., Blasco, G., and Villar, L., 1994, Geología del extremo sur del Salar de Pocitos, provincia de Salta, República Argentina: 7° Congreso Geológico Chileno Actas, no. 1, p. 220–224.

Manuscript Accepted by the Society September 4, 1998

Printed in U.S.A.

Geological Society of America
Special Paper 336
1999

Paleomagnetic constraints on the evolution of Paleozoic suspect terranes from southern South America

Augusto E. Rapalini
Laboratorio de Paleomagnetismo Daniel Valencio, Departamento de Ciencias Geológicas, FCEyN, Universidad de Buenos Aires, Pabellón 2, Ciudad Universidaria, 1428 Buenos Aires, Argentina
Ricardo A. Astini
Cátedra de Estratigrafía y Geología Histórica, Facultad de Ciencias Exactas, Físicas y Naturales, Universidad Nacional de Córdoba, Av. Velez Sarsfield 299, 5000, Córdoba, Argentina
Carlos M. Conti
Laboratorio de Paleomagnetismo Daniel Valencio, Departamento de Ciencias Geológicas, FCEyN, Universidad de Buenos Aires, Pabellón 2, Ciudad Universidaria, 1428 Buenos Aires, Argentina

ABSTRACT

The Paleozoic evolution of southwestern Gondwana apparently involved the accretion of several large suspect terranes at different times and under different tectonic circumstances. The recent acquisition of paleomagnetic data from Paleozoic rocks of this region, albeit still scarce, permit establishment of some constrains on the tectonic and paleogeographic evolution of some of these terranes. A reliable paleomagnetic pole from the Early Cambrian Cerro Totora Formation in northern Precordillera confirms that the Argentine Precordillera is an exotic terrane that was part of or adjacent to Laurentia, very likely at the Ouachita Embayment, in the Early Cambrian. Paleomagnetic poles from an Early Ordovician magmatic belt in northwest Argentina permit the identification of a peri-Gondwanic–rotated terrane, called the Eastern Puna–Famatina terrane, with volcanic arc affinities. Early Paleozoic paleomagnetic poles from the Western Puna (Sierra de Almeida, Chile) were previously interpreted as evidence of a para-autochthonous nature of the Arequipa-Antofalla Massif that rotated successively clockwise and counterclockwise. New preliminary paleomagnetic data in the same region (Salar del Rincón, Argentina) are not easily reconciled with the previous data, unless a Late Ordovician pre-tectonic remagnetization is assumed for the latter. The paleomagnetic and geologic data permit an alternate interpretation based on the fact that the Arequipa and Antofalla blocks can be considered as different terranes. In this case the paleomagnetic data obtained so far is representative of the Antofalla block alone and suggests at least 1,000-km latitudinal displacement of this block with respect to its present position in the South American margin since the Late Cambrian–Early Ordovician. Accretion of this terrane to the Gondwana margin must have occurred in latest Ordovician. Paleozoic paleomagnetic poles from Patagonia indicate that this terrane shares a common apparent polar wander path (APWP) with Gondwana since the Early Devonian. This suggests that no significant movement occurred between these two crustal blocks since that time. Many questions regarding the complex tectonic evolution of southwestern South America that can potentially be addressed by paleomagnetism remain unanswered.

Rapalini, A. E., Astini, R. A., and Conti, C. M., 1999, Paleomagnetic constraints on the evolution of Paleozoic suspect terranes from southern South America, *in* Ramos, V. A., and Keppie, J. D., eds., Laurentia-Gondwana Connections before Pangea: Boulder, Colorado, Geological Society of America Special Paper 336.

INTRODUCTION

Paleomagnetism has proved to be an essential tool in deciphering the tectonic evolution of continental margins in which accretion of displaced and exotic terranes was an important process (Beck 1989; Van der Voo, 1993). The paleomagnetic technique permits the determination of quantitative paleolatitudinal and paleoazimuthal controls in the evolution of terranes prior to and during their accretion, as well as determination of the age of this process. Sometimes, paleomagnetism can be decisive in ruling out or supporting some tectonic models, and the paleogeography of most exotic terranes is very difficult to ascertain without the aid of paleomagnetic data.

The Paleozoic evolution of the western continental margin of southern South America was a complex one, apparently involving the accretion and displacement of several large terranes in pre–Late Paleozoic times (e.g., Ramos, 1984, 1988; Ramos et al., 1986; Mpodozis and Ramos, 1989; Astini et al., 1995). Although not fully known in detail, the assembly of present-day South America probably began in Late Proterozoic times (Ramos, 1988) when a few cratonic nuclei were assembled into the Gondwana supercontinent (see Rogers et al., 1994). However, large areas of southwestern South America are not part of these more ancient, cratonic blocks. Instead, they were apparently accreted later on, at different times and under different tectonic conditions, displaying a pattern of younger ages of accretion toward the present western continental margin. Although a broad picture of successive accretions during latest Proterozoic-Paleozoic times have been obtained, mainly from geologic and biostratigraphic considerations, the details of this evolution are still poorly known. Probably this has been due in part to the scarcity of paleomagnetic data from Paleozoic rocks in southwestern South America. Yet, in recent years, a few significant paleomagnetic results have been obtained that place important constraints on the tectonic evolution of the southwestern continental margin of Gondwana during the Paleozoic. These results, together with their tectonic and paleogeographic implications, are briefly reviewed in this contribution.

ARGENTINE PRECORDILLERA

The Argentine Precordillera (AP in Fig.1) is the most clearly exotic Paleozoic terrane of this margin (see Astini and Thomas, this volume). Its present morphostructural features are due to Andean tectonics (Jordan et al., 1983). It is a thrust-fold belt located to the east of the main Andes, the uplift of which is due to subduction of the oceanic Nazca plate under the South American continent. The north-south extent of the Paleozoic block, however, probably exceeds 800 km, reaching the San Rafael block in southern Mendoza (Astini et al., 1995; Bordonaro et al., 1996). The Argentine Precordillera consists of a crustal block of Greenvillian age (Mahlburg Kay et al., 1996) with a thick cover of folded Paleozoic and Tertiary rocks (Baldis and Bordonaro, 1981; Allmendinger et al., 1990). The development of a thick carbonate platform between the Early Cambrian and Early Ordovician with typical Laurentian faunas (see Benedetto et al., 1995; Astini et al., 1995) is its most outstanding stratigraphic fea-

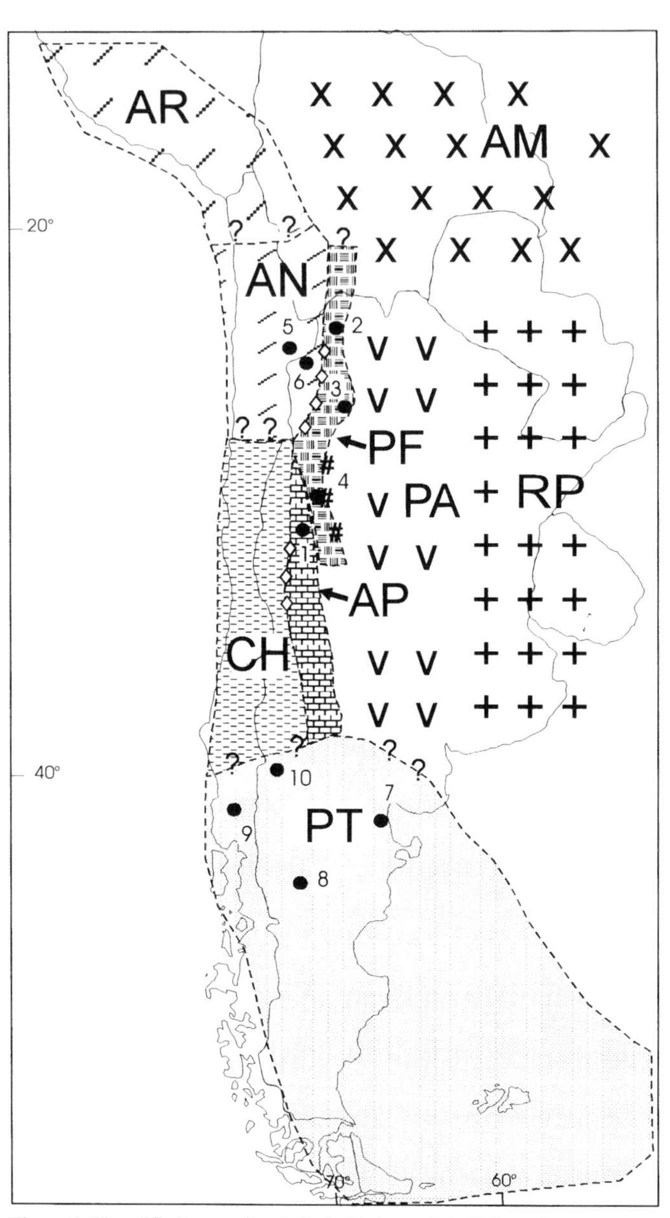

Figure 1. Simplified map of main Late Proterozoic to Paleozoic suspect terranes in southwestern South America. AR = Arequipa Massif; AN = Antofalla block; CH = Chilenia; AP = Argentine Precordillera; PF = Eastern Puna–Famatina; PT = Patagonia; PA = Pampia; RP = Rio de la Plata craton; AM = Amazonia craton. Question marks indicate terrane boundaries with hypothetical locations; open diamonds: exposed ophiolites; # symbols: mylonitic zones; solid circles: location of paleomagnetic studies discussed in the text. Numbers: 1 refers to Cerro Totora; 2, Eastern Puna; 3, Cuchiyaco; 4, Famatina; 5, Sierra de Almeida; 6, Quebrada de Lari; 7, Sierra Grande; 8, Sierra de Tepuel; 9, Chilean granites; 10, Choiyoi volcanics. Simplified from Ramos (1988), Conti et al. (1996), and Bahlburg and Hervé (1997). Minor suspect terranes on the Chilean coast, such as Mejillonia (Bahlburg and Hervé 1997), Pichidangui (Forsythe et al., 1987) and Madre de Dios (Forsythe and Mpodozis, 1979), are not shown.

ture. Despite this being known for several decades (Poulsen, 1958; Borrello, 1965), it was not until much later (Ramos et al., 1986; Dalla Salda et al., 1992; Astini et al., 1995) that the possibility of the Argentine Precordillera being an accreted terrane was considered seriously.

The 1995 Penrose Conference on the possible Laurentian origin of the Argentine Precordillera (Dalziel et al., 1996) concluded that the Argentine Precordillera is most likely an exotic fragment derived from Laurentia and accreted to Gondwana sometime during the Ordovician. The main evidence for this is the majority of Cambrian fossil faunas nearly identical to those from continental shelf successions of Eastern Laurentia and unique in South America, the very similar Cambrian-Ordovician stratigraphic successions to those of the southern Appalachians, and the isotopic data from the basement rocks with much closer affinities with the Appalachian basement than with adjacent basement rocks in South America. The evolution of the faunal provincialism and the stratigraphy (Benedetto et al., 1995) indicate clear Gondwanan affinities for the Argentine Precordillera by the Late Ordovician. This suggests that the accretion of the Argentine Precordillera took place sometime during the Ordovician. Despite this evidence, no consensus has been reached on the exact timing of separation from Laurentia and accretion to Gondwana, the tectonic mechanism involved in this process or the exact position and kinematics of rifting (Dalla Salda et al., 1992; Astini et al., 1995; Thomas and Astini, 1996; Dalziel, 1997).

Until very recently, no paleomagnetic data were available to test and constrain the tectonic models mentioned before. This was due, in part to the few paleomagnetic studies carried out on this region, but also to an apparently widespread and pervasive Permian remagnetizing event that inhibited determination of the primary magnetic components from many of the Lower Paleozoic rocks (Rapalini and Tarling, 1993; Truco and Rapalini, 1996).

A successful paleomagnetic study was recently carried out on the Lower Cambrian Cerro Totora Formation (Rapalini and Astini, 1998), exposed in the northern Precordillera (location 1 in Fig. 1). The Cerro Totora Formation consists in a mixed succession of evaporites, red siliciclastics, and carbonates (Astini and Vaccari, 1996) with trilobites of the Bonnia-Olenellus zone (Vaccari, 1990). This succession has been interpreted as typical synrift deposits (Astini and Vaccari, 1996), possibly accumulated during the initial rifting of the Argentine Precordillera from the Ouachita embayment of south-central North America (Thomas and Astini, 1996).

Rapalini and Astini (1998) reported a paleomagnetic study from 12 sites (80 samples) located on red siltstones (9 sites), dolomitic limestones (2 sites), and green shales (1 site) of the Cerro Totora Formation. Following standard demagnetization techniques within-site consistent characteristic remanent magnetization directions were isolated from the nine hematite-bearing siltstone sites and one site of the magnetite-bearing dolomitic limestone. A positive fold test indicated a pre-tectonic (pre–Late Tertiary) magnetization. A virtual geomagnetic pole was computed for each site ($N = 10$), the mean of which is the paleomagnetic pole of the Cerro Totora Formation (CT in Table 1). The position of CT is not consistent with any expected pole position for South America in the Phanerozoic, suggesting that a post-Cambrian remagnetization of the rocks is unlikely. CT also disagrees with the Cambrian path for the already assembled Gondwana (Fig. 2A) (Meert and Van der Voo, 1996). However, if the Argentine Precordillera is placed in the Ouachita embayment, according to the hypothesis of Thomas and Astini (1996), CT becomes perfectly consistent in position and age with the Cambrian segment of the Laurentian APWP (Fig. 2B).

This paleomagnetic result provides strong evidence that the Argentine Precordillera is an exotic terrane derived from Laurentia and most likely from the southern Appalachians. This also implies that all early Paleozoic or older terranes located to the west (i.e., the basement of Chilenia) (Ramos et al., 1986) are necessarily allochthonous. Since there is no younger early Paleozoic paleomagnetic pole yet available for the Argentine Precordillera, its kinematic history between the Early Cambrian and the Late Ordovician remains unconstrained by paleomagnetic data.

EASTERN PUNA–FAMATINA TERRANE

An outstanding geologic feature of northwest Argentina is a north-trending belt of Lower Ordovician magmatic rocks (Fig.1) that comprises the Famatina system and the "Faja Eruptiva" of Eastern Puna (Puna Oriental). The age of this belt is very well defined by means of fossil assemblages found in intercalated sediments as Tremadoc–early Llanvirn (e.g., Coira, 1973; Koukharsky and Mirré, 1974; Coira and Koukharsky, 1991; Aceñolaza, 1992; Vaccari et al., 1992), as well as radiometric datings on intrusive bodies in the Eastern Puna (Omarini et al., 1984; Lork and Bahlburg, 1993) between 476 and 467 Ma. This volcanism has been assigned to a magmatic arc on the basis of its geochemical signature and geotectonic setting (Coira et al., 1982; Ramos, 1986; Mannheim, 1993; Toselli et al., 1996). This volcanic arc was apparently emplaced on oceanic or quasi-oceanic crust in northern Puna (Ramos, 1986) and Famatina (Toselli et al., 1996), whereas in southern Puna there is evidence of underlying continental crust (Ramos, 1986).

Conti et al. (1996) presented paleomagnetic data from Lower Ordovician rocks at four localities (16 sites) on this belt, including a reassessment of previously published data by Valencio et al. (1980) on the Suri Formation in the Famatina system (locations 2–4 in Fig.1). Conti et al. (1996) found consistent pretectonic magnetizations on different units of the same age along this belt, and computed three paleomagnetic poles of Early Ordovician age for the Puna Oriental (Eastern Puna), Cafayate area (Cuchiyaco Granodiorite), and the Famatina system (Table 1). The same magnetization direction was found in different lithologies carried by different magnetic minerals. The poles derived from the paleomagnetic directions are consistent between them, but disagree with the Early Ordovician mean pole position for Gondwana (Grunow, 1995). The disagreement is found in

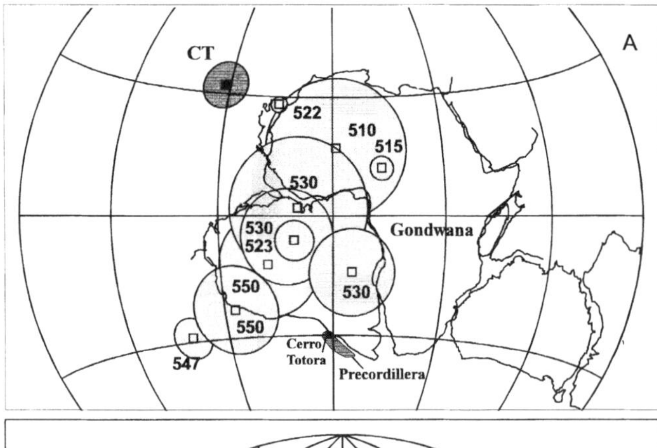

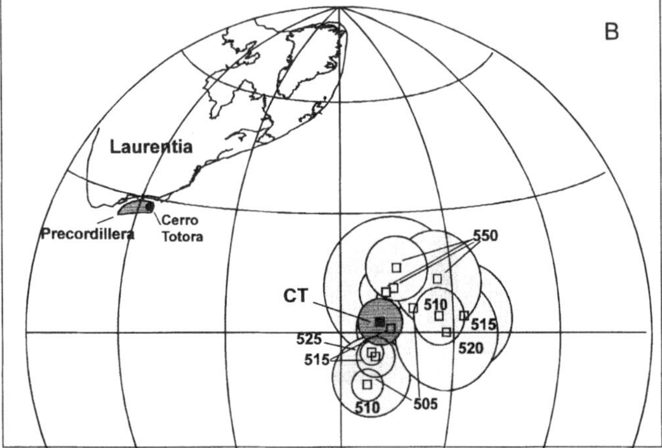

Figure 2. A, Early Cambrian paleomagnetic pole from the Cerro Totora Formation (CT) in African coordinates after Gondwana reconstruction (de Wit et al., 1988) and the Gondwana 550–510-Ma APWP (Meert and Van der Voo, 1996). The Precordillera has been kept attached to South America in its present location. Numbers indicate the most likely age (in Ma) of the Gondwana poles. B, Position of Cerro Totora Formation pole (CT), after placing the Argentine Precordillera against the southeastern Laurentian margin following the model of Thomas and Astini (1996), and the 550–505 APWP of Laurentia (Torsvik et al., 1996). Numbers indicate the most likely age (in Ma) of each Laurentian paleomagnetic pole. From Rapalini and Astini (1998).

declination only (Fig. 3), with no paleolatitudinal discordance. A clockwise rotation of 52.6° ± 11.1° was computed by Conti et al. (1996). These paleomagnetic results led these authors to propose that the Lower Ordovician volcano-sedimentary belt extending from La Rioja province (Famatina) to Jujuy province (eastern Puna), and called Puna Oriental–Famatina (PF in Fig. 1) is a rotated terrane that was accreted to Gondwana in Middle Ordovician times. Geologic data (Coira et al., 1982; Allmendinger et al., 1983; Astini et al., 1995, Blasco et al., 1996; Bahlburg and Hervé, 1997) support the existence of an oceanic realm between the Arequipa-Antofalla and the Argentine Precordillera terranes on one side and Puna Oriental–Famatina on the other during Ordovician times. Conversely, the eastern boundary of this terrane is clearly defined by a thick mylonitic belt (Lopez and Toselli, 1993) in the south that marks the limit with the Proterozoic Pampia terrane (PA in Fig. 1). The northeastern boundary is not so clearly defined by a mylonitic zone, as by the tectonic discontinuity between Puna and Cordillera Oriental provinces (Allmendinger et al., 1983; Mon and Hongn, 1987; Bahlburg and Hervé, 1997). The age of rotation and accretion cannot be determined on the basis of the available paleomagnetic data; however, the most likely age on geologic considerations corresponds to the Middle Ordovician, both in Puna Oriental (Bahlburg and Hervé, 1997) and in Famatina (Rapela et al., 1992; Toselli et al., 1996).

The similar paleolatitudes indicated by the paleomagnetic data for this terrane and the southwestern continental margin of Gondwana in Early Ordovician times suggest that the Puna Oriental–Famatina was a peri-Gondwanic magmatic arc. A similar paleogeographic reconstruction was obtained for the Famatina system by Benedetto and Sánchez (1996) and Astini and Benedetto (1996) on the basis of biogeographic considerations, and by Toselli et al. (1996) based on the geochemical signature of the magmatic rocks.

The Puna Oriental–Famatina volcanic arc was probably only one of several short-lived magmatic arcs that developed in the southern Iapetus during Ordovician time (Van der Voo, 1993; Mac Niocaill et al., 1997; Dalziel, 1997). According to Astini et al. (1995), the Famatina volcanic arc (and its probable northward extension in eastern Puna) was produced by the consumption of oceanic crust associated with the approach of the Precordillera terrane toward the Gondwana margin. A rotation of this arc in Middle Ordovician time produced its accretion to the Gondwana margin.

WESTERN PUNA (AREQUIPA?–ANTOFALLA) TERRANE

The existence of Precambrian crystalline rocks along the southern coast of Perú, generally referred to as the Arequipa Massif (AR in Fig. 1), has long posed the possibility of an exotic origin. The rocks belonging to this massif gave Rb/Sr radiometric ages of 1,910 ± 36 and 1,918 ± 33 Ma (Dalmayrac et al., 1977; Shackleton et al., 1979). This Proterozoic block was extended southward in many models to include outcrops of metamorphic rocks in Antofalla and Belén in northern Chile, with ages ranging from 1,460 to 777 Ma (Pacci et al., 1980; Damm et al., 1990), leading to a larger Arequipa-Antofalla block (Ramos, 1988) (AR-AN in Fig. 1). Different early Paleozoic tectonic evolutions (see Forsythe et al., 1993) have been inferred for the northern (Arequipa) and the southern (Antofalla) parts of this terrane. In the north, no evidence has been found for the existence of ocean floor of early Paleozoic age between the Precambrian coastal rocks and the basement of the Amazon basin (Dalmayrac et al., 1980; Sempere, 1995). However, Ordovician ophiolites exist in the Western Puna of Argentina to the east of the Antofalla metamorphic rocks (Allmendinger et al., 1983; Blasco et al., 1996), suggesting the existence of an ocean between this area and Gondwana in Ordovician time. Recent radiometric datings by Basei et al.(1996) cast

TABLE 1. SUMMARY OF PALEOMAGNETIC POLES FROM PALEOZOIC SUSPECT TERRANES OF SOUTHWESTERN SOUTH AMERICA

Terrane	Formation	Paleomagnetic Pole (PP) Latitude (°)	Longitude (°)	A95 (dp/dm) (°)	Paleomagnetic Results*	PP Age	Ref†
Precordillera	Cerro Totora F. (CT)	37.0	314.1	5.8	AF/Th demag, +FT N = 10 (n = 65), IRM	Early Cambrian	1
Puna Oriental-Famatina	Acoyte F., Chiquero F., (PO)	-17.7	357.9	13.0	AF/Th demag, + FT, IRM, N = 5 (n = 21).	Early Ordovician	2
	Cuchiyaco Gr. (CY)	-30.0	0.5	10.0	AF/Th demag, IRM, N = 5 (n = 23)	Early Ordovician	2
	Suri F. (FS)	-13.0	356.3	6.0	AF/Th demag, N = 6 (n = 18)	Early Ordovician	2
	Overall mean (PF)	-21.8	358.7	7.5	+FT, Consistency test N = 16 (n = 61)	Early Ordovician	2
Western Puna	Choschas Pluton + Pre-Sil lavas (CHO)	58.1	309.2	(3.8/7.5)	AF/Th demag, both polarities, n = 24	Late Cambrian–Early Ordovician	3
	Late Ordovician plutons (OP)	0.0	271.0	(12.4/13.7)	AF/Th demag, +FT, both polarities, n = 20	Late Ordovician	3
	Pre-Sil lavas (roof pendant?) (OL)	-25.2	260.5	(13.7/15.0)	AF/Th demag, both polarities, n = 16	Silurian?	3
	Las Vicuñas F. (FV)	-45.1	283.4	10.6	AF/Th demag, +FT, N = 4 (n = 12)	Late Ordovician?	4
Patagonia	Sierra Grande F. (lower member) (SG1)	3.4	238.0	(8.0/14.0)	AF/Th demag, both polarities, N = 4 (n = 14)	Silurian??	5
	Sierra Grande F. (upper member) (SG2)	-42.0	283.5	20.0	Th/Ch demag, +FT, N= 2 (n = 10)	Early Devonian	5
	Tepuel Group (TP)	-31.7	316.1	(15.0/16.0)	AF/Th demag, +FT, IRM, detrital rem, N = 16 (n = 62)	Middle Carboniferous	6
	Chilean Intrusives (CI)	-57.4	323.5	19.0	AF/Th demag, both polarities, N = 7 (n = 58)	Late Carboniferous–Early Permian	7
	Sierra Grande F. (SG3)	-77.3	310.7	(7.0/8.0)	AF/Th demag, IRM, FT syntectonic, N = 13 (n = 88)	Late Permian?	8
	Choiyoi F. (CH)	-21.0	232.3	8.0	AF/Th demag, +FT, N = 33 (n = 85)	Late Permian	9

*AF/Th/Ch demag mean complete demagnetization by alternating field, thermal cleaning, or chemical leaching; +FT = positive fold test; N = number of sites; n = number of samples; IRM = isothermal remanence experiments..
†1 = Rapalini and Astini, 1998; 2 = Conti et al., 1996; 3 = Forsythe et al., 1993; 4 = this chapter; 5 = Rapalini and Vilas, 1991; 6 = Rapalini et al., 1994; 7 = Beck et al., 1991; 8 = Rapalini, 1998; 9 = Rapalini et al., 1989.

doubts on the Proterozoic ages of some of the Chilean (Antofalla) outcrops. This, as well as evidence for substantial differences in age and composition of the basement of the Andes to the north and south of 20°–21°S (Wörner et al., 1992; Aitchenson et al., 1995), led Bahlburg and Hervé (1997) to postulate that the Arequipa-Antofalla block be subdivided into two blocks: an Early to Middle Proterozoic Arequipa block in the north and a Late Proterozoic–Early Paleozoic Antofalla block in the south (Fig. 1).

Forsythe et al. (1993) recently presented the first early Paleozoic paleomagnetic data from the western Puna in the Sierra de Almeida (northern Chile) (location 5 in Fig. 1). These data are summarized in Table 1 (poles CHO, OP, and OL). The lower Paleozoic rocks they studied belong to a magmatic belt known as the Faja Eruptiva de la Puna Occidental (Western Puna Eruptive belt) (Palma et al., 1986). Three paleomagnetic poles were obtained from these rocks. The oldest one corresponds to the Late Cambrian (502 ± 10 Ma) Choschas Granite and the Cambrian-Ordovician volcanics of the Cordon de Lila Complex (CHO in Fig. 6A,B). This pole presented an inconclusive tilt test but has no resemblance to expected younger poles for Gondwana. A second pole (OP in Table 1 and Fig. 6) was obtained from the Tucucaro, Tilopozo, and Alto del Inca plutons with Rb/Sr ages of 441 ± 8, 452 ± 4, and 468 ± 100, 433 ± 22 Ma, respectively. The remanence directions from these plutons were corrected for tilt of a pre-Devonian unconformity, resulting in a positive tilt test. A third pole (OL in Table 1 and Fig. 6) was obtained for the Cambrian-Ordovician lavas of the Cordón de Lila Complex at a different locality where they may be a roof pendant of a Silurian pluton. Forsythe et al. (1993) suggested a remagnetized Silurian age for this pole.

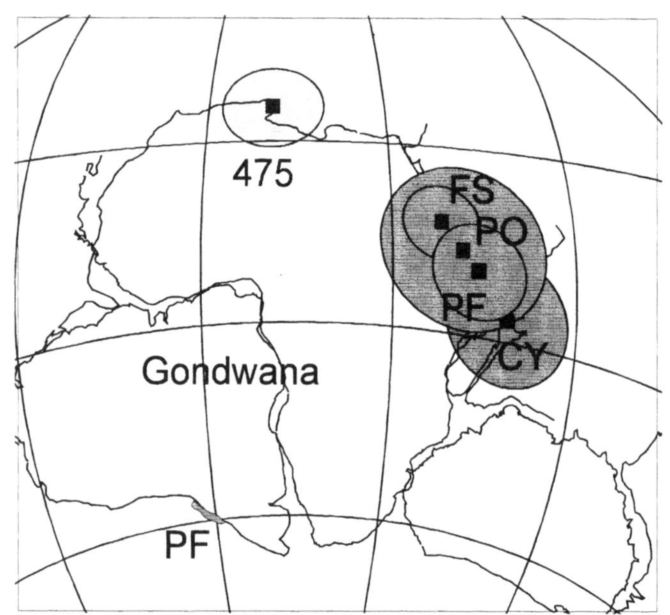

Figure 3. Comparison of the Early Ordovician paleomagnetic poles from the Eastern Puna–Famatina terrane with the 475-Ma mean paleomagnetic pole for Gondwana (Grunow, 1995) in a Gondwana reconstruction (de Wit et al., 1988). Details and symbols of paleomagnetic poles as in Table 1. PF = Eastern Puna–Famatina terrane. Modified from Conti et al. (1996).

A new paleomagnetic study was carried out recently by Conti (1995) on the Lower Tremadocian Las Vicuñas Formation (Moya et al., 1993). This is a 200-m-thick volcaniclastic sequence exposed at Quebrada de Lari (location 6 in Figs. 1 and 4) in the western Puna of Argentina. The outcrops of this unit are relatively near the Sierra de Almeida in Chile (see Fig. 1) and also belong to the Western Puna Eruptive belt. Thirty-one samples from seven sites were collected from the Las Vicuñas Formation. Application of thermal demagnetization proved ineffective to isolate the magnetic components because of chemical changes caused by the experimental heating. However, AF demagnetization (Fig. 5) permitted the definition of the characteristic remanent magnetization in 12 samples from four sites, presumably carried by magnetite. This remanence was generally isolated at demagnetizing fields ranging from 10 to 80 mT. A secondary component coincident with the present dipole direction was found also in most samples carried by goethite (Conti, 1995). Comparison of remanence directions (Table 2) in situ and after tectonic correction shows a better grouping after full tectonic correction. The pre-Silurian rocks of the area were subjected to at least two main deformational phases: the Late Ordovician Ocloyic and the Late Cenozoic Andean (Moya et al., 1993). The latter is represented by open northwest-trending folds. All sampling sites were located on one limb of the Andean fold (Fig. 4); therefore, the positive result of the fold test suggests a pre-Ocloyic magnetization. However, the small number of samples and the fact that the site with a substantial different structural attitude is represented by a single sample make interpretation of results highly speculative. Nevertheless, the in situ remanence directions do not correspond to those expected for Recent times, which is taken as strong evidence of at least a pre-Andean magnetization. A preliminary paleomagnetic pole for the Vicuñas Formation (FV) is shown in Figure 6 (see also Tables 1 and 2).

Forsythe et al. (1993) interpreted their data from Sierra de

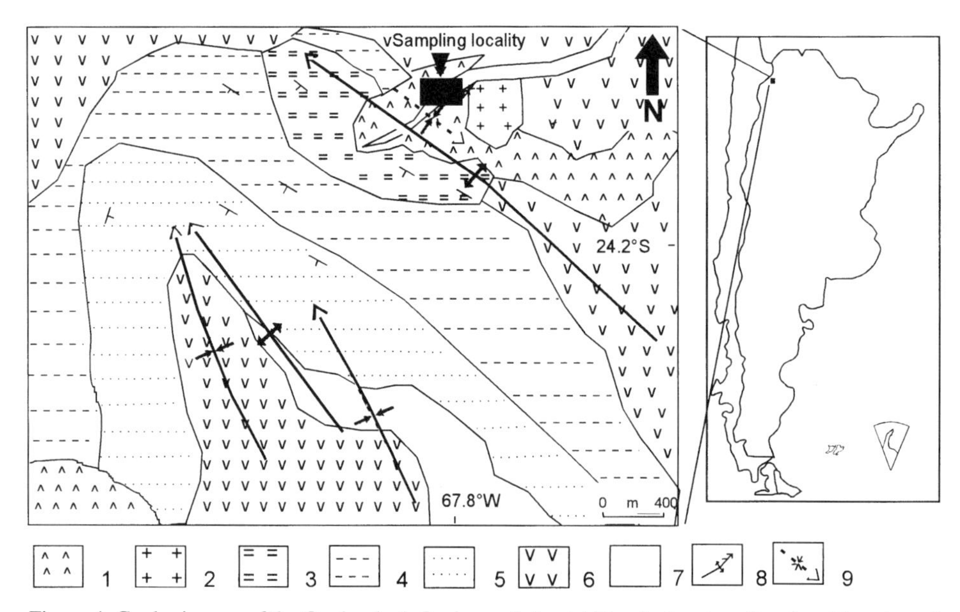

Figure 4. Geologic map of the Quebrada de Lari near Salar del Rincón (western Puna) and location of the sampling locality of the Early Ordovician Vicuñas Formation. Numbers: 1, Vicuñas Formation (Tremadoc); 2, rhyolitic porphyry (Ordovician?); 3, Salar del Rincón Formation (Silurian); 4, Cerro Oscuro Formation (Carboniferous); 5, Arizaro Formation (Permian); 6, Cenozoic volcanics; 7, Recent sediments; 8, Tertiary fold; 9, pre-Silurian fold. Taken from Moya et al. (1993).

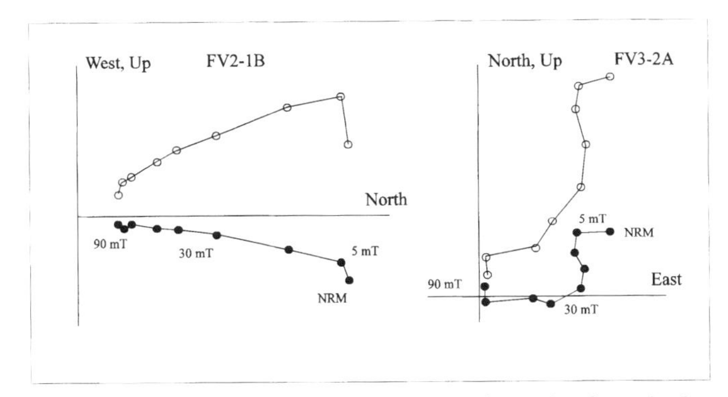

Figure 5. Characteristic magnetic behavior of samples from the Las Vicuñas Formation submitted to AF demagnetization. Closed (open) symbols correspond to vectorial representations on the horizontal (vertical) planes.

Almeida as representative of the entire Arequipa-Antofalla block. It is clear from Figure 6 that some of the poles are not consistent with coeval poles from Gondwana. In particular, the Late Cambrian–Early Ordovician pole (CHO) shows an important departure from the expected position according to the recently presented Early Ordovician mean pole for Gondwana (Grunow, 1995). Forsythe et al. (1993) proposed a counterclockwise rotation of around 50° for the Arequipa-Antofalla block after the Early Ordovician to account for this discrepancy. They also suggested no significant discordance in paleolatitude. The other two poles (OP and OL) were assigned to the Late Ordovician and Early Silurian, respectively. Due to the fact that the Middle Ordovician–Early Devonian segment of the Gondwana APWP is ill-defined (Van der Voo, 1993), Forsythe et al. (1993) suggested that the only significant discordance of the paleomagnetic poles from the Sierra de Almeida with respect to the Gondwana path was for the Late Cambrian–Early Ordovician. These data were used to propose a model of a para-autochthonous evolution of the Arequipa-Antofalla block, in which this block rotated clockwise, probably in the latest Proterozoic, around an Euler pole located in northern Perú. This would lead to the opening of a basin with development of oceanic crust in the southern part, consistent with the geologic observations of ophiolitic rocks only in northwest Argentina. Closure of the basin could not be deciphered from the paleomagnetic data due to difficulty in determining whether the Late Ordovician and Silurian poles were discordant compared with the Gondwana APWP. Geologic considerations pointed to a Late Ordovician (Ocloyic) closure of the basin.

If the position of the Late Cambrian–Early Ordovician (CHO) pole is compared rigorously respect to the ~475-Ma mean pole for Gondwana (Grunow, 1995), it shows significant discordance both in paleoazimuth and paleolatitude. The corresponding values are: for rotation, 49.4° ± 8.6° (counterclockwise), for paleolatitude discordance, 17.7° ± 8.1°. This suggests at least 10° lower paleolatitude for the Sierra de Almeida in Early Ordovician times than expected. Unless this is considered as an artifact of the paleomagnetic data, these data suggest that the Sierra de Almeida is part of a displaced terrane. Although the model presented by Forsythe et al. (1993) can still be sustained, mainly on geologic grounds, an alternate interpretation of the paleomagnetic data is possible.

Considering that the Antofalla block is possibly a separate terrane from the Arequipa Massif (Bahlburgh and Hervé, 1997), the paleomagnetic data obtained by Forsythe et al. (1993) and Conti (1995) correspond only to the Antofalla block. This block apparently has much younger crust (Neoproterozoic?) than the Arequipa Massif and is bounded to the east by Ordovician ophiolitic rocks with mid-ocean ridge basalt geochemical signature (Blasco et al., 1996). This is consistent with the consumption of oceanic crust during Ordovician times to the east of the Antofalla block. These observations suggest that the Antofalla block had an independent evolution from the Arequipa Massif in early Paleozoic times. The paleomagnetic data (CHO pole) suggest that in the Early Ordovician it was located at least 1,000 km (in lower paleolatitudes) from its present position on the South American margin. During Ordovician time this terrane moved toward the continental margin and was accreted in the latest Ordovician–Early Silurian, causing the Ocloyic tectonic event (Bahlburg and Hervé, 1997).

The ill-defined nature of the Late Ordovician–Silurian segment of the Gondwana APWP does not permit the exact time of accretion of the Antofalla block to be determined from the paleomagnetic observations. However, the OP pole, of probable Late Ordovician age, falls west (present-day coordinates) from the generally assumed Middle–Late Ordovician APWP segment of Gondwana. This might suggest that some relative displacement

TABLE 2. PALEOMAGNETIC DATA FROM LAS VICUNAS FORMATION

Site	n	Dec. in Situ (°)	Inc. in Situ (°)	α95 (°)	Bed. Str. (°)	Bed. Dip (°)	Dec. Corr. (°)	Inc. Corr (°)	α95 (°)
FV2	2	57.0	-39.5	....	341	47	354.7	-78.9	
FV3	5	61.0	-42.7	17.0	338	42	22.6	-82.8	
FV4	4	49.0	-44.8	15.0	338	42	344.7	-76.6	
FV6	1	46.5	-7.6	....	327	70	16.2	-74.0	
All sites	4 (12)	52.9	-34.0	21.2			3.3	-78.5	5.9

Dec. = declination; Inc. = inclination; Bed. Str. = bedding strike; Bed. dip = bedding dip; Dec. Corr. = declination corrected for bedding; Inc. Corr. = inclination corrected for bedding.

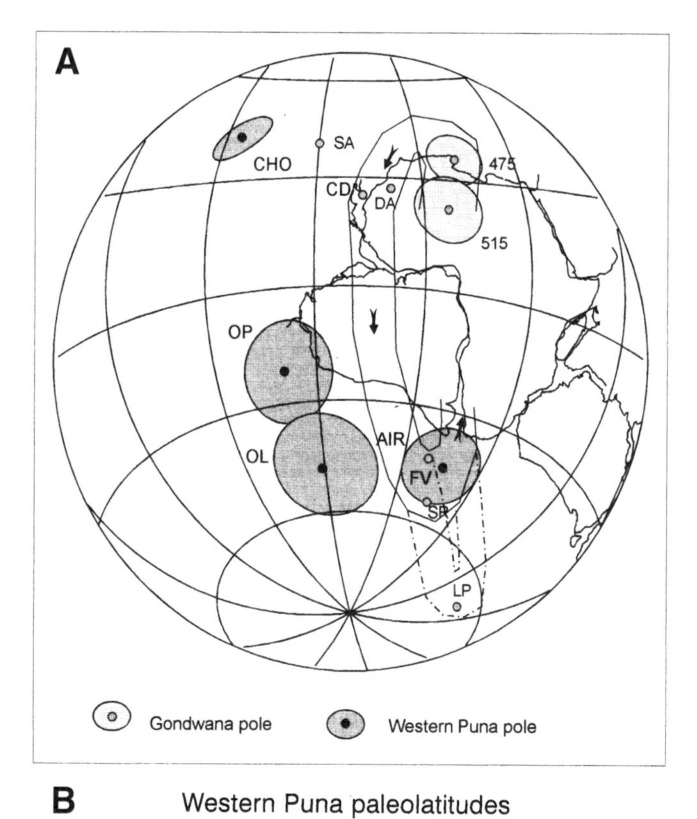

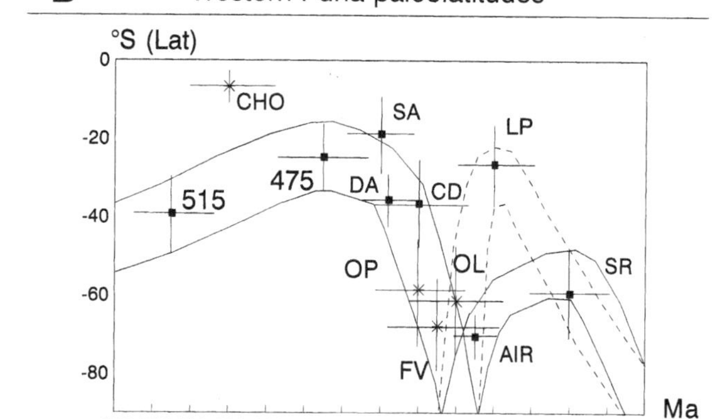

Figure 6. A, Early Paleozoic paleomagnetic poles for the Western Puna and their 95% confidence ovals in a Gondwana reconstruction (according to de Wit et al., 1988). The 515- and 475-Ma mean paleomagnetic poles for Gondwana (Grunow, 1995) and the hypothetical trace of the Ordovician–Devonian apparent polar wander path are also shown. An alternate Silurian segment according to the Lipeon pole (Conti et al., 1995) is shown as dotted lines. Individual poles for Gondwana (SA = Salala Ring Complex; CD = Cedarberg Formation; DA = Damara Granites; AIR = Air Ring Complex; SR = Snowy River Volcanics; from selections by Grunow, 1995, and Conti et al., 1995) are shown without their confidence circles. Details of Western Puna poles are presented in Table 1. B, Paleolatitude vs. geologic time plot for a site in the Western Puna (24°S, 68°W) according to the Gondwana and Western Puna paleomagnetic poles shown in (A) with their estimated errors. Note the discordant paleolatitude obtained from the Choschas pole with respect to the expected paleolatitude from the Gondwana poles. Both possible paleolatitudinal tracks for the Silurian are shown. Letter symbols as in Figure 6A.

(rotation?) took place between the Antofalla and Gondwana land masses in the Late Ordovician. The position of the FV pole is problematic if a Tremadocian age is assumed for it, as it is totally inconsistent with the CHO pole from Sierra de Almeida. The fact that FV is compatible with a latest Ordovician–Early Silurian pole position for Gondwana (Van der Voo, 1993; Conti et al., 1995; Berquó and Ernesto, 1997) suggests that its magnetization is of that age and that it could be related with the beginning of the Ocloyic orogeny.

Another argument presented by Forsythe et al. (1993) in favor of the para-autochthonous nature of the Arequipa-Antofalla block is that the early Paleozoic poles obtained in their study show a trend similar to that of the Gondwana APWP for those times, suggesting that both landmasses moved together. However, Van der Voo (1994) has noted that similar trends are found for various continents for this time interval. He suggested that the "loop" of the Gondwana APWP is due to an episode of true polar wander. In such a case, similar trends would be expected for all continental blocks, regardless of their respective drift histories.

Tosdal et al. (1994) presented isotopic data that suggest that the Arequipa Massif is closely related to the Amazonia craton. These data were obtained from outcrops north of 20°S and therefore pertain to the Arequipa block. The paleomagnetic data, however, were obtained from rocks south of 23°S. Therefore, the boundary between the Arequipa and Antofalla blocks should lie somewhere between 20° and 23°S. It is interesting that a clear change in the character of the Andean basement was found at around 20°–21°S by Wörner et al. (1992) and Aitchenson et al. (1995) from the geochemistry of Cenozoic lavas. Furthermore, Götze et al. (1994) have recently found a major gravimetric high in western Puna extending south of 22°S, which they related to a fundamental change in the basement characteristics at that latitude. They also reported a much higher rigidity of the basement between 20° and 22°S with respect to that between 22° and 26°S.

The hypothesis suggested above should be considered as an alternate interpretation to that presented by Forsythe et al. (1993). However, new and more reliable early Paleozoic paleomagnetic data from both the Arequipa and the Antofalla blocks are needed to determine which, if either, of these hypotheses is correct.

PATAGONIA

Patagonia is the largest of the Paleozoic suspect terranes of southern South America (Fig.1). In some models this terrane has been portrayed as a single tectonic block comprising the Malvinas Plateau (Ramos, 1988; Mahlburgh Kay et al., 1989). The peculiar geologic features of Patagonia have attracted the attention of many geologists since the first decades of this century (Keidel, 1925; Windhausen, 1931). In the last two decades, several hypotheses have proposed an allochthonous nature for Patagonia (Dalmayrac et al., 1980; Martinez, 1980; Ramos, 1984, 1988; Sellés Martinez, 1988). Instead, other models suggest an essentially authochtonous character of this block, at least since the early Paleozoic (Varela et al., 1991; Dalla Salda et al., 1992).

The most favored of the "allochthonous" models that considers Patagonia as a single block is that proposed by Ramos (1984). This model suggests that an ocean was consumed between the northern boundary of Patagonia and southern Gondwana by subduction beneath Patagonia until a collision took place in the Middle or Late Paleozoic. Deformation in Sierra de la Ventana area was interpreted as the result of this collision. This model has remained controversial for several reasons: (1) there is a lack of a suture with obducted ocean floor between both continental blocks; (2) there is no biostratigraphic or palaeoclimatic evidence suggesting different palaeogeographies for Patagonia and South America in the Late Palaeozoic; (3) there are comparable Late Proterozoic to early Paleozoic basement rocks in northern Patagonia and the Pampean Ranges (PA in Fig. 1) in central Argentina (e.g., Varela et al., 1991); and (4) the location of the hypothetical boundary in western Argentina and Chile is unclear.

It is clear that high-quality palaeomagnetic data from Paleozoic rocks in Patagonia could solve the controversy surrounding its tectonic history. However, suitable Paleozoic sedimentary or volcanic exposures are scarce in Patagonia with most rocks of this age being intrusive or metamorphic basement. Nevertheless, some paleomagnetic data have been obtained from Paleozoic rocks of Patagonia in recent years (Rapalini et al., 1989, 1994; Beck et al., 1991; Rapalini and Vilas, 1991; Rapalini, 1998). The paleomagnetic poles obtained from Paleozoic rocks in Patagonia are summarized in Table 1. Six paleomagnetic poles of possible Paleozoic age from Patagonia have been published. Three of them (SG1, SG2, SG3) correspond to the Silurian–Early Devonian Sierra Grande Formation, exposed in northeast Patagonia (location 7 in Fig. 1). Another pole (TP) belongs to the middle–late Carboniferous Tepuel Group in central western Patagonia (location 8 in Fig. 1). The CI pole was determined from late Carboniferous–Early Permian (305–285 Ma) intrusive rocks exposed in the Chilean forearc (location 9 in Fig. 1). The CH pole was determined from Late Permian volcanics exposed in northwest Patagonia (location 10 in Fig. 1). Rapalini (1998) has recently discussed the reliability of these poles. The TP and CH poles were considered highly reliable; SG2, SG3, and CI, moderately reliable; while SG1 was qualified as being of low reliability.

The positions of these poles with respect to the middle–late Paleozoic APWP of Gondwana is presented in Figure 7. Poles SG2 of Early Devonian age, TP of middle Carboniferous age, and CI of late Carboniferous–Early Permian age are consistent with the Gondwana path. Pole SG3 is a recently obtained pole (Rapalini, 1998) for the deformation of the Sierra Grande sedimentary rocks. The position is consistent with the Gondwana path for late Early to Late Permian times, suggesting folding and remagnetization of these rocks at that time. The discordant position of CH is due to a local tectonic rotation of the sampling area (Rapalini et al., 1989). CH shows, however, concordant paleolatitudes with that expected from Late Permian poles for Gondwana. The single discordant pole from Patagonia is SG1. While this could be taken as evidence for an accretional origin of Patagonia, the low reliability of this pole suggests that no tectonic interpretation should be based on it (Rapalini, 1998). All poles other than SG1 indicate that both Patagonia and Gondwana shared the same apparent polar wander path since the Early Devonian, suggesting no significant movement of Patagonia with respect to Gondwana since that time.

The paleomagnetic data summarized above suggest that, if Patagonia is an accreted terrane, its collision with Gondwana occurred before the Devonian. However, both the Sierra Grande sediments in northeast Patagonia (Rapalini, 1998) and the sedimentary successions exposed at Sierra de la Ventana, north of the Patagonian margin (Japas, 1987; von Gosen et al., 1990; Cobbold et al., 1991; Tomezzoli and Vilas, 1996) were deformed in Permian times. This renders unlikely any tectonic model involving a continent-continent collision to account for the deformational event around the northern boundary of Patagonia, unless deformation can be explained occurring well over 100 Ma after collision started. Tectonic models that propose that Patagonia was part of Gondwana since the early Paleozoic (e.g., Varela et al., 1991; Dalla Salda et al., 1992), seem to be more consistent with the paleomagnetic data obtained to date. However, determination of the pre-Devonian paleogeographic evolution of Patagonia will require acquisition of new paleomagnetic poles.

CONCLUDING REMARKS

The tectonic evolution of southern South America in the

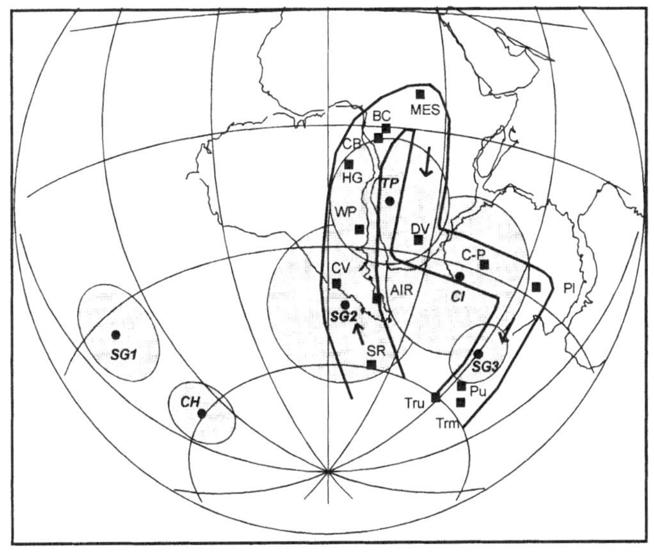

Figure 7. Paleozoic paleomagnetic poles from Patagonia with their 95% confidence ovals and the Early Devonian–Triassic apparent polar wander path of Gondwana. The Gondwana path is based on individual high-quality poles from Early Devonian to Late Carboniferous (see Rapalini, 1998) and mean Western Gondwana poles from Late Carboniferous to Triassic-Jurassic boundary (Van der Voo, 1993). See discussion in the text. Modified from Rapalini (1998).

Paleozoic probably involved the accretion of several terranes. The recent acquisition of paleomagnetic data on Paleozoic rocks from this region, albeit still scarce, provides constraints on the tectonic and paleogeographic evolution of some of these terranes.

Recently obtained paleomagnetic results from Early Cambrian sediments of the Argentine Precordillera confirm that this terrane was part of Laurentia at that time. Although not constrained by paleomagnetism, it is very likely that by the late Ordovician it was already accreted to the western Gondwana margin based on paleobiogeographic affinities (Benedetto et al., 1995) and continuity of the Gondwanan Hirnantian glacial event throughout the proto-Andean region of South America (Buggisch and Astini, 1993). This implies that all pre-Silurian rocks situated to the west of the Argentine Precordillera are also allochthonous.

Paleomagnetic data from geologic units of Early Ordovician age along a magmatic belt of this age in northwest Argentina led to the definition of an accreted terrane with volcanic arc characteristics called the Puna Oriental–Famatina terrane. This terrane was most likely a peri-Gondwanic, short-lived volcanic arc that was accreted to the Gondwana margin in Middle Ordovician times.

Paleomagnetic results from early Paleozoic rocks in the western Puna magmatic belt were previously interpreted as representative of a para-authochtonous single Arequipa-Antofalla block. This block was supposed to rotate successively clockwise and counterclockwise to rift apart from and then collide with the Gondwana margin. The paleomagnetic and geologic data permit an alternate interpretation based on the fact that the Arequipa and Antofalla blocks can be considered separate terranes. In this case, the paleomagnetic data obtained so far is representative of the Antofalla block alone and suggests at least 1,000 km of latitudinal displacement of this block with respect to its present position on the South American margin for the Late Cambrian–Early Ordovician. Accretion of this terrane to the Gondwana margin by latest Ordovician is consistent with both paleomagnetic and geologic data.

Paleomagnetic poles of different qualities from Patagonia indicate that this terrane shares a common apparent polar wander path with Gondwana since the Early Devonian. This suggests that no significant relative movements occurred between these two crustal blocks since those times.

Despite the progress made in recent years, many questions concerning the tectonic evolution of the southwestern continental margin of Gondwana in Paleozoic times still remain. Some of these questions that could potentially be answered with further paleomagnetic investigations include: Which were the kinematics for the transfer of the Argentine Precordillera from Laurentia to Gondwana? Was the Puna Oriental–Famatina terrane a single (or multiple) volcanic arc? Were the Antofalla and Arequipa blocks separate terranes in the early Paleozoic? If so, what was the evolution of Arequipa? What was the paleogeographic evolution of Chilenia before its collision with Gondwana? Were some of these terranes related prior to their accretion to Gondwana? Was Patagonia a single tectonic block in Paleozoic times and what was its paleogeographic evolution before the Devonian?

Many more unanswered questions remain. It is hoped that further paleomagnetic studies in southern South America will yield some answers in the coming years.

ACKNOWLEDGMENTS

This work is the product of several years of paleomagnetic study. Collaboration and discussion with V. Ramos, J. Vilas, D. Tarling, S. Flint, P. Turner, B. Coira, M. Koukharsky, C. Vasquez, O. Bordonaro, and many colleagues at the Laboratorio Daniel Valencio in Buenos Aires are gratefully acknowledged. The final manuscript benefited from the critical reviews by Bob Butler and Rubén Somoza. The University of Buenos Aires, (particularly, UBACyT grants EX-135 and EX-002 supported the final stages of the study, Consejo Nacional de Investigaciones Científicas y Técnicas (Argentina), Fundación Antorchas, and the University of Plymouth (United Kingdom), contributed different kinds of support. This is a contribution to the International Geological Correlation Program Project 376 (Laurentia-Gondwana Connections before Pangea).

REFERENCES CITED

Aceñolaza, F. G., 1992, El sistema Ordovícico de Latinoamérica, *in* Gutierrez Marco, J. C., Saavedra, J., and Rabano. I., eds., El Paleozoico Inferior de Ibero-América: Public: Madrid, Universidad de Extremadura, Spain, p. 85–118.

Aitchenson, S. J., Harmon, R. S., Moorbath, S., Schneider, A., Soler, P., Soria-Escalante, E., Steele, G., Swainbank, I., and Wörner, G., 1995, Pb isotopes define basement domains of the Altiplano, central Andes: Geology, v. 23, p. 555–558.

Allmendinger, R. W., Ramos, V. A., Jordan, T. E., Palma, M., and Isacks, B. L., 1983, Paleogeography and Andean structural geometry, NW Argentina: Tectonics, v. 2, no. 1, p. 1–16.

Allmendinger, R. W., Figueroa, D., Snyder, D., Beer, J., Mpodozis, C. and Isacks, B. L., 1990, Foreland shortening and crustal balancing in the Andes at 30°S latitude: Tectonics, v. 9, no. 4, p. 789–809.

Astini, R. A., and Benedetto, J. L., 1996, Paleoenvironmental features and basin evolution of a complex volcanic-arc region in the pre-Andean western Gondwana: The Famatina belt: 3rd International Symposium on Andean Geodynamics, Extended Abstracts, p. 755–758.

Astini, R. A., and Vaccari, N. E., 1996, Sucesión evaporítica del Cámbrico inferior de la Precordillera: significado geológico: Revista de la Asociación Geológica Argentina, v. 51, p. 97–106.

Astini, R. A., Benedetto, J. L., and Vaccari, N. E., 1995, The early Paleozoic evolution of the Argentine Precordillera as a Laurentian rifted, drifted and collided terrane: a geodynamic model: Geological Society of America Bulletin, v. 107, p. 253–273.

Bahlburg, H., and Hervé, F., 1997, Geodynamic evolution and tectonostratigraphic terranes of northwestern Argentina and northern Chile: Geological Society of America Bulletin, v. 109, p. 869–884.

Baldis, B. A., and Bordonaro, O., 1981, Evolución de facies carbonáticas en la cuenca cámbrica de la Precordillera de San Juan: Actas 8º Congreso Geológico Argentino, San Luis, v. 1, p. 385–397.

Basei, M. A., Charrier, R., and Hervé, F., 1996, New ages (U-Pb, Rb-Sr, K-Ar) from supposed pre-Cambrian units in northern Chile: 3rd International Symposium on Andean Geodynamics, Extended Abstracts, p. 763–766.

Beck, M. E., Jr., 1989, Paleomagnetism of continental North America, implications for displacement of crustal blocks within the Western Cordillera, Baja California to British Columbia: Geological Society of America Bulletin, v. 101, p. 471–492.

Beck, M. E., García, A., Burmester, R. F., Munizaga, F., Hervé, F., and Drake, R., 1991, Paleomagnetism and geochronology of late Paleozoic granitic rocks from the Lake District of southern Chile: implications for accretionary tectonics: Geology, v. 19, p. 332–335.

Benedetto, J. L., and Sánchez, T. M., 1996, Paleobiogeography of brachiopod and molluscan faunas along the South American margin of Gondwana during the Ordovician, *in* Baldis, B. A., and Aceñolaza, F. G., eds., Early Paleozoic evolution in NW Gondwana: San Miguel de Tucumán, Argentina, Serie Correlación Geológica, v. 12, p. 23–38.

Benedetto, J. L., Vaccari, N. E., Carrera, M. G., and Sánchez, T. M., 1995, The evolution of faunal provincialism in the Argentine Precordillera during the Ordovician: new evidence and paleontological implications, *in* Cooper, J. D., Droser, M. L., and Finney, S. C., eds., Ordovician odyssey: SEPM Pacific Section, v. 77, p. 181–184.

Berquó, T. S., and Ernesto, M., 1997, Paleomagnetism of the early Paleozoic sediments from the Caacupé Group, eastern Paraguay: 8th Scientific Assembly, Uppsala, International Association of Geomagnetism and Aeronomy (IAGA 97), Abstracts, p. 59.

Blasco, G., Villar, L., and Zappettini, E. O., 1996, El complejo ofiolítico desmembrado de la Puna argentina. Provincias de Jujuy, Salta y Catamarca: Actas, 13º Congreso Geológico Argentino and 3º Congreso de Exploración de Hidrocarburos, Buenos Aires, v. 3, p. 653–667.

Bordonaro, O., Keller, M., and Lehnert, O., 1996, El Ordovícico de Ponón Trehué en la Provincia de Mendoza (Argentina): redefiniciones estratigráficas: Actas, 13º Congreso Geológico Argentino and 3º Congreso de Exploración de Hidrocarburos, Buenos Aires, v. 1, p. 541–550.

Borrello, A. V., 1965, Sobre la presencia del Cámbrico Inferior olenellidiano en la Sierra de Zonda, Precordillera de San Juan: Ameghiniana, v. 3, p. 313–318.

Buggisch, W., and Astini, R. A., 1993, The late Ordovician Ice age: new evidence from the Argentine Precordillera, *in* Findlay, R. H., Unrug, R., Banks, M. R., and Veevers, J. J., eds., Gondwana Eight; Assembly, evolution and dispersal: Rotterdam, Netherlands, Balkema, p. 439–447.

Cobbold, P. R., Gapais, D., and Rossello, E. A., 1991, Partitioning of transpressive motions within a sigmoidal foldbelt: the Variscan Sierras Australes, Argentina: Journal of Structural Geology, v. 13, no. 7, p. 743–758.

Coira, B., 1973, Resultados preliminares sobre la petrología del ciclo eruptivo concomitante con la sedimentación de la Formación Acoyte en la zona de Abra Pampa, prov. de Jujuy, Revista de la Asociación Geológica Argentina, v. 27, no. 1, p. 85–90.

Coira, B., and Koukharsky, M., 1991, Lavas en almohadilla, Ordovícico en el Cordón de Escaya, Puna Septentrional, Argentina: Actas, 6º Congreso Geológico Chileno, v. 1, p. 674.

Coira, B., Davidson, J., Mpodozis, C., and Ramos, V., 1982, Tectonic and magmatic evolution of the Andes of northern Argentina and Chile: Earth Science Reviews, v. 18, p. 89–95.

Conti, C. M., 1995, Paleomagnetismo del Paleozoico Inferior y Medio del Noroeste Argentino y sus implicancias en la evolución tectónica y geodinámica del márgen oeste del Gondwana [Ph.D. thesis]: Buenos Aires, Argentina, Universidad de Buenos Aires, 301 p.

Conti, C. M., Rapalini, A. E., and Vilas, J. F., 1995, Palaeomagnetism of the Silurian Lipeon Formation, NW Argentina, and the Gondwana apparent polar wander path: Geophysical Journal International, v. 121, p. 848–862.

Conti, C. M., Rapalini, A. E., Coira, B., and Koukharsky, M., 1996, Paleomagnetic evidence of an early Paleozoic rotated terrane in NW Argentina. A clue for Gondwana-Laurentia interaction?: Geology, v. 24, p. 953–956

Dalla Salda, L., Cingolani, C., and Varela, R., 1992, Early Paleozoic orogenic belt of the Andes in southwestern South America: result of Laurentia-Gondwana collision?: Geology, v. 20, p. 617–620.

Dalmayrac, B., Lancelot, J. R., and Leyreloup, A., 1977, Two billion year granulites in the late Precambrian metamorphic basement along the southern Peruvian coast: Science, v. 198, p. 49–51.

Dalmayrac, B., Laubacher, G., Marocco, R., Martinez C., and Tomasi, P., 1980, La Chaine Hercynienne d'Amerique du Sud. Structure et evolution d'une orogene intracratonique: Geologische Rundschau, v. 69, no. 1, p. 1–21.

Dalziel, I. W. D., 1997, Neoproterozoic-Paleozoic geography and tectonics: review, hypothesis, environmental speculation: Geological Society of America Bulletin, v. 109, p. 16–42.

Dalziel, I. W. D., Dalla Salda, L. H., Cingolani C., and Palmer, A. R., 1996, The Argentine Precordillera, a Laurentian terrane; Penrose Conference Report: GSA Today, February, p. 16–18.

Damm, K. W., Pichowiak, S., Harmon, R. S., Todt, W., Keley, S., Omarini, R., and Niemeyer, H., 1990, Pre-Mesozoic evolution of the central Andes: the basement revisited, *in* Mahlburgh Kay, S., and Rapela, C. W., eds., Plutonism from Antarctica to Alaska: Geological Society of America Special Paper, v. 241, p. 101–126.

de Wit, M., Jeffery, M., Bergh, H., and Nicolaysen, L., 1988, Geological map of Gondwana, with explanations and references: Tulsa, Oklahoma, American Association of Petroleum Geologists, scale 1:10,000,000.

Forsythe, R. D., and Mpodozis, C., 1979, El archipiélago Madre de Dios, Patagonia Occidental, Magallanes: rasgos generales de la estratigrafía y estructura del "basamento" pre-Jurásico Superior: Revista Geológica de Chile, v. 7, p. 13–29.

Forsythe, R. D., Kent, D. V., Mpodozis, C., and Davidson, J., 1987, Palaeomagnetism of Permian and Triassic rocks, central Chilean Andes, *in* Elliot, D. H., Collinson, J. W., and McKenzie, G. D., eds., Gondwana Six: American Geophysical Union Geophysical Monograph Series, p. 241–252.

Forsythe, R. D., Davidson, J., Mpodozis, C., and Jesinkey, C., 1993, Lower Paleozoic relative motion of the Arequipa block and Gondwana; paleomagnetic evidence from Sierra de Almeida of northern Chile: Tectonics, v. 12, p. 219–236.

Götze, H. J., Lahmeyer, B., Schmidt, S., and Strunk, S., 1994, The lithospheric structure of the central Andes (20–26°S) as inferred from interpretation of regional gravity, *in* Reutter, K. J., Scheuber, E., and Wigger, P. J., eds., Tectonics of the southern central Andes: Berlin, Springer-Verlag, p. 7–21.

Grunow, A. M., 1995, Implications for Gondwana of new Ordovician paleomagnetic data from igneous rocks in southern Victoria Land, east Antarctica: Journal of Geophysical Research, v. 100, p. 12589–12603.

Japas, M. S., 1987, Análisis cuantitativo de la deformación en el Sector Oriental de las Sierras Australes de Buenos Aires y su implicancia geodinámica [Ph.D. thesis]: Buenos Aires, Argentina, Universidad de Buenos Aires, 359 p.

Jordan, T. E., Isacks, B. L., Ramos V. A., and Allmendinger, R. W., 1983, Mountain building in the central Andes: Episodes, v. 3, p. 20–26.

Keidel, J., 1925, Sobre el desarrollo paleogeográfico de las grandes unidades geológicas de la Argentina: Sociedad Argentina de Estratigrafía y Geografía, GAEA, Anales, v. 4, p. 251–312.

Koukharsky, M., and Mirré, J. C., 1974, Nuevas evidencias del vulcanismo ordovícico en la Puna: Revista de la Asociación Geológica Argentina, v. 29, no. 1, p. 128–134.

Lopez, J., and Toselli, A., 1993, La faja milonítica de TIPA: faldeo oriental del Sistema de Famatina, Argentina, Actas, 12º Congreso Geológico Argentino and 2º Congreso de Exploración de Hidrocarburos, Mendoza, v. 3, p. 39–42.

Lork, A., and Bahlburgh, H., 1993, Precise U-Pb ages of monazites from the Faja Eruptiva de la Puna Oriental and the Cordillera Oriental, NW Argentina: Actas, 12º Congreso Geológico Argentino and 2º Congreso de Exploración de Hidrocarburos, Mendoza, v. 4, p. 1–6.

Mac Niocaill, C., van der Pluijm, B. A., and Van der Voo, R., 1997, Ordovician paleogeography and the evolution of the Iapetus ocean: Geology, v. 25, p. 159–162.

Mahlburg Kay, S., Ramos, V. A., Mpodozis, C., and Sruoga, P., 1989, Late Paleozoic to Jurassic silicic magmatism at the Gondwana margin: analogy to the Middle Proterozoic in North America?: Geology, v. 17, p. 324–328.

Mahlburg Kay, S., Orrell, S. and Abbruzzi, J. M., 1996, Zircon and whole rock Nd-Pb evidence for a Grenville age and a Laurentian origin for the basement of the Precordillera in Argentina: Journal of Geology, v. 104, p. 637–648.

Mannheim, R., 1993, Génesis de las volcanitas eopaleozoicas del Sistema de Famatina, Actas, 12º Congreso Geológico Argentino and 2º Congreso de Exploración de Hidrocarburos, Mendoza, v. 4, p. 147–155.

Martinez, C., 1980, Structure et evolution de la Chaine Hercynienne et de la Chaine Andine dans le nord de la Cordillere des Andes de Bolivie: Travaux et Documents de L'Ostrom, no. 119, 352 p.

Meert, J. G., and Van der Voo, R., 1996, Paleomagnetic and $^{40}AR/^{39}Ar$ study of the Sinyai Dolerite, Kenya: implications for Gondwana Assembly: Journal of Geology, v. 104, p. 131–142.

Mon, R., and Hongn, F., 1987, Estructura del Ordovícico de la Puna: Revista de la Asociación Geológica Argentina, v. 31, p. 65–72.

Moya, M., Malanca, S., Hongn, F. and Bahlburgh, H., 1993, El Tremadoc temprano en la Puna Occidental Argentina: Actas, 12º Congreso Geológico Argentino and 2º Congreso de Exploración de Hidrocarburos; Mendoza, v. 2, p. 20–30.

Mpodozis, C., and Ramos, V. A., 1989, The Andes of Chile and Argentina, *in* Ericksen, G. E., Cañas Pinochet, M. T., and Reinemud, J. A., eds., Geology of the Andes and its relation to hydrocarbon and mineral resources: Circumpacific Council for Energy and Mineral Resources Earth Sciences Series, v. 11, p. 59–90.

Omarini, R., Viramonte, J., Cordani, U., Salfity, J., and Kawashita, K., 1984, Estudio geocronológico Rb-Sr de la Faja Eruptiva de la Puna en el sector de San Antonio de los Cobres, Provincia de Salta: Actas, 9º Congreso Geológico Argentino, S. C. Bariloche, v. 3, p. 146–158.

Pacci, D., Munizaga, F., Hervé, F., Kawashita, K., and Cordani, U., 1980, Acerca de la edad Rb-Sr precámbrica de rocas de la Formación Esquistos de Belén, Departamento de Parinacota, Chile: Revista Geológica de Chile, v. 11, p. 43–58.

Palma, M. A., Parica, P. D., and Ramos, V. A., 1986, El granito de Archibarca: su edad y significado tectónico, provincia de Catamarca: Revista de la Asociación Geológica Argentina, v. 41, p. 414–419.

Poulsen, V., 1958, Contribution to the Middle Cambrian paleontology and stratigraphy of Argentina: Copenhague Matematisc fysiske Meddelelser, Der kongelige Danske Videnskabernes Selsbak, v. 31, p. 1–22.

Ramos, V. A., 1984, Patagonia: un continente paleozoico a la deriva? Actas, 9º Congreso Geológico Argentino, S. C. Bariloche, v. 2, p. 311–325.

Ramos, V. A., 1986, El diastrofismo Oclóyico: un ejemplo de tectónica de colisión durante el Eopaleozoico en el Noroeste Argentino: Revista del Instituto Geológico Minero, Jujuy, v. 6, p. 13–28.

Ramos, V. A., 1988, Late Proterozoic–early Paleozoic of South America—a collisional history: Episodes, v. 11, no. 3, p. 168–174.

Ramos, V. A., Jordan, T. E., Allmendinger, R. W., Mpodozis, C., Kay, S. M., Cortés J. M., and Palma, M., 1986, Paleozoic terranes of the central Argentine—Chilean Andes: Tectonics, v. 5, no. 6, p. 855–880.

Rapalini, A. E., 1998, Syntectonic magnetization of the mid-Paleozoic Sierra Grande Formation. Further constraints for the tectonic evolution of Patagonia: Journal of the Geological Society of London, v. 155, p. 105–114.

Rapalini, A. E., and Astini, R. A., 1998, Paleomagnetic confirmation of the Laurentian origin of the Argentine Precordillera: Earth and Planetary Science Letters, v. 155, p. 1–14.

Rapalini, A. E., and Tarling, D. H., 1993, Multiple magnetizations in the Cambro-Ordovician carbonate platform of the Argentine Precordillera and their tectonic implications: Tectonophysics, v. 227, p. 49–62.

Rapalini, A. E., and Vilas, J.F., 1991, Preliminary paleomagnetic data from the Sierra Grande Formation: tectonic consequences of the first Middle Paleozoic paleopoles from Patagonia: Journal of South American Earth Sciences, v. 4, no. 1–2, p. 25–41.

Rapalini, A. E., Vilas, J. F., Bobbio, M. L., and Valencio, D. A., 1989, Geodynamic interpretations from paleomagnetic data of late Paleozoic rocks in the southern Andes, *in* Hillhouse, J. W., ed., Deep structure and past kinematics of accreted terranes: American Geophysical Union Geophysical Monograph Series, v. 5, no. 50, p. 41–57.

Rapalini, A. E., Tarling, D. H., Turner, P., Flint, S., and Vilas, J. F., 1994, Palaeomagnetism of the Carboniferous Tepuel Group, central Patagonia, Argentina: Tectonics, v. 13, no. 5, p. 1277–1294.

Rapela, C. W., Coira, B., Toselli, A., and Saavedra, J., 1992, El magmatismo del Paleozoico Inferior en el sudoeste de Gondwana, *in* Gutierrez Marco, J. C., Saavedra, J., and Rabano. I., eds., El Paleozoico Inferior de Ibero-América: Madrid, Publicacion Universidad de Extremadura, Spain, p. 21–68.

Rogers, J. J. W., Unrug, R., and Sultan, M., 1994, Tectonic assembly of Gondwana: Journal of Geodynamics, v. 19, p. 1–34.

Sellés Martinez, J., 1988, The relationship between Patagonian and South American blocks: frontal collision or transpression?: 7th Gondwana Symposium, Sao Paulo, Brazil, Abstracts, p. 139.

Sempere, T., 1995, Phanerozoic evolution of Bolivia and adjacent regions, *in* Tankard, A. J., Suárez Soruco, R., and Welsink, H. J., eds., Petroleum basins of South America: American Association of Petroleum Geologists Memoir, v. 62, p. 207–230.

Shackleton, R. M., Ries, A. C., Coward, M. P., and Cobbold, P. R., 1979, Structure, metamorphism and geochronology of the Arequipa Massif of coastal Perú: Geological Society of London Memoir, v. 12, p. 1–21.

Thomas, W., and Astini, R. A., 1996, The Argentine Precordillera: a traveler from the Ouachita embayment of North American Laurentia: Science, v. 273, p. 752–757.

Tomezzoli, R. N., and Vilas, J .F., 1996, Paleomagnetismo del Grupo Pillahiuncó en Sierra de la Ventana (Estancias Las Julianas y San Carlos): Actas, 13º Congreso Geológico Argentino and 3º Congreso de Exploración de Hidrocarburos, Buenos Aires, v. 2, p. 481–488.

Torsvik, T., Smethurst, M., Meert, J., Van der Voo, R., McKerrow, S., Brasier, M., Sturt, B., and Walderhaug, H., 1996, Continental break-up and collision in the Neoproterozoic and Paleozoic—a tale of Baltica and Laurentia: Earth Science Review, v. 40, p. 229–258.

Tosdal, R. M., Munizaga, F., Williams, W. C., and Bettencourt, J. S., 1994, Middle Proterozoic crystalline basement in the central Andes, western Bolivia and northern Chile: a U-Pb and Pb isotopic perspective: Actas, 7º Congreso Geológico Chileno, v. 2, p. 1464–1467.

Toselli, A. J., Durand, F. R., Rossi de Toselli, J. N., and Saavedra, J., 1996, Esquema de evolución geotectónica y magmática eopaleozoica del Sistema de Famatina y sectores de Sierras Pampeanas: Actas, 13º Congreso Geológico Argentino and 3º Congreso de Exploración de Hidrocarburos, Buenos Aires, v. 5, p. 443–462.

Truco, S., and Rapalini, A. E., 1996, New evidence of a widespread Permian remagnetizing event in the central Andean zone of Argentina: 3rd International Symposium on Andean Geodynamics, St. Malo, France, Extended Abstracts, p. 799–802.

Vaccari, N. E., 1990, Primer hallazgo de trilobites del Cámbrico inferior en la Provincia de la Rioja (Precordillera Septentrional): Revista de la Asociación Geológica Argentina, v. 43, p. 558–561.

Vaccari, N. E., Benedetto, J. L., Waisfeld, B. G., and Sánchez, T., 1992, La fauna de *Neseuretus* en la Formación Suri (oeste de Argentina): edad y relaciones paleobiogeográficas: Revista Española de Paleontología, v. 8, no. 2, p. 185–190.

Valencio, D. A., Vilas, J. F., and Mendia, J. E., 1980, Palaeomagnetism and K-Ar ages of Lower Ordovician and Upper Silurian–Lower Devonian rocks from north-west Argentina: Geophysical Journal of the Royal Astronomical Society, v. 62, p. 27–39.

Van der Voo, R., 1993, Paleomagnetism of the Atlantic, Tethys and Iapetus Oceans: Cambridge, United Kingdom, Cambridge University Press, 411 p.

Van der Voo, R, 1994, True polar wander during the middle Paleozoic?: Earth and Planetary Science Letters, v. 122, p. 239–243.

Varela R., Dalla Salda, L., Cingolani, C., and Gomez, V., 1991, Estructura, petrología y geocronología del basamento de la región del Limay, provincias de Río Negro y Neuquén, Argentina: Revista Geológica de Chile, v. 18, p. 147–163.

von Gosen, W., Buggisch, W., and Dimieri, L. V., 1990, Structural and metamorphic evolution of the Sierras Australes (Buenos Aires Province/Argentina): Geologische Rundschau, v. 79, no. 3, p. 797–821.

Windhausen, A., 1931, Geología Argentina. II parte. Geología Histórica y Regional del territorio argentino: Buenos Aires, Argentina. Editorial Peusser, 645 p.

Wörner, G., Moorbath, S., and Harmon, R. S., 1992, Andean Cenozoic volcanic centers reflect basement isotopic domains: Geology, v. 20, p. 1103–1106.

Manuscript Accepted by the Society September 4, 1998

Geological Society of America
Special Paper 336
1999

Pb isotope evidence for Colombia–southern México connections in the Proterozoic

Joaquin Ruiz
Department of Geosciences, University of Arizona, Tucson, Arizona 85721, United States
Richard M. Tosdal
U.S. Geological Survey, Menlo Park 94025, United States
Pedro A. Restrepo*
Department of Geosciences, University of Arizona, Tucson, Arizona 85721, United States
Gustavo Murillo-Muñetón*
Department of Geology, University of Southern California, Los Angeles, California 90089-0740, United States

ABSTRACT

Identical present-day lead isotope compositions of whole rocks from the Mesoproterozoic Guichicovi and Oaxaca Complexes in southern México and Grenville rocks from the Mesoproterozoic Santa Marta and Garzón Massifs in the Colombian Andes indicate these now widely separated Grenville-age rocks shared a U/Pb history that was distinct from Grenville-age rocks from northern and east-central México, and Texas and parts of the Appalachians in the United States. The southern México–Colombian rocks are characterized by a large range of $^{206}Pb/^{204}Pb$ values (17.5–24.6) and a corresponding range of $^{207}Pb/^{204}Pb$ values (15.53–16.31) that scatter along two 1,250 ± 50 Ma reference isochrons that lie above and below the average crustal growth curve. Variable Th/U values, indicated by a wide range of $^{208}Pb/^{204}Pb$ values (36.8–45.1), characterize the complexes, with rocks of the Guichicovi Complex and the Garzón Massif sharing low time-integrated Th/U values whereas the Santa Marta Massif has a tendency toward elevated Th/U values. These Pb isotopic similarities provide additional constraints to reconstructions of the pre-Pangea collisions between Gondwana and Laurentia. The Pb isotopic data, coupled with paleontologic data from unconformably overlying Early Paleozoic rocks, suggest that these Grenville-age basement blocks shared a common history in the late Proterozoic, and that they were part of Gondwana during the early Paleozoic. The southern México rocks, the Oaxaca and Guichicovi Complexes, are considered to have been transferred from that part of northwestern Gondwana now occupied by Colombia to the southern tip of Laurentia during Paleozoic orogenic events related to the opening and closing of the proto–Atlantic Ocean. This model requires a major suture separating México Grenville-age rock, which are considered to be offset pieces of the Grenville Province of Laurentia, from Gondwanan crust in southern México; the suture must lie north of the Guichicovi-Oaxaca Complexes. The Pb isotopic data also require that the Mesoproterozoic megaterrane Oaxaquia consists of several discrete blocks assembled in the late Proterozoic(?) or Paleozoic.

*Present addresses: Restrepo: Conoco Inc., Advance Exploration Organization, 600 N. Dairy Ashford, Houston, Texas 77252-2197; Murillo-Muñetón: Department of Geology/Geophysics, Texas A&M University, College Station, Texas 77843.

Ruiz, J., Tosdal, R. M., Restrepo, P. A., and Murillo-Muñetón, G., 1999, Pb isotope evidence for Colombia–southern México connections in the Proterozoic, *in* Ramos, V. A., and Keppie, J. D., eds., Laurentia-Gondwana Connections before Pangea: Boulder, Colorado, Geological Society of America Special Paper 336.

INTRODUCTION

Continental reconstructions of the Americas based on age, isotopic, lithostratigraphic, and faunal correlations, as well as paleomagnetic data (Bond et al., 1984; Hoffman, 1991; Dalziel, 1991, 1997; Moores, 1991; Dalla Salda et al., 1992a,b; Park, 1992; Keppie, 1993; Keppie et al., 1996) have focused on possible tectonic interactions between western South America and northeastern North America in the Proterozoic and early Paleozoic. Of critical importance in these reconstructions are small continental fragments, referred to as suspect terranes (Coney et al., 1980) or tectonic tracers (Dalziel et al., 1994), that lie between more coherent cratonic cores. Along the southern margin of Laurentia, two of these small terranes, the Paleozoic Acatlan Complex and Mesoproterozoic (~1 Ga) Oaxaca Complex, plus an offset piece, the Guichicovi Complex in southern México, have long presented problems with plate tectonic reconstructions. This plate tectonic misfit led Yañez et al. (1991) and Restrepo-Pace et al. (1997) to test the possibility that the southern México Grenville-age rocks were related to rocks of similar age in the Colombian Andes rather than offset pieces of the Grenville province of eastern Laurentia, as is considered most likely for Grenville-age rocks in northern and east-central México south of the Ouachita suture (James and Henry, 1993; Lawlor et al., 1999, and references therein). Despite these studies, and others such as those of Keppie and Ortega-Gutiérrez (1995), the question still remains whether the Grenville-age rocks of southern México are transferred terranes from South America, as suggested by Ruiz et al. (1988a,b) and Yañez et al. (1991) based on age and fossil data, or whether all Grenville-age rocks in México are part of one coherent terrane, named Oaxaquia by Ortega-Gutiérrez et al. (1995), as suggested by similar structural fabrics and metamorphic histories of outcropping Grenville-age rocks. We examine this question herein using Pb isotope compositions of whole rocks from Proterozoic rocks of the Garzón and Santa Marta Massifs in Colombia and of the Guichicovi and Oaxaca Complexes in southern México (Figs. 1–3) to constrain potential tectonic links between México's Grenville-age terranes, and those of Texas, the Appalachian Mountains in eastern North America, and Colombia.

REGIONAL SETTING

Colombian Grenville

The northern termination of the Andes of South America in Colombia branches into three north-northeastern–trending ranges separated by narrow valleys. The ranges, underlain by upthrust basement rocks, are known as the Santa Marta, Santander, and Garzón Massifs (Figs. 1 and 2). In general, the massifs consist of high-grade metamorphic rocks of Precambrian age, low-grade metamorphosed pelitic rocks of Early Paleozoic age, and marine sedimentary rocks of Cambrian and Ordovician, Devonian, and Permo-Carboniferous age. The presence of a Proterozoic basement in these ranges has long been recognized (MacDonald and Hurley, 1969; Goldsmith et al., 1971; Ward et al., 1973; Tschanz et al., 1969, 1974), and K/Ar and Rb/Sr ages of rocks from the Garzón Massif led Alvarez and Cordani (1980), Alvarez (1981, 1984), Kroonenberg (1982a), and Priem et al. (1989) to suggest that sparse basement outcrops in the Colombian Andes were part of a much larger Grenville-age metamorphic belt. U-Pb and Sm-Nd data reported by Restrepo-Pace et al. (1997) have confirmed the presence of the postulated Grenville-age belt in the Colombian Andes.

The Garzón Massif (Fig. 2), the most extensive and best studied block, is bounded on the east by the east-verging Borde Llanero fault system, which places Precambrian crystalline rocks over Tertiary clastic rocks of the Andean foreland. On the west, the massif is bounded by the west-verging Garzón-Suaza thrust fault, which places the basement rocks over Mesozoic and Cenozoic sedimentary rocks filling the upper Magdalena Valley. Triassic and Jurassic plutons intruded the basement complex along the western margin. The Garzón Massif metamorphic rocks extend west beneath the Magdalena Valley fill into the eastern margin of the central Andean range. Outcrops are limited to road cuts and river gorges, since the area is covered with lush vegetation.

Banded felsic and mafic granulites are the dominant rock in the Garzón Massif (Radelli, 1962; Kroonenberg, 1982a,b). In detail, these rocks are layered charnockites and pelitic gneisses alternating with mafic granulites, marbles, and calc-silicate layers and cross-cut by aplite dikes (Kroonenberg, 1982b). Amphibolites and orthopyroxene hornblendite and metaultramafic lenses are also present. Orthopyroxene and perthitic and/or anthiperthitic feldspars indicate granulite-grade metamorphic conditions. The absence of cordierite in pelitic rocks and the presence of olivine plus plagioclase in mafic rocks suggest intermediate pressures during metamorphism (Kroonenberg, 1982b). Retrograde hornblende rims orthopyroxene in some samples.

Subordinate rocks in the Garzón Massif are pegmatitic and augen orthogneisses known as the Guapotón and Mancagua gneisses. These gneisses are gray to pink colored, hornblende- and biotite-bearing granitic pegmatitic gneisses that are concordantly foliated with the hosting Garzón Group rocks. They are regarded as being syntectonic granitoids (Kroonenberg 1982a,b).

Unconformably overlying the metamorphic rocks is a Cambrian shelf sequence (Harrington and Kay, 1951; Rushton, 1962; Bridger, 1982). The Cambrian sequence is 1,500 m thick, consisting of limestone, dolomite, serpentinized diabase, and, toward the top, sandstone and shale beds; *Ehmania* (Harrington and Kay, 1951) and *Paradoxides* (Rushton, 1962) trilobites have been recovered from this sequence (Fig. 3).

The second block, the Santander Massif located north of the Garzón Massif (Figs. 1 and 2), has one of the most complete Late Proterozoic and Phanerozoic rock records of the northern Andes (Restrepo-Pace et al., 1997). The Santander Massif is bounded on the west by the left-lateral Bucaramanga–Santa Marta fault and on the east by the east-vergent Pamplona-Cubugón-

Mercedes thrust system. The Bucaramanga gneisses, the oldest rocks, are quartzofeldspathic gneisses with subordinate interlayered amphibolitic gneisses and diopside-tremolite-epidote–bearing calc-silicate rocks and the Silgará schists. The Bucaramanga gneiss commonly contains quartz-plagioclase-biotite-andalusite-cordierite ± sillimanite or quartz-plagioclase-biotite-sillimanite-K-feldspar ± garnet, indicating high metamorphic grade (Ward et al., 1973). The contact between the Bucaramanga gneisses and the low to medium metamorphic grade Silgará schists is enigmatic and currently appears to be defined by the biotite-sillimanite isograd. Silurian(?), Devonian, and Permo-Carboniferous marine sedimentary rocks, a thick Jurassic continental

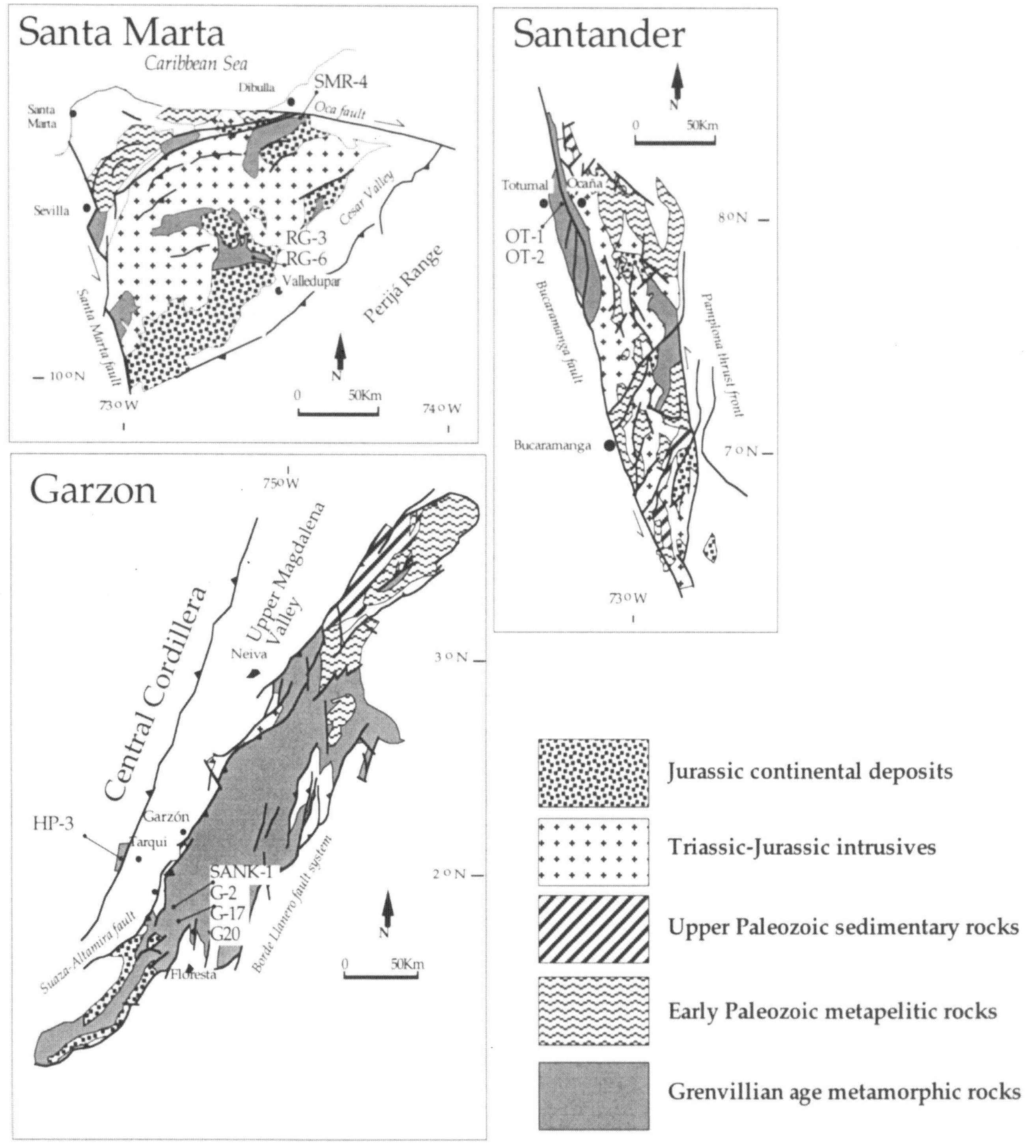

Figure 1. Regional location and geologic maps of Colombian Grenville-age massifs (from Restrepo-Pace et al., 1997).

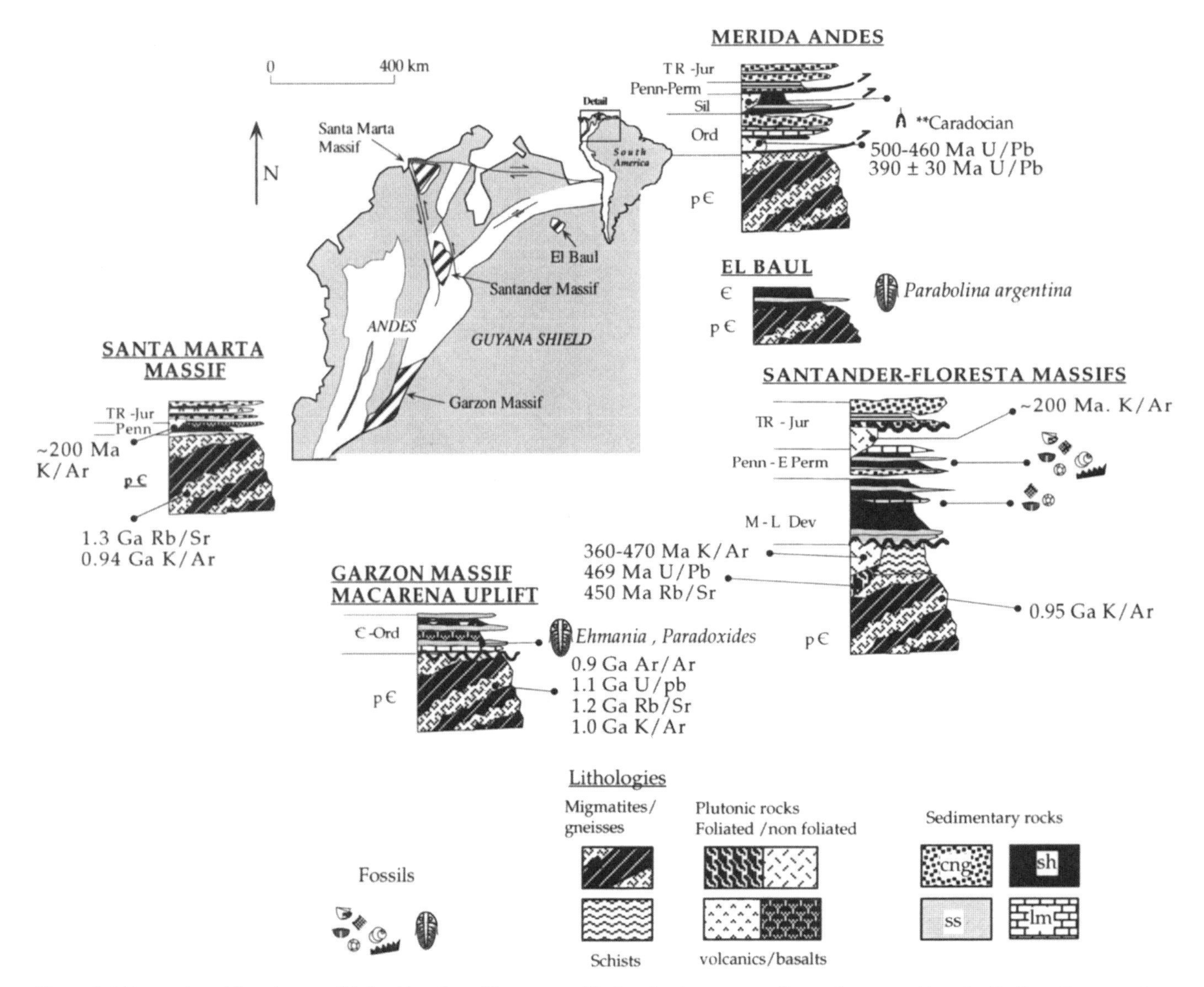

Figure 2. Lithostratigraphic columns of Colombian Grenville-age massifs showing important radiometric ages and key fossils (from Restrepo-Pace et al., 1997).

molasse, Cretaceous marine sedimentary rocks, and Tertiary continental sedimentary rocks unconformably overlie the metamorphic rocks. Thermal effects related to intrusion of Triassic and Jurassic calc-alkaline plutons into the basement complex have strongly perturbed many pre-Jurassic isotopic systems (Ward et al., 1973; Restrepo-Pace et al., 1997).

The third basement block in Colombia, the Santa Marta Massif (Figs. 1 and 2), is an isolated triangular-shaped uplifted block bounded on the north by the right-lateral Oca fault, on the west by the left-lateral Santa Marta–Bucaramanga fault and on the southeast by the Cesar Valley. Three northeast-southwest–trending tectonostratigraphic belts form the massif (Tschantz et al., 1969, 1974). The southeastern belt consists of Proterozoic granulite facies migmatites composed of interlayered pelitic and garnet-orthopyroxene-biotite quartzofeldspathic granulites, orthopyroxene-clinopyroxene metabasites, and hornblende-clinopyroxene mafic gneisses and anorthosites (Radelli, 1961; Tschantz et al., 1969).

Southern México Grenville

The complex pre-Mesozoic geology of México consists of a variety of fault-bounded terranes that were assembled in the Mesozoic (Fig. 3) (Ortega-Gutiérrez, 1991; Coney and Campa, 1983; Sedlock et al., 1993; Murillo-Muñetón, 1994). The oldest rocks in these terranes are high metamorphic grade Mesoproterozoic gneisses that form the Oaxaca (Ortega-Gutiérrez, 1981) and Guichicovi Complexes (Murillo-Muñetón, 1994) in southern México, the Novillo and Huiznopala gneisses in east-central and northern México (Patchett and Ruiz, 1987; Ruiz et al., 1988a,b; Lawlor et al., 1999), and the lower grade, but still Mesoproterozoic, rocks of Sierra del Cuervo and El Carrizalillo in northern

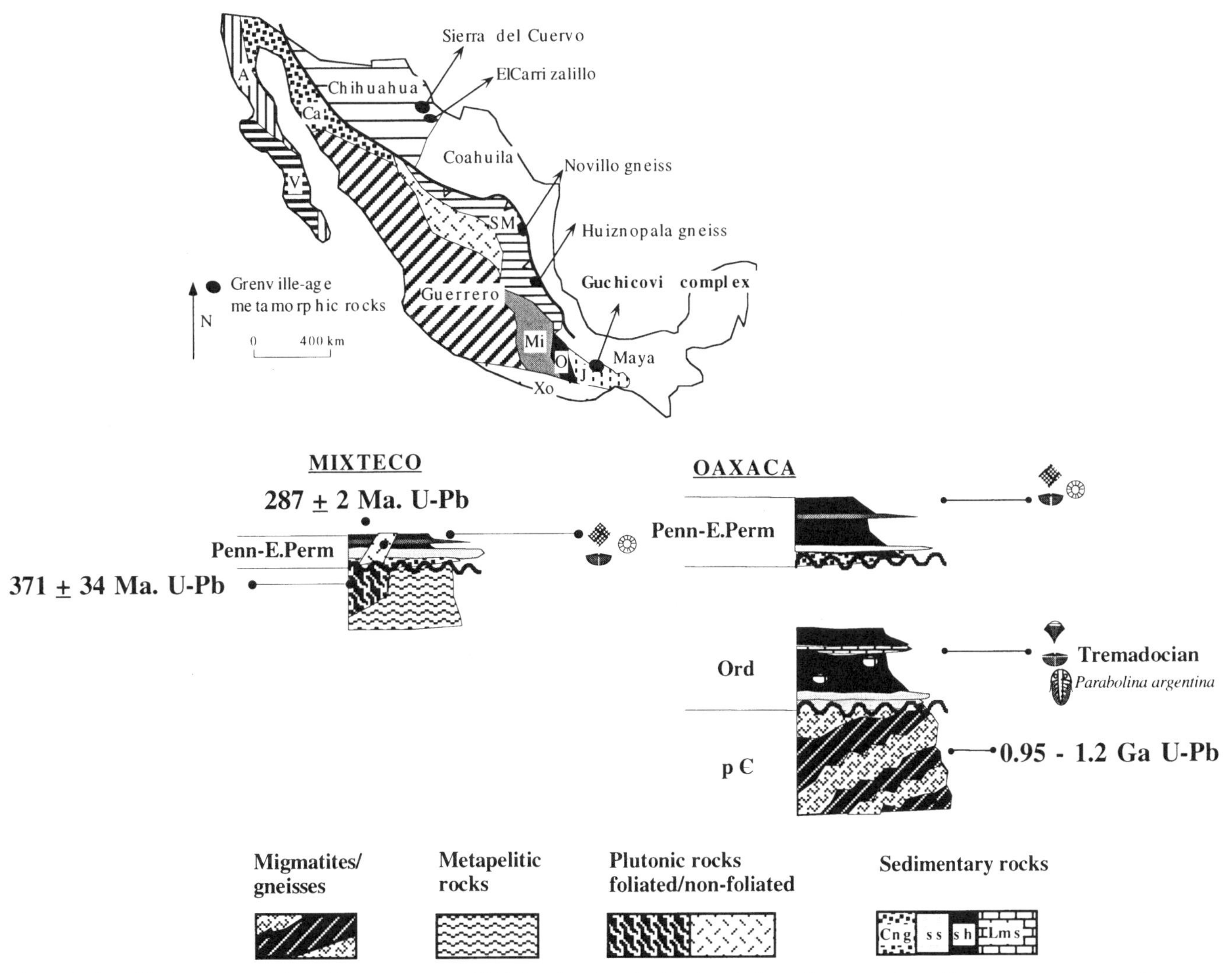

Figure 3. Simplified tectono-stratigraphic terrane map of México showing distribution of Grenville-age rocks and important ages and fossils typical of the region.

México. Many geologic features are shared by these high-grade metamorphic complexes. These include similar foliation trends, metamorphic grade and cooling histories, and lithologies. Based on these apparently common geologic features, Ortega-Gutiérrez et al. (1995) proposed that the outcropping metamorphic complexes may be all connected in the subsurface, thereby forming a large Grenville-age terrane, which they named Oaxaquia. Pb isotopic data (Keppie and Ortega-Gutiérrez, 1995; Lawlor et al., 1999) have been used to support this inference, although different studies reached different conclusions as to a Gondwanan (Keppie and Ortega-Gutiérrez, 1995) or Appalachian parentage (Lawlor et al., 1999) for some or all of these out-of-place rocks south of the Ouachita suture (James and Henry, 1993; Keppie et al., 1996). However, these studies lacked Pb isotopic data from Grenville-age rocks in southern México and Colombia, so their bearing on the question of the integrity of Oaxaquia and the origin of the southernmost México Grenville-age rocks is limited.

The Oaxaca and Guichicovi Complexes in southern México are thought to have originally formed a contiguous Grenville-age granulite facies metamorphic terrane, since disrupted by strike-slip faulting in the Mesozoic (Murillo-Muñetón, 1994). The strike-slip fault, with up to 300 km of dextral displacement, now marks the contact between the Oaxaca and Juarez terranes (Fig. 3). Tectonic models of México have suggested that Grenville-age metamorphic complexes of southern México are separated by a major suture from similar-age rocks in northern and east-central México (Yañez et al., 1991). Furthermore, the models suggested that the complexes may have been transferred from South America to their present position after a collision in the Paleozoic. The transfer model is based on the observation that the Oaxaca and Guichicovi Complexes are bounded by major faults, which place Paleozoic metasedimentary rocks to the west of the Proterozoic rocks. The westward position of Paleozoic sedimentary rocks relative to Proterozoic metamorphic rocks is reversed from what characterizes the Appalachian margin, leading to speculations that the Acatlan and Oaxacan Complexes were part of the South American margin during Paleozoic Wilson Cycle collisions (Yañez et al, 1991). Additional indication of the allochthonous character of these blocks is provided by the Tremadocian trilobites in Paleozoic rocks that unconformably overlie the Oaxacan granulitic basement (Robison and Pantoja-Alor, 1968). These fauna more closely resemble those present along Gondwana than those in equivalent-age rocks in Laurentia (Palmer, 1973). Conversely, paleomagnetic data presented by Ballard et al. (1989) are interpreted to indicate that the Oaxaca Complex was derived from a paleolatitude similar to that of the Adirondack-Ontario region, although this interpretation is not unique and the data could be consistent with a provenance along the northwestern Gondwanan margin (Keppie and Ortega-Gutiérrez, 1995).

The Guichicovi Complex in southern México (Fig. 3) consists dominantly of layered mafic and felsic granulites, amphibolites, and quartzofeldspathic gneisses with or without clinopyroxene, garnet, biotite, and graphite. Protoliths are interpreted to be a sequence of sedimentary rocks, principally of arkosic composition, intruded by premetamorphic intermediate and mafic intrusion (Murillo-Muñetón, 1994). The common amphibolite may have mafic lava or marl protoliths. Calc-silicate gneiss, marble, and quartzite are less common rock types. Similar rocks make up the Oaxaca Complex in which, in addition to the rocks already described, large exposures of anorthosite are also known. Both the Guichicovi and Oaxaca Complexes underwent granulite facies metamorphism at intermediate crustal depths (Mora and Valley, 1985; Murillo-Muñetón, 1994; Murillo-Muñetón et al., 1994). The U-Pb closure temperature in zircons during the granulite facies metamorphism occurred between 1,060 and 1,100 Ma in the Oaxaca Complex (Silver et al., 1994). Granulite facies metamorphism is interpreted to have continued to a slightly younger time in the Guichicovi Complex, as shown by U-Pb geochronology presented below.

Model Nd ages for the Mexican Grenville-age rocks are around 1.5 Ga (Ruiz et al, 1988a). These values are similar to those of most Grenville-age rocks in Texas and the eastern United States (Ruiz et al., 1988b) and similar also to the Grenville-age exposures of Colombia (Restrepo-Pace et al., 1997).

ANALYTIC METHODS

Samples from Colombia have been previously discussed by Restrepo-Pace et al. (1997), and all are well characterized by U-Pb, $^{40}Ar/^{39}Ar$, and Sm/Nd isotopic work. Fifteen of these samples (4 from the Santa Marta Massif and 11 from the Garzón Massif) along with 19 samples from the Guichicovi Complex described by Murillo-Muñetón (1994) were analyzed at the U.S. Geological Survey (USGS) in Menlo Park, California, for their Pb isotopic compositions. In addition, two samples from the Guichicovi Complex were dated using U-Pb methods, also at the USGS in Menlo Park.

Zircons were separated from granulitic gneisses using standard gravimetric and magnetic techniques, and then hand-picked and inspected to ensure purity. Zircon dissolution was modified from Krogh (1982) and Mattinson (1987), and mass spectrometry utilized a Finigan-Mat 262 multiple-collector mass spectrometer.

Whole-rock dissolution techniques, Pb isotope column chromatography, and error analysis are described by Wooden et al. (1992) and Arribas and Tosdal (1994). Measured Pb isotopic compositions were corrected for 0.125% fractionation per atomic mass unit, based upon replicate analyses of NBS 981 and 982. Laboratory procedural blanks were 1 ng Pb or less and the precision of the isotopic measurements was generally better than 0.05% at the 2σ confidence level. Lead isotope compositions are reproducible to the ±0.08, ±0.1, and ±0.14% (2σ), respectively, for $^{206}Pb/^{204}Pb$, $^{207}Pb/^{204}Pb$, and $^{208}Pb/^{204}Pb$.

U-Pb GEOCHRONOLOGY OF GUICHICOVI COMPLEX

The two samples of the Guichicovi Complex dated by U-Pb methods are a garnet-hornblende granitic gneiss (CHOX-58) and a garnet-clinopyroxene-hornblende quartzofeldspathic gneiss (MIXTE-22) (Murillo-Muñetón, 1994). Protoliths for both rocks are thought be feldspathic sedimentary rocks, most likely of arkosic composition. Both contained abundant zircon of uniform crystal habit typified by a purplish amber color. There was no evidence for a mixed population of zircons. These rock were dated because their high metamorphic grade and abundant zircon of metamorphic origin would constrain the timing of granulite metamorphism.

Eight fractions of zircons from the two samples were analyzed (Table 1). Most of the U-Pb ages overlap concordia within the limits of their analytic precision (Fig. 4). Two concordant zircon fractions from CHOX-58 (fractions 58a and 58b in Fig. 4 inset) indicate an age of 986 ± 4 Ma for granulite facies metamorphism. The third fraction from this sample is slightly discordant and younger. These concordant U-Pb ages indicates that no older inherited zircons were present in the metasedimentary gneiss. This observation is remarkable in view of the inferred arkosic protolith and suggests either that detrital zircons were not present to any great amounts (an unlikely scenario) or that granulite facies metamorphism recrystallized any older zircons—the more likely interpretation. Inferred peak metamorphic temperatures as high as 755 ± 25°C, depending on the mineral thermometer, were attained in the Guichicovi Complex (Murillo-Muñetón, 1994; Murillo-Muñetón and Anderson, 1994), and these temperatures are sufficient to facilitate zircon recrystallization and diffusion of radiogenic Pb out of zircons during protracted high-grade metamorphism (Chiarenzelli and McLelland, 1993).

In contrast to CHOX-58, inaccurate measuring statistics and the resulting poor analytic data hinder interpretation of the U-Pb data for MIXTE-22. Nevertheless, U-Pb ages of most zircon fractions overlap concordia within their uncertainties at ages (977–984 Ma) that are slightly younger than the 986 ± 4-Ma age interpreted for CHOX-58. The oldest fraction overlaps within analytic uncertainty (fraction 22b in Fig. 4 inset) the 986 ± 4-Ma interpreted age of CHOX-52. U-Pb ages for the youngest zircon

TABLE 1: U-PB GEOCHRONOLOGIC DATA FOR ZIRCONS FROM GRANULITE-FACIES GNEISSIC ROCKS IN THE GUICHICOVI COMPLEX OF THE LA MIXTEQUITA MASSIF, SOUTHERN MÉXICO

Fraction†	Weight (mg)	^{206}Pb* (ppm)	^{238}U (ppm)	Observed ratios†† 206Pb/204Pb	207Pb/206Pb	208Pb/206Pb	Atomic ratios§# 206Pb*/238U	207Pb*/235U	207Pb*/206Pb*	Ages, Ma 206Pb*/238U	207Pb*/235U	207Pb*/206Pb*
Garnet-hornblende granitic gneiss: *CHOX-58; 16°57; 95°14.6'*												
M>163	2.4	30.0	206.9	3086	0.07659	0.0858	0.16553(0.1)	1.6449(0.7)	0.07207(0.6)	987.4	987.6	988±13
N>163	2.0	29.3	203.7	2232	0.07830	0.0912	0.16523(0.2)	1.6407(0.4)	0.07202(0.3)	985.8	986.0	986±7
N<80	2.5	28.0	196.5	5181	0.07474	0.0872	0.16410(0.1)	1.6309(0.3)	0.07208(0.3)	979.5	982.2	988±6
Garnet-pyroxene quartzofeldspathic gneiss: *MIXTE-22; 17°03.9'; 95°10.1'*												
N>163	4.4	8.9	62.5	6369	0.07397	0.0996	0.16484(0.1)	1.6325(0.7)	0.07183(0.7)	983.6	982.8	981±13
N<80	3.4	10.4	73.6	7042	0.07408	0.1054	0.16387(0.2)	1.6301(1.0)	0.07215(1.0)	978.2	981.9	990±20
M>163	5.3	10.0	70.0	4184	0.07554	0.1056	0.16459(0.2)	1.6392(1.7)	0.07223(1.5)	982.2	985.4	992±31
M<80	2.0	10.3	72.8	3546	0.07583	0.1091	0.16361(0.2)	1.6223(1.4)	0.07192(1.3)	976.8	978.9	984±26
M<80	4.8	6.3	57.5	5333	0.07473	0.1037	0.12607(0.1)	1.2542(0.4)	0.07215(0.4)	765.4	825.3	990±6

* Denotes radiogenic Pb. Sample dissolution and ion exchange chemistry modified from Krogh (1973) and Mattinson (1987).
† N-Nonmagnetic and M-Magnetic at 1.8 ° and 0.5° side slope on a Franz Isodynamic separator. Sizes are in microns.
†† Observed ratios collected on Farrady cups on Finigan-Mat MAT 262 multiple collector mass spectrometer at the U.S. Geological Survey in Menlo Park. Uncertainties in the $^{208}Pb/^{206}Pb$ and $^{207}Pb/^{206}Pb$ rations are less than 0.1% and the uncertainty in the $^{206}Pb/^{204}Pb$ ratios are less than 20%.
§ Observed ratios were corrected for 0.125% per unit mass fractionation based on replicate analyses of NBS 981 and 983, for laboratory blank that has averaged <0.2 ng Pb, and for common Pb based upon a 1000 Ma age using average crustal growth curve of Stacey and Kramers (1975).
\# Atomic ratios calculated using the following constants: $^{238}U/^{235}U$=137.88; ^{235}U=0.98485 X $10^{-9}yr^{-1}$; ^{238}U=0.155125 X $10^{-9}yr^{-1}$. Uncertainties (2σ) in percent (%) are shown in parentheses.

fraction (fraction 22a in Fig. 4) are considerably younger, and lie on a discordia whose lower intercept (–27 ± 54 Ma) suggests modern Pb loss or continuous diffusive Pb loss. Based on the distribution of the U-Pb ages, we consider Pb loss to be the likely explanation for all scatter along concordia of the U-Pb ages.

As zircons from MIXTE-22 are interpreted to have variable amounts of Pb loss, it is also possible that the 986 ± 4-Ma age for granulite facies metamorphism derived from CHOX-58 is also a minimum age. This possibility is rejected because of the agreement of $^{207}Pb^*/^{206}Pb^*$ ages (between 981 and 990 Ma) for the various zircon fractions from both samples and the concordant U-Pb age (Table 1). Granulite facies metamorphism in the Guichicovi Complex was, therefore, occurring at 986 ± 4 Ma. The data do not allow us to constrain when metamorphism began nor how long it lasted.

Pb ISOTOPE GEOCHEMISTRY OF THE GRENVILLE BELT

Present-day Pb isotopic compositions for Grenville-age whole rocks from Colombia are similar to those determined in rocks from the Guichicovi Complex in southern México (Table 2; Fig. 5). The range of Pb isotopic composition for the Oaxaca Complex reported by Martiny et al. (1997) forms a small field that lies at slightly lower $^{206}Pb/^{204}Pb$ (17.1–17.25) than the Pb isotopic range for the Guichicovi Complex (Fig. 5). Together, the Pb isotopic data indicate that the southern México and Colombian Grenville-age exposed rocks shared a common U/Pb evolution, and likely, a common tectonic history, although much detailed chronologic and geologic work is required to confirm this linkage.

As a whole, the Pb isotopic compositions of southern México and Colombia metamorphic complexes are characterized by a large range in $^{206}Pb/^{204}Pb$ (17.1–25.6) and a corresponding range in $^{207}Pb/^{204}Pb$ (15.5–15.9) (Fig. 5B). Within the Colombian samples, the Garzón Massif shows a large range in $^{206}Pb/^{204}Pb$ and $^{207}Pb/^{204}Pb$, whereas the Santa Marta Massif is characterized by a limited range of $^{206}Pb/^{204}Pb$ (17.1–17.9) and $^{207}Pb/^{204}Pb$ (15.47–15.55) (Table 2). The Guichicovi Complex from México likewise displays the entire range of $^{206}Pb/^{204}Pb$ (17.1–25.6) and $^{207}Pb/^{204}Pb$, with two graphite-bearing quartzose rocks (ISTEH-33 and ISTEH-83) having extremely high $^{206}Pb/^{204}Pb$ (39.8 and 48.0) and $^{207}Pb/^{204}Pb$ (17.44 and 17.76) but correspondingly low $^{208}Pb/^{204}Pb$ (40.7 and 45.9) (Table 2). Their Pb isotopic compositions are probably controlled by trace minerals within the metasedimentary rocks that most likely retained low Th/U and high U/Pb values of their protoliths. The extreme U/Pb values may reflect a relative enrichment of U and consequent low Th/U values of sedimentary rocks rich in carbonaceous material (Kesler et al., 1994).

The $^{206}Pb/^{204}Pb$ and $^{207}Pb/^{204}Pb$ isotopic compositions from Colombia and southern México metamorphic complexes are bounded by two subparallel arrays (Fig. 5B). One array, lying above the average crustal growth curve of Stacey and

Kramers (1975), is characterized by a large range in Pb isotopic compositions ($^{206}Pb/^{204}Pb > 18.5$) and consequently a range of U/Pb values. The second array lies below the average crustal growth curve and has only a narrow range of $^{206}Pb/^{204}Pb$ (between 17.1 and 18.0) and lower $^{207}Pb/^{204}Pb$ values (Fig. 5B). Whole-rock Pb isotopic data from metaigneous rocks in the Oaxaca Complex have Pb isotopic ranges that lie at the low $^{206}Pb/^{204}Pb$ (17.14–17.25) end of this lower array (Fig. 5B). Several samples have Pb isotopic compositions between the two arrays. Neither array defines a statistically meaningful isochron, although the Pb isotopic values do scatter about 1,250 ± 50 Ma reference isochrons (Fig. 5). There must have been a degree of Pb isotopic heterogeneity in these rocks in the Mesoproterozoic, an observation that is expected in view of the mixture of igneous and immature sedimentary protoliths of the analyzed samples. Conversely, it is also possible that the Pb isotopic compositions were not completely homogenized during high-grade metamorphism in the Mesoproterozoic.

In contrast to the coherence of the $^{206}Pb/^{204}Pb$ and $^{207}Pb/^{204}Pb$ of the Colombian massifs and southern México metamorphic rocks, $^{208}Pb/^{204}Pb$ shows differences between and within regions (Fig. 5A). In particular, there is a much wider range of $^{208}Pb/^{204}Pb$ at a given $^{206}Pb/^{204}Pb$ for the Colombian massifs than for the Guichicovi and Oaxaca Complexes. The Guichicovi Complex data plot as a field that extends parallel to and toward the right of the average crustal growth curve suggests that Th/U in these rocks is ≤4, the approximate crustal value. The Garzón Massif follows a similar pattern, with the exception of one sample (G-3 in Table 2) which has elevated $^{208}Pb/^{204}Pb$ (41.15) at a low $^{206}Pb/^{204}Pb$ value (18.3). A time-integrated Th/U > 4 must characterize this rock. In contrast, the Santa Marta Massif has a steep array formed by a large range of $^{208}Pb/^{204}Pb$ (36.65–39.65) over a limited range of $^{206}Pb/^{204}Pb$ (17.1–17.9) that extends above the average crustal growth curve, suggestive of Th/U ≥ 4. Pb isotopic data for the Oaxaca Complex plots in the low $^{208}Pb/^{204}Pb$ part of the field for the Santa Marta Massif.

LINKAGES BETWEEN MEXICAN, COLOMBIAN, AND APPALACHIAN GRENVILLE TERRANES

Whether the Mexican Grenville-age rocks form one microplate, Oaxaquia, and the time at which this microplate was assembled, are central questions for Late Proterozoic and early Paleozoic reconstructions of the Americas (Ortega-Gutiérrez et al., 1995). Radiogenic isotopes provide one means for constraining these models, with the Pb and Nd isotopic systems being the most commonly used. For México, the common Nd model age for Grenville-age rocks of around 1.5 Ga effectively precludes this isotopic system from use as a tracer of displaced blocks of Grenville-age crust. Fortunately, the Pb isotopic variations presented herein seem to distinguish similarities and difference between the granulite exposures.

México has Grenville-age rocks at Sierra del Cuervo, Carrizalillo, Novillo, and Huiznopala in northern and east-central México

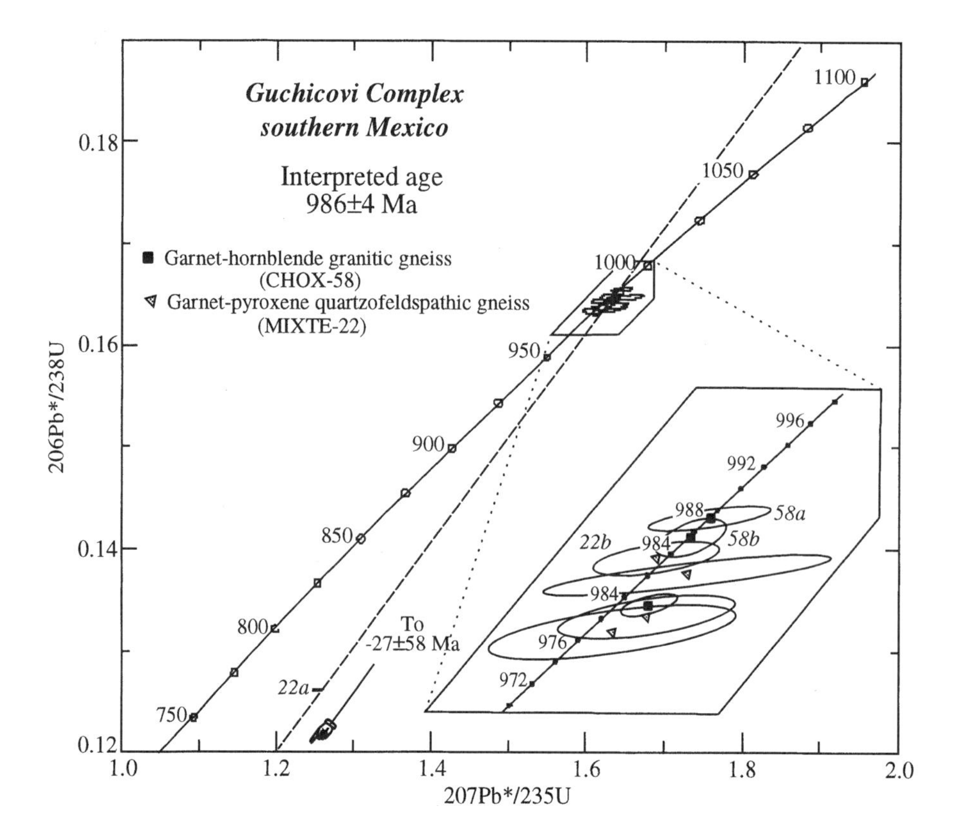

Figure 4. U-Pb concordia diagrams for granulitic rocks from the Guichicovi Complex, southern México.

TABLE 2: PRESENT-DAY WHOLE ROCK PB ISOTOPIC COMPOSITIONS FOR PROTEROZOIC CRYSTALLINE ROCKS IN THE COLOMBIAN ANDES. AND FROM THE GUCHICOVI COMPLEX, OAXACA, MEXICO[1].

Sample	Rock type	$^{206}Pb/^{204}Pb$	$^{207}Pb/^{204}Pb$	$^{208}Pb/^{204}Pb$
Columbia				
Santa Marta massif				
RG-1	Granulite	17.419	15.502	36.651
RG-3	Biotite granulite	17.645	15.529	38.687
RG-6	Biotite granulite	17.905	15.550	39.658
RG-8	Granulite	17.107	15.471	36.885
Garzón massif				
GSnAn-4	Quartz-kspar-bt gneiss	20.827	15.870	39.864
SANKR-1	Hornblende augen granite	17.904	15.568	36.868
HP-3	Hornblende quartz-kspar gneiss	24.403	16.156	40.530
HP-5	Hornblende quartz-kspar gneiss	19.167	15.714	36.866
G-1	Amphibolitic gneiss	18.494	15.664	37.103
G-2	Amphibolitic gneiss	18.296	15.633	41.146
G-7	Dioritic orthogneiss	18.405	15.643	37.800
G-9	Dioritic orthogneiss	18.390	15.634	37.336
G-11	Granitic gneiss	17.917	15.576	37.746
G-11a	Granulite	17.894	15.575	37.752
G-12	Granulite	18.567	15.665	37.855
Guchicovi gneiss				
MIXTE-9	Granite gneiss	22.713	15.976	42.174
MIXTE-10	Pyroxene-hornblende-feldspar gneiss	21.123	15.859	40.150
MIXTE-14	Granitic gneiss	20.013	15.827	38.471
MIXTE-15	Hornblende-two pyroxene granulite	17.671	15.571	36.818
MIXTE-19	Clinopyroxene-plagioclase gneiss	24.830	16.174	41.597
MIXTE-20	Hornblende granitic gneiss	22.546	15.973	40.052
MIXTE-22	Garnet-px-hnbd-quartz-feld. gneiss	17.527	15.528	36.776
MIXTE-23	Retrograde hornblende-biotite gneiss	19.709	15.755	38.513
MIXTE-26	Hornblende-two pyroxene granulite	18.412	15.617	37.314
MIXTE-28	Hornblende-quartz-feldspar gneiss	17.710	15.533	37.319
MIXTE-46	Retro. diorite granulitic orthogneiss	18.493	15.631	38.201
MIXTE-50	Garnet granitic gneiss	19.856	15.761	40.040
MIXTE-60	Clinopyroxene granitic gneiss	25.595	16.307	45.119
MIXTE-62	Garnet-hnbd-two pyroxene granulite	19.141	15.706	38.032
MIXTE-63	Altered garnet mafic granulite	18.622	15.594	37.882
MIXTE-65	Granitic granulite (charnockite)	17.595	15.528	36.945
CHOX-58	Garnet-hornblende granitic gneiss	17.607	15.535	37.049
ISTEH-33	Graphite-2 pyroxene felsic granulite	39.794	17.440	45.698
ISTEH-83	Graphite-amph. quartzite	48.006	17.758	40.936

SYMBOLS: alter = weakly hydrothermally altered; bt=biotite; feld = feldspar; hnbd=hornblende; kspar-K-feldspar; px-pyroxene; retro - retrograde metamorphosed.

1. See Restrepo-Pace (1997) and Murillo-Muñetón (1994) for more detailed sample description and locations.

(Fig. 3) (Mauger et al, 1983; Patchett and Ruiz, 1987; Ruiz et al., 1988a; Lawlor et al., 1999), and in the Oaxaca and Guichicovi Complexes in southern México (Fig. 3) (Patchett and Ruiz, 1987; Ruiz et al., 1988; Murillo-Muñetón et al., 1994). The metamorphic grade, cooling age, and Nd model ages for all these rocks are very similar, excluding the metamorphic grade at El Carrizalillo and Sierra del Cuervo, which are not granulite facies metamorphic rocks but rather are characterized by lower greenschist and amphibolite-facies mineral assemblages (Mauger et al., 1983). Nevertheless, the overall similarities of Grenville-age rocks were used to suggest that these blocks were only the outcropping parts of a megaterrane that may be continuous in the subsurface (Ortega-Gutiérrez et al., 1995). Using Pb isotopes, Keppie and Ortega-Gutiérrez (1995) concluded that Oaxaquia was a coherent crustal terrane. Their data, however, do show variations in the U/Pb history for Grenville-age rocks that are suggestive of multiple base-

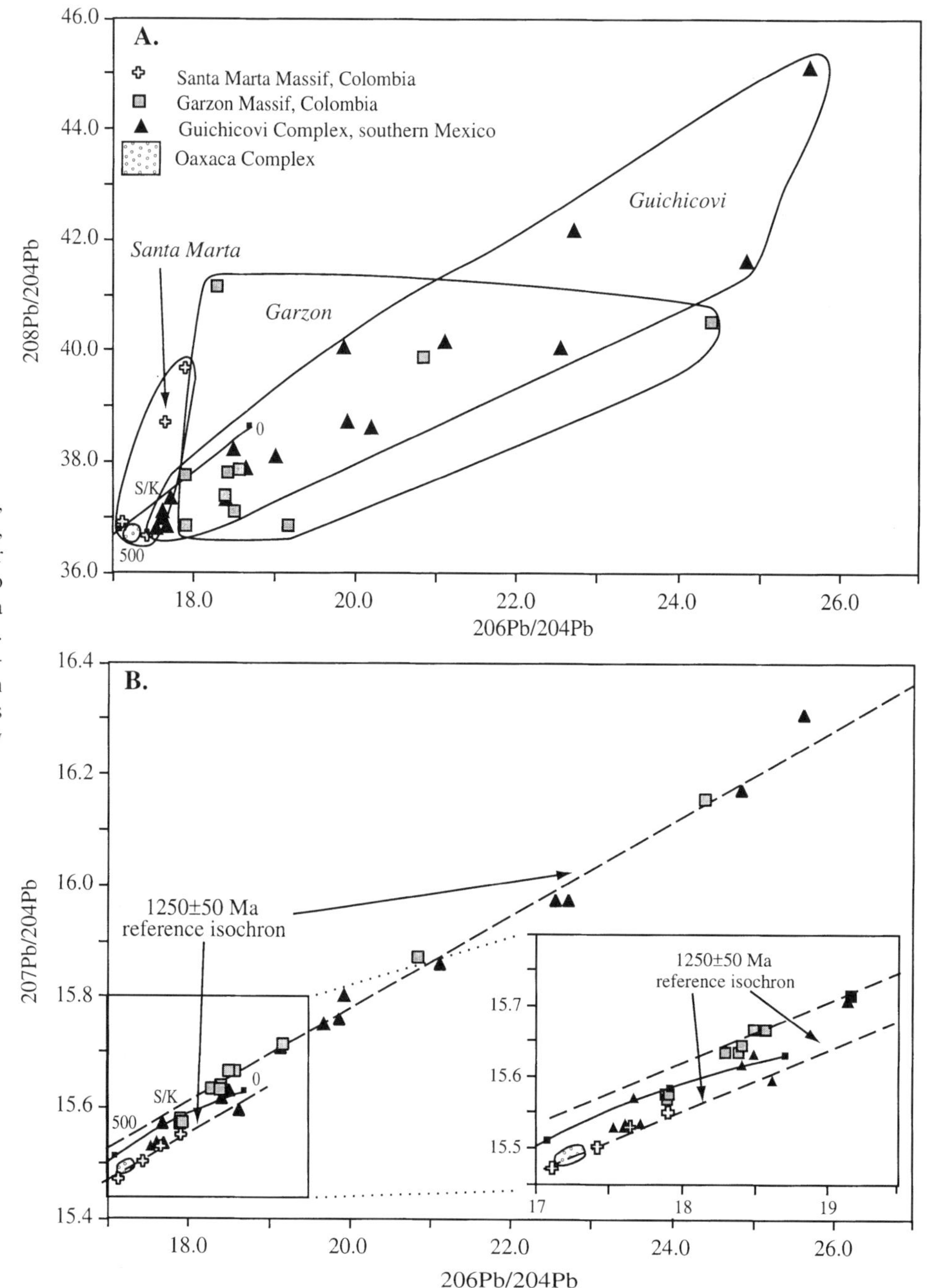

Figure 5. Present-day whole rock. A, $^{208}Pb/^{204}Pb$ vs. $^{206}Pb/^{204}Pb$ diagram and B, $^{207}Pb/^{204}Pb$ vs. $^{206}Pb/^{204}Pb$ diagram for Grenville-age rocks from southern México and Colombia. Field data for the Oaxaca Complex are from Martiny et al. (1997). Uncertainty in the Pb isotopic compositions are <0.1% (2σ) and are smaller than the plotted symbols. Growth curve (S/K) is the average crustal growth curve of Stacey and Kramers (1975).

ment blocks of potentially different origins. Furthermore, their conclusions were limited by a lack of Pb isotopic data from Grenville-age rocks in southern México and Colombia.

Pb isotopic compositions for Colombian massifs and the southern México complexes and the Guichicovi and Oaxaca Complexes are indistinguishable (Fig. 6), an isotopic coherency that is consistent with the regions being related geologically as proposed by Yañez et al. (1991) and Restrepo-Pace et al. (1997). Within the Colombian–southern México Pb isotopic data, it is interesting to note that Pb isotopic compositions of the northernmost Colombia massif, the Santa Marta Massif, are similar to those of the northernmost southern México complex, the Oaxaca Complex, whereas those of the Garzón Massif, the southernmost Colombian massif, are more similar to the Guichicovi Complex, the southernmost southern México complex. Are these Pb isotopic and geographic similarities simply coincidental? Or do they suggest the possibility that within the context of the Colombian–southern México connection advocated herein (see also Keppie and Ortega-Gutiérrez, 1995), the Santa Marta Massif–Oaxaca Complex and the Garzón Massif–Guichicovi Complex were related spatially and Pb isotopically prior to being rifted apart? Although possible within the available data, considerable more data are required to confirm or disprove this hypothesis.

Grenville-age rocks of northern and east-central México, south of the Ouachita suture, such as the Huiznopala gneiss and the source of cobbles in the Paleozoic Las Uvas conglomerate in the Coahuila terrane, also have similar Pb isotopic compositions (Fig. 6) (Lopez, 1997; Lopez et al., 1997; Lawlor et al., 1999) that bear no relation to the Pb isotopic compositions of Precambrian rocks north of the Ouachita suture in Texas (James and Henry, 1993). For east-central México rocks, Lawlor et al. (1999) conclude that their Pb isotopic histories permit, but do not require, linkages with Grenville-age terranes of eastern Laurentia, specifically in the southern Appalachian and the Adirondack Mountains in the northern Appalachians (Fig. 6). Looking south, east-central Mexican Grenville-age rocks have Pb isotopic compositions that slightly overlap Pb isotopic fields for rocks of similar age in southern México and Colombia (Fig. 6). Pb isotopic compositions for nonacid-washed samples from the Huiznopala gneiss would lie at higher $^{207}Pb/^{204}Pb$ for a given $^{207}Pb/^{204}Pb$ than shown in Figure 6, and define a field more similar to the fields for southern México and Colombia (K. L. Cameron, writ-

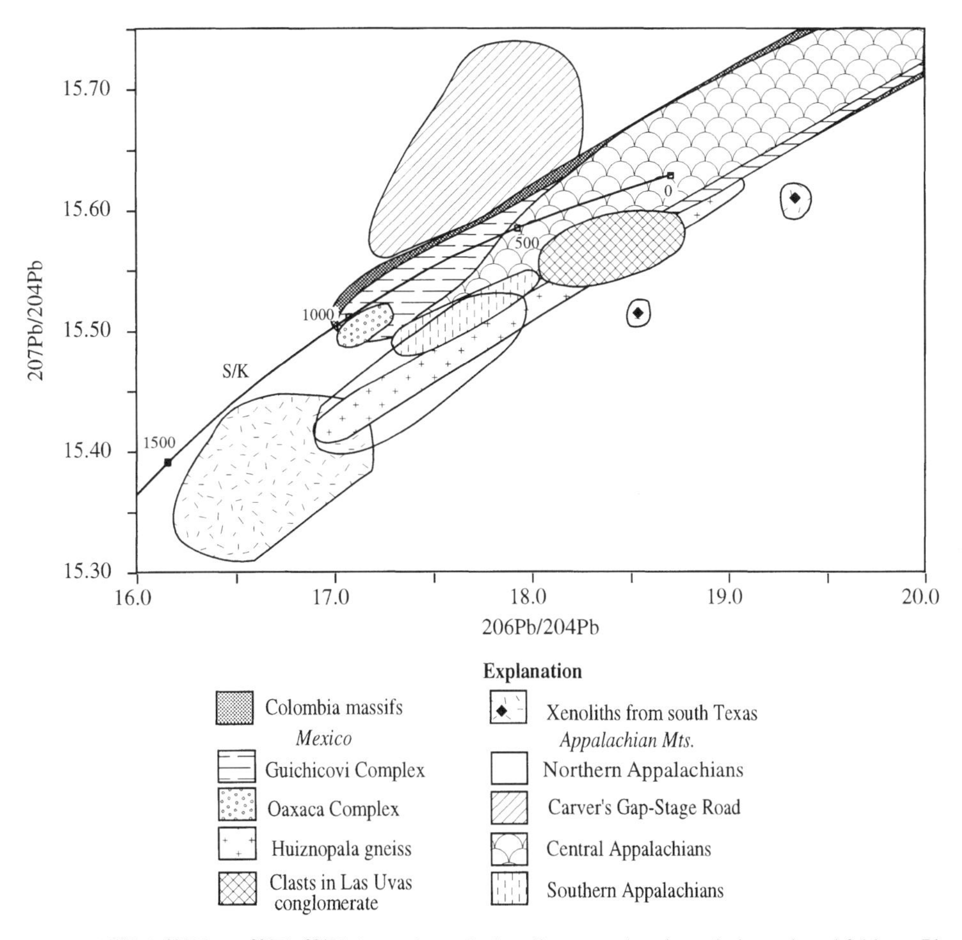

Figure 6. $^{207}Pb/^{204}Pb$ vs. $^{206}Pb/^{204}Pb$ isotopic evolution diagrams showing whole-rock and feldspar Pb isotopic compositions for Grenville-age rocks in México, Colombia, and the eastern United States. Growth curve (S/K) is the average crustal growth curve of Stacey and Kramers (1975). Sources of data: Colombia and southern México (this study; Martiny et al., 1997); Adirondack Mountains in the northern Appalachians (DeWolf and Metzger, 1994); central and southern Appalachians (see below for specific basement terranes) and Carver's Gap Gneiss–Stage Road Layered Gneiss in the eastern United States (Sinha et al., 1996); cobbles in the Las Uvas conglomerate (Lopez, 1997; Lopez et al., 1997); Huiznopala gneiss (Lawlor et al., 1999); and Texas and northern México (Zartman and Wasserburg, 1969; Cameron et al., 1992; Ward and Cameron, 1996; K. L. Cameron, written communication, 1996; see also James and Henry, 1993). Grenville-age basement terranes of Sinha et al. (1996) included within the central Appalachians are the Baltimore Gneiss, State Farm Gneiss, part of the Blue Ridge, and rocks in the Pine Mountain Complex in the southern Appalachians. Grenville-age basement terranes of Sinha et al. (996) included within the southern Appalachians are the Sauratown Mountains, Corbin Gneiss, Tallulah Falls, and from the Honeybrook Uplands in the central Appalachians.

ten communication, 1997). However, it is important to recognize that the Huiznopala gneiss has undergone low-grade retrograde metamorphism and possesses a penetrative cataclastic fabric. Because of the low temperature history of these rocks, Lawlor et al. (1999) conclude that nonacid-washed whole-rock samples are a mix of Pb isotopic signatures, and that the Pb isotopic compositions of acid-washed whole-rock samples better reflect their primary Pb isotopic compositions. As such, the Pb isotopic field of the Huiznopala gneiss shown for comparison in Figure 6 consists of Pb isotopic composition for acid-washed whole-rock samples. Similar degrees of retrograde recrystallization and low-temperature fabric development are not present in the granulite facies rocks in southern México and Colombia, and this effect on their Pb isotopic compositions can be ignored. With this caveat in mind, the east-central México rocks are characterized by a lower $^{207}Pb/^{204}Pb$ at a given $^{206}Pb/^{204}Pb$ than are the southern México–Colombian rocks (Fig. 6). As such, these two regions possess slight differences in their respective U/Pb histories. This difference leads to the conclusion that the east-central México complexes are not obviously related to the southern México complexes. If true, then a suture must lie between them.

On a broader context, Pb isotopic compositions of Grenville-age crystalline rocks in the southern and central Appalachians of the eastern United States also vary by geographic location (Sinha et al., 1996). In detail, these basement terranes can be subdivided into three Pb isotopically and geographically distinct terranes. The westernmost consists of the Carver's Gap Gneiss and the Stage Road Layered Gneiss within the Blue Ridge. Pb isotopic compositions of these rocks are characterized by elevated $^{207}Pb/^{204}Pb$ lying above the average crustal growth curve (Fig. 6). Sinha et al. (1996) concluded that they are fragments of a terrane, or terranes, accreted to the cratonic core of North America during the Grenville Orogeny, an interpretation that is supported by the similarity of their Pb isotopic compositions with those typical of Proterozoic rocks along the present west coast of Gondwana (Tosdal, 1996) and also forming the basement to the allochthonous Avalonia Terrane (Ayuso and Bevier, 1991), also derived from northern South America (Nance and Murphy, 1994; 1996; Keppie et al., 1996). As all other Grenville-age rocks in the southern and central Appalachians lie east of the Carver's Gap Gneiss–Stage Road Layered Gneiss, they are also presumably allochthonous to the cratonic core of eastern North America. One group of central Appalachian complexes (Baltimore Gneiss, State Farm Gneiss, part of the Blue Ridge, and rocks in the Pine

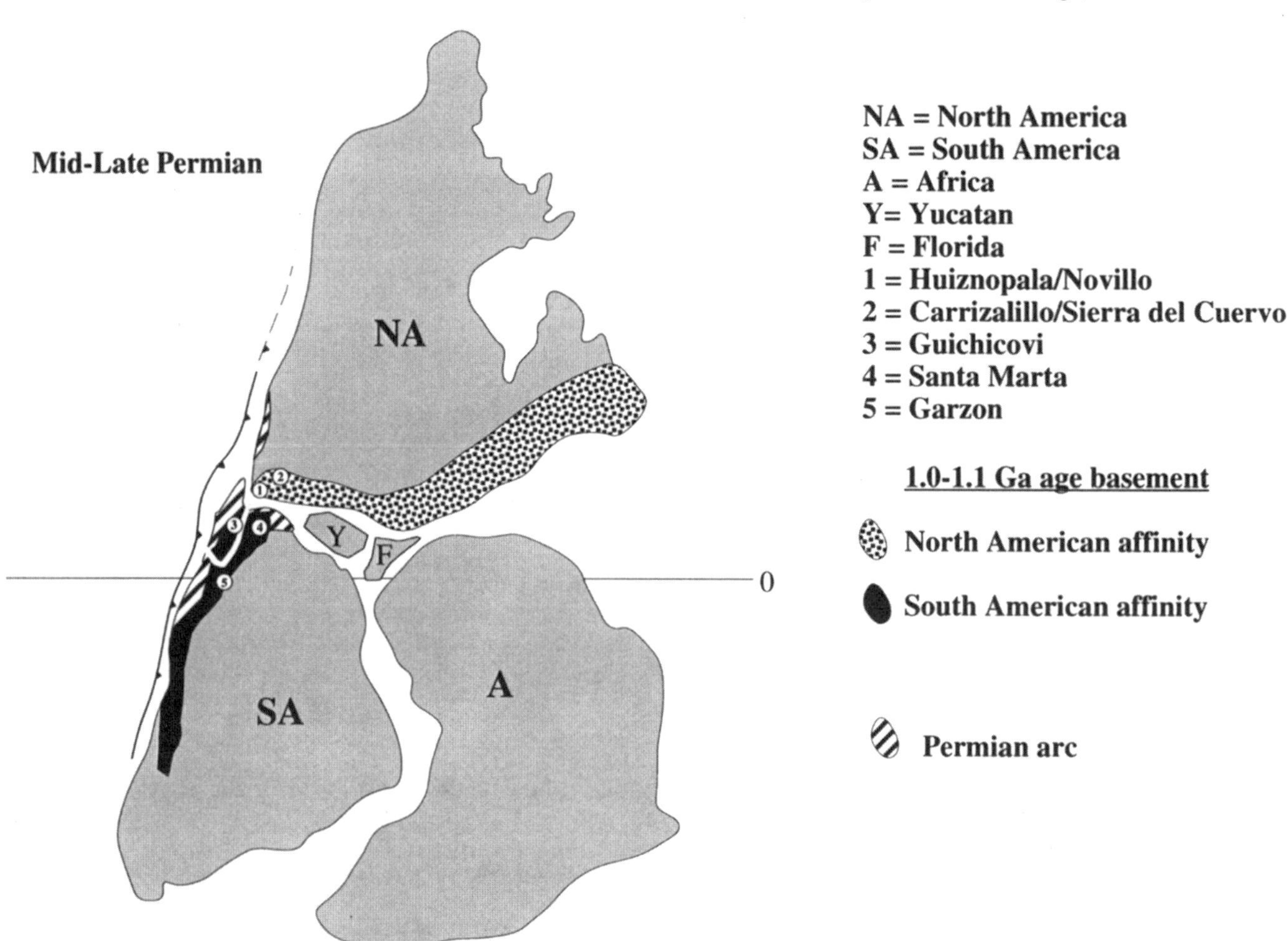

Figure 7. Reconstruction of Gondwana and Laurentia during mid-Late Permian showing the location of Grenville-age rocks now found in México.

Mountain Complex in the southern Appalachians) have Pb isotopic compositions (Sinha et al., 1996) similar to those determined for southern México and Colombia (Fig. 6). Of particular note is the extremely large range of $^{206}Pb/^{204}Pb$, which could simply reflect a predominance of sedimentary protoliths with a wide range of U/Pb values. Such an extended range of $^{206}Pb/^{204}Pb$ is a distinctive feature that is different from other Grenville-age complexes in the Appalachians and also east-central México. The similarity in Pb isotopic characteristics of central Appalachian basement terranes implies common U/Pb histories that could permit them to have been related at some time. In a similar fashion, other southern Appalachian Grenville-age complexes have Pb isotopic compositions similar to those of east-central México and to the Adirondack Mountains in the northern Appalachian (see discussion in Lawlor et al., 1999). Although suggestive, the different Pb isotopic correlations do not prove genetic linkages between the central Appalachian terranes and southern México and Colombia. However, the complicated plate tectonic history between break-up of Rodinia and amalgamation of Pangea at the end of the Paleozoic do provide ample opportunity to accrete blocks of either continental fragment (Laurentia or Gondwana-Amazonia) to the other as Laurentia moved past Gondwana (Dalziel, 1991, 1997; Bahlburg, 1993; Keppie et al., 1996). We emphasize that knowing the precise paleogeographic reconstructions of these blocks during Grenville times is obviously complex and requires much additional geologic, isotopic, and geochronologic data from all Grenville-age basement rocks of eastern North America, México, and Colombia.

Figure 7 shows a reconstruction that attempts to place the suture between Gondwana and Laurentian Grenville-age crust in México. The reconstruction is constrained by the piercing point in southern México and Colombia afforded by the Permo-Triassic arc present in large parts of Cordillera in North and South America. The reconstruction shows all blocks in the region that are Permian or older, and excludes segments of the younger crust that makes up large parts of western Colombia and México. The reconstruction also shows the location of various Grenville-age blocks of México. Noteworthy in this model is that Grenville-age rocks of southern México were part of Gondwana and must have been left behind after the proto-Atlantic opened.

CONCLUSIONS

Pb isotopic data argue that southern México Grenville-age rocks shared a similar U/Pb evolution with Colombian Grenville-age massifs, and possibly to parts of the central Appalachians. This U/Pb history was different from northern and east-central Mexican Grenville-age terranes, and to other parts of the Appalachians. This conclusion has several important tectonic implications. (1) The Oaxaquia terrane has not been a coherent megaterrane since the Mesoproterozoic, but rather it is a composite terrane amalgamated at various times in the late Proterozoic(?) and Paleozoic. (2) The Grenville-age rocks of southern México were likely transferred from their original position along the margin of Gondwana from somewhere near the present location of Colombia during collisions as part of the Wilson Cycle or perhaps during earlier events. (3) There must be an suture separating Gondwanan rocks in southern México from the rest of México. This suture is located somewhere north of the Guichicovi-Oaxaca Complexes, and might preserve a record of collisional tectonism along the margin of Gondwana during latest Proterozoic or earlier Paleozoic collisions as Laurentia moved by Gondwana. (4) The similarities of the U/Pb history of several central and southern Appalachian Grenville-age basement terranes to those for México and Colombia suggest potential correlations not widely recognized in attempts to reconstruct pre-Pangea interactions between Gondwana and Laurentia. If some Appalachian Grenville-age rocks are indeed related to southern México-Colombian rocks, then these Appalachian basement terranes must also have been transferred to Laurentia during latest Proterozoic or earlier Paleozoic collisions.

ACKNOWLEDGMENTS

This work was partly supported by National Science Foundation Grants EAR-935061 (to J.R.) and EAR-9219347 (to J.L. Anderson, University of Southern California), and by the W. M. Keck Foundation, the National Hispanic Scholarship Fund, the Foss Foundation for Mineralogic Research, the Graduate Student Fund of the Department of Geological Sciences of the University of Southern California, and the Instituto Mexicano del Petroleo. This manuscript was written while one of us (R.M.T.) was a visiting scientist at the Department of Geological Sciences, Cornell University, which is thanked for its logistical support. We thank Robert Ayuso, Ken Cameron, Peter Coney, William Dickinson, and Peter Schaff for their careful reviews of previous drafts.

REFERENCES CITED

Alvarez, J., 1981, Determinación de edad Rb/Sr en rocas del Macizo de Garzón, Cordillera Oriental de Colombia: Geología Norandina, v. 4, p. 31–38.

Alvarez, J., 1984, Una edad K/Ar del Macizo de Garzón, Departamento del Huila (Colombia): Geología Norandina, v. 7, p. 35–38.

Alvarez, J., and Cordani, U. G., 1980, Precambrian basement within the septentrional Andes: age and geological evolution: Abstracts, XXVI International Geologic Congress, Paris, v. 1, no. 10, p. 18–19.

Arribas, A., Jr., and Tosdal, R. M., 1994, Isotopic composition of Pb in ore deposits of the Betic Cordillera, Spain: origin and relationship to other European deposits: Economic Geology, 89, p. 1074–1093.

Ayuso, R. A., and Bevier, M. L., 1991, Regional differences in Pb isotopic compositions of feldspars in plutonic rocks of the northern Appalachian Mountains, U.S.A. and Canada: a geochemical method of terrane correlation: Tectonics, v. 10, p. 191–212.

Bahlburg, H., 1993, The hypothetical Pacific continent revisited: new evidence from the mid-Paleozoic basins of northern Chile: Geology, v. 21, p. 909–912.

Ballard, M. M., Van der Voo, R., and Urrutia-Fucugauchi, J., 1989, Paleomagnetic results from the Grenvillian-aged rocks from Oaxaca, Mexico: evidence for a displaced terrane: Precambrian Research, v. 42, p. 343–352.

Bond, G. C., Nickeson, P. A., and Kominz, M. A., 1984, Breakup of a supercontinent between 625 Ma and 555 Ma: new evidence and implications for continental histories: Earth and Planetary Science Letters, v. 70, p. 325–345.

Bridger, C. S., 1982, El Paleozoico inferior de Colombia: Bogotá, Universidad Nacional de Colombia, 222 p.

Cameron, K. L., Robinson, J. V., Niemeyer, S., Nimz, G. J., Kuentz, D. C., Harmon, R. S., Bohlen, S. R., and Collerson, K. D., 1992, Contrasting styles of pre-Cenozoic and mid-Tertiary crustal evolution in northern Mexico: evidence from deep crustal xenoliths from La Olivina: Journal of Geophysical Research, v. 97, B12, p. 17353–17376.

Chiarenzelli, J. R., and McLelland, J. M., 1993, Granulite facies metamorphism, palaeo-isotherms and disturbance of the U-Pb systematics of zircon in anorogenic plutonic rocks from the Adirondack Highlands: Journal of Metamorphic Geology, v. 11, p. 59–70.

Coney, P. J., and Campa, M., 1983, Lithotectonic terrane map of Mexico: U.S. Geological Survey Miscellaneous Studies Map MF 1974-D, scale 1:1250000.

Coney, P. J., Jones, D. L., and Monger, J. W. H., 1980, Cordilleran suspect terranes: Nature, v. 228, p. 329–333.

Dalla Salda, L. H., Cingolani, C. A., and Varela, R., 1992a, Early Paleozoic belt of the Andes in southwestern South America: result of Laurentia-Gondwana collision?: Geology, v. 20, p. 617–620.

Dalla Salda, L. H., Dalziel, I. W. D., Cingolani, C. A., and Varela, R., 1992b, Did the Taconic Appalachians continue into southern South America?: Geology, v. 20, p. 1059–1062.

Dalziel, I. W. D., 1991, Pacific margins of Laurentia and east Antarctica–Australia as a conjugate rift pair: evidence and implications for an Eocambrian supercontinent: Geology, v. 19, p. 598–601.

Dalziel, I. W. D., 1997, Neoproterozoic-Paleozoic geography and tectonics: review, hypothesis, and environmental speculation: Geological Society of America Bulletin, v. 109, p. 16–42.

Dalziel, I. W. D., Dalla Salda, L. H., and Gahagan, L. M., 1994, Paleozoic Laurentia-Gondwana interaction and the origin of the Appalachian-Andean mountain system: Geological Society of America Bulletin, v. 106, p. 243–252.

DeWolf, C. P., and Mezger, K., 1994, Lead isotope analyses of leached feldspars: constraints on the early crustal history of the Grenville Orogen: Geochimica et Cosmochimica Acta, v. 58, p. 5537–5550.

Goldsmith, R., Marvin, R. F., and Menhert, H. H., 1971, Radiometric ages of the Santander massif, eastern Cordillera, Colombian Andes: U.S. Geological Survey Professional Paper 750D, p. 44–49.

Harrington, J. H., and Kay, M., 1951, Cambrian and Ordovician faunas of eastern Colombia: Journal of Paleontology, v. 25, p. 655–668.

Hoffman P. A., 1991, Did the birth of North America turn Gondwana inside out?: Science, v. 252, p. 1409–1411.

James, E. W., and Henry, C. D., 1993, Southeastern extent of the North American craton in Texas and northern Chihuahua as revealed by Pb isotopes: Geological Society of America Bulletin, v. 105, p. 116–126.

Keppie, J. D., 1993, Transfer of the northeastern Appalachians (Meguma, Avalon, Gander, and Exploits terranes) from Gondwana to Laurentia during Middle Paleozoic continental collision: Proceedings of the First Circum-Pacific and Circum-Atlantic Terrane Conference, Guanajuato, México, Instituto de Geología Universidad Nacional Autónoma de México, p. 71–73.

Keppie, J. D., and Ortega-Gutiérrez, F., 1995, Provenance of Mexican Terranes: isotopic constraints: International Geology Reviews, v. 37, p. 813–824.

Keppie, J. D., Dostal, J., Murphy, J. B., and Nance, R. D., 1996, Terrane transfer between eastern Laurentia and western Gondwana in the early Paleozoic: constraints on global reconstructions, *in* Nance, R. D., and Thompson, M. D., eds., Avalonian and related peri-Gondwanan terranes of the Circum-North Atlantic: Geological Society of America Special Paper 304, p. 369–380.

Kesler, S. E., Cummings, G. L., Krstic, D., and Appold, M. S., 1994, Lead isotope geochemistry of Mississippi Valley–type deposits of the southern Appalachians: Economic Geology, v. 89, p. 307–321.

Krogh, T. E., 1982, A low-contamination method of hydrothermal decomposition of zircon and extraction of U and Pb for isotope age determinations: Geochimica et Cosmochimica Acta, v. 37, p. 485–494.

Kroonenberg, S., 1982a, A Grenvillian granulite belt in the Colombian Andes and its relations to the Guiana Shield: Geologia Mijnbouw, v. 61, p. 325–333.

Kroonenberg, S., 1982b, Litología, metamorfismo y orígen de las granulitas del macizo de Garzón, Cordillera Oriental (Colombia): Geología Norandina, v. 6, p. 39–46.

Lawlor, P. J., Ortega-Gutiérrez, F., Cameron, K. L., Ochoa-Camarillo, H., Lopez, R., and Sampson, D. E., 1999, U/Pb geochronology, geochemistry, and provenance of the Grenvillian Huiznopala gneiss of eastern Mexico: Precambrian Research (in press).

Lopez, R., 1997, The Pre-Jurassic geotectonic evolution of the coahuila terrane, northeastern Mexico: Grenville basement, a Late Paleozoic arc, Triassic plutonism, and the events south of the Ouachita suture [Ph.D. thesis]: Santa Cruz, University of California.

Lopez, R., Cameron, K. L., and Jones, N. W., 1997, Evidence from U-Pb zircons ages for the Grenvillian reworking of 1.8 Ga crust south of the Ouachita Suture in the Coahuila Terrane, northern Mexico [abs.]: Eos (Transactions, American Geophysical Union), v. 78, no. 46, p. F785.

MacDonald, W. D., and Hurley, P. M., 1969, Precambrian gneisses from northern Colombia, South America: Geological Society of America Bulletin, v. 80, p. 1867–1872.

Martiny, B., Martinez-Serrano, R., and Ayuso, R. A., 1997, Pb isotope geochemistry of Tertiary igneous rocks and continental crustal complexes, southern Mexico [abs.]: Eos (Transactions, American Geophysical Union), v. 78, no. 46, p. F844.

Mattinson, J. M., 1987, U-Pb ages of zircons: a basic examination of error propagation: Chemical Geology, Isotope Geology Section, v. 66, p. 151–162.

Mauger, R. L., McDowell, F. W., and Blount, J. G., 1983, Grenville-age Precambrian rocks of the Los Filtros area near Aldama, Chihuahua, *in* Clark K. F., and Goodal, P. C., eds., Geology and mineral resources of north-central Chihuahua Field Guidebook: El Paso, Texas, Geological Society of America, p. 165–168.

Moores, E. M., 1991, Southwest U.S.–East Antarctic (SWEAT) connection: a hypothesis: Geology, v. 19, p. 425–428.

Mora, C. I., and Valley, J. W., 1985, Ternary feldspar thermometry in granulites from the Oaxacan Complex, Mexico: Contributions to Mineralogy and Petrology, v. 89, p. 215–225.

Murillo-Muñetón, 1994, Petrologic and geochronologic study of the Grenville-age granulites and post-granulite plutons from the La Mixtequita area, State of Oaxaca, southern Mexico, and their tectonic significance [M.S. thesis]: Los Angeles, University of Southern California, 163 p.

Murillo-Muñetón, G., and Anderson, J. L., 1994, Thermobarometry of Grenville(?) granulites of the La Mixtequita Massif, southern Mexico: Geological Society of America Abstracts with Programs, v. 26, no. 2, p. 76.

Murillo-Muñetón, G., Anderson, J. L., and Tosdal, R. M., 1994, A new Grenville-age granulite terrane in southern Mexico: Geological Society of America Abstracts with Programs, v. 26, no. 7, p. A48.

Ortega-Gutiérrez, F., 1991, Metamorphic belts of southern Mexico and their tectonic significance: Geofisica Internacional, v. 20-3, p. 177–202.

Ortega-Gutiérrez, F., Ruiz, J., and Centeno-Garcia, E., 1995, Oaxaquia—a Proterozoic microcontinent in Mexico: Geology, v. 24, p. 136–198.

Nance, R. D., and Murphy, J. B., 1994, Contrasting basement isotopic signatures and the palinspastic restoration of peripheral orogens: examples for the Neoproterozoic Avalonian-Cadomian belt: Geology, v. 22, p. 617–620.

Nance, R. D., and Murphy, J. B., 1996, Basement isotopic signatures and Neoproterozoic paleogeography of Avalonian-Cadomian and related terranes in the circum-North Atlantic, *in* Nance, R. D., and Thompson, M. D., eds., Avalonian and related peri-Gondwanan terranes of the Circum-North Atlantic: Geological Society of America Special Paper 304, p. 333–346.

Palmer, A. R., 1973, Cambrian trilobites, *in* Hallam, A., ed., Atlas of paleobiogeography: New York, Elsevier Scientific, p. 7.

Park, R. G., 1992, Plate kinematic history of Baltica during the Middle to Late Proterozoic: a model: Geology, v. 20, p. 725–728.

Patchett, P. J., and Ruiz, J., 1987, Nd isotopic ages of crust formation and metamorphism in the Precambrian of eastern and southern México: Contributions to Mineralogy and Petrology, v. 96, p. 523–528.

Priem, H. N. A., Kroonenberg, S. B., Boelrijk, N. A. I. M., and Hebeda, E. H.,

1989, Rb-Sr and K-Ar evidence for the presence of a 1.6 Ga basement underlying the 1.2 Ga Garzón–Santa Marta granulite belt in the Colombian Andes: Precambrian Research, v. 42, p. 315–324.

Radelli, L., 1961, Introducción al estudio de la geología y de la petrografía del Macizo de Santa Marta: Geología Colombiana–Universidad Nacional de Colombia, no. 2, p. 41–115.

Radelli, L., 1962, Introducción al estudio de la Petrografía del Macizo de Garzón: Geología Colombiana–Universidad Nacional de Colombia, no. 3, p. 17–46.

Restrepo-Pace, P. A., Ruiz, J., Gehrels, G., and Cosca, M., 1997, Geochronology and Nd isotopic data of Grenville-age rocks in the Colombian Andes: new constraints for Late Proterozoic–Early Paleozoic paleocontinental reconstructions of the Americas: Earth and Planetary Science Letters, v. 150, p. 427–441.

Robison, R. A., and Pantoja-Alor, J., 1968, Tremadocian trilobites from the Nochixtlán region, Oaxaca, México: Journal of Paleontology, v. 42, p. 767–800.

Ruiz, J., Patchett, P. J., and Ortega-Gutiérrez, F., 1988a, Proterozoic and Phanerozoic basement terranes of Mexico from Nd isotopic studies: Geological Society of America Bulletin, v. 100, p. 274–281.

Ruiz, J., Patchett, P. J., and Arculus, R. J., 1988b, Nd-Sr isotope composition of lower crustal xenoliths—evidence for the origin of felsic volcanic rocks in Mexico: Contributions to Mineralogy and Petrology, v. 99, p. 36–43.

Rushton, A. W. A., 1962, Paradoxides from Colombia: Geological Magazine, v. 100, p. 255–257.

Sedlock R. L., Ortega-Gutiérrez, F., and Speed, R. C., 1993, Tectonostratigraphic terranes and tectonic evolution of Mexico: Geological Society of America Special Paper 278, 153 p.

Silver, L. T., Anderson, T. H., and Ortega-Gutiérrez, F., 1994, The "thousand million year" orogeny in eastern and southern Mexico: Geological Society of America Abstracts with Programs, v. 26, no. 7, p. A48.

Sinha, A. K., Hogan, J. P., and Parks, J., 1996, Lead isotope mapping of crustal reservoirs within the Grenville superterrane: I. Central and southern Appalachians, *in* Basu, A., and Hart, S., eds., Earth processes: reading the isotopic code: Washington, D.C., American Geophysical Union Monograph 95, p. 293–305.

Stacey, J. S., and Kramers, J. D., 1975, Approximation of terrestrial lead isotope evolution by a two-stage model: Earth and Planetary Science Letters, v. 26, p. 207–221.

Tosdal, R. M., 1996, The Amazon-Laurentian connection as viewed from the Middle Proterozoic rocks in the central Andes, western Bolivia and northern Chile: Tectonics, v. 15, p. 827–842.

Tschantz, C. M., Jimeno, A., and Cruz, J., 1969, Geology of the Santa Marta area (Colombia): Instituto Nacional Investigacion Minera, Informe 1829, 288 p.

Tschantz, C. M., Marvin, R. F., Cruz, J., Mehnert, H., and Cebulla, G., 1974, Geologic evolution of the Sierra Nevada de Santa Marta area, Colombia: Geological Society of America Bulletin, v. 85, p. 273–284.

Ward, D. E., Goldsmith, R., Cruz, J., and Restrepo, H., 1973, Geología de los cuadrángulos H-12 Bucaramanga y H-13 Pamplona, Departamento de Santander: Boletín de Geología Ignea, v. 21, 132 p.

Ward, R., and Cameron, K. L., 1996, Petrology and geochemistry of granulite-facies xenoliths, west Texas: evidence for Proterozoic age deep crust: Geological Society of America Abstracts with Programs, v. 28, no. 1, p. 68.

Wooden, J. L., Czamanske, G. K., Bouse, R. M., Likhachev, A. P., Kunilov, V. E., and Lyul'lp, V., 1992, Pb isotope data indicate a complex mantle origin for the Noril'sk Talnakh ores, Siberia: Economic Geology, v. 87, p. 1153–1165.

Yañez, P., Ruiz, J., Patchett, J. P., Ortega-Gutiérrez, F., and Gehrels, G., 1991, Isotopic studies of the Acatlán Complex, southern México: implications for Paleozoic North American tectonics: Geological Society of America Bulletin, v. 103, p. 817–828.

Zartman, R. E., and Wasserburg, G. J., 1969, The isotopic composition of lead in potassium feldspars from some 1.0-b.y.-old North American igneous rocks: Geochimica et Cosmochimica Acta, v. 33, p. 901–942.

Manuscript Accepted by the Society September 4, 1998

Printed in U.S.A.

Geological Society of America
Special Paper 336
1999

Middle American Precambrian basement: A missing piece of the reconstructed 1-Ga orogen

J. Duncan Keppie and F. Ortega-Gutiérrez
Instituto de Geología, Universidad Nacional Autónoma de México, 04510 México D.F., México

ABSTRACT

The geologic record of the ~1-Ga basement of Middle America (México and Central America) is compared with those from the Grenvillian orogen of southern and eastern North America, the Sveconorwegian orogen in southern Norway and Sweden, the basement massifs of the northern Andes of Colombia and Perú, and the western Amazon craton. These comparisons suggest that closest correlatives of Oaxacan Complex in México crop out in the Adirondack Highlands and the Colombian massifs because they all contain metasediments and a ~1.18–1.06-Ga anorthosite-mangerite-charnockite-granite (AMCG) suite. Peak P-T conditions for the various ~1-Ga regions indicate that the Oaxacan Complex is most comparable to the Adirondacks, the Colombian massifs, and the Norwegian internal belt, but the latter is dissimilar in being remobilized ~1.75–1.55-Ga Gothian basement. Similarly, comparison of the temperature-time paths for the various areas of the orogen indicates closest correlatives of Oaxaquia are the Adirondack Highlands, the internal parts of the Grenville orogen in Labrador and the Garzon Massif in Colombia. Comparing the age and lithologies of the oldest strata overlying the ~1-Ga basement, the Late Cambrian–Early Ordovician cover of the Oaxacan Complex is most similar to that in the Colombian massifs, and both have a Gondwanan faunal assemblage. The close correlation between the Adirondack Highlands, Oaxaquia, and the Colombian massifs suggests that they mostly originated as arcs in the Grenville ocean between Laurentia, Baltica, and Amazonia. The arcs were caught between colliding cratons and depressed into the roots of the Grenville orogen. The subsequent break-up of Rodinia left the Adirondack Highlands with Laurentia, and Oaxaquia and the Colombian massifs with Gondwana as indicated by Cambro-Ordovician faunal affinites and the similarity of their Paleozoic cover to that on the Amazon craton. The generally younger unconformity on the Gondwana side of Iapetus may be interpreted in terms of asymmetrical listric rifting with Gondwana on the upper plate.

INTRODUCTION

The Grenville orogeny is inferred to have resulted from the final amalgamation of the supercontinent Rodinia. Since Hoffman (1991) used the ~1-Ga orogens to reconstruct a single Grenvillian-aged belt and asked the question "Did the breakout of Laurentia turn Gondwanaland inside-out?" there have been several papers comparing the geologic records of various parts of the reconstructed orogen (Fig. 1) (e.g., Wasteneys et al., 1995; Sadowski and Bettencourt, 1996; Romer, 1996). Most of this work has dealt with the main parts of the reconstructed orogen, e.g., the Grenville, Sveconorwegian, Sunsas, Irumide-Kibaran, and Namaqua-Natal belts. The smaller fragments of Grenvillian rocks in Middle America (México and Central America) and northwestern South America (Figs. 2 and 3), such as Oaxaquia in México, the Ch'ortis block comprising most of Honduras and

Keppie, J. D., and Ortega-Gutiérrez, F., 1999, Middle American Precambrian basement: A missing piece of the reconstructed 1-Ga orogen, *in* Ramos, V. A., and Keppie, J. D., eds., Laurentia-Gondwana Connections before Pangea: Boulder, Colorado, Geological Society of America Special Paper 336.

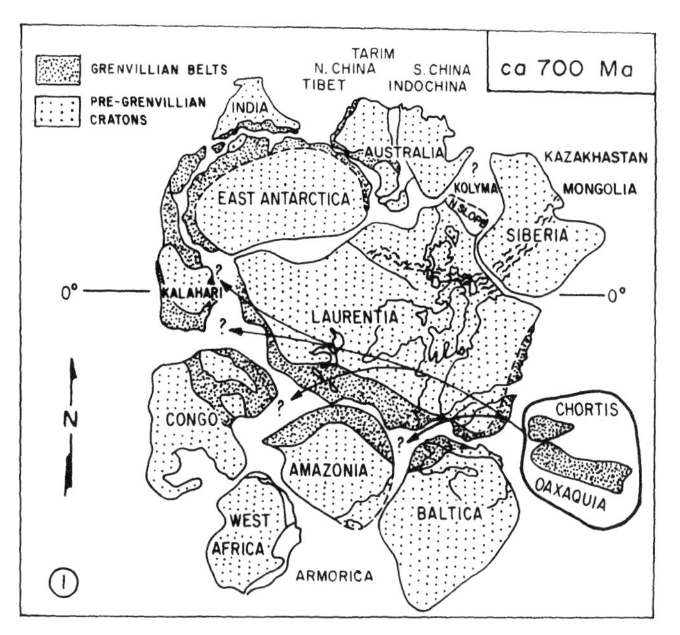

Figure 1. The reconstructed Grenville Orogen of Hoffman (1991) showing potential locations for the Oaxaquia and Chortis blocks.

southern Guatemala, and the Santander, Garzón, Cesar, Sierra Nevada de Santa Marta, and Antioquía massifs of Colombia were generally not considered. With the recognition that ~1-Ga rocks underlie much of México (Ortega-Gutíerrez et al., 1995) and that the ~1-Ga rocks forming the basement of the Chortis block may represent its southern extension (Manton, 1996), it is clear that this region represents a Grenvillian fragment of considerable size: ~2,000 by 500 km. This Middle American basement has been variously considered to be: (1) a southern extension of the Laurentian Grenvillian orogen based on similarities of the geologic records (de Cserna, 1971; Shurbet and Cebull, 1987); (2) an allochthonous Grenvillian terrane originating near the Adirondacks and Ontario based on paleomagnetic data (Ballard et al., 1989); and (3) an allochthonous terrane of Gondwanan affinity using Cambro-Ordovician faunal provinciality data (Robison and Pantoja-Alor, 1968; Rowley and Pindell, 1989) and Pb isotopic data from Mesozoic and Cenozoic rocks (Keppie and Ortega-Gutiérrez, 1995).

In an attempt to resolve these diverse opinions, a more comprehensive comparison of the geologic records of Oaxaquia with potential correlatives in eastern-southern Laurentia and western Gondwana is undertaken here. Lithologic data are accompanied by comparison of pressure-temperature-time (P-T-t) curves using the concept that cooling histories of different terranes are generally distinct (cf. van der Pliujm et al., 1994). Conversely, it is

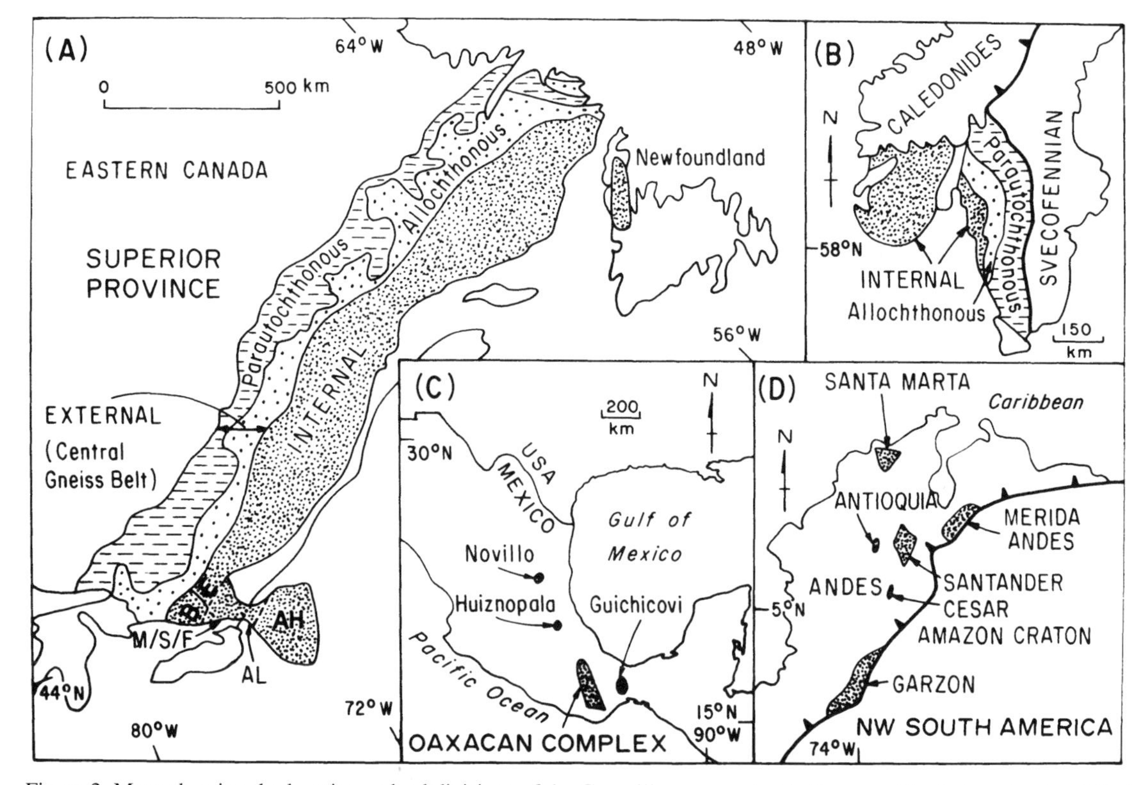

Figure 2. Maps showing the location and subdivisions of the Grenville orogen. A, Northeastern North America (modified from Davidson, 1995): B/E = Bancroft/Elzevir Terranes; M/S/F = Mazinaw/Sharbot Lake/Frontenac Terranes; AL = Adirondack Lowlands; AH = Adirondack Highlands. B, Scandinavia (modified from Romer, 1996). C, México (modified from Ortega-Gutiérrez et al., 1995). D, Colombia (modified from Restrepo-Pace et al., 1997).

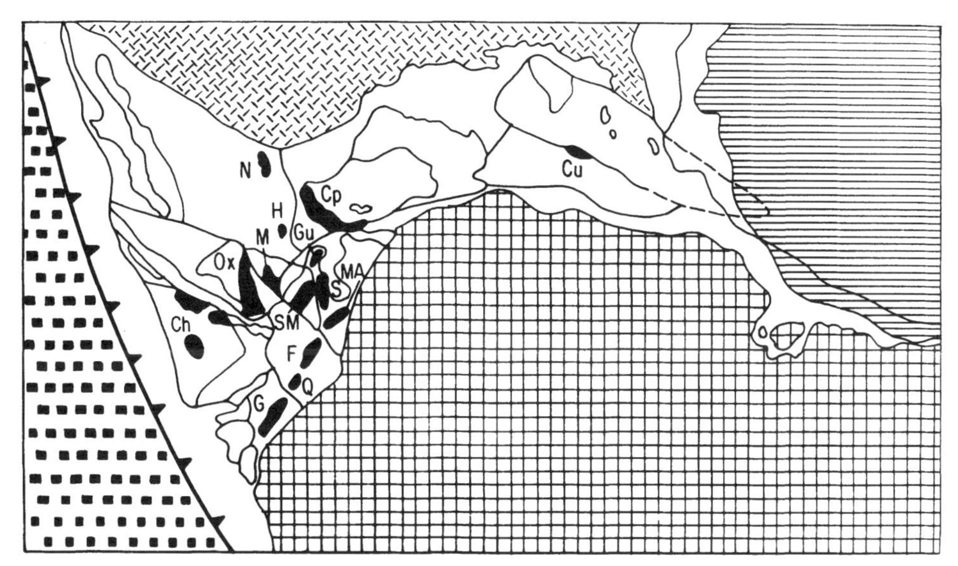

Figure 3. Pangea reconstruction (modified after Pindell, 1985) showing locations of 1-Ga basement in Middle America, and northern South America. Ch = Chortis; Cp = Chiapas; Cu = Cuba; F = Floresta; G = Garzón; Gu = Guajira; H = Huiznopala; M = Mixtequita (Guichicovi Complex); MA = Mérida Andes; N = Novillo; Ox = Oaxacan Complex; Q = Quetame (Cesar); S = Santander; SM = Santa Marta.

inferred that parts of the same terrane would share the same cooling history. Furthermore, rocks metamorphosed at various depths in an orogen probably show different relative cooling histories. In the upper levels of an orogen, where internal parts are generally thrust over external areas cooling histories may reveal progressively older to younger cooling ages, respectively. The opposite pattern might be expected in deep levels of an orogen as isostatic uplift and exhumation progressively expose more central and deeper levels.

Closure temperatures for individual minerals may vary with composition, cooling rate, grain size, and activation energy, but under average metamorphic conditions they may be approximated as: U-Pb in garnet, >800°C; in zircon, ~750°–650°C; in monazite, ~650°C; in titanite, ~625°C; in rutile, ~400°C (Ghent et al., 1988; Parrish, 1990; Mezger et al., 1991; Heaman and Parrish, 1991); and $^{40}Ar/^{39}Ar$ in hornblende (slow cooling), ~480°C, in muscovite, ~385°C; in white micas in phyllites, ~325°C; in biotite, ~300°C; and in K-feldspar, ~150°C (Harrison, 1981; Robbins, 1972; Harrison et al., 1985; Reynolds et al., 1987; Kirschner et al., 1996). Where such data are unavailable, other types of geochronologic data were used: K-Ar mineral ages are assumed to record closure temperatures similar to $^{40}Ar/^{39}Ar$ mineral ages, and Rb-Sr in whole rocks is assumed to record closure at ~700°C (Giletti, 1991).

REVIEW OF GRENVILLIAN GEOLOGIC RECORDS

México

Outcrops of ~1-Ga rocks are most extensive in southern México (10,000 km^2) where they make up the Oaxacan Complex (Fig.2A). This complex consists of metapelite, quartzofeldspathic gniess, calc-silicate, amphibolite, and marble intruded by anorthosite, charnockite, and garnetiferous orthogneiss, all of which are metamorphosed to granulite facies (Fig.4) (Anderson and Silver, 1971; Ortega-Gutiérrez, 1984). The oldest dated intrusion is ~1,113 Ma (Silver et al., 1994), which provides a younger age limit on the paragneisses. Nd isotopic data for igneous rocks of the Oaxacan Complex gave T_{DM} ages between 1.47 and 1.6-Ga, suggesting that their source could be significantly older (Ruiz et al., 1988). Typical metamorphic assemblages are: cpx – opx – plag ± gt – qzt – hb – bt in mafic gneisses, qtz – Kf – plag – gt ± sill – bi – sp – cor in pelitic granulites, and cc – scp – cpx ± woll – qtz in calc-silicates (Ortega-Gutiérrez, 1984). Peak temperatures and pressures using coexisting garnet and pyroxene are estimated to be 700°–750°C and 7.2–8.2 kb (Fig. 5) (Mora et al., 1986). Intense deformation accompanying peak metamorphism is recorded by thrusts and subhorizontal foliation with strong north- or northwest-trending stretching lineations parallel to tight isoclinal fold axes. Local kinematic indicators show top to the south movement. Subsequent north-northwest–trending upright folds were accompanied by mylonization in the steeply dipping limbs with recrystallization of biotite-chlorite-actinolite/hornblende recording upper greenschist–lower amphibolite facies conditions. K-Ar ages for hornblende, muscovite, biotite, and K-feldspar from cross-cutting, posttectonic pegmatites are ~927, ~925, ~875, and ~775 Ma, respectively (Fig.6) (Fries et al., 1962; Fries and Rincon-Orta, 1965). The Oaxacan Complex is unconformably overlain by Tremadocian rocks of the Tiñu Formation (Robison and Pantoja-Alor, 1968).

Bodies of ~1-Ga meta-anorthosite, garnetiferous orthogneiss, and charnockite with similar north-northwest structural trends are predominant in the small inliers at Novillo, Huiznopala, and Mixtequita (Guichicovi Complex) (Figs. 2 and 3) (Fries and Rincón-

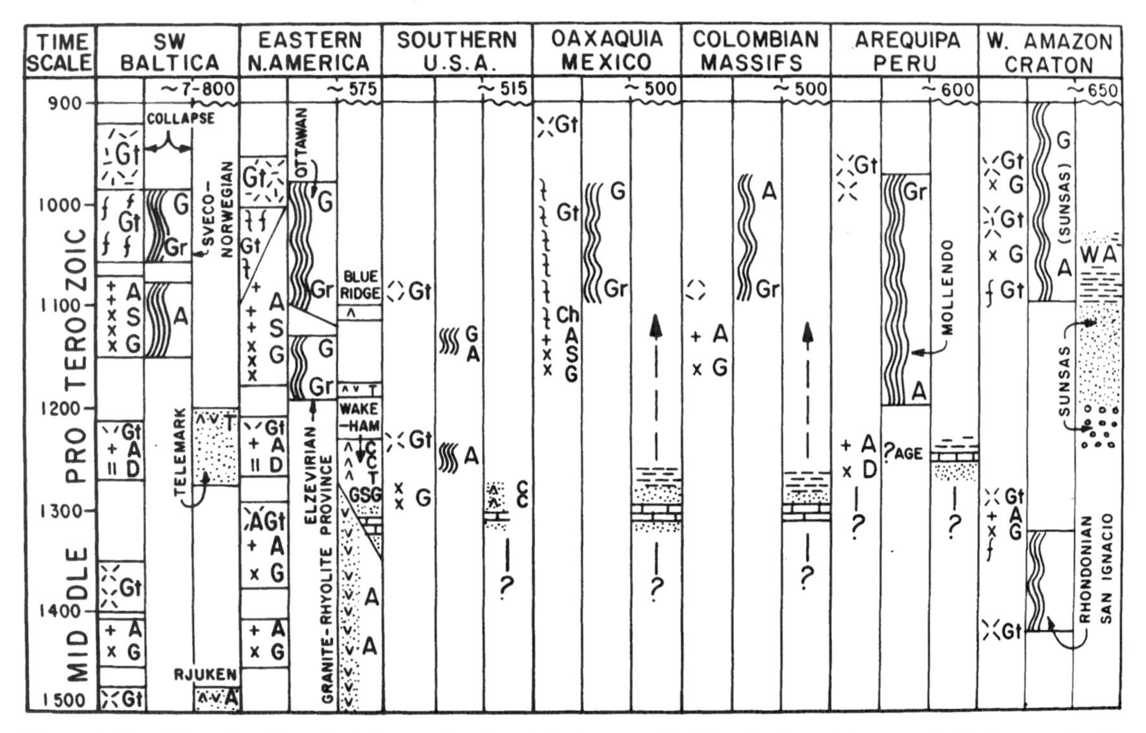

Figure 4. Correlation chart for selected areas showing plutonic record (left-hand column: Gt = granitoid; A = anorthosite; Ch = charnockite; S = syenite; G = gabbro; D = dikes), deformation (wavy lines) and metamorphism (middle column: Gr = granulite facies; A = amphibolite facies; G = greenschist facies), and stratigraphy (right-hand column: C = calc-alkaline; T = tholeiitic; A = alkaline volcanic rocks). References for data in vertical columns: *Southwestern Baltica* (data from Johansson et al., 1991; Kullerud and Dahlgren, 1993; Andrésson and Dallmeyer, 1995; Romer and Smeds, 1996; Romer, 1996, and references therein); *Eastern N. America* (data from Bohlen et al., 1985; Dallmeyer, 1987; Rivers et al., 1989; Anovitz and Essene, 1990; McLelland and Chiarenzelli, 1990; Wardle et al., 1990; Prevec et al., 1990; Gower et al., 1991; van der Pluijm et al., 1994; Davidson, 1995 and references therein); *Southern U.S.A.* (data from Mosher, 1993, 1996; Roback, 1996; Rougie et al., 1996; Thomas, 1991); *Oaxaquia, México* (data from Ortega-Gutiérrez et al., 1995, 1997, and references therein; Weber and Köhler, 1997); *Columbian massifs* (data from Trumpy, 1943; Ganser, 1955; Jenks et al., 1956; Radelli, 1962; Arnold, 1966; Renzoni, 1968; Ceidel, 1968; Irving, 1971; Ward et al., 1973; Pulido-González, 1979; Maya-Sánchez, 1992; Restrepo-Pace et al., 1997, and references therein); *Arequipa Massif, Peru* (data from Shackleton et al., 1979; Wasteneys et al., 1995, and references therein); *W. Amazon Craton* (data from Litherland et al., 1986; Sadowski and Bettencourt, 1996, and references therein).

Orta, 1965; Ortega-Gutiérrez, 1978; Ramírez-Ramírez, 1992; Murillo-Muñetón, 1994; Ortega-Gutiérrez et al., 1995). Peak metamorphism is characterized by the assemblage cpx–gt–qtz–plag–rut–ilm in mafic gneisses at Novillo and Mixtequita (Fig.5). The assemblage orthopyroxene-gedrite-phlogopite-titaniferous hornblende-quartz-plagioclase in Huiznopala gneiss suggests rather high metamorphic temperatures. Thermobarometric studies carried out on the Guichicovi and Huiznopala gneisses (Murillo-Muñetón, 1994; Lawlor et al., 1999) yield temperatures similar to those in the Oaxacan Complex (Fig. 5). On the other hand, mineral assemblages suggest that the Huiznopala gneisses formed at pressures between those at Novillo and Oaxacan Complexes. Thermobarometric studies of garnet and pyroxene in the Novillo gneisses reveal a zonation produced during retrograde metamorphism falling from 730° to 775°C at 8.9–9.7 kb to 595° to 642°C at 7.4–8.3 kb (Orozco, 1991). Zircon from Novillo granulites gave an U-Pb age of 1,018 Ma (Silver et al., 1994), and a K-Ar age for biotite yielded ~775 Ma (Fig. 6) (Fries and Rincón-Orta, 1965). Volcanic arc charnockite from Huiznopala has yielded an igneous U/Pb zircon age of ~1,212 Ma (Ortega-Gutiérrez et al., 1997). Granite gneiss in Mixtequita (Guichicovi Complex) has an U/Pb zircon upper intercept age of 1,238 ± 56 Ma interpreted as the time of intrusion (Weber and Köhler, 1997). Metamorphic minerals from the Huiznopala and Guichicovi gneisses yielded U-Pb zircon ages of 987 ± 4 and ~990–980 Ma, and K-Ar hornblende and biotite ages of ~925 (Guichicovi) and ~825–880 Ma, respectively (Murillo-Muñeton, 1994; Ortega-Gutíerrez et al., 1995, 1997; Weber and Köhler, 1997). The apparent northward younging of cooling ages (Oaxaca-Huiznopala-Novillo) may be related to progressive isostatic uplift of deeper roots of the orogen. The Novillo gneisses are unconformably overlain by Silurian rocks containing brachiopods of South American affinity (Stewart et al., 1993, this volume), whereas rocks not older than Jurassic overlie the Guichicovi Complex.

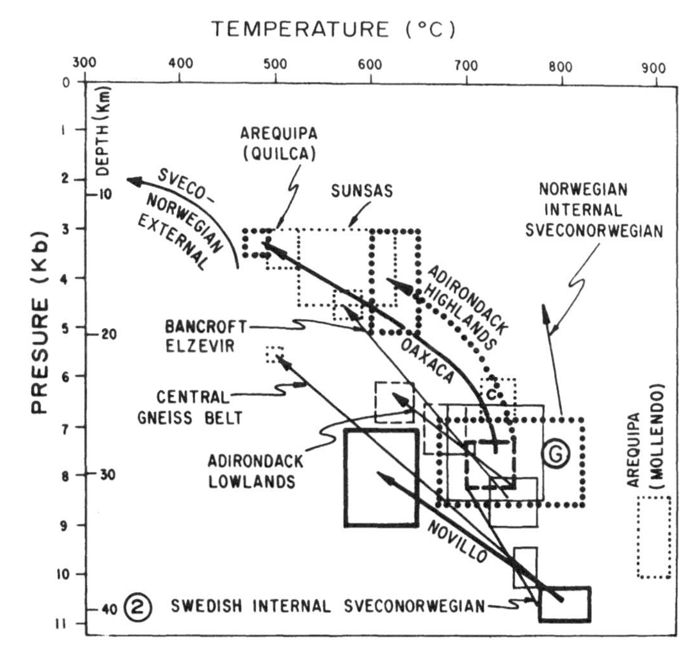

Figure 5. Pressure (P) – Temperature (T) plot of selected Grenvillian-age areas showing P-T paths. Circled letter G = Guichicovi; C = Colombian massifs.

Texas

Proterozoic gneisses, mafic and pelitic schists, amphibolites, minor marble, serpentinite, and mafic igneous bodies in Texas have been subdivided into several domains, which are inferred to represent the southern margin of Laurentia, its miogeocline, and an accreted arc terrane (Mosher, 1993, 1996). The rocks record multiple deformations at metamorphic grades reaching amphibolite facies in the south and decreasing northward into greenschist facies. U-Pb isotopic data indicates two orogenic events (Roback, 1996; Rougvie et al., 1996): (1) a regional metamorphic event interpreted as arc-continent collision took place at ~1,256–1,253 Ma, and was followed by widespread granitoid magmatism over the next ~20 Ma; and (2) another regional metamorphic event that peaked at ~1,147–1,128 Ma, followed by intrusion of late syn- to posttectonic plutons at ~1,090–1,070 Ma, and cooling through biotite blocking temperatures by ~1,042 Ma. The oldest overlying rocks of the Llano uplift are Middle Cambrian (Thomas, 1991).

Grenville orogen

The Grenville orogen in eastern Canada and the United States has been subdivided into three belts (from Grenville Front toward the interior): exterior thrust or parautochthonous Gneiss belt, exterior allochthonous Metasedimentary belt, and interior allochthonous Granulite belt (Wynne-Edwards, 1972; Davidson, 1986; Gower et al., 1991). The inboard Gneiss belt includes imbricated Archean–Early Proterozoic craton composed mainly of upper amphibolite to granulite facies igneous rocks that reached peak metamorphic conditions of ~780°–830°C at ~10–11 kb decreasing to ~500°C at ~5.5 kb (Anovitz and Essene, 1990). Within this belt, Grenvillian plutonic rocks are extremely rare. U-Pb ages for monazite and titanite, and $^{40}Ar/^{39}Ar$ ages of hornblende yield ~1,110–1,074, ≥~1,074, and ~995–948 Ma, respectively, interpreted as cooling ages (van der Pluijm et al., 1994).

The central Metasedimentary belt consists of greenschist to granulite facies marble and other metasedimentary and metavolcanic rocks intruded by calc-alkaline granitoids and plutons of the gabbro-anorthosite-monzonite-granite suite. This belt has been subdivided into a number of terranes (from northwest to southeast): Bancroft, Elzevir, Mazinaw, Sharbot Lake, and Frontenac–Mont Laurier–Adirondack Lowlands (Davidson, 1986, 1995). The Bancroft and Frontenac Terranes contain shallow marine sediments deposited between ~1,425 and 1,350 Ma, whereas the Elzevir Terrane is made up of ~1,300–1,240-Ma bimodal volcanic rocks with some shallow marine sediments (Lumbers et al., 1990). The Elzevir Terrane is composed of marbles, siliclastics, and 1.29–1.25-Ga tholeiitic–calc-alkaline volcanic and plutonic rocks interpreted to have been erupted in a backarc basin that was subsequently telescoped by collision between Laurentia and the Adirondack Highlands during the ~1,190–1,160-Ma Elzevirian Orogeny (Starmer, 1996). Granulite facies metamorphism peaked at ~750°C and 8–8.5 kb, decreasing to ~575°C at ~4.5 kb (Bancroft and Elzevir) or ~625°C at ~6.5 kb (Adirondack Lowlands) (Anovitz and Essene, 1990). The Bancroft and Elzevir Terranes record U-Pb zircon, monazite, and titanite ages of ~1,045, ~1,041, and ~1,024– 1,045 Ma, respectively, and a $^{40}Ar/^{39}Ar$ hornblende plateau age of ~959 Ma (van der Pluijm et al., 1994). Ages in the Adirondack Lowlands range from ~1,168–1,127 (U-Pb in garnet) to ~1,171–1,137 Ma (U-Pb in monazite), and ~1,156–1,103 (U-Pb in titanite) to ~1,008–953 Ma (U-Pb in rutile) (van der Pluijm et al., 1994).

The Granulite belt, represented by the Adirondack Highlands–Morin Terrane, is characterized by upper amphibolite–granulite facies metasediments (marble, pelitic and psammitic gneiss) intruded by ~1,350–1,300-Ma magmatic arc tonalites and 1.17–1.12 Ga anorthosites and charnockitic granitoids (McLelland and Chiarenzelli, 1990; Daly and McLelland, 1991). Peak metamorphism in the Adirondack Highlands reached ~675°–825°C at 6.5–8 kb, falling to ~600°–650°C at 3–5 kb (Bohlen et al., 1985). Corresponding ages range from ~1,154 to1,013 Ma (U-Pb in garnet) through ~1,033 Ma (U-Pb in monazite), ~1,050–982 Ma (U-Pb in titanite) and ~950–900 Ma ($^{40}Ar/^{39}Ar$ in hornblende) to ~911–885 Ma (U-Pb in rutile) (van der Pluijm et al., 1994). These rocks are unconformably overlain by Neoproterozoic–Early Cambrian rift facies rocks of the Appalachians.

Grenvillian massifs within the Appalachians have been subdivided into several terranes: New England, Blue Ridge, Mars Hill, and Goochland (Bartholomew and Lewis, 1992). They correlate the massifs of New England with the Adirondack Highlands, and infer that the granulites facies metavolcanic and

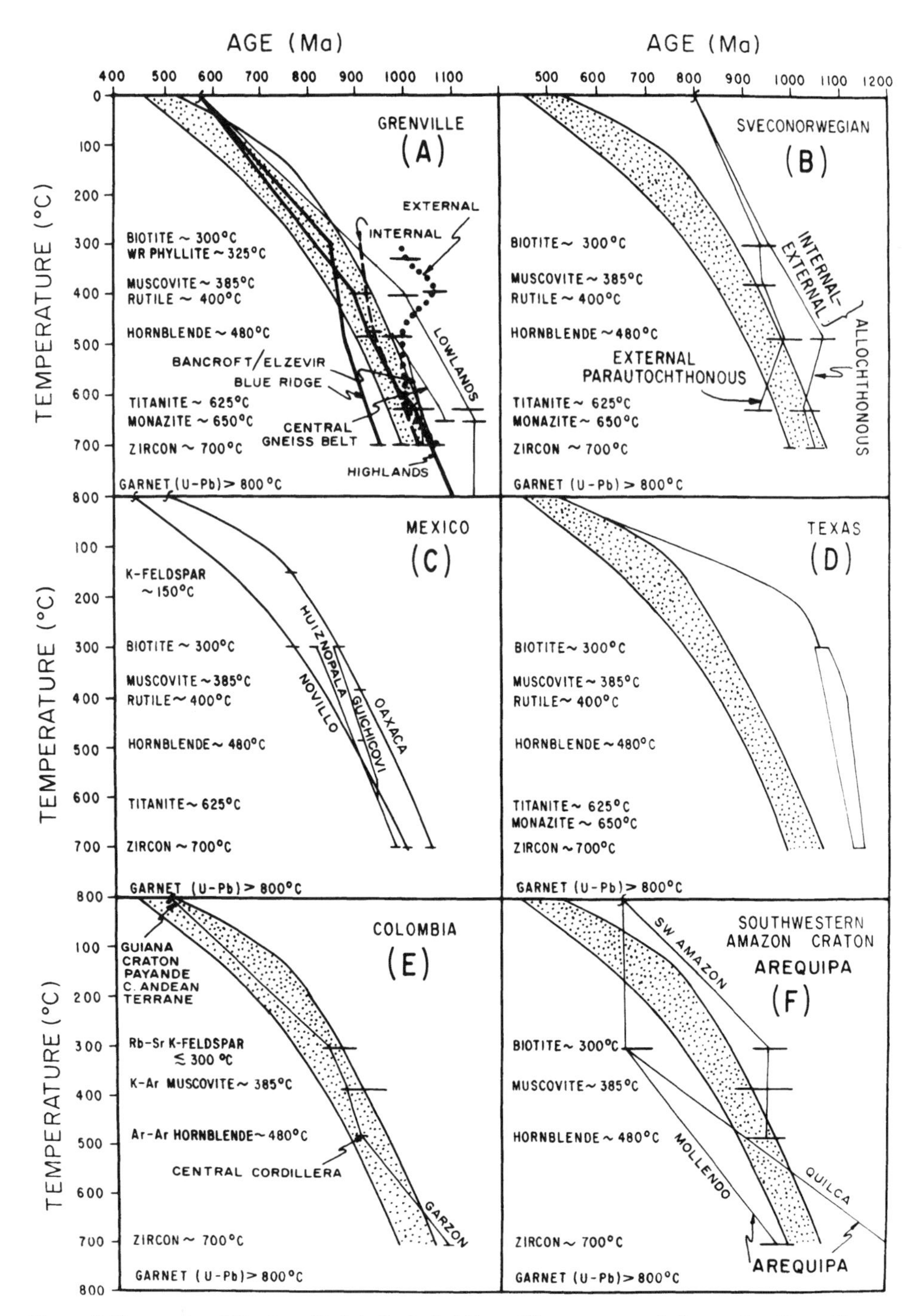

Figure 6. Temperature (T) – time (t) plot of selected Grenvillian-age areas. A, Laurentian Grenville orogen; B, Sveconorwegian orogen; C, México: Oaxacan Complex, Novillo, Huiznopala, Guichicovi; D, Texas; E, Garzón Massif, Colombia; F, southwestern Amazon craton and Arequipa Massif. Shaded area on each plot represents field for Oaxaquia. S-shaped line represents age of unconformity at the base of overlying units.

metasedimentary rocks intruded by charnockite and anorthosite of the Blue Ridge Terrane represent a 1,130–1,030-Ma volcanic arc formed on the southeastern margin of the Adirondacks. Peak metamorphic conditions in the Blue Ridge Terrane reached ~750°–800°C and ~8 kb and occurred at ~980–915 Ma (Herz and Force, 1984; Herz, 1984; Sinha and Bartholomew, 1984) with cooling through argon closure temperatures in hornblende and biotite at 880 and 850 Ma, respectively (Sutter et al., 1980). The 1.8–1.2-Ga Mars Hill metasedimentary terrane is interpreted as a fragment of an older craton accreted during the Grenville orogeny (Bartholomew and Lewis, 1992). On the other hand, the Goochland Terrane has been variously interpreted as a displaced piece of autochthonous Laurentian Grenville basement or an exotic Gondwanan terrane accreted during Paleozoic tectonism (Hatcher, 1989; Keppie et al., 1996).

The exterior allochthonous belt in eastern Labrador is represented by the Mealy Mountains Terrane, which consists mainly of igneous protoliths (anorthosite, granitoids, gabbro, troctolite) locally intruding pelitic gneisses (Gower, 1996). The Mealy Mountain Terrane is inferred to have rifted away from Laurentia at ~1,710 Ma followed by ~1,680–1,660-Ma arc magmatism terminating by accretion to Laurentia at ~1,650 Ma, succeeded by bimodal and granitoid magmatism until ~1,600 Ma. The Pinware Terrane forms part of the interior allochthonous belt inferred to represent arc magmatism developed on the southern margin of Laurentia (Gower et al., 1991). It comprises >1,500-Ma quartzite, calc-silicate units, pelitic schist and gneiss, and banded mafic rocks intruded by ~1,510–1,450-Ma granitoid rocks (Gower, 1996). These events were followed by granitoid magmatism at ~1,150–1,110 and ~1,040–1,026 Ma and metamorphism at ~1,003 Ma (Connelly et al., 1994), cooling through U-Pb titanite and rutile closure temperatures at ~1,030–968 and ~930–920 Ma, respectively (Schärer et al., 1986; Gower et al, 1991; Gower, 1996), and $^{40}Ar/^{39}Ar$ hornblende and biotite closure temperatures at ~946–926 and ~914–911 Ma, respectively (Dallmeyer, 1987). Comparable $^{40}Ar/^{39}Ar$ hornblende and biotite closure temperatures in the external belt are ~1,011–991 and ~1,018–985 Ma, respectively (Owen et al., 1988). In Newfoundland, the Grenvillian rocks are unconformably overlain by Neoproterozoic–Early Cambrian rift facies rocks of the Appalachians.

Sveconorwegian orogen

Romer and Smeds (1996) extrapolated the three-fold subdivision of the Canadian Grenville orogen into southern Norway and Sweden. Here, the western edge of the Baltic Shield consists of >1,800-Ma Svecofennian crust intruded by ~1,800–1,650-Ma Trans-Scandinavian Igneous belt bounded by the Protogine zone, a Gothian shear zone reactivated during the Sveconorwegian orogeny (Larson and Berglund, 1992). To the west lie the <1,750-Ma Gothian terranes that were accreted to Baltica between ~1,700–1,600 Ma (Starmer, 1996). The external belt in southern Sweden comprises granitoid gneisses, amphibolites, and supracrustals overlain by ~1,640–1,610- and ~1,150(?)-Ma supracrustals in the east, and ~1,760-Ma island arc rocks intruded by Gothian and Sveconorwegian plutons in the west (Ahäll and Daly, 1989; Ahäll et al., 1995). The internal belt in southern Norway has been subdivided into several terranes made up of <1,700-Ma pelitic-semipelitic gneisses, quartzites, carbonates, and dacitic-andesitic gneisses, in places overlain by ~1,500-Ma Rjukan Group rhyolites and basalts, Seljord Group quartzites, and ~1,150-Ma Bandak Group rhyolites, basalts, and sediments, intruded in the extreme west by ~1,010–935-Ma anorthosites (Falkum, 1985; Demaiffe and Michot, 1985; Menuge, 1988; Starmer, 1985, 1996). These rocks were variably deformed and metamorphosed by the ~1,090–980-Ma Sveconorwegian orogeny, succeeded by ~930–920-Ma posttectonic intrusions (Romer, 1996). Thrusting of the internal over external belts resulted in progressive cooling, first in the internal belt through closure temperatures for U-Pb in zircon at ~1,060–1,030-Ma and in titanite at ~1,040–1,000 Ma (Hansen et al., 1989; Eliasson and Schöberg, 1991; Johansson et al., 1993), K/Ar in hornblende at ~1,080–1,040 Ma, and $^{40}Ar/^{39}Ar$ in muscovite and biotite at ~999 ± 6 and ~970–885 Ma, respectively (Page et al., 1996), and then in the external belt through closure temperatures of U-Pb in titanite at ~950–930 Ma (Johansson, 1991; Connelly et al., 1995), $^{40}Ar/^{39}Ar$ in hornblende, muscovite, and biotite at ~1,010–960, ~963–905, and 975–900 Ma, respectively (Andréasson and Dallmeyer, 1995; Page et al., 1996). These rocks were exhumed before deposition of the ~800–700-Ma Visingsö Group (Bonhomme and Welin, 1983; Wikström and Karis, 1993).

Columbian massifs

Several massifs in the Colombian Andes have basement rocks of ~1-Ga age: Sierra Nevada de Santa Marta, Santander, Cesar, Garzon, and Antioquía (Maya-Sánchez, 1992). The basement rocks of the Garzón Massif consist of granulite, charnockite, marble, granitic gneiss, migmatite, and amphibolite. Single abraded zircons from an augen gneiss plot near the upper end of a chord with an upper intercept of 1,098 ± 9 Ma (Restrepo-Pace et al., 1997). Unfortunately, hornblende and biotite from the Garzón Massif yield discordant $^{40}Ar/^{39}Ar$ spectra indicative of excess argon (Restrepo-Pace et al., 1997). However, hornblende from nearby orthogneiss in the central Cordillera yields a plateau age of 911 ± 2 Ma (Restrepo-Pace et al., 1997). Rb-Sr whole-rock ages gave ~1,596–601 Ma with large errors (Alvarez, 1981; Priem et al., 1989), K-Ar hornblende ages ranging from ~1,000 ± 25 to ~925 ± 50 Ma (Alvarez, 1981), and a K-Ar phlogopite age of ~912 ± 35 Ma (Priem et al., 1989). Given the excess argon encountered by Restrepo-Pace et al. (1997), these K-Ar ages must be viewed with caution.

In the Santander Massif, upper amphibolite facies, pelitic and calc-silicate gneisses, and amphibolite (Bucaramanga Gneiss) have yielded a K-Ar hornblende age of ~945 ± 40 Ma (Goldsmith et al., 1971). $^{40}Ar/^{39}Ar$ spectra from hornblende in the Santander Massif are highly discordant (Restrepo-Pace et al., 1997), suggesting that these K-Ar ages be viewed with caution. These gneisses are overlain by greenschist–lower amphibolite facies slate, phyllite, and graywacke of the Silgará Formation. Correlative rocks in the Sierra Macarena (Guejar Formation) contain abundant Cambro-Ordovician trilobites, brachiopods, and graptolites (Irving, 1971) and pass westward from platformal through basinal in the Cordillera Oriental to volcanic in the cental Cordillera. These rocks are unconformably overlain by shale, quartzite, and siltstone of the 700–1000-m-thick Devonian Floresta Formation containing bryozoa (Irving, 1971; Ward et al., 1973). Similar Devonian rocks occur along the entire eastern

Colombian Andes and into the Mérida Andes of Venezuela and were deformed at very low grades before being overlain by Carboniferous-Permian rocks (Ceidel, 1968).

In the Santa Marta Massif, granulite facies gneisses associated with anorthosites and anorthositic gneisses (Los Mangos granulite) intruded by nelsonite have yielded ~1,300 ± 100 and 752 ± 70 Ma (Rb-Sr whole rocks) (MacDonald et al., 1969; Tschanz et al., 1974; D. Faure, unpublished data, 1978), and a ~940 ± 30-Ma K-Ar hornblende age (Tschanz et al., 1974). Discordant U-Pb zircon data give $^{207}Pb/^{206}Pb$ ages ranging between 1,543 and 1,037 Ma; however, the detrital character of the zircons indicates inheritance and precludes determination of a crystallization age (Restrepo-Pace et al., 1997). Discordant $^{40}Ar/^{39}Ar$ biotite spectra recorded by Restrepo-Pace et al. (1997) indicate disturbance of the isotopic system and for this reason urge caution in using these K-Ar ages. A granite intrusive on the Guajira Peninsula yielded a ~1,250-Ma U-Pb zircon age (Irving, 1975).

Although no thermobarometric studies have been performed on the Colombian Grenvillian massifs, common mineral assemblages have been described: opx – cpx – plag – gt – cd. The apparent absence of the higher pressure assemblage cpx – gt – qtz – plag may indicate pressures lower than <7 kb at temperatures of 750°–800°C in the Colombian massifs.

Arequipa Massif

The basement of the Arequipa Massif is composed of granulite facies paragneisses, dioritic gneisses, and migmatites (Shackleton et al., 1979; Wasteneys et al., 1995). Peak metamorphic assemblages consisting of hypersthene-sillimanite-garnet indicate ~8.5–10-kb pressures at ~900°C, which in places are extensively retrograded. U-Pb zircon geochronology indicates peak metamorphism took place at ~1,198 +6/-4 and ~970 ± 23 Ma (Wasteneys et al., 1995). K-Ar ages on biotite have yielded ~679 ± 12 and ~642 ± 16 Ma (Stewart et al., 1974). These gneisses are unconformably overlain by greenschist facies tilloids, psammites, carbonates, phyllites, and slates of the Marcona Formation (Shackleton et al., 1979). The tilloids have been correlated with the Port Askaig Tillite in Scotland (Caldas, 1979), a correlation that implies they are Neoproterozoic in age.

Southwestern Amazon craton

The western margin of the Amazon craton is made up of the ~1,500–1,250-Ma Rondonian/San Ignacio orogen bordered by the ~1,250–900-Ma Sunsas orogen (Litherland et al., 1986). The Sunsas, Aguepei, and Vibosi Groups were deposited unconformably on the >~1,280-Ma Rondonian basement and consist of 800–(?)6000 m of fluviatile-deltaic, basal conglomerate, quartzite, mudstone and siltstone, and upper quartzitic sandstone. On the Amazon craton, these units are undeformed, but the Sunsas orogeny produced polyphase deformation at amphibolite facies (garnet-staurolite-sillimanite) along the southwestern margin. The Sunsas tectonothermal activity is represented in the Rondonian basement by reactivation and retrograde metamorphism accompanied by intrusion of syn- to posttectonic granitoids and resetting of K-Ar ages. Rb-Sr whole-rock analyses of a syntectonic layered sill gave an age of ~993 ± 139 Ma, whereas a posttectonic granite yielded an age of ~1,005 ± 12 Ma. These ages are similar to K-Ar mica ages from syn- to posttectonic granitoids, which generally range from ~1,000 to 950 Ma (Litherland et al., 1986). Reset K-Ar ages in basement have ranges of ~960–900 Ma in hornblende, ~950–850 Ma in muscovite, and ~1,150–913 Ma in biotite (Litherland et al., 1968). Rocks of the Sunsas orogen are unconformably overlain by a ~2,500-m-thick, upward sequence consisting of a local volcanic unit overlain by tillite, conglomerate, stromatolitic limestone, sandstone, and shale (Litherland et al., 1986). The tillite and stromatolites suggest a Vendian age, whereas *Phycodes pedum* trace fossils in the sandstone unit indicate an earliest Cambrian age. Deformation of these rocks varies from gentle folding to polyphase folding and slaty cleavage in the shales and local northeast-vergent thrust zones bordering basement contacts. These units are unconformably overlain by ~350 m of Lower-Middle Silurian fossiliferous (*Clarkeia* fauna) marine sandstones, siltstones, and conglomerates succeeded upward by >500 m of Lower-Middle Devonian fossiliferous marine conglomerate, sandstone, siltstone, and mudstone.

DISCUSSION

Lithologies of the Oaxacan Complex are most similar to those of the Adirondack Highlands, the Blue Ridge Terrane, and the Colombian massifs: they all contain metasediments and a ~1.18–1.06-Ga anorthosite-mangerite-charnockite-granite (AMCG) suite (Figs. 2 and 3). Arc magmatism of several different ages is common to all three regions: ~1.35–1.3-Ga tonalite of the Adirondack Highlands, ~1.2-Ga charnockite in the Huiznopala gneisses of México, and 1.1–1.0-Ga calc-alkaline volcanic suites of the Blue Ridge Terrane. Peak metamorphism seems to be slightly younger in the Blue Ridge Terrane. The lack of Archean–Early Proterozoic basement in the Oaxacan Complex and Huiznopala Gneiss suggests that it cannot have been derived from an Early Proterozoic Laurentia that makes up most of the Grenville and Sveconorwegian provinces. The Texas Grenvillian rocks are unlike the Oaxacan Complex because they lack the AMCG suite and the orogenic activity is older (>1,090 Ma) and at lower grades of metamorphism. Similarly, the exposed Sunsas orogen lacks the AMCG suite and reaches only amphibolite facies metamorphism.

Examination of the peak P-T conditions for the various ~1-Ga regions (Fig. 5) indicates that the Oaxacan Complex is comparable with the Adirondacks, the Colombian massifs, and the Norwegian internal belt, with the exception that the latter contains remobilized ~1.75–1.55-Ga Gothian basement (Fig. 4). The high-pressure peak metamorphism at Novillo is similar to that recorded in parts of the Central Gneiss belt in Canada and the Swedish internal belt, but in both the latter areas the protoliths are older than Novillo, i.e., Archean–Early Proterozoic. The high-temperature peak metamorphism in the Arequipa Massif (sillimanite-hypersthene) appears to be unique.

Comparison of the temperature-time paths (Fig. 6) indicates that, along the eastern margin of Laurentia, the cooling of the Grenville orogeny is diachronous not only across, but along, the belt, becoming progressively younger from Texas (~1,150–1,030 Ma) through the Canadian Grenville (1,100–1,080 Ma), to the Sveconorwegian (1,080–1,060 Ma). This trend is broken in the Mexican basement, where the Grenvillian orogeny occurred between ~1,070 and 980 Ma, confirming those studies that suggest that it is not continuous with the basement of Texas. Cooling curves for the various areas of the orogen indicate closest comparisons between Oaxaquia and the Adirondack Highlands, the internal parts of the Grenville orogen in Labrador, and the Garzon Massif in Colombia. Earlier cooling of the internal versus external parts of the orogen, consistent with overthrusting in the upper levels of the orogen, is most clearly seen in the Sveconorwegian orogen, but may also be discerned in the southwestern Grenville Province where the Adirondack Lowlands cooled before the Bancroft-Elzevir and Central Gneiss belt (Fig. 6). On the other hand, later cooling of deepest parts of the orogen is visible in México, and the Adirondack Highlands versus the inboard terranes, and, assuming continuity beneath the Andes, the Arequipa compared with the southwestern Amazon craton.

Comparing the age and lithologies of the oldest strata overlying the ~1-Ga basement (Fig. 4), the Late Cambrian–Early Ordovician cover of the Oaxacan Complex is most similar to that in the Colombian massifs. Although it is only slightly older (Middle Cambrian) in Texas and Chihuahua, the faunal provinciality is dissimilar: Laurentian in Texas/Chihuahua versus mainly Gondwanan in México. Elsewhere, rocks of the Laurentian Grenville orogen are overlain by late Neoproterozoic rocks and older Neoproterozoic rocks in Scandinavia. Similarly in South America, the oldest rocks overlying the ~1-Ga basement are Neoproterozoic.

Taking all these data into consideration, the Oaxacan Complex is most similar to the Colombian massifs, although the Adirondack Highlands is a close second. Both of these regions represent allochthonous terranes. The Colombian massifs lie within the northern Andes, making it difficult to determine their exact position at ~1-Ga. Similarly, the Adirondack Highlands appears to have originated as an arc in the Grenville ocean between Laurentia, Baltica, and Amazonia (Starmer, 1996). By analogy, it may be inferred that Oaxaquia, the Colombian massifs, and the Adirondack Highlands originated in the Grenvillian ocean between Laurentia, Baltica, and Gondwana, and were caught in the center as these cratons collided to be depressed into the roots of the Grenville orogen (Fig. 7). Subsequent break-up of Rodinia left the Adirondack

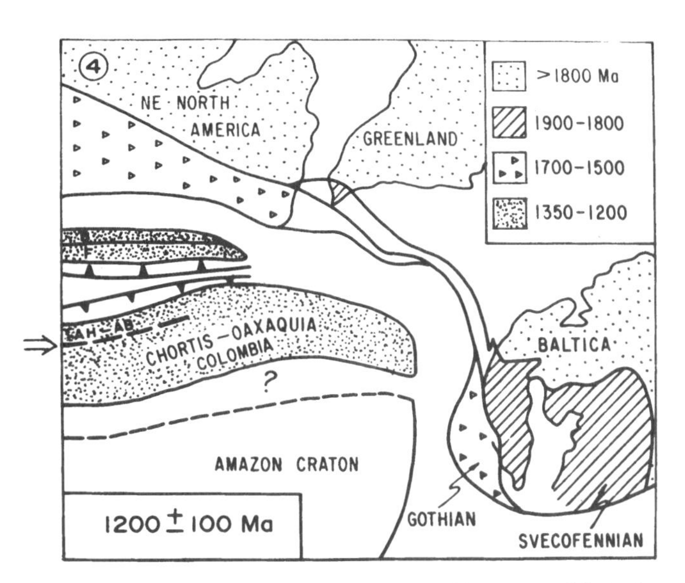

Figure 7. Palinspastic model for ~1,200 ± 100 Ma (modified after Starmer, 1996).

Highlands with Laurentia, and Oaxaquia and the Colombian massifs with Gondwana as indicated by Cambro-Ordovician and Silurian faunal affinites and the similarity of their Paleozoic cover to that on the Amazon craton. The generally younger unconformity on the Gondwana side of Iapetus may be interpreted in terms of asymmetric listric rifting with Gondwana on the upper plate.

The foregoing geologic correlations are confirmed by the similarity in the lead isotopic signatures of the Mexican Guichicovi Complex, the Colombian Santa Marta and Garzon Massifs, and the Appalachian Blue Ridge Massifs (Ruiz et al., this volume). In contrast, the Huiznopala gneisses are slightly less radiogenic, a feature Ruiz et al. (this volume) correlates with the Laurentian Grenville of Texas, and separate northern and southern México along the trans-Mexican volcanic belt into Laurentian and Gondwanan Grenville that were brought together during the late Paleozoic assembly of Pangea. However, this is neither consistent with the Gondwanan faunal affinities in Novillo (Stewart et al., this volume), nor with the geologic and P-T-t data present here. An alternate explanation of the Pb isotopic data may be that there are discrete Grenvillian terranes each with their own Pb signature. Further isotopic work is necessary to analyze the various possibilities.

ACKNOWLEDGMENTS

This work was funded by the Insituto de Geología, Universidad Nacional Autónoma de México, DGAPA, Project IN10195, and Consejo Nacional de Ciencia y Tecnología Project 0225P-T9506. We are grateful to T. H. Anderson and Z. de Cserna for their thoughtful reviews of the manuscript, and Luis Burgos for drafting the figures.

REFERENCES CITED

Ahäll, K.-I., and Daly, J. S., 1989, Age, tectonic setting and provenance of Östfold-Marstrand belt supracrustals: westward crustal growth of the Baltic shield at 1760 Ma: Precambrian Research, v. 45, p. 45–61.

Ahäll, K.-I., Persson, P.-O., and Skiöld, T., 1995, Westward accretion of the Baltic Shield: implications from the 1.6 Ga Amäl-Horred Belt, SW Sweden: Precambrian Research, v. 70, p. 235–251.

Alvarez, J., 1981, Determinación de edad Rb/Sr en rocas del Macizo de Garzón, Cordillera Oriental de Colombia: Geología Norandina, Colombia, no. 4, p. 31–38.

Anderson, T. H., and Silver, L. T., 1971, Age of granulite metamorphism during the Oaxacan orogeny, México: Geological Society of America Abstracts with Programs, v. 3, p. A492.

Anovitz, L. M., and Essene, E. J., 1990, Thermobarometry and pressure-temperature paths in the Grenville Province of Ontario: Journal of Petrology, v. 31, p. 197–241.

Andréasson, P. G., and Dallmeyer, R. D., 1995, Tectonothermal evolution of high-alumina rocks within the Protogine Zone, southern Sweden: Journal of Metamorphic Geology, v. 13, p. 461–474.

Arnold, H. C., 1966, Upper Paleozoic Sabaneta-Palmarito sequence of Merida Andes, Venezuela: American Association of Petroleum Geologists Bulletin, v. 50, p. 2366–2387.

Ballard, M. M., van der Voo, R., and Urrutia-Fucugauchi, J., 1989, Paleomagnetic results from Grenvillian-aged rocks from Oaxaca, México: evidence for a displaced terrane: Precambrian Research, v. 42, p. 343–352.

Bartholomew, M. J., and Lewis, S. E., 1992, Appalachian Grenville Massifs: pre-Appalchian translational tectonics, *in* Mason, R., ed., Basement tectonics: International Basement Tectonics Association Publication 7, p. 363–374.

Bohlen, S. R., Valley, J. W., and Essene, E. J., 1985, Metamorphism in the Adirondacks, I. Petrology, pressure and temperature: Journal of Petrology, v. 26, p. 971–992.

Bonhomme, M. G., and Welin, E., 1983, Rb-Sr and K-Ar isotopic data on shale and siltstone from the Visingsö Group. Lake Vättern basin, Sweden: Geologiska Föreningens i Stockholm Förhandlingar, v. 105, p. 363–366.

Caldas, J., 1979, Evidencias de una glaciación Precambriana en la Costa Sur del Peru: Arica-Chile, Segundo Congreso Geológico Chileno, p. J29–38.

Ceidel, F., 1968, El Grupo Girón, una molasa mesozoica de la Cordillera Oriental: Boletín Geológico, Ministerio Minerales y Petróleos, Servicio Geológico Nacional, v. 16, nos. 1-3, p. 6–95.

Connelly, J. N., Rivers, T., and James, D. T., 1995, Thermotectonic evolution of the Grenville Province of western Labrador: Tectonics, v. 14, p. 202–217.

Dallmeyer, R. D., 1987, $^{40}Ar/^{39}Ar$ mineral age record of variably superimposed Proterozoic tectonothermal events in the Grenville Orogen, central Labrador: Canadian Journal of Earth Sciences, v. 24, p. 314–333.

Daly, J. S., and McLelland, J. M., 1991, Juvenile Middle Proterozoic crust in the Adirondack Highlands, Grenville Province, northeastern North America: Geology, v. 19, p. 119–122.

Davidson, A., 1986, New interpretations in the southwest Grenville Province, *in* Moore, J. M., Davidson, A., and Baer, A. J., eds., The Grenville Province: Geological Association of Canada Special Paper 31, p. 61–74.

Davidson, A., 1995, A review of the Grenville orogen in its North American type area: Journal of Australian Geology and Geophysics, v. 16, p. 3–24.

De Cserna, Z., 1971, Precambrian sedimentation, tectonics, and magmatism in México: Geologische Rundschau, v. 60, p. 1488–1513.

Demaiffe, D., and Michot, J., 1985, Isotope geochronology of the Proterozoic crustal segment of southern Norway: a review, *in* Tobi, A. C., and Touret, J. L. R., eds., The deep Proterozoic crust in the North Atlantic Provinces: NATO Advanced Study Institute Series C, v. 158, p. 411–433.

Eliasson, T., and Schöberg, H., 1991, U-Pb dating of the post-kinematic Sveconorwegian Bohus granite, SW Sweden: evidence of restitic zircon: Precambrian Research, v. 51, p. 337–350.

Falkum, T., 1985, Geotectonic evolution of southern Scandinavia in light of a late Proterozoic plate collision, *in* Tobi, A. C., and Touret, J. L. R., eds., The deep Proterozoic crust in the North Atlantic Provinces: NATO Advanced Study Institute Series C, v. 158, p. 309–322.

Fries, C., Jr., and Rincón-Orta, C., 1965, Nuevas aportaciones geocronológicas y técnicas empleadas en el Laboratorio de Geocronometría: Universidad Nacional Autónoma de México, Instituto de Geología Boletín, v. 73, p. 57–133.

Fries, C., Jr., Schmitter, E., Damon, P. E., Livingston, D. E., and Erickson, R., 1962, Edad de las rocas metamórficas em las cañones de La Peregrina y de Caballeros, parte centro-occidental de Tamaulipas: Universidad Nacional Autónoma de México, Instituto de Geología Boletín, v. 64, p. 55–69.

Ganser, A., 1955, Ein Beitrag zur Geologie und Petrographie der Sierra Nevada de Santa Marta (Kolumbien, Sudamerika): Schwizer Mineralogie Petrographie Mitteilungen, v. 35, p. 209–279.

Ghent, E. D., Stout, M. Z., and Parrish, R. R., 1988, Determination of metamorphic pressure-temperature-time (P-T-t) paths: Mineralogical Association of Canada Short Course on Heat, Metamorphism and Tectonics, v. 14, p. 155–188.

Giletti, B. J., 1991, Rb and Sr diffusion in alkali feldspars, with implications for cooling histories of rocks: Geochimica et Cosmochimica Acta, v. 55, p. 1331–1343.

Goldsmith, R., Marvin, R. F., and Menhart, H. H., 1971, Radiometric ages of the Santander massif, eastern Cordillera, Colombian Andes: U.S. Geological Survey Professional Paper 750D, p. 44–49.

Gower, C. F., 1996, The evolution of the Grenville Province in eastern Labrador, Canada, *in* Brewer, T. S., ed., Precambrian crustal evolution in the North Atlantic region: Geological Society of London Special Publication 112, p. 197–218.

Gower, C. F., Heaman, L. M., Loveridge, W. D., Schärer, U., and Tucker, R. D., 1991, Grenvillian magmatism in the eastern Grenville Province, Canada: Precambrian Research, v. 51, p. 315–336.

Hansen, B. T., Persson, P.-O., Sollner, F., and Lindh, A., 1989, The influence of disturbed U-Pb systems in zircons from metamorphic rocks in southwest Sweden: Lithos, v. 23, p. 123–136.

Harrison, T. M., 1981, Diffusion of ^{40}Ar in hornblende: Contributions to Mineralogy and Petrology, v. 78, p. 324–331.

Harrison, T. M., Duncan, I., and McDougall, I., 1985: Diffusion of ^{40}Ar in biotite: temperature, pressure, and compositional effects: Geochimica et Cosmochimica Acta, v. 49, p. 2461–2468.

Hatcher, R. D., 1989, Tectonic synthesis of the U.S. Appalachians: Boulder, Colorado, Geological Society of America, Geology of North America, v. F-12, p. 511–536.

Heaman, L., and Parrish, R. R., 1991, U-Pb geochronology of accessory minerals: Mineralogical Association of Canada Short Course Handbook, v. 19, p. 59–102.

Herz, N., 1984, Rock suites in Grenvillian terrane of the Roseland district, Virginia, Part 2. Igneous and metamorphic petrology, *in* Bartholomew, M. J., ed., The Grenville event in the Appalachians and related topics: Geological Society of America Special Paper 194, p. 200–214.

Herz, N., and Force, E. R., 1984, Rock suites in Grenvillian terrane of the Roseland district, Virginia, Part 1. Lithologic relations, *in* Bartholomew, M. J., ed., The Grenville event in the Appalachians and related topics: Geological Society of America Special Paper 194, p. 187–199.

Hoffman, P. E., 1991, Did the breakout of Laurentia turn Gondwanaland inside out?: Science, v. 252, p. 1409–1412.

Irving, E. M., 1971, La evolución estructural de los Andes mas septentrionales de Colombia: Boletín Geológico, Ministerio Minerales y Petróleos Instituto Nacional Investigación Geología y Minerales, v. 19, no. 2, 90 p.

Jenks, W. F., ed., 1965. Handbook of South American geology: Geological Society of America Memoir 65, 378 p.

Johansson, A., Lindh, A., and Möller, C., 1991, Late Sveconorwegian (Grenville) high-pressure granulite facies metamorphism in southwest Sweden: Journal of Metamorphic Petrology, v. 9, p. 283–292.

Johansson, A., Meier, M., Oberli, F., and Wikman, H., 1993, The early evolution of the southwest Swedish gneiss province: geochronological and isotopic evidence from southernmost Sweden: Precambrian Research, v. 64, p. 361–368.

Keppie, J. D., and Ortega-Gutiérrez, F., 1995, Provenance of Mexican terranes: isotopic constraints: International Geology Review, v. 37, p. 813–824.

Keppie, J. D., Dostal, J., Murphy, J. B., and Nance, R. D., 1996, Terrane transfer between eastern Laurentia and western Gondwana in the early Paleozoic: constraints on global reconstructions, *in* Nance, R. D., and Thompson, M. D., eds., Avalonian and related peri-Gondwanan terranes of the circum-North Atlantic: Geological Society of America Special Paper 304, p. 369–380.

Kirschner, D. L., Cosca, M. A., Masson, H., and Hunziker, J. C., 1996, Staircase $^{40}Ar/^{39}Ar$ spectra of fine-grained white mica: timing and duration of deformation and empirical constraints on argon diffusion: Geology, v. 24, p. 47–750.

Kullerud, L., and Dahlgren, S. H., 1993, Sm-Nd geochronology of Sveconorwegian granulite facies mineral assemblages on the Bramble Shear Belt, south Sweden: Precambrian Research, v. 64, p. 389–402.

Larson, S. A., and Berglund, J., 1992, A chronological subdivision of the Trans-Scandinavian igneous belt—three magmatic episodes?: Geologiska Föreningens i Stockholm Förhandlingar, v. 114, p. 459–461.

Lawlor, P. J., Ortega-Gutiérrez, F., Cameron, K. L., Ochoa-Camarillo, H., and Sampson, D. E., 1999, U/Pb geochronology, geochemistry, and provenance of the Grenvillian Huiznopala Gneiss of eastern México: Precambrian Research, v. 94, p. 73–100.

Litherland, M., Annells, R. N., Appleton, J. D., Berrangé, J. P., Bloomfield, K., Burton, C. C. J., Darbyshire, D. P. F., Fletcher, C. J. N., Hawkins, M. P., Klinck, B. A., Llanos, A., Mitchell, W. I., O'Connor, E. A., Pitfield, P. E. J., Power, G., and Webb, B. C., 1986, The geology and mineral resources of the Bolivian Precambrian shield. British Geological Survey Overseas Memoir 9, 153 p.

Lumbers, S. B., Heaman, L. M., Vertolli, V. M., and Wu, T.-W., 1990, Nature and timing of middle Proterozoic magmatism in the Central Metasedimentary Belt, Grenville Province, Ontario, *in* Gower, C. F., Rivers, T., and Ryan, A. B., eds., Mid-Proterozoic Laurentia-Baltica: Geological Association of Canada Special Paper 38, p. 243–276.

MacDonald, W. D., and Hurley, P. M., 1969, Precambrian gneisses from northern Colombia, South America: Geological Society of America Bulletin, v. 80, p. 1867–1872.

Manton, W. I., 1996, The Grenville of Honduras: Geological Society of America Abstracts with Programs, v. 28, no. 7, p. A493.

Maya-Sánchez, M., 1992, Catálogo de dataciones isotópicas en Colombia: Instituto de Investigaciones en Geocienicas, Mineralogía y Química, v. 32, no. 1-3, p. 127–187.

McLelland, J. M., and Chiarenzelli, J. R., 1990, Geochronological studies in the Adirondack Mountains and the implications of a Middle Proterozoic tonalite suite, *in* Gower, C. F., Rivers, T., and Ryan, A. B., eds., Mid-Proterozoic Laurentia-Baltica: Geological Association of Canada Special Paper 38, p. 175–194.

Menuge, J. F., 1988, The petrogenesis of massif anorthosites: a Nd and Sr isotopic investigation of the Proterozoic of Rogaland/Vest Agder, SW Norway: Contributions to Mineralogy and Petrology, v. 98, p. 363–373.

Mezger, K., Rawnsley, C. M., Bohlen, S. R., and Hanson, G. N., 1991, U-Pb garnet, sphene, monazite and rutile ages: implications for duration of high-grade metamorphism and cooling histories, Adirondack Mts., New York: Journal of Geology, v. 99, p. 415–428.

Mora, C. I., Valley, J. W., and Ortega-Gutiérrez, F., 1986, The temperature and pressure conditions of Grenville-age granulite facies metamorphism of the Oaxacan Complex, southern México: Universidad Nacional Autónoma de México, Instituto de Geología Revista, v. 5, p. 222–242.

Mosher, S., 1993, Western extensions of Grenville age rocks, Texas, *in* Reed, J. C., Bickford, M. E., Houston, R. S., Link, P. K., Rankin, D. W., Sims, P. K., and van Schmus, W. R., eds., Boulder, Colorado, Geological Society of America, Geology of North America, v. C-2, Precambrian, conterminous U.S., p. 365–378.

Mosher, S., 1996, Grenville orogenesis along the southern margin of Laurentia: Geological Society of America Abstracts with Programs, v. 28, no. 1, p. A55.

Murillo-Muñetón, G., 1994, Petrologic and geochronologic study of Grenvillian-age granulites and post-granulitic plutons from La Mixtequita area, State of Oaxaca in southern México: [M. S. thesis]: University of Southern California, 163 p.

Orozco, M. T., 1991, Geotermobarometría de granulitas precámbricas del basamento de la Sierra Madre Oriental: Convención sobre la Evolución Geológica de México Memoria, p. 138–143.

Ortega-Gutiérrez, F., 1978, El Gneiss Novillo y rocas metamórficas asociadas en los cañones del Novillo y de La Peregrina, área de Ciudad Victoria, Tamaulipas: Universidad Nacional Autónoma de México, Instituto de Geología Revista, v. 2, p. 19–30.

Ortega-Gutiérrez, F., 1984, Evidence of Precambrian evaporites in the Oaxacan granulite complex of southern México: Precambrian Research, v. 23, p. 377–393.

Ortega-Gutiérrez, F., Ruiz, J., and Centeno-Garcia, E., 1995, Oaxaquia—a Proterozoic microcontinent accreted to North America during the late Paleozoic: Geology, v. 23, p. 1127–1130.

Ortega-Gutiérrez, F., Lawlor, P., Cameron, K. L., and Ochoa-Camarillo, H., 1997, New studies of the Grenvillean Huiznopala Gneiss, Molango area, State of Hidalgo, México—preliminary results: Instituto de Investigaciones en Ciencias de la Tierra de la Universidad del Estado de Hidalgo e Instituto de Geología de la Universidad Nacional Autónoma de México, II Convención sobre la Evolución Geológica de México y recursos asociados, Pachuca, Hgo., Libro-guía de las excursiones geológicas, Excursión 2, p. 19–25.

Owen, J. V., Dallmeyer, R. D., Gower, C. F., and Rivers, T., 1988, Metamorphic conditions and $^{40}Ar/^{39}Ar$ geochronologic contrasts across the Grenville front zone, coastal Labrador, Canada: Lithos, v. 21, p. 13–35.

Page, L. M., Stephens, M. B., and Wahlgren C.-H., 1996, $^{40}Ar/^{39}Ar$ geochronological constraints on the tectonothermal evolution of the eastern segment of the Sveconorwegian Orogen, south-central Sweden, *in* Tobi, A. C., and Touret, J. L. R., eds., The deep Proterozoic crust in the North Atlantic Provinces: NATO Advanced Study Institute Series C, v. 158, p. 315–330.

Parrish, R. R., 1990, U-Pb dating of monazite and its application to geological problems: Canadian Journal of Earth Sciences, v. 27, p. 1431–1450.

Pindell, J. L., 1985, Alleghanian reconstruction and subsequent evolution of the Gulf of México, Bahamas, and Proto-Caribbean: Tectonics, v. 4, p. 1–39.

Prevec, S. A., McNutt, R. H., and Dickin, A. P., 1990, Sr and Nd isotopic and petrological evidence for the age and origin of the White Bear complex and associated units from the Grenville Province in eastern Labrador, *in* Gower, C. F., Rivers, T., and Ryan, B., eds., Mid-Proterozoic Laurentia-Baltica: Geological Association of Canada Special Paper 38, p. 65–78.

Priem, H. N. A., Kroonenberg, S. B., Boelrijk, N. A. I. M., and Hebeda, E. H., 1989, Rb-Sr and K-Ar evidence for the presence of a 1.6 Ga basement underlying the 1.2 Ga Garzón-Santa marta granulite belt in the Colombian Andes: Precambrian Research, v. 42, p. 315–324.

Pulido-González, O., 1979, Geología de las Planchas 135 San Gil y 151 Charala; Departamento de Santander: Boletín Geológico del Ministerio de Minas y Energía, Instituto Nacional Investigaciones Geológicas y Minerales, v. 23, no. 2, p. 40–78.

Radelli, L., 1962, Acerca de la geología de la serranía de Perijá entre Codazzi y Villanueva: Geología Colombiana, no. 1, p. 23–41.

Ramírez-Ramírez, C., 1992, Pre-Mesozoic geology of the Huizachal-Peregrina anticlinorium, Ciudad Victoria, Tamaulipas and adjacent parts of eastern México [Ph. D. thesis]: Austin, University of Texas, 316 p.

Renzoni, G., 1968, Geología del Macizo de Quetame: Geología Colombiana, Universidad Nacional Colombia, v. 5, p. 75–127.

Restrepo-Pace, P. A., Ruiz, J., Gehrels, G., and Cosca, M., 1997, Geochronology and Nd isotopic data of Grenville-age rocks in the Colombian Andes: new constraints for Late Proterozoic–early Paleozoic paleocontinental reconstructions of the Americas: Earth and Planetary Science Letters, v. 150, p. 427–441.

Reynolds, P. H., Elias, P., Muecke, G. K., and Grist, A. M., 1987, Thermal history of the southwestern Meguma Zone, Nova Scotia, from an $^{40}Ar/^{39}Ar$ and fission track study of intrusive rocks: Canadian Journal of Earth Sciences,

v. 24, p. 1952–1965.
Rivers, T., Martignole, J., Gower, C. F., and Davidson, A., 1989, New tectonic subdivision of the Grenville Province, southeast Canadian Shield: Tectonics, v. 8, p. 63–84.
Roback, R. C., 1996, Mesoproteroozic poly-metamorphism and magmatism in the Lllano Uplift, Central Texas: Geological Society of America Abstracts with Programs, v. 28, no. 7, p. A377.
Robbins, G. A., 1972, Radiogenic argon diffusion in muscovite under hydrothermal conditions [M.S. thesis]: Providence, Rhode Island, Brown University, 23 p.
Robison, R., and Pantoja-Alor, J., 1968, Tremadocian trilobites from Nóchixtlan region, Oaxaca, México: Journal of Paleontology, v. 42. p. 767–800.
Romer, R. L., 1996, Contiguous Laurentia and Baltica before the Grenvillian-Sveconorwegian orogeny?: Terra Nova, v. 8, p. 173–181.
Romer, R. L., and Smeds, S.-A., 1996, U-Pb columbite ages of pegmatites from Sveconorwegian terranes in southwestern Sweden: Precambrian Research, v. 76, p. 15–30.
Rougvie, J. R., Carlson, W. D., Connelly, J. N. Roback, R. C., and Copeland, P., 1996, Late thermal evolution of Proterozoic rocks in the northeastern Llano Uplift, Central Texas: Geological Society of America Abstracts with Programs, v. 28, no. 7, p. A376–377.
Rowley, D. B., and Pindell, J. L., 1989, End Paleozoic–early Mesozoic western Pangean reconstruction and its implications for the distribution of Precambrian and Paleozoic rocks around Meso-America: Precambrian Research, v. 42, p. 411–444.
Ruiz, J., Patchett, P. J., and Ortega-Gutiérrez, F., 1988, Proterozoic and Phanerozoic basement terranes of México from Nd isotopic studies: Geological Society of America Bulletin, v. 100, p. 274–281.
Sadowski, G. R., and Bettencourt, J. S., 1996, Mesoproterozoic tectonic correlations between eastern Laurentia and the western border of the Amazon Craton: Precambrian Research, v. 76, p. 213–227.
Schärer, U., Krogh, T. E., and Gower, C. F., 1986, Age and evolution of the Grenville Province from U-Pb systematics in accessory minerals: Contributions to Mineralogy and Petrology, v. 94, p. 438–451.
Shackleton, R. M., Ries, A. C., Coward, M. P., and Cobbold, P. R., 1979, Structure, metamorphism and geochronology of the Arequipa Massif of coastal Peru: Journal Geological Society of London, v. 136, p. 195–214.
Shurbet, D. H., and Cebull, S. E., 1987, Tectonic interpretation of the westernmost part of the Ouachita-Marathon (Hercynian) orogenic belt, west Texas–México: Geology, v. 15, p. 458–461.
Silver, L. T., Anderson, T. H., and Ortega-Gutíerrez, F., 1994, The "thousand" year old orogeny of southern and eastern México: Geological Society of America Abstracts with Programs, v. 26, p. A48.
Sinha, A. K., and Bartholomew, M. J., 1984, Evolution of the Grenville terrane in the central Virginia Appalachians, *in* Bartholomew, M. J., ed., The Grenville event in the Appalachians and related topics: Geological Society of America Special Paper 194, p. 175–186.
Starmer, I. C., 1985, The evolution of the south Norwegian Proterozoic as revealed by the major and mega-tectonics of the Kongsberg and Bamble Sectors, *in* Tobi, A. C., and Touret, J. L. R., eds., The deep Proterozoic crust in the North Atlantic Provinces: NATO Advanced Study Institute Series C, v. 158, p. 259–290.
Starmer, I. C., 1996, Accretion, rifting, rotation and collision in the North Atlantic supercontinent, 1700–950 Ma, *in* Brewer, T. S., ed., Precambrian crustal evolution in the North Atlantic region: Geological Society Special Publication 112, p. 219–248.
Stewart, J. W., Evernden, J. F., and Snelling, N. J., 1974, Age determinations from Andean Peru: a reconnaissance survey: Geological Society of America Bulletin, v. 85, p. 1107–1116.
Stewart, J. H., Blodget, R. B., Boucot, A. J., and Carter, J. L., 1993, Middle Paleozoic exotic terrane near Ciudad Victoria, northeastern México, and the southern Margin of Paleozoic North America, *in* Ortega-Gutiérrez, F., Coney, P., Centeno-García, E., and Gómez-Caballero, A., eds., Guanajuato, México, *in* Proceedings, 1st Circum-Pacific and Circum-Atlantic Terrane Conference, p. 147–149.
Sutter, J. F., Crawford, M. L., and Crawford, W. A., 1980, $^{40}Ar/^{39}Ar$ age spectra of coexisting hornblende and biotite from the Piedmont of Pennsylvania: their bearing on the metamorphic and tectonic history: Geological Society of America Abstracts with Programs, v. 12, p. A85.
Thomas, W. A., 1991, The Appalchian-Ouachita rifted margin of southeastern North America. Geological Society of America Bulletin, v. 103, p. 415–431.
Trumpy, D., 1943, Pre-Cretaceous of Colombia: Geological Society of America Bulletin, v. 54, p. 1281–1304.
Tschanz, C. M., Marvin, R. F., Cruz, J., Mehnart, H., and Cebulla, G., 1974, Geologic evolution of the Sierra Nevada de Santa Marta area, Colombia: Geological Society of America Bulletin, v. 85, p. 273–284.
Van der Pluijm, B. A., Mezger, K., Cosca, M. A., and Essene, E. J., 1994, Determining the significance of high-grade shear zones by using temperature-time paths, with examples from the Grenville orogen: Geology, v. 22, p. 743–746.
Ward, D. E., Goldsmidt, R., Restrepo, A. H., and Cruz, B. J., 1973, Geología de los Cuadrángulos H-12 Bucaramanga y H-13 Pamplona, Departamento de Santander, Boletín Geológico (Bogotá), v. 21 (1-3), 125 p.
Wardle, R. J., Ryan, B., Philippe, S., and Schärer, U., 1990, Proterozoic crustal development, Goose Bay region, Grenville province, Labrador, Canada, *in* Gower, C. F., Rivers, T., and Ryan, A. B., eds., Mid-Proterozoic Laurentia-Baltica: Geological Association of Canada Special Paper 38, p. 197–214.
Wasteneys, H. A., Clark, A. H., Farrar, E., and Langridge, R. J., 1995, Grenvillian granulite-facies metamorphism in the Arequipa Massif, Peru: a Laurentia-Gondwana link: Earth and Planetary Science Letters, v. 132, p. 63–73.
Weber, B., and Köhler, H., 1997, La evolución magmática y metamórfica del complejo Guichicovi de edad grenvilleana, Estado de Oaxaca: II Convención sobre la Evolución Geológica de México, Resúmenes, p. 88–89.
Wikström, A., and Karis, L., 1993, Note on the basement-cover relationships of the Visingsö Group in the northern part of the Lake Vättern basin, south Sweden: Geologiska Föreningens i Stockholm Förhandlingar, v. 115, p. 311–313.
Wynne-Edwards, H. R., 1972, The Grenville Province, *in* Price, R. A., and Douglas, R. J. W. eds., Variations in tectonic styles in Canada: Geological Association of Canada Special Paper 11 , p. 263–334.

MANUSCRIPT ACCEPTED BY THE SOCIETY SEPTEMBER 4, 1998

Printed in U.S.A.

Geological Society of America
Special Paper 336
1999

Review of Paleozoic stratigraphy of México and its role in the Gondwana-Laurentia connections

José Luis Sánchez-Zavala, Elena Centeno-García, and Fernando Ortega-Gutiérrez
Instituto de Geología, Universidad Nacional Autónoma de México, Ciudad Universitaria, 04510 México D.F., México

ABSTRACT

México is made up of several allochthonous terranes floored by Proterozoic metamorphic basements. Their Paleozoic sedimentary cover, though preserved as isolated dispersed exposures, is important for reconstructing the paleogeographic and tectonic evolution of México during this time. Sedimentary rocks of northwestern México (Chihuahua, Caborca, and Cortéz terranes) have been related to the evolution of the actual western and southern margins of Laurentia. Early- to mid-Paleozoic stratigraphy and faunas of central and southern México (Oaxaquia) are different from those of North America, suggesting that these terranes evolved away from Laurentia until Carboniferous time. Faunal affinities with Venezuela suggest that Oaxaquia probably evolved near Gondwana. Metamorphic rocks of the Mixteco terrane of southern México record a major orogenic event of middle Paleozoic age. During Carboniferous time, terranes of southern and central México apparently approach Laurentia, as suggested by the faunal distribution. Two types of magmatic activity were recorded in the Permian: a basaltic-andesitic sequence in northern Oaxaquia block, and a rhyolitic province in the Chihuahua, southern Oaxaquia, Mixteco, and Maya terranes. Oceanic arc sequences of Pacific affinity seem to be recorded in the Cortéz and Juchatengo terranes. Terranes of northwestern México (Caborca and Cortéz) were apparently displaced to their actual position during mid-Jurassic time.

INTRODUCTION

Most of México is formed by several tectonostratigraphic terranes (Campa and Coney, 1987; Sedlock et al., 1993) (Fig. 1). Its location, as part of a relatively narrow and elongate crustal bridge between the North and South American cratons and between two oceanic basins of contrasting history, makes its tectonic evolution complex. During the last decades there has been considerable debate regarding the affinity of the terranes of México. Whether they are related to the North or the South American cratons, or whether they are of Cordilleran or Appalachian affinity has not been well constrained. Most of the paleogeographic reconstructions of Pangea show an overlap of 20–60% of México and South America, suggesting a different geographic position for most of southern México during Paleozoic time (Bullard et al., 1965; Pindell, 1985; Scotese and McKerrow, 1990). Although the tectonic evolution of the terranes that form México is not well understood as yet, a summary of the new advances in their Paleozoic stratigraphy and a discussion of their evolution are the topics of this chapter.

Figure 1 shows the distribution of the tectonostratigraphic terranes that contain Paleozoic sedimentary sequences (modified from Coney and Campa,1987; Ortega et al., 1995). The largest exposures of Paleozoic rocks are found in the northwestern part of the country and are associated with the North American craton (Chihuahua terrane), Caborca, and Cortéz terranes. In other areas Paleozoic strata are found as isolated exposures, mostly along the eastern margin in the Oaxaquia and Coahuila terranes.

Sánchez-Zavala, J. L., Centeno-García, E., and Ortega-Gutiérrez, F., 1999, Review of Paleozoic stratigraphy of México and its role in the Gondwana-Laurentia connections, *in* Ramos, V. A., and Keppie, J. D., eds., Laurentia-Gondwana Connections before Pangea: Boulder, Colorado, Geological Society of America Special Paper 336.

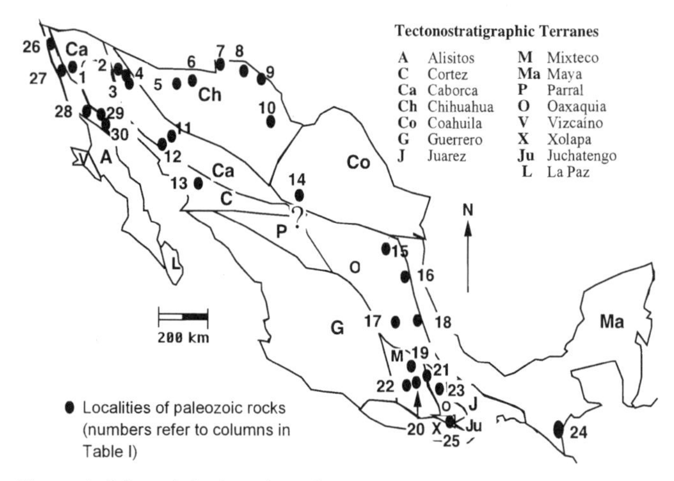

Figure 1. Map of the location of the Paleozoic sedimentary rocks and tectonostratigraphic terranes of México (modified from Campa and Coney, 1983).

In southern México, Paleozoic sedimentary rocks are exposed in the Mixteco, southern Oaxaquia, Juchatengo, and part of Maya terranes (Fig. 1).

CRYSTALLINE BASEMENTS

The oldest sialic basements of México are exposed in the Caborca (Longoria and Perez, 1979; Anderson et al., 1979), and Chihuahua terranes where they are interpreted as the southern extension of the North American craton. These complexes are older than 1.7 Ga (Anderson and Silver, 1981). In contrast, the Oaxaquia block and the southeastern parts of the Chihuahua terrane are floored by Middle Proterozoic rocks that include Novillo Gneiss, Los Filtros, Oaxaca, and Guichicovi complexes, respectively (Fig. 1). All have metamorphic ages between 911 and 1,080 Ma, and neodymium model ages from 1.4 to 1.8 Ga (Anderson and Silver, 1971; Ortega-Gutiérrez et al., 1977; Patchett and Ruiz, 1987; Murillo-Muñetón, 1994), which are similar to the those of Grenville Province of North America. Ortega-Gutiérrez et al. (1995) proposed that the Proterozoic basements of the Oaxaquia block, part of Juárez, and the Maya and Coahuila terranes might have evolved together as a large piece of continent, termed Oaxaquia. These basements, the Novillo, Huiznopala, Oaxaca, and Guichicovi complexes, share the following similarities: (1) massif type anorthosite-charnockite complexes, (2) protoliths rich in sedimentary rocks of shallow-marine platform or continental rift–related facies, (3) common granulite metamorphic facies, (4) a U-Pb zircon age of about 1.0 Ga, and (5) apparently common cooling histories. These evidences strongly suggests a coherent geologic history for this continental block (Ortega-Gutiérrez et al., 1995; Keppie and Ortega-Gutiérrez, 1995). Therefore, descriptions of Precambrian terranes are referred to herein as Northern (northern Sierra Madre terrane), Central (central Sierra Madre terrane), and Southern (Zapoteco terrane) Oaxaquia.

The nature of the basements of the Coahuila and Cortéz terranes is unknown. The oldest units exposed in both terranes are low-grade metamorphic rocks of Paleozoic age (Coney and Campa, 1987), known only from wells and from fragments in sedimentary cover.

The basement of the Mixteco terrane (Campa and Coney, 1983), located west of Oaxaquia (Ortega-Gutiérrez et al., 1995) (Fig. 1), is constituted by low- to high-grade metamorphosed igneous and sedimentary rocks (Acatlán complex) of pre-Mississippian age (Ortega-Gutiérrez, 1978). The evolution of the Acatlán complex is very significant to the pre-Mesozoic North America geology because of its geographic position, its early to middle Paleozoic age, polymetamorphic high- to low-pressure character, its protolith character (ocean floor, trench turbidites, reworked Proterozoic crust), and its structural relationship with the Oaxaca complex (Ortega-Gutiérrez, 1993).

Basement of the Maya terrane is variable in composition and age. There are reports of Cambrian metamorphic rocks in the Yucatan Peninsula (Dallmeyer, 1982). Precambrian rocks in central Oaxaca (Guichicovi complex) are as old as 1.2 Ga in age (Weber and Köhler, 1997), and exposures of the Chuacus schist on the Chiapas-Guatemala border yield Precambrian (1.0 Ga) ages as well (Gomberg et al., 1968; McBirney and Bass, 1969).

Below, we summarize the stratigraphy of the Paleozoic sedimentary sequences of México. We have chosen to divide this review into temporal periods. The references cited in the text are purposely selective rather than exhaustive. Detailed descriptions of the main Paleozoic formations of México are available through the Geological Society as Data Repository Item #200021 (DR #200021).

PALEOZOIC SEDIMENTARY COVER

Late Proterozoic to Cambrian

Sedimentary Late Proterozoic to Cambrian rocks are restricted to the northwest areas of México, in the Chihuahua and Caborca terranes (Figs. 1, 2, and 3). In the Chiuahua terrane the sedimentary units (Figs. 1 and 2) are mostly quartzitic sandstones and arkoses that change upward to limestone with chert and some dolostone in the Chihuahua terrane. They were defined as the Bolsa, Capote, Esperanza and Abrigo Formations, and range in age from Early to mid-Late Cambrian (Fig.2; DR #200021) (Bridges, 1970; Aponte, 1974; Hayes, 1975). This sequence was deposited in shallow marine environments and has been interpreted as the southern extension of the western North America miogeocline (Bridges, 1970; Stewart et al., 1990).

Late Proterozoic rocks of the Caborca terrane consist of arkosic and quartzitic sandstone that changes upward to limestone and includes: the Arpa, Caborca, Clemente, Pitiquito, Papalote, Gamuza and Tecolote Formations (Arellano, 1956; Cooper and Arellano, 1946; Anderson et al., 1979; Longoria and Pérez, 1979; Stewart et al., 1984). They contain stromatolites of Middle to Late (Riphean) Proterozoic age, and have been interpreted as shallow marine deposits (Cevallos and Weber, 1980;

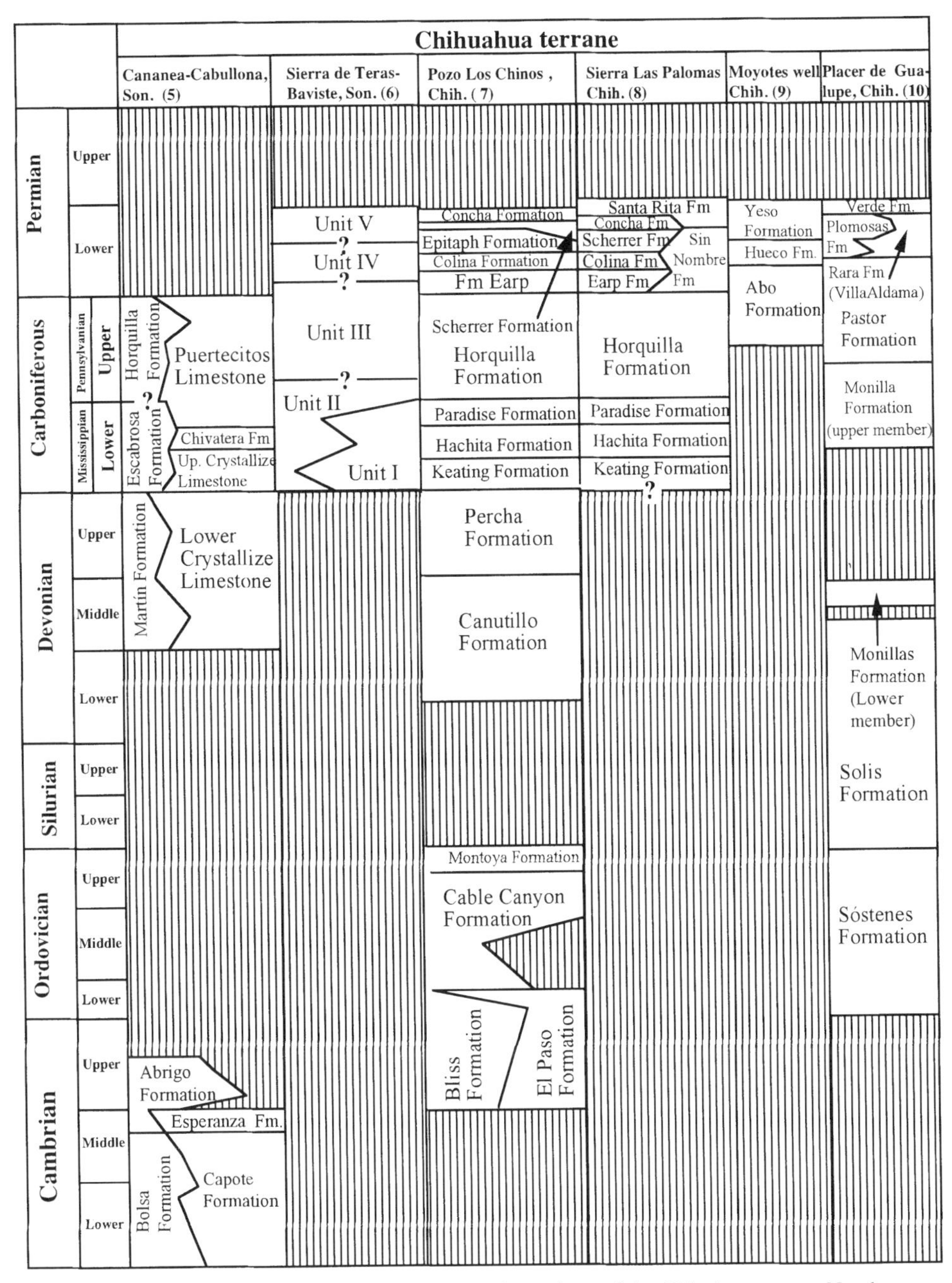

Figure 2. Correlation chart for the major Paleozoic formations of the Chiuahua terrane. Numbers are referred to in Figure 1.

Weber and Cevallos, 1980). They lie unconformably on the ~1.7-Ga Bamori complex (Damon et al., 1962; Anderson et al., 1979; Stewart et al., 1984). The sequence changes transitionally to interbedded limestone, dolomite, and some quartzitic sandstone of La Ciénega and Puerto Blanco Formations (Stewart et al., 1984). Basaltic lava flows and volcanic conglomerates interbedded with quartzitic sandstone lie between the La Ciénega and Puerto Blanco Formations, near the transitional contact between the Precambrian and Cambrian (Stewart et al., 1984). The Puerto Blanco Formation is made up of quartzitic sandstone and siltstone, limestone, and conglomerate. The lava flows are alkaline basalts similar to those from modern rift settings (Anderson, 1993). The Playa San Felipe Group of Baja California is made up of sandstone, siltstone, limestone and alkaline basalts, and it has been correlated with the Puerto Blanco Formation (Fig. 3; DR #200021) (Anderson, 1993). Similar basaltic horizons are exposed in several localities along the western North America miogeocline (Poole et al., 1992).

The Cambrian sequence continues with quartzites of the Proveedora Formation, and limestone and siltstone of the Buelna,

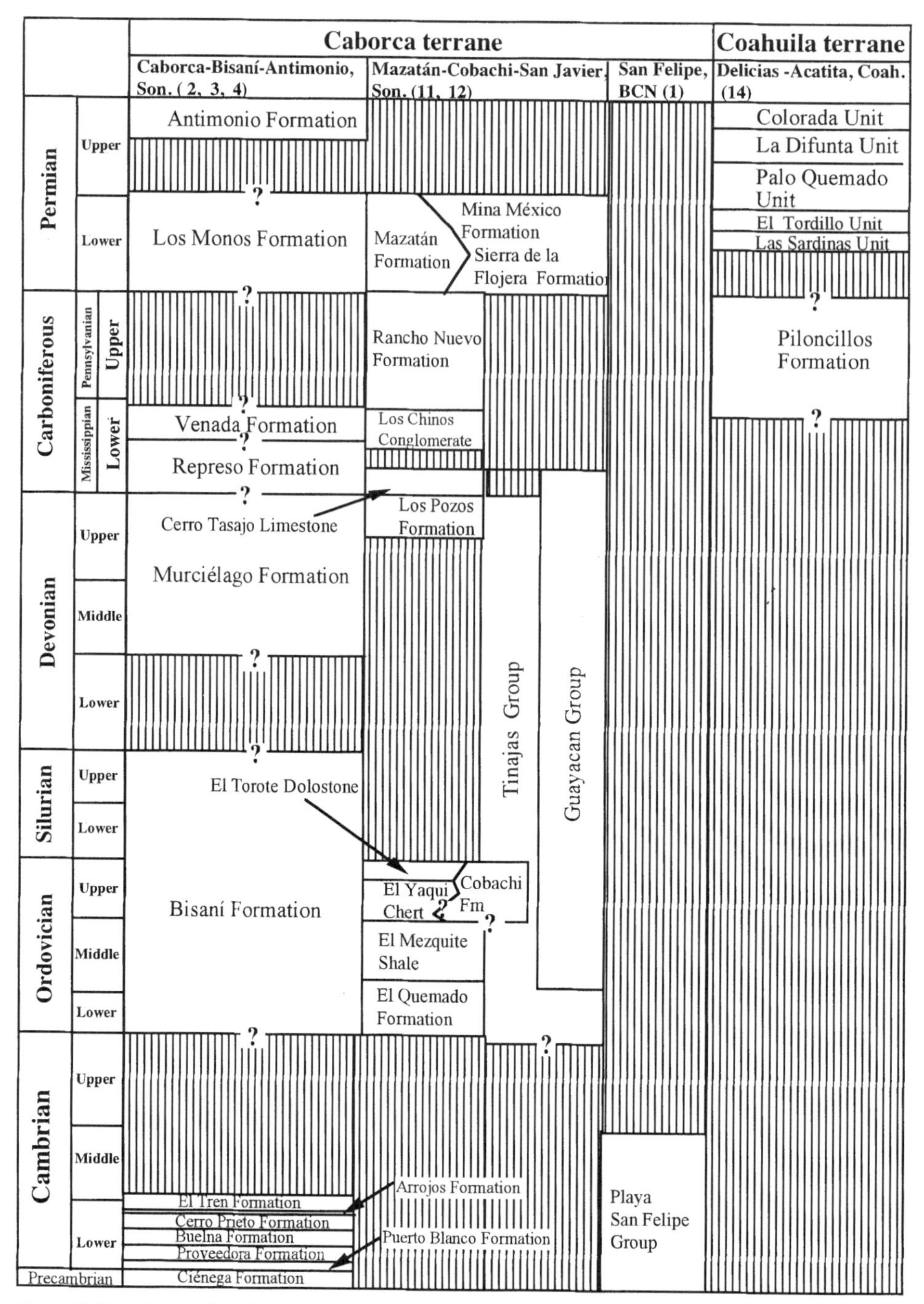

Figure 3. Correlation chart for the Paleozoic formations of the Caborca and Coahuila terranes. Numbers are referred to in Figure 1.

Cerro Prieto, Arrojos and El Tren Formations (Fig. 3) (Stewart et al., 1984). Clastic and calcareous rocks of northern Baja California (Playa San Felipe Group) have been considered of Late Proterozoic to Cambrian age, owing to lithologic similarities with the Proveedora and Puerto Blanco Formations in Sonora and rocks in Nevada of the same age (Anderson, 1993). However, fossils have not been reported from this area.

The Late Proterozoic to Cambrian rocks of the Caborca terrane were deposited on a shallow marine cratonal-platform environment (Stewart et al., 1990), and have been interpreted as either the continuation, or laterally displaced parts, of the western North American margin (Stewart et al., 1990). Stratigraphy and lithologic characteristics are similar to those in the Precambrian-Paleozoic sequences of Nevada, which are made up of sandstone, carbonates, and basaltic lava flows (Poole et al., 1992). Isotope ages of the Proterozoic magmatism in the Cordillera of western North America range from 918 to 730 Ma (Poole et al., 1992). The isotope ages do not agree with paleontological dating, since fossil faunas from the same units are interpreted to be Late Proterozoic to Cambrian (650–570 Ma) (Poole et al., 1992). Although accurate dating has not been carried out in Sonora, the sedimentary sequence between the basaltic lava flows and the layers with archaeocyathids is thin, suggesting that the lavas might be uppermost Late Proterozoic or Early Cambrian (Anderson et al., 1979; Stewart, 1988). If this younger age is proved, the rifting event in Sonora would thus be much younger than in other parts of the North American Cordillera.

Cambrian to Ordovician

Exposures of Cambro-Ordovician sedimentary rocks are rare in México (DR #200021). They are limited to some outcrops of Ordovician to Silurian rocks in northwest México (Caborca terrane), Late Cambrian to Late Ordovician sediments on the margin of the North America craton (Chihuahua terrane), and small outcrops of Tremadocian rocks in southern México (southern Oaxaquia) (Figs. 1–6).

Cambro-Ordovician strata of the North American craton (Chihuahua terrane) are made up of limestone, shale with chert, and dolostone. They are primarily Lower Ordovician in age, but extend up to the Lower Devonian (Sóstenes and Solis Formations) in central areas of the Chihuahua terrane. Drilling in northern parts of the terrane has recorded Late Cambrian to Late Ordovician sediments of the Bliss, El Paso, Cable Canyon and Montoya Formations (Reynolds, 1972) that were deposited in shallow marine environments. Their stratigraphy and faunal content are similar to those in Arizona and New México, suggesting continuity with the western North America miogeocline (Stewart, 1988).

Limestone and dolomite are exposed in the northern parts of the Caborca terrane in central Sonora. They contain conodonts with a wide range in age, from Ordovician to Silurian (Brunner, 1975). These rocks have been described as the Bisaní Formation and were deposited in shallow marine environments (Cooper and Arellano, 1946). Contacts with other units are not exposed; however, they are inferred to be unconformable with the underlying El Tren Formation (mid-Cambrian) and overlying Murciélago Formation (Devonian), based on regional relationships. In contrast, rocks of the southern Caborca terrane, in central Sonora (Mazatán, Cobachi, and San Javier areas) are made up of shales, arkoses, and conglomerates, with some limestone. Units of the Cobachi area are the Cobachi Formation and Guayacan Group, which have fossils of graptolites that suggest mid-Ordovician and Ordovician to Mississippian ages, respectively. The Tinajas Group is made up of quartzite, limestone, and dolostone and is Ordovician-Devonian in age (Keith and Noll, 1987). The Guayacan Group consists of deep marine siliciclastic sequences, and ranges in age from mid-Ordovician to Devonian (Noll, 1981). Formations of the Mazatan area (Fig. 3) are El Quemado, El Mezquite, El Yaqui, and El Torote (Poole and Madrid, 1988). They contain fossils that range from Early to Late Ordovician age and were deposited on a platform-shelf environment (Poole et al., 1995). These units, made up shale, sandstone, and chert, have been interpreted as part of the eugeocline sequence (Stewart et al., 1990). Both mio- and eugeocline sequences are covered unconformably by Devonian sediments (Poole and Madrid, 1988).

Sedimentary rocks of the northwestern Cortéz terrane (Baja California Peninsula) are made up of quartzite, calcareous rocks and chert (Rancho San Marcos group) that contains conodonts of Late Ordovician age (Fig. 3) (Lothringer, 1993). Metamorphosed volcanic and sedimentary units from the Cortéz terrane (Fig. 4) (El Fuerte Formation) are interpreted to be early Paleozoic in age (Mullan, 1978; Gastil and Miller, 1983), based on their regional relationships (Fig. 4; DR #200021).

Ordovician (Tremadocian) sedimentary rocks of Oaxaquia are located in the south, and they rest unconformably on the Grenvillian-age Oaxacan complex (Fig. 5) (Robison and Pantoja-Alor, 1968; Pantoja-Alor, 1970). They belong to the Tiñú Formation, that is made up of interbedded limestone and black shale containing abundant trilobites of Tremadocian age (DR #200021) (Robison and Pantoja-Alor, 1968). It was deposited in an external platform-slope or mid-fan environment, and it is tectonically overlain by Mississippian rocks of the Santiago Formation (Pantoja-Alor, 1970; Centeno-García et al., 1997). Rocks that were originally described as Cambrian and Ordovician in northern Oaxaquia (La Presa Quartzite and Victoria limestone) (Carrillo-Bravo, 1961) are now known to be part of the Precambrian basement and Silurian sediments, respectively (Fig. 6) (Ramírez-Ramírez, 1978; Stewart et al., 1993; Boucot et al., 1997).

SILURIAN TO DEVONIAN

The geologic evolution of México from Silurian to Devonian is complex. Sedimentary rocks of Silurian-Devonian time are exposed in the Chihuahua, Caborca, and Cortéz terranes and northern Oaxaquia (Figs. 2–4, 6). Carbonate rocks and shale from the Chihuahua terrane (Martin Formation, Lower Crystallize Limestone, Solis, Canutillo, Monillas, and Percha Formations) (Cooper and Arellano, 1946; Bridges, 1964; Reynolds, 1972) were deposited on shallow marine settings. As with most of the Paleozoic sequences of Chihuahua, these rocks are correlated with those from Arizona, New Mexico, and Texas, and form part of the southern platform deposits of Laurentia (DR #200021) (Stewart et al., 1990). The Siluro-Devonian sequence of the Caborca terrane is made up of limestone and shales deposited in shallow marine environments (Bisaní, and Murciélago Formations) (Cooper and Arellano, 1946; Mulchay and Velasco, 1954). Upper and lower contact relationships between the Murciélago Formation and other units are unknown. However, lower contact with the Bisaní Formation is interpreted to be discordant (Cooper and Arellano, 1946). The Bisaní and Murciélago Formations have been correlated with the sequences of the western Cordillera in Nevada (Stewart, 1988).

Middle Paleozoic rocks of northern Oaxaquia comprise the Cañón de Caballeros and La Yerba Formations (Fig. 6; DR #200021). These rocks are described in detail by J. Stewart et al.

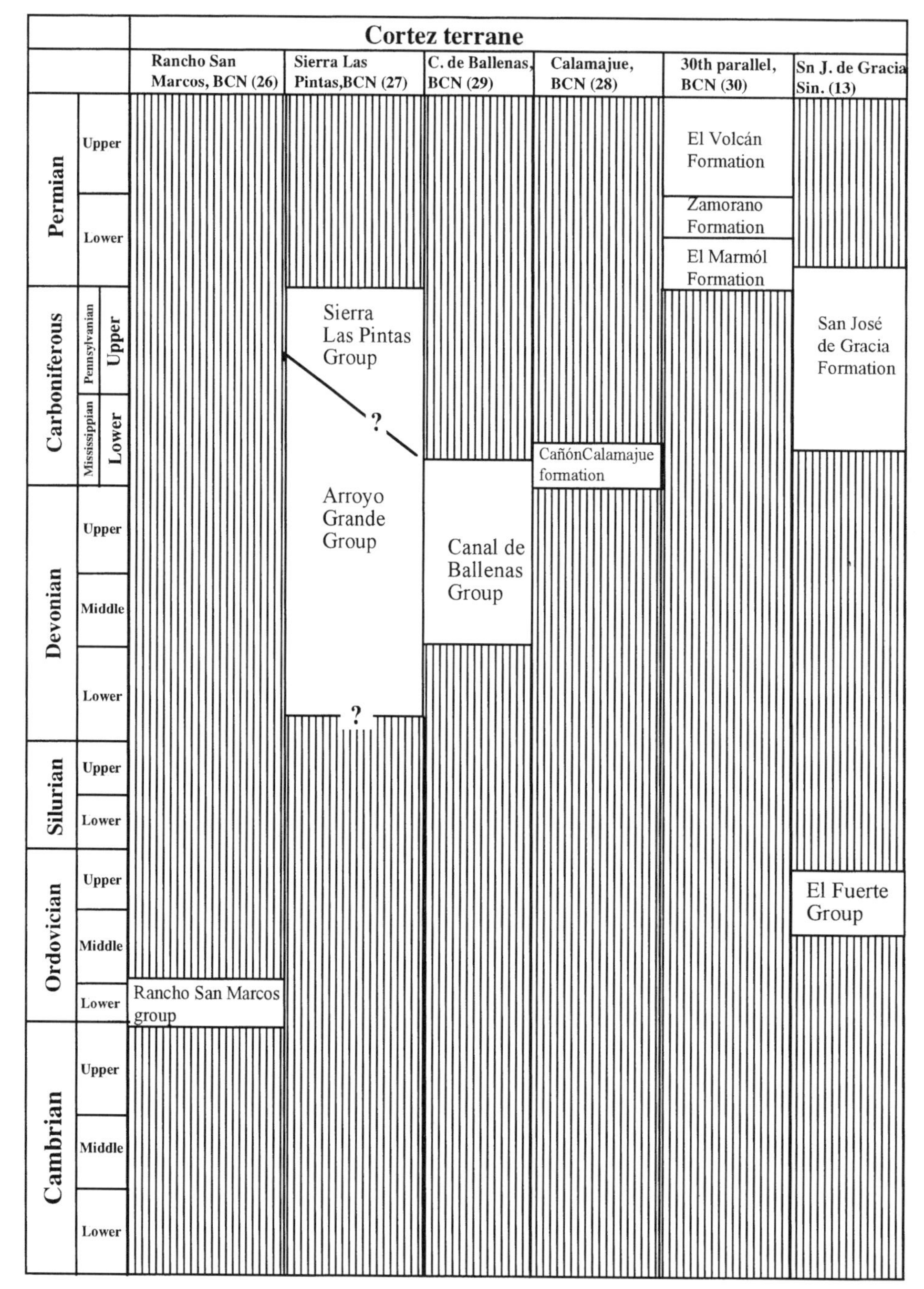

Figure 4. Correlation chart for the Paleozoic units of the Cortéz terrane. Numbers are referred to in Figure 1.

in this volume. The Cañón de Caballeros Formation consists of limestone and interstratified shale and quartzitic sandstone (Carrillo-Bravo, 1959, 1961). New studies of fossil brachiopods from the Cañón de Caballeros Formation indicate a Late Silurian age (Stewart et al., 1993; Boucot et al., 1997). La Yerba Formation consists of siliceous rocks, shale, sandstone, and limestone, and is considered to be part of the Cañón de Caballeros Formation (Stewart et al., 1993). Rocks that were described as novaculite of the La Yerba Formation (Carrillo-Bravo, 1961) are now known to be rhyolitic volcanic rocks of Mississippian age (Gursky and Ramírez-Ramírez, 1986).

Middle Paleozoic rocks from northern Oaxaquia have traditionally been considered to be a southern continuation of the Ouachita orogenic belt of Texas and adjacent areas (Flawn et al., 1961). However, the Cañón de Caballeros and the La Yerba Formations contain fauna with Old World Realm (Rhenish-Bohemian region; Boucot, 1975; Boucot et al., 1997). This is distinctly different from fauna of comparable age in North America (Stewart

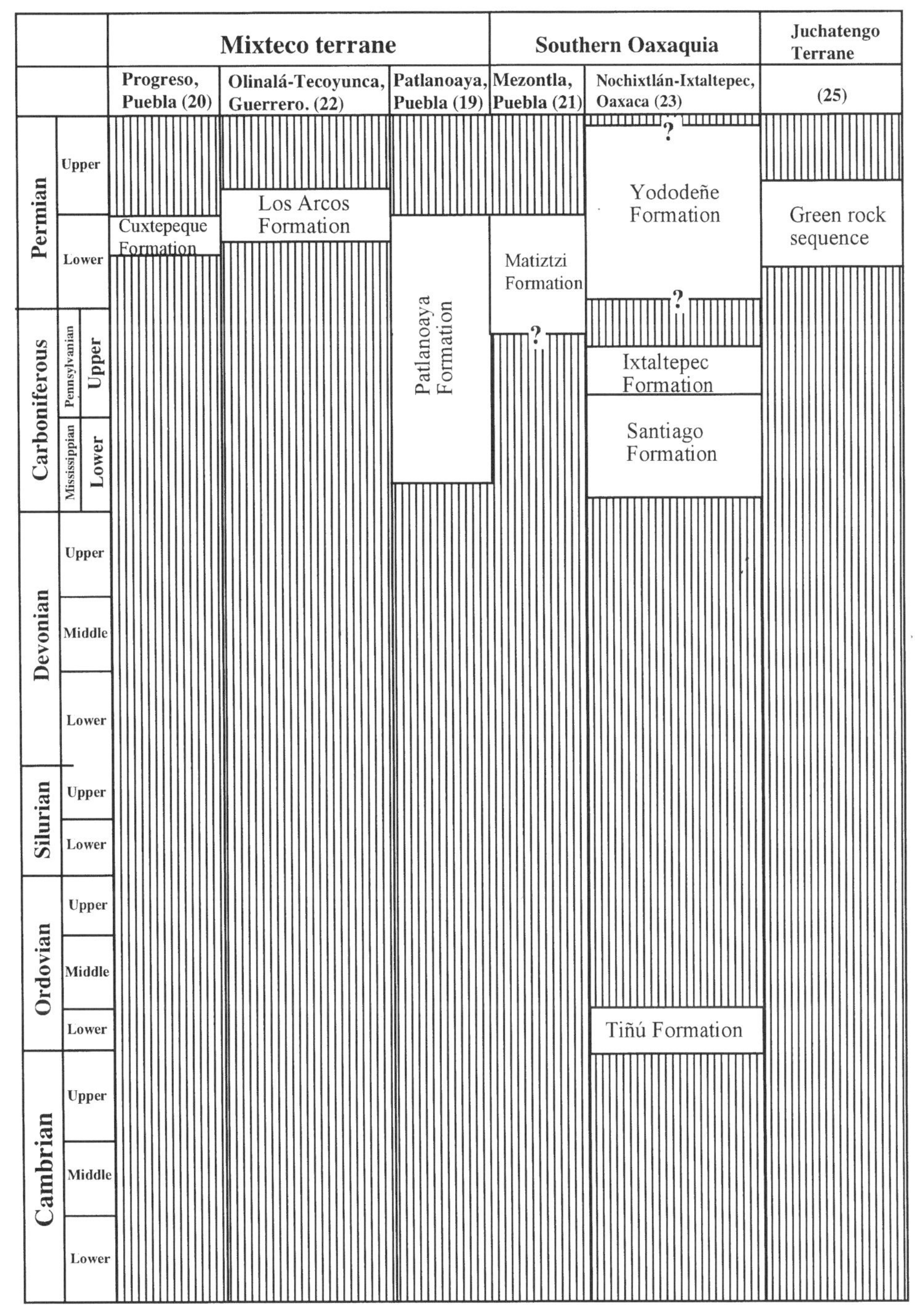

Figure 5. Correlation chart for the Paleozoic formations of southern Oaxaquia and the Mixteco terrane. Numbers are referred to in Figure 1.

et al., 1993). The La Yerba fauna has more similarities with faunas from the Mérida Andes of Venezuela (Boucot et al., 1997).

The Cortéz terrane contains mid- to Late Devonian rocks (Figs. 1 and 4; DR #200021). The terrane is made up of volcaniclastic rocks interbedded with shale, limestone, and some basaltic pillow lavas (Arroyo Grande and Canal de Ballenas Groups) (Campbell and Crocker, 1993; Leier-Engelhardt, 1993), and contains conodonts and some fossil corals.

Other rocks of early and middle Paleozoic times are metamorphic, such as the Acatlán complex that forms the basement of the Mixteco terrane (Campa and Coney, 1984; Sedlock et al., 1993). Its evolution is important for understanding the paleogeography of early and middle Paleozoic in México. The Acatlán complex includes marine sedimentary sequences, syn-collisional granitoids, high-level syn-orogenic to late tectonic granitoids, oceanic mafic-ultramafic complexes, and tectonic inclusions

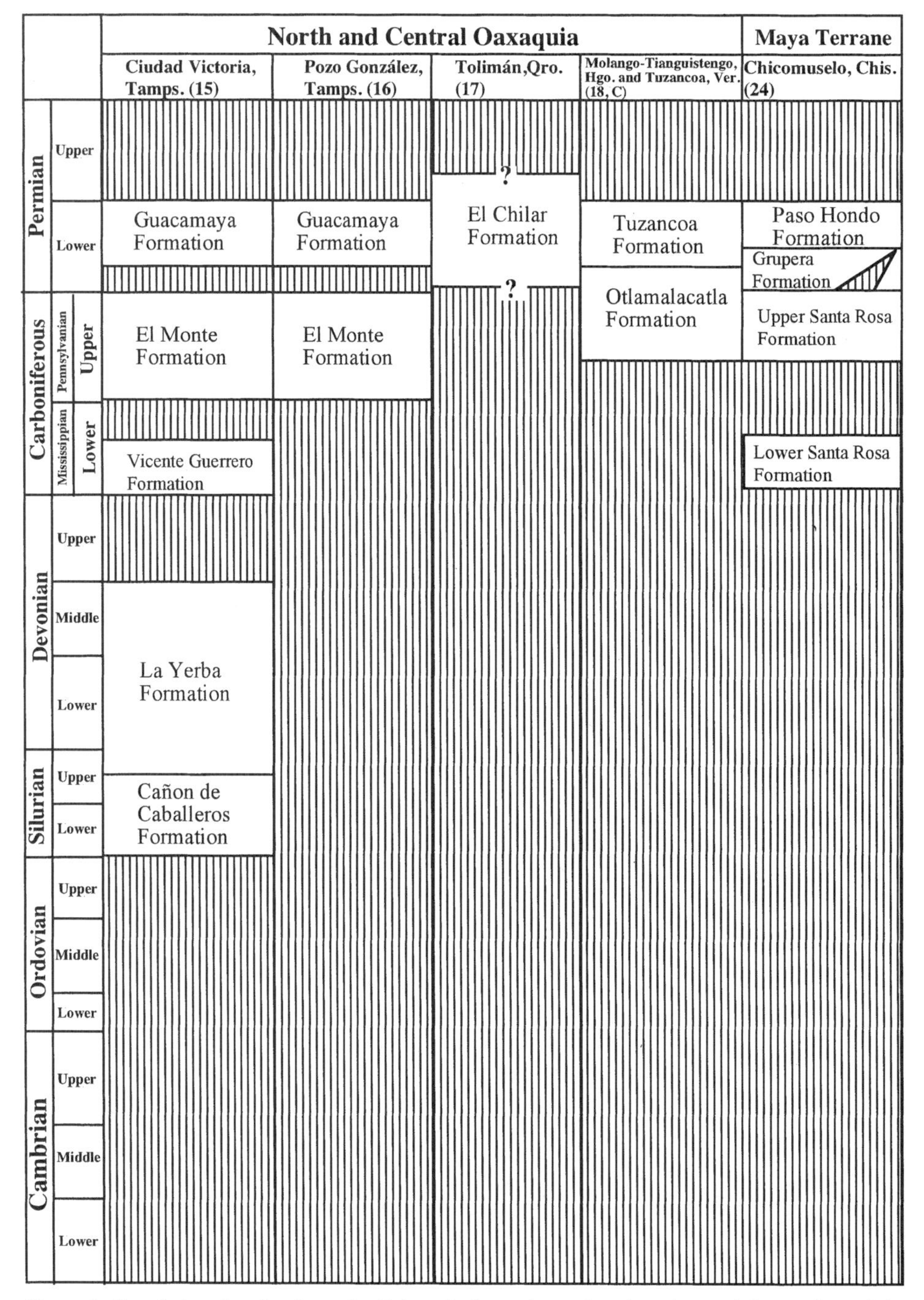

Figure 6. Correlation chart for the major Paleozoic formations of north and central Oaxaquia, and the Maya terrane. Numbers are referred to in Figure 1.

within flysch lithologies (Ortega-Gutiérrez, 1978, 1993; Ortega-Gutiérrez et al., 1997).

MISSISSIPPIAN TO PERMIAN

Carboniferous to Permian rocks are the most abundant of all the Paleozoic sequences of México (Figs. 1–6; DR #200021). However, their exposures are isolated in each terrane and only a few are overlapping major tectonic contacts.

Late Paleozoic rocks deposited on the North America Craton (Chihuahua terrane) are Carboniferous only at the western part in Sonora, and extend up to the Permian in Chihuahua State. Overall, composition of Carboniferous rocks is homogeneous throughout the Chihuahua terrane. However, these rocks have been defined with variable nomenclature in each locality

(Figs. 1 and 2). The Carboniferous rocks are defined as the Escabrosa and Horquilla Formations; the Upper Crystallize Limestone, Chivatera Formation, and Puertecitos Limestone; and Units I, II, and II in Sonora (DR #200021) (Mulchay and Velasco, 1954; Viveros, 1965; Tovar, 1968a; Aponte, 1974). They are similar in lithology and faunal content with the Keating, Hachita, Paradise, Horquilla Abo, Monilla, Pastor, and Rara Formations of Chihuahua (Fig. 2; DR #200021) (Diaz and Navarro, 1964; Bridges, 1964; Tovar, 1968b). Most of them are made up of limestone, calcareous shale, and sandstone, with some chert nodules that were deposited in shallow marine environments. The faunal content is abundant, and includes a wide range of corals, brachiopods, fusulinids, crinoids, and bivalves. These rocks are more terrigenous toward the west-northwest (Chihuahua State). Pennsylvanian rocks are more clastic than the Mississippian sandstone and conglomerate in central-north Chihuahua State (Horquilla Formation). Overall, the percentage of clastic material increases toward the west throughout the Chihuahua terrane.

Deposition in shallow marine environments continued in the Early Permian in the Chihuahua terrane. Dominant lithologies are limestone, dolostone, shale, and sandstone with abundant invertebrate fossil fauna (Units IV and V, Earp, Colina, Scherrer, Epitaph, Concha, Abo, Hueco, Rara, Villa Aldama, Plomosas, and Verde Formations) (Diaz and Navarro, 1964; Tovar, 1968b; Patterson, 1978). Locally there are sequences of evaporites in northern Chihuahua (Yeso Formation). The ages of these sequences range from Wolfcampian to Leonardian. Permo-Carboniferous rocks of the Chihuahua terrane are interpreted to be the southern extent of the continental platform of western North America (Stewart et al., 1990).

Late Paleozoic rocks of the Caborca terrane can be grouped into two belts (Figs. 1 and 3; DR #200021): (1) The first belt is a belt of calcareous shallow marine rocks of an age that extends from Carboniferous to Permian time. These rocks are made up of very fossiliferous limestone and dolostone, and are formed by the Represo and Venada Formations of Kinderhookian-Osagean and Late Meramecian ages, respectively (Cooper and Arellano, 1946). They are interpreted as miogeoclinal deposits (Stewart et al., 1990). Contact relationships of the Venada and Represo Formations with other Paleozoic units are not exposed (Cooper and Arellano, 1946). This belt also includes limestone of the Monos Formation of Early Permian age, and part of the El Antimonio Formation of Late Permian age (González-León, 1996). The El Antimonio Formation is made up of siltstone and debris-flow deposits suggestive of deeper environments of deposition (González-León, 1996). This belt is interpreted as miogeoclinal deposits and trends northwest-southeast along the Sonora State (Fig. 3). (2) The second belt is a belt of sedimentary rocks of deeper water facies that lies westward of the miogeoclinal belt. It is made up of limestone, interbedded shale and sandstone, and some conglomerate (Cerro Tasajo limestone, Los Chinos conglomerate, and Rancho Nuevo and Mazatán Formations). These units range in age from Early Carboniferous to Early Permian, and are interpreted as eugeoclinal deep marine sediments (Poole and Madrid, 1988).

There are two major tectonic events affecting the eugeoclinal rocks. The oldest, an Early Mississippian thrusting event that verges toward the south is found in the pre-Mississippian rocks of southern Caborca terrane (Stewart et al., 1990). These are unconformably overlain by Upper Mississippian and Pennsylvanian sediments (Los Chinos conglomerate and Rancho Nuevo Formation) (Stewart et al., 1990). Apparently, the second and youngest tectonic event emplaced the Sierra de La Flojera Turbidites, Mina Vieja, and Mazatán Formations over the miogeoclinal sequences by Late Permian time (Stewart et al., 1990). These major regional disconformities have not been identified in other areas of México. This is also the only area in which the Paleozoic-Mesozoic transition has been reported in México.

The Cortéz terrane contains few exposures of Permo-Carboniferous rocks (Figs. 1 and 4; DR #200021). They are made by strongly deformed limestone at the base, followed by quartzitic sandstone, shale, and thick sequences of banded chert in southeastern Cortéz terrane (San José de Gracia Formation) (Carrillo-Martínez, 1971). This unit contains fossils of Late Mississippian to Late Pennsylvanian age (Carrillo-Martínez, 1971), and is considered to be allochthonous with respect to the Caborca terrane. Contact relationships with the metamorphosed El Fuerte sequence that contains volcanic and volcaniclastic rocks has not been well understood. Other units are a sequence of calcareous shale and sandstone interbedded with hyaloclastites, and pillowed and massive basaltic flows (Sierra Las Pintas Group). This group is exposed in the northwestern parts of the Cortéz terrane (Baja California). A sequence of low-graded metamorphosed shale, sandstone, and chert is exposed in the area (Cañón de Calamajue Formation). Fossils that belong to this unit range in age from Early Devonian to Early Mississippian (Campbell and Crocker, 1993; Leier-Engelhardt, 1993). Permian rocks in the northern Cortéz terrane (Fig. 4) are mostly deep marine shale, quartzite, and limestone (El Mármol, Zamorano, and El Volcán Formations). They were deposited in a continental slope basin, and are interpreted to be part of the Sonoran eugeoclinal sequences. The age of deformation of late Paleozoic rocks of the Cortéz terrane is still uncertain.

Exposures of late Paleozoic rocks in the Coahuila terrane (Figs. 1 and 3) are made up of volcaniclastic sequences (shale, sandstone, and conglomerate), interbedded with layers of calcareous debris flow and volcanic rocks that form blocks several meters in diameter (Piloncillos Las Sardinas, El Tordillo, Palo Quemado, La Difunta, and Colorada units) (King, 1944; Wardlaw et al., 1979). They contain abundant crinoids, fusulinids, trilobites, ammonoids, and other fossils of Leonardian to Guadalupian age (DR #200021). These units represent sediments associated with a submarine volcanic arc (McKee and Jones, 1988).

There are several exposures of late Paleozoic rocks in the Oaxaquia block. In northern Oaxaquia (Ciudad Victoria), they are formed by the Vicente Guerrero, Del Monte, and Guacamaya

Formations of Early Mississippian, mid-Late Pennsylvanian, and Early Permian age, respectively (Figs. 1 and 6; DR #200021). Rocks in the Ciudad Victoria area are described by Stewart et al. in detail in this volume. They are mostly alternating sandstone, shale, siltstone, and some conglomerate, and contain some calcareous intervals (Stewart et al., 1993). Upper parts of the Vicente Guerrero Formation contain a rhyolitic flow that yielded early Mississippian U/Pb age (334 ± 39 Ma) (Stewart et al., this volume). These units are interpreted as a "flysch" of orogenic origin (Carrillo-Bravo, 1965).

Similar nomenclature to the Ciudad Victoria area was used to described Permian rocks in central Oaxaquia (Molango area) (Carrillo-Bravo, 1965). However, recent studies have shown that the stratigraphy differs from stratigraphy in Ciudad Victoria. Paleozoic rocks in the Molango area are made of large volumes of volcanic and volcaniclastic rocks (Figs. 1 and 6) (Rosales-Lagarde et al., 1997). This unit is made of andesitic brecciated and massive lava flows, volcanic conglomerates, and tuffs, with some interbedded limestone, shale, and volcanic sandstone. It changes upward to alternating volcanic shale, sandstone, and conglomerate deposited as turbidites (Otlamalacatla and Tuzancoa Formations) (Rosales-Lagarde et al., 1997). Fossil crinoids, fusulinids, and brachiopods are dated as Wolfcampian-Leonardian in age, but there are some brachiopods that might be Pennsylvanian in age (Carrillo-Bravo, 1961; Moreno-Cano and Patiño-Ruiz, 1981). This sequence is similar to the Permian rocks from the Coahuila terrane in north-central México, and to volcaniclastic rocks of the Tolimán area in central Oaxaquia (El Chilar Formation) (Figs. 1 and 6).

Late Paleozoic rocks of southern Oaxaquia are the Santiago and Ixtaltepec Formations; the Matzitzi Formation, deposited over the contact between Oaxaquia and the Mixteco terrane; and the Cuxtepec, Patlanoaya, and Los Arcos Formations, located in the Mixteco terrane (Figs. 1 and 5) (Villaseñor-Martínez et al., 1987; Corona-Esquivel, 1981; Silva-Pineda, 1970; Weber and Cevallos, 1994).

The Santiago Formation is made up of limestone, shales, and sandstones that contain Early Mississippian faunas. This unit was deposited in a shallow marine environment and rests tectonically over the Tremadocian Tiñu Formation. The Ixtaltepec Formation consists of interbedded shale and sandstone, with some limestone beds. It contains abundant Lower-Middle Pennsylvanian invertebrates, which indicates a shallow marine environment (Figs. 1 and 5; DR #200021) (Pantoja-Alor, 1970; Sour-Tovar and Quiroz, 1991). These units are intruded by rhyolitic and andesitic sills. The sequence continues to redbeds of the Yododeñe Formation of unknown age. The latter is considered to be Permian in age because of stratigraphic position (Pantoja-Alor, 1970). Late Paleozoic rocks of southern Oaxaquia were affected by extensional low-angle sliding, folding, and lateral shearing before mid-Cretaceous time (Centeno-García et al., 1997).

The Matzitzi Formation was originally described as a continental clastic unit of Pennsylvanian age that rests unconformably on the Precambrian Oaxacan Complex (Silva-Pineda, 1970). However, Weber and Cevallos (1994) have extended the age range to Permian (Leonardian). This unit is made up of intercalated conglomerates, shale, and sandstone with abundant fossil flora. Recent mapping of the area has shown that the Matzitzi Formation rests unconformably on the Acatlán complex (Mixteco terrane), and it contains at least one thick ignimbrite horizon (Figs. 1 and 5; DR #200021) (Centeno-García et al., 1997).

Late Paleozoic rocks in the Mixteco terrane are mostly marine. The Patlanoaya sequence, of Mississippian (Osagean) to Permian age, is made up of basal conglomerates, shale, sandstone, and some limestone layers deposited on a shallow marine setting. The Los Arcos Formation is made up of interbedded limestone, shale, and scarce sandstone and calcareous sandstone (Corona-Esquivel, 1981). It was considered to be unconformably overlain by the Las Lluvias Ignimbrite (Corona-Esquivel, 1981). However, abundant fragments of tuff in the sandstone, similar to the Las Lluvias ignimbrite, suggest that the ignimbrite might be contemporaneous to the Los Arcos Formation (Fig. 5) (Centeno-García et al., 1997).

The Juchatengo terrane consists of gabbroic and plagiogranitic dikes and stocks, basalts and volcaniclastic rocks, and interbedded black shale, chert, siltstone, sandstone, and minor limestone from which a possible Middle Permian ammonoid was recovered (Grajales-Nishimura, 1988). The Juchatengo terrane has been interpreted as remains of a back arc/ocean floor basin (Grajales-Nishimura, 1988). The boundary between the Oaxaquia and Juchatengo terranes is not exposed. However, the Juchatengo terrane is also inferred to have accreted to the southern margin of Oaxaquia by the Permian because both terranes are intruded by a suite of Permian-Triassic calc-alkaline plutons (Figs. 1 and 5) (Grajales-Nishimura, 1988; Torres-Vargas et al., 1993).

The Maya terrane is covered by a Mississippian to Permian sedimentary sequence. The Pennsylvanian-Mississippian Santa Rosa Formation (Chiapas State), is divided in two units (Figs. 1 and 6), and is made up of calcareous shale, and sandstone that contains abundant shallow marine fossils (Malpica, 1977). The Santa Rosa Formation is unconformably overlain by shale, limestone, and sandstone of the Grupera Formation (Malpica, 1977). The sequence continues transitionally to limestone and dolostone of the Permian Paso Hondo Formation (Malpica, 1977). Abundant crinoids, fusulinids, trilobites, brachiopods, and other fossils have dated these formations as Wolfcampian and Leonardian, respectively (DR #200021) (Malpica, 1977).

An important characteristic of the upper Paleozoic sequences of México is that volcanic and intrusive rocks are more abundant than in lower Paleozoic units. This magmatism can broadly be grouped into three distinctive suites: basic (basaltic-andesitic) submarine province of central México, which includes the magmatism of the Coahuila terrane, northern and central Oaxaquia; oceanic tholeiitic basalts of the Juchatengo terrane; and felsic (ignimbritic) magmatism of southern and northern México, which is exposed in southern Oaxaquia and the Mixteco and Chihuahua terranes, and also found in the Maya terrane in Belize.

OVERVIEW OF THE PALEOZOIC TECTONIC EVOLUTION

Global paleogeographic reconstructions for the Paleozoic have México positioned in a wide range of locations (Scotese and McKerrow,1990; Yañez et al., 1991; Sedlock et al., 1993; Stewart et al., 1993; Keppie, 1977; Ortega-Gutiérrez, 1993; Dalziel, 1997; Dalziel et al., 1994). However, we consider that the paleogeographic models summarized here are the most representative. They are mostly based on faunal and lithologic affinities, paleomagnetism, and the nature of the tectonic environment where they formed.

Precambrian to Cambrian evolution of northwestern México has been related to the evolution of the continental margin of western North America (Stewart et al., 1993). Sediments along this margin are associated with rifting from Middle Proterozoic to Middle Cambrian. This passive margin probably originated by the break-up of Rodinia (Poole et al., 1992; Dalziel, 1997). Early stages of rifting are of Middle to Late Proterozoic age in central parts of the western U.S. Cordillera. However, if magmatism in Sonora is found to be Proterozoic-Cambrian, it could suggest younger ages for the rifting than those proposed by Dalziel (1997) (Fig. 7).

Cambro-Ordovician tectonic evolution of northwestern México (Fig. 8) has been linked to Laurentia (Western Cordillera of North America) specifically with Nevada (Stewart et al., 1988). Sedimentary rocks of the Caborca terrane are interpreted to be displaced parts of the North American miogeosyncline of Nevada or the continuation of the western North America miogeosyncline (Anderson et al., 1979; Stewart et al., 1990). If an early Paleozoic age for the El Fuerte Formation (Cortéz terrane) is confirmed, this sequence could represent the remnants of an accreted island arc (Fig. 8). Sediments of the Chihuahua terrane are the continuation of the southern continental platform of Arizona and New Mexico (Stewart et al., 1990) (Fig. 8). Dalziel (1997) proposed a paleo transform that ran from Laurentia to Gondwana (Ellsworth-Mohave-Sonora transform, or EMST), that might have changed to subduction during Cambrian-Ordovician time. However, the lack of magmatism in the sequences of the Chihuahua terrane suggests that there was no subduction along the southwestern margin of Laurentia during that time (Fig. 8).

Paleogeographic evolution of Oaxaquia during Cambrian-Ordovician time is not well constrained. Paleozoic reconstructions have located the terrane around the Venezuelan and Colombian Andes (Keppie, 1977), southern North America (Scotese and McKerrow, 1990), near the Chilean Andes (Dalziel et al., 1994), or adjacent to the Amazon craton (Keppie and Ortega-Gutiérrez, 1995; Keppie et al., 1996). Tiñú faunas are not conclusive for paleogeographic reconstructions, but show more affinities with South America than with Europe, North America, and Africa (Whittington and Hughes, 1972; Sour-Tovar, 1990). Isotopic signatures and environment of deposition of the Tiñú Formation indicate that southern Oaxaquia might have been associated to a major cratonic land mass with a Precambrian core, such as Gondwana (Fig. 8) (Centeno-García et al., 1997).

Silurian to Devonian evolution of the Chihuahua terrane remains associated with the southern North America platform. Thus, the Oaxaquia block might have been either part of a Paleozoic exotic microcontinental block or a rifted fragment of a non–North American Paleozoic continent, perhaps Gondwana (Fig. 9) (Stewart et al., 1993; Boucot et al., 1997). Devonian volcanic and volcaniclastic sequences of the Cortéz terrane might have formed in an oceanic arc setting, but their petrogenesis has not been determined (Fig. 9).

The Oaxaquia block probably remained linked to Gondwana during Silurian-Devonian? time (Fig. 9). This is suggested by the Late Silurian brachiopods of northern Oaxaquia (Ciudad Victoria), which show more affinity to brachiopods from the Venezuelan Andes (Gondwana) (Boucot et al., 1997).

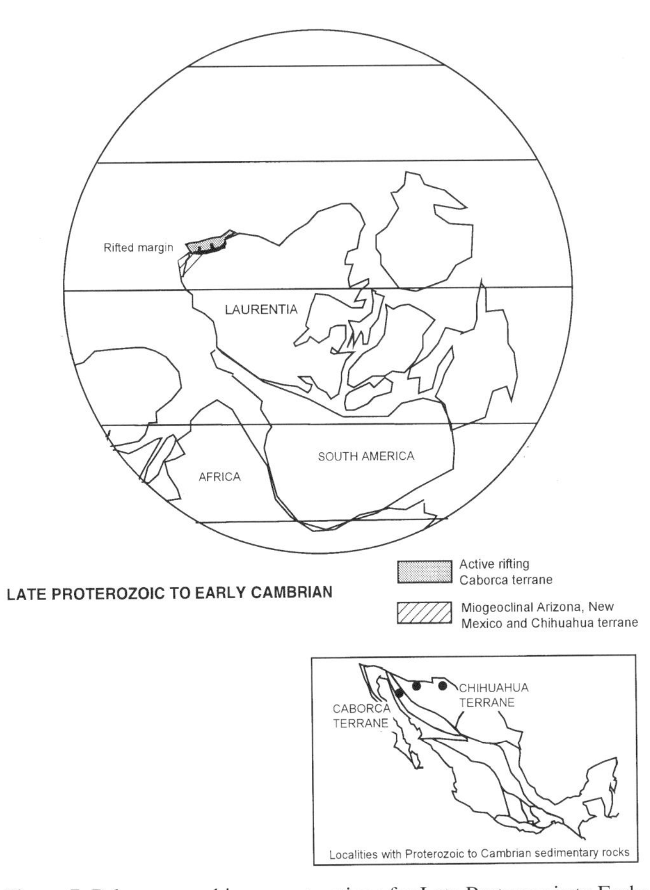

Figure 7. Paleogeographic reconstructions for Late Proterozoic to Early Cambrian time. Continent positions were modified from Dalziel (1997). Basaltic magmatism and sedimentary facies of the Caborca terrane suggest that rifting of the western–North America margin might have continued up to the Cambrian in Sonora.

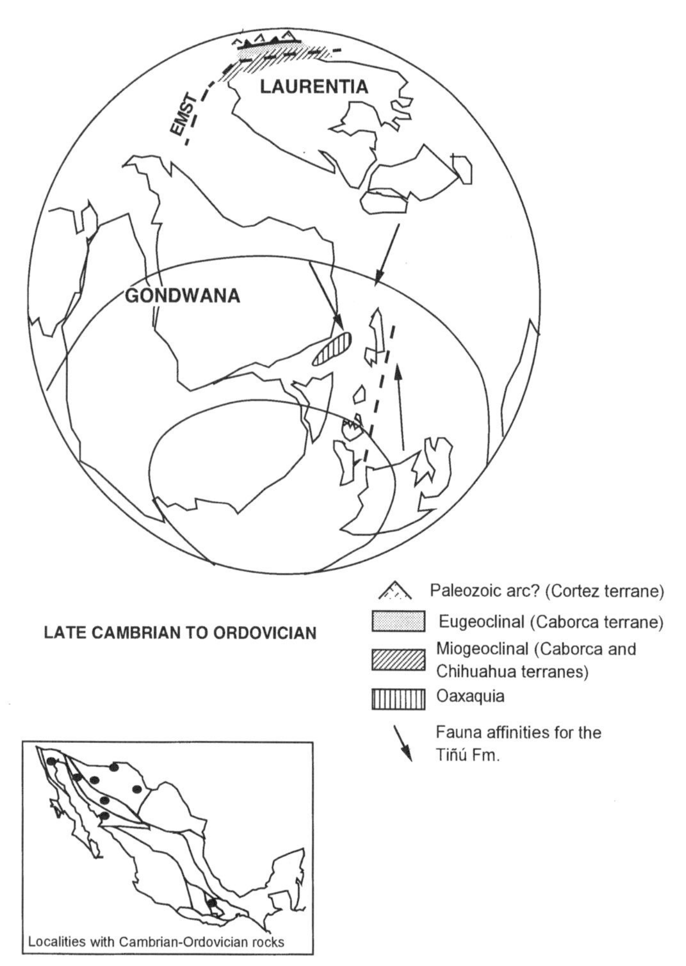

Figure 8. Cambrian to Ordovician Paleogeographic reconstruction (modified from Dalziel, 1997). Sedimentary characteristics of the Tiñu Formation suggest a Gondwana position for the Oaxaquia block. Sedimentary rocks of northwestern México (Caborca terrane) might have evolved north of their position (Stewart et al., 1990). They probably represent a passive margin sequence. Volcanic rocks of the Cortéz terrane might be remnants of an oceanic arc that was active during Paleozoic time. EMST = Ellsworth-Mohave-Sonora transform.

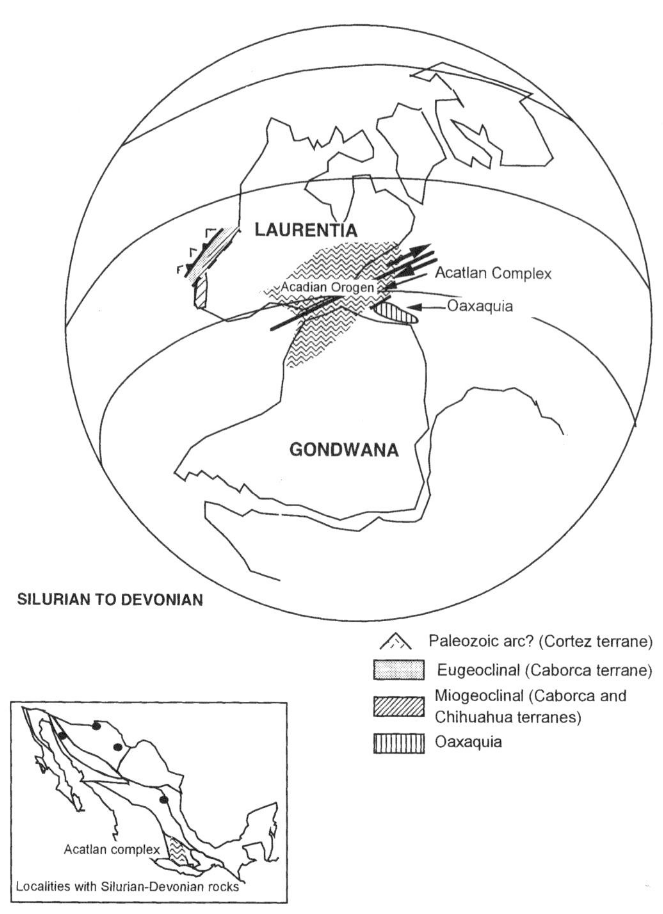

Figure 9. Silurian-Devonian reconstruction (modified from Kent and Van der Voo, 1990; Van der Voo, 1993). The Caborca and Cortéz terrane were still evolving in the Western Cordillera of North America. Rocks of the Acatlán complex were deformed during the Ordovician-Silurian and another phase in the Middle Devonian (Ortega-Gutiérrez et al., 1997). Oaxaquia might have been related to Gondwana.

The early history that characterized the formation of the Acatlán complex (Mixteco terrane) is poorly known. Two main tectonic events have been recorded in the Acatlán complex: one in the Ordovician-Silurian and another in the Middle Devonian. They have been interpreted as part of the collisional and transpressional interactions between Gondwana and Laurentia from Ordovician to Devonian time (Fig. 9) (Kent and Van der Voo, 1990; Van der Voo, 1993; Ortega-Gutiérrez et al., 1997).

While the Chihuahua sequences remain as shallow marine deposits, the tectonic evolution of the Caborca terrane was controlled by collision and accretion during the Permo-Caboniferous. It is evidenced by two major unconformities that suggest the accretion of the eugeoclinal sequences. Whether the deformation of eugeoclinal rocks, of the Caborca terrane is associated with the accretion of the Cortéz terrane (Fig. 10) is unknown. Relationship between the development of the Marathon-Ouachita thrust belt and sedimentary sequences of northern México has not been well established. Metamorphic rocks at the U.S.-México border area (103° longitude) suggest that a deformational belt of early Paleozoic age extended down into México.

The similarity of early Mississippian faunas in Ciudad Victoria to those in ancestral North America suggests a close approach of the Oaxaquia block to Laurentia by early Mississippian time (Stewart et al., 1993; Ortega-Gutiérrez et al., 1995) (Fig. 10). Reconstruction of México in the western margin of Pangea during the late Paleozoic by Pindell (1985) placed a collisional belt along Oaxaquia. However, the widespread magmatism suggests a different tectonic setting. The origin and tectonic significance of the three magmatic provinces are still unknown. Torres et al. (1993) reported a belt of Permian granitoids emplaced along the suture between the Maya terrane and Oaxaquia that shows trace element compositions typical of subduction zones (Fig. 10). Thus, the

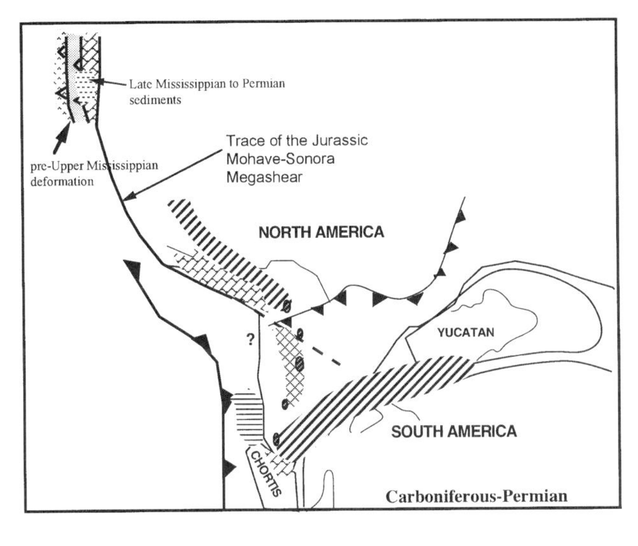

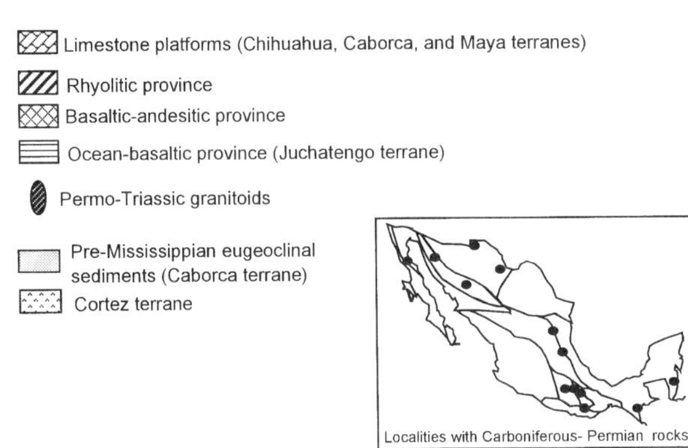

Figure 10. Reconstruction showing the distribution of Late Carboniferous to Permian magmatic provinces and the evolution of the Caborca and Cortéz terranes in the Cordillera (reconstruction modified from Rowley and Pindell, 1989).

volcanic rocks of the Molango area might represent the subaereal activity of this late Paleozoic volcanic arc suggested by Torres et al. (1993) (Fig. 10) (Rosales-Lagarde et al., 1997). Most of southern Oaxaquia, the Mixteco, and the Maya terranes were covered by shallow marine or continental sediments as a platform or passive margin-type environment. However, the opening of a narrow deep oceanic basin between Oaxaquia and the Mixteco terrane has been suggested by Vachard et al. (1997).

ACKNOWLEDGMENTS

Special thanks to Allison Palmer, Reinhard Weber, and Robert Lopez for their constructive reviews. We also thank Duncan Keppie for his useful comments. This paper represents a contribution to International Geological Correlation Project 376. We are grateful for funding supplied by the Universidad Nacional Autónoma de México (Project PAPIIT-IN101095).

REFERENCES CITED

Anderson, P. V., 1993, Prebatholithic stratigraphy of the San Felipe area, Baja California Norte, *in* Gastil, G., and Miller, R. H., The prebatholithic stratigraphy of peninsular California: Geological Society of America Special Paper 279, p. 1–10.

Anderson, T. H., and Silver, L. T., 1971, Age of granulite metamorphism during the Oaxacan orogeny, México: Geological Society of America Abstracts with Programs, v. p. A3–4.

Anderson, T. H., and Silver L. T., 1981, An overview of Precambrian rocks in Sonora, México: Universidad Nacional Autónoma de México, Instituto de Geología, Revista, v. 5, p. 131–139.

Anderson, T. H., Eells, J. L., and Silver, L. T., 1979, Precambrian and Paleozoic rocks of the Caborca region, Sonora, México, *in* Anderson, T. H., and Roldán-Quintana, J., eds., Geology of northern Sonora: Geological Society of America Field Trip Guidebook, p. 1–22.

Aponte, B. M., 1974, Estratigrafía del Paleozoico (Cámbrico-Pensivánico), del Centro de Sonora [Tesis Profesional]: México, D.F., Instituto Politécnico Nacional, Escuela Superior de Ingeniería y Arquitectura, 48 p.

Arellano, A. R. V., 1956, Relaciones del Cámbrico de Caborca, especialmente con la base del Paleozoico, *in* Rodges, J., ed., El Sistema Cámbrico, su paleogeografía y el problema de su base; pt II: Australia, America: México, D.F., Symposium. XX Congreso Geológico Internacional, v. 2; p. 509–527.

Boucot, A. J., 1975, Evolution and extinction rate controls: Amsterdam, Elsevier, Developments in Paleontology and Stratigraphy, v. 1, 427 p.

Boucot, A. J., Blodgett, R. B., and Stewart, J. H., 1997, European province Late Silurian brachiopods from the Ciudad Victoria area, Tamaulipas, northeastern México, *in* Klapper, G., Murphy, M. A., and Talent, J. A., Paleozoic sequence stratigraphy, biostratigraphy, and biogeography: studies in honor of J. Grenville (Jess) Johnson: Geological Society of America Special Paper 321, p. 273–293.

Bridges, L. W., 1964, Stratigraphy of Mina Plomosas–Placer de Guadalupe area, *in* Geology of Mina Plomosas–Placer de Guadalupe area, Chihuahua, México, Field Guidebook: West Texas Geological Society Publication 64-50, p. 50–93.

Bridges, L. W., 1970, Paleozoic history of the southern Chihuahua tectonic belt, *in* The geological framework of the Chihuahua tectonic belt:. West Texas Geological Society Symposium, p. 67–74.

Brunner, P., 1975, Estudio Estratigráfico del Devónico en el área del Bísani, Caborca, Sonora: Instituto Mexicana del Petróleo Revista, v. 7, no. 1, p. 16–45.

Bullard, E. C., Everett, J. E., and Smith, A. G., 1965, The fit of the continents around the Atlantic: Philosophical Transactions of the Royal Society of London, ser. A, v. 258, p. 41–51.

Campa, M. F., and Coney, P., 1983, Tectono-stratigraphic terranes and mineral resource distributions of México: Canadian Journal of Earth Sciences, v. 20, p. 1040–1051.

Campbell, M., and Crocker, J., 1993, Geology west of the Canal de Ballenas, Baja California, México, *in* Gastil, G., and Miller, R. H.: The prebatholithic stratigraphy of peninsular California: Geological Society of America Special Paper 279, p. 61–76.

Carrillo, B. J., 1959, Notas sobre el Paleozoico de la región de Ciudad Victoria, Tamps.: Asociación Mexicana de Geólogos Petroleros Boletín, v. 11, p. 673–680.

Carrillo-Bravo, B. J., 1961, Geología del Anticlinorio Huizachal-Peregrina al NW de Ciudad Victoria, Tamps.: Asociación Mexicana de Geólogos Petroleros Boletín, v. 13, p. 1–98.

Carrillo-Bravo, J.,1965, Estudio Geológico de una parte del anticlinorio de Huayacocotla: Asociación Mexicana de Geólogos Petroleros Boletín, v. 17, p. 73–96.

Carrillo-Martínez, M., 1971, Geología de la Hoja San Jose de Gracia, Sin [Tesis Profesional]: México, D.F., Universidad Nacional Autónoma de México, Facultad de Ingeniería, 56 p.

Centeno-García, E., Sánchez-Zavala, J. L., Patchett, J., Sour-Tovar, F., and

Ortega-Gutiérrez, F., 1997, Stratigraphy, Nd sediment provenance, and paleogeography of Paleozoic sequences in southern México, *in* Terrane Dynamics–97, International Conference on Terrane Geology: University of Canterbury, Royal Society of New Zealand, Institute of Geological and Nuclear Sciences, Abstracts, p. 42–45.

Cevallos, S., and Weber, R., 1980, Arquitectura, estructura y ambiente de depósito de algunos estromatolitos del Precámbrico sedimentario de Caborca, Sonora, UNAM: Revista Instituto de Geología, v. 4, no. 2, p. 97–103.

Coney, P., and Campa, M. F., 1987, Lithotectonic terrane map of México (west of the 91st meridian): U.S. Geological Survey Miscellaneous Field Studies Map MF-1874-D, scale 1:2,500,000.

Cooper, G. A., and Arellano, A. R. V., 1946, Stratigraphy near Caborca, northwest Sonora, México: American Association of Petroleum Geologists Bulletin, v. 30, p. 606–619.

Corona-Esquivel, R. J., 1981 (1984), Estratigrafía de la región de Olinalá-Tecocoyunca, noreste del Estado de Guerrero: Universidad Nacional Autónoma de México, Instituto de Geología Revista, v. 5, p. 1–16.

Dallmeyer, R. D., 1982, Pre-Mesozoic basement of the southeastern Gulf of México: Geological Society of America Abstracts with Programs, v. 14, p. A471.

Dalziel, I. W. D., 1997, Overview: Neoproterozoic-Paleozoic geography and tectonics: review, hypothesis, environmental speculations: Geological Society of America Bulletin, v. 109, p. 16–42.

Dalziel, I. W. D., Dalla-Salda, L. H., and Gahagan, L. M., 1994, Paleozoic Laurentia-Gondwana interaction and the origin of the Appalachian-Andean mountain system: Geological Society of America Bulletin, v. 106, p. 243–252.

Damon, P. E., Livingston, D. E., Mauger, R. L., Giletti, B. J., and Pantoja-Alor, J., 1962, Edad del Precámbrico "Anterior" y de otras rocas del zócalo de la región de Caborca-Altar de la parte noroccidental del Estado de Sonora: Universidad Nacional Autónoma de México, Instituto de Geología, Boletín, v. 64, p. 11–44.

Diaz, T. G., and Navarro, G., 1964, Litología y correlación estratigráfica del Paleozoico Superior en la región de Palomas, Chihuahua, México: Asociación Mexicana de Geólogos Petroleros, Boletín v. 16, p. 107–120.

Flawn, P. T., Goldstein, A., Jr., King, P. B., and Weaver, C. E., 1961, The Ouachita system: Austin, University of Texas Bureau of Economic Geology Publication 6120, 401 p.

Gastil, G., and Miller, R. H., 1983, Prebatholithic terranes of southern and peninsular California, U.S.A. and México, Status report, *in* Stevens, C. H., ed., Pre-Jurassic rocks, western North American suspect terranes: Society of Economic Mineralogists and Paleontologists, Pacific Section, p. 49–61.

Gomberg, D. N., Banks, P. D., and McBirney A. R., 1968, Preliminary zircon ages from Central Cordillera: Science, v. 162, p. 121–122.

Gonzalez-Leon, C., 1996, Stratigraphy and paleogeographic setting of the Antimonio Formation, *in* Gonzalez-Leon, C., and Stanley, G., ed., US–México cooperative research, *in* International Workshop on the Geology of Sonora, Estacion Regional del Noroeste, Publicaciones Ocasionales No. 1: Instituto de Geología UNAM, Memoir, p. 23–32.

Grajales-Nishimura, J. M., 1988, Geology, geochronology, geochemistry, and tectonic implications of the Juchatengo green rock sequence, State Oaxaca, southern México [M.S. thesis]: Tucson, University of Arizona, 145 p.

Gursky, H. J., and Ramírez-Ramírez, C., 1986, Notas preliminares sobre el reconocimiento de vulcanitas ácidas en el Cañón de Caballeros (Núcleo del Anticlinorio Huizachal-Peregrina, Tamaulipas, México), *in* Barbarín, J. M., and Gursky, H. J., eds., Aspectos geológicos del noreste de México: Universidad Autónoma de Nuevo León, Actas de la Facultad de Ciencias de la Tierra, Linares, v. 1, p. 11–22.

Hayes, P. T., 1975, Cambrian and Ordovician rocks of southern Arizona and New Mexico and westernmost Texas: U.S. Geological Survey Professional Paper 873, 98 p.

Keith, B., and Noll, J. H., 1987, Preliminary geologic map of the Cerro Cobachi area, Sonora, México: U.S. Geological Survey Miscellaneous Fields Studies Map, MF-1980, scale 1:20,000.

Kent, D. V., and Van der Voo, R., 1990, Palaeozoic palaeogeography from palaeomagnetism of the Atlantic-bordering continents, *in* McKerrow, W. S., and Scotese, C. R., eds., Palaeozoic palaeogeography and biogeography: Geological Society of London Memoir 12, p. 49–56.

Keppie, J. D., 1977, Plate tectonics interpretation of Paleozoic world maps: Nova Scotia Department of Mines Paper 77-3, 45 p.

Keppie, J. D., and Ortega-Gutiérrez, F.., 1995, Provenance of Mexican terranes: isotopic constraints: International Geology Review, v. 37, p. 813–824.

Keppie, J. D., Dostal, J., Murphy, J. B., and Nance, R. D., 1996, Terrane transfer between eastern Laurentia and western Gondwana in the early Paleozoic: constrains on global reconstructions, *in* Nance, R. D., and Thompson, M. D., eds., Avalonian and related peri-Gondwanan terranes of the circum–North Atlantic: Boulder, Colorado, Geological Society of America Special Paper 304, p. 369–380.

King, R. E., 1944, Geology and paleontology of the Permian area northwest of the Las Delicias, southwestern Coahuila México: Geological Society of America Special Paper 52, 130 p.

Leier-Engelhardt, P., 1993, Middle Paleozoic strata of the Sierra de Las Pintas, northeastern Baja California Norte, México, *in* Gastil, G., and Miller, R. H., The prebatholithic stratigraphy of peninsular California: Geological Society of America Special Paper 279, p. 23–42.

Longoria, J. F., and Perez, V. A., 1979, Bosquejo Geológico de los cerros Chino y Rajón, cuadrángulo Pitiquito-La Primavera (NW de Sonora): Universidad Autónoma de Sonora, Departamento de Geología Boletín, v. 1, p. 119–144.

Lothringer, C. J., 1993, Alloctonous Ordovician strata of Rancho San Marcos, Baja California Norte, México, *in* Gastil, G., and Miller, R. H.: The prebatholithic stratigraphy of peninsular California: Geological Society of America Special Paper 279, p. 11–22.

Malpica, C. R., 1977, Estudio Estratigráfico del Paleozoico de la Cuenca de Chicomuselo, Chiapas: Instituto Mexicana del Petróleo, Subdirección Técnica de Exploración, Informe Técnico Especial, Proyecto C-1003, parte II, 73 p.

McBirney, A. R., and Bass, M. N., 1969, Structural relations of pre-Mesozoic rocks of northern Central America: American Association of Petroleum Geologists Memoir 11, p. 269–280.

McKee, J. W., and Jones, N. W., 1988, Las Delicias Basin-A record of Late Paleozoic arc volcanism in northeastern México: Geology, v. 16, p. 37–40.

Moreno-Cano, L. C., and Patiño-Ruiz, J., 1981, Estudio Paleozoico en la región de Calnalí, Hidalgo (Sierra Madre Oriental) [Tesis Profesional]: México, D.F., Instituto Politécnico Nacional, Escuela Superior de Ingeniería y Arquitectura, 30 p.

Mulchay, R. B., and. Velasco, J. R., 1954, Sedimentary rocks at Cananea, Sonora, México, and tentative correlation with the section at Bisbee and the Swisshelm Mountains, Arizona: Transactions, Society of Mining Engineers, American Institute of Mining, Metallurgical and Petroleum Engineers, v. 199, p. 628–632.

Mullan, H. S., 1978, Evolution of the Nevadan orogen in northwestern México: Geological Society of America Bulletin, v. 89, p. 1175–1188.

Murillo-Muñetón, G., 1994, Petrologic and geochronology study of Grenvillian-age granulites and post-granulites plutons from La Mixtequita area, state of Oaxaca, southern México [M.S. thesis]: Los Angeles, University of Southern California, 163 p.

Noll, J. H., Jr., 1981, Geology of the Picacho Colorado area, northern Sierra Cobachi, central Sonora, México [M.S. thesis]: Flagstaff, Northern Arizona University, 165 p.

Ortega-Gutiérrez, F., 1978, Estratigrafía del Complejo Acatlán en la Mixteca Baja, Estados de Puebla y Oaxaca: Universidad Nacional Autónoma de México, Revista del Instituto de Geología, v. 2, p. 112–131.

Ortega-Gutiérrez, F., 1993, Tectonostratigraphic analysis and significance of the Paleozoic Acatlán complex of southern México, *in* Ortega-Gutiérrez, F., Centeno-García, E., Morán-Zenteno, D. J., and Gómez-Caballero, A., eds., Terrane geology of southern México, 1st circum-Pacific and circum-Atlantic terrane conference, Guanajuato, México, Guidebook, Field Trip B: Universidad Nacional Autónoma de México, Instituto de Geología

p. 54–60.

Ortega-Gutiérrez, F., Anderson, T. H., and Silver, L. T., 1977, Lithologies and geochronology of the Precambrian craton of southern México: Geological Society of America Abstracts with Programs, v. 9, p. A1121–A1122.

Ortega-Guitiérez, F., Ruiz, J., and Centeno-García, E., 1995, Oaxaquia—a Proterozoic microcontinent accreted to North America during the late Paleozoic: Geology, v. 23, p. 1127–1130.

Ortega-Gutiérrez, F., Elías-Herrera, M., and Sánchez-Zavala, J. L., 1997, Tectonic evolution of the Acatlán complex: new insights, *in* II Convención sobre la evolución geológica de México y recursos asociados, Pachuca, Hidalgo, Simposia y Coloquio: Universidad Autónoma de Hidalgo, Instituto de Investigaciones en Ciencias de la Tierra, and Universidad Nacional Autónoma de México, Instituto de Geología, p. 19–21.

Pantoja-Alor, A. J., 1970, Rocas sedimentarias paleozoicas de la región centro septentrional de Oaxaca: Sociedad Geológica Mexicana, Libreto Guía de la Excursión México-Oaxaca, p. 67–84.

Patchett, P. J., and Ruiz, J., 1987, Nd isotopic ages of crust formation and metamorphism in the Precambrian of eastern and southern México: Contributions to Mineralogy and Petrology, v. 96, p. 523–528.

Patterson, W. D., 1978, Geology of Permian rocks near Ascencion northern Chihuahua, México [M.S. thesis]: El Paso, Texas, El Paso University, 70 p.

Pindell, J. L., 1985, Alleghanian reconstruction and subsequent evolution of the Gulf of México, Bahamas, and Proto-Caribbean: Tectonics, v. 4, p. 1–39.

Poole, F. G., and Madrid, R. J., 1988, Allocthonous Paleozoic eugeoclinal rocks of the Barita de Sonora mine area, central Sonora, México, *in* Rodríguez-Torres, R., ed., El Paleozoico de la región central del Estado de Sonora: Libreto Guía de la Excursión Geológica para el Segundo Simposio sobre la Geología y Minería del Estado de Sonora, Excursiones de Campo, Universidad Nacional Autónoma de México, Hermosillo, Sonora, p. 32–41.

Poole, F. G., Stewart, J. H., Palmer, A. R., Sanberg, C. A., Madrid, R. J., Ketner, K. B., Carter, C., and Morales-Ramírez, J. M., 1992, Latest Precambrian to latest Devonian time-development of a continental margin, *in* Burchfiel, B. C., Lipman, P. W., and Zoback, M. L., eds., The Cordilleran orogen—Conterminous U.S.: Boulder, Colorado, Geological Society of America, Geology of North America, v. G-3, p. 9–56.

Poole, F. G., Stewart, J. H., Repetski, J. E., Ross, R. J., Ketner, K. B., Amaya-Martínez, R., and Morales-Ramírez, J. M., 1995, Ordovician carbonate-shelf rocks of Sonora, México, *in* Cooper, J. D., Droser, M. L., and Finney, S. C., eds., Ordovician odyssey: Short Papers for the Seventh International Symposium on the Ordovician system, Las Vegas, Nevada: SEPM, Pacific Section, book 77, p. 267–275.

Ramírez-Ramírez, R. C., 1978, Reinterpretación Tectónica del Esquisto Granjeno de Ciudad Victoria, Tamaulipas: Universidad Nacional Autónoma de México, Instituto de Geología, Revista, v. 2, p. 31–36.

Reynolds, M. S., 1972, Informe Final Pozo Los Chinos No. 1, Petróleos Mexicanos, Superintendencia General de Exploración, Zona Noreste, Informe Geológico, NE-M-1194, 105 p. (unpublished)

Robison, R., and Pantoja-Alor, J., 1968, Tremadocian trilobites from the Nochixtlán region, Oaxaca, México: Journal of Paleontology, v. 42, no. 3, p. 767–800.

Rosales-Lagarde, L., Centeno-García, E., Ochoa-Camarillo, H., and Sour-Tovar, F., 1997, Permian volcanism in eastern México—preliminary report, *in* II Convención Geológica Sobre la Evolución Geológica de México y recursos asociados, Pachuca, Hidalgo, Libro Guía de las excursiones geológicas, Excursión 1: Universidad Autónoma de Hidalgo, Instituto de Investigaciones en Ciencias de la Tierra y Universidad Nacional Autónoma de México, Instituto de Geología, p. 27–32.

Rowley, D. B., and Pindell, J. L., 1989, End Paleozoic–early Mesozoic western Pangean reconstruction and its implications for the distribution of Precambrian and Paleozic rocks around Meso-America: Precambrian Research, v. 42, p. 411–444.

Scotese, C. R., and McKerrow, W. S., 1990, Revised world maps and introduction, *in* McKerrow, W. S., and Scotese, C. R., eds., Paleozoic paleogeography and biogeography: Geological Society of London Memoir 12, p. 1–21.

Sedlock, R. L., Ortega-Gutíerrez, F., and Speed, R. C., 1993, Tectonostratigraphic terranes and tectonic evolution of México: Geological Society of America Special Paper 278, 153 p.

Silva-Pineda, A., 1970, Plantas del Pensilvánico de la región de Tehuacán, Puebla: Universidad Nacional Autónoma de México, Instituto de Geología Boletín, v. 29, 109 p.

Sour-Tovar, F., 1990, Comunidades Cámbrico-Ordovícicas del área de Santiango Ixtalpepec, Oaxaca (Formación Tiñú): Implicaciones paleoambientales y paleogeográficas: Sociedad Paleontológica Mexicana Revista, v. 3, p. 7–24.

Sour-Tovar, F., and Quiroz, B. S. A., 1991, Icnofósiles paleozoicos de Nochixtlán, Oaxaca: Sociedad Paleontológica Mexicana, III Congreso Nacional de Paleontología, México, D.F., v. 3, p. 131.

Stewart, J. H., 1988, Latest Proterozoic and Paleozoic southern margin of North America and the accretion of México: Geology, v. 16, p. 186–189.

Stewart, J. H., McMenamin, M. A. S., and Morales-Ramírez, J. M., 1984, Upper Proterozoic and Cambrian rocks in the Caborca region, Sonora, México—Physical stratigraphy, biostratigraphy, paleocurrent studies, and regional relations: U.S. Geological Survey Professional Paper 1309, 36 p.

Stewart, J. H., Poole, F. G., Ketner, K. B., Madrid, R. J., Roldán-Quintana, J., and Amaya-Martínez, R., 1990, Tectonics and stratigraphy of the Paleozoic and Triassic southern margin of North America, Sonora, México, *in* Gehrels, G. E., and Spencer, J. E., eds., Geologic excursions through the Sonoran Desert region, Arizona and Sonora: Arizona Geological Survey Special Paper 7, p. 183–202.

Stewart, J. H., Blodgett, R. B., Boucot, A. J., and Carter, J. L., 1993, Middle Paleozoic exotic terrane near Ciudad Victoria, northeastern México, and the southern margin of Paleozoic North America, *in* Ortega-Gutiérrez, F., Centeno-García, E., Morán-Zenteno, D. J., and Gómez-Caballero, A., eds., Pre-Mesozoic basement of NE México, lower crust and mantle xenoliths of central México, and northern Guerrero terrane, 1st circum-Pacific and circum-Atlantic terrane conference, Guanajuato, México, Guidebook, Field Trip A: Universidad Nacional Autónoma de México, Instituto de Geología, p. 55–57.

Torres-Vargas, R., Ruiz, J., Murillo-Muñetón, G., and Grajales-Nishimura, J. M., 1993, The Paleozoic magmatism in México—evidence for the shift from circum-Atlantic to circum-Pacific tectonism, *in* Ortega-Gutiérrez, F., Centeno-García, E., Morán-Zenteno, D. J., and Gómez-Caballero, A., eds., 1st circum-Pacific and circum-Atlantic terrane conference, Guanajuato, México, Proceedings, Universidad Nacional Autónoma de México, Instituto de Geología, p. 154–155.

Tovar, R. J., 1968a, Medición a detalle de la sección paleozoica expuesta en la Sierra de Teras Bavispe, Sonora: Petróleos Mexicanos, Superintendencia General de Exploración, D.F.N.E. Distrito Chihuahua, Informe Geológico, NE-M-1073, s/p. (unpublished).

Tovar, R. J., 1968b, Medición a detalle de la sección paleozoica expuesta en la Sierra de Las Palomas, Chihuahua: Petróleos Mexicanos, Superintendencia General de Exploración, D.F.N.E. Distrito Chihuahua, Informe Geológico, NE-M-1069, s/p. (unpublished).

Vachard, D. U., Grajales, M., Flores de Dios, A., Torres, R., and Buitrón, B., 1997, Patlanoaya and juchatengo: two key sequences for understanding the late Paleozoic geological history of México, *in* II Convención sobre la evolución geológica de México y recursos asociados, Memoria: Pachuca, Hidalgo, Universidad Autónoma de Hidalgo, Universidad Nacional Autónoma de México.

Van der Voo, R., 1993, Paleomagnetism of the Atlantic Tethys and Iapetus oceans: Cambridge University Press, 411 p.

Villaseñor-Martínez, A. B., Martínez-Cortes, A., and Contreras y Montero, B., 1987, Biostratigrafía del Paleozoico Superior de San Salvador Patlanoaya, Puebla, México: Sociedad Mexicana de Paleontología Boletín v. 1, p. 396–417.

Viveros, M. A., 1965, Estudio Geológico de la Sierra de Cabullona Municipio de Agua [Tesis Profesional]: México, D.F., Universidad Nacional Autónoma de México, Facultad de Ingeniería, 82 p.

Wardlaw, B., Furnish, W. M., and Nestell, M. K., 1979, Geology and paleontology of the Permian beds near Las Delicias, Coahuila, México: Geological

Society of America Bulletin, v. 90, p. 111–116.
Weber, R., and Ceballos, S., 1980, El significado sedimentario de los estromatolitos del Precámbrico sedimentario de la región de Caborca, Sonora: UNAM, Revista Instituto de Geología, v. 4, no. 2, p. 104–110.
Weber, R., and Ceballos, S., 1994, Perfil actual y perspectivas de la paleobotánica en México: Sociedad Botánica de México Boletín, v. 55, p. 141–148.
Weber, B., and, Köhler, H., 1997, La evolución magmática y metamórfica del Complejo Guichicovi de edad Grenvilleana, estado de Oaxaca, *in* II Convención sobre la evolución Geológica de México, Pachuca, Hidalgo, Resúmenes: Universidad Autónoma del Estado de Hidalgo, Instituto de Investigaciones en Ciencias de la Tierra, Universidad Nacional Autónoma de México, Instituto de Geología, p. 88–89.
Whittington, H. B., and Hughes, C. P., 1972, Ordovician geography and faunal provinces deduced from trilobite distribution: Philosophical Transactions of the Royal Society of London, ser. B, v. 263, p. 235–278.
Yañez, P., Ruiz, J., Patchett, P. J., Ortega-Gutiérrez, F., and Gehrels, G., 1991, Isotopic studies of the Acatlán complex, southern México: Geological Society of America Bulletin, v. 103, p. 817–828.

Manuscript Accepted by the Society September 4, 1998

Geological Society of America
Special Paper 336
1999

Exotic Paleozoic strata of Gondwanan provenance near Ciudad Victoria, Tamaulipas, México

John H. Stewart
U.S. Geological Survey, MS-901, 345 Middlefield Road, Menlo Park, California 94025, United States
Robert B. Blodgett
Department of Zoology, Oregon State University, Cordley Hall 3029, Corvallis, Oregon 97331, United States
Arthur J. Boucot
Department of Zoology, Oregon State University, Cordley Hall 3029, Corvallis, Oregon 97331, United States
John L. Carter
The Carnegie Museum of Natural History, 4400 Forbes Avenue, Pittsburgh, Pennsylvania 15213, United States
Robert López
Earth Science Department, University of California, Santa Cruz, California 95064, United States

ABSTRACT

Outcrops in Cañones de la Peregrina and de Caballeros near Ciudad Victoria, Tamaulipas, México, consist of the Grenville-age (1,018 Ma) Novillo gneiss. The second overlying strata are composed, in ascending order, of: (1) the Cañón de Caballeros Formation (Silurian) consisting of conglomerate, silicic volcanic arenite, limestone, and sandstone and siltstone; (2) the Vicente Guerrero Formation (Lower Mississippian), consisting of quartz and lithic arenite, siltstone, and shale; (3) the El Aserradero Rhyolite (Mississippian, a 334 ± 39-Ma U-Pb zircon age); (4) the unconformably overlying Del Monte Formation (Lower and Middle Pennsylvanian), consisting of bioclastic grainstone, sandstone, and limy sandstone; (5) the Guacamaya Formation (Lower Permian), consisting of turbiditic siltstone and sandstone rich in volcanic detritus; and (6) unconformably overlying Jurassic and Cretaceous conglomerate, limestone, siltstone, and sandstone. The Cañón de Caballeros and the Vicente Guerrero Formation are each about 100 m thick; the Del Monte Formation, about 200 m thick; and the Guacamaya Formation, about 1,260 m thick.

An extensive fauna of brachiopods, gastropods, trilobites, and corals from the Silurian Cañón de Caballeros Formation has been assigned to the European Province of the North Atlantic Region, a non–North American fauna. The Silurian fauna is similar to that in the allochthonous West Avalonia terrane in eastern coastal North America, to fauna in Venezuela, and to fauna in northwest Africa, Europe, the British Isles, and Scandinavia. The Silurian fauna in the Ciudad Victoria area is most similar to fauna in the Cordillera de Mérida in Venezuela. The original distribution of the fauna of the European Province appears to have been along the South American–northwest Africa margin of the supercontinent Gondwana, and at that time, adjacent Baltica, and separate from the Laurentian (ancestral North American) continent. This pattern suggests that the Ciudad Victoria Silurian strata were derived from somewhere along this margin of Gondwana (most likely near Venezuela) and were transported tectonically to their present position near ancestral North America. The

Stewart, J. H., Blodgett, R. B., Boucot, A. J., Carter, J. L., and López, R., 1999, Exotic Paleozoic strata of Gondwanan provenance near Ciudad Victoria, Tamaulipas, México, *in* Ramos, V. A., and Keppie, J. D., eds., Laurentia-Gondwana Connections before Pangea: Boulder, Colorado, Geological Society of America Special Paper 336.

emplacement of the Ciudad Victoria rocks against ancestral North America probably took place in the late Paleozoic. Mississippian igneous and deformational events noted in the Paleozoic succession near Ciudad Victoria and in the southern United States may indicate the initial events of this tectonic activity. The rocks of the Ciudad Victoria area were apparently rifted from South America and left in their position near ancestral North America when South America separated from Laurentia during the Mesozoic break-up of the supercontinent Pangea.

INTRODUCTION

Paleozoic rocks exposed in Cañones de la Peregrina and de Caballeros near Ciudad Victoria, State of Tamaulipas, northeastern Mexico (Fig. 1), are significant because they contain a non–North American Silurian fauna (Boucot et al., 1997) that appears to be significantly out of place relative to its present position. Originally, these rocks were considered a continuation of rocks of the Marathon region of Texas (Flawn et al., 1961; Anderson and Schmidt, 1983), but this correlation is no longer valid (Stewart, 1988; Stewart et al., 1993). This chapter describes the Paleozoic stratigraphy of the Ciudad Victoria area, and interprets this stratigraphy in terms of regional paleobiogeographic and tectonic patterns.

The study focuses on the stratigraphy and paleontology of Silurian and Mississippian rocks in Cañón de la Peregrina and Cañón de Caballeros northwest of Ciudad Victoria (Fig. 1). Field work was carried out by J. H. Stewart, R. B. Blodgett, and A. J. Boucot during November 4–24, 1993, and paleontologic studies by Blodgett, Boucot, and J. L. Carter during 1993–1994. Geochronologic studies were performed by Robert Lopez from 1993–1996. A description of the Silurian fauna based on this study has already been published by Boucot et al. (1997).

Access to Cañones de la Peregrina and de Caballeros (Fig. 2) is by dirt roads that can be negotiated by a four-wheel-drive vehicle or a pickup truck. Cañón de la Peregrina and all but the upper part of Cañón de Caballeros is accessible by a high-clearance van. Dense vegetation makes foot access to many parts of the area difficult, and field traverses are greatly facilitated by trails used for farming and cattle raising. Due to the dense vegetation, outcrops in much of the area are poor, and a clear understanding of geologic relations is difficult. We concur with Heim's (1940) statement about the Sierra Madre Oriental near Ciudad Victoria: "No jungle known to the writer presents so many obstacles to geological work."

PREVIOUS WORK

In 1925, Parker A. Robertson, a geologist with the Mexican Gulf Oil Company, discovered fossils in Cañón de la Peregrina, 13 km west northwest of Ciudad Victoria. These fossils were identified as Early Mississippian in age by Girty (1926). Heim (1940), reporting on work for the Shell Group of oil companies in 1925, also described these same fossiliferous beds, as well as other rocks that he also assigned to the Paleozoic. Aside from these initial reports, internal oil company reports, a report of Permian fusulinids from Cañón de la Peregrina (Muir, 1936, p. 8), and field trips to the area (Humphrey and Díaz, 1953; Díaz, 1956), little work was done on the Paleozoic rocks near Ciudad Victoria until the extensive studies of Carrillo-Bravo (1959, 1961, 1963). He (Carrillo-Bravo, 1959, 1961, 1963) mapped in detail the geology of major drainage in the eastern Sierra Madre Oriental, including Cañones de la Peregrina and de Caballeros that contain the most complete exposures of Paleozoic rocks in the Ciudad Victoria area. Carrillo-Bravo (1959, 1961, 1963) further indicated the presence of Cambrian(?), Ordovician(?),

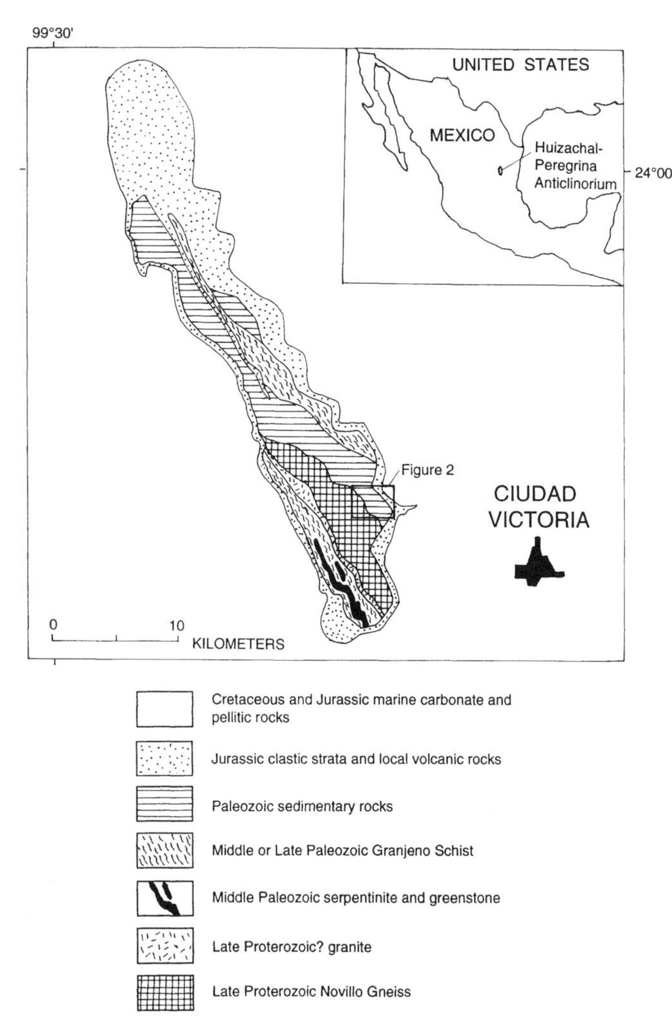

Figure 1. Generalized map of Huizachal-Peregrina anticlinorium.

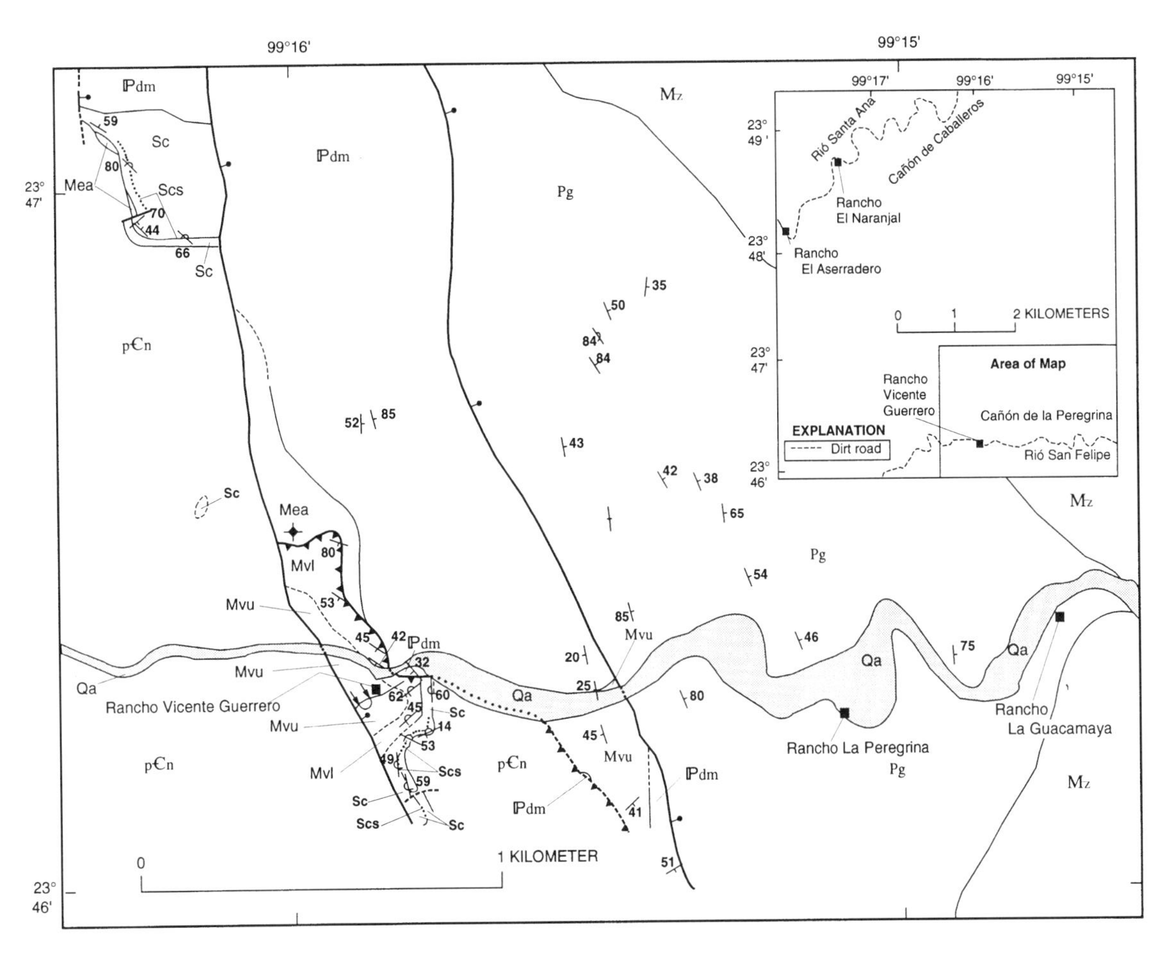

EXPLANATION

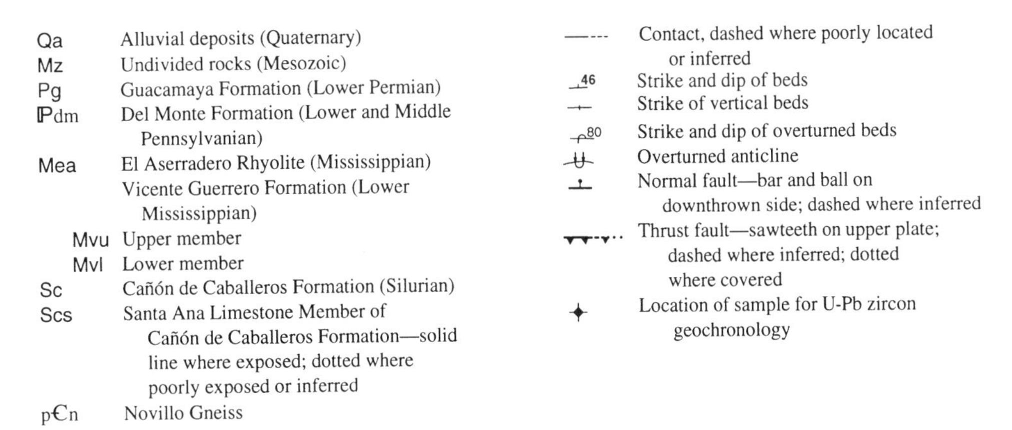

Figure 2. Geologic map of Cañón de Peregrina area.

Silurian, Devonian, Mississippian, Pennsylvanian, and Permian rocks, but as described below, his recognition of Cambrian(?), Ordovician(?), and Devonian rocks is considered erroneous. Subsequent to Carrillo-Bravo's work, reports on Paleozoic rocks in the area include description of Pennsylvanian goniatites (Murray et al., 1960), petrographic studies of Paleozoic rocks (Tellez-Giron, 1970), field studies and a field guide to Paleozoic rocks by students of Pan American University (Conklin et al., 1974), a study of a Silurian to Devonian brachiopod (Boucot, 1975), a report on rhyolitic volcanic rock (Gursky and Ramirez-Ramirez, 1986), a study of Permian rocks in Cañón de la Peregrina (Gursky and Michalzik, 1989), a study of the pre-Mesozoic geology of the region (Ramirez-Ramirez, 1978, 1981, 1992), field guides (Russell, 1981, Ortega-Gutiérrez et al., 1993), and a study of the Paleozoic stratigraphy by Gursky (1996). Stewart et al (1993) published a preliminary appraisal of the mid-Paleozoic stratigraphy, paleontology, and tectonics of the Ciudad Victoria area. In that appraisal, some rocks in the area were assigned to the Devonian, but this age assignment has proven to be incorrect.

REGIONAL SETTING

Paleozoic rocks in Cañones de la Peregrina and de Caballeros lie in a large anticlinorium (Huizachal-Peregrina anticlinorium) along the easternmost part of the Sierra Madre Oriental (Fig. 1). The anticline exposes the three main pre-Mesozoic units: the Proterozoic Novillo Gneiss, the mid-Paleozoic (metamorphic age) Granjeno Schist, and relatively unmetamorphosed Paleozoic rocks (Figs. 1–3). The Precambrian Novillo Gneiss (Carrillo-Bravo, 1961; Fries et al., 1962; Denison et al., 1971; Cserna et al., 1977; Ortega-Gutiérrez, 1978; Garrison et al., 1980; Ruiz et al., 1988; Ramirez-Ramirez, 1992) has been dated by K-Ar methods (Fries et al., 1962; Denison et al., 1971) and by Sm-Nd methods (Ruiz et al., 1988) as between 860 and 930 Ma, by Rb-Sr methods as 1,140 ± 80 Ma (Garrison et al., 1980), and by U-Pb zircon methods as 1,018 ± 3 Ma (Silver et al., 1994). The Novillo

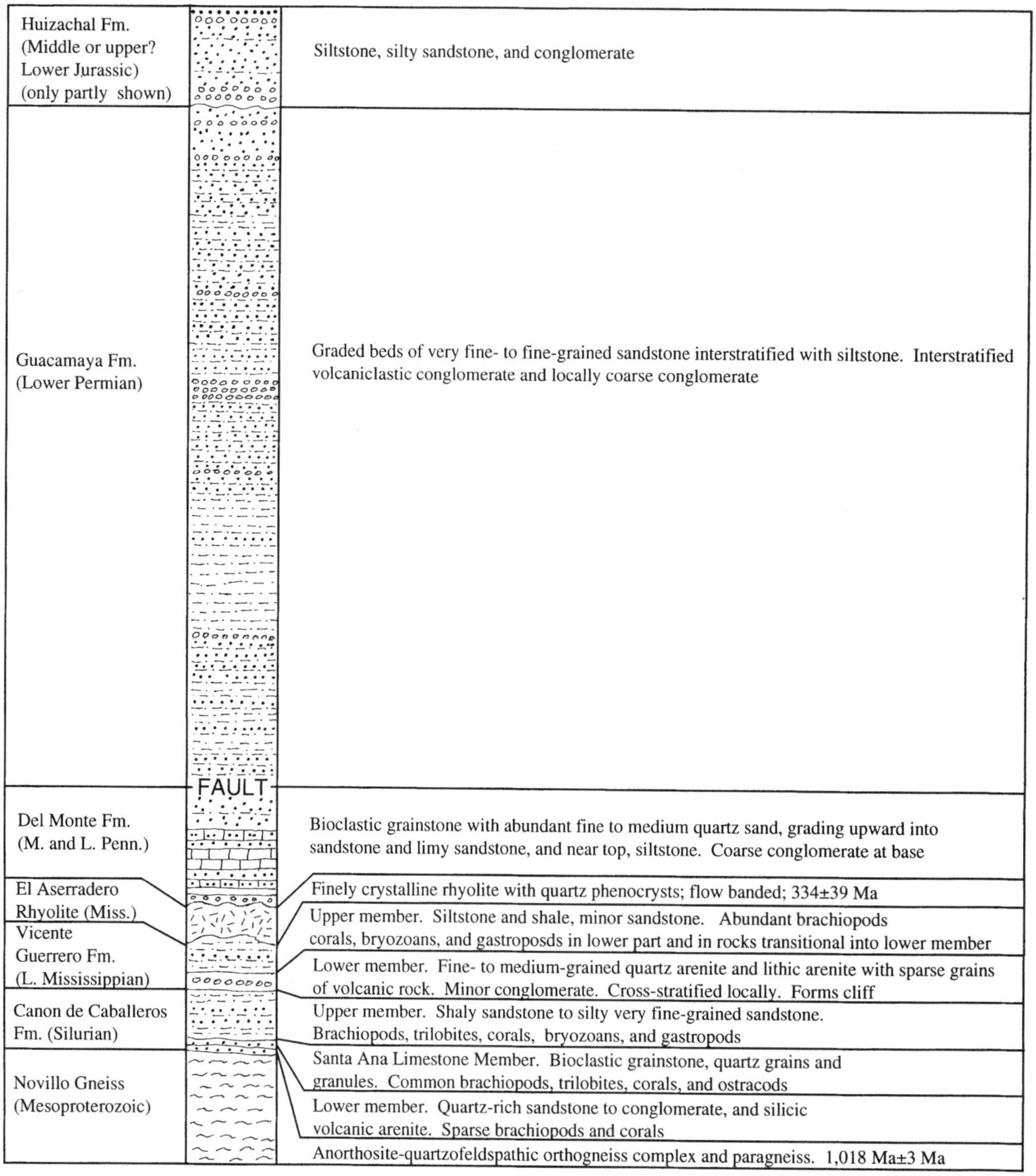

Unit	Description
Huizachal Fm. (Middle or upper? Lower Jurassic) (only partly shown)	Siltstone, silty sandstone, and conglomerate
Guacamaya Fm. (Lower Permian)	Graded beds of very fine- to fine-grained sandstone interstratified with siltstone. Interstratified volcaniclastic conglomerate and locally coarse conglomerate
FAULT	
Del Monte Fm. (M. and L. Penn.)	Bioclastic grainstone with abundant fine to medium quartz sand, grading upward into sandstone and limy sandstone, and near top, siltstone. Coarse conglomerate at base
El Aserradero Rhyolite (Miss.)	Finely crystalline rhyolite with quartz phenocrysts; flow banded; 334±39 Ma
Vicente Guerrero Fm. (L. Mississippian)	Upper member. Siltstone and shale, minor sandstone. Abundant brachiopods corals, bryozoans, and gastroposds in lower part and in rocks transitional into lower member
	Lower member. Fine- to medium-grained quartz arenite and lithic arenite with sparse grains of volcanic rock. Minor conglomerate. Cross-stratified locally. Forms cliff
Canon de Caballeros Fm. (Silurian)	Upper member. Shaly sandstone to silty very fine-grained sandstone. Brachiopods, trilobites, corals, bryozoans, and gastropods
	Santa Ana Limestone Member. Bioclastic grainstone, quartz grains and granules. Common brachiopods, trilobites, corals, and ostracods
	Lower member. Quartz-rich sandstone to conglomerate, and silicic volcanic arenite. Sparse brachiopods and corals
Novillo Gneiss (Mesoproterozoic)	Anorthosite-quartzofeldspathic orthogneiss complex and paragneiss. 1,018 Ma±3 Ma

Figure 3. Stratigraphic column of Mesoproterozoic to Jurassic rocks in Cañones de Peregrina and de Caballeros. Thickness and lithologic column of Permian rocks from Gursky and Michalzik (1989). Age of Novillo Gneiss from Silver et al. (1994) and of El Aserradero Rhyolite from this report.

Gneiss consists of two main parts (Ortega-Gutiérrez, 1978): an anorthositic-quartzofeldspathic orthogneissic complex that includes massive anorthosite and a paragneiss sequence of quartzo-feldspathic gneiss with some layers of marble and graphitic schist. In tectonic contact with the Novillo Gneiss or separated from it by a granite, is the Granjeno Schist (Carrillo-Bravo, 1961; Cserna et al., 1977; Cserna and Ortega-Gutiérrez, 1978; Garrison, 1978; Garrison et al., 1980; Ramirez-Ramirez, 1992; Ortega-Gutiérrez et al., 1993), consisting of low-grade mica-schist interbedded with abundant greenstone, metachert, and rare carbonate rocks. It includes large bodies of serpentinite with relic textures and structures that indicate it was originally a layered ultramafic complex (Ortega-Gutiérrez et al., 1993). Isotopic ages on the Granjeno Schist (Ramirez-Ramirez, 1992, p. 113) have a range of from 257–330 Ma and suggest a mid-Paleozoic metamorphic age for the unit. Ortega-Gutiérrez et al. (1993) indicated that a seven-point Rb-Sr isochron of 330 Ma for the Granjeno Schist may the best age of the unit. Pillow basalt, olistolithic blocks (Ramirez-Ramirez, 1992), and metachert suggest that the protolith of the Granjeno Schist was originally a deposit in a deep ocean basin adjacent to a convergent continental margin (Ramirez-Ramirez, 1992; Ortega-Gutiérrez et al., 1993). Relatively unmetamorphosed Paleozoic rocks in the Huizachal-Peregrina anticlinorium (the main subject of this chapter) are the third major sequence of pre-Mesozoic rocks in the Huizachal-Peregrina anticlinorium. These Paleozoic rocks are herein shown to rest unconformably on the Novillo Gneiss, but their relation to the Granjeno Schist is unknown.

The Novillo Gneiss, Granjeno Schist, and the Paleozoic rocks in the Huizachal-Peregrina anticlinorium are unconformably overlain by folded but conspicuously less deformed Mesozoic rocks. The oldest part of this sequence is divided into two formations referred to by Carrillo-Bravo (1961) as the Huizachal Formation (below) and the La Joya Formation (above). The same two units were defined by Mixon et al. (1959) as the La Boca Formation and La Joya Formation of the Huizachal Group. Silva-Pineda (1979) and Michalzik (1991) used the nomenclature of Carrillo-Bravo (1961), and we follow their proposal here. The Huizachal Formation (La Boca Formation of Mixon et al., 1959) consists of a sequence, locally over 2000 m thick, of reddish brown and greenish gray siltstone, silty sandstone, sandstone, and conglomerate (Carrillo-Bravo, 1961; Belcher, 1979). It was originally dated as Upper Triassic on the basis of fossil plants (Mixon et al., 1959), but study of pollen from the same plant localities indicates an Early or Middle Jurassic age (Rueda-Gaxiola et al., 1991). In addition, a mammal-like reptile (Clark and Hopson, 1985) from the Huizachal Formation (La Boca Formation) is certainly younger than Late Triassic, and may be post–Early Jurassic in age. This and other vertebrates from the Huizachal Formation (La Boca Formation) are considered to be "late(?) Early or Mid-Jurassic" by Fastovsky et al. (1987) or Middle Jurassic by Fastovsky et al. (1995). The La Joya Formation unconformably overlies the Huizachal Formation (La Boca Formation) and consists of reddish brown and greenish gray conglomerate, limestone, siltstone, and sandstone. The La Joya Formation is transitional upward into Upper Jurassic carbonate rocks, and is considered to be of mainly pre-Oxfordian, or perhaps pre-late Oxfordian, age (Michalzik, 1991). La Joya Formation is overlain by a thick sequence of Upper Jurassic and Cretaceous strata consisting largely of carbonate and evaporitic rocks (Carrillo-Bravo, 1961).

STRATIGRAPHY

Introduction

Carrillo-Bravo's (1961) study of the Paleozoic rocks in the Huizachal-Peregrina anticlinorium led him to propose seven formation names; La Presa Quartzite and Naranjal Conglomerate, both of Cambrian(?) age; Victoria Limestone of Ordovician(?) age; Cañón de Caballeros Formation, of Silurian age; La Yerba Formation, of Devonian age; Vicente Guerrero Formation, of Mississippian age; Del Monte Formation, of Pennsylvanian age; and Guacamaya Formation, of Permian age. Of these formations, only four (the Cañón de Caballeros, Vicente Guerrero, Del Monte, and Guacamaya Formations) are here considered to be valid. The La Presa Quartzite is now included as a part of a paragneiss sequence in the Proterozoic Novillo Gneiss (Ramirez-Ramirez, 1992; Gursky, 1996). The Naranjal Conglomerate is considered to be late Paleozoic by Ramirez-Ramirez (1992) and Gursky (1996), and may, at least, in part, be equivalent to a conglomerate at the base of the Pennsylvanian Del Monte Formation. Restudy by Boucot and Blodgett of the original faunal collections from the Victoria Limestone indicates that it is Silurian in age, and that this limestone is actually the same as a widespread limestone unit (the Santa Ana Limestone Member) in the lower part of the Silurian Cañón de Caballeros Formation. The La Yerba Formation was considered by Carrillo-Bravo (1961) to be Devonian and to consist of novaculite (a dense light colored chert) and siliciclastic strata. However, Gursky and Ramirez-Ramirez (1986) determined that the so-called novaculite is actually a volcanic rock (rhyolite to rhyodacite) that is dated herein as Mississippian. Fauna that were originally reported by Carrillo-Bravo (1961), Boucot (1975), Stewart et al. (1993) as Devonian appear, after a restudy by Boucot and Blodgett, to be either Silurian or Mississippian in age. Our field studies indicate that all of the sedimentary rocks mapped as Devonian by Carrillo-Bravo (1961), students of Pan-American University (Conklin et al., 1974), and Gursky (1996) are actually Mississippian in age and part of the Vicente Guerrero Formation.

Cañón de Caballeros Formation (Silurian)

The Cañón de Caballeros Formation was named by Carrillo-Bravo (1961) for outcrops in Cañón de Caballeros about 800 m southwest of Rancho El Naranjal. The outcrops lie along, or on the northwest bank of, Río Santa Ana in an area where the river is close to the road at lat 23°48.45′N and long 99°17.45′W.

The Cañón de Caballeros Formation is exposed at the type

locality in Cañón de Caballeros and in two areas in Cañón de la Peregrina (Fig. 2). The formation is divided into three members: a lower member, a Santa Ana Limestone Member (new name), and an upper member (Fig. 3).

Lower member. The lower member is exposed in Cañón de la Peregrina where it consists of light brown, quartz-rich, fine- to very coarse-grained sandstone to conglomerate and overlying, and possibly interstratified, silicic volcanic arenite. The sandstone to conglomerate contains granules and pebbles as large as 4 cm of white or transparent quartz. The silicic volcanic arenite is fine to coarse grained and typically composed of about 40% finely crystalline to microcrystalline lithic grains; 35% embayed angular to rounded subhedral to euhedral quartz; 15% angular to subangular subhedral to anhedral potassium feldspar; sparse siltstone clasts; and about 10% matrix. A few of the finely crystalline to microcrystalline lithic grains can be identified as rhyolite or felsite, and most of these finely crystalline to microcrystalline grains are probably of volcanic material. The euhedral to subhedral, unabraded, embayed grains of quartz indicate a nearby silicic volcanic source, presumably an unconsolidated or relatively unconsolidated tuff that was essentially coeval with deposition of the Cañón de Caballeros Formation.

The lower member is considered to lie depositionally on the Novillo Gneiss, although exposures were not complete enough to observe this relation in the field directly. However, a depositional contact is indicated by the presence of coarse conglomerate in the lower member and by mapped relations in Cañón de la Peregrina that show the lower member and the Santa Ana Limestone Member to be consistently adjacent to outcrops of the Novillo Gneiss. Such consistency would be unlikely if the contact were a fault. In addition, the same stratigraphy was observed in Cañón de Caballeros at the type locality of the Cañón de Caballeros Formation. Here the lower member is covered, but the Santa Ana Limestone Member nevertheless is only about 6 m stratigraphically above the presumed depositional contact with exposed Novillo Gneiss in a position similar to that in Cañón de la Peregrina. We interpret that the lower member is depositionally above the Novillo Gneiss in Cañón de Caballeros, although Carrillo-Bravo (1961) shows a fault between the Cañón de Caballeros Formation and the Novillo Gneiss, and Ramirez-Ramirez (1992, plate 1) considered the outcrop of gneiss to be a block in Pennsylvanian sediments.

Fossils at one locality in sandstone of the lower member include a rostroconch, stricklandiid brachiopod, *Delthyris*? sp., fenestellid bryozoa, *Atrypa*? "*reticularis*," tabulate and rugose corals, and unidentified brachiopods and ?trilobite fragments (Boucot et al., 1997). An early to mid-Wenlock age is suggested by the stricklandiid brachiopod (Boucot et al., 1997).

Santa Ana Limestone Member. The name Santa Ana Limestone Member of the Cañón de Caballeros Formation is here given to a distinctive thin limestone unit recognized in both Cañónes de la Peregrina and de Caballeros. It is named for Río Santa Ana, and the type locality is the same as that of the Cañón de Caballeros Formation along Río Santa Ana in Cañón de Caballeros at lat 23°48.45′N and long. 99°17.45′W. At the type locality of the Cañón de Caballeros Formation, as mentioned above, the Novillo Gneiss is exposed, but the lower member of the Cañón de Caballeros Formation (probably 6 m thick) is covered. The type Santa Ana Limestone Member is above this covered interval. Carrillo-Bravo (1961, Fig. 5, opposite p. 27) published a photograph of what we here call the type section of the Santa Ana Limestone Member. The member is about 2.5 m thick and composed of a distinctive bioclastic grainstone commonly with abundant coarse grains and granules of polycrystalline and monocrystalline quartz apparently derived from metamorphic rocks and/or quartz veins such as those common in the Novillo gneiss.

In Cañón de la Peregrina, the Santa Ana Limestone Member appears to be disrupted structurally, and is present as discontinuous outcrops in the lower part of the Cañón de Caballeros Formation. In Cañón de la Peregrina, it is similar or nearly identical in thickness and lithology to the member in Cañón de Caballeros.

The Santa Ana Limestone Member contains a Silurian fauna of brachiopods, corals, ostracods, and trilobites (Boucot et al., 1997). These fossils include the brachiopods *Sphaerirhynchia* sp., *Delthyris* sp., *Dalejina* sp., *Strophoprion* sp., *Orbiculoidea* sp., *Atrypa* "*reticularis*," *Isorthis* sp., *Eospirifer* sp., *Leptaena* "*rhomboidalis*," *Resserella* sp., *Howellella* sp., "*Rhynchonella*" aff. *stricklandii*, *Pentamerus* sp., *Cyrtia* sp., stropheodontid and athyrid brachiopods, bellerophontid gastropods, dalmanitid trilobites, rugose corals, favositid corals, pelmatozoan columnals, and pterineoids. These fossils indicate a Wenlock to Ludlow age (Boucot et al., 1997). The presence of *Sphaerirhynchia* indicates a position no older than late Wenlock, and the Santa Ana Member is younger than the early to middle Wenlock age suggested for the lower member. However, the *Pentamerus*, which is present in Cañón de Caballeros, is most consistent with a late Llandovery age, because this genus is known only from the late Llandovery in the world, except for interior and Appalachian parts of North America where it ranges into the early and middle Wenlock. *Pentamerus* in Cañón de Caballeros may be Wenlock in age also.

Upper member. The upper member is exposed at the type section of the Cañón de Caballeros Formation and in Cañón de la Peregrina. In Cañón de Caballeros, the upper member consists of thin to thick beds of shaly dark gray to olive-gray siltstone to silty very fine-grained sandstone, and a few beds of fine- to medium-grained sandstone. The fine- to medium-grained sandstone contain abundant finely crystalline to microcrystalline lithic grains, some of which can be identified as rhyolitic or felsitic rock. Many of the finely crystalline to microcrystalline grains are considered to be volcanic material. The strata are structurally contorted locally and generally poorly exposed. The exact thickness of the member is not precisely known in Cañón de Caballeros, but appears to be about 100 m. In Cañón de Caballeros, a short covered interval separates the upper member and siltstone containing Mississippian brachiopods and bryozoa.

In Cañón de la Peregrina, the upper member consists of yellow-brown, olive-gray, and medium gray siltstone with a few layers of fine- to medium-grained sandstone. In outcrops about 0.3 km southeast of Rancho Vicente Guerrero, dark gray to yellow-brown chert or silicified siltstone is present in what appears to be the lowermost part of the upper member of the Cañón de Caballeros Formation. The upper member contains two Silurian faunas. The older one includes 23 genera of brachiopods, as well as trilobites, bryozoa, corals, and gastropods (Boucot et al., 1997). The presence of *Macropleura* and *Sphaerirhynchia* suggest a late Wenlock to Ludlow age, as does *Amphistrophiella* (*Sulcastrophiella*) *carrillobravoi* (Boucot et al., 1997) known previously as *Amphistrophia* sp. B (Boucot et al., 1972) from the late Wenlock–Ludlow of the Cordillera de Mérida (Boucot et al., 1972). The younger fauna includes nine genera of brachiopods. The presence of the coarsely plicate *Delthyris* from the younger fauna indicates a pre–late Pridoli age and *Baturria* indicates a late Ludlow to Pridoli age. The original *Baturria* collection (USGS Ca 1775) from Cañón de la Peregrina was assigned to the Pridoli to Gedinnian (latest Silurian and earliest Devonian) by Boucot (1975) largely because of the presence of a *Delthyris* that was wrongly assigned to *D.* (*Quadrifarius*). Reexamination of the fossil indicates that the identification is in error, and that the fossil is merely a large, pre–late Pridolian type of typical *Delthyris* that can range into the early Pridolian. As mentioned by Boucot (1975), the *Baturria* from Cañón de la Peregrina is most similar to species of *Baturria* of Ludlow and Pridoli age (Late Silurian) from Spain (Carls, 1974), rather than to species of Gedinnian age (Early Devonian). The large size of *Dalejina* present in the *Baturria*-bearing beds also serves to distinguish this part of the upper member from the lower part of the member. The strata that contain *Baturria* are the youngest part of the upper member based on the age of the fossils, but lithologically these strata are indistinguishable from other parts of the member.

As described below, the contact of the Cañón de Caballeros Formation and the Vicente Guerrero Formation appears to be conformable. Locally, the Cañón de Caballeros Formation is overlain unconformably by the Lower Pennsylvanian Del Monte Formation.

The Cañón de Caballeros Formation is considered to be a shallow-water photic zone marine deposit. This interpretation is based on the presence of an abundant, well-preserved shallow-water fauna of brachiopods, trilobites, corals, bryozoa, and gastropods. The fauna, particularly delicate bryozoa, indicate little transport of the fossil material.

Vicente Guerrero Formation (Lower Mississippian). The type locality of the Vicente Guerrero Formation (Carrillo-Bravo, 1961) is directly east of the Rancho Vicente Guerrero. Here, the formation consists of two members (Fig. 3). The lower member is about 30 m thick and consists of medium to light gray, fairly well-sorted, fine- to medium-grained sandstone. The sandstone consists of quartz and lithic arenite containing finely crystalline to microcystalline lithic rocks, some of which can be identified as rhyolite, felsite, and other fine-grained volcanic rocks. The sandstone is mostly thin to thick bedded, but small-scale, low-angle, cross-strata in sets 3–6 cm thick were observed on a few surfaces. The quartzitic sandstone also contains minor 5–15-cm-thick layers of conglomerate with subrounded clasts mostly 0.5–3 cm in diameter, but locally as large as 6 cm. These clasts consist of white quartz and minor amounts of dark gray chert or silicified siltstone. Siltstone clasts and slump structures are present in some sandstone beds. The lower member is resistant and forms massive outcrops along rivers and locally on the sides of canyons. The upper member consists of dark gray siltstone and shale and minor amounts of olive-gray, silty, fine-grained sandstone. Where exposed along the road directly west of Rancho Vicente Guerrero, the upper member is highly contorted structurally. The upper member is at least 50 m thick.

The Vicente Guerrero Formation is also recognized in Cañón de la Peregrina about 1 km east of Vicente Guerrero Ranch and in Cañón de Caballeros (Carrillo-Bravo, 1961). We recognized the lower member of the formation in the type locality and a hundred meters east of Rancho El Aserradero in Cañón de Caballeros, but elsewhere we were not sure we could distinguish the two members.

Brachiopods and to a lesser extent, gastropods, corals, and bryozoa, are abundant in the Vicente Guerrero Formation. Most of these fossils appear to be present near the transitional contact between the lower and upper members of the formation. The fossils are present both in sandstone and in siltstone, commonly where the two lithologies are interlayered. Based on new collections, plus the materials in the collections of the U.S. Geological Survey and National Museum of Natural History in Washington, D.C. and the Field Museum of Natural History in Chicago, at least 20 species of articulate brachiopods are identified as present in the Vicente Guerrero Formation. Seven of these cannot be identified to species and four clearly represent new ones. The nine remaining species consist of the following: *Rugosochonetes* cf. *R. multicosta* (Winchell, 1863), ?*Tolmatchoffia winchelli* (Girty, 1903), *Rotaia subtrigona* (Meek & Worthen, 1860), *Actinoconchus* cf. *A. lamellosus* (Léveillé, 1835), *Cleiothyridina* cf. *C. tenuilineata* (Rowley, 1900), *Camarophorella mutabilis* Hyde, 1908, *Punctospirifer subtexta* (White, 1860), *Syringothyris typa* Winchell, 1863, and *Beecheria chouteauensis* (Weller, 1914). The *Rugosochonetes*, *Cleiothyridina*, *Punctospirifer*, *Syringothyris*, and *Beecheria* clearly limit the age of this assemblage to early Osagean or that of the Burlington Limestone, probably lower Burlington, of the upper Mississippi Valley region of the United States. Only *Rotaia subtrigona* ranges above the Burlington into the Keokuk.

The contact of the Vicente Guerrero Formation and the underlying Cañón de Caballeros Formation was not observed in detail. It was mapped on a hill a few hundred meters east of Rancho Vicente Guerrero, and here the contact appears to be conformable, although exposures are poor and the structure probably complex in detail. In Cañón de Caballeros, the upper contact of the Cañón de Caballeros Formation lies in a covered interval.

As described below, the Vicente Guerrero Formation is overlain along a significant angular unconformity by the Lower Pennsylvanian Del Monte Formation.

The Vicente Guerrero Formation is considered to be a shallow-water marine deposit. This interpretation is supported by the presence of a brachiopod, coral, and gastropod fauna. Some sandstone beds in the lower member of the Vicente Guerrero Formation are massive, contain rip-up clasts and local slump structures, and conceivably could be relatively deep-water gravity-flow deposits. However, no graded bedding or Bouma divisions were noted, and, in the type area of the formation, one sandstone bed contained small-scale cross-strata in several sets stacked one on the other. Such cross-strata are more suggestive of shallow-water traction-current deposits than sediment gravity flows. In addition, the fauna in the Vicente Guerrero Formation is abundant and well preserved. No indication of significant abrasion of the shelly material is evident.

El Aserradero Rhyolite (Mississippian). In Cañón de Caballeros, Gursky and Ramirez-Ramirez (1986) and Ramirez-Ramirez (1992) recognized that rocks, originally considered to be a Devonian novaculite (Carrillo-Bravo, 1961) in the La Yerba Formation, are rhyolite or rhyodacite. These rocks are now recognized in both Cañones de la Peregrina and de Caballeros (Gursky, 1996; and this report). In addition to the supposed novaculite, the La Yerba Formation as defined by Carrillo-Bravo (1961) also contains siliciclastic rocks. These siliciclastic rocks of the original La Yerba Formation now appear to everywhere belong to the Mississippian Vicente Guerrero Formation. We here propose to abandon the name La Yerba Formation and to use the name El Aserradero Rhyolite for the rhyolite and rhyodacite exposed in both Cañones de la Peregrina and de Caballeros. This name was originally applied to the rhyolite by Gursky and Ramirez-Ramirez (1986) and by Gursky (1996), although the name was not formally proposed by them. We also propose that the type locality of the El Aserradero Rhyolite is in the area 0.1–0.2 km S60°E of Rancho El Aserradero, on the south side of the canyon of the Río Santa Ana about 50–150 m above the river (Gursky and Ramirez-Ramirez, 1986). The type locality is at lat 23°48.1′W, long 99°17.5′N.

The El Aserradero Rhyolite (Fig. 3) is light gray, dense, and very fine grained. It is flow banded and locally contains breccia. The rhyolite may locally be an intrusive unit because, in a few areas, it appears, to be within outcrops of the Novillo Gneiss. Elsewhere it may be mostly a lava flow. Two types of rhyolite are recognized petrographically (Gursky and Ramirez-Ramirez, 1986). One type is finely crystalline to slightly porphyritic and composed of quartz, feldspar, and sparse mica (mostly muscovite). The other type contains phenocrysts of quartz, alkali feldspar, and very rare plagioclase, biotite, and muscovite.

Field relations examined by us do not clearly indicate the age of the El Aserradero Rhyolite relative to the Silurian Cañón de Caballeros or the Mississippian Vicente Guerrero Formation. However, we did not observe any field areas where the rhyolite is clearly interstratified with these formations, and we think, on the basis of field relations, that the rhyolite is younger than the Early Mississippian Vicente Guerrero Formation. A pre–Lower Pennsylvanian age is indicated by the presence of rhyolite clasts in the Lower and Middle Pennsylvanian Del Monte Formation. We observed such clasts in conglomerate float from strata near the base of the Del Monte Formation at one locality about 0.5 km northeast of Rancho Vicente Guerrero, and Ramirez-Ramirez (1992) has also reported rhyolite detritus in the Del Monte Formation. A Mississippian age for the El Aserradero Rhyolite is confirmed by a lower intercept U-Pb zircon age of 334 ± 39 Ma for the rhyolite (see the section on Geochronology)

Del Monte Formation (Lower Pennsylvanian). The Del Monte Formation was named by Carrillo-Bravo (1961) for outcrops in Cañón de la Peregrina 0.5–1 km north of Vicente Guerrero. It also crops out in Cañón de Caballeros and in a canyon 11 km north of Cañón de Caballeros (Carrillo-Bravo, 1961). The lower part of the formation consists of bioclastic grainstone with abundant fine to medium quartz sand. The grainstone forms 0.5–2-m-thick graded beds and bold, cliffy, outcrops. Higher in the formation, limy sandstone and sandstone become more abundant, and even higher, limy siltstone, silty limestone, and siltstone are dominant. Conglomerate is described locally at the base and within the formation (Carrillo-Bravo, 1961, Conklin et al., 1974; Ramirez-Ramirez, 1992). The basal conglomerate contains clasts as large as 20 cm of sandstone and black siltstone. Possibly this basal conglomerate is the same unit that Carrillo-Bravo (1961) referred to as the Naranjal Conglomerate in Cañón de Caballeros. The Del Monte Formation is reported by Carrillo-Bravo (1961) to be about 200 m thick and more than 220 m thick by Gursky (1996). The Del Monte Formation is considered to be Early Pennsylvanian in age on the basis of fusulinids, corals, and goniatites (Carrillo-Bravo, 1961; Murray et al., 1960), and as young as Middle Pennsylvanian (Desmoinesian) on the basis of conodonts (N. Savage, written communication, 1995).

The Del Monte Formation rests unconformably on older rocks. In Peregrina and Cañón de Caballeros, it rests either on the Lower Mississippian Vicente Guerrero Formation or on the Silurian Cañón de Caballeros Formation (Fig. 2) (Carrillo-Bravo, 1961). These relations suggest pre–Lower Pennsylvanian deformation of the Silurian and Lower Mississippian rocks.

The Del Monte Formation is a turbiditic unit characterized by thick, graded beds of lime grainstone with quartz sand deposited by sand flows or debris flows. Local coarse conglomerates are considered by Ramirez-Ramirez (1992) to be proximal debris flows.

Guacamaya Formation (Lower Permian). The Guacamaya Formation is the youngest and thickest Paleozoic formation exposed in the Peregrina-Huizachal anticlinorium. The type locality is in Cañón de la Peregrina between Rancho La Guacamaya and Rancho La Peregrina (Carrillo-Bravo, 1961). It crops out in both Cañones de la Peregrina and de Caballeros and in canyons that are 7 and 11 km north of Cañón de Caballeros (Carrillo-Bravo, 1961). A measured thickness of 1,260 m was obtained by Gursky and Michalzik (1989) in Cañón de la Peregrina. The formation consists mostly of 2–20-cm-thick, graded beds of very fine- to fine-grained sandstone interstratified with siltstone beds of comparable thickness (Gursky and Michalzik,

1989). Bouma divisions are evident locally. The section also includes fine- to coarse-grained volcaniclastic sandstone and locally coarse conglomerate. In Cañón de la Peregrina, the formation is in fault contact with the Del Monte Formation (Fig. 2) and unconformably overlain by the La Joya Formation of Middle Jurassic age. The Guacamaya Formation is Early Permian in age on the basis of fusulinids (Carrillo-Bravo, 1961; Tellez-Giron, 1970). In Cañón de Caballeros, Carrillo-Bravo (1961) reported that the Guacamaya Formation is in fault contact with older rocks and unconformably overlain by the La Joya Formation of Middle Jurassic age and, a few kilometers north of Cañón de Caballeros, by the Huizachal Formation (La Boca Formation of Mixon et al., 1959) of late(?) Early or Middle Jurassic age.

The Guacamaya Formation is a deep-water flysch deposit (Gursky and Michalzik, 1989) characterized mainly by fine-grained distal turbidite layers. Local coarse layers indicate more proximal sources and higher energy environments.

STRUCTURE

Our mapping was confined to the eastern part of Cañón de la Peregrina (Fig. 2). In the western part of this area, the Silurian Cañón de Caballeros Formation, and presumably its contact with the Novillo Gneiss, are steeply dipping to overturned. The two major outcrops of the Cañón de Caballeros Formation are separated by a major north–northwest-trending, down-to-the-east, normal fault. Directly east of Rancho Vicente Guerrero, the Novillo Gneiss, the Cañón de Caballeros Formation, and the Vicente Guerrero Formation are in a thrust plate over the Del Monte Formation. The thrust can be seen along the south bank of the Río San Felipe in the type locality of the Vicente Guerrero Formation about 100 m east of Rancho Vicente Guerrero. The upper plate is a large northeast-trending overturned syncline in the Novillo Gneiss, Cañón de Caballeros Formation, and the Vicente Guerrero Formation. The overturned limb is to the southeast, which suggests northwest structural transport of the upper plate. Small-scale folds exposed in a road cut in the upper member of the Cañón de Caballeros Formation about 100 m east of Rancho Vicente Guerrero also indicate westward structural movement.

The Del Monte Formation rests unconformably on the Cañón de Caballeros and Vicente Guerrero Formations and generally appears less structurally contorted than these older rocks, even though it forms the lower plate of the thrust described above. In the northwestern part of the map area (Fig. 2), the Del Monte Formation appears to rest depositionally on the Cañón de Caballeros Formation, whereas in the west-central part of the map area, it apparently rests unconformably on the El Aserradero Rhyolite. As mapped the Del Monte Formation is everywhere separated by a fault from the Guacamaya Formation. The Guacamaya Formation generally dips moderately to steeply east to northeast, although locally it contains northwest-trending folds (Carrillo-Bravo, 1961, Ramirez-Ramirez, 1992). The Del Monte and Guacamaya Formations appear similar in the degree of structural deformation, although the contact between the two units appears to be a fault everywhere, and the depositional relations between the two units cannot be directly observed.

The Paleozoic section in the Huizachal-Peregrina anticlinorium is unconformably overlain by folded, but less deformed, Mesozoic rocks (Carrillo-Bravo, 1961; Ramirez-Ramirez, 1992)

GEOCHRONOLOGY

A sample for U-Pb zircon dating was collected from the Aserradero Rhyolite at a locality 0.6 km west northwest of Vicente Guerrero Ranch on the north side of the Cañón de la Peregrina (Fig. 2).

Approximately 200 zircons were separated from 32 kg of crushed rock. The grains were separated into fractions (Fractions A–D in Fig. 4; Table 1) according to their magnetic susceptibility, shape, size, and abundance of inclusions. All fractions were air-abraded to about two-thirds of their original size, dissolved in HF, and passed through 150-µl ion-exchange columns following the methods of Krogh (1973) and Mattinson (1987). The zircons are pale pink, core-free, and contain few inclusions. Morphologies range from euhedral four-sided prismatic grains with and without pyramidal tips to euhedral multifaceted bipyramidal grains with a 2:1 aspect ratio.

The U-Pb data are discordant with a considerable range in U-Pb ages (Fig. 4; Table 1), and the discordance is most likely the result of analyzing mixed populations of inherited basement zircons and Paleozoic igneous zircons. The discordant array defines a lower intercept on concordia at 334 ± 39 Ma and an upper intercept of 1,086 ± 94 Ma (Fig. 4) The lower intercept is taken as the estimate of the El Aserradero Rhyolite crystallization age, and this age agrees well with the presumed stratigraphic position of the rhyolite.

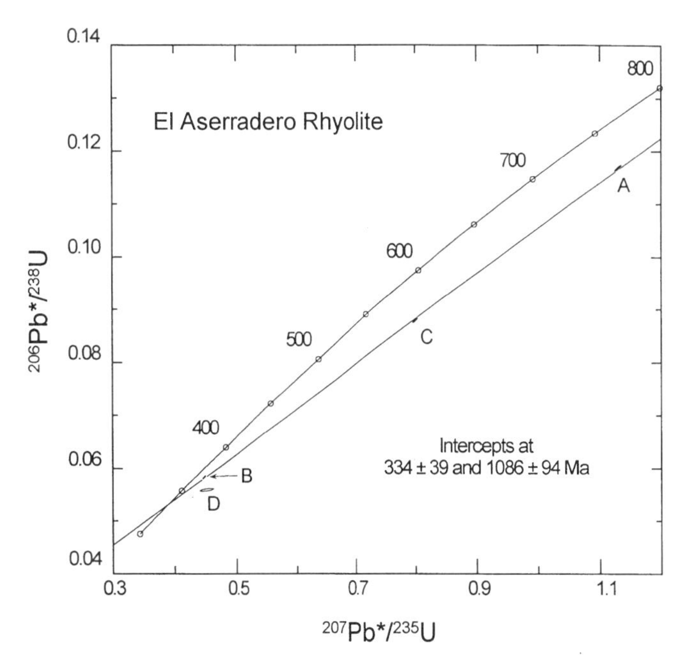

Figure 4. Concordia diagram for U-Pb age of El Aserradero Rhyolite. Letters A through D indicate individual zircon fractions.

TABLE 1. U-Pb GEOCHRONOLOGIC DATA FOR ZIRCONS FROM EL ASERRADERO RHYOLITE, TAMAULIPAS, MEXICO

Sample # Fraction§	Weight (µg)	$^{206}Pb^*$ (ppm)	^{238}U (ppm)	Observed Ratios§ $^{206}Pb/^{204}Pb$	Observed Ratios§ $^{207}Pb/^{206}Pb$	Observed Ratios§ $^{208}Pb/^{206}Pb$	Atomic Ratios§ $^{206}Pb^*/^{238}U$	Atomic Ratios§ $^{207}Pb^*/^{235}U$	Atomic Ratios§ $^{207}Pb^*/^{206}Pb^*$	Ages** $^{206}Pb^*/^{238}U$	Ages** $^{207}Pb^*/^{235}U$	Ages** $^{207}Pb^*/^{206}Pb^*$
A 90µm nm2 abr	130	26.8	265	9,477	0.07158	0.09833	0.11683	1.1300	0.07015	712.3	767.7	932.7 ± 1.2
B 50µm m1 abr	100	11.2	223	4,490	0.05880	0.13977	0.05832	0.4473	0.05563	365.4	375.4	437.5 ± 1.2
C 200µm m2 abr	15	24.3	321	8,613	0.06758	0.12171	0.0879	0.7997	0.06598	543.1	596.7	805.9 ± 1.6
D 30µm m4 abr	25	5.8	119	66	0.27638	0.67499	0.05602	0.4525	0.05858	351.4	379.0	551.5 ± 38

†m = magnetic; nm = nonmagnetic on the Frantz isodynamic separator at a 10 degree side slope and 1.7 amperes; µm = micrometer; abr = abraded (approximately 8 hours at 4 psi, 12 hours at 2 psi). Dissolutions and chemistry were made following the methods of Krogh (1973), using Parrish (1987)-type microcapsules. $^{206}Pb^*$ = radiogenic Pb ppm.
§Two sigma uncertainties on the $^{207}Pb/^{206}Pb$ and $^{208}Pb/^{206}Pb$ ratios are generally <0.05%. Two sigma uncertainties in $^{206}Pb/^{204}Pb$ ratio vary from 0.1% to 0.7%. Estimated uncertainties of the $^{206}Pb/^{238}U$ age are ± 0.4% based on replicate analysis of a single zircon fraction.
**Decay constants used: $^{238}U = 1.55125 \times 10^{-10}y^{-1}$; $^{235}U = 9.8485 \times 10^{-10}y^{-1}$; $^{238}U/^{235}U = 137.88$. $^{207}Pb^*/^{206}Pb^*$ age uncertainties are 2 sigma and from the data reduction program PBDAT of K. Ludwig (1991). Observed ratios are not corrected. In calculating ages, the measured ratios are adjusted for mass fractionation of 0.1% per a.m.u. for Pb and 5 to 10 pg of Pb blank. Initial Pb corrections are based on a feldspar separate with 208:207:206:204 of 18.23:15.53:37.75:1. Isotopic data were measured on a VG 54-30 sector multicollector mass spectrometer with a pulse counting Daly detector at the University of California, Santa Cruz.

Fraction D is anomalous because of its low $^{206}Pb/^{204}Pb$ value. The low value resulted either from analytic problems or from mixing of zircon with apatite, thorite, or some other mineral in the sample.

The upper intercept age of 1,086 ± 94 Ma is quite reasonable for the basement in the Ciudad Victoria area. This age within uncertainty overlaps the published U-Pb zircon age of 1,018 ± 18 Ma for the Novillo Gneiss near Ciudad Victoria (Silver et al., 1994). Furthermore, it fits nicely into the very narrow range of published U-Pb zircon ages (1,018–1,080 Ma) of the Grenville exposures in Mexico (Ortega-Gutiérrez et al., 1995a,b).

PALEOBIOGEOGRAPHY

As proposed by Boucot (1990), the Silurian marine faunas of the world are divided into two realms—the North Silurian and the Malvinokaffric. The North Silurian Realm is further divided into the North Atlantic Region and the Uralian-Cordilleran Region, and finally the North Atlantic Region is divided into the North American Province and the European Province. The Silurian fauna of Cañones de la Peregrina and de Caballeros is assigned to the European Province of the North Atlantic Region (Boucot et al., 1997). This assignment is based on the presence of *Mclearnitesella* and abundant *Isorthis* (*Ovalella*) in the Wenlock-Ludlow–age portion of the upper member of the Cañón de Caballeros Formation. These genera are present in the European Province in Europe and South America (Boucot, 1990; Boucot et al., 1997), but unknown from the North American Province (Boucot, 1990). In addition, *Amphistrophiella (Sulcastrophiella) carrillobravoi* (Boucot et al., 1997) which is present in the upper member of the Cañón de Caballeros Formation, is known elsewhere only from the Cordillera de Mérida of Venezuela (Boucot et al., 1972). The fauna in the Cordillera de Mérida is also assigned to the European Province, although possibly representing another unit in the North Atlantic Region, separate from both the North American and European Provinces. Nevertheless, the fauna from the Cordillera de Mérida has most affinity with the European Province. The younger Ludlow-Pridoli fauna from the Cañón de Caballeros Formation contains *Baturria*, *Isorthis* (*Ovalella*), and a large *Delthyris*. The *Baturria* and *Isorthis* (*Ovalella*) are present in the Silurian of Europe (Carls, 1974), and are unknown in the North American Province (Boucot, 1990). The large *Delthyris* is unlike the smaller ones occurring in the North American Province, indicating a European Province assignment.

In addition to the Ciudad Victoria area, faunas of the European Province are known from coastal Massachusetts, Maine, and New Brunswick in the United States and Canada, in the Cordillera de Mérida in Venezuela, in North Africa, and in Europe and adjacent regions (Fig. 5; Appendix 1). Faunas of the North Atlantic Region also are present in Argentina near the border with Chile (Isaacson et al., 1976), but the faunas at this locality are not distinctive enough to be assigned to either the European or North American Province.

We consider that areas with a Silurian European fauna were once close together so that faunal interchange was possible. To show this paleogeography, we use the mid-Silurian reconstruction of Torsvik et al. (1996), which is based mainly on paleomagnetic data. We have, however, modified their reconstruction in one major way—Baltica–East Avalonia–West Avalonia is placed close to Gondwana (Fig. 6) and is separated from Laurentia (ancestral North America) by a major ocean. In the reconstruction of Torsvik et al. (1996), Baltica–East Avalonia–West Avalonia is close to North America, separated from Gondwana by a major ocean. Our modification is consistent with the paleomagnetic data that allow for longitudinal changes in the position of major blocks, and results in a relatively coherent distribution of areas

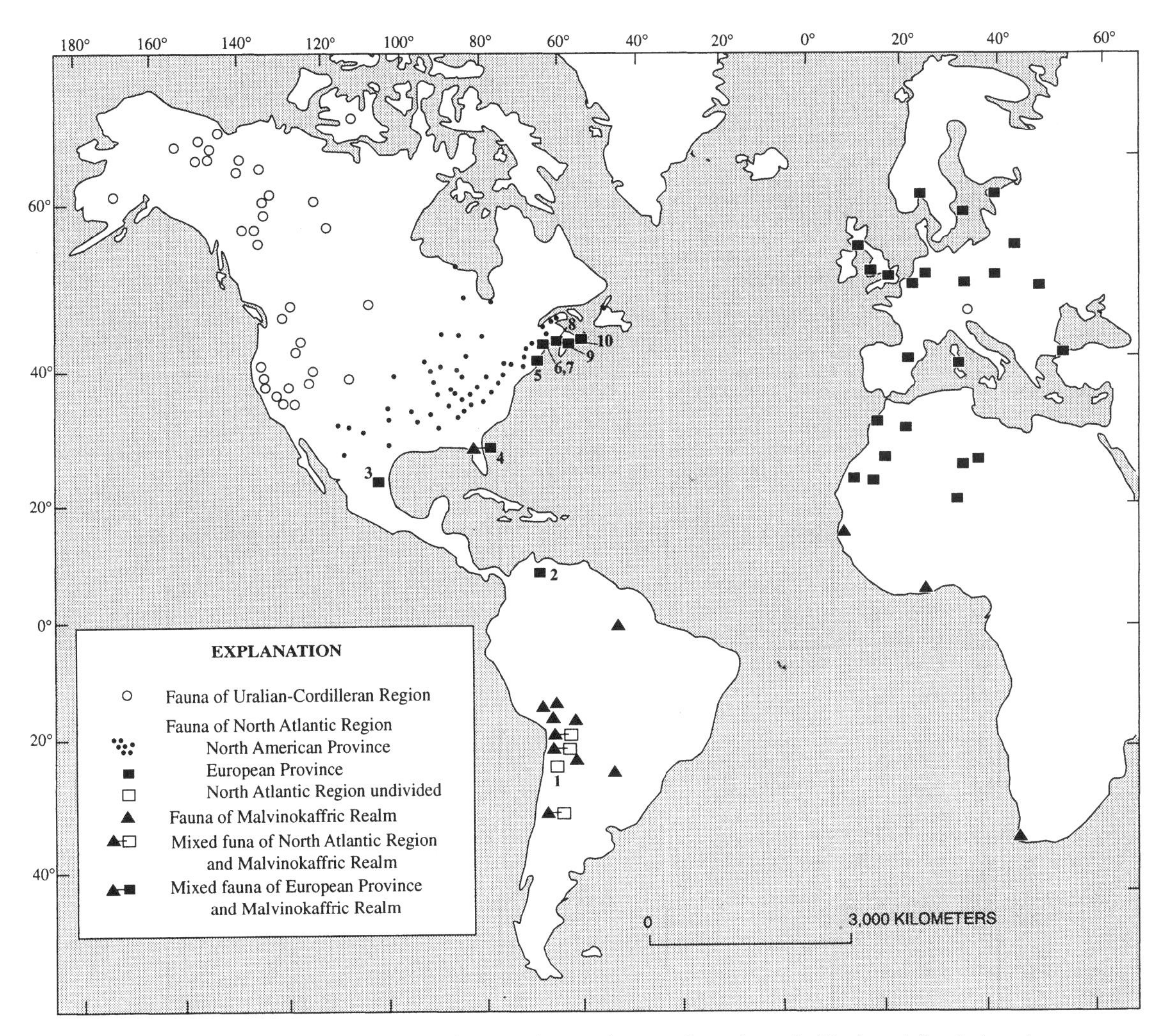

Figure 5. Distribution of Silurian fauna realms, regions, and provinces in North and South America, Europe, and part of Africa. Numbers refer to localities in North America described in Table 2. In western Argentina and Bolivia in South America (near locality 1) fauna are a mixture of the Malvinokaffric Realm and North Atlantic Region types (Isaacson et al., 1976; A. J. Boucot, unpublished data, 1994). In the subsurface in Florida (locality 3), Silurian chitinozoans are of Malvinokaffric Realm types and bivalves are of European Province types (Laufeld, 1979, Pojeta et al., 1976; Boucot, 1990). Figure locations based on many sources including Boucot (1990), Berry and Boucot (1970, 1972a,b, 1973), Ziegler et al. (1974), Isaacson et al. (1976), and Boucot (unpublished data, 1994).

containing the Silurian European faunal province. In our reconstruction (Fig. 6), the Sierra Madre terrane, which includes the Ciudad Victoria area, is placed near what is now Venezuela in South America. This position is preferred because of the similar Silurian fauna in the two areas. The brachiopod *Amphistrophiella* (*Sulcastrophiella*) *carrillobravoi* n.sp. is known only from the Ciudad Victoria area and the Cordillera de Mérida, Venezuela.

Placing an ocean between Baltica–East Avalonia–West Avalonia and North America is consistent with Silurian and Devonian biogeographic information, but is inconsistent with some geologic information that is interpreted to indicate that Baltica–East Avalonia–West Avalonia was in contact with the North American continent by Silurian time. For the Silurian, the paleontologic information indicates that Baltica–East Avalonia–West Avalonia and Laurentia (ancestral North America) have different faunas and no intermixing of specific genera and species of the two provinces. By later Early Devonian time, however, some local intermixing of North America and European faunas is noted, and by later Middle Devonian time this intermixing is widespread (Boucot, 1993). This paleontologic information is consistent with the idea that Laurentia and Baltica–East Avalonia–West Avalonia were in contact for the first time in the later Early or Middle Devonian. This paleontologic information, on the other hand, is inconsistent with other information such as: (1) the timing of structural events (Keppie, 1993), (2) the possible overlap of Silurian strata and volcanic rocks from West Avalonia onto Laurentia (Chandler et al., 1987), (3) a possible Grenville (Laurentia) provenance of Silurian rocks in West Avalonia

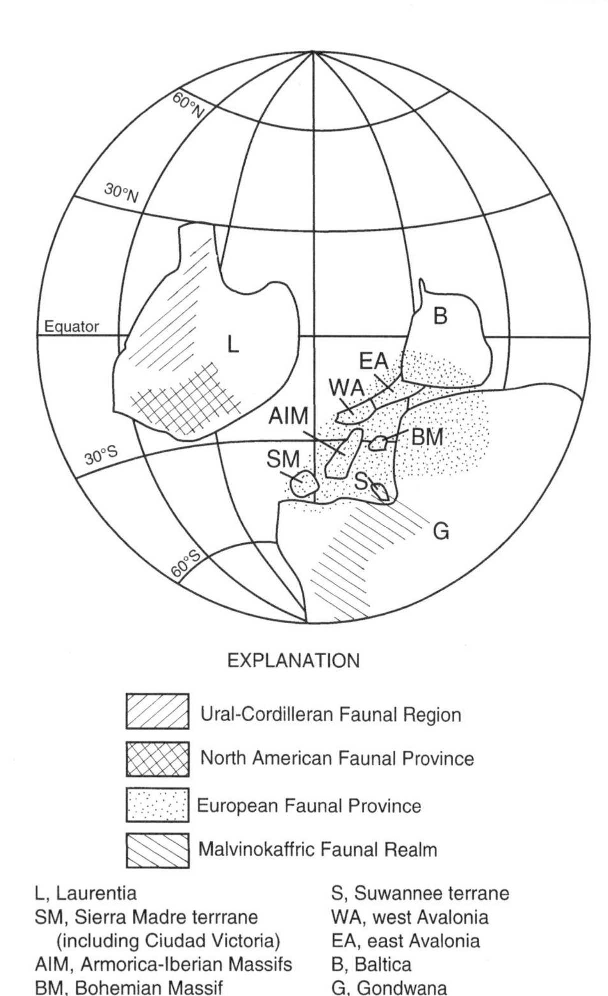

Figure 6. Interpretive distribution of Silurian faunal regions, provinces, and realms. Distribution of major continents and blocks modified from paleomagnetic reconstruction of Torsvik et al. (1996). In the reconstruction of Torsvik et al. (1996), Baltica–East Avalonia–West Avalonia are shown adjacent to Laurentia, whereas in the reconstruction shown here these regions are shown at the same paleolatitude, but near Gondwana. The paleolatitude position of the Bohemian Massif is an average of an Upper Ordovician and an Upper Silurian magnetic pole (Torsvik et al., 1996). The position of the Armorica-Iberian Massifs are unknown paleomagnetically for the Silurian, but is drawn in a likely position similar to that shown by Torsvik et al. (1996). The Sierra Madre terrane (including Ciudad Victoria) is placed near present-day Venezuela based on the similar Silurian fauna of the Cordillera de Mérida in Venezuela and of Ciudad Victoria (see text).

(Murphy et al., 1995), and (4) problematical (in our view) paleontologic information (Keppie et al., 1996). This information is interpreted to indicate that Baltica–East Avalonia–West Avalonia is in contact with Laurentia by Silurian time. We know of no easy way to reconcile the two interpretations (Baltica–East Avalonia–West Avalonia in contact with Laurentia, vs. separate from Laurentia), and suspect that one of the two hypotheses is incorrect. The only simple interpretation that conceivably could reconcile the two hypotheses is that all of the areas with a European fauna were close together, similar to the distribution of the European fauna shown in Figure 6, and that part of the Baltica–East Avalonia–West Avalonia area was close enough to Laurentia to share structural, sedimentary, and volcanic events, but was separated from Laurentia by a relatively small seaway that somehow prohibited faunal interchange. In this hypothesis, no major ocean existed between Gondwana and Laurentia during the Silurian. We think this latter hypothesis is unlikely.

The Mississippian fauna from the Vicente Guerrero Formation of Ciudad Victoria, in contrast to the Silurian fauna, has strong affinities to North America. In fact, the most remarkable aspects of the Mississippian fauna is its similarity to the fauna of the Borden Delta (Waverly Group) of central and eastern Ohio. The Waverly assemblage of large syringothyridids, large elongate multicostate productoids, and representatives of the genera *Rugosochonetes*, *Actinoconchus*, *Cleiothyridina*, *Tylothyris*, and *Punctospirifer* might be designated the Waverly Delta biofacies. The unusual athyridid *Camarophorella* is also present in the Waverly of Ohio and in the Vicente Guerrero Formation but is, in general, a very rare genus.

The North American affinities of the Mississippian assemblage from Ciudad Victoria cannot be doubted, but does this fauna also have affinities with other Mexican or South American faunas? Mississippian fauna from Oaxaca, Mexico, has clear affinities to North America (Sour-Tovar et al., 1996), but comparisons with the Ciudad Victoria fauna have not been made. In South America, Carboniferous brachiopods are reported from Bolivia, Peru, Chile, Argentina, and Brazil. Several Lower Carboniferous brachiopod species have been described from western Argentina by Amos (1957, 1958), but they have no close affinities with the present assemblage. Dutro and Isaacson (1991) reported on a low diversity Lower Carboniferous brachiopod fauna from northern Chile with no apparent affinities to the present one. The Upper Carboniferous brachiopods of Brazil are well known but are not germane to the Lower Carboniferous fauna of Ciudad Victoria. The remaining faunas from Mexico and South America are known only from general reports or undescribed museum collections. Therefore, the question regarding the affinities of the Ciudad Victoria fauna with those elsewhere in Mexico or with South America cannot be answered at the present time. However, the strong affinities of the Lower Mississippian fauna of Ciudad Victoria with those in North America suggests a free interchange of fauna between, and a presumed geographic closeness of, the two areas.

TECTONOSTRATIGRAPHIC TERRANES AND REGIONAL CORRELATIONS

The Ciudad Victoria area has been assigned by Campa and Coney (1983) to the Sierra Madre terrane that is one of many terranes containing Precambrian or lower and middle Paleozoic (pre-Permian) rocks in the United States, Mexico, Cuba, Central America, and South America (Fig. 7; Appendix 2). The distribution, boundary relations, and tectonic history of these terranes are poorly known. Some so-called terranes may be continuations of

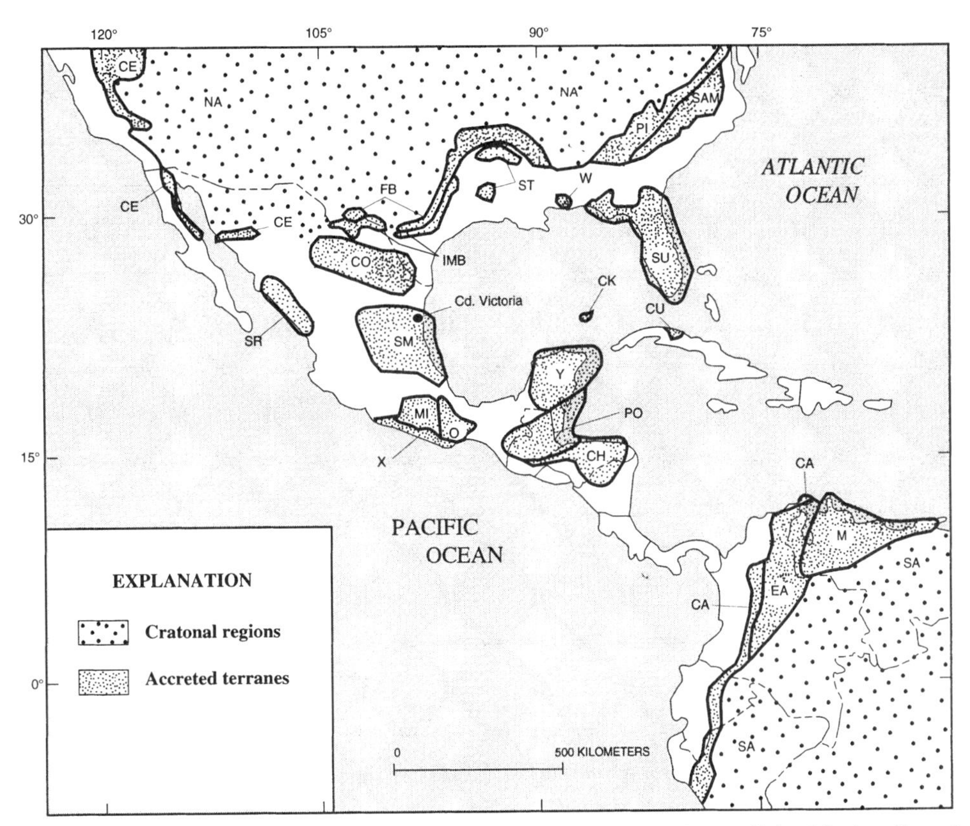

Figure 7. Terranes containing pre-Permian rocks in southern United States, Cuba, Mexico, Central America, and northern South America. See Appendix 2 for summary description of individual terranes.

adjacent terranes whereas other terranes may be composite. Ortega-Gutiérrez et al. (1995b), in particular, considered that all of the Sierra Madre and Oaxaca terranes, as well as adjacent areas, are a single microcontinental block, rather than several separate terranes, and have referred to this block as "Oaxaquia." Regardless of how they are distinguished, these terranes, or larger microcontinental blocks, have had a complex history involving interaction between cratonal and marginal areas of ancestral North America, South America, and Africa. Many lie within the zone where Central America and part of Mexico overlap with South America in the Pangean reconstruction (Pindell, 1985; Pindell and Barrett, 1990), and must have had a different location during the time of the Pangean supercontinent.

Some of the terranes (Sierra Madre, Oaxaca, Polochic, Chortis, central Cuba, eastern Andean) (Appendix 2), contain Grenville-age basement rocks. In addition, a few terranes (Mixteca, Xolapa, and Mérida) (Appendix 2) contain intrusive or metamorphic rocks with either primary, inherited, or detrital zircons of Grenville-age that apparently indicate either a Grenville basement or inherited or detrital zircons derived from such a basement. The Grenville-age rocks may have all had a common origin in the Grenville orogeny that is interpreted by Hoffman (1991) to be the product of a continental collision between what is now the western part of Amazonian Shield in South America, and eastern and southern North America. The Grenville-related terranes, in this interpretation, are fragments of this collision zone, and during subsequent events may have been attached to either North or South America and reached their final location by multiple tectonic events during the Paleozoic, Mesozoic, and Cenozoic (Pindell, 1985; Dalziel, 1992; Dalziel et al., 1994; Ortega-Gutiérrez et al., 1995b).

The Suwannee terrane has a Pan-African basement (Appendix 2) and has been interpreted as lying between Africa, North America, and South America in the Pangean reconstruction (Pindell, 1985). Northern Yucatan (Krogh et al., 1993) and Catoche Knoll (Schlager et al., 1984) also appears to have a Pan-African basement, and thus may have had a similar position.

The Paleozoic rocks in the Ciudad Victoria area were originally correlated with rocks of the Frontal belt (Fig. 7) of the Ouachita orogenic belt in the southern United States, particularly with the Marathon region of Texas (Flawn and Diaz, 1959; Flawn et al., 1961; Carrillo-Bravo, 1961; Anderson and Schmidt, 1983). This correlation now seems untenable. The original interpretation was based heavily on the supposed recognition of a Devonian novaculite in the Ciudad Victoria area (Carrillo-Bravo, 1961) and the correlation of this novaculite with the Caballos Novaculite of the Marathon Region. The Caballos Novaculite is a highly conspicuous and distinctive unit of the Marathon region. As described originally by Gursky and Ramirez-Ramirez (1986) and Ramirez-Ramirez (1992), the so-called novaculite of the Ciudad Victoria area is actually a rhyolite (Aserradero Rhyolite noted herein). This revelation puts in question the correlation of the Ciudad Victoria rocks with those in the Marathon Region (Stewart, 1988).

Other lithologic and stratigraphic differences between Ciudad Victoria and Marathon region are also evident: (1) The oldest Paleozoic rocks in the Ciudad Victoria area are Silurian in age and rest depositionally on Proterozoic Grenville-age basement rocks, whereas in the Marathon region, rocks as old as Cambrian are recognized and no basement rocks are exposed; (2) Silurian rocks in the Ciudad Victoria area are largely siltstone, sandstone, and conglomerate, whereas Silurian rocks in the Marathon region form the lower part of the Caballos Novaculite (Barrick, 1987; McBride, 1989; Noble, 1990, 1993, 1994); (3) Devonian rocks are not recognized in the Ciudad Victoria area, whereas Devonian rocks form part of the Caballos Novaculite in the Marathon region (McBride, 1989; Noble, 1990, 1993); (4) Lower Mississippian (Early Osagean) strata form the Vicente Guerrero Formation in the Ciudad Victoria area, but rocks of such age are not recognized in the Marathon region (Noble, 1990, 1993); (5) Lower Pennsylvanian strata form the relatively thick calcarenite-rich Del Monte Formation in the Ciudad Victoria region, whereas comparable-age strata in the Marathon region are relatively thick and mostly siliciclastic (McBride, 1989); and (6) Lower Permian strata form thick flysch deposits of the Guacamaya Formation of the Ciudad Victoria area, whereas during Early Permian time deposition in the flysch basins of the Marathon region ceased following the final events of the Ouachita orogeny.

In the Sierra Madre terrane, outcrops of Paleozoic rocks are sparse outside of the Ciudad Victoria area. Metamorphic rocks in Aramberri, about 30 km west northwest of the Ciudad Victoria area appear to be correlative to the Granjeno Schist, based on similar isotopic ages (Denison et al., 1971). In the Molango area and nearby regions, about 300 km south of Ciudad Victoria, Grenville-age basement rocks (Ortega-Gutiérrez et al., 1995a, 1997) and Paleozoic sedimentary and volcanic rocks crop out (Carrillo-Bravo, 1965; Moreno-Cano and Patiño-Ruiz, 1981; Ochoa-Camerillo, 1997; Rosales-Lagarde et al., 1997; Centeno-García and Rosales-Lagarde, 1997). The oldest Paleozoic rocks known in the Molango area are Pennsylvanian and Permian in age (Centeno-García and Rosales-Lagarde, 1997) and consist of sedimentary and volcanic rocks. Part of the sequence in the Molango area lithologically resembles the Permian Guacamaya Formation of the Ciudad Victoria area.

Lithologic and stratigraphic similarities of the Paleozoic rocks in the Ciudad Victoria area with other terranes in the United States, Mexico, and Central America are generally not evident. The shallow-water Silurian and Lower Mississippian siliciclastic rocks in the Ciudad Victoria area are lithologically unlike carbonate-rich strata of the Silurian or Mississippian cratonal-platform or miogeoclinal belts near the margins of Paleozoic North America (Laurentia). Neither do the Ciudad Victoria rocks lithologically resemble offshelf deeper water chert-argillite-greenstone assemblages that flanked Paleozoic North America. On the other hand, the low-grade metamorphic rocks of the Granjeno Schist in the Ciudad Victoria and Aramberri areas of the Sierra Madre terrane may be analogous to the metamorphic rocks in the interior metamorphic belt of the Ouachita orogenic belt. Correlations of rocks in the Sierra Madre terrane with those of the Coahuila terrane to the north are difficult because Paleozoic rocks in the Coahuila terrane are mainly in the subsurface. However, local outcrops indicate a Pennsylvanian and Permian magmatic arc system (Appendix 2) considered to be part of the same arc that produced the volcanoclastic rocks in the Ciudad Victoria area (Torres et al., 1992; Centeno-García et al., 1995; Rosales-Lagarde et al., 1997). In addition, plutonic and volcanic rocks in the Coahuila terrane yield isotopic ages similar to that of the Mississippian El Aserradero Rhyolite in the Ciudad Victoria area (Appendix 2). Correlation of rocks of the Ciudad Victoria area with those of the Sonabari and Rusias terranes in Sinaloa does not seem likely because the Sinaloa terranes are not known to contain Silurian strata, and chert-rich successions (Gastil et al., 1991), which are not present in the Ciudad Victoria area, are common. Correlation of the Paleozoic rocks of Ciudad Victoria rocks with those of Oaxaca, Mexico, is not clear either. The Oaxaca terrane contains Grenville-age crystalline basement rocks, as does the Ciudad Victoria area, but the unconformably overlying Paleozoic strata consist of a thin carbonate and siliciclastic rocks of Cambrian and Early Ordovician age, and a thick, largely siliciclastic Mississippian succession (Appendix 2). Although considered here to be different terranes, the Ciudad Victoria, Molango, and Oaxaca rocks could, nevertheless, all be parts of a single microcontinent, as proposed by Ortega-Gutiérrez and other (1995b) and the variation in Paleozoic stratigraphy due to variations in regional patterns of deposition. The Mixteca, Xolapa, and Chortis terranes in southern Mexico are composed primarily of igneous and metamorphic rocks, that are dissimilar to those at Ciudad Victoria. As mentioned above, the Yucatan terrane in Mexico has a Pan-African basement and is thus unlike the Ciudad Victoria area. The Polochic terrane in Guatemala, Belize, and southernmost Mexico contains Grenville-related rocks, but lower Paleozoic sedimentary rocks are unrecognized, and the lithology of the Pennsylvanian and Permian rocks does not match that of rocks of comparable age in the Ciudad Victoria area (Appendix 2).

As described previously, the Silurian fauna in the Ciudad Victoria area has the closest ties to Silurian fauna in the Cordillera de Mérida (Mérida terrane), Venezuela (Fig. 7; Appendix 2). The Ciudad Victoria area has a Grenville-age basement and the Mérida terrane contains paragneiss with Grenville-age zircon (Appendix 2). Grenville-age basement rocks are definitely present in the eastern Andean terrane (Appendix 2), directly west of the Mérida terrane. In both the Ciudad Victoria and the Cordillera de Mérida areas, Silurian strata consist mostly of siliciclastic rocks, and both areas are characterized by a diverse history of sedimentation and igneous and metamorphic activity. However, the stratigraphic, structural, and igneous history of the two areas does not correspond in detail (Appendix 2), and if the two are related, they must represent different parts of what was originally a stratigraphically and orogenically variable region.

As outlined above, the Ciudad Victoria area does not correspond in detail to the stratigraphic, structural, or igneous history of other terranes in the southern United States, Mexico, central America, or northern South America. Possibly, the Ciudad Victoria area was part of a separate structural block, or part of a larger microcontinental block (Oaxaquia), as proposed by Ortega-Gutiérrez et al. (1995b). If it was a separate block, or part of a larger microcontinent, its geologic history might have differed significantly from other terranes. Nevertheless, if the Ciudad Victoria area was separate from other terranes, it apparently was close enough to South America, or some other area with a Silurian European fauna, for faunal interchange.

PALEOZOIC IGNEOUS, METAMORPHIC, AND DEFORMATIONAL EVENTS

The Ciudad Victoria area is characterized by (1) a Silurian volcanic event; (2) a Mississippian igneous and deformational event; (3) a middle or late Paleozoic, possibly Mississippian, metamorphic event; (4) a Permian volcanic event; and (5) a post–Lower Permian and pre-Jurassic deformational event. Only the first three of these events is considered in detail here.

The Silurian volcanic event is recognized by the presence of abundant unabraded silicic volcanic material in sandstone in the lower member of the Silurian Cañón de Caballeros Formation. The inferred Silurian volcanic activity is approximately 417–443 Ma according to the time scale of Tucker and McKerrow (1995). It corresponds approximately in time, but not in stratigraphic, structural, or tectonic setting, with igneous and metamorphic rocks in the following areas (Fig. 8; Appendix 2): (1) schist, eclogites, and a granitoid in the Acatlán complex of the Mixteca terrane, México; (2) a poorly described 410-Ma age for a rhyolite in a drill hole in the Yucatan terrane, México; (3) granitoids in Belize (Potochic terrane); (4) gneiss and metaigneous rocks in the eastern Andean terrane, Colombia and Venezuela; and (5) igneous rocks in the Mérida terrane. In addition, the Pennsylvanian Haymond Formation in the Frontal belt in Texas contains isotopically dated igneous and metamorphic boulders (including metarhyolite) of Ordovician to possibly Mississippian age that could have had a source in igneous rocks to the south in the Coahuila and Sierra Madre terranes, including inferred Silurian volcanic rocks and Mississippian El Aserradero Rhyolite in the Ciudad Victoria area (Appendix 2).

The Mississippian igneous and deformational event is marked by the El Aserradero Rhyolite that has a U-Pb age of 334 ± 39 Ma (Mississippian) and by deformation of the Silurian Cañón de Caballeros Formation and the Lower Mississippian Vicente Guerrero Formation prior to deposition of the Lower and Middle Pennsylvanian Del Monte Formation. The Mississippian igneous and deformational event in the Ciudad Victoria area corresponds in age with the initial occurrences of volcanic detritus in sedimentary units in ancestral North American and with the presence of detrital and volcanic rocks in the Frontal belt of Ouachita orogenic belt and intrusive rocks in the Coahuila terrane (Fig. 8; Appendix 2). These relations suggests that, by Mississippian time, the Ciudad Victoria area (Sierra Madre terrane) and the Coahuila terrane were close enough to North America to share similar volcanic events and perhaps to allow for the transport of volcanic detritus to the North American terrane. As has been described previously, this position close to ancestral North America is supported by the similarity of the Mississippian fauna in the Ciudad Victoria area and that in North America. The Mississippian volcanic event in the Ciudad Victoria area overlaps in time with igneous events in many other terranes in México, Central America, and South America (Fig. 8; Appendix 2). The stratigraphic, structural, and tectonic settings for these events, however, does not correspond closely, if at all, with events in the Ciudad Victoria area.

A middle or late Paleozoic, possibly Mississippian, metamorphic event produced the Granjeno Schist. The metamorphic age of the Granjeno Schist is 257–330 Ma, based on K-Ar and Rb-Sr dating (Ramirez-Ramirez, 1992, Sedlock et al., 1993), although a seven-point Rb-Sr isochron of 330 Ma (Mississippian) appears to be the best determination of the metamorphic age of the Granjeno Schist (Ortega-Gutíerrez et al., 1993). The metamorphic age of the Granjeno Schist corresponds well with the age of the El Aserradero Rhyolite. However, as pointed out by Ramirez-

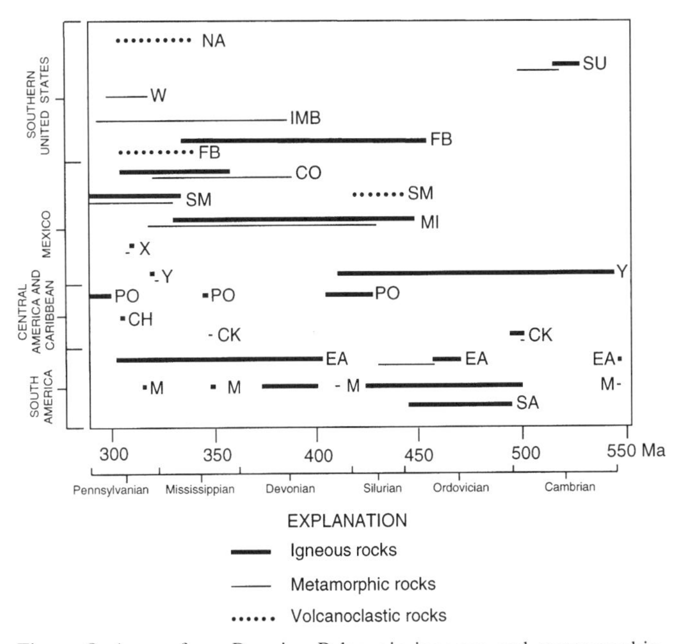

Figure 8. Ages of pre-Permian Paleozoic igneous and metamorphic rocks and volcanoclastic sedimentary rocks in southern United States, México, Central America and Caribbean region, and northern South America. Based mainly on isotopic dating. Bar is general span of individual ages from all types of isotopic dating and from stratigraphic information. Specific data and sources of information shown in Appendix 2. Ages of geologic periods after Harland et al. (1989) and Tucker and McKerrow (1995). Symbols for terranes: NA = North America; SU = Suwannee; W = Wiggins; IMB = Interior metamorphic belt of Ouachita orogenic belt; FB = Frontal belt of Ouachita orogenic belt; CO = Coahuila; SM = Sierra Madre; MI = Mixteca; X = Xolapa; Y = Yucatan; PO = Polochic; CH = Chortis; CK = Catoche Knoll; EA = Eastern Andean; M = Mérida; SA = South America.

Ramirez (1992), the Granjeno Schist appears to have been emplaced into the Ciudad Victoria area by a younger, probably late Paleozoic, event. In support of this view, Ramirez-Ramirez (1992) has noted that the Novillo Gneiss and the Granjeno Schist are in contact with one another, yet K-Ar ages on the Novillo Gneiss do not appear to be significantly reset by the event that metamorphosed the Granjeno Schist. This relation suggests that the Granjeno Schist was emplaced cold into its present position at some time after metamorphism (Ramirez-Ramirez, 1992). In addition, the essentially unmetamorphosed Paleozoic stratigraphic section in the Ciudad Victoria area lies close to outcrops of the metamorphic Granjeno Schist, a situation that would be unlikely if they were close together at the time of metamorphism. Finally, Ramirez-Ramirez (1992) has noted that folds presumably formed during metamorphism have different orientations in different blocks, suggesting that these blocks are related to deformation younger than metamorphism and are probably related to post-metamorphic emplacement of the Granjeno Schist.

The metamorphic event that produced the Granjeno Schist may be broadly similar to the metamorphic event, or events, that produced the interior zone of the Ouachita orogenic belt and metamorphic rocks in the Aramberri area, 30 km west-northwest of the Ciudad Victoria area (Appendix 2).

Volcanic detrital material, in addition to that in Silurian rocks, is also present in the Ciudad Victoria area as a minor component of arenites and grainstones in the Lower Mississippian Vicente Guerrero Formation and the Lower and Middle Pennsylvanian Del Monte Formation. This volcanic material could have had a source in older volcanic rocks, or from reworked Silurian silicic volcanic sandstone, and may not indicate volcanic activity coeval with the Vicente Guerrero or Del Monte Formations. However, such Mississippian and Pennsylvanian activity cannot be ruled out. The Permian Guacamaya Formation contains locally abundant detrital volcanic material (Gursky and Michalzik, 1989) that is indicative of coeval volcanic activity.

INTERPRETATION OF THE TECTONIC HISTORY OF THE CIUDAD VICTORIA AREA

The tectonic history of the Ciudad Victoria area includes the following events:

1. The Grenville-age Novillo Gneiss may have formed during the Grenville or related orogenies in what is now the eastern United States and eastern Canada (Ballard et al., 1989; Ortega-Gutiérrez et al., 1995b) or in South America (Keppie and Ortega-Gutiérrez, 1995). These orogenic events have been related (Hoffman, 1991; Dalziel, 1992; Dalziel et al., 1994; Keppie and Ortega-Gutiérrez, 1995) to collision between the northeastern part of ancestral North America and the northwestern part of ancestral South America.

2. Erosion or nondeposition characterizes Cambrian and Ordovician time. Orogenic activity is also possible at this time, although we found no evidence of this.

3. By Silurian time, thin, shallow-marine, largely clastic deposits (Cañón de Caballeros Formation) were laid down unconformably over the Novillo Gneiss. The lower member of the Cañón de Caballeros Formation contains abundant volcanic debris indicative of nearby volcanic activity. The Cañón de Caballeros Formation contains a North Silurian Realm, North Atlantic Region, European Province fauna. One genus and species of brachiopod is known only from the Ciudad Victoria area and the Cordillera de Mérida, Venezuela, suggesting that the Ciudad Victoria area may have originally been located near Venezuela.

4. Erosion or nondeposition took place during Devonian time. In addition, if the interpretation is accepted that what is now the Ciudad Victoria area was originally near the Mérida terrane in Venezuela in the Silurian and remained part of South America during the Devonian, and that the Eastern Andean terrane was also part of South America during the Devonian, then what is now the Ciudad Victoria area may have migrated close to North America by Devonian time. This interpretation follows because the Devonian strata in the Eastern Andean terrane in northern South America have fauna closely related to the Devonian North American fauna (Appendix 2).

5. Shallow-marine clastic deposits (Vicente Guerrero Formation) were deposited in Early Mississippian time. These deposits have a North American fauna, and indicate that by this time the Ciudad Victoria area lay close enough to North America for faunal interchange, as it may have also during the Devonian.

6. A Mississippian igneous and deformational event is marked by the emplacement of the Aserradero Rhyolite (344 ± 39 Ma) and by folding and erosion of Mississippian and older rocks before deposition of the unconformably overlying Lower and Middle Pennsylvanian Del Monte Formation. The igneous and structural activity can be related to the close approach of the Ciudad Victoria area to ancestral North America (Laurentia), or some other crustal block, perhaps during the initial stages of the Appalachian orogeny. The presence of Mississippian volcanic material in the Ouachita orogenic belt (Appendix 2), together with indications that orogenic activity started in the Ouachita belt in the Mississippian (Viele and Thomas, 1989), supports the idea that the Ciudad Victoria area was part of the approaching plate or plates that produced the Appalachian orogeny. This activity appears to be part of widespread igneous and deformational events during the middle and late Paleozoic throughout the region from North America to South America (Fig. 8).

7. In the middle or late Paleozoic, metamorphic events (Appendix 2 dated as 257–330 Ma, but mostly likely about 330 Ma, Mississippian), produced the Granjeno Schist. As described above, the Granjeno Schist, and associated serpentinite probably were emplaced as a structural plate into its present position after metamorphism. Emplacement occurred prior to deposition of Jurassic sedimentary rocks that unconformably overlie the Paleozoic rocks in the Ciudad Victoria area and probably in the late Paleozoic (Ramirez-Ramirez, 1992), well after the time of the metamorphism of the Granjeno Schist.

8. Deep-water marine deposits (Del Monte Formation)

formed in Early and Middle Pennsylvanian time. These deposits may represent synorogenic deposits during encroachment of the Ciudad Victoria area against ancestral North America during the Appalachian orogeny, or may indicate that orogenic activity slackened, and a deep sedimentary basin developed in the Ciudad Victoria area after initial orogenic events in the Mississippian. During this time, the Coahuila terrane, if it is indeed a terrane separate from the Ciudad Victoria area, may have lain between Ciudad Victoria and the Ouachita orogenic belt.

9. Amalgamation of the Ciudad Victoria area with the Pangean supercontinent occurred by Early Permian time when tectonic activity in the Ouachita orogenic belt ceased (Viele and Thomas, 1989).

10. A Lower Permian magmatic arc (indicated by abundant volcanic material in the Guacamaya Formation) developed along the margin of the newly formed Pangean supercontinent, or during the final stages in the formation of this continent (Torres et al., 1992; Centeno-García et al., 1995; Rosales-Lagarde et al., 1997).

11. Deposition of Lower Jurassic coarse conglomeratic rocks in rift-basins (Belcher, 1979) unconformably over folded Lower Permian and older rocks.

12. Break-up of the Pangean supercontinent started by Jurassic time and resulted in the opening of the Gulf of Mexico and the Atlantic Ocean. Deposition of thick clastic and carbonate deposits of Jurassic and Cretaceous age followed in the Ciudad Victoria area. The Paleozoic terranes of México, Central America, and Caribbean regions were dispersed by complex tectonic activity during the late Mesozoic and Cenozoic (Pindell, 1985; Rowley and Pindell, 1989; Pindel and Barrett, 1990)

ACKNOWLEDGMENTS

We thank Wolfgang Stinnesbeck of the Geologishes Institut, Universitat Karlsruhe, Karlsruhe, Germany, formerly with the Universidad de Nuevo León, Linares, Mexico, for his help in meeting the residents of Cañón de la Peregrina, for his geologic discussions, and for loaning us collections of Paleozoic fossils from Cañón de la Peregrina. Hans-Jürgen Gursky, Geologisches-Paläontologisches Institut, Darmstadt, Germany, provided information on fossil localities and access to the area. Carmelo Fernández was our guide in the field and his knowledge of trails in Cañón de la Peregrina was invaluable. We also thank the people of the Ejido de la Peregrina for letting us work on their land and for their friendly support. Finally, we thank Zoltan de Cserna, Elena Centeno-García, C. M. González León, and S. D. Ludington for their helpful review of the manuscript. Funds for field work in México were supplied by National Geographic Society Grant 4981-93.

APPENDIX 1

Silurian fauna localities of the European Province and the Malvinokaffric Realm in North America, and of the European Province and undivided North Atlantic Region in South America (Localities shown in Figure 5)

1. Cerro Rincón, Argentina. About lat 24°13′S, long 67°30′W. Strata with Silurian fauna of North Atlantic Region (Isaacson et al., 1976), but not distinguishable as either the European or North American Province. Silurian fossils in Bolivia (Isaacson et al., 1976; A. J. Boucot, unpublished information, 1994) to the north of Cerro Rincón contain faunas of North Atlantic Region in the lower few meters and of the Malvinokaffric Realm in the remainder of the succession.

2. Cordillera de Mérida, Venezuela. Centered on about lat 7°45′N, long 71°15′W. Strata with Silurian fauna of European Province, North Atlantic Region, rest on cool climate Middle and Upper Ordovician strata and are overlain by Pennsylvanian and Permian rocks (Boucot et al., 1972).

3. Ciudad Victoria, lat 23°46′N, long 99°16′N. Silurian fauna of European Province, North Atlantic Region (Boucot et. al., 1997; this report).

4. Subsurface Florida, centered about lat 30°N, long 83°W. A mixing of fauna of different biogeographic areas. Contains Silurian chitinozoans of Malvinokaffric Realm (Laufeld, 1979) and Silurian bivalves of European Province, North Atlantic Region (Pojeta et al., 1976).

5. Georgetown, Massachusetts (near Rowley). Coastal volcanic belt, West Avalonia terrane. About lat 42°43.3′N, long 70°59.3′W. In Newbury Volcanic Series (or complex). Silurian fauna of European Province, North Atlantic Region. (Berry and Boucot, 1970, p. 189–190). Berry and Boucot stated that the locality has Pridoli and early Gedinnian beds correlative to the Pembroke Formation of the Eastport Volcanic Series.

6. North Haven in Penobscot Bay, Maine. About lat 44°7.3′N, long 68° 49.2′W, (Berry and Boucot, 1970, p. 116). Ames Knob Formation; Late Llandoverian to Pridolian or Gedinnian age. Silurian fauna of European Province, North Atlantic Region. Coastal volcanic belt.

7. Eastport, Maine. Silurian fauna of European Province, North Atlantic Region, and Devonian fauna of Rhenish-Bohemian region, Old World Realm. About lat 44°54′N, long. 67°01′W. Coastal volcanic belt. Some information in Berry and Boucot 1970) under Eastport Formation, Quoddy Formation, Denny's Formation, and Pembroke Formation.

8. Long Reach to Sussex, New Brunswick. Coastal volcanic belt, West Avalonia terrane. Centered about lat 45°30′N, long 66°00′W, Long Reach and Jones Creek Formations. Silurian fauna of European Province, North Atlantic Region. (Berry and Boucot, 1970, p. 180, 168).

9. Cobequid Mountains, Nova Scotia, centered about 45 km southeast of Springhill, about lat 45°30′N, long 63°50′W. West Avalonia terrane. Silurian fauna of European Province, North Atlantic Region, and Devonian fauna of Old World Realm (Rhenish-Bohemian Region). In eastern part of area, strata are like those at Arisaig (loc. 10, below). In western part of area, strata are like Long Reach (loc. 8, above) (A. J. Boucot, unpublished data, 1994).

10. Arisaig and Knoydart, Nova Scotia, approx. lat 45°45.4′N, long. 62°10.0′W, and lat 45°42.7′N, long 62°14.7′W, respectively. West Avalonia terrane. Silurian fauna of European Province, North Atlantic Region and Lower Devonian fauna of Rhenish-Bohemian Region, Old World Realm. (Boucot et al. (1974).

APPENDIX 2

Pre-Permian terranes with emphasis on distribution and age of post-Proterozoic and pre-Permian igneous, metamorphic, and volcanoclastic rocks in the southern United States, Mexico, Central America, Cuba, and northern South America (Ages are those listed in publications and have not been recalculated using new age constants).

UNITED STATES AND NORTHERNMOST MEXICO

NA, North America. Cratonal Precambrian crystalline basement rocks ranging in age from over 3,000 to about 900 Ma, Paleozoic sedimentary cover rocks, and flanking Precambrian and Paleozoic shelf or miogeoclinal deposits. Volcanic detrital material is present in lower Upper Mississippian and Lower Pennsylvanian sandstones in the Black Warrior basin of Mississippi and Alabama (Mack et al., 1983). Widespread radioactive markers in Mississippian shales of western Texas are attributed to ejecta from explosive volcanic events (Decatur and Rosenfeld, 1982).

PI, Piedmont and associated terranes, southern Appalachian Mountains. Mainly late Precambrian and early Paleozoic metaclastic rocks lying on Grenvillian basement (Williams and Hatcher, 1982). In the Northern Appalachians, correlative terranes are regarded as deformed and metamorphosed rocks at the eastern edge of the North American miogeocline (Williams and Hatcher, 1982; Hatcher, 1989)

SAM, Southern Appalachians. Includes rocks of the Belair, Kiokee, Carolina Slate, and Charlotte belts, and parts of the Raleigh and Kings Mountain belts (Rankin et al., 1989). Consists of Cambrian mafic to felsic metavolcanic and metaigneous rocks, metamudstone, metawacke, quartzite, paragneiss; Silurian, Devonian, and Carboniferous undeformed granitic and gabbroic rocks (Rankin et al., 1989, and references therein). Considered to be related to West Avalonia terrane of northeastern United States and eastern Canada, which was emplaced against Laurentia during multiple events starting in the Ordovician and lasting into the Devonian and Mississippian, or during a single Devonian and Mississippian event (Osberg et al., 1989; Hatcher, 1989).

SU, Suwannee terrane in the subsurface of Alabama, Georgia, and Florida (Rankin et al., 1989, and references therein; Thomas et al., 1989, and references therein). Consists of calc-alkaline felsic volcanic-plutonic rocks, undeformed granite, a high-grade metamorphic sequence, and fossiliferous Lower Ordovician to Middle Devonian platform sedimentary rocks. Undeformed granite is dated isotopically as 525–530 Ma (Rankin et al., 1989; Dallmeyer et al., 1987). High-grade metamorphic rocks dated isotopically as 500–520 Ma (Dallmeyer, 1986). Mixed faunal assemblage of Malvinokaffric Realm and European Province (see Appendix 1).

W, Wiggins terrane. Pre-Mesozoic rocks in subsurface, composed of phyllite, chlorite schist, quartzite, gneiss, amphibolite, and granite (Thomas et al., 1989, and references cited therein). K-Ar age of 299 ± 9 Ma on muscovite-sericite from phyllite. Granite ages range in age from 284 ± 14 (whole-rock K-Ar) to 272 ± 10 Ma (Rb-Sr). Gneiss yields $^{40}Ar/^{39}Ar$ incremental release spectra that define ages of 300–310 Ma, and phyllite whole-rock spectra that indicate age of 315–320 Ma.

ST, Sabine Uplift and Texarkana Platform. The Sabine Uplift consists of subsurface Carboniferous flysch, pre-Desmoinesian rhyolite porphyry, Pennsylvanian shallow-water carbonates, and Pennsylvanian and Permian deeper water clastics and minor carbonates (Nicholas and Waddell, 1989). The Texarkana Platform consists mostly of Pennsylvanian shallow-water fossiliferous limestone and fine-grained clastic rocks (Nicholas and Waddel, 1989; Viele and Thomas, 1989). Both areas are considered by Viele and Thomas (1989) to be part of a plate outboard of the Ouachita orogenic belt. A pre-Desmoinesian (pre–upper Middle Pennsylvanian) rhyolite porphyry and tuff is present in a drill holes in the Sabine Uplift, eastern Texas, and yielded an apparently erroneous Rb-Sr age of 255 ± 15 Ma (Nicholas and Waddell, 1989).

IMB, Interior metamorphic belt of Ouachita orogenic belt (Flawn et al., 1961; Viele and Thomas, 1989). Phyllite, slate, marble, quartzite, minor metavolcanic rock, and brecciated granite. Metamorphic ages mostly range from 300 to 370 Ma. Specific localities and ages follow: (1) In McCurtain County, Oklahoma, K-Ar dates from schist from drill cores give 307 ± 6 and 296 ± 6 Ma on muscovite, and the following dates on phyllites and one schistose metasandstone: 318 ± 6, 317 ± 6, 301 ± 6, 311 ± 6, 315 ± 6, 356 ± 7, 360 ± 7, 369 ± 7, 368 ± 7, 378 ± 8, 322 ± 6, and 325 ± 7 Ma (Denison et al., 1977). (2) In Collin County, Texas, K-Ar whole-rock age on diabase from drill core (Humble No. 1, Miller) are 384 ± 8 and 389 ± 8 Ma (Denison et al., 1977). (3) In Waco Uplift, Texas, K-Ar dates from drill core (Shell No. 1, Barrett) are 304 ± 6 Ma on talc from marble, and 337 ± 6 and 336 ± 6 Ma on biotite from granitic rock (Denison et al., 1977). (4) In Milam County, Texas, K-Ar dates on muscovite from phyllite in drill cores (Davis No. 1, Coffee) are 315 ± 6 and 310 ± 6 Ma (Denison et al., 1977). (5) In Bexar, Medina, and Maverick Counties, Texas, south of the Llano Uplift, deep wells penetrated dacite, andesite, and basalt flows, and granite and granodiorite (Flawn et al., 1961; Niem, 1977). These volcanic rocks, according to Flawn et al. (1961), could correlate with volcanic rocks in the Stanley Group. (6) In the Devils River uplift, Texas, K-Ar whole-rock dates from drill cores are as follows: metarhyolite, 383 ± 8 and 385 ± 8 Ma; metadacite, 372 ± 7 and 373 ± 7 Ma; and muscovite from phyllite, 327 ± 7 and 304 ± 6 Ma (Denison et al., 1977). The ages on the volcanic rocks are apparently metamorphic ages because Rb-Sr ages on the same rocks are Neoproterozoic in age. In addition, Nicholas and Rozendal (1975) have reported that K-Ar ages from drill cores in the Devils River uplift range in age from 256 to 431 Ma and show no relation to depth.

FB, Frontal belt of Ouachita orogenic belt. Pre-orogenic off-shelf (Cambrian to Devonian) and synorogenic deep-water clastic deposits (Mississippian to Pennsylvanian) of Ouachita orogenic belt (Viele and Thomas, 1989). In the Ouachita Mountains of Arkansas, Oklahoma, and east Texas, tuffs as much as 40 m thick are present in the basal part of the Upper Mississippian Stanley Group (Niem, 1977; Decatur and Rosenfeld, 1982; Loomis et al., 1994); and granite boulders in the Ordovician Blakely Sandstone yield Rb-Sr ages (perhaps metamorphic or alteration ages) of 489 ± 55, 432 ± 75, and 351 ± 47 Ma (Denison et al., 1977) and U-Pb zircon ages of 1,284 ± 12, 1,350 ± 30, and 1,407 ± 13 Ma (Bowring, 1986). In the Marathon region, Texas, tuffaceous material is present in sandstone of the upper one-third of the Upper Mississippian and Pennsylvanian Tesnus Formation (McBride, 1989) and igneous and metamorphic boulders in the Pennsylvanian Haymond Formation give Rb-Sr ages of 369–457 Ma (Denison et al., 1969).

CE, Cordilleran eugeoclinal rocks. Offshelf, deep-water, chert, argillite, shale, quartzite, limestone, and greenstone emplaced over Cordilleran miogeoclinal rocks in the Antler orogeny (Late Devonian and Early Mississippian) and the Sonoma orogeny (Late Permian to Early Triassic) in the western United States and in the Sonora orogeny (Late Permian to Triassic) in northwest Mexico (Miller et al., 1992; Stewart et al., 1990; Poole et al., 1992, 1995). Includes Paleozoic deep-water strata in Baja California, Mexico (Gastil and Miller, 1993). Locally within the northwesternmost part of Figure 7, includes allochthonous terranes outboard of the offshelf deposits described here.

MEXICO AND CENTRAL AMERICA

SI, Sonabari and Rusias terranes of Campa and Coney (1983). Consists of relatively small outcrops in the Pacific Coastal Plain of Sinaloa. The El Fuerte area, in northernmost Sinaloa, includes metamorphosed chert-argillite-limestone, schist, and metavolcanic and other metamorphic rocks (Mullan, 1978). Thin turbidite limestone beds associated with chert near El Fuerte collected by F. G. Poole and W. R. Page yielded conodonts identified as Ordovician by A. G. Harris (oral communication, 1997). In addition, a metamorphic rock near El Fuerte has a Triassic protolith age (Anderson and Schmidt, 1983), but other rocks in the El Fuerte are undated. The San José de Gracia area contains mainly bedded chert, with minor interlayered argillite and calcarenite with mid-Pennsylvanian to Early Permian conodonts (Gastil et al., 1991). Other sequences contain chert, shale, limestone, and quartzite and limestone containing Early Mississippian to late Pennsylvanian fusulinids (Gastil et al., 1991; López-Ramos, 1981, p. 5). Rocks in the Mazatlán area consist of flysch deposits of argillite, sandstone, conglomeratic sandstone, and limestone and contain Paleozoic bryozoa (López-Ramos, 1981, p. 5).

Possible extensions of the Sonabari and Rusias terranes are the prebatholithic metasedimentary and metaigneous rocks in Baja California

Sur (an uncertain K-Ar age of 335 Ma; data summarized by Sedlock et al., 1993, p. 48), and the muscovite-amphibole schist (K-Ar age of 326 ± 26 and 311 Ma; data summarized by Sedlock et al., 1993, p. 60) in the Santa María del Oro area, Durango.

CO, Coahuila terrane (Coahuiltecano terrane of Sedlock et al., 1993). Mostly subsurface terrane of strongly deformed, fine-grain, low-grade metamorphic rocks cut by Triassic granitoids (Flawn et al., 1961; Handschy et al., 1987; Sedlock et al., 1993). The western part of the terrane contains Pennsylvanian to Permian mass-gravity marine deposits and volcaniclastic deposits indicative of a closeby magmatic arc (McKee et al., 1988). Terrane is bordered on the south by the hypothetical Mojave-Sonora megashear, along which 700–800 km of left-lateral offset has been proposed (Anderson and Schmidt, 1983). The Coahuila terrane may be a separate block or originally may have continued southward into the Sierra Madre terrane. Isotopic ages are (1) Rb-Sr model ages of 370 ± 4.5 and 387 ± 4 Ma on cobbles of quartz-biotite-muscovite-chlorite schist from Cretaceous conglomerate in Potero Colorado (McKee et al., 1990); (2) U-Pb ages of 331 ± 4 Ma on dacitic pluton and 303 ± 5 Ma on a volcanic rock at Las Delicias, Coahuila (R. Lopez, written communication, 1997); (3) Rb-Sr model age of 358 ± 70 Ma on granite gneiss from drill core (Pemex No. 1, La Perla well) (Denison et al., 1969); (4) U-Pb zircon upper-intercept ages of about 350 Ma on two intermediate-composition granulite xenoliths (Rudnick and Cameron, 1991; Cameron et al., 1992; Cameron and Jones, 1993), and an unpublished three-point isochron on zircon from xenoliths has an upper intercept age of 321 Ma (K. L. Cameron, written communication, 1994); and (5) Rb-Sr age of 350 ± 15 Ma on metasedimentary clasts in Cretaceous conglomerate in northern Chihuahua (Bridges, 1971).

SM, Sierra Madre terrane of Campa and Cone (1983) and subsurface rocks of the Gulf Coastal Plain. Northern part of the Oaxaquia terrane of Ortega-Gutiérrez and approximately the same as the Guachichil terrane of Sedlock et al. (1993). The Sierra Madre terrane is mostly covered by Mesozoic or Cenozoic sedimentary or volcanic rocks. Pre-Mesozoic rocks crop out in the Ciudad Victoria, the Aramberri, and the Molango areas. In the Ciudad Victoria area (Carrillo-Bravo, 1961; this report), Grenville-age basement rocks are unconformably overlain by: (1) Silurian shallow-water rocks (including silicic volcanic arenite), (2) Lower Mississippian shallow-water rocks, (3) a Mississippian rhyolite (334 ± 39 Ma; this report), and (4) unconformably overlying relatively deep-water Pennsylvanian and Lower Permian strata. The Permian strata contain abundant volcanoclastic detrital rocks. Structurally emplaced against these strata are low-grade metasedimentary and metavolcanic rocks (Granjeno Schist), and serpentinite. Sedlock et al. (1993, p. 26) lists 14 Rb-Sr and K-Ar ages on the Granjeno Schist (see also Ramirez-Ramirez, 1992, p. 117) that range from 257–330 Ma. Ortega-Gutiérrez et al. have (1993) indicated that a seven-point Rb-Sr isochron that gives an age of 330 Ma is probably significant and may be the age of metamorphism. In the Aramberri area, Nuevo Leon, graphic schist have a metamorphic age of 293–294 Ma (Denison et al., 1971) and are probably equivalent to the Granjeno Schist of the Ciudad Victoria area. The Molango area, Hidalgo, (Carrillo-Bravo, 1965; Moreno-Cano and Patiño-Ruiz, 1981; Ochoa-Camerillo, 1997; Ortega-Gutiérrez et al., 1997; Rosales-Lagarde et al, 1997; Centeno-García and Rosales-Lagarde, 1997) contains Grenville-age gneiss, and unconformably overlying Upper Pennsylvanian and Lower Permian sedimentary and volcanic rocks. The Upper Pennsylvanian and Lower Permian rocks consist of shale, sandstone, limestone, lava flows, tuffs, volcanic breccia, volcaniclastic sandstone, and conglomerate. Brachiopods, trilobites, and crinoids are common shale and limestone. Some shale and sandstone is lithologically similar to that in the Guacamaya Formation of the Ciudad Victoria area.

The Sierra Madre terrane is considered to extend into the subsurface of the Gulf Coastal Plain of eastern Mexico where it consists of Precambrian gneiss, Paleozoic(?) metasandstone and quartzite, and Upper Paleozoic(?) schist and metasedimentary rocks (López-Ramos, 1972). Sedlock et al. (1993, p. 30) lists a K-Ar? age of 320 Ma on an igneous rock from southern Tamaulipas and Pozo Rica region. The subsurface areas of the Sierra Madre terrane also contain Mesozoic conglomeratic rocks and large areas of Triassic(?) plutonic rocks. The Sierra Madre terrane, as shown in Figure 7, also extends to the west and southwest of Ciudad Victoria into areas where Grenville-age basement is indicated by isotopic studies of xenoliths (Ruiz et al., 1988).

MI, Mixteca terrane (Campa and Coney, 1983). Same as Mixteco terrane of Sedlock et al. (1993). The oldest rocks in the Mixteca terrane comprise the Acatlán complex (description and isotopic ages from Yañez et al., 1991, and Sedlock et al., 1993, and references therein) consisting of:

1. The Petlalcingo Subgroup of schist, amphibolite, quartzite, and phyllite, and other metamorphic rocks. Isotopic ages are U-Pb zircon age of 356 ± 140 Ma on migmatite/paragneiss, Sm-Nd model age of 429 ± 50 on schist, Rb-Sr age of 386 ± 6 Ma on schist, Sm-Nd modal age of 349 ± 27 Ma on schist, K-Ar age of 346 ± 28 on schist, and K-Ar age of 328 ± 26 Ma on schist.
2. The Xayacatlan Formation of serpentinized peridotites, eclogitized and amphibolitized metabasites, pelitic schist, and quartzite, which are interpreted as a dismembered ophiolite; isotopic ages on the Xayacatlan Formation are Sm-Nd model age of 416 ± 12 Ma on schist, Sm-Nd model age of 388 ± 44 Ma on eclogite, Rb-Sr age of 386 ± 6 Ma, Rb-Sr age of 332 ± 4 Ma on eclogite, and Rb-Sr age of 318 ± 4 Ma on schist.
3. The Esperanza granitoids (mostly polymetamorphosed mylonite gneiss). Isotopic ages are as follows: U-Pb zircon lower intercept of 371 ± 34 Ma, U-Pb zircon age of 425 ± 13 Ma, Rb-Sr whole-rock age of 428 ± 24 and 448 ± 175 Ma, Rb-Sr whole rock and white mica age of 330 ± 5 Ma.
4. The Upper Devonian(?) Tecomate Formation consisting of arkosic metaclastic rocks, calcareous metapelites, and limestone. The clastic rocks in the Tecomate Formation contain clasts of the Esperanza granitoids.
5. The Totoltepec stock with a U-Pb age of 287 ± 2 Ma (Yañez et al., 1991).

The Esperanza granitoids are interpreted as the product of Early-Late Devonian collision of the Mixteca and Oaxaca terranes (Sedlock et al., 1993) and perhaps related to Acadian orogenesis in eastern North America (Sedlock et al., 1993, and references therein). The Carboniferous event (Totoltepec stock) is interpreted as the result of collision of Gondwana and North America (Yañez et al., 1991). The U-Pb study of the Esperanza granitoid indicates either that the granitoid is a Grenville-age intrusion and underwent Devonian resetting or that the lower intercept is the age of intrusion and deformation, and the upper intercept is the result of inheritance from Grenville-age rocks (Yañez et al., 1991).

In the northern part of the Mixteca terrane, the Acatlán complex is unconformably overlain by Early Mississippian marine strata and apparently also by sandstone, siltstone, and conglomerate of Pennsylvanian and probably Permian age.

O, Oaxaca terrane (Campa and Coney, 1983). Same as Zapoteco terrane of Sedlock et al. (1993) and a part of Oaxaquia, a microcontinental block as defined by Ortega-Gutiérrez et al. (1995b). Consists of Grenville-age granulite and anorthosite crystalline basement rocks (Silver et al., 1994; Ortega-Gutiérrez et al., 1995b) that are unconformably overlain by a sedimentary succession that consists of Cambrian and Ordovician limestone and shale and Mississippian and Pennsylvanian shale and sandstone (Pantoja-Alor, 1970; Pantoja-Alor and Robison, 1967). Taxa from Cambrian and Ordovician strata have Gondwana affinity (Robison and Pantoja-Alor, 1968; Ortega-Gutiérrez et al., 1995b). U-Pb dating on zircon from gneiss and granite, with unstated relation to basement and sedimentary rocks described above, give lower intercept ages of 353 ± 250 and 331 ± 359 Ma, respectively (Herrmann et al., 1994, p. 472). These dates have unacceptable large uncertainties, but are similar to 371 ± 35 Ma (U-Pb zircon) age of the Esperanza pluton in the Acatlán complex (Yañez et al., 1991).

X, Xolapa terrane (Campa and Coney, 1983). Same as Chatino terrane of Sedlock et al. (1993). High-grade metamorphic to migmatitic orthogneiss and paragneiss intruded by undeformed Tertiary plutonic rocks. Based on U-Pb zircon dating, these rocks have a Tertiary metamorphic age but the protoliths either received detritus from a continental region of Grenville age or the Xolapa terrane has a Grenville-age basement (Sedlock et al., 1993, and references therein; Herrmann et al., 1994). Rb-Sr whole rock age of 311 ± 30 Ma on gabbro and monzonite in Tierra Caliente complex and 308 ± 5 Ma for metasedimentary rocks. Information cited by Sedlock et al. (1993, p. 39). Also see Herrmann et al. (1994, Fig. 2).

Y, Yucatan terrane. Sparse subsurface information indicates that Paleozoic or Paleozoic(?) rocks in the subsurface of Yucatan consist of metavolcanic rock, quartzite, and schist (Sedlock et al., 1993, and references therein). From the Yucatan No. 1 drill hole, López-Ramos (1975, p. 267) indicated an age of 410 Ma (no method of dating given) on rhyolite and the probable existence of a metamorphic event at about 300 Ma (no method of dating given). López-Ramos (1981, p. 275) from the same drill hole indicates an Rb-Sr age of 410 Ma for the rhyolite and a 330-Ma age for a possible metamorphic event. He indicates that the 300-Ma date is only slightly older than the 280-Ma age of the Bladen Volcanics of Maya Mountains of Belize. U-Pb ages on zircons from breccia related to the Chicxulub crater in the northern part of Zucatan (Yucatan No. 6 drill hole) indicate a source rock of 544.5 ± 5.0 Ma and a lesser source rock of 418 ± 6 Ma (Krogh et al., 1993). In addition, a single near-concordant 320 ± 31-Ma shocked zircon grain in Colorado, USA, that comes from ejecta considered also to be from the Chicxulub impact site, suggests a younger igneous event in Yucatan. The 544.5 ± 5.0-Ma age from the Yucatan No. 6 drill hole suggests a Pan-African basement age (Ortega-Gutiérrez et al., 1995b) for the Yucatan terrane.

PO, Polochic terrane. This name is proposed here to describe metamorphic and sedimentary rocks exposed in Guatemala, Belize, and southernmost Mexico. The oldest rocks in the Polochic terrane are the Chuacús Group, composed of gneiss, schist, marble, mylonized metagranite, and minor volcanic and greenstone and quartzite. Zircon from gneiss in the Chuacús Group yielded an U-Pb age of 1,075 ± 25 Ma and indicates either the age of the source terrane of the original sediments or the age of principal metamorphism (Gomberg et al., 1968). This Grenville-age component distinguishes the terrane from Yucatan to the north that, in part at least, has a Pan-African-age basement. The next younger rocks are Late Silurian plutons in Belize that have yielded a U-Pb zircon age of 418 ± 3.6 Ma, a minimum U-Pb age of 404 Ma, a U-Pb age of about 410–420 Ma, and an Rb-Sr age of 428 ± 41 Ma (Steiner and Walker, 1996, and references therein). A U-Pb age of 345 ± 20 Ma has been obtained from the Rabinal granite in Guatemala (Gomberg et al., 1968). These plutons are known, or presumed, to intrude the Chuarcús Group. The Chuarcús Group, or the granitoids that intrude it, are unconformably(?) overlying by the Pennsylvanian to Lower Permian Santa Rosa Group (Donnelly and López-Ramos, 1990; Sedlock et al., 1993). The Santa Rosa Group consists mostly of marine conglomerate, sandstone (flysch), volcanic rocks, and a younger unit of limestone and shale. In Belize, the Santa Rosa Group includes the Bladen Volcanics that consist of 1,500–1,800 m of rhyolitic and dacitic flows, ash-flow tuffs, and tuff breccias (Bateson and Hill, 1977; Donnelly and López-Ramos, 1990). The isotopic age of the Bladen Volcanics is listed by Bass and Zartman (1969; López-Ramos, 1975) as 300 Ma (no method of dating given), by López-Ramos (1981, p. 277) as 280 Ma (no method of dating given), and by Cole and Andrews-Jones (1979) as 232 Ma (K-Ar method). The Santa Rosa Group is overlain by the mid-Permian Chochal Formation that consists of limestone, dolomite, and shale.

CH, Chortis terrane, southern Guatemala, Honduras, and northern Nicaragua. This terrane, lying south of the Motogua fault in Guatemala, include poorly dated gneiss, schist, and other metamorphic rocks and a younger low-grade succession of phyllite (Horne et al., 1990; Sedlock et al., 1993). One area of Grenville-age basement rocks (U-Pb zircon age of 1.0 Ga; Manton, 1996) are known in northern Honduras. An Rb-Sr date of 305 ± 12 Ma on a metaigneous complex (Horne et al., 1976) that intrudes the metamorphic rocks suggests a pre-mid–Pennsylvanian protolith age for some of the metamorphic rocks. The terrane may include rocks of diverse origin and structural setting. Rocks of the terrane are unconformably overlain by Mesozoic sedimentary and volcanic rocks.

CARIBBEAN AREA

CU, Central Cuba, marbles and siliciclastic metasedimentary rocks and granite (Renne et al., 1989). $^{40}Ar/^{39}Ar$ plateau age for phlogopite from a marble is 903.5 ± 7.1 Ma and U-Pb zircon data from the granite indicate an inherited zircon component of about 900 Ma, considered to represent a Grenville-age event. No evidence of a Pan–African age thermal overprint was noted.

CK, Catoche Knoll. Drill cores from the Gulf of Mexico Deep Sea Drilling Project Leg 77 on Catoche Knoll and a nearby knoll yield $^{40}Ar/^{39}Ar$ ages of 496 ± 8 and 501 ± 9 Ma on amphibolite, 500 ± 8 Ma on phyllite, and 348 ± 8 Ma on gneiss (Schlager et al., 1984). These dates from the Gulf of Mexico indicate a Pan–African-age basement similar to that in the Suwannne and Yucatan terranes, but different from the Grenville-age basement in Cuba.

SOUTH AMERICA

CA, Central Andean terrane. A complex terrane, considered by Restrepo and Toussaint (1988) and Restrepo-Pace (1992) to be a single terrane, but divided by Etayo-Serna et al. (1983) into six separate terranes. Characterized by a variety of metasedimentary and metavolcanic rocks in polymetamorphic complexes that, based on isotopic dating, were metamorphosed and intruded by igneous rocks in the Mesoproterozoic and Devonian and metamorphosed again in the Permian to Triassic (Forero-Suarez, 1990; Restrepo and Toussaint, 1988; McCourt, et al., 1984). Pre-Cretaceous sedimentary rocks are apparently absent in the Central Andean terrane. McCourt et al. (1984) considered that during the Devonian(?) the terrane represented a magmatic arc related to a subduction zone along the western margin of South America. Restrepo (1982) has summarized data that contains the following dates: K-Ar age of 343 ± 12 on muscovite from gneiss, K-Ar age of 482 ± 50 Ma on amphibolite, K-Ar whole-rock age of 326 ± 82 Ma on pillow lava, and K-Ar whole-rock age of 312 ± 15 Ma on metadiabase.

EA, Eastern Andean terrane. A complex terrane considered to be part of a single terrane by Restrepo and Toussaint (1988) and Restrepo-Pace (1992), but divided by Etayo-Serna et al. (1983) into four terranes. Local basement rocks consist of Mesoproterozoic metamorphic and igneous rocks, including Grenville-age granulite (MacDonald and Hurley, 1969; Alvarez, 1981; Kroonenberg, 1982; Etayo-Serna et al., 1983; Restrepo and Toussaint, 1988; Priem et al., 1989; Case et al., 1990; Restrepo, 1995). Basement rocks are unconformably overlain by low- to medium-grade metamorphic rocks locally intruded by granitoids of Ordovician to Devonian age (Forero-Suarez, 1990). In the Santander massif these metamorphic and igneous rocks are 477 ± 16 Ma (mid-Ordovician) on the basis of U-Pb dating of zircon from a synkinematic pluton (Restrepo, 1995) and mostly Ordovician and Silurian in age on basis of K-Ar and Rb-Sr isotopic dating. In the Santander massif, Goldsmith et al. (1971) have reported K-Ar age of 413 ± 30 Ma (421 Ma recalculated) for hornblende from a metadiorite, K-Ar age of 432 ± 8 (439 Ma recalculated) for muscovite from pegmatite in gneiss, K-Ar age of 439 ± 12 Ma (448 Ma recalculated) for muscovite from pegmatite in gneiss, Rb-Sr whole-rock age of 450 ± 80 Ma (439 Ma recalculated) for a granitic gneiss, K-Ar age of 457 ± 13 Ma (465 Ma recalculated) for muscovite from a pegmatite in gneiss; Boinet et al. (1985) has reported a K-Ar date of 456 ± 22.8 Ma on gabbro and 349 ± 17.5 Ma on quartz monzonite; Etayo-Serna et al. (1983) reported K-Ar ages of 546 ± 48 Ma on a quartz monzonite, a K-Ar age of

394 ± 23 Ma on a monzonite, and an Rb-Sr age of 471 ± 22 on a granitic stock.

The predominantly Ordovician and Silurian igneous and metamorphic rocks of the Eastern Andean terrane are unconformably overlain by unmetamorphosed sedimentary rocks of Devonian and Carboniferous age (Restrepo and Toussaint, 1988; Benedetto, 1984). The Devonian strata are exposed in widely spaced outcrops in the Eastern Andean terrane and consist of a few hundred meters to as much as 1000 m of marine sandstone, siltstone, conglomerate, and sparse limestone (Benedetto, 1984). The Sierra de Perijá in the eastern part of the Eastern Andean terrane contains a Devonian fauna of the eastern North Americas Realm. This fauna is also characteristic of the eastern part of the United States and Canada (Benedetto, 1984; Boucot, 1988). Carboniferous strata in the Eastern Andean terrane are as much as 1000 m thick and consist of a nonmarine lower part composed of sandstone, conglomerate, sandstone with fragments of plants, and an upper part of marine shales and limestone (Benedetto, 1984).

Silurian strata are absent in the Eastern Andean terrane, except at one locality where low-grade metamorphic rocks (phyllite, schist, and quartzite) contain spores and acritarchs indicative of a Silurian age (Grösser and Prossl, 1990; Pimentel de Bellizzia et al, 1992).

A group of igneous rocks mostly younger than those in the Santander massif are recognized in the Sierra de Perijá in the northern part of the Eastern Andean terrane. Here, U-Pb zircon ages from the Lajas granite suggest crystallization age from about 310 to 385 Ma (Dasch, 1982). Rb-Sr dates on this same granite are 334 ± 12 and 380 ± 18 Ma (Espejo et al., 1980). Rb-Sr dates on the Río Antray granite are 312 ± 18 and 402 ± 48 Ma (Espejo et al., 1980). Martín-Bellizzia (1968) has indicated ages (method of dating unknown) of 304 ± 80 Ma on a granite gneiss, 302 ± 42 Ma on the granite of El Palmar, and 370 ± 20 Ma, and 320 ± 20 Ma on the granite of El Tetumo.

M, Mérida terrane of Richards and Coney (1990). Includes the deformed pericratonal belt of Benedetto (1982) and Benedetto and Ramírez-Puig (1982). Consists of rocks in the Cordilleran de Mérida, metasedimentary rocks of the El Baúl massif (Feo-Codecido et al., 1984; Pimental de Bellizzia, 1992), subsurface rocks in the Llanos basins (Feo-Codecido et al., 1984) and poorly dated metamorphic rocks in the Cordillera de la Costa (Urbani, 1989). The Mérida terrane is considered a continuation of the Eastern Andean terrane by Restrepo and Toussaint (1988) and Restrepo-Pace (1992), but a separate terrane by Richards and Coney (1990). A belt of igneous rocks of mostly pre-Devonian (possibly mostly Ordovician) age extends in a northwest-trending belt from the Eastern Andean terrane into the Mérida terrane (Forero-Suarez, 1990), suggesting that the two terranes, if they are different, may have been juxtaposed sometime before the Devonian.

The Cordillera de Mérida in the Mérida terrane is a complex polydeformed and polycomponent structural block (Case et al., 1990). In the southern Cordillera de Mérida, low-grade metamorphic rocks of Precambrian(?) to as young as possibly Cambrian age are overlain unconformably by, or possibly everywhere in fault contact with, unmetamorphosed Upper Ordovician and Silurian strata. These Upper Ordovician and Silurian strata are 350 m thick and consist mostly of siltstone, mudstone, sandstone, calcareous siltstone, and clayey limestone (Boucot et al., 1972; Benedetto and Ramírez-Puig, 1982; Pimentel de Bellizzia, 1992; Benedetto et al., 1992). The fauna in the Silurian part of this section (Boucot et al., 1972) is related to the Silurian European Province, North Atlantic Region (Boucot, 1990), and is similar to the fauna at Ciudad Victoria (this report). In the central and northern part of the Cordillera de Mérida, Precambrian paragneiss is considered to have been derived from source rocks that are about 1050–1200 m.y. old. The paragneiss is inferred to be intruded by approximately 600-Ma-old granitic rocks (Burkley, 1976; Maurrasse, 1990). Permian-Carboniferous, and perhaps older, slates, as well as sandstone and conglomerate are widespread in the Cordillera de Mérida. The relation of these younger, or perhaps in part equivalent, sequences to the Upper Ordovician and Silurian rocks is uncertain.

Plutonic rocks, mostly of Ordovician and Silurian age, are widespread in the Cordillera de Mérida and are in part older than and in part younger than the sedimentary succession. Burkley (1976) reported U-Pb ages on zircons from 15 plutons. A large majority of these ages are in the range of 500–425 Ma. Two plutons are about 390 Ma and one is about 620 Ma. An augen gneiss is about 350 Ma. K-Ar ages of 460 ± 15, 400 ± 7, 380 ± 20, 348 ± 15, and 297 Ma on plutonic bodies are reported by Martín-Bellizzia (1968) and Schubert (1969). Other isotopic ages in the Cordillera de Mérida are Rb-Sr isochron ages of 430–490 Ma from granitoids, gneiss, and schist (Cordani et al., 1985); Rb-Sr ages of 289 to 293 Ma on sedimentary rocks (Burkley, 1976); Rb-Sr age of 410 ± 40 Ma on mica schist (Bass and Shagam, 1960); and an Rb-Sr age of 373 ± 20 Ma on a pegmatite (Martín-Bellizzia, 1968).

In the Cordillera de Mérida, Benedetto and Ramírez-Puig (1982) emphasized an Ordovician tectonic event, based on the presence of conglomerate of Ordovician age, and a Late Silurian or Devonian orogenic event based on the intrusion of isotopically dated Devonian granitic rocks into paleontologically dated Silurian sedimentary rocks.

In the eastern part of the Mérida terrane, subsurface granitic and metamorphic rocks in the Llanos basins have yielded the following isotopic ages (Feo-Codecido et al. (1984): Rb-Sr age of 463 ± 45 Ma on granite pegmatite and 433 ± 50 Ma on unknown rock type; and K-Ar ages of 412 ± 21 Ma on gneiss and schist, 411 ± 12 and 406 ± 8 Ma on schist, 347 ± 10 on pelitic hornfels, and 330 and 321 Ma on granite. In the Cordillera de la Costa, Districto Federal, Venezuela, Urbani (1989) has reported Rb-Sr ages of 424 and 321 Ma on gneiss.

SA, South America. Consists of the South American craton composed of Precambrian metamorphic and igneous rocks ranging in age from greater than 2,000 to about 900 Ma, and thin Neoproterozoic(?) and Paleozoic, and younger, cover strata along the margins (Kroonenberg, 1982; Teixeira et al., 1989; Benedetto and Ramírez-Puig, 1982; Etayo-Serna et al., 1983; Mojica and Villarroel, 1990). In the northwestern part of the South American craton (San José del Guaviare, Colombia), Pinson et al. (1962) obtained the following Paleozoic isotopic ages on a complex of syenites, pegmatites, and fine-grained syenite aplites: K-Ar ages of 445 ± 22, 460 ± 23, 485 ± 25, and 485 ± 25 Ma, and an Rb-Sr age of 495 ± 25 Ma.

REFERENCES CITED

Alvarez, J., 1981, Determinación de la edad Rb/Sr en rocas del Macizo de Garzón, Cordillera Oriental de Colombia: Geología Norandina, v. 4, p. 31–38.

Amos, A. J., 1957, New syringothyrid brachiopods from Mendoza, Argentina: Journal of Paleontology, v. 31, p. 99–104.

Amos, A. J., 1958, Some Lower Carboniferous brachiopods from the Volcán Formation, San Juan, Argentina: Journal of Paleontology, v. 32 (5), p. 838–845.

Anderson, T. H., and Schmidt, V. A., 1983, The evolution of Middle America and the Gulf of Mexico–Caribbean Sea region during Mesozoic time: Geological Society of America Bulletin, v. 94, p. 941–966.

Ballard, M. M., Van der Voo, R., and Urrutia-Fucugauchi, J., 1989, Paleomagnetic results from the Grenvillian-aged rocks from Oaxaca, Mexico; evidence for a displaced terrane: Precambrian Research, v. 42, p. 343–352.

Barrick, J. E., 1987, Conodont biostratigraphy of the Caballos Novaculite (Early Devonian–Early Mississippian), northwestern Marathon uplift, west Texas, *in* Austin, R. L., ed., Conodonts: investigative techniques and applications: Chichester, United Kingdom, Ellis Horwood Ltd., for the British Micropalaeontology Society, p. 120–135.

Bass, M. N., and Zartman, R. E., 1969, The basement of Yucatan Peninsula [abs.]: Eos (Transactions, American Geophysical Union) v. 50, p. 313.

Bass, M. N., and Shagam, R., 1960, Edades Rb-Sr de las rocas cristalinas de los Andes Meridenos, Venezuela: Venezuela Dirección Geología, Boletín Geología, Publicación Especial 3, v. 1, p. 377–381.

Bateson, J. H., and Hall, I. H. S., 1977, The geology of the Maya Mountains, Belize: Institute of Geological Sciences, Overseas Memoir 3, 43 p.

Belcher, R. C., 1979, Depositional environments, paleomagnetic and tectonic significance of the Huizachal Redbeds (lower Mesozoic), northeastern Mexico (Ph.D. thesis): Austin, University of Texas, 286 p.

Benedetto, J. L., 1982, Las Unidades tecto-estratigráficas Paleozoicas del norte de Sudamérica, Apalaches del sur y noreste de Africa: Quinto Congreso Latinoamericano de Geología, Buenos Aires, Argentina, Actas I, p. 469–487.

Benedetto, J. L., 1984, Les brachiopodes devoniens de la Sierra de Perijá (Venezuela); systematique et implications paleogeographiques: Université de Bretagne Occidentale, Biostratigraphie du Paleozoique 1, 191 p.

Benedetto, J. L., and Ramírez-Puig, E., 1982, La secuencia sedimentaria Precámbrico-Paleozoico Inferior pericratónica del extremo norte de Sudamérica y sus relaciones con las cuencas del norte de Africa: Quinto Congreso Latinoamericano de Geología, Buenos Aires, Argentina, Actas II, p. 411–425.

Benedetto, J. L, Sanchez, T. M., and Brussa, E. D., 1992, Las Cuencas Silúricas de América Latina, *in* Gutiérrez-Marco, J. C., Saavedra, J., and Rábano, I, eds., Paleozoico Inferior de Ibero-América: Universidad de Extremadura, p. 119–148.

Berry, W. B. N., and Boucot, A. J., 1970, Correlation of the North American Silurian rocks: Geological Society of America Special Paper 102, 289 p.

Berry, W. B. N., and Boucot, A. J., 1972a, Correlation of the South American Silurian rocks: Geological Society of America Special Paper 133, 59 p.

Berry, W. B. N., and Boucot, A. J., 1972b, Correlation of the southeast Asian and Near Eastern Silurian rocks: Geological Society of America Special Paper 137, 6 p.

Berry, W. B. N., and Boucot, A. J., 1973, Correlation of the African Silurian rocks: Geological Society of America Special Paper 147, 83 p.

Boinet, T., Bourgois, J., Bellon, H., and Toussaint, Jean-François, 1985, Age et répartition du magmatisme Prémésozoïque des Andes de Colombie: Comptes Rendus de l'Académie des Sciences, Paris, v. 300, series II, no. 10, p. 445–450.

Boucot, A. J., 1975, Evolution and extinction rate controls: Amsterdam, Elsevier Scientific, Developments in Paleontology and Stratigraphy, vol. 1, 427 p.

Boucot, A. J., 1988, Devonian biogeography: an update, *in* Proceedings, 2nd International Symposium Devonian System: Canadian Society of Petroleum Geologists Memoir 14, v. 3, p. 211–227.

Boucot, A. J., 1990, Silurian biogeography, *in* McKerrow, W. S., and Scotese, C. R., eds., Palaeozoic palaeogeography and biogeography: Geological Society of London Memoir 12, p. 191–196.

Boucot, A. J., 1993, Comments on Cambrian-to-Carboniferous biogeography and its implications for the Acadian orogeny, *in* Roy, D. C., and Skehan, J. W., eds., The Acadian orogeny: recent studies in New England, Maritime Canada, and the autochthonous foreland: Geological Society of America Special Paper 275, p. 41–49.

Boucot, A. J., Johnson, J. G., and Shagam, R., 1972, Braquiópodos Silúricos de Los Andes merideños de Venezuela *in* Memoria, Cuarto Congreso Geológico Venezolano, v. 2: Venezuela, Ministerio de Minas e Hidrocarburos, Dirección de Geología, Boletín de Geología, Publicación Especial 5, p. 585–727.

Boucot, A. J., Dewey, J. F., Dineley, D. L., Fletcher, R., Fyson, W. K., Griffin, J. G., Hickox, C. F., McKerrow, W. S., and Ziegler, A. M., 1974, Geology of the Arisaig area, Antigonish County, Nova Scotia: Geological Society of America Special Paper 139, 191 p.

Boucot, A. J., Blodgett, R. B., and Stewart, J. H., 1997, European Province Late Silurian brachiopods from the Ciudad Victoria area, Tamaulipas, northeastern Mexico, *in* Klapper, G., Murphy, M. A., and Talent, J. A., eds., Paleozoic sequence stratigraphy, biostratigraphy, and biogeography: Studies in honor of J. Granville ("Jess") Johnson: Geological Society of America Special Paper 321, p. 273–293.

Bowring, S. A., 1986, U-Pb zircon ages of granitic boulders in the Ordovician Blakely Sandstone, Arkansas, and implications for their provenance, *in* Guidebook, Economic geology of central Arkansas, Society of Economic Geologists Field Trip No. 2, February 28–March 1, 1986: Little Rock, Arkansas Geological Commission, p. 31.

Bridges, L. W. D., 1971, Paleozoic history of the southern Chihuahua tectonic belt, *in* Seewald, K., and Sundeen, D., 1971, The geologic framework of the Chihuahua tectonic belt: West Texas Geological Society Publication 71-59, p. 67–74.

Burkley, L. A., 1976, Geochronology of the central Venezuelan Andes [Ph.D. thesis]; Cleveland, Ohio, Case Western Reserve University, 150 p.

Cameron, K. L., and Jones, N., 1993, A reconnaissance Nd-Sr isotopic study of pre-Cenozoic igneous and metaigneous rocks of the Coachuila terrane, northeastern Mexico, *in* Ortega-Gutiérrez, F., Coney, P. J., Centeno-García, E., and Gómez-Caballero, A., eds., Proceedings, 1st Circum-Pacific and Circum-Atlantic Terrane Conference: Guanajuato, México, Universidad Nacional Autónoma de México, Instituto de Geología, p. 24–27.

Cameron, K. L., Robinson, J. V., Niemeyer, S., Nimz, G. J., Kuentz, D. C., Harmon, R. S., Bohlenk, S. R., and Collerson, K. D., 1992, Contrasting styles of pre-Cenozoic and mid-Tertiary crustal evolution in northern Mexico: evidence from deep crustal xenoliths from La Olivina: Journal of Geophysical Research, v. 97, no. B12, p. 17353–17376.

Campa, M. F., and Coney, P. J., 1983, Tectono-stratigraphic terranes and mineral resources of Mexico: Canadian Journal of Earth Sciences, v. 20, p. 1040–1051.

Carls, P., 1974, Die Proschizophoriinae (Brachiopoda; Siluri-Devon) der Östlichen Iberischen Ketten (Spanien): Senckenbergiana Lethaea, v. 55, no 1/5, p. 153–227.

Carrillo-Bravo, J., 1959, Notas sobre el Paleozoico de la región de Ciudad Victoria, Tamaulipas: Asociación Mexicana de Geólogos Petroleros, Boletín, v. 11, nos. 11 and 12, p. 673–681.

Carrillo-Bravo, J., 1961, Geología del anticlinoria Huizachal-Peregrina al NW de Ciudad Victoria, Tamaulipas: Asociación Mexicana de Geólogos Petroleros, Boletín, v. 13, p. 1–98.

Carrillo-Bravo, J., 1963, Geology of the Huizachal-Peregrina anticlinorium northwest of Ciudad Victoria, Tamaulipas, *in* Geology of Peregrina Canyon and Sierra del Abra, Mexico; Annual Field Trip Guidebook: Corpus Christi Texas, Corpus Christi Geological Society, p. 11–23.

Carrillo-Bravo, J., 1965, Estudio geológico de una parte del anticlinorio de Huayacocotla: Asociación Mexicana de Geólogos Petroleros, Boletín, v. 17, no. 5-6, p. 73–96.

Case, J. E, Shagam, R., Giegengack, R. F., 1990, Geology of the northern Andes: An overview, *in* Dengo, G., and Case, J. E. eds., 1990, The Caribbean region: Boulder, Colorado, Geological Society of America, Geology of North America, v. H, p. 177–200.

Centeno-García, E., Ochoa-Camarillo, H., and Sour-Tovar, F., 1995, Permian volcanism in eastern Mexico, preliminary report: Geological Society of America Abstracts with Programs, v. 26, no. 6, p. A73.

Centeno-García, E., and Rosales-Lagarde, L., 1997, Itinerario de la excursión al anticlinorio de Huayacocotla en la región de Molango, Estado de Hidalgo, México—tercer día, *in* Gómez-Caballero, A., and Alcayde-Orraca, M., eds., Guia de las excursiones geológicas: Pachuca, México, Instituto de Investigaciones en Ciencias de la Tierra of the Univeridad Autónoma del Estado de Hidalgo and Instituto de Geología of the Universidad Nacional Autónoma de México, II Convención sobre la evolución de México y recursos asociados, p. 41–43.

Chandler, F. W., Sullivan, R. W., and Currie, K. L., 1987, The age of the Springdale Group, western Newfoundland, and correlative rocks—evidence for a Llandovery overlap assemblage in the Canadian Appalachians: Transactions of the Royal Society of Edinburgh: Earth Sciences, v. 78, part 1, p. 41–49.

Clark, J. M., and Hopson, J. A., 1985, Distinctive mammal-like reptile from Mexico and its bearing on the phylogeny of the Tritylodontidae: Nature, v. 315, p. 398–400.

Cole, G. L., and Andrews-Jones, D. A., 1979, Geological and economic evaluation including exploration results of the 1978 field season: Belmopan, Belize Ministry of Geology, Belize Annual Report, no. 4, 37 p.

Conklin, J., Corpstein, P., Fralick, K., Mikels, J., Morales, N., and Rodgers, M., eds., 1974, Geology of Huizachal-Peregrina anticlinorium, Field Conference Guidebook; Southwestern Association of Student Geological Societies: Edinburg, Texas Pan American Geological Society, 134 p.

Cordani, U., García, J. R., Pimentel de Bellizzia, N., and Etchart, H., 1985, Comentario sobre dataciones geocronológicas en la región de los Andes centrales, *in* Memoria, 6th Congreso Geológico de Venezuela, Caracas, 1985: Venezuela Sociedad Venezolana de Geólogos, v. 3, p. 1571–1585.

Cserna, Z. de, and Ortega-Gutiérrez, F., 1978, Reinterpretactión tectónica del Equisto Granjeno de Ciudad Victoria, Tamaulipas; Contestactión: Universidad Nacional Autónoma de México, Instituto de Geología, Revista, v. 2, p. 212–215.

Cserna, Z. de, Graf, J. L., Jr., and Ortega-Gutiérrez, Fernando, 1977, Alóctono del Paleozoico inferior en la región de Ciudad Victoria, Estado de Tamaulipas: Universidad Nacional Autónoma de México, Instituto de Geología, Revista, v. 1, p. 33–43.

Dallmeyer, R. D., 1986, Contrasting accreted terranes in the southern Appalachians and Gulf Coast subsurface: Geological Society of America Abstracts with Programs, v. 18, p. 578–579.

Dallmeyer, R. D., Caen-Vachette, M., and Villeneuve, M., 1987, Emplacement age of post-tectonic granites in southern Guinea (west Africa) and the peninsular Florida subsurface: implications for origins of southern Appalachian exotic terranes: Geological Society of America Bulletin, v. 99, p. 87–93.

Dalziel, I. W. D., 1992, On the organization of American plates in the Neoproterozoic and the breakout of Laurentia: GSA Today, v. 2, no. 11, p. 237, and 240–241

Dalziel, I. W. D., Dalla Salda, L. H., and Gahagan, L. M., 1994, Paleozoic Laurentia-Gondwana interaction and the origin of the Appalachian-Andean mountain system: Geological Society of America, v. 106, p. 243–252.

Dasch, L. E., 1982, U-Pb geochronology of the Sierra de Perijá, Venezuela [M.S. thesis]: Cleveland, Ohio, Case Western Reserve University, 164 p.

Decatur, S. H., and Rosenfeld, J. H., 1982, Mississippian radioactive marker beds and their possible relationship to the Ouachita orogeny: Geological Society of America Abstracts with Programs, v. 14, p. 109.

Denison, R. E., Kenny, G. S., Burke, W. H., Jr., and Hetherington, E. A., Jr., 1969, Isotopic ages from igneous and metamorphic boulders from the Haymond Formation (Pennsylvanian), Marathon basin, Texas, and their significance: Geological Society of America Bulletin, v. 80, p. 245–256.

Denison, R. E., Burke, W. H., Jr., Hetherington, E. A., Jr., and Otto, J. B., 1971, Basement rock framework of parts of Texas, southern New Mexico and northern Mexico, *in* Seewald, K., and Sundeen, D. eds., The geologic framework of the Chihuahua tectonic belt: Midland, Texas, West Texas Geological Society Publication 71-59, p. 3–14.

Denison, R. E., Burke, W. H., Otto, J. B., and Hetherington, E. A., 1977, Age of igneous and metamorphic activity affecting the Ouachita fold belt, *in* Stone, C. G., ed., Symposium on the geology of the Ouachita Mountains, vol. 1: Little Rock, Arkansas Geological Commission, p. 25–40.

Díaz, T., 1956, Ruta: Ciudad Victoria, Tamps.-Cañón de Peregrina, *in* Estratigrafía del Cenozoico y del Mesozoico a lo largo de la carretera entre Reynosa, Tamps. y México, D. F., Tectónica de la Sierra Madre Oriental, Volcanismo en el Valle de México: Congreso Geológico Internacional, México City, Mexico, Excursiones A-14 y C-6, p. 63–68.

Donnelly, T. W., and López-Ramos, E., 1990, The Maya block and Motagua suture zone, *in* Dengo, G., and Case, J. E. eds., 1990, The Caribbean region: Boulder, Colorado, Geological Society of America, Geology of North America, v. H, p. 37–55.

Dutro, J. T., Jr., and Isaacson, P. E., 1991, Lower Carboniferous brachiopods from Sierra de Almeida, northern Chile, *in* Mackinnon, D. I., Lee, D. E., and Campbell, J. D., eds., Brachiopods through time: Rotterdam, Netherlands, Balkema, p. 327–332

Espejo C. A., Etchart, H. L., Cordani, U. G., and Kawashita, K., 1980, Geocronología de intrusivas ácidas en la Sierra de Perijá, Venezuela: Venezuela Ministerio de Energía y Minas, Boletín de Geología, v. 14, p. 245–254.

Etayo-Serna, F., Barrero-Lozano, D., Lozano-Quiroga, H., González-Iregui, H., Orrego-López, A., Ballesteros-Torres, I., Forero-Onofre, H., Ramírez-Quiroga, C., Zambrano-Ortiz, F., Duque-Caro, H., Vargas-Higuera, R., Alvarez-Agudelo, J., Ropaín-Ulloz, C., Cardozo-Puentes, E., Galvis-García, N., Sarmiento-Rojas, L., Albers, J. P., Case, J. E., Greenwood, W. R., Singer, D. A., Berger, B. R., Cox, D. P., Hodges, C. A., and Bowen, R. W., 1983, Mapa de terrenos geológicos de Colombia: Bogotá, Colombia, Instituto Nacional de Investigaciones Geológico-Mineras (INGEOMINAS), 157 p.

Fastovsky, D. E., Clark, J. M., and Hopson, J. A., 1987, Preliminary report of a vertebrate fauna from an unusual paleoenvironmental setting, Huizachal Group, Early or Mid-Jurassic, Tamaulipas, Mexico, *in* Currie, P. J., and Koster, E. H., 4th Symposium on Mesozoic Terrestrial Ecosystems, Short Papers: Occasional Paper of the Tyrrell Museum of Paleontology, no. 3, p. 82–87.

Fastovsky, D. E., Clark, J. M., Strater, N. H., Montellano, M., Hernandez, R., and Hopson, J. A., 1995, Depositional environments of a Middle Jurassic terrestrial vertebrate assemblage, Huizachal Canyon, Mexico: Journal of Vertebrate Paleontology, v. 15, p. 561–575.

Feo-Codecido, G., Smith, F. D., Jr., Aboud, N., and Di Giacomo, E., 1984, Basement and Paleozoic rocks of the Venezuelan Llanos basins, *in* Bonini, W. E., Hargraves, R. B., and Shagam, R., eds., 1984, The Caribbean–South American plate boundary and regional tectonics: Geological Society of America Memoir 162, p. 175–187.

Flawn, P. T., and Diaz, T., 1959, Problems of the Paleozoic tectonics in north-central and north-eastern Mexico: American Association of Petroleum Geologists Bulletin, v. 43, p. 224–230.

Flawn, P. T., Goldstein, A., Jr., King, P. B., and Weaver, C. E., 1961, The Ouachita system: Austin, University of Texas, Bureau of Economic Geology Publication Number 6120, 401 p.

Forero-Suarez, A., 1990, The basement of the Eastern Cordillera, Colombia: an allochthonous terrane in northwestern South America: Journal of South American Earth Sciences, v. 3, no. 2/3, p. 141–151.

Fries, C., Jr., Schmitter, E., Damon, P. E., Livingston, D. E., and Erickson, R., 1962, Edad de las rocas metamórficas en los Cañones de la Peregrina y de Caballeros, parte centro-occidental de Tamaulipas: Universidad Nacional Autónoma de México, Instituto de Geología, Boletín, v. 64, p. 4, p. 55–69.

Garrison, J. R., Jr., 1978, Reinterpretation of isotopic age data from the Granjeno Schist, Ciudad Victoria, Tamaulipas: Universidad Nacional Autónoma de México, Instituto de Geología, Revista, v. 2, p. 87–89.

Garrison, J. R., Jr., Ramírez-Ramírez, C., and Long, L. E., 1980, Rb-Sr isotopic study of the ages and provenance of Precambrian granulite and Paleozoic greenschist near Ciudad Victoria, Mexico, *in* Pelger, R. H., Jr., ed., The origin of the Gulf of Mexico and the early opening of the central north Atlantic ocean: Baton Rouge, Louisiana State University, p. 37–49.

Gastil, R. G., and Miller, R. H., eds., 1993, The Prebatholithic stratigraphy of Peninsular California: Geological Society of America Special Paper 279, 163 p.

Gastil, R. G., Miller, R., Anderson, P., Crocker, J., Campbell, M., Buch, P., Lothringer, C., Leier-Engelhardt, P., Delattre, M., and Hoobs, J., 1991, The relation between the Paleozoic strata on opposite sides of the Gulf of California: Geological Society of America Special Paper 254, p. 7–17.

Girty, G. H., 1926, A new area of Carboniferous rocks in Mexico: Science, v. 63, no. 1628, p. 286–287.

Goldsmith, R., Marvin, R. F., and Mehnert, H. H., 1971, Radiometric ages in the Santander massif, eastern Cordillera, Colombian Andes: Geological Survey Research 1971, U.S. Geological Survey Professional Paper 750-D, p. D44–D-49.

Gomberg, D. M., Banks, P. O., and McBirney, A. R., 1968, Guatamala: preliminary zircon ages from central cordillera: Science, v. 162, p. 121–122.

Grösser, J. R., and Prossl, K. F., 1990, First evidence of Silurian in Colombia: palinostratigraphical data for the Metamorphic Group Quétame, Cordillera Oriental: XII Geowissenschaftliche Lateinamerikanische Kolloquium, München, p. 47.

Gursky, H.-J., 1996, Paleozoic stratigraphy of the Peregrina Canyon area, Sierra

Madre Oriental, NE México: Zentralblatt für Geologie und Paläontologie, Teil I, H.7/8, p. 973–989.

Gursky, H.-J., and Michalzik, D., 1989, Lower Permian turbidites in the northern Sierra Madre Oriental, México: Zentralblatt für Geologie und Paläontologie, Teil I, H. 5/6, p. 821–838.

Gursky, H.-J., and Ramirez-Ramirez, C., 1986, Notas preliminares sobre el descubrimiento de volcanitas ácidas en El Cañón de Caballeros (Núcleo del Anticlinorio Huizachal-Peregrina, Tamaulipas, México, *in* Barbarín, J. M., and Gursky, H.-J., eds. Aspectos geológicos del noreste de México: Actas de la Facultad de Ciencias de la Tierra, Universidad Autónoma de Nuevo León, Linares, Toma 1, no. 1, p. 11–22.

Handschy, J. W., Keller, G. R., and Smith, K. J., 1987, The Ouachita system in northern México: Tectonics, v. 6, no. 3, p. 323–330.

Harland, W. B, Armstrong, R. L., Cox, A. V., Craig, L. E., Smith, A. G., and Smith, D. G., 1989, A geologic time scale: Cambridge, United Kingdom, Cambridge University Press, 263 p.

Hatcher, R. D, Jr., 1989, Tectonic synthesis of the U.S. Appalachians, *in* Hatcher, R. D., Jr., Thomas, W. A., and Viele, G. W., eds., The Appalachian-Ouachita orogen in the United States: Boulder, Colorado, Geological Society of America, Geology of North America, v. F-2 p. 511–535

Heim, A., 1940, The front ranges of the Sierra Madre Oriental, México, from Ciudad Victoria to Tamazunchale: Eclogae Geologicae Helvetiae, v. 33, no. 2, p. 313–352.

Herrmann, U. R., Nelson, B. K., and Ratschbacker, L., 1994, The origin of a terrane: U/Pb zircon geochronology and tectonic evolution of the Xolapa complex (southern México): Tectonics, v. 13, no. 2, p. 455–474.

Hoffman, P. F., 1991, Did the breakout of Laurentia turn Gondwanaland inside-out?: Science, v. 252, p. 1409–1411.

Horne, G. S., Clark, G. S., and Pushkar, P., 1976, Pre-Cretaceous rocks in northwestern Honduras: basement terrane in Sierra de Omoa: American Association of Petroleum Geologists Bulletin, v. 60, no. 4, p. 566–583.

Horne, G. S., Finch, R. C., and Donnelly, T. W., 1990, The Chortis block, *in* Dengo, G., and Case, J. E. eds., 1990, The Caribbean region: Boulder, Colorado, Geological Society of America, Geology of North America, v. H, p. 55–76.

Humphrey, W. E., and Díaz, T., 1953, Excursión al Cañón de la Peregrina, Ciudad Victoria, Tamaulipas: Asociación Mexicana de Geólogos Petroleros, Primera Convención Nacional, Resumen 29, México, D. F.

Isaacson, P. E., Antelo, B., and Boucot, A. J., 1976, Implications of a Llandovery (Early Silurian) brachiopod fauna from Salta Province, Argentina: Journal of Paleontology, v. 50, no. 6, p. 1103–1112.

Keppie, J. D., 1993, Synthesis of Palaeozoic deformational events and terrane accretion in the Canadian Appalachians: Geologische Rundschau, v. 82, no. 3, p. 381–431.

Keppie, J. D., and Ortega-Gutiérrez, F., 1995, Provenance of Mexican terranes: Isotopic constraints: International Geology Review, v. 37, p. 813–824.

Keppie, J. D., Dostal, J., Murphy, J. B., and Nance, R. D., 1996, Terrane transfer between eastern Laurentia and western Gondwana in the early Paleozoic: constraints on global reconstructions, *in* Nance, R. D., and Thompson, M. D., eds., Avalonian and related peri-Gondwanan terranes in the Circum–North Atlantic: Geological Society of America Special Paper 304, p. 369–380.

Krogh, T. E., 1973, A low-contamination method for hydrothermal dissolution of zircons and extraction of U and Pb for isotopic age determinations: Geochimica Cosmochimica Acta, 37, p. 485–494.

Krogh, T. E., Kamo, S. L., Sharpton, V. L., Marin, L. E., and Hildebrand, A. R., 1993, U-Pb ages of single shocked zircons linking distal K/T ejecta to the Chicxulub crater: Nature, v. 366, p. 731–734.

Kroonenberg, S., 1982, A Grenvillian granulite belt in the Colombian Andes and its relation to the Guiana shield: Geologie en Mijnbouw, v. 61, no. 4, p. 325–333.

Laufeld, S., 1979, Biogeography of Ordovician, Silurian and Devonian chitinozoans, *in* Gray, J., and Boucot, A. J., eds., Historical biogeography, plate tectonics, and the changing environments: Corvallis, Oregon State University Press, p. 75–90.

Loomis, J., Weaver, B., and Blatt, H., 1994, Geochemistry of Mississippian tuffs from the Ouachita Mountains, and implications for the tectonics of the Ouachita orogen: Geological Society of America Bulletin, v. 106, p. 1158–1171.

López-Ramos, E., 1972, Estudio del basamento ígneo y metamórfico de las zonas Norte y Poza Rica (entre Nautla, Ver. y Jiménez, Tamps.): Asociación Geólogos Mexicanos Petroleros Boletín, v. 26, 266–323.

López-Ramos, E., 1975, Geological summary of the Yucatan Peninsula, *in* Nairn, A. E. M., and Stelhi, F. G., eds., The ocean basins and margins, v. 3: New York, Plenum Press, p. 257–285.

López-Ramos, E., 1981, Geología de México: v. III, México City, published privately, 446. p.

MacDonald, W. D., and Hurley, P. M., 1969, Precambrian gneisses from northern Colombia: Geological Society of America Bulletin, v. 80, p. 1867–1872.

Mack, G. H., Thomas, W. A., and Horsey, C. A., 1983, Composition of Carboniferous sandstones and tectonic framework of southern Appalchian-Ouachita orogen: Journal of Sedimentary Petrology, v. 53, p. 931–946.

Manton, W. I., 1996, The Grenville of Honduras: Geological Society of America, Abstracts with Programs, v. 28, no. 7, p. A493.

Martín-Bellizzia, C., 1968, Edades isotópicas de rocas Venezolanas: República de Venezuela, Ministerio de Minas e Hidrocarburos, Dirección de Geología, Boletín de Geología, v. 10, no. 19, p. 356–380

Mattinson, J. M., 1987, U-Pb ages of zircons: a basic examination of error propagation: Chemical Geology, v. 66, p. 151–162.

Maurrasse, F. J-M. R., 1990, Stratigraphic correlation for the circum-Caribbean region, *in* Dengo, G., and Case, J E. eds., 1990, The Caribbean region: Boulder, Colorado, Geological Society of America, Geology of North America, v. H, plate 4.

McBride, E. F., 1989, Stratigraphy and sedimentary history of pre-Permian Paleozoic rocks of the Marathon uplift, *in* Hatcher, R. D., Jr., Thomas, W. A., and Viele, G. W., eds., The Appalachian-Ouachita orogen in the United States: Boulder, Colorado, Geological Society of America, Geology of North America, v. F-2, p. 603–620.

McCourt, W. J., Aspden, J. A., and Brook, M., 1984, New geological and geochronological data from the Colombian Andes; continental growth by multiple accretion: Geological Society of London Journal, v. 141, p. 831–845.

McKee, J. W., Jones, N. W., and Anderson, T. H., 1988, Las Delicias basin: a record of late Paleozoic arc volcanism in northeastern México: Geology, v. 16, p. 37–40.

McKee, J. W., Jones, N. W., and Leon, L. E., 1990, Stratigraphy and provenance of strata along the San Marcos fault, central Coahuila, México: Geological Society of America Bulletin, v. 102, no. 5, p. 593–614.

Michalzik, D., 1991, Facies sequence of Triassic-Jurassic red beds in the Sierra Madre Oriental (NE México) and its relation to the early opening of the Gulf of México: Sedimentary Geology, v. 71, p. 243–259.

Miller, E. L, Miller, M. M., Stevens, C. H., Wright, J. E., and Madrid, R., 1992, Late Paleozoic tectonic evolution of the western U.S. Cordillera, *in* Burchfiel, B. C., Lipman, P. W., and Zoback, M. L., 1992, The Cordilleran orogen: conterminous U.S.: Boulder, Colorado, Geological Society of America, Geology of North America, v. G-3, p. 57–106.

Mixon, R. B., Murray, G. E, and Díaz, T., 1959, Age and correlation of Huizachal Group (Mesozoic), State of Tamaulipas, México: American Association of Petroleum Geologists Bulletin, v. 43, p. 757–771.

Mojica, J., and Villarroel, C., 1990, Sobre la distribución y facies del Paleozoico Inferior sedimentario en el extremo NW de Sudamérica: Bogota, Colombia, Universidad Nacional de Colombia, Facultad de Ciencias, Geología Colombiana No. 17, p. 219–226.

Moreno-Cano, L. A., and Patiño-Ruiz, J., 1981, Estudio del Paleozoico en la region de Calnali, Hidalgo (en la Sierra Madre Oriental) [thesis]: México City, México, Instituto Politécnico Nacional, 54 p.

Muir, J. M., 1936, Geology of the Tampico region, México: Tulsa, Oklahoma, American Association of Petroleum Geologists, 280 p.

Mullan, H. S., 1978, Evolution of part of the Nevada orogen in northwest México: Geological Society of America, v. 89, p. 1175–1188.

Murphy, J. B., Keppie, J. D., Dostal, J., Waldron, J. W. F., and Cude, M. P., 1995, Geochemical and isotopic characteristics of Early Silurian clastic sequences in Antigonish Highlands, Nova Scotia, Canada: constraints on the accretion of Avalonia in the Appalachian-Caledonide orogen: Canadian Journal of Earth Sciences, v. 33, no. 3, p. 379–388.

Murray, G. E., Furnish, W. M., and Carrillo B. J., 1960, Carboniferous goniatites from Caballeros Canyon, State of Tamaulipas, Mexico: Journal of Paleontology, v. 34, no. 4, p. 731–737.

Nicholas, R. L., and Rozendal, R. A., 1975, Subsurface positive elements within the Ouachita foldbelt in Texas and their relationship to the Paleozoic cratonic margin: American Association Petroleum Geologists Bulletin, v. 59, p. 193–216.

Nicholas, R. L., and Waddell, D. E., 1989, The Ouachita system in the subsurface of Texas, Arkansas, and Louisiana, *in* Hatcher, R. D., Jr., Thomas, W. A., and Viele, G. W., eds., The Appalachian-Ouachita orogen in the United States: Boulder, Colorado, Geology of North America, v. F-2, Geological Society of America, p. 661–672.

Niem, A. R., 1977, Mississippian pyroclastic flow and ash-flow deposits in the deep-marine Ouachita flysch basin, Oklahoma and Arkansas: Geological Society of America Bulletin, v. 88, p. 49–61.

Noble, P. J., 1990, Radiolarian biostratigraphy: evidence for an Early Mississippian basin-wide hiatus at the Caballos-Tesnus boundary, Marathon basin, Texas, *in* Laroche, T. M., and Higgins, L., eds., Marathon thrust belt: structure, stratigraphy, and hydrocarbon potential; SEPM Field Seminar: Midland, West Texas Geological Society and Permian Basin Section, p. 83–91.

Noble, P. J., 1993, Paleoceanographic and tectonic implications of a regionally extensive Early Mississippian hiatus in the Ouachita system, southern mid-continental United States: Geology, v. 21, p. 315–318.

Noble, P. J.,1994, Silurian radiolarian zonation for the Caballos Novaculite, Marathon uplift, west Texas: Bulletin of American Paleontology, v. 106, no. 345, 55 p.

Ochoa-Camarillo, H., 1997, Geología del anticlinorio Huayacocotla en la región de Molango, Hgo., México, *in* Gómez-Caballero, A., and Alcayde-Orraca, M., eds., Guía de las excursiones geológicas: Pachuca, México, Instituto de Investigaciones en Ciencias de la Tierra de la Univeridad Autónoma del Estado de Hidalgo and Instituto de Geología de la Universidad Nacional Autónoma de México, II Convención sobre la evolución de México y recursos asociados, p. 1–17.

Ortega-Gutiérrez, F., 1978, El Gneis Novillo y rocas metamórficas asociadas en los Cañones del Novillo y de la Peregrina, área de Ciudad Victoria, Tamaulipas: Universidad Nacional Autónima de México, Instituto de Geología, Revista, v. 2, no. 1, p. 19–30.

Ortega-Gutiérrez, F., Centeno-García, E., Morán-Zenteno, D. J., and Gómez-Caballero, A. eds., 1993, Pre-Mesozoic basement of NE Mexico, lower crust and mantle xenoliths of central Mexico, and northern Guerrero terrane, *in* Field Trip Guidebook A, Proceedings, 1st Circum-Pacific and Circum-Atlantic Terrane Conference: Guanajuato, México, Universidad Nacional Autónoma de México, Instituto de Geología, 96 p.

Ortega-Gutiérrez, F., Lopez, R., Cameron, K. L., Ochoa-Camarillo, H., and Sanchez-Zavala, J. L., 1995a, Grenvillian Huiznopala Gneiss and its correlation with other Grenville-age high-grade terranes in eastern Mexico: Geological Society of America Abstracts with Programs, v. 27, p. A398.

Ortega-Gutiérrez, F., Ruiz, J., and Centeno-Garcia, E., 1995b, Oaxaquia, a Proterozoic microcontinent accreted to North America during the late Paleozoic: Geology, v. 23, p. 1127–1130.

Ortega-Gutiérrez, F., Lawlor, P., Cameron, K. L., and Ochoa-Camarillo, H., 1997, New studies of the Grenvillian Huiznopala Gneiss, Molango area, State of Hidalgo, Mexico—preliminary results, *in* Gómez-Caballero, A., and Alcayde-Orraca, M., eds., Guía de las excursiones geológicas: Pachuca, México, Instituto de Investigaciones en Ciencias de la Tierra de la Univeridad Autónoma del Estado de Hidalgo e Instituto de Geología de la Universidad Nacional Autónoma de México, II Convención sobre la evolución de México y recursos asociados, p. 19-25

Osberg, P. H., Tull, J. F., Robinson, P., Hon, R., and Butler, J. R., 1989, The Acadian orogen, *in* Hatcher, R. D., Jr., Thomas, W. A., and Viele, G. W., eds., The Appalachian-Ouachita orogen in the United States: Boulder, Colorado, Geological Society of America, Geology of North America, v. F-2, p. 179–232.

Pantoja-Alor, J., 1970, Rocas sedimentarias paleozoicas de la región centro-septentrional de Oaxaca: Oaxaca, Sociedad Geológica Mexicana, Excursion México, Oaxaca, p. 67–84.

Pantoja-Alor, J., and Robison, R. A., 1967, Paleozoic sedimentary rocks in Oaxaca, Mexico: Science, v. 157, p. 1033–1035.

Pimentel de Bellizzia, N., Bellizzia-G., A, and Ulloa, C., 1992, Paleozoico Inferior: una síntesis del Noroeste de América del Sur (Venezuela, Colombia y Ecuador), *in* Gutiérrez-Marco, J. C., Saavedra, J., and Rábano, I, eds., Paleozoico Inferior de Ibero-América: Universidad de Extremadura, p. 203–224.

Pindell, J. L., 1985, Alleghenian reconstruction and subsequent evolution of the Gulf of Mexico, Bahamas, and proto-Caribbean: Tectonics, v. 4, p. 1–39.

Pindell, J. L., and Barrett, S. F., 1990, Geological evolution of the Caribbean region; A plate-tectonic perspective, *in* Dengo, G., and Case, J. E. eds., 1990, The Caribbean region: Boulder, Colorado, Geological Society of America, Geology of North America, v. H, p. 405–432.

Pinson, W. H., Jr., Hurley, P. M., Mencher, E., and Fairbairn, H. W., 1962, K-Ar and Rb-Sr ages of biotites from Colombia, South America: Geological Society of America Bulletin, v. 73, p. 907–910.

Pojeta, J., Jr., Kríz, J., and Berdan, J. M., 1976, Silurian-Devonian pelecypods and Paleozoic stratigraphy of subsurface rocks in Florida and Georgia and related Silurian pelecypods from Bolivia and Turkey: U.S. Geological Survey Professional Paper 879, p. 32.

Poole, F. G., Stewart, J. H., Palmer, A. R., Sandberg, C. A., Madrid, R. J., Ross, R. J, Jr., Hintze, L. F., Miller, M. M., and Wrucke, C. T., 1992, Latest Precambrian to latest Devonian time, development of a continental margin, *in* Burchfiel, B. C., Lipman, P. W., and Zoback, M. L., eds., 1992, The Cordilleran orogen: conterminous U.S.: Boulder, Colorado, Geological Society of America, Geology of North America, v. G-3, p. 9–56

Poole, F. G., Stewart, J. H., Berry, W. B. N., Harris, A. G., Repetski, J. E., Madrid, R. J., Ketner, K. B., Carter, C., and Morales-Ramirez, J. M., 1995, Ordovician ocean-basin rocks of Sonora, Mexico, *in* Cooper, J. D., Droser, M. L., and Finney, S. C., eds., Ordovician odyssey, Short Papers for the 7th International Symposium on the Ordovician System: Las Vegas, Nevada, SEPM, Pacific Section, no. 77, p. 277–284.

Priem, H. N. A., Kroonenberg, S. B., Boelrijk, N. A. I. M., and Hebeda, E. H., 1989, Rb-Sr and K-Ar evidence for the presence of a 1.6 Ga basement underlying the 1.2 Ga Garzón–Santa Marta granulite belt in the Colombian Andes: Precambrian Research, v. 42, p. 315–324.

Ramirez-Ramirez, C., 1978, Reinterpretación tectónica del Esquisto Granjeno de Ciudad Victoria, Tamaulipas: Universidad Nacional Autónoma de México, Instituto de Geología, Revista, v. 2, p. 31–36.

Ramirez-Ramirez, C., 1981, Pre-Mesozoic geology of Huizachal-Peregrina anticlinorium near Ciudad Victoria, northeastern Mexico: Geological Society of America Abstracts with Programs, v. 13, no. 2, p. 102.

Ramirez-Ramirez, C., 1992, Pre-Mesozoic geology of Huizachal-Peregrina anticlinorium, Ciudad Victoria, Tamaulipas, and adjacent parts of eastern Mexico [Ph.D. thesis]: Austin, University of Texas, 317 p.

Rankin, D. W., Drake, A. A., Jr., Glover, L, III, Goldsmith, R., Hall, L. M., Murray, D. P., Ratcliffe, N. M., Read, J. F., Secor, D. T., Jr., and Stanley, R. S., 1989, Pre-orogenic terranes, *in* Hatcher, R. D., Jr., Thomas, W. A., and Viele, G. W., eds., The Appalachian-Ouachita orogen in the United States: Boulder, Colorado, Geological Society of America, Geology of North America, V. F-2, p. 7–100.

Renne, P. R., Mattinson, J. M., Hatten, C. W., Somin, M., Onstott, T. C., Millán, G., and Linares, E., 1989, $^{40}Ar/^{39}Ar$ and U-Pb evidence for Late Proterozoic (Grenville-age) continental crust in north-central Cuba and regional tectonic implications: Precambrian Research, v. 42, p. 325–341.

Restrepo. J. J., 1982, Compilación de edades radiométricas de Colombia: Departamentos Andinos hasta 1982: Universidad Nacional de Colombia, Seccional Medellín, Departmento de Ciencias de la Tierra, Boletín de Ciencias de la Tierra, p. 201–248.

Restrepo, J. J., and Toussaint, J. F., 1988, Terranes and continental accretion in the Colombian Andes: Episodes, v. 11, no. 3, p. 189–193.

Restrepo, P. A., 1995, Late Precambrian to early Mesozoic tectonic evolution of the Colombian Andes, based on new geochronological, geochemical, and isotopic data [Ph.D. thesis]: Tucson, University of Arizona, 195 p.

Restrepo-Pace, P.A., 1992, Petrotectonic characterization of the Central Andean Terrane, Colombia: Journal of South American Earth Sciences, v. 5, no. 1, p. 97–116.

Richards, D. R., and Coney, P. J., 1990, Tectonic evolution and terranes of the Andes: a regional view: Geological Society of America Abstracts with Programs, v. 22, no. 7, p. A327–A328.

Robison, R., and Pantoja-Alor, J., 1968, Tremadocian trilobites from Nochixtlan region, Oaxaca, Mexico: Journal of Paleontology, v. 42, p. 767–800.

Rosales-Lagarde, L., Centeno-García, E., Ochoa-Camarillo, H., and Sour-Tovar, F., 1997, Permian volcanism in eastern Mexico—preliminary report, *in* Gómez-Caballero, A., and Alcayde-Orraca, M., eds., Guía de las excursiones geológicas: Pachuca, México, Instituto de Investigaciones en Ciencias de la Tierra de la Univeridad Autónoma del Estado de Hidalgo e Instituto de Geología de la Universidad Nacional Autónoma de México, II Convención sobre la evolución de México y recursos asociados, p. 27–32.

Rowley, D. B., and Pindell, J. L., 1989, End Paleozoic–Early Mesozoic western Pangean reconstruction and its implications for the distribution of Precambrian and Paleozoic rocks around Meso-America: Precambrian Research, v. 42, p. 411–444.

Rudnick, R. L., and Cameron, K. L., 1991, Age diversity of the deep crust in northern Mexico: Geology, v. 19, p. 1197–1200.

Rueda-Gaxiola, J., Lopez-Ocampa, E., Dueñas, M. A., and Rodriguez-Benitez, J. L., 1991, Las fosas de Huizachal–Peregrina y de Huayacocotla: dos partes de un graben relacionado con el origen del Golfo de México: Universidad Nacional Autónoma de México, Instituto de Geología, Evolución Geológica México, Pachuca, Hidalgo, Memoria, p. 189–192.

Ruiz, J, Patchett, P. J., and Ortega-Gutiérrez, F., 1988, Proterozoic and Phanerozoic basement terranes of Mexico and Nd isotopic studies: Geological Society of America Bulletin, v. 100, p. 274–281.

Russell, J. L., 1981, Geology of Peregrina and Novillo Canyons, Ciudad Victoria, Mexico: 1981 Gulf Coast Association of Geological Societies Convention Field Trip: Corpus Christi, Texas, Corpus Christi Geological Society, 23 p.

Schlager, W., Buffler, R. T., Angstadt, D., Bowdler, J. L., Cotillon, P. H., Dallmeyer, R. D., Halley, R. B., Kinoshita, H., Magoon, L. B., III, McNulty, C. L., Patton, J. W., Risciotto, K. A., Premoli-Silva, I., Avello-Suarez, O., Testarmata, M. M., Tyson, R. V., and Watkins, D. K., 1984, Deep Sea Drilling Project, Leg. 77, southeastern Gulf of Mexico: Geological Society of America Bulletin, v. 95, p. 226–236.

Schubert, C. 1969, Geologic structure of a part of the Barinas mountain front, Venezuelan Andes: Geological Society of America Bulletin, v. 80, p. 443–458.

Sedlock, R. L., Ortega-Gutiérrez, F., and Speed, R. C., 1993, Tectonostratigraphic terranes and tectonic evolution of Mexico: Geological Society of America Special Paper 278, 153 p.

Silva-Pineda, A., 1979, La flora triásica de México: Universidad Nacional Autónoma de México, Instituto de Geología, Revista, v. 3, no. 2, p. 138–145.

Silver, L. T., Anderson, T. H., and Ortega-Guitierrez, F., 1994, The "thousand million year" orogeny in eastern and southern Mexico: Geological Society of America Abstracts with Programs, v. 26, no. 7, p. A48.

Sour-Tovar, F., Quiroz-Barroso, S. A., and Navarro-Santillan, D., 1996, Carboniferous invertebrates from Oaxaca, southern Mexico: mid-continent paleogeographic extension: Geological Society of America Abstracts with Programs, v. 28, no. 7, p. A365.

Steiner, M. B., and Walker, J. D., 1996, Late Silurian plutons in Yucatan: Journal of Geophysical Research, v. 101, no. B8, p. 17727–17735.

Stewart, J. H., 1988, Latest Proterozoic and Paleozoic southern margin of North America and the accretion of Mexico: Geology, v. 16, p. 186–189.

Stewart, J. H., Poole, F. G., Ketner, K. B., Madrid, R. J., Roldán-Quintana, J., and Amaya-Martínez, R., 1990, Tectonics and stratigraphy of the Paleozoic and Triassic southern margin of North America, Sonora, Mexico, *in* Gehrels, G. E., and Spencer, J. E., eds., Geologic excursions through the Sonoran Desert region, Arizona and Sonora: Arizona Geological Survey Special Paper 7, p. 183–202.

Stewart, J. H., Blodgett, R. B., Boucot, A. J., and Carter, J. L., 1993, Middle Paleozoic terrane near Ciudad Victoria, northeastern Mexico, and the southern margin of Paleozoic North America, *in* Ortega-Gutiérrez, F., Coney, P. J., Centeno-García, E., and Gómez-Cabellero, A., eds., Proceedings 1st Circum-Pacific and Circum-Atlantic Terrane Conference: Guanajuato, México, Universidad Nacional Autónoma de México, Instituto de Geología, p. 147–149.

Teixeira, W., Tassinari, C. C. G., Cordani, U. G., and Kawashita, K., 1989, A review of the geochronology of the Amazonian craton: tectonic implications: Precambrian Research, v. 42, p. 213–227.

Tellez-Giron, C., 1970, Microfacies y microfósiles Paleozoicos del área de Ciudad Victoria, Tamps., N.E. de México: Instituto Mexicano del Petróleo, Serie Monográfica No. 1, Publicación No. 70 AG/068.

Torres, V. R., Ruiz, J., Grajales, M., and Murillo, G., 1992, Permian magmatism in eastern and southern Mexico and its tectonic implications: Geological Society of America Abstracts with Programs, v. 24, p. 64.

Thomas, W. A., Chowns, T. M., Daniels, D. L., Neathery, T. L., Glover, L., III, and Gleason, R. J., 1989, The subsurface Appalachians beneath the Atlantic and Gulf Coastal Plains, *in* Hatcher, R. D., Jr., Thomas, W. A., and Viele, G. W., eds., The Appalachian–Ouachita orogen in the United States: Boulder, Colorado, Geological Society of America, Geology of North America, v. F-2, p. 445–458

Torsvik, T. H., Smethurst, M. A., Meert, J. G., Van der Voo, R., McKerrow, W. S., Brasier, M. D., Sturt, B. A., and Walderbaug, H. J., 1996, Continental break-up and collision in the Neoproterozoic and Palaeozoic—a tale of Baltica and Laurentia: Earth Science Reviews, v. 40, no. 3-4, p. 229–258.

Tucker, R. D., and McKerrow, W. S., 1995, Early Paleozoic chronology: a review in light of new U-Pb zircon ages from Newfoundland and Britan: Canadian Journal of Earth Science, v. 32, p. 368–379.

Urbani, F., 1989, Observaciones sobre la Edad del gneis de Sebastopol y el paragneiss de La Mariposa, districto Federal: Caracas, Universidad Central de Venezuela, Facultad de Ingeniería; GEOS, no. 29, Memorias de las Jornadas 50th Aniversario de la Escuela de Geología, Minas y Geofísica, p. 278–280.

Viele, G. W., and Thomas, W. A., 1989, Tectonic synthesis of the Ouachita orogenic belt, *in* Hatcher, R. D., Jr., Thomas, W. A., and Viele, G. W., eds., The Appalachian-Ouachita orogen in the United States: Boulder, Colorado, Geological Society of America, Geology of North America, v. F-2, p. 695–728.

Williams, H., and Hatcher, R. D., Jr., 1982, Suspect terranes and accretionary history of the Appalachian orogen: Geology, v. 10, p. 530–536.

Yañez, P., Ruiz, J., Patchett, P. J., Ortega-Gutiérrez, F., and Gehrels, G. E., 1991, Isotopic studies of the Acatlán complex, southern Mexico: implications for Paleozoic North American tectonics: Geological Society of America Bulletin, v. 103, p. 817–828.

Ziegler, A. M., Rickards, R. B., and McKerrow, W. S., 1974, Correlation of the Silurian rocks of the British Isles: Geological Society of America Special Paper 154, 154 p.

Manuscript Accepted by the Society on September 4, 1998

Printed in USA

Geological Society of America
Special Paper 336
1999

Neoproterozoic–early Paleozoic evolution of Avalonia

J. Brendan Murphy
Department of Geology, St. Francis Xavier University, Antigonish, Nova Scotia, Canada B2G 2W5
J. Duncan Keppie
Instituto de Geología, Universidad Nacional Autónoma de México, 04510 México, D.F., México
J. Dostal
Department of Geology, St. Mary's University, Halifax, Nova Scotia, Canada, B3H 3C3
R. Damian Nance
Department of Geology, Ohio University, Athens, Ohio 45701, United States

ABSTRACT

Avalonia is a terrane that occupied a position peripheral to Gondwana in the Neoproterozoic and early Paleozoic, and was transferred to Laurentia by the Late Ordovician. There is fragmentary evidence of early arc (ca. 780 to 660 Ma) magmatism, although it is unclear whether this represents continuous or episodic subduction. This was followed by the main arc phase from 630 to 590 Ma, which produced calc-alkalic volcanic and plutonic rocks and volcaniclastic rocks deposited in backarc basins. Pb-Pb isotopic data support previous interpretations that Avalonia was peripheral to the Amazon craton at this time. Neodymium isotopic signatures suggest that subduction polarity was to the northwest (present coordinates), which implies a 180° rotation of Avalonia prior to accretion with Laurentia. From 590 to 540 Ma, a transition from an arc to a transtensional intracontinental environment occurred diachronously from New England to the British Isles. Comparison with Cenozoic analogs suggest this can be attributed to collision of the margin with an oceanic ridge, and progressive generation of a continental transform fault as the triple point migrated along the Gondwanan margin, in a manner similar to the generation of the modern San Andreas fault system. This model also explains the change in kinematics along major faults active at this time, from sinistral to dextral and structural inversion of backarc basins.

INTRODUCTION

Avalonia occurs along the southeastern flank of the Appalachian orogen, outcropping discontinuously in fault-bounded blocks from the Avalon Peninsula-type area in southeastern Newfoundland to southern New England (Fig. 1) (Williams, 1979; O'Brien et al., 1983). Its Neoproterozoic evolution is exotic to Laurentia and is dominated by arc-related tectonothermal activity thought to represent subduction beneath the Gondwanan margin (Keppie et al., 1991; Murphy and Nance, 1991). Early Mesozoic reconstructions suggest former continuity with Neoproterozoic sequences in the British Isles and France (Williams, 1979). In detail, however, the Neoproterozoic arc-related sequences vary considerably in age and are believed to represent a collage of suspect terranes that were amalgamated prior to the deposition of a platformal Early Cambrian sequence that contains Acado-Baltic (Avalonian) fauna (Keppie, 1985, 1993).

The Neoproterozoic evolution of Avalonia provides fundamental constraints on paleocontinental reconstructions during the Neoproterozoic and early Paleozoic. Paleontologic, paleomagnetic, and isotopic data indicate that Avalonia originated as one of a number of terranes along the South American and the northwest African periphery of Gondwana, and became attached to Laurentia by the Late Ordovician (e.g., Cawood et al., 1994;

Murphy, J. B., Keppie, J. D., Dostal, J., and Nance, R. D., 1999, Neoproterozoic–early Paleozoic evolution of Avalonia, *in* Ramos, V. A., and Keppie, J. D., eds., Laurentia-Gondwana Connections before Pangea: Boulder, Colorado, Geological Society of America Special Paper 336.

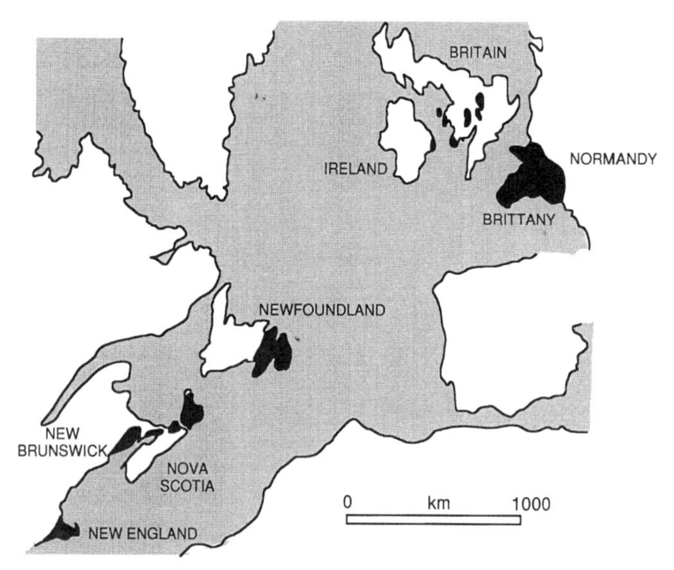

Figure 1. Pre-Mesozoic continental reconstruction showing the position of Avalonia and Cadomia (Armorican Massif) in the northern Appalachians and western Europe (after Krogh et al., 1988, and Nance et al., 1991).

McKerrow et al., 1991; Golonka et al., 1994; Hodych and Buchan, 1994; Nance and Thompson, 1996; Keppie et al., 1996; Van Staal et al., 1996; Keppie et al., in press). Isotopic data from Avalonia indicates a Neoproterozoic position peripheral to the Amazon craton and from Cadomia, suggests a position off the west African craton (Keppie and Krogh, 1990; Nance and Murphy, 1994; Van Staal et al., 1996).

In the late Neoproterozoic (between ca. 590 and 540 Ma), subduction beneath the Avalonian portion of the Gondwanan margin terminated (e.g., Murphy and Nance, 1989; Keppie et al., 1991). The style of Neoproterozoic orogenic activity indicates that termination of subduction was not due to continental collision (Murphy and Nance, 1989; Strachan et al., 1990); rather, it is attributed to the progressive development of a continental transform margin. This observation implies that Avalonia continued to face an open ocean between 590 and 530 Ma, in contrast to the fate of oceans consumed by collisional orogenesis that typify much of Pan-African tectonic activity (Murphy and Nance, 1989; Strachan et al., 1990; Keppie et al., 1991).

In this chapter, we review and synthesize the most recent data from Avalonia, which we believe provides new insights into the Neoproterozoic–Early Paleozoic arc to platform transition. In addition, the enhanced geochronologic database now affords more direct comparison with Cenozoic analogs.

GENERAL GEOLOGY

Several stages of Neoproterozoic tectonothermal activity have been recognized within Avalonia. There is fragmentary evidence for an early arc phase (ca. 780–660 Ma), and main arc activity occurred from 630 to 590 Ma. From 590 to 540 Ma, however, the record is highly variable, with some parts of Avalonia recording little or no magmatism (e.g., New England, mainland Nova Scotia), whereas others have voluminous magmatism (e.g., Cape Breton, Newfoundland). By about 550 Ma, intracontinental magmatism had become established in most regions of Avalonia, with the exception of the British Isles.

Pre–630-Ma magmatism

A generalized Neoproterozoic–Early Paleozoic tectonostratigraphy for Avalonia is shown in Figure 2. In Atlantic Canada, vestiges of ca. 780–660 Ma arc-related events may represent early stages of Neoproterozoic arc activity or a separate pre-Avalonian tectonothermal episode. Representatives occur in mainland Nova Scotia (the 734-Ma calc-alkalic Economy River Gneiss: Doiget al., 1993), in Cape Breton Island (the ca. 676-Ma arc-related Stirling belt, Bevier et al., 1993; 700–630-Ma backarc volcanism in the Creignish Hills: Keppie et al., 1998), and in southern Newfoundland (the ca. 763-Ma rift ophiolites of the Burin Group volcanics: Krogh et al., 1988; the 683-Ma Connaigre Bay Group calc alkalic rhyolite: Swinden and Hunt, 199; and the 686-Ma calc alkalic Roti granite: O'Brien et al., 1990, 1992). Similar evidence for an early arc-related event is found in correlative Avalonian rocks in southeastern Ireland (ca. 700-Ma calc alkalic orthogneiss: R. Doig, personal communication, 1993; Gibbons and Horak, 1996), in Britain (the ca. 670-Ma calc alkalic Malvern Plutonic complex: Tucker and Pharoah, 1991, Strachan et al., 1996a), and in the Cadomian Armorican Massif of France (the 746 ± 17-Ma Port Morvan orthogneiss: Egal et al., 1996).

630–590-Ma magmatism

The main phase of arc-related magmatism occurred between 630 and 590 Ma (e.g., Barr and White, 1988; Pe-Piper and Piper, 1989; Murphy et al., 1990; O'Brien et al., 1990; Dostal et al., 1990; Strachan et al., 1996b; Keppie et al., 1998). Coeval development of sedimentary basins such as the Boston Bay Group in New England (Thompson et al., 1996) and the Georgeville Group in mainland Nova Scotia (Murphy and Keppie, 1987) are attributed to oblique convergence that resulted in strike slip motion and intra-arc rifting in areas of localized extension (Fig. 2). However, in many individual regions within Avalonia, arc-related magmatism occurs within a relatively narrow time interval during the 630–590-Ma period. This suggests that the site of magmatism within the Avalonian arc varies in time and location (e.g., Strachan et al., 1996a; Murphy et al., 1997), presumably because of changes in the nature of subduction along the Gondwanan margin during this time period.

In New England, subduction-related magmatism commenced at ca. 630 Ma, and the transition from arc to extensional magmatism probably occurred by ca. 590 Ma (Thompson et al., 1996; Mancuso et al., 1996). In southern New Brunswick, arc-related volcanic rocks of the Broad River Group and co-genetic plutonic rocks are dated at ca. 630–600 Ma (Bevier and Barr,

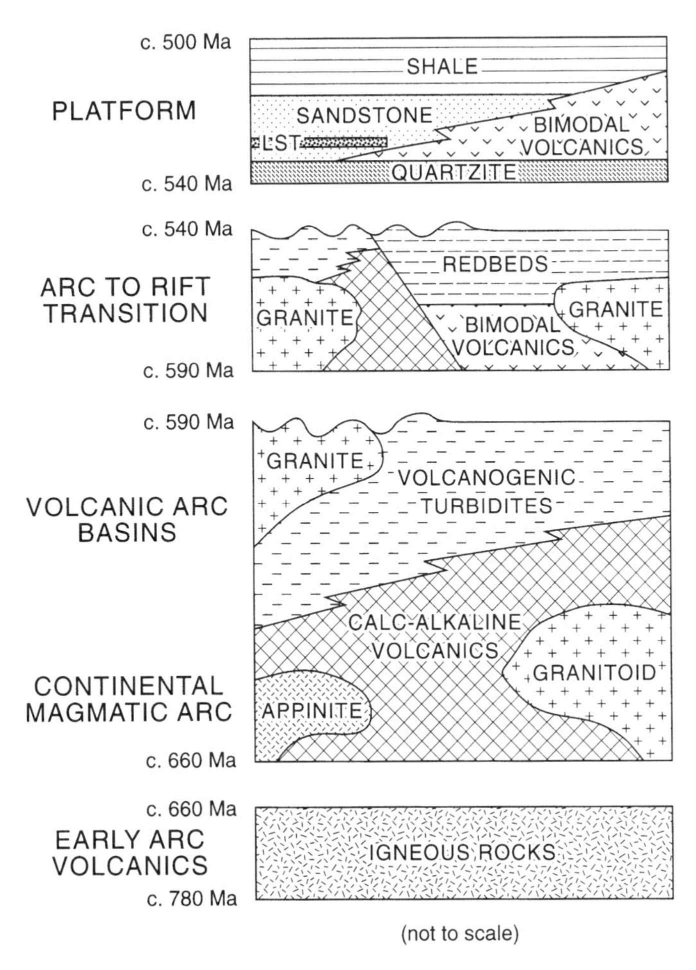

Figure 2. Interpretive tectonostratigraphic column for the Neoproterozoic rocks of Avalonia and Cadomia (see text for sources).

1990; Bevier et al., 1993; Barr et al., 1994; Barr and White, 1996). In mainland Nova Scotia, the Jeffers Group of the Cobequid Highlands was deposited from ca. 628 to 610 Ma, whereas the partially correlative Georgeville Group ranges from ca. 618 to 610 Ma (Murphy et al., 1997). These groups both contain calc-alkalic arc-related volcanics interbedded with rift-related tholeiitic rocks, which are overlain by a thick sequence of turbidites and minor volcanic rocks (Pe-Piper and Piper, 1987, 1989; Murphy and Keppie, 1987; Murphy et al., 1990, 1991). The rift-related volcanic rocks are thought to signal development of the basin into which the turbidites were deposited (Murphy and Nance, 1989).

In southern Cape Breton Island, ca. 630–600 Ma plutonic and volcanic rocks (East Bay Hills, Coxheath Hills, and Pringle Mountain groups) (Bevier et al., 1993) are similar to those of southern New Brunswick. These rocks show a continuous range from mafic to felsic compositions and are predominantly calc-alkalic (Barr et al., 1996).

The Avalon Peninsula of Newfoundland has two distinct lithotectonic assemblages, separated by the Paradise Sound fault (Barr and Kerr, 1997) along which the ca. 760 Ma ophiolitic complex of the Burin Group occurs. According to Keppie et al. (1991), these relationships indicate that the fault may represent the vestiges of a suture between two portions of Avalonia that were amalgamated prior to the deposition of an early Paleozoic overstep sequence. To the east of the Paradise Sound fault, the age and general geology are broadly similar to those represented in the Antigonish and Cobequid Highlands of mainland Nova Scotia (O'Brien et al., 1996). A thick succession of arc and arc-rift tholeiitic to calc alkalic mafic to felsic volcanic and plutonic rocks (Harbour Main Group and correlatives) is succeeded by a 4–5 km sequence of volcaniclastic turbidites and tuffs (Connecting Point Group), the deposition of which is thought to continue to ca. 580 Ma (O'Brien et al., 1996; Barr and Kerr, 1997). To the west of the Paradise Sound fault, arc-related rocks are represented by the calc alkaline Love Cove and Marystown Groups and correlatives including the ca. 615-Ma Cap de Miquelon Group, which occurs on Miquelon Island and consists of metasedimentary and mafic metavolcanic rocks whose geochemistry is typical of lavas emplaced in continental arcs and active tectonic margins (Rabu et al., 1996).

In Britain and southeastern Ireland, the main phase of arc activity occurred between 620 and 590 Ma. Representatives of this activity include the ca. 615-Ma granitoid complexes in Anglesey and the Welsh mainland (Tucker and Pharoah, 1991; Horak, 1993), and the ca. 616–610-Ma Glinton and Orton volcanics and co-genetic plutonic rocks in the English Midlands (Noble and Tucker, 1992).

The Neoproterozoic evolution of the Cadomian belt records the amalgamation of volcanic arc and backarc complexes into a composite terrane along the Gondwanan margin (Strachan et al., 1996b). In the northern portion of the belt (Tregor–La Hague terrane), a suite of calc-alkalic gabbros, diorites, and granite plutons ranges in age from 615–560 Ma (Dallmeyer et al., 1994). In the central and southern Armorican massif (St. Brieuc, St. Malo, and Mancellian terranes), arc-related metavolcanic and metasedimentary rocks of the Brioverian succession and co-genetic plutons probably range in age from ca. 610 to 585 Ma (e.g., Guerrot and Peucat, 1990; Egal et al., 1996).

590–540-Ma magmatism

As noted by Barr and Kerr (1997), igneous rocks in the 590–540 Ma interval are relatively sparsely represented and show varying petrologic characteristics from one part of Avalonia to another. This implies along-strike variation in tectonic setting. Rocks of this age are virtually absent in New England but occur in southern New Brunswick, mainland Nova Scotia, southeastern and central Cape Breton Island, and on the western Avalon Peninsula of Newfoundland. Voluminous latest Neoproterozoic (ca. 560–550 Ma) plutonic and volcanic rocks occur in southern New Brunswick and are attributed to an intracontinental extensional regime (Barr and White, 1996). In the Antigonish and Cobequid Highlands, ca. 580–570 Ma magmatism occurs in small, isolated plutons as postorogenic A-type granitic bodies (Pe-Piper et al., 1996; Murphy et al., 1998). Emplacement of these granite bodies is attributed to coeval transcurrent activity (Murphy et al., 1998). This time interval coincides with structural inversion of the Cobequid-Antigonish backarc basin associated with a change from sinistral to dextral strike-slip motion.

In Cape Breton Island, 580–550-Ma magmatism varies from immature arc in southern Cape Breton Island to mature arc in central Cape Breton Island (Barr, 1993; Dostal et al., 1996; Barr and Kerr, 1997). In the western Avalon Peninsula, the geochemistry of ca. 575–565-Ma volcanic and plutonic rocks indicate that they are derived from an immature volcanic arc (Barr and Kerr, 1997). In the eastern Avalon Peninsula, volcanic and plutonic rocks of this age are bimodal with alkalic affinities. The granites have A-type chemistry (Tuach, 1991) and are attributed to an intracontinental extensional environment. In the British Isles, an important phase of late arc activity occurred from ca. 570 to 550 Ma, producing igneous rocks along the Welsh borderlands and the English Midlands (e.g., the Warren House volcanics, the Uriconian Volcanic Group) (Tucker and Pharoah, 1991) and the development of the Anglesey blueschists (Dallmeyer and Gibbons, 1987; Gibbons and Horak, 1996).

In the North Armorican Massif, arc-related activity continued until at least ca. 560 Ma in the Tregor–La Hague and St Brieuc terranes (e.g., Dallmeyer et al., 1994; D'Lemos et al., 1992a; D'Lemos and Brown, 1993; Strachan et al., 1996b). However, in the St. Malo and Mancellian terranes, regional deformation at 570 Ma was succeeded by progressive metamorphism, crustal anatexis and granite emplacement peaking at ca. 540 Ma which occurred in a sinistral transpressive regime (e.g., D'Lemos et al., 1992b).

In summary, ca. 630–590 Ma Avalonian rocks from New England to the British were dominated by subduction-related activity in which several backarc basins were developed, although it is clear that, in any one locality, arc-magmatism occurred during a narrow interval within that period. Inversion of basins in New England and Atlantic Canada appears to have occurred about 600 Ma, and is related to a change from sinistral to dextral strike slip motion along major faults. From 590 to 550 Ma, the tectonic environment was transitional, such that arc activity and intracontinental extension occurred more or less synchronously in different parts of the belt. The switch from arc to rift intracontinental magmatism appears to have occurred diachronously from southeast to the northwest: New England (ca. 590 Ma), southern New Brunswick (600–560 Ma), mainland Nova Scotia (610–580 Ma), Cape Breton Island (550–540 Ma), Britain (550–540 Ma), Cadomia (post–540 Ma).

POLARITY OF SUBDUCTION

The polarity of Avalonian subduction in the Neoproterozoic is crucial to interpretations of its tectonic history along the Gondwanan margin. Recent precise age dating has allowed across-strike comparison of coeval rocks, from which the polarity of subduction may be deduced. Using trace element and isotopic data, Dostal et al. (1996) concluded that variations in ε_{Nd} and rare earth elements (REE) in 580–550-Ma igneous rocks in Cape Breton Island are compatible with northwesterly directed subduction (present coordinates). Conversely, Barr (1993) suggested that subduction of 630–600 Ma rocks in Cape Breton Island was directed to the south on the basis of K_2O-SiO_2 relationships. Taken together, this would imply a flip in the polarity of subduction between 600 and 580 Ma. The possibility of mobility of potassium during secondary alteration, however, makes this parameter less reliable than Nd isotopic data, which are relatively insensitive to alteration (e.g., De Paolo, 1981, 1988).

In order to gain a continuous across-strike comparison for 630–600 Ma rocks, data from volcanic rocks in the East Bay Hill and the Coxheath Hills in Cape Breton Island are presented here to supplement published data from other parts of Maritime Canada (Table 1). These rocks are calc-alkalic and consist of interbedded basalts and rhyolites (Thicke, 1988; Dostal et al., 1990, 1992). Mantle-normalized trace element data for the basaltic rocks in both belts are characterized by depletion in Nb and Ti, and enrichment in Th, features typical of subduction-related settings (Fig. 3). The associated felsic rocks of the Coxheath belt have elevated abundances of incompatible elements, and negative Eu and Ti anomalies that indicate extensive fractionation (Fig. 3). However, the relatively flat slope of the heavy REE elements in the felsic rocks contrasts with the monotone decreasing slope in basaltic rocks, and suggests that the felsic rocks were formed by crustal anatexis (Fig. 3C). The ε_{Nd}^t values for the rhyolites (t = 613 Ma, calculated for the age of extrusion) varies from +1.8 to +0.74, and depleted mantle model ages (T_{DM}) range from 1,028 Ma to 1,160 Ma (Table 1). The data fall within the envelope for Avalonian crustally derived rocks (Fig. 4) derived by drawing ε_{Nd}-time growth lines (Murphy et al., 1996; Keppie et al., 1997). The data are similar to those of coeval Avalonian felsic rocks elsewhere in Cape Breton Island (Dostal et al., 1990; Barr and Hegner, 1992; Barr, 1993), the Antigonish Highlands (Murphy et al., 1996) and in southern New Brunswick (Dostal and McCutcheon, 1990; Barr and White, 1996), which are also interpreted to have been derived by crustal anatexis.

The ε_{Nd}^t values for the basalts (t = 623 Ma, calculated for the age of extrusion) vary from +2.01 to +0.38 in the East Bay block and +0.72 to –1.01 in the Coxheath block (Table 2). These values are similar to coeval mafic rocks elsewhere in Cape Breton and in southern New Brunswick (e.g., Whalen et al., 1994), but are considerably lower than coeval mafic rocks in the Antigonish Highlands. Because the Antigonish Highlands basalts are thought to have been emplaced in an extensional rift/backarc setting (Murphy et al., 1990) and therefore thinner continental crust, the lower ε_{Nd}^t values in Cape Breton and New Brunswick may reflect a greater degree of crustal contamination.

TABLE 1. MAJOR AND TRACE ELEMENT ANALYSES OF REPRESENTATIVE VOLCANIC ROCKS FROM EAST BAY HILLS AND COXHEATH BLOCKS

	East Bay Hills Block			Coxheath Block					
Sample	20	21	24	29	221	74	28	27	26
(wt. %)									
SiO_2	51.14	52.51	53.98	46.08	49.66	52.50	68.43	72.01	73.30
TiO_2	1.24	0.98	1.04	1.16	1.28	1.01	0.45	0.43	0.41
Al_2O_3	18.86	19.45	18.52	17.63	19.59	17.05	16.33	13.11	14.10
Fe_2O_3*	11.71	8.38	7.95	10.41	9.84	9.33	2.37	2.66	2.66
MnO	0.16	0.08	0.15	0.20	0.12	0.18	0.03	0.04	0.02
MgO	4.47	3.58	6.04	3.61	3.80	4.94	0.47	0.86	0.54
CaO	3.60	7.13	1.81	6.67	5.01	5.54	1.00	1.80	0.35
Na_2O	3.52	3.53	6.30	3.30	3.14	3.47	5.47	2.79	3.40
K_2O	2.06	1.22	0.42	3.16	2.11	2.83	3.18	3.94	3.45
P_2O_5	0.25	0.21	0.19	0.23	0.22	0.22	0.07	0.07	0.05
LOI	3.13	2.70	3.74	7.65	4.83	2.51	1.84	1.42	1.45
Total	100.14	99.77	100.14	100.10	99.60	99.58	99.64	99.13	99.73
(Mg)	0.43	0.46	0.60	0.41	0.43	0.51	0.28	0.39	0.29
(ppm)									
Cr	52	42	25	45	17	72	3	26	9
Ni	79	38	67	33	22	46	....	2	3
V	257	212	151	320	244	217	21	31	25
Zn	137	63	86	89	67	176	35	28	34
Rb	109	45	16	118	81	88	104	134	120
Ba	387	396	122	318	519	766	475	700	470
Sr	363	521	176	143	516	483	151	215	110
Ta	0.36	0.71	0.35	0.21	0.32	0.43	0.99	0.99	0.86
Nb	5.3	5.8	4.3	4.5	5.4	8.9	14.2	10.0	11.4
Hf	2.65	2.98	3.35	2.12	2.86	3.72	7.33	5.12	4.77
Zr	105	118	143	82	117	161	342	228	249
Y	18	17	19	17	21	20	33	24	27
Th	4.24	5.32	3.59	2.89	3.29	13.32	15.24	16.64	13.57
La	17.80	18.90	11.55	14.75	21.64	34.06	21.54	31.84	22.46
Ce	40.15	40.02	28.20	33.77	45.29	68.62	42.54	68.70	55.07
Pr	5.08	4.77	3.65	4.31	5.48	7.96	5.86	7.76	6.47
Nd	21.17	19.31	16.30	18.10	22.39	30.00	22.91	28.37	23.99
Sm	4.52	3.99	3.81	3.91	4.84	5.70	4.90	5.36	4.56
Eu	1.25	1.16	1.12	1.10	1.58	1.29	0.90	0.93	0.75
Gd	4.01	3.60	3.81	3.82	4.54	4.90	4.72	4.55	3.93
Tb	0.57	0.51	0.59	0.55	0.66	0.69	0.78	0.68	0.66
Dy	3.67	3.33	3.84	3.55	4.25	4.20	5.64	4.81	4.56
Ho	0.69	0.64	0.75	0.66	0.81	0.79	1.21	0.87	0.94
Er	1.94	1.71	2.16	1.85	2.25	2.24	3.93	2.62	3.07
Tm	0.29	0.25	0.31	0.26	0.32	0.31	0.61	0.43	0.49
Yb	1.93	1.66	1.91	1.67	2.15	2.13	3.98	2.83	3.35
Lu	0.28	0.23	0.27	0.24	0.31	0.30	0.63	0.43	0.53

(Mg) = $Mg/(Mg+Fe^t)$

DePaolo (1981) observed that ε_{Nd}^t values are increasingly negative with distance from the trench, reflecting increasing involvement of ancient crust. A compilation of this data for 550–580 and 600–630 Ma across Cape Breton Island and New Brunswick shows similarly decreasing ε_{Nd}^t to the northwest in both regions (Fig. 5). This suggests that subduction was northwest-directed in Maritime Canada, during both the 630–600 Ma and subsequent 580–550 Ma magmatism, and that the polarity flip between these two periods of magmatism proposed for Cape Breton Island did not occur. In Cape Breton Island, the northwestern migration of calc-alkaline magmatism and inversion of the Antigonish-Cobequid backarc basin at ca. 600 Ma was accompanied by a reversal in kinematics from sinistral to dextral along major fault zones (Nance and Murphy, 1990; Murphy et al., 1991) and is coincident with shallowing of the subduction zone (Keppie et al., 1998). This shallowing of the subduction zone is supported by the more gradual northward increase in ε_{Nd}^t in the 580–550 Ma arc rocks compared to the 630–600 Ma arc rocks both in Cape Breton and in southern New Brunswick (Fig. 5A,B).

TABLE 2. NEODYMIUM ISOTOPIC DATA FOR VOLCANIC ROCKS FROM EAST BAY HILLS AND COXHEATH BLOCKS

Sample	Type	Sm (ppm)	Nd (ppm)	$^{147}Sm/^{144}Nd$	$^{143}Sm/^{144}Nd_m$	Age (Ma)	$^{143}Sm/^{144}Nd_i$	$\varepsilon_{Nd}{}^t$	T_{DM}
East Bay Hills Block									
20	Basalt	6.4	29.1	0.132	0.51243	623	0.51189	+1.14	1157
24	Basalt	5.2	21.6	0.146	0.51253	623	0.51194	+2.01	1166
21	Basalt	4.6	21.9	0.128	0.51238	623	0.51185	+0.38	1197
Coxheath Block									
221	Basalt	6.6	29.6	0.136	0.51243	613	0.51189	+0.72	1213
74	Basalt	7.1	37.2	0.116	0.51226	613	0.51180	-1.01	1227
29	Basalt	5.3	23.4	0.137	0.51239	613	0.51184	-0.18	1302
26	Rhyolite	6.3	31.7	0.121	0.51242	613	0.51194	+1.70	1048
27	Rhyolite	6.1	31.3	0.118	0.51241	613	0.51194	+1.80	1028
28	Rhyolite	5.3	24.9	0.128	0.51240	613	0.51189	+0.74	1160

Age (Ma) = assumed crystallization age; $^{143}Sm/^{144}Nd_m$ and $^{143}Sm/^{144}Nd_i$ = measured (m) and initial (i) isotopic ratios, respectively; T_{DM} = Nd model age (in Ma) for fractionation from depleted mantle (DM) to the crust; $\varepsilon_{Nd}{}^t$ = the fractional difference between the $^{143}Sm/^{144}Nd$ of rock and the bulk earth at the time of crystallization.

NEOPROTEROZOIC POSITION OF AVALONIA; ADDITIONAL EVIDENCE FOR AN AMAZONIAN CONNECTION

The episodic history of arc-related magmatism from ca. 780 to 540 Ma in some or all of Avalonia, in addition to paleomagnetic data (e.g., Van der Voo, 1988), provides strong evidence that Avalonian tectonothermal activity occurred along the periphery of Gondwana and was not directly related to coeval rifting along the Laurentian margin (e.g., Williams, 1979). Preliminary U-Pb geochronologic data (Keppie et al., 1998) and Nd isotopic data (Nance and Murphy, 1994, 1996; Keppie et al., 1996) suggest that Avalonia was positioned along the periphery of the Amazon craton. Comparisons among Pb-Pb isotopic data of Avalonia (e.g., Ayuso et al., 1996) and the Amazon craton (Keppie and Ortega-Gutierrez, 1995) support this interpretation.

Concordant detrital zircons from Avalonia in mainland Nova Scotia yield age ranges of 612–694, 977–1223, 1332, 1506–1545, 1606–1630, 1827–2000, and 2600–3000, Ma (Keppie et al., 1998) (Fig. 6). Geochemical and isotopic data from the sedimentary rocks in which these zircons were analyzed indicate that these rocks were locally derived from typical Avalonian basement (Murphy and MacDonald, 1993). These detrital zircon ages can be sourced only within the Amazon craton, supporting Nd isotopic evidence that Avalonia lay adjacent to the Amazon craton.

The Pb-Pb data provide information on the nature of the composition of continental basement because they reflect time-integrated average compositions of large portions of crustal sources (e.g., Doe and Zartman, 1979). As noted by Ayuso et al. (1996) in Cape Breton Island, initial Pb isotopic values of feldspars may reflect the composition and age of the continental source. Plotted on Pb-Pb isotopic diagrams (Fig. 7), these data fall between the orogene and upper crust fields, reflecting the lack of a mantle component. However, the data also lie within the evolutionary envelope of the Amazon craton and are compatible with an Amazonian source for these rocks.

The peripheral position of the Avalonian arc relative to the Amazonian portion of Gondwana requires the predominant polarity of subduction would have been to the south during the Neoproterozoic. The northwesterly direction of subduction in Atlantic Canada implied by the $\varepsilon_{Nd}{}^t$ data therefore suggests a rotation of more than 90° of this portion of Avalonia prior to its accretion to Laurentia. This is consistent with the model of Keppie et al. (1996, 1997), in which this rotation is inferred to have taken place during dextral relative motion between Gondwana and Laurentia in the early Paleozoic.

RECENT ANALOGS

Although the Mesozoic-Cenozoic history of western North America has been suggested as a recent analog for peri-Gondwanan tectonic activity (e.g.. Strong et al., 1978; Nance et al., 1991), a detailed comparison has been hindered by a lack of precise geochronologic dating of magmatic events. Neoproterozoic palinspastic restoration of these terranes along the Gondwanan margin (Keppie et al., 1996; Nance and Murphy, 1996) suggest that subduction occurred along most of the margin during 630–600 Ma, the main phase of arc activity. However, the mechanism for termination of subduction is unclear. Using the reconstruction of Bond et al. (1984), Murphy and Nance (1989) attributed termination to generation of a transform fault associated with the opening of the Iapetus ocean. However, this model cannot explain the diachronous termination of subduction within Avalonia. Intracontinental tectonics had developed by ca. 590 Ma in New England, and 580 Ma in mainland Nova Scotia, whereas convergent margin tectonics continued until at least 550 Ma in Cape Breton Island, 550 Ma in the British Isles and 540 Ma in Cadomia.

The Mesozoic-Cenozoic history of southwestern North America offers a plausible analog for such activity. Arc-related tectonics resulted in the Antler (Mississipian), Sonoma (Permian-

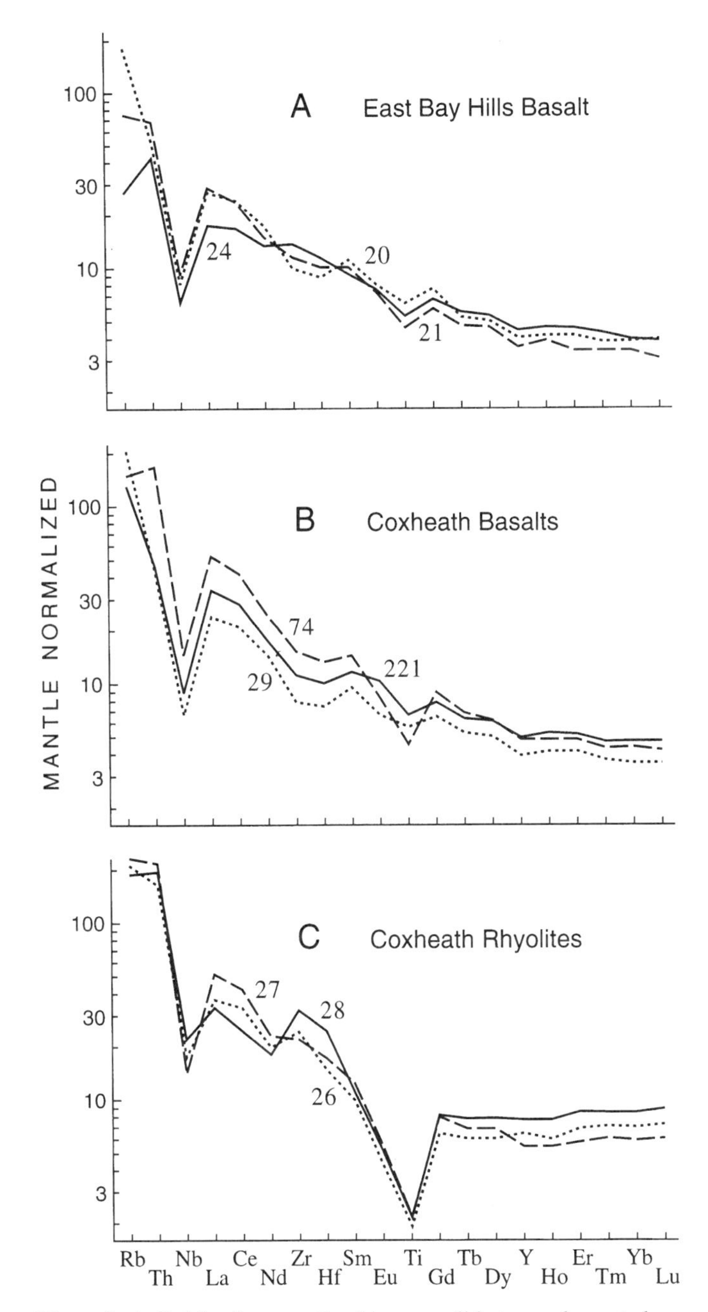

Figure 3. A–C, Mantle-normalized incompatible trace element abundances in representative basalts and rhyolites from the East Bay Hills and Coxheath blocks (Cape Breton Island). Normalizing values after Sun and McDonough (1989).

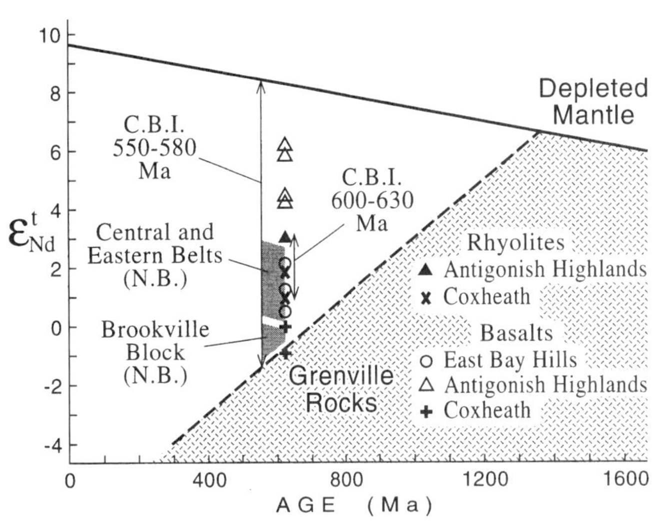

Figure 4. Summary $\varepsilon_{Nd}{}^{t}$ vs. time (Ma) diagram of isotopic data from ca. 630–550 Ma rocks of the East Bay Hills and Coxheath blocks (see table 1) and the Antigonish Highlands (data from Murphy et al., 1996). Arrows show the range of data from Barr and Hegner (1992) and Dostal et al. (1996). For comparison with Avalonian and Grenvillian fields, the data from southern New Brunswick (Whalen et al., 1994) are also shown.

Triassic), and Nevadan (Jurassic) orogenic events (Burchfiel et al., 1992), and voluminous magmatism at various times in the Jurassic-Cretaceous (Burchfiel et al., 1992). Between the Late Cretaceous and Eocene, the Laramide orogeny (Dickinson and Snyder, 1979) was characterized by an almost complete absence of magmatism, (the "Paleocene magmatic gap"), a feature that has been attributed to flat-slab subduction extending 1,200 km into the continental interior or collision of a microcontinent Baja-BC (e.g., Livaccari et al., 1981; Severinghaus and Atwater, 1990; Maxson and Tikoff, 1996).

Geodynamic modeling by Zhong and Gurnis (1995) suggests this flat-slab subduction was a consequence of the relatively rapid westward migration of the North American convergent margin by oceanward continental motion or trench migration, a scenario that is commonly interpreted to have accompanied the Laramide orogenic event. In northern Nevada, this period was followed by renewed magmatism that coincided with the steepening of the subducting Farallon slab associated with slower convergence, the development of the Kula-Farallon ridge and northward passage of the trailing edge of Baja-BC (Gans et al., 1989; Coney 1979; Maxson and Tikoff, 1996). Ongoing subduction throughout much of the early Tertiary was succeeded by the progressive generation of the San Andreas fault over the past 30 m.y. (Atwater, 1970; Severinghaus and Atwater, 1990). This was related to the collision of the East Pacific Ridge spreading center with the trench, and the northward migration of the resulting triple point. Thus, the termination of subduction and initiation of intracontinental tectonics was diachronous along the margin and occurred along a relatively restricted segment of the continental margin. To the north of the Mendocino triple point, subduction of the Juan de Fuca plate continues to the present day and results in the arc activity represented by the Cascades. To the south of Baja, continued subduction of the Cocos plate results in the present-day arc activity in Mexico.

Along the Gondwanan margin in the Neoproterozoic, changes in the profile in the subduction zone as modeled by Zhong and Gurnis (1995) alone may explain gaps in magmatic record between 660 and 630 Ma and the relatively short duration of magmatism in any one portion of Avalonia during peak arc activity (e.g., Strachan et al. 1996b). Termination of subduction and the diachronous gen-

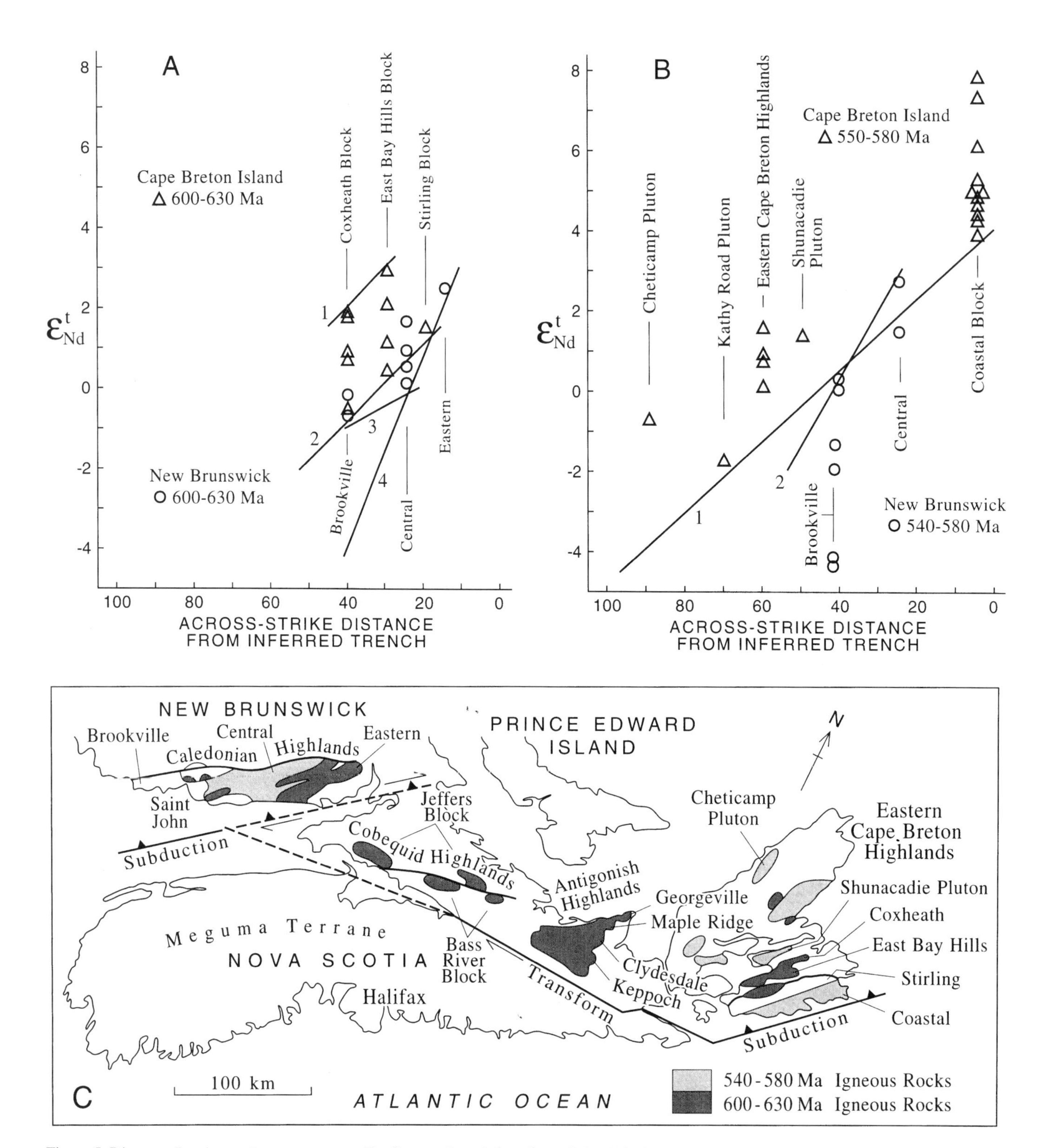

Figure 5. Diagram showing ε_{Nd}^{t} versus across strike distance from inferred trench for 630–600 Ma (A) 580–540 Ma rocks and (B) in Cape Breton Island and southern New Brunswick. Isotopic data from Table 1, Barr and Hegner (1992), Dostal et al. (1996), Murphy et al. (1996) and Whalen et al. (1994). In A, lines 1 and 2 represent the upper and lower limits for ε_{Nd}^{t} evolution in Cape Breton, and line 3 is the lower limit for ε_{Nd}^{t} evolution in New Brunswick. In B, line 1 is the lower limit for ε_{Nd}^{t} evolution in Cape Breton, and line 2 is the upper limit for ε_{Nd}^{t} evolution in New Brunswick. C, The locations of the regions from which the samples were derived.

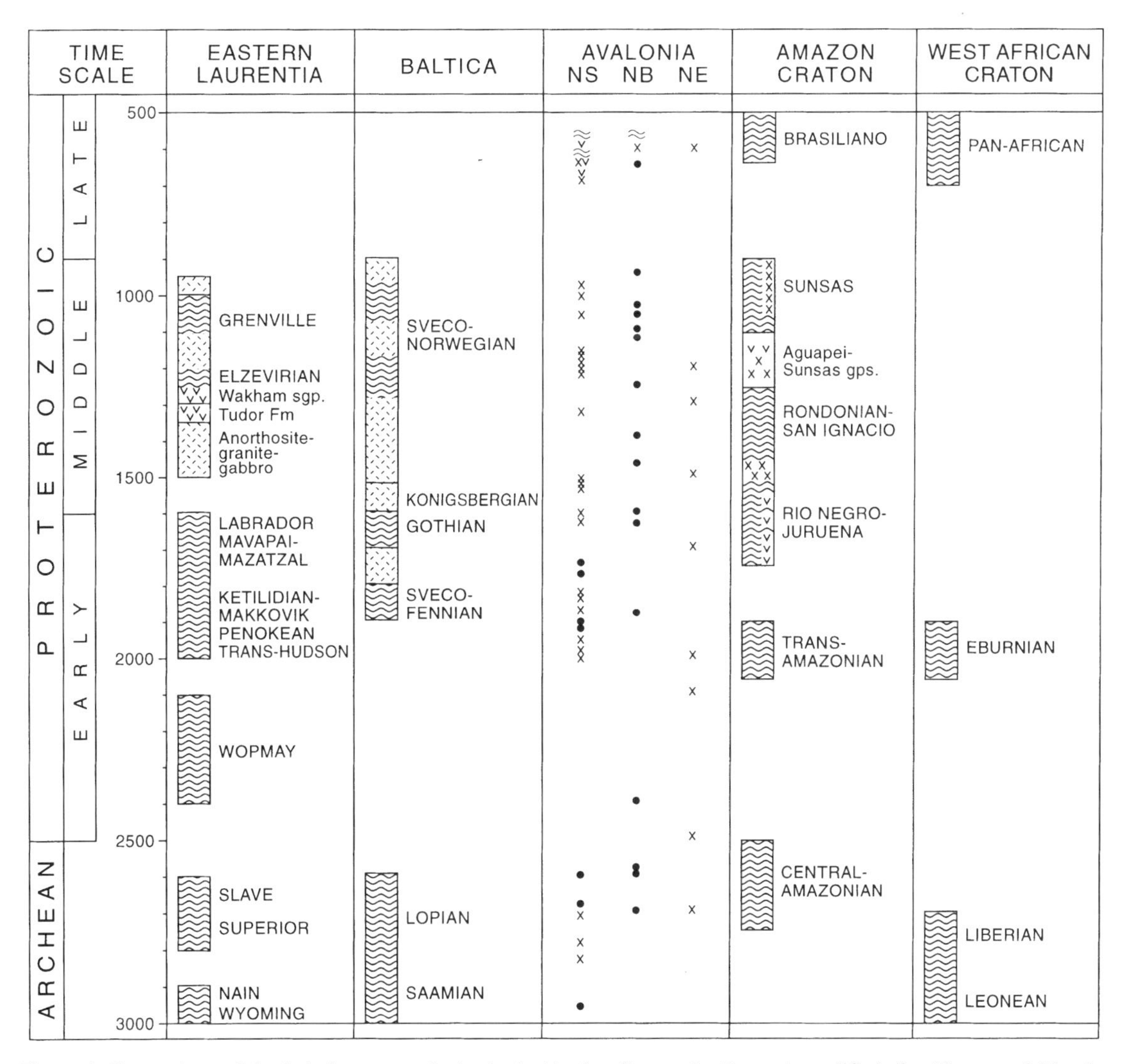

Figure 6. Comparison of detrital zircon ages in Avalonia (Avalon Composite Terrane) modified after Nance and Murphy (1994) and Keppie et al. (1998). Data compiled from New Brunswick (Bevier et al., 1990) and New England (Karabinos and Gromet, 1993) with tectonothermal events in Baltica (Gower et al., 1990; Goodwin, 1991, Starmer, 1993), eastern Laurentia (Hoffman, 1989; Connelly and Heaman, 1993; Martignole et al., 1994; Gower and Tucker, 1994), the Amazon craton (Texiera et al., 1989; Sadowski and Bettencourt, 1996, and references therein) and northwest Africa (Rocci et al., 1991). Waves denote periods of orogenesis; dashes denote plutons, v = volcanic rocks; x = concordant U-Pb zircon ages; • = discordant $^{207}Pb/^{206}Pb$ zircon ages.

eration of intracontinental magmatism may reflect the migration of a triple point along the Gondwanan margin from New England to Cape Breton Island, possibly reflecting collision with a spreading center (Figs. 8 and 9). Migration of this triple point may have displaced outboard terranes eastward along the margin-Cadomia might represent such a displaced terrane. Subduction continued later in adjacent terranes to the west and east in a manner similar to the modern Cascades and the Mexican Cordillera.

CONCLUSIONS

The tectonothermal evolution of Avalonia provides important constraints for Neoproterozoic and early Paleozoic paleocontinental reconstructions because of its peripheral position along the Gondwanan margin. There are at least three main phases in its tectonic history. (1) Vestiges of an early arc phase from ca. 780 to 660 Ma are preserved in many parts of the belt; however, it is unclear whether they represent episodic or continuous subduction. (2) The main arc phase occurs between 630 and 590 Ma and produced voluminous calc alkalic volcanic and plutonic rocks in addition to intra-arc and backarc basins in which volcaniclastic sedimentary rocks were deposited. In each individual location along the belt, however, the duration of arc volcanism was limited suggesting changes in the profile of the subducting slab during evolution. Nd isotopic data indicate that subduction polarity was to the northwest in Atlantic Canada (present coordinates). U-Pb geochronology and

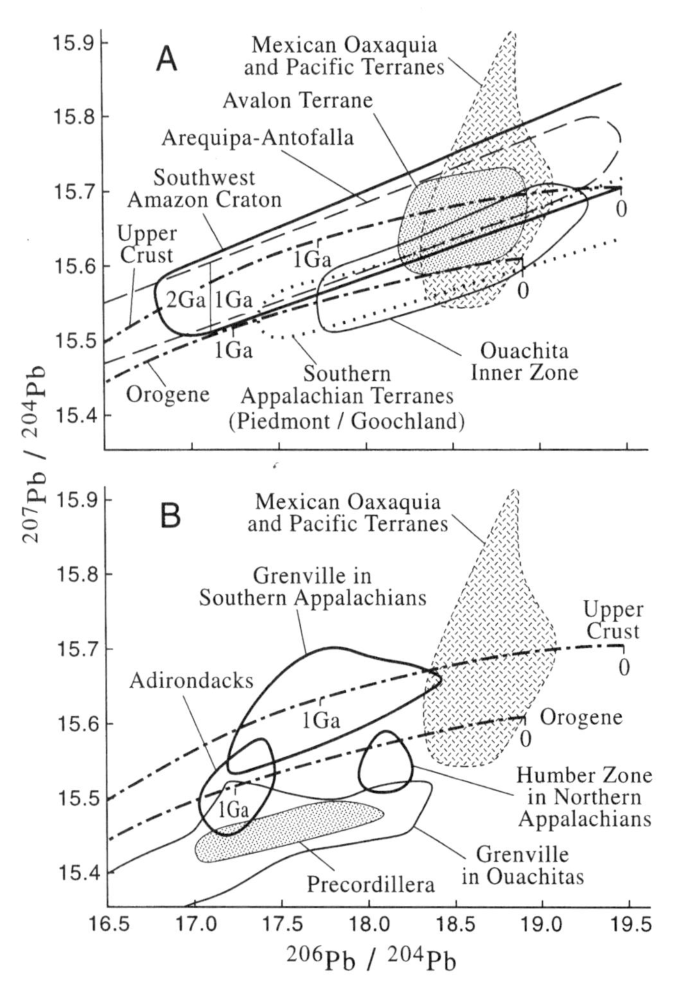

Figure 7. A and B, Comparison between $^{207}Pb/^{204}Pb$ vs. $^{206}Pb/^{204}$ Pb isotopic compositions of Avalonia (compiled from Ayuso and Bevier, 1991; Whalen et al., 1994; Ayuso et al., 1996) with potential source locations. Data from the Mexican and Ouxaquia terranes is from Keppie and Ortega-Gutierrez (1995); Grenvillian rocks in the Ouachitas and the Oauchita inner zone from James and Henry (1993a,b); the southern Appalachians, Piedmont and Goochland terranes from Sinha et al. (1996); the Adirondacks from Zartman (1969); the Humber Zone in the northern Appalachians from Ayuso and Bevier (1991); the Precordillera of western Argentina from Kay et al. (1996); the Arequipa-Antofalla terrane of southern Peru–northern Chile from Aitcheson et al. (1995) and the southwest Amazon craton from Tosdal et al. (1994).

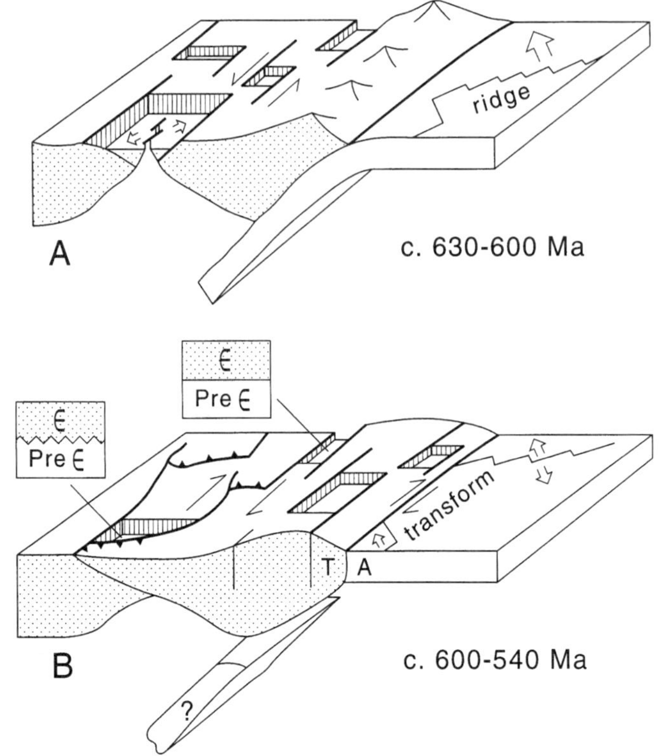

Figure 8. A, Schematic plate tectonic model for the main arc phase from 630 to 590 Ma in which oblique subduction results in calc alkalic magmatism along the length of the belt and sinistral motion along major faults resulting in the local opening of volcanic arc basins locally floored by oceanic crust. B, The collision of an oceanic ridge results in the progressive development of a continental transform fault, diachronous termination of arc magmatism, and development of intracontinental extension, dextral motion along major faults, and structural inversion of volcanic arc basins.

Nd-Sm and Pb-Pb isotopic studies support a position peripheral to the Amazonian craton during the main phase of arc activity. (3) The third phase, from 590 to 540 Ma represents a transition from an arc to a transform environment. Available kinematic data suggest that this transition was accompanied by a switch from sinistral to dextral kinematics and structural inversion. Recent U-Pb data suggest that this occurred diachronously. Intracontinental magmatism occurred in New England and mainland Nova Scotia by ca. 580 Ma, whereas arc magmatism continued much later in Cape Breton Island, the British Isles, and in correlative rocks in Cadomia, an adjacent peri-Gondwanan terrane.

The enhanced geochronologic database enables more realistic comparisons with Cenozoic analogs. The Neoproterozoic tectonothermal evolution of Avalonia is similar to that of western North America, in which episodes of arc magmatism were succeeded by the generation of the San Andreas transform associated with the collision of an oceanic spreading ridge. A similar setting is invoked for Avalonia and would explain the diachronous nature of the arc-transform transition and the switch from sinistral to dextral kinematics and associated structural inversions.

ACKNOWLEDGEMENTS

We are grateful to Robin Strachan and Chris Hepburn for thoughtful, constructive reviews. This project was supported by the Natural Sciences and Engineering Research Council general and lithoprobe grants to (J. B. M. and J. D.), the University Council Research Grants at St. Francis Xavier University, and the James Chair of Pure and Applied Sciences at St. Francis Xavier University, (to J. D. K. and R. D. N.). This is a contribution to International Geological Correlation Program Project 376.

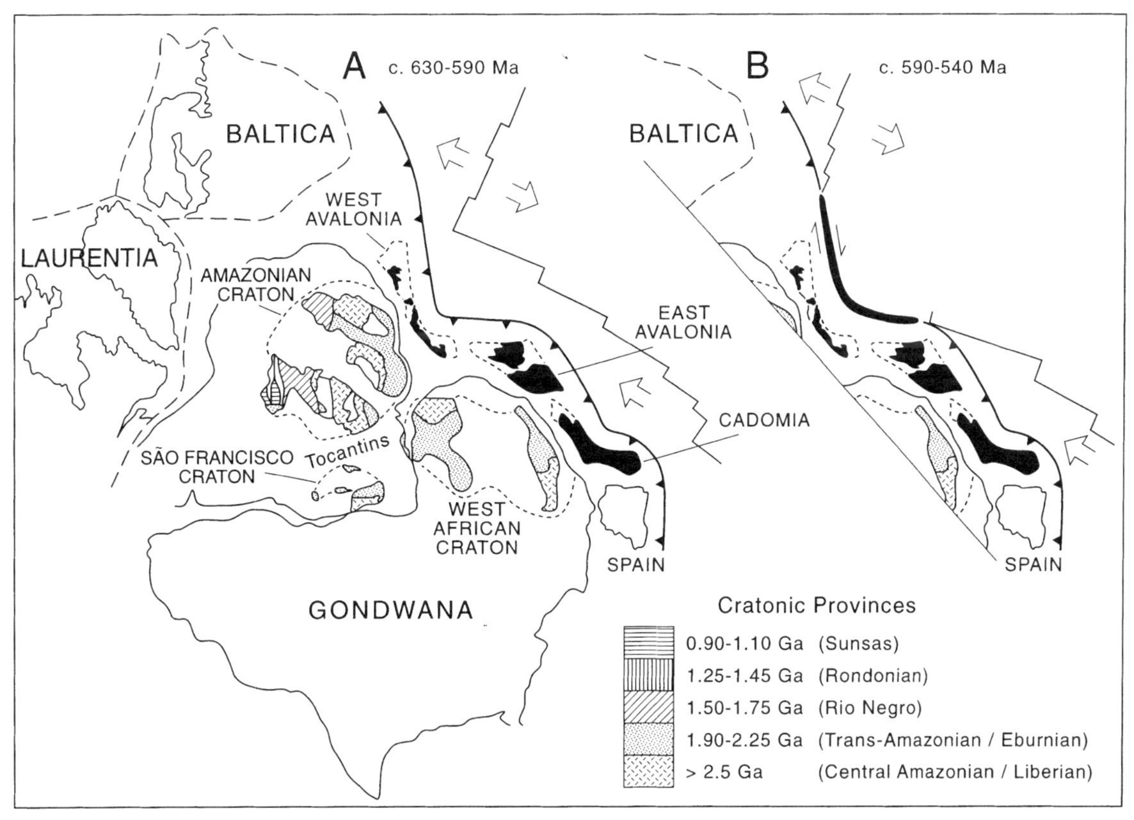

Figure 9. Global reconstruction (modified after Dalziel, 1992) of Avalonia and Cadomia along the periphery of Gondwana (modified after Nance and Murphy, 1994, 1996; Keppie et al., 1996) showing (A) subduction along the length of Avalonia and Cadomia from 630 to 590 Ma and (B) the diachronous termination of subduction and generation of a continental transform fault related to collision of an oceanic ridge. Note that subduction-related magmatism is continuing in Britain and Cadomia, in contrast to the Appalachian tract.

REFERENCES CITED

Aitcheson, S. J., Harmon, R. S., Moorbath, S., Schneider, A., Soler, P., Soria-Escalante, E., Steele, G., Swainbank, Y., and Worner, G., 1995, Pb isotopes define basement terranes of the Altiplano, central Andes: Geology, v. 33, p. 555–558.

Atwater, T., 1970, Implications of plate tectonics for the Cenozoic tectonic evolution of western North America: Geological Society of America Bulletin, v. 81, p. 3518–3536.

Ayuso, R. A., and Bevier, M. L., 1991, Regional differences in Pb isotopic compositions of feldspars in plutonic rocks of the northeastern Appalachian mountains, U.S.A., and Canada: geochemcial method of terrane correlation: Tectonics, v. 10, p. 191–212.

Ayuso, R. A., Barr, S. M., and Longstaffe, F. J., 1996, Pb and O isotopic constraints on the source of granitic rocks from Cape Breton Island, Nova Scotia: American Journal of Science, v. 276, p. 789–817.

Barr, S. M., 1993, Geochemical and tectonic setting of late Precambrian volcanic and plutonic rocks in southeastern Cape Breton Island, Nova Scotia: Canadian Journal of Earth Sciences, v. 30, p. 1147–1154.

Barr, S. M., and Hegner, E. 1992, Nd isotopic compositions of felsic igneous rocks in Cape Breton Island, Nova Scotia: Canadian Journal of Earth Sciences, v. 29, p. 650–657.

Barr, S. M., and Kerr, A., 1997, Late Precambrian plutons in the Avalon terrane of New Brunswick, Nova Scotia and Newfoundland, *in* Sinha, A. K., Whalen, J. B., and Hogan, J. P., eds., The nature of magmatism in the Appalachian orogen: Geological Society of America Memoir 191, p. 45–74.

Barr, S. M., and White, C. E. 1988, Petrochemistry of contrasting late Precambrian volcanic and plutonic associations, Caledonian Highlands, southern New Brunswick: Maritime Sediments and Atlantic Geology, v. 24, p. 353–372.

Barr S. M., and White, C. E., 1996, Contrasts in late Precambrian-early Paleozoic tectonothermal history between Avalon composite terrane sensu stricto and other possible peri-Gondwanan terranes in southern New Brunswick and Cape Breton Island, Canada, *in* Nance, R. D., and Thompson, M. D., eds., Avalonian and related peri-Gondwanan terranes of the circum North Atlantic. Geological Society of America Special Paper 304, p. 95–108.

Barr, S. M., Bevier, M. L., White, C. E., and Doig, R., 1994, Magmatic history of the Avalon terrane in southern New Brunswick, Canada, based on U-Pb (zircon) geochronology: Journal of Geology, v. 102, p. 399–409.

Bevier, M. L., and Barr, S. M., 1990, U-Pb constraints on the stratigraphy and tectonic history of the Avalon terrane, New Brunswick, Canada: Journal of Geology, v. 98, p. 53–63.

Bevier, M. L., Barr, S. M., and White, C. E., 1990, Late Precambrian U-Pb ages for the Brookville Gneiss, southern New Brunswick. Journal of Geology, v. 98, p. 955–968.

Bevier M. L., Barr, S. M., White, C. E., and Macdonald, A. S., 1993, U-Pb geochronologic constraints of the volcanic evolution of the Mira (Avalon) terrane, southeastern Cape Breton Island, Nova Scotia: Canadian Journal of Earth Sciences, v. 30, p. 1–10.

Bond, G. C., Nickerson, P. A., and Kominz, M. A., 1984, Breakup of a supercontinent between 625 Ma and 555 Ma: new evidence and implications for continental histories: Earth and Planetary Science Letters, v. 70, p. 325–345.

Burchfiel B. C., Cowan, D. S., and Davis, G. A., 1992, Tectonic overview of the Cordilleran orogen in the western United States, *in* Burchfiel, B. C., Lipman, P. W., and Zoback, M. L., eds., The Cordilleran orogen: conterminous U.S.: Boulder, Colorado, Geological Society of America, Geology of North America, v. G-3, p. 407–430.

Cawood, P. A., Dunning, G. R., Lux, D., and van Gool, J. A. M., 1994, Timing of peak metamorphism and deformation along the Appalachian margin of Laurentia in Newfoundland: Geology, v. 22, p. 399–402.

Coney, P. J., 1979, Tertiary evolution of Cordilleran metamorphic core complexes, *in* Armentrout, J. W., et al., eds., Cenozoic paleogeography of the western United States: Society of Economic Paleontologists and Mineralogists Pacific Section, Symposium 3, p. 15–28.

Connelly, J. N., and Heaman, L. M., 1993, U-Pb geochronological constraints on the tectonic evolution of the Grenville province, western Labrador: Precambrian Research, v. 63, p. 123–142.

Dallmeyer, R. D., and Gibbons, W., 1987, The age of blueschist metamorphism in Anglesey, North Wales: evidence from $^{40}Ar/^{39}Ar$ mineral ages of the Penmynydd Schists: Geological Society of London Journal, v. 144, p. 843–850.

Dallmeyer, R. D., D'Lemos, R. D., and Strachan, R. A., 1994, Timing of Cadomian and Variscan tectonothermal activity, La Hague and Alderney, North Armorican Massif: evidence from $^{40}Ar/^{39}Ar$ mineral ages: Geological Journal, v. 29, p. 29–44.

Dalziel, I. W. D., 1992, On the organization of American plates and the breakout of Laurentia: GSA Today, v. 2, p. 237, 240–241.

DePaolo, D. J., 1981, Neodymium isotopes in the Colorado Front Range and crust-mantle evolution in the Proterozoic: Nature, v. 29, p. 193–196.

DePaolo, D. J., 1988, Neodymium isotope geochemistry: an introduction: New York, Springer-Verlag, 187 p.

Dickinson, W. R., and Synder, W. S., 1979, Geometry of subducted slabs related to the San Andreas transform: Journal of Geology, v. 87, p. 609–627.

D'Lemos, R. D., and Brown, M.,1993, Sm-Nd isotope characteristics of late Cadomian granite magmatism in northern France and the Channel Islands: Geological Magazine, v. 130, p. 797–804.

D'Lemos, R. D., Dallmeyer, R. D., and Strachan, R. A., 1992a, $^{40}Ar/^{39}Ar$ dating of plutonic rocks from Jersey, Channel Islands: Proceedings of the Ussher Society, v. 8, p. 50–53.

D'Lemos, R. D., Brown, M., and Strachan, R. A., 1992b, Granite generation, ascent and emplacement within a transpressional orogen: Geological Society of London Journal, v. 149, p. 487–490.

Doe, B. R., and Zartman, R. E., 1979, Plumbotectonics. I: The Phanerozoic, *in* Barnes, H., ed., Geochemistry of hydrothermal ore deposits: New York, John Wiley, p. 22–70.

Doig, R., Murphy, J. B., and Nance, R. D., 1993, Tectonic significance of the Late Proterozoic Economy River Gneiss, Cobequid Highlands, Avalon composite terrane, Nova Scotia: Canadian Journal of Earth Sciences, v. 30, p. 474–479.

Dostal, J., and McCutcheon, S. R., 1990, Geochemistry of Late Proterozoic basaltic rocks from southeastern New Brunswick, Canada: Precambrian Research, v. 47, p. 83–98.

Dostal, J., Keppie, J. D., and Murphy, J. B. 1990, Geochemistry of Late Proterozoic basaltic rocks from southeastern Cape Breton Island, Nova Scotia: Canadian Journal of Earth Sciences, v. 27, p. 619–631.

Dostal, J., Keppie, J. D., and Zhai, M., 1992, Geochemistry of mineralized and barren Late Proterozoic felsic volcanic rocks in southeastern Cape Breton Island, Nova Scotia: Precambrian Research, v. 56, p. 33–49.

Dostal, J., Keppie, J. D., Cousens, B. L., and Murphy, J. B. 1996, 550–580 Ma magmatism in Cape Breton Island, (Nova Scotia, Canada): the product of NW-dipping subduction during the final stage of assembly of Gondwana: Precambrian Research, v. 76, p. 96–113.

Egal E., Guerrot, C., Le Goff, D., Thieblemont, D., and Chantraine, J., 1996, The Cadomian orogeny revisited in northern France, *in* Nance, R. D., and Thompson, M. D., eds., Avalonian and related peri-Gondwanan terranes of the circum–North Atlantic. Geological Society of America Special Paper 304, p. 281–318.

Gans, P. B., Mahood, G. A., and Schermer, E., 1989, Synextensional magmatism in the Basin and Range province: a case study from the eastern Great Basin: Geological Society of America Special Paper 233, 53 p.

Gibbons, W., and Horak, J., 1996, The evolution of the Neoproterozoic Avalonian subduction system: evidence from the British Isles, *in* Nance, R. D., and Thompson, M. D., eds., Avalonian and related peri-Gondwanan terranes of the circum–North Atlantic. Geological Society of America Special Paper 304, p. 269–280.

Goodwin, A. M., 1991, Precambrian Geology: London, Academic Press, 666 p.

Golonka, J., Ross, M. I., and Scotese, C. R.,1994, Phanerozoic, paleogeographic and paleoclimatic modeling maps, in Beauchamp, B., Embry, A., and Glass, D., eds., Pangea. Canadian Society of Petroleum Geologists Memoir 17, p. 1–47.

Gower, C. F., and Tucker, R. D., 1994, The distribution of pre-1400 Ma crust in the Grenville province: implications for rifting in Laurentia-Baltica during eon 14: Geology, v. 22, p. 827–830.

Gower, C. F., Ryan, A. B., and Rivers, T., 1990, Mid-Proterozoic Laurentia-Baltica: an overview of its geological evolution and a summary of the contributions made by this volume, *in*, Gower, C. F., Rivers, T., and Ryan, A. B., eds., Mid-Proterozoic Laurentia-Baltica: Geological Association of Canada Special Paper 38, p. 1–22.

Guerrot, C., and Peucat, J. J., 1990, U-Pb geochronology in the Late Proterozoic Cadomian belt of NW France, *in* D'Lemos, R. D., Strachan, R. A., and Topley, C. G., eds., The Cadomian orogeny. Geological Society of London Special Publication 51, p. 383–393

Hodych, J. P., and Buchan, K. L., 1994. Paleomagnetism of the Early Silurian Cape Marys sills of the Avalon Peninsula of Newfoundland [abs.]: Eos (Transactions, American Geophysical Union), v. 75, p. 16.

Hoffman, P. F., 1989, Precambrian geology and tectonic history of North America, *in* Bally, A. M., and Palmer, A. R., eds., The geology of North America; an overview: Boulder, Colorado, Geology of North America, v. A, p. 447–512.

Horak, J., 1993, The late Precambrian Coedana and Sarn complexes, northwest Wales—A geochemical and petrological study [Ph.D. thesis]: Cardiff University of Wales, 415 p.

James, E. W., and Henry, C. D., 1993a, Southeastern extent of the North American craton in Texas and northern Chihuahua as revealed by Pb isotopes: Geological Society of America Bulletin, v. 105, p. 116–126.

James, E. W., and Henry, C. D., 1993b, Lead isotopes of ore deposits in Trans-Pecos Texas and northeastern Chihuahua, Mexico: basement igneous and sedimentary sources of metals: Economic Geology, v. 88, p. 934–947.

Karabinos, P., and Gromet, L. P., 1993, Application of single-grain zircon evaporation analyses to detrital grain studies and age discrimination in igneous suites: Geochimica et Cosmochimica Acta, v. 57, p. 4257–4267.

Kay, S. M., Orrell, S., and Abbruzzi, J. M., 1996, Zircon and whole rock Nd-Pb evidence for a Grenvillian age and Laurentian origin for the basement of the Precordilleran terrane in Argentina: Journal of Geology, v. 104, p. 637–648.

Keppie J. D., 1985. The Appalachian college; *In* Gee, D. G., and Sturt, B., eds., The Caledonide orogen, Scandinavia, and related area. New York, Wiley, p. 1217–1226.

Keppie, J. D., 1993, Synthesis of Paleozoic deformational events and terrane accretion in the Canadian Appalachians: Geologische Rundschau, v. 82, p. 381–431.

Keppie J. D., and Krogh, T. E.,1990, Detrital zircon ages from late Precambrian conglomerate, Avalon composite terrane, Antigonish Highlands, Nova Scotia: Geological Society of America Abstracts with Programs, v. 22, no. 2, p. 27–28.

Keppie, J. D., and Ortega-Gutierrez, F., 1995, Provenance of Mexican terranes: isotopic constraints: International Geology Reviews 37, p. 813–824.

Keppie, J. D., Nance, R. D., Murphy, J. B., and Dostal, J. 1991, The Avalon terrane, *in* Dallmeyer, R. D., and Lecorche, J. P. eds., The Western African orogens and Circum Atlantic correlatives, Springer-Verlag, p. 315–333.

Keppie, J. D., Dostal, J., Murphy, J. B., and Nance, R. D., 1996. Terrane transfer between eastern Laurentia and western Gondwana in the early Paleozoic: constraints on global reconstructions, *in* Nance, R. D., and Thompson, M. D. eds., Avalonian and related peri-Gondwanan terranes of the circum North Atlantic: Geological Society of America Special Paper 304,

p. 369–380.

Keppie, J. D., Dostal, J., Murphy, J. B., and Cousens, B. L., 1997, Paleozoic within-plate volcanic rocks in Nova Scotia (Canada) reinterpreted: isotopic constraints on magmatic source and paleocontinental reconstructions: Geological Magazine, v. 134, p. 425–437.

Keppie, J. D., Davis, D. W., and Krogh, T. E., 1998, U-Pb geochronological constraints on Precambrian stratified units in the Avalon composite terrane of Nova Scotia, Canada: Tectonic implications: Canadian Journal of Earth Sciences, v. 35, p. 222–236.

Krogh, T. E., Strong, D. F., O'Brien, S. J., and Papezik, V. S. 1988, Precise U-Pb dates from the Avalon terrane in Newfoundland: Canadian Journal of Earth Sciences, v. 25, p. 442–453.

Livaccari, R. F., Burke, K., and Sengör, A. M. C., 1981, Was the Laramide orogeny related to subduction of an oceanic plateau?: Nature, v. 289, p. 276–279.

Mancuso, C. I., Gates, A. E., and Puffer, J. H., 1996, Geochemical and petrologic evidence of Avalonian arc to rift transition from granitoids in southeastern Rhode Island, *in* Nance, R. D., and Thompson, M. D., eds., Avalonian and related peri-Gondwanan terranes of the circum North Atlantic. Geological Society of America Special Paper 304, p. 95–108.

Martignole, J., Machado, N., and Indares, A., 1994, The Wakeham terrane: a Mesoproterozoic terrestrial rift in the eastern part of the Grenville province: Precambrian Research, v. 68, p. 291–306.

Maxson, J., and Tikoff, B., 1996, Hit-and-run collision model for the Laramide orogeny, western United States: Geology, v. 24, p. 968–972.

McKerrow, W. S., Dewey, J. F., and Scotese, C. R., 1991, The Ordovician and Silurian development of the Iapetus Ocean: Special Papers in Paleontology, v. 44, p. 165–178.

Murphy, J. B., and Keppie, J. D., 1987, The stratigraphy of the late Precambrian Georgeville Group, Antigonish Highlands, Nova Scotia: Maritime Sediments and Atlantic Geology, Avalon Symposium volume, v. 23, p. 49–61.

Murphy, J B.. and MacDonald, D. A., 1993, Geochemistry of late Proterozoic arc-related volcaniclastic turbidite sequences, Antigonish Highlands, Nova Scotia: Canadian Journal of Earth Sciences, v. 30, p. 2273–2282.

Murphy, J. B, and Nance, R. D., 1989, A model for the evolution of the Avalonian-Cadomian belt, Geology v. 17, p. 735–738.

Murphy, J. B, and Nance, R. D. 1991, Supercontinent model for the contrasting character of Late Proterozoic orogenic belts, Geology v. 19, p. 469–472.

Murphy, J. B., Keppie, J. D., Dostal, J., and Hynes, A. J. 1990, Late Precambrian Georgeville Group: a volcanic arc rift succession in the Avalon terrane of Nova Scotia, *in* D'Lemos, R. D., Strachan, R. A., and Topley, C. G., eds., The Cadomian orogeny: Geological Society of London Special Publication 51, p. 383–393.

Murphy, J. B. Keppie, J. D. and Hynes, A. J. 1991, Geology of the Antigonish Highlands, Geological Survey of Canada Paper 89-10, 114 p.

Murphy, J B., Keppie, J D., Dostal, J., and Cousens, B L., 1996, Repeated lower crustal melting beneath the Antigonish Highlands, Avalon Composite Terrane, Nova Scotia: Nd isotopic evidence and tectonic implications, *in* Nance, R. D., and Thompson, M. D., eds., Avalonian and related peri-Gondwanan terranes of the circum North Atlantic. Geological Society of America Special Paper 304, p. 109–120.

Murphy, J. B., Keppie, J. D., Davis, D., and Krogh, T. E., 1997, Tectonic significance of new U-Pb age data for Neoproterozoic igneous units in the Avalonian rocks of northern mainland Nova Scotia, Canada: Geological Magazine, v. 134, p. 113–120.

Murphy, J. B., Anderson, A. J., and Archibald, D A., 1998, Alkali feldspar granite and associated pegmatites in an arc-wrench environment: the petrology of the Late Proterozoic Georgeville Pluton, Antigonish Highlands, Avalon Composite Terrane, Nova Scotia, Canada: Canadian Journal of Earth Sciences, v. 35, p. 110–120.

Nance, R. D., and Murphy, J. B. 1994, Contrasting basement signatures and the palinspastic restoration of peripheral orogens: example from the Neoproterozoic Avalonian-Cadomian belt: Geology, v. 22, p. 617–620.

Nance, R. D., and Murphy, J. B., 1996, Basement isotopic signatures and Neoproterozoic paleogeography of Avalonian-Cadomian and related terranes in the circum North Atlantic, *in* Nance, R. D., and Thompson, M. D., eds., Avalonian and related peri-Gondwanan terranes of the circum North Atlantic, Geological Society of America Special Paper 304, p. 333–346.

Nance, R. D., and Thompson, M. D. 1996, eds., Avalonian and related peri-Gondwanan terranes of the circum North Atlantic: Geological Society of America Special Paper 304, 390 p.

Nance, R. D., Murphy, J. B., Strachan, R. A., D'Lemos, R. D., and Taylor, G. K., 1991, Tectonostratigraphic evolution of the Avalonian-Cadomian belt and the breakup of a late Precambrian supercontinent: an interpretive review, *in* Stern, R. J., and Kroner, A., eds., Tectonic regimes and crustal evolution in the Late Proterozoic: Precambrian Research, v. 53, p. 41–78.

Noble, S. R., and Tucker, R. D., 1992, Late Ordovician closure of the Tornquist sea: isotopic evidence from SE England: Geological Association of Canada: Mineralogical Association of Canada Program with Abstracts, v. 17, p. A85.

O'Brien, S. J., Wardle, R. J., and King, A. F., 1983, The Avalon zone: a Pan-African terrane in the Appalachian orogen of Canada: Geological Journal, v. 18, p. 195–222.

O'Brien, S. J., Strong, D. F., and King, A. F. 1990. The Avalon zone type area: southeastern Newfoundland Appalachians, *in* Strachan, R. A., and Taylor, G. K. eds., Avalonian and Cadomian geology of the North Atlantic: Glasgow and London, Blackie, p. 166–194.

O'Brien, S. J., O'Driscoll, C. F., and Tucker, R. D., 1992, A reinterpretation of the geology of parts of the Hermitage Peninsula, southwestern Avalon zone, Newfoundland: Newfoundland Department of Mines and Energy Geological Survey Branch Report 92-1, p. 185–194.

O'Brien, S. J., O'Brien, B. H., Dunning, G. R., and Tucker, R. D., 1996, Late Proterozoic Avalonian and related peri-Gondwanan rocks of the Newfoundland Appalachians, *in* Nance, R. D., and Thompson, M. D., eds., Avalonian and related peri-Gondwanan terranes of the circum North Atlantic. Geological Society of America Special Paper 304, p. 9–28.

Pe-Piper, G., and Piper, D. J. W., 1987, The pre-Carboniferous rocks of the Cobequid Hills, Avalon zone, Nova Scotia: Maritime Sediments and Atlantic Geology, v. 23, p. 41–49.

Pe-Piper, G., and Piper, D. J. W., 1989, The Late Hadrynian Jeffers Group, Cobequid Highlands, Avalon zone of Nova Scotia: a back arc volcanic complex: Geological Society of America Bulletin, v. 101, p. 364–376.

Pe-Piper, G., Piper, D. J. W., and Koukouvelas, I., 1996, Precambrian plutons of the Cobequid Highlands, Nova Scotia, Canada, *in* Nance, R. D., and Thompson, M. D., eds., Avalonian and related peri-Gondwanan terranes of the circum North Atlantic, Geological Society of America Special Paper 304, p. 121–132.

Rabu, D., Guerrot, C., Tegyey, M., Murphy, J. B., Keppie, J. D., 1996, St. Pierre–Micquelon: a new piece in the Avalonian puzzle, *in* Nance, R. D., and Thompson, M. D., eds., Avalonian and related peri-Gondwanan terranes of the North Atlantic. Geological Society of America Special Paper 304, p. 66–94.

Sadowski, G. R., and Bettencourt, J. S., 1996, Mesoproterozoic tectonic correlations between eastern Laurentia and the western border of the Amazon craton: Precambrian Research, v. 76, p. 213–227.

Severinghaus, J., and Atwater, T., 1990, Cenozoic geometry and thermal state of the subducting slabs beneath North America, *in* Wernicke, B. P., ed., Basin and Range extensional tectonics near the latitude of Las Vegas, Nevada, Geological Society of America Memoir 176, p. 1–22.

Sinha, A. K., Hogan, J .P., and Parks, J., 1996, Lead isotope mapping of crustal reservoirs within the Grenville superterrane: I. central and southern Appalachians, *in* Basu, A., and Hart, S. R., eds., Earth processes: reading the isotopic code: American Geophysical Union monograph 95, p. 293–305.

Starmer, I. C., 1993, The Sveconorwegian Orogeny in southern Norway, relative to deep crustal structures and events in the North Atlantic Proterozoic supercontinent: Norwegian Geol. Tidsskr., v. 73, p. 109–132.

Strachan, R. A., Roach, R. A., and Treloar, P. J., 1990, Cadomian terranes in the North Armorican massif, *in* Strachan, R. A., and Taylor, G. K., eds., Avalonian and Cadomian geology of the North Atlantic: Glasgow and London, Blackie, p. 65–92.

Strachan, R. A., Nance, R. D., Dallmeyer, R. D., D'Lemos, R. S., and Murphy, J. B., 1996a, Late Precambrian tectonothermal evolution of the Malverns Complex, U.K.: Journal of the Geological Society of London, v. 153, p. 589–600.

Strachan, R. A., D'Lemos, R. D., and Dallmeyer, R. D., 1996b, Neoproterozoic evolution of an active plate margin, *in* Nance, R. D., and Thompson, M. D., eds., Avalonian and related peri-Gondwanan terranes of the circum North Atlantic, Geological Society of America Special Paper 304, p. 333–346.

Strong, D. F., O'Brien, S. J., Strong, P. G., Taylor, S. W., and Wilton, D. H. C., 1978, Aborted Proterozoic rifting in Newfoundland: Canadian Journal of Earth Sciences, v. 15, p. 117–131.

Sun, S.-S., and McDonough, W. F., 1989, Chemical and isotopic systematics of ocean basalts: implications for mantle composition and processes, *in* Saunders, A. D., and Tarney, J., eds., Magmatism in the ocean basins: Geological Society of London Special Publication 42, p. 313–345.

Swinden S., and Hunt, P. A., 1991, A U-Pb age from the Connaigre Bay Group, southwestern Newfoundland: implications for regional correlations and metallogenesis. Radiometric age and isotopic studies: Report 4, Geological Survey of Canada, Paper 90-2, p. 3–10.

Texiera, W., Tassinari, C. C. G., Cordani, U. G., and Kawashita, K., 1989, A review of the geochronology of the Amazon craton: tectonic implications: Precambrian Research, v. 42, p. 213–227.

Thicke, M. J.,1988, The geology of Late Hadrynian metavolcanic and granitoid rocks of the Coxheath Hills-northeast East Bay Hills areas, Cape Breton Island, Nova Scotia: [M.Sc. thesis]: Wolfville, Nova Scotia, Canada, Acadia University.

Thompson M. D., Hermes, O. D., Bowring, S. A., Isachsen, C. E., Besancon, J. R., and Kelly, K. L., 1996, Tectonostratigraphic implications of Late U-Pb zircon ages in the Avalon Zone of southeastern New England, *in* Nance, R. D., and Thompson, M. D., eds., Avalonian and related peri-Gondwanan terranes of the circum North Atlantic, Geological Society of America Special Paper 304 , p. 333–346.

Tosdal, R. M., Munizaga, F., Williams, W. C., and Bettencourt, J. S., 1994, Middle Proterozoic crystalline basement in the central Andes, western Bolivia and northern Chile: a U-Pb and Pb isotope perspective: 7 Congreso Geologico Chileno 1994, Universidad de Concepcionm Actas v. II, p. 1464–1467.

Tuach, J., 1991, Geology and geochemistry of the Cross Hills plutonic suite, Fortune Bay, Newfoundland (1M/10): Newfoundland Department of Mines and Energy, Geological Survey Branch Report 91-2, 73 p.

Tucker, R. D., and Pharoah, T. C., 1991, U-Pb zircon ages for Late Precambrian igneous rocks in southern Britain: Geological Society of London Journal, v. 148, p. 435–443.

Van der Voo, R., 1988, Paleozoic paleogeography of North America, Gondwana, and intervening displaced terranes: comparisons of paleomagnetism within paleoclimatology and biogeographical patterns: Geological Society of America Bulletin, v. 100, p. 311–324.

Van Staal, C. R., Sullivan, R. W., and Whalen, J. B., 1996, Provenance and tectonic history of the Gander zone in the Caledonide-Appalachian orogen: implications for the assembly of Avalon, *in* Nance, R. D., and Thompson, M. D., eds., Avalonian and related peri-Gondwanan terranes of the circum-Atlantic: Geological Society of America Special Paper 304, p. 347–368.

Whalen, J. B., Jenner, G. A., Currie, K. L., Barr, S. M., Longstaffe, F. J., and Hegner, E., 1994, Geochemical and isotopic characteristics of granitoids of the Avalon zone, southern New Brunswick: possible evidence of repeated delamination events: Journal of Geology, v. 102, p. 269–82.

Williams, H., 1979, The Appalachian orogen in Canada: Canadian Journal of Earth Sciences, v. 16, p. 792–807.

Zartman, R. E., 1969, Lead isotopes in igneous rocks of the Grenville Province as a possible clue to the presence of older crust: Geological Association of Canada Special Paper 5, p. 193–205.

Zhong, S., and Gurnis, M., 1995, Mantle convection with plates and mobile, faulted plate margins: Science, v. 267, p. 838–843.

Manuscript Accepted by the Society September 4, 1998

Printed in U.S.A.

Geological Society of America
Special Paper 336
1999

Odyssey of terranes in the Iapetus and Rheic oceans during the Paleozoic

J. Duncan Keppie
*Instituto de Geología, Universidad Nacional Autónoma de México,
Ciudad Universitaria, Delegacion Coyoacan, 04510 México D.F., México*
Victor A. Ramos
*Universidad de Buenos Aires, Departamento de Ciencias Geológicas, Ciudad Universitaria,
Pabellon II, (1428) Nuñez-Buenos Aires, Argentina*

ABSTRACT

Geologic, paleomagnetic, and faunal data indicate that Paleozoic terranes bordering the Iapetus and Rheic oceans may be classified as native or exotic with respect to adjacent cratons. Native terranes include the Notre Dame–Shelburne Falls arc lying along the eastern margin of Laurentia, Cadomia situated adjacent to North Africa, and the Famatina, Puna and Arequipa-Antofalla Terranes occurring adjacent to western South America. Exotic terranes include the western South American terranes, Cuyania and Chilenia, which are inferred to have been derived from southern Laurentia; and the North American terranes, Oaxaquia, Florida, Carolina, Avalonia, and associated terranes, which have a Gondwanan provenance in northern South America–West Africa. Early Cambrian opening of Iapetus was followed in Late Cambrian–Early Ordovician by extensional subduction that produced volcanic arcs and backarc basins on both margins (Notre Dame–Shelburne Falls and Famatinian arcs), which changed to compressional arcs collapsing the backarc basins during the mid-Late Ordovician. During the Ordovician, Cuyania was transferred from southern Laurentia to western Gondwana, and Carolina/Piedmont and Avalonia moved from northwestern Gondwana to eastern Laurentia. This exchange indicates that the southern part of Iapetus was preferentially subducted beneath South America as the central part was mainly subducted beneath eastern Laurentia, and suggests that southern and central Iapetus were separated by a major transform fault. After the Late Ordovician demise of Iapetus, the Rheic ocean persisted between Laurentia and Gondwana, and terranes appear to have remained adjacent to their neighboring craton until the terminal amalgamation of Pangea in the Permo-Carboniferous. During the Mesozoic break-up of Pangea, the Oaxaquia, Chortis, Florida, and Cadomia Terranes were left adjacent to Laurentia and Baltica.

INTRODUCTION

This chapter attempts to depict the movement of terranes, especially those of exotic provenance, in the Iapetus and Rheic oceans during the Paleozoic using the recent global reconstructions of Dalziel (1995, 1997) as base maps (Figs. 1–9). The paleomagnetic data used to produce these base maps is given in Dalziel (1995, 1997). Alternative global maps produced by several other workers, such as van der Voo (1993) and Golonka et al. (1995), using the same paleomagnetic data generally place the major continents at similar paleolatitudes but at different paleolongitudes. This difference results in variations in the widths of the various oceans. Thus, the wide Ordovician Iapetus proposed by van der Voo (1993) and Golonka et al. (1995) contrasts with the narrow Iapetus between eastern Laurentia and western South America depicted by Dalziel (1997). Given the exchange of

Keppie, J. D., and Ramos, V. A., 1999, Odyssey of terranes in the Iapetus and Rheic oceans during the Paleozoic, *in* Ramos, V. A., and Keppie, J. D., eds., Laurentia-Gondwana Connections before Pangea: Boulder, Colorado, Geological Society of America Special Paper 336.

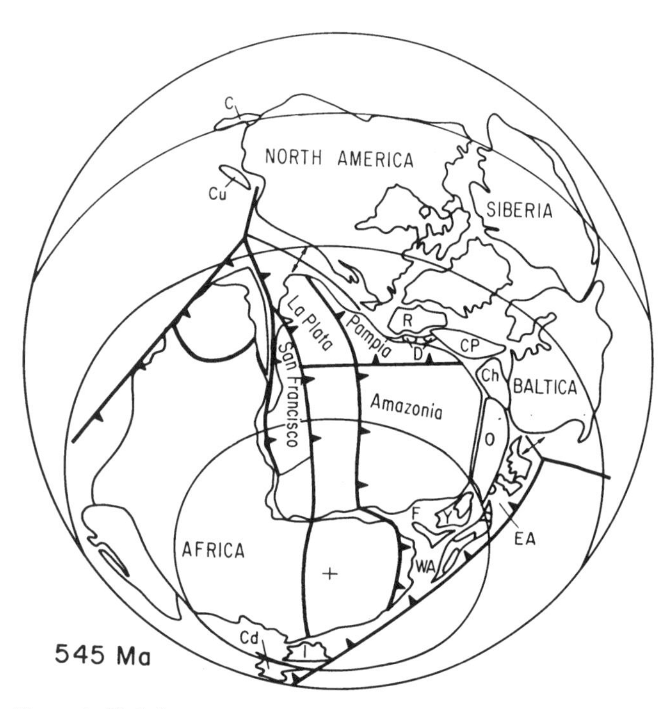

Figure 1. Global reconstruction for the Precambrian-Cambrian boundary (~545 Ma). C = Chilenia; Cd = Cadomia; Ch = Chortis; CP = Carolina, Piedmont, and Goochland; Cu = Cuyania; D = Dalradia; EA = East Avalonia; I = Iberia; F = Florida; O = Oaxaquia; R = Rockall (a promontory of eastern Laurentia; Dalziel, 1994); WA = West Avalonia; Y = Yucatan.

terranes between Laurentia and South America in the Ordovician, we prefer the latter reconstruction. Similarly, van der Voo (1993) and Golonka et al. (1995) collide North and South America/Africa in the Devonian, whereas Dalziel (1995) prefers a wide Rheic ocean. In this case, the occurrence of the same faunal province in Laurentia and northern South America would favor the former (Keppie et al., 1996). However, for consistency in the preparation of the base maps, we used the series produced for us by Lisa Gahagan and Ian Dalziel, because the main purpose of this chapter and of the IGCP (International Geological Correlation Program) Project 376, Laurentia-Gondwana Connections before Pangea, is to determine the provenence of terranes presently located in eastern-southern Laurentia and western-northern Gondwana.

To date, most of the work has concentrated on defining individual terranes and attempting to determine their provenance, e.g., Avalonia and the Precordillera (Keppie et al., 1998; Astini et al., 1995). In certain situations, these different workers show different terranes being derived from the same location. The first attempt to palinspastically locate many of these terranes showed their locations through time but not their dimensions (Keppie et al., 1996). This chapter shows both their locations and dimensions relative to the major cratons (Figs. 1–9).

Brief summaries of the geologic records of the various terranes are given below, including evidence for provenance and times of accretion. They may be grouped into native terranes, which presently lie close to their latest Precambrian location, and exotic terranes, which have been derived from a location a great distance from their present position. Native terranes include the Notre Dame–Shelburne Falls arc lying along the eastern margin of Laurentia, Cadomia adjacent to North Africa, and Famatina, Puna, and Arequipa-Antofalla adjacent to western South America. Exotic terranes include the western South America terranes Cuyania and Chilenia, which are inferred to have been derived from southern Laurentia; and the North American terranes Oaxaquia, Florida, Carolina, Avalonia, and other associated terranes, which have Gondwanan provenance.

TERRANES

Dalradia

Dalradia extends across the southern Highlands of Scotland and northern Ireland and consists of Neoproterozoic rocks that were deformed and metamorphosed twice: first, prior to intrusion of a 590 ± 2-Ma granite, and subsequently, in the Early Ordovician associated with accretion to the Laurentian margin (Bluck and Dempster, 1991). The absence of a Neoproterozoic tectonothermal event in the adjacent Laurentian margin led Bluck and Dempster (1991) to suggest a Gondwanan provenance for Dalradia, which Dalziel (1994) suggested lay in the Arica embayment. On the other hand, Soper (1994) concluded that Dalradia is part of the Laurentian passive margin.

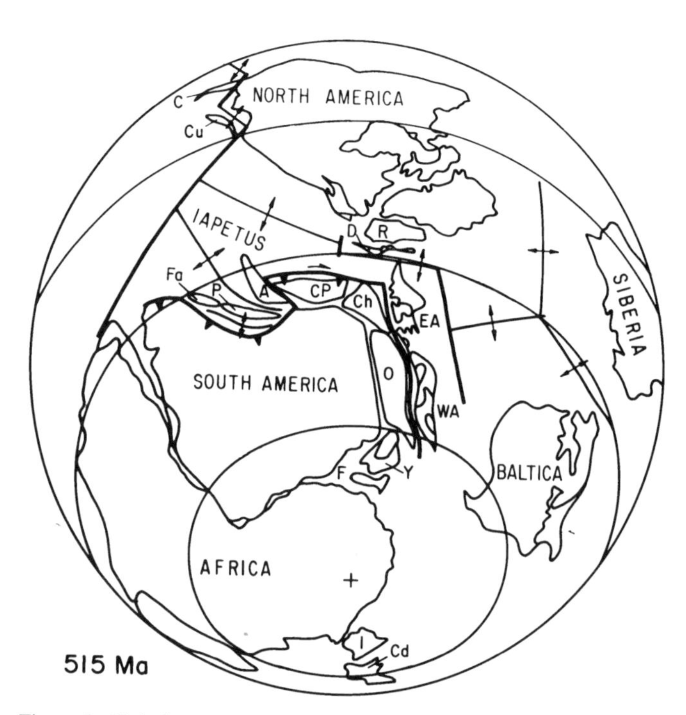

Figure 2. Global reconstruction for the mid-Late Cambrian (~515 Ma). A = Arequipa-Antofalla; Fa = Famatina; P = Puna (for other abbreviations, see Fig. 1).

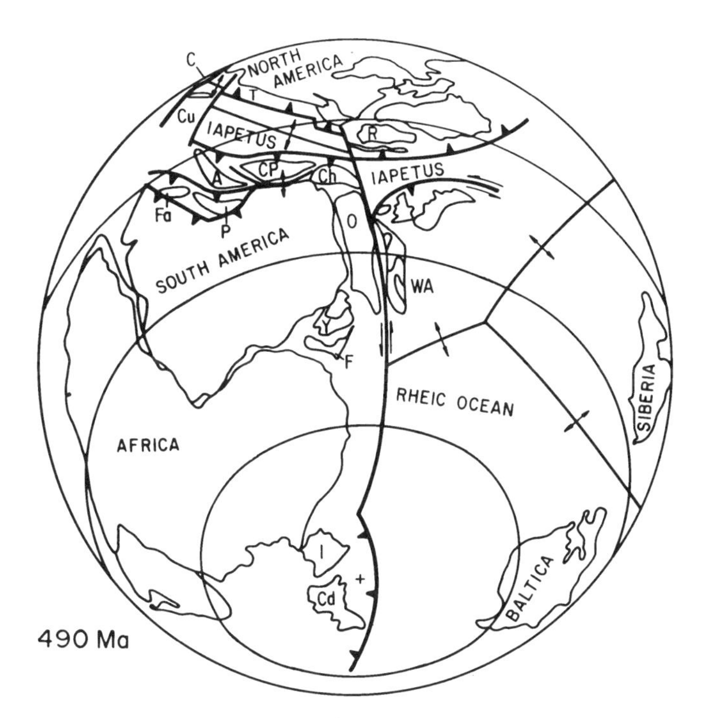

Figure 3. Global reconstruction for the Early Ordovician (Tremadoc-Arenig, ~490 Ma). For abbreviations, see Figures 1 and 2. T = Notre Dame–Shelburne Falls arc terrane.

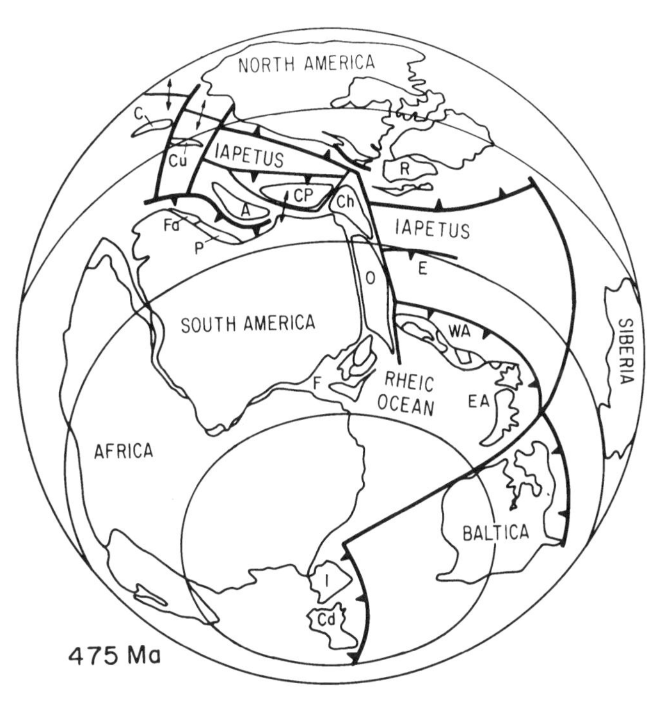

Figure 4. Global reconstruction for the Early-Middle Ordovician (~475 Ma). For abbreviations, see Figures 1–3. E = Exploits terrance.

Cadomia (Iberian, Armorican and Bohemian massifs)

Cadomia includes Neoproterozoic Iberian, Armorican, and Bohemian Massifs in western-central Europe. It is underlain by a ca. 2-Ga basement overlain by ~700–540 Ma calc-alkaline magmatic arc complexes and periarc basins (Chantraine et al., 1994; Chaloupsky et al., 1995; Strachan et al., 1996). These units are unconformably overlain by lower Paleozoic sedimentary and volcanic rocks that contain Cambrian Tethyan faunas and Archaeocyathids, overlain by the distinctive Ordovician Armorican quartzite, common to the northern Gondwanan margin of North Africa (Doré, 1994; Robardet et al., 1994; Chaloupsky et al., 1995). Paleomagnetic data indicate that these massifs were carried with the northern margin of Africa from high paleolatitudes in the Ordovician to subequatorial positions during the Carboniferous (Bachtadse et al., 1995), where they collided with Laurentia/Avalonia and Baltica to produce the Variscan orogen.

Notre Dame–Shelburne Falls Arc

During the Ordovician, subduction along Laurentian margin of Iapetus produced the Notre Dame–Shelburne Falls arc that stretched from northern Ireland through the Notre Dame terrane of Atlantic Canada into the Shelburne Falls arc in New England (Karabinos et al., 1998) and possibly into the southern Appalachians. Its peri-Laurentian location is consistent with paleomagnetic data (Mac Niocaill et al., 1997). Collision of this arc with Laurentia during the Ordovician produced the Taconian orogeny (Keppie, 1993).

Avalonia (+ Meguma, Gander/Wexford/Lake District, and Exploits)

Paleozoic Avalonia extends along the eastern side of the northern Appalachians and into the southern Caledonides. It is characterized by an uppermost Neoproterozoic–lower Paleozoic overstep sequence containing a unique Cambrian Avalonian fauna and an Upper Silurian Rhenish-Bohemian fauna (Keppie et al., 1996, 1997; Murphy et al., this volume). This overstep sequence is underlain by Neoproterozoic miogeoclinal rocks (limited to the western portion of the terrane) overlain by a magmatic arc sequence that may be subdivided into a fragmentary, early extensional arc stage (780–630 Ma), a main extensional arc phase (630–610 Ma), a late phase that shows a diachronous transition from a transpressional arc to a transtensional intracontinental environment (610–550 Ma). Polarity of subduction from 700 to 550 Ma was toward the northwest (present coordinates) in western Avalonia, which implies that a sizable ocean existed to the southeast (Keppie et al., 1998).

Avalonia is bordered by Cambro-Ordovician continental rise prisms: Meguma continental rise bordering the Rheic Ocean and Gander/Wexford/Lake District continental rise on the Iapetus margin bordered by a Cambro-Ordovician arc (Exploits arc), outboard of which lay an intra-Iapetan oceanic arc bearing the Celtic island fauna (Exploits terrane) (Keppie et al., 1996, 1997, 1998; van Staal et al., 1996). Neoproterozoic-Cambrian paleomagnetic data, Nd isotopic data, and Cambrian

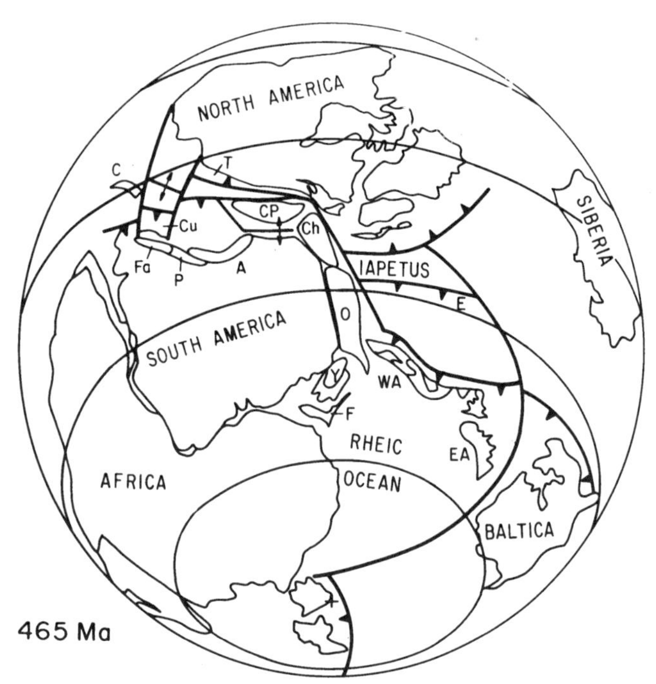

Figure 5. Global reconstruction for the Middle Ordovician (~465 Ma). For abbreviations, see Figures 1–4.

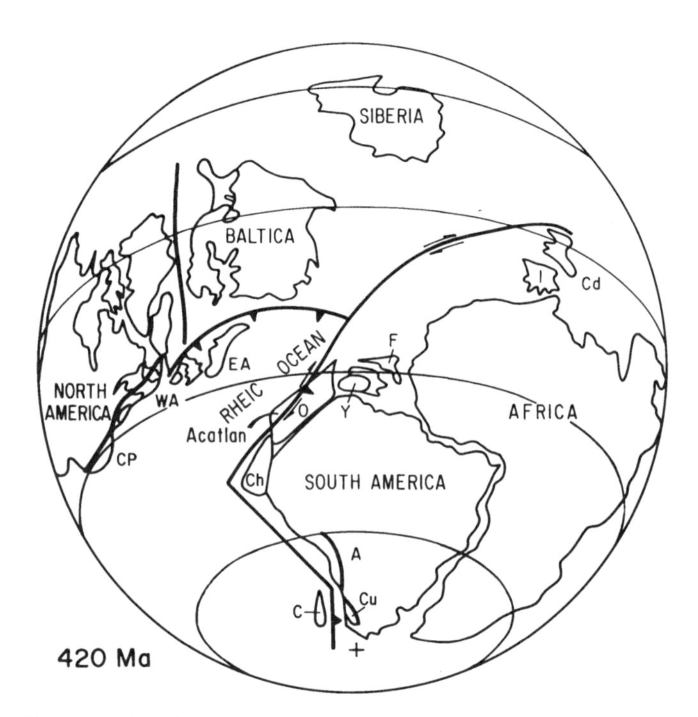

Figure 6. Global reconstruction for the mid-Late Silurian (~420 Ma). For abbreviations, see Figures 1 and 2.

faunal affinities in Avalonia suggest Gondwanan connections (Theokritoff, 1979; van der Voo, 1993; Keppie et al., 1997). Detrital zircons in Avalonia, Gander, and Exploits Terranes indicate sources in Amazonia, whereas those in the Meguma Terrane show provenance from northwest Africa (Keppie et al., 1998). Paleomagnetic data are consistent with a Neoproterozoic-Cambrian location off northwestern Gondwana (van der Voo, 1993). Such a provenance appears to be inconsistent with a large, long-lived, Neoproterozoic ocean existing southeast of western Avalonia (present coordinates), and has led to the inference that it has been rotated through 180° during the early Paleozoic (Keppie et al., 1998). Such a rotated reconstruction would place Avalonia adjacent to Amazonia with the inferred large ocean lying to the north (present coordinates).

Late Ordovician–Early Silurian accretion of Avalonia to Laurentia is indicated by shared deformation (Keppie, 1993) followed by the presence of a Llandovery overstep sequence extending from Laurentia across the Appalachian central mobile belt onto Avalonia and the Meguma Terrane (Chandler et al., 1987), and the first appearance of Laurentian-Baltic geochemical signatures in Lower Silurian sediments of Avalonia (Murphy et al., 1996). These data are consistent with paleomagnetic data that indicate the gradual convergence of Laurentia and Avalonia throughout the Ordovician (Mac Niocaill et al., 1997).

Carolina (+ Goochland and Piedmont Terranes)

The Carolina Terrane is characterized as a composite arc consisting of (1) an older, ~630–580-Ma juvenile oceanic arc sequence that is inferred to have been deformed in the Virgilina orogeny before deposition of (2) a younger, ~580–540-Ma mature arc sequence that is, in turn, overlain by (3) Middle Cambrian sedimentary rocks containing cool-water trilobites similar to those from Armorica and Baltica (Samson et al., 1990; Hibbard and Samson, 1995). The Goochland Terrane is represented by a ~1-Ga granulite facies metamorphic sequence that has been inferred to be a part of both Laurentian Grenville and an exotic terrane (Glover, 1989; Rankin et al., 1989). Hibbard and Samson (1995) have suggested that the switch from juvenile to mature isotopic arc signature is related to underthrusting of the Goochland Terrane beneath the Carolina Terrane, although the two Carolina arc sequences may represent distinct terranes (J. P. Hibbard, written communication, 1998). The Piedmont Terrane consists of a Cambro-Ordovician composite volcanic arc, forearc, and accretionary complex that appears to record a latest Neoproterozoic-Cambrian (~530–510 Ma) Potomac Orogeny (Hibbard and Samson, 1995). The time of amalgamation of the Piedmont and Carolina Terranes is uncertain, and rocks of the Piedmont Terrane have been correlated with both Laurentia and Gondwana. However, correlatives of the Potomac Orogeny are relatively rare and may be found only in the Pampean Orogeny of western South America and the Ross-Delamerian Orogeny in Antarctica and Australia, thus favoring a peri-Gondwanan origin. Thus, the Piedmont is inferred to be a continuation of the Pampean orogenic belt. Accretion of the Carolina terrane to Laurentia has been related to both the Middle Ordovician Blountian orogeny and to the Carboniferous Alleghanian orogeny (Hatcher, 1989).

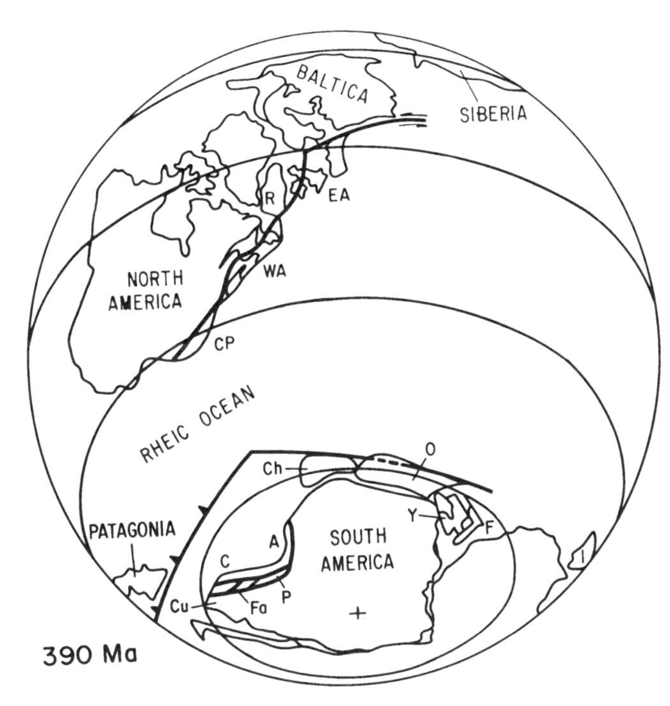

Figure 7. Global reconstruction for the Early-Middle Devonian (~390 Ma). For abbreviations, see Figures 1 and 2.

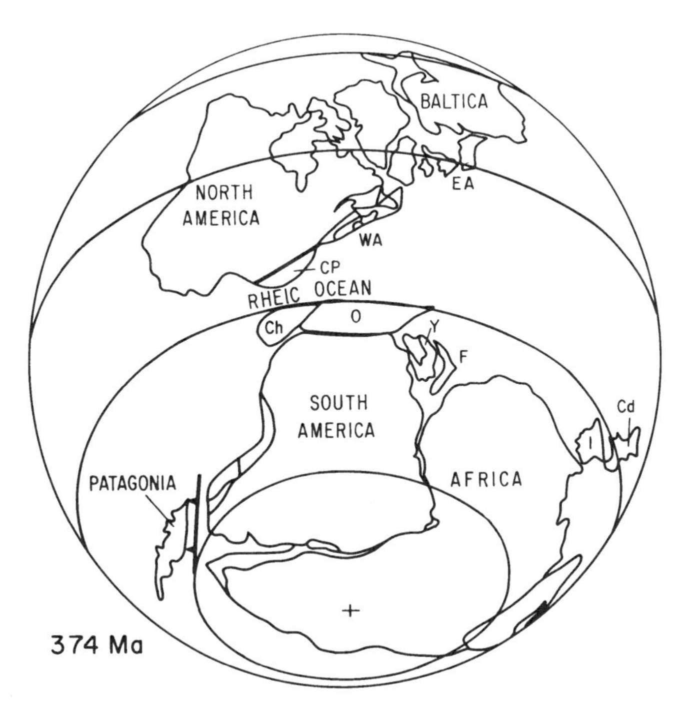

Figure 8. Global reconstruction for the mid-Late Devonian (~374 Ma). For abbreviations, see Figures 1 and 2.

Florida

Peninsular Florida is underlain by a late Neoproterozoic-Cambrian (535–510 Ma) basement unconformably overlain by undeformed Ordovician-Devonian sedimentary rocks containing high-latitude trilobite and acritarch fauna of Gondwanan affinity, which have generally been correlated with the Bove basin of West Africa (Whittington and Hughes, 1974; Cramer and Diez, 1974; Dallmeyer, 1987; Dallmeyer et al., 1987). This assignment is consistent with paleomagnetic data (Opdyke et al., 1987). It is inferred to have been accreted to Laurentia in the Permo-Carboniferous during final amalgamation of Pangea.

Yucatan

Zircons from plutonic rocks recovered from boreholes in the Yucatan Peninsula have yielded late Neoproterozoic ages (Krogh et al., 1993), whereas Late Silurian ages were given by plutonic rocks in the Maya Mountains (Steiner and Walker, 1996). In the southern part of the Yucatan block, metamorphic rocks of the undated Chaucus Group are intruded by late Paleozoic and younger granitoids (Sedlock et al., 1993). These units are inferred to be unconformably overlain by Carboniferous and Permian volcanic and sedimentary rocks. The Neoproterozoic plutons in the Yucatan block suggest affinities with the Pan-African and Brasiliano orogens of Gondwana. The Yucatan block is inferred to have been caught in the collision between North and South America. Paleomagnetic data from the Chiapas Massif indicate a ~60° anticlockwise rotation of the Yucatan block during Late Triassic–Middle Jurassic opening of the Gulf of Mexico (Molina-Garza et al., 1992).

Oaxaquia (+ Chortis and Acatlan)

The backbone of Mexico is underlain by ~1-Ga granulites of the Oaxacan Complex. Comparison of lithologic, Pb isotope, and P-T-t data suggest that the Oaxacan Complex originated between eastern Laurentia (Adirondacks) and northwestern South America (Keppie and Ortega-Gutiérrez, 1995, this volume). Paleomagnetic data from the Oaxacan Complex are consistent with derivation from eastern Laurentia (in the vicinity of the Adirondacks and Ontario) and northwestern Gondwana (Ballard et al., 1989). The Oaxacan Complex is unconformably overlain by latest Cambrian–earliest Ordovician platformal rocks containing trilobites of Gondwanan (Bolivian) affinity (Robison and Pantoja-Alor, 1968; Keppie and Ortega-Gutiérrez, this volume). These rocks are overlain by Carboniferous and ?Permian shallow marine-continental sedimentary rocks. Reports of ~1-Ga basement in the Chortis block suggest it represents a southern continuation of Oaxaquia (Manton, 1996).

The western side of Oaxaquia is occupied by an ophiolitic sequence (Acatlan Complex) of unknown age, which was subducted and metamorphosed to eclogite facies during the Silurian, and dextrally transported in the late Paleozoic (Ortega-Gutiérrez et al., 1997). Like the Yucatan, Oaxaquia (including Chortis and Acatlan) is inferred to have been trapped between North and South America during the formation of Pangea, and to have been left on the Laurentian margin during Mesozoic separation.

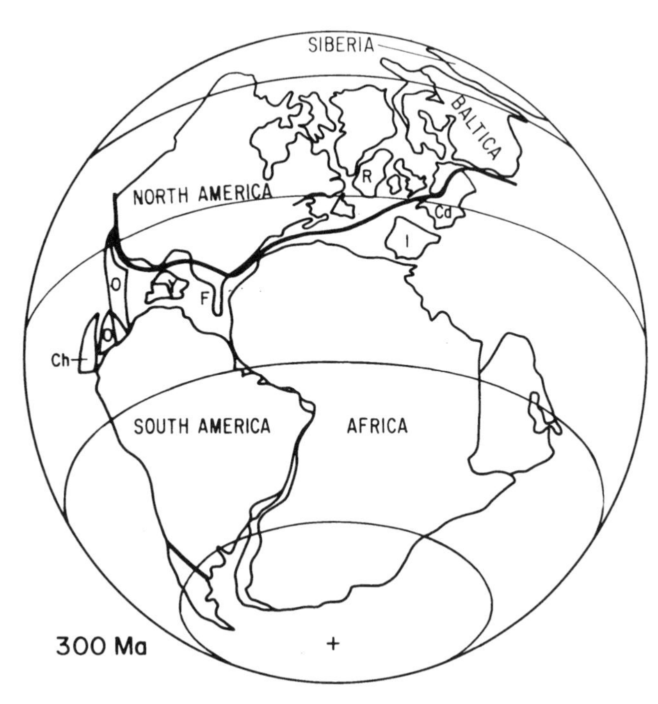

Figure 9. Global reconstruction for the Carboniferous-Permian boundary (~300 Ma). For abbreviations, see Figures 1 and 2.

Arequipa-Antofalla

The basement of the exotic Arequipa-Antofalla Terrane of Ramos (1988) has yielded ages of ~2 and ~1 Ga in the Peruvian segment and in parts of northern Chile (Shackelton et al., 1979, Wasteneys et al., 1995, Damm et al., 1990). Analysis of the metamorphic facies in this basement in northern Chile and northwestern Argentina showed an additional metamorphic peak at ~500 Ma, which, combined with the homogeneity of neodymium model ages, led Lucassen et al. (1996) to correlate this basement with typical Gondwanan basement in the Sierras Pampeanas. The early Paleozoic history of northwestern Argentina and northern Chile shows that the basement of Arequipa-Antofalla underwent partial remobilization during Ordovician times, when a magmatic arc was developed in these rocks (Coira et al., 1982; Ramos, 1988). The subduction-related rocks gave way to collisional suites at ~450 Ma (Davidson et al., 1983). Deformation starting in the Middle Ordovician and final collision at the end of the Ordovician was postulated by Bahlburg and Hervé (1997). Available paleomagnetic data for this terrane indicate considerable rotation relative to adjacent South America but at similar paleolatitudes in Early Ordovician times, followed by Middle Ordovician accretion to the Puna and other peri-Gondwanan terranes (Forsythe et al., 1989).

Puna

The Puna Terrane consists of an Early Ordovician arc-backarc overlain by a Middle Ordovician foreland basin containing Celtic-Gondwanan fauna that was deformed by the Late Ordovician Oclóyic orogeny (Ramos, 1996; Bahlburg and Hervé, 1997). Time of deformation in the Puna is younger than the 476–467-Ma granitoids (Bahlburg and Hervé, 1997), and it is assumed to be ~460 Ma. Preliminary Early Ordovician paleomagnetic data indicate slight separation of the Puna from western South America, but at similar paleolatitudes to Avalonia (Conti et al., 1996). The lead isotopic data of large basement xenoliths in Cenozoic volcanic rocks have Gondwanan signatures distinct from those of typical Laurentian basement (Coira and Caffe, 1995).

Famatina

The Famatina Terrane consists of late Proterozoic–(?)Early Cambrian turbiditic deposits overlain by Cambro-Ordovician volcanic arc lavas (Turner, 1960; Mannheim, 1993; Toselli et al., 1996; Aceñolaza et al., 1996). Arenigian and Llanvirnian sedimentary rocks contain an abundant fauna of trilobites and brachiopods with Gondwanan affinities (Mángano and Buatois, 1996). These sequences are intruded by a series of subduction-related granitoids (Toselli et al., 1996) and postcollisional leucogranites (Mannheim, 1993). Ordovician paleomagnetic data indicate that the Famatina Terrane lay at paleolatitudes close to 30° (Conti et al., 1996), similar to the paleolatitudes obtained in the Exploits arc by van der Voo et al. (1991). Collision and accretion of this terrane to western Gondwana is inferred to have occurred at ~450 Ma, based on the change between subduction-related igneous rocks and postcollisional granites.

Cuyania

This composite terrane formed by the amalgamation of the basement of present Precordillera and Pie de Palo Terranes. Both basements are characterized by juvenile crust of Grenvillian age, amalgamated at ~1,050–950 Ma, during an episode of island arc collision (Vujovich and Ramos, 1993; Ramos et al., 1996). Isotopic signatures and geochronology of the Precordilleran basement are exotic and distinct to surrounding Gondwanan terranes, but are typical of present southeastern Laurentia, such as the Llano uplift of Texas (Kay et al., 1996). The thick carbonate platform deposits of Cambrian-Ordovician age of the Precordillera were inferred to be exotic to the protomargin of Gondwana, which is dominated by clastic platforms (Ramos et al., 1986). The rift-drift transition of these carbonate rocks, known since the early work of Bond et al. (1984), is completely different from the rest of the early Paleozoic Gondwana margin (Ramos, 1997). Cambrian–Early Ordovician faunal affinities, such as in trilobites, characteristic of inner shelf facies, show unique and strong linkages with Laurentia (Dalziel et al. 1996). Middle to Late Cambrian agnostids in present western Precordillera indicate an open sea and suggest an early separation between Laurentia and Precordillera (Bordonaro and Liñán, 1994). Middle to Late Ordovician faunas have switched from Laurentian to Gondwanan affinities (Astini, 1996). Paleoclimatic indicators record an origin of Precordillera in subtropical environments in the Cambrian, shifting to subglacial conditions by the end of the Ordovician (Astini et al., 1996). Paleomagnetic data on the Early Cambrian rocks of the Precordillera (Rapalini et al., 1998, this volume) indicate paleolatitudes of approximate 20°, similar to the Ouachita embayment of Laurentia, as inferred by Thomas and Astini (1996).

Time of docking is constrained by the age of the Middle-Late Ordovician foreland basin deposits and the deformation of Western Sierras Pampeanas (Astini et al., 1996; Ramos et al., 1996). The 565 ± 45-Ma U-Pb zircon age of ophiolitic rocks along the western margin of the Precordillera suggests that Cuyania separated as a microcontinent from Laurentia in the latest Neoproterozoic or Cambrian (Davis et al., 1997), rather than the "Texas plateau" model proposed by Dalziel (1997).

Chilenia

The basement of this large terrane underlies the main Andes of Argentina and Chile, and it is exposed in a series of minor erosional windows or it is preserved as roof pendants in some of the large granitic batholiths (Ramos et al., 1986). The large amount of upper Paleozoic granitoids developed at both slopes of the Andes of Argentina and Chile was taken as evidence for the continental nature of the basement of Chilenia by Nasi et al. (1985); this interpretation was later confirmed by the isotopic ratio of the granites (Mpodozis and Kay, 1992). The early Paleozoic cover is scarce and different from Precordillera, although isolated patches of Silurian carbonates are present. The microflora described in these rocks have no clear provincialism that may be linked either with Laurentia or Gondwana. On the other hand, the new 1,069 ± 36-Ma age obtained for the Chilenia Terrane (Ramos and Basei, 1997), combined with the absence of Brasiliano deformation, suggests Laurentian affinities for the Chilenia Terrane. Accretion to Gondwana is postulated during Middle-Late Devonian times (Ramos et al., 1986) or Late Silurian–Early Devonian times, based on the age of deformation and metamorphism along the suture. Final amalgamation occurred in Early Carboniferous times, where continental to shallow marine deposits overstep both terranes.

Reconstructions

Using paleomagnetic data, faunal provinciality, times of accretion, and the geologic records of the various terranes, an attempt has been made to locate the terranes on the series of Paleozoic global reconstructions by Dalziel (1995, 1997) (Figs. 1–9). This is an iterative process between the different time frames because a definitive location on one time frame influences its position on other time frames. For example, the transfer of Avalonia and Carolina from Gondwana to Laurentia, and Cuyania from southern Laurentia to western South America during the Ordovician, suggests proximity between northwestern South America and eastern Laurentia at this time.

CONCLUSIONS

Examination of the global reconstructions (Figs. 1–9) reveals several different geographically distinct domains: (1) an exotic southern domain in central Argentina and Chile in which the Cuyania and Chilenia Terranes traveled from southern Laurentia to southwestern Gondwana during the Ordovician, Silurian, and possibly Devonian; (2) a native central domain in northern Argentina and Peru, where the Famatina, Puna, and Arequipa-Antofalla Terranes originated in adjacent South America to be separated slightly in the Cambrian–Early Ordovician, only to be reaccreted during the mid–Late Ordovician; (3) a southern North America exotic domain where Gondwanan terranes (Oaxaquia, Chortis, and Florida) were trapped in the terminal, Permo-Carboniferous Pangean collision between Gondwana and Laurentia, only to be left on the Laurentian margin during the Mesozoic opening of the Gulf of Mexico; (4) a central Laurentian native domain (Notre Dame–Shelburne Falls arc) inboard of an exotic domain where the Carolina, Avalonia, and neighboring terranes were transferred from Peru, northern South America, and northwestern Africa during the Ordovician; and (5) a native European domain in which Cadomia originated adjacent to high-latitude North Africa, moving to lower latitudes in the early Paleozoic, only to be caught in the collision between Africa and Laurentia-Baltica during the late Paleozoic Variscan orogeny, and to be left on the Laurentia-Baltic margin during the opening of Tethys.

These observations suggest two different plate regimes existed in Iapetus during the early Paleozoic (Figs. 1–5). Southern Iapetus was underlain by a plate that was being subducted under western South America during which Cuyania and Chilenia traveled from west to east (present coordinates). This convergence was extensional in Cambrian–Early Ordovician times in central South America, producing backarc basins between the native Famatina, Puna, and Arequipa-Antofalla Terranes and changed to compressional in the Late Ordovician and Silurian when they were reaccreted to western Gondwana. The accretion of the Cuyania Terrane, derived from Laurentia, implies that the mid-Iapetus ridge was subducted beneath South America. Central Iapetus was underlain by a plate that was being subducted on both margins, producing the Notre Dame–Shelburne Falls and Famatina-Piedmont arcs. However, the transfer of Carolina, Avalonia, and neighboring terranes from Gondwana to Laurentia suggests that Iapetus was preferentially subducted beneath eastern Laurentia. The sinistral transpression in the northern Appalachians may be attributed to escape tectonics north of the South American promontory (Keppie, 1993). After the Late Ordovician demise of Iapetus, the Rheic ocean persisted between Laurentia and Gondwana, and terranes appear to have remained adjacent to their place of origin until the terminal amalgamation of Pangea in the Permo-Carboniferous (Figs. 6–9). Estimates of the width of the late Paleozoic Rheic ocean vary considerably from zero to several thousand kilometers. Thus, in the Devonian, Kent and van der Voo (1990), van der Voo (1993), and Golonka et al. (1995) have shown collision between eastern Laurentia and either South America or

Africa, whereas Dalziel (1995) has depicted a sizable Rheic ocean until collision in the Late Devonian. These differences are predominantly longitudinal preferences in placing the major cratonic regions. Faunal provinciality data favor proximity between eastern Laurentia and northwestern Gondwana (Keppie et al., 1996). Remnants of the Rheic ocean may occur in the Acatlan Complex of Mexico and the ophiolites of Cadomia. During the Mesozoic break-up of Pangea, the Oaxaquia, Chortis, Florida, and Cadomia Terranes were left adjacent to Laurentia and Baltica.

ACKNOWLEDGMENTS

We thank Lisa Gahagan and Ian Dalziel for their assistance in preparing the global base maps, Luis Burgos for drafting the figures for publication, and Zoltan de Cserna and James Hibbard for thoughtful reviews. This work was funded by Project IN101095 of the Instituto de Geología, Universidad Autónoma de México, and by CONACYT Project 0255P-T9506 (both to J.D.K.), for which we are grateful.

REFERENCES CITED

Aceñolaza, F. G., Miller, H., and Toselli, A. J., 1996, Geología del Sistema de Famatina: Münchner Geologische Hefte, v. A 19, p. 1–410.

Astini, R. A., 1996, Las fases diastróficas del Paleozoico medio en la Precordillera del oeste Argentino, *in* XIII° Congreso Geológico Argentino y III° Congreso Exploración de Hidrocarburos: Actas (Buenos Aires), V, p. 509–526.

Astini, R. A., Benedetto, J. L., and Vaccari, N. E., 1995, The early Paleozoic evolution of the Argentina Precordillera as a Laurentian rifted, drifted, and collided terrane: a geodynamic model: Geological Society of America Bulletin, v. 107, p. 253–273.

Astini, R., Ramos, V. A., Benedetto, J. L., Vaccari, N. E., and Cañas, F. L., 1996, La Precordillera: un terreno exótico a Gondwana, *in* XIII° Congreso Geológico Argentino y III° Congreso Exploración de Hidrocarburos: Actas (Buenos Aires), V, p. 293–324.

Bachtadse, V., Torsvik, T. H., Tait, J. A., and Soffel, H. C., 1995, Paleomagnetic constraints on the paleogeographic evolution of Europe during the Paleozoic, *in* Dallmeyer, R. D., Franke, W., and Weber, K., eds., Pre-Permian geology of central and eastern Europe: Heidelberg, Germany, Springer-Verlag, p. 567–578.

Bahlburg, H., and Hervé, F., 1997, Geodynamic evolution and tectonostratigraphic terranes of northwestern Argentina and northern Chile: Geological Society of America Bulletin, v. 109, p. 869–884.

Ballard, M. M., Van der Voo, R., and Urrutia-Fucagauchi, J., 1989, Paleomagnetic results from Grenvillian-aged rocks from Oaxaca, Mexico: evidence for a displaced terrane: Precambrian Research, v. 42, p. 343–352.

Bluck, B. J., and Dempster, T. J., 1991, Exotic metamorphic terranes in the Caledonides: tectonic history of the Dalradian block, Scotland: Geology, v. 19, p. 1133–1136.

Bond, G. C., Nickelson, P. A., and Kominz, M. A., 1984, Breakup of a supercontinent between 625 Ma and 55 Ma: new evidence and implications for continental histories: Earth and Planetary Science Letters, v. 70, p. 325–345.

Bordonaro, O., and Liñán, E., 1994, Some Middle Cambrian agnostoids from the Precordillera Argentina: Revista Española de Paleontología, v. 9(1), p. 105–114.

Chaloupsky, J., Chlupác, I., Masek, J., Waldhausrová, J., and Cháb, J., 1995, Teplá-Barrandian Zone (Bohemicum): stratigraphy, *in* Dallmeyer, R. D., Franke, W., and Weber, K., eds., Pre-Permian geology of central and eastern Europe: Heidelberg, Germany, Springer-Verlag, p. 379–391.

Chandler, F. W., Loveridge, D., and Currie, K. L., 1987, The age of the Springdale Group, western Newfoundland, and correlative rocks—evidence for a Llandovery overlap sequence in the Canadian Appalachians: Transactions of the Royal Society of Edinburgh: Earth Sciences, v. 78, p. 41–49.

Chantraine, J., Chauvel, J. J., and Rabu, D., 1994, The Cadomian orogeny in the Armorican Massif: lithostratigraphy, *in* Keppie, J. D., ed., Pre-Mesozoic geology of France and related areas: Heidelberg, Germany, Springer-Verlag, p. 81–95.

Coira, B., and Caffe, P., 1995, Xenoliths hosted in Andean Cenozoic volcanic rocks as samples of northern Puna crystalline basement, *in* Laurentian-Gondwanan connections before Pangea Field Conference, Program with Abstracts 11-12, San Salvador de Jujuy, Argentina, Instituto de Geologia y Mineria, Universidad Nacional de Jujuy.

Coira, B. L., Davidson, J. D., Mpodozis, C., and Ramos, V. A., 1982, Tectonic and magmatic evolution of the Andes of northern Argentina and Chile: Earth Science Review, v. 18 (3–4), p. 303–332.

Conti, C. M., Rapalini, A. E., Coira, B., and Koukharsky, M., 1996, Paleomagnetic evidence of an early Paleozoic rotated terrane in northwest Argentina: a clue for Gondwana-Laurentia interaction?: Geology, v. 24, p. 953–956.

Cramer, F. H., and Diez, M. del C. R., 1974, Early Paleozoic palynomorph provinces and paleoclimate: Society of Economic Paleontologists and Mineralogists Special Publication 21, p. 177–188.

Dallmeyer, R. D., 1987, $^{40}Ar/^{39}Ar$ age of detrital muscovite with Lower Ordovician sandstone in the coastal plain basement of Florida: implications for west African linkages: Geology, v. 15, p. 998–1001.

Dallmeyer, R. D., Caen-Vachette, M., and Villeneuve, M., 1987, Emplacement age of post-tectonic granites in southern Guinea (west Africa) and the peninsula Florida subsurface: implications for origins of southern Appalachian exotic terranes: Geological Society of America Bulletin, v. 99, p. 87–93.

Dalziel, I. W. D., 1994, Precambrian Scotland as a Laurentia-Gondwana link: origin and significance of cratonic promontories: Geology, v. 22, p. 589–592.

Dalziel, I. W. D., 1995, Earth before Pangea: Scientific American, January, p. 58–63.

Dalziel, I. W. D., 1997, Overview: Neoproterozoic-Paleozoic geography and tectonics: review, hypothesis, environmental speculation: Geological Society of America Bulletin, v. 109, p. 16–42.

Dalziel, I. W. D., Dalla Salda, L., Cingolani, C., and Palmer, P., 1996, The Argentine Precordillera: a Laurentian terrane?: GSA Today, v. 6(2), p. 16–18.

Damm, K. W., Pichowiak, S., Harmon, R. S., Todt, W., Kelley, S., Omarini, R., and Niemeyer, H., 1990, Pre-Mesozoic evolution of the central Andes; the basement revisited, *in* Kay, S. M., and Rapela, C. W., eds., Plutonism from Antarctica to Alaska: Geological Society of America Special Paper 241, v. 101–126.

Davidson, J., Mpodozis, C., and Rivano, S., 1983, El Paleozoico de la Sierra de Almeida, al oeste de Monturaqui, Alta Cordillera de Antofagasta, Chile: Revista Geológica de Chile, v. 12, p. 3–23.

Davis, J. S., McClelland, W. C., Roeske, S. M., Gerbi, C., and Moores, E. M., 1997, The early to middle Paleozoic tectonic history of the SW Precordillera terrane and Laurentia-Gondwana interactions: Geological Society of America Abstracts with Programs, v. 29, no. 6, p. A380.

Doré, F., 1994, The Variscan orogeny in the Armorican Massif: Cambrian of the Armorican Massif, *in* Keppie, J. D., ed., Pre-Mesozoic geology of France and related areas: Heidelberg, Germany, Springer-Verlag, p. 136–141.

Forsythe, R. D., Davidson, J., Mpodozis, C., and Jesinkey, C., 1989, Lower Paleozoic relative motion of the Arequipa block and Gondwana; paleomagnetic evidence from Sierra de Almeida of northern Chile: Tectonics, v. 12, p. 219–236.

Glover, L. G., III, 1989, Tectonics of the Virginia Blue Ridge and Piedmont, *in* Field Trip Guidebook T385: Washington, D.C., American Geophysical Union, 59 p.

Golonka, J., Ross, M. I., and Scotese, C. R., 1995, Phanerozoic paleogeographic and paleoclimatic modelling maps, *in* Pangea: global environments and resources: Canadian Society of Petroleum Geologists Memoir 17, p. 1–47.

Hatcher, R. D., Jr., 1989, Tectonic synthesis of the U.S. Appalachians, *in* Hatcher, R., Thomas, W., and Viele, G., eds., The Appalachian-Ouachita orogen in the United States: Boulder, Colorado, Geological Society of America, Geology of North America, v. F-2, p. 511–535.

Hibbard, J. P., and Samson, S. D., 1995, Orogenesis exotic to the Iapetan cycle in the southern Appalachians, *in* Hibbard, J. P., van Staal, C. R., and Cawood, P. A., eds., Current perspectives in the Appalachian-Caledonian orogen: Geological Society of Canada Special Paper 41, p. 191–205.

Karabinos, P., Samson, S., Hepburn, C., and Stoll, H., 1998, Taconian orogeny in the New England Appalachians: collision between Laurentia and the Shelburne Falls arc: Geology, v. 26, p. 215–218.

Kay, S. M., Orrell, S., and Abbruzzi, J. M., 1996, Zircon and whole rock Nd-Pb isotopic evidence for a Grenville age and a Laurentian origin for the Precordillera terrane in Argentina: Journal of Geology, v. 104, p. 637–648.

Keppie, J. D., 1993, Synthesis of Paleozoic deformational events and terrane accretion in the Canadian Appalachians: Geologische Rundschau, v. 82, p. 381–431.

Keppie, J. D., and Ortega-Gutiérrez, F., 1995, Provenance of Mexican terranes: isotopic constraints: International Geology Review, v. 37, p. 813–824.

Keppie, J. D., Dostal, J., Murphy, J. B., and Nance, R. D., 1996, Terrane transfer between eastern Laurentia and western Gondwana in the early Paleozoic: constraints on global reconstructions, *in* Nance, R. D., and Thompson, M. D., eds., Avalonian and related peri-Gondwanan terranes of the circum-North Atlantic: Geological Society of America Special Paper 304, p. 369–380.

Keppie, J. D., Dostal, J., Murphy, J. B., and Cousens, B. L., 1997, Paleozoic within-plate volcanic rocks in Nova Scotia (Canada) reinterpreted: isotopic constraints on magmatic source and paleocontinental reconstructions: Geological Magazine, v. 134, p. 425–447.

Keppie, J. D., Davis, D. W., and Krogh, T. E., 1998, U-Pb geochronological constraints on Precambrian stratified units in the Avalon Composite Terrane of Nova Scotia, Canada: tectonic implications: Canadian Journal of Earth Sciences, v. 35, p. 222–236.

Krogh, T. E., Kamo, S. L., Sharpton, B., Marin, L., and Hildebrand, A. R., 1993, U-Pb ages of single shocked zircons linking distal K/T ejecta to the Chicxulub crater: Nature, v. 366, p. 731–734.

Lucassen F., Wilke, H. G., Viramonte, J., Becchio, R., Franz, G., Laber, A., Wemmer, K., and Vroon, P., 1996, The Paleozoic basement of the central Andes (18–26°S) a metamorphic view, *in* III° International Symposium on Andean Geodynamics, St. Maló, Abstracts, p. 779–782.

Mac Niocaill, C., van der Pluijm, B. A., and van der Voo, R., 1997, Ordovician paleogeography and the evolution of the Iapetus ocean: Geology, v. 25, p. 159–162.

Mángano, M. G., Buatois, L. A., 1996, Estratigrafía, sedimentología y evolución paleoambiental de la Formación Suri en la subcuenca de Chaschuil, Ordovícico del Sistema de Famatina, *in* Aceñolaza, F. G., Miller, H., and Toselli, A. J., eds., Geología del Sistema de Famatina: Münchner Geologische Hefte, A19, p. 51–76.

Mannheim, R., 1993, Genese der vulkanite und subvulkanite des altpaläozischen Famatina-systems, NW-Argentinien, und seine geodynamische entwicklung: Münchner Geologische Hefte, v. 9, 130 p.

Manton, W. I., 1996, The Grenville of Honduras: Geological Society of America Abstracts with Programs, v. 28, no. 7, p. A493.

Molina-Garza, R. S., Van der Voo, R., and Urrutia-Fucugauchi, J., 1992, Paleomagnetism of the Chiapas massif, southern Mexico: evidence for rotation of the Maya block and implications for the opening of the Gulf of Mexico: Geological Society of America Bulletin, v. 104, p. 1156–1168.

Mpodozis, C., and Kay, S. M., 1992, Late Paleozoic to Triassic evolution of the Gondwana margin: evidence from Chilean frontal cordilleran batholiths (28°–31°S): Geological Society of America Bulletin, v. 104, p. 999–1014.

Murphy, J. B., Keppie, J. D., Dostal, J., Waldron, J. W. F., and Cude, M. P., 1996, Geochemical and isotopic characteristics of Early Silurian clastic sequences in Antigonish Highlands, Nova Scotia, Canada: constraints on the accretion of Avalonia in the Appalachian-Caledonide Orogen: Canadian Journal of Earth Sciences, v. 33, p. 379–388.

Nasi, C., Mpodozis, C., Cornejo, P., Moscoso, R., and Maksaev, V., 1985, El batolito de Elqui-Limarí (Paleozoico superior–Triásico): Características petrográficas, geoquímicas y significado tectónico: Revista Geológica de Chile, v. 25–26, p. 77–111.

Opdyke, N. D., Jones, D. S., McFadden, B. J., Smith, D. L., Mueller, P. A., and Shuster, R. D., 1987, Florida as an exotic terrane: paleomagnetic and geochronologic investigation of lower Paleozoic rocks from the subsurface of Florida: Geology, v. 15, p. 900–903.

Ortega-Gutiérrez, F., Elias-Herrara, M., and Sánchez-Zavala, J. L., 1997, Tectonic evolution of the Acatlan Complex: new insights, *in* II Convencion sobre la evolucion geologica de Mexico y recursos asociados, Simpois y Coloquio: México, D.F., Instituto de Geología, Universidad Nacional Autónoma de México, p. 19–20.

Ramos, V. A., 1988, Tectonics of the Late Proterozoic–Early Paleozoic: a collisional history of southern South America: Episodes (Ottawa), v. 11(3), p. 168–174.

Ramos, V. A., 1996, Suspect Laurentian affinities of southern South America: International Geological Congress, 30th, Beijing, Abstracts v. 1, p. 9.

Ramos, V. A., 1997, The early Paleozoic margin of South America: a conjugate passive margin of Laurentia?: XIV° Reunión de Geología del Oeste Peninsular: Vila Real, Portugal, Comunicacoes, p. 5–10.

Ramos, V. A., and Basei, M., 1997, The basement of Chilenia: an exotic continental terrane to Gondwana during the early Paleozoic: Symposium on Terrane Dynamics '97: Canterbury, New Zealand, Department of Geological Sciences, University of Canterbury, p. 140–143.

Ramos, V. A., Jordan, T. E., Allmendinger, R. W., Mpodozis, C., Kay, S. M., Cortés, J. M., and Palma, M. A., 1986, Paleozoic terranes of the central Argentine-Chilean Andes: Tectonics, v. 5, p. 855–880.

Ramos, V. A., Vujovich, G. I., and Dallmeyer, R. D., 1996, Los klippes y ventanas tectónicas de la estructura preándica de la Sierra de Pie de Palo (San Juan): edad e implicaciones tectónicas: XIII° Congreso Geológico Argentino y III° Congreso Exploración de Hidrocarburos, Actas (Buenos Aires), V, p. 377–392.

Rankin, D. W., Drake, A. A., Jr., Glover, L., IV, Goldsmith, R., Hall, L. M., Murray, D. T., Radcliffe, N. M., Read, J. F., Secor, D. T., and Stanley, R. F., 1989, Pre-orogenic terranes, *in* Hatcher, R., Thomas, W., and Viele, G., eds., The Appalachian-Ouachita orogen in the United States: Boulder, Colorado, Geological Society of America, Geology of North America, v. F-2, p. 7–99.

Rapalini, A. E., and Astini, R. A., 1998, Paleomagnetic confirmation of the Laurentian origin of the Argentine Precordillera: Earth and Planetary Science Letters, v. 155, p. 1–14.

Robardet, M., Bonjour, J. L., Paris, F., Morzadec, P., and Racheboeuf, P. R., 1994, The Variscan orogeny in the Armorican Massif: Ordovician, Silurian, and Devonian of the Medio-North-Armorican domian, *in* Keppie, J. D., ed., Pre-Mesozoic geology of France and related areas: Heidelberg, Germany, Springer-Verlag, p. 142–151.

Robison, R., and Pantoja-Alor, J., 1968, Tremadocian trilobites from Nochixtlan region, Oaxaca, Mexico: Journal of Paleontology, v. 42, p. 767–800.

Samson, S. L., Palmer, A. R., Robison, R. A., and Secor, D. T., 1990, Biogeographical significance of Cambrian trilobites from the Carolina slate belt: Geological Society of America Bulletin, v. 102, p. 1459–1470.

Sedlock, R. L., Ortega-Gutiérrez, F., and Speed, R. C., 1993, Tectonostratigraphic terranes and tectonic evolution of Mexico: Geological Society of America Special Paper 278, 153 p.

Shackelton, R. M., Ries, A. C., Coward, M. P., and Cobbold, P. R., 1979, Structure, metamorphism and geochronology of the Arequipa Massif of coastal Peru: Quarterly Journal Geological Society of London, v. 136, p. 195–214.

Soper, N. J., 1994, Was Scotland a Vendian RRR junction? Geological Society of London Journal, v. 151, p. 579–582.

Steiner, M. B., and Walker, J. D., 1996, Late Silurian plutons in Yucatan: Journal of Geophysical Research, v. 101, p. 17727–17735.

Strachan, R. A., D'Lemos, R. S., and Dallmeyer, R. D., 1996, Neoproterozoic evolution of an active plate margin: North Armorican Massif, France, *in* Nance, R. D., and Thompson, M. D., eds., Avalonian and related peri-Gondwanan terranes of the circum-North Atlantic: Geological Society of America Special Paper 304, p. 325–332.

Theokritoff, G., 1979, Early Cambrian provincialism and biogeographic boundaries in the North Atlantic region: Lethaia, v. 12, p. 281–295.

Thomas, W. A., and Astini, R. A., 1996, The Argentine Precordillera: a traveller from the Ouachita embayment of North American Laurentia: Science, v. 273, p. 752–757.

Toselli, A., Saavedra, J., and Rossi de Toselli, J. N., 1996, Interpretación geotectónica del magmatismo del sistema de Famatina, *in* Aceñolaza, F. G., Miller, H., and Toselli, A. J., eds., Geología del Sistema de Famatina: Münchner Geologische Hefte, A 19, p. 283–292.

Turner, J. C. M., 1960, Las Sierras transpampeanas como unidad estructural: Primeras Jornadas Geológicas Argentinas (Salta, 1960), Actas (Buenos Aires), II, p. 387–402.

Van der Voo, R., 1993, Paleomagnetism of the Atlantic, Tethys and Iapetus Oceans: Cambridge, United Kingdom, University Press, 411 p.

Van der Voo, R., Johnson, R. J. E., van der Pluijm, B. A., and Knutson, L. C., 1991, Paleomagnetism of some vestiges of Iapetus: paleomagnetism of the Ordovician Robert's Arm, Summerford and Chanceport groups, central Newfoundland: Geological Society of America Bulletin, v. 103, p. 1564–1575.

Van Staal, C., Sullivan, R. W., and Whalen, J. B., 1996, Provenance and tectonic history of the Gander Zone in the Caledonide-Appalachian orogen: implications for the assembly of Avalon, *in* Nance, R. D., and Thompson, M. D., eds., Avalonian and related peri-Gondwanan terranes of the circum-North Atlantic: Geological Society of America Special Paper 304, p. 347–368.

Vujovich, G. I., and Ramos, V. A., 1993, The western Sierras Pampeanas island arc terranes, *in* Proceedings, 1st Circum-Pacific and Circum-Atlantic Terrane Conference (Guanajuato): México, D.F., Universidad Nacional Autónoma de México, Instituto de Geología, p. 166–169.

Wasteneys, H. A., Clark, A. H., Farrar, E., and Langridge, R. J., 1995, Grenvillian granulite facies metamorphism in the Arequipa Massif, Peru: a Laurentia Gondwana link: Earth and Planetary Science Letters, v. 132, p. 63–73.

Whittington, H. B., and Hughes, C. P., 1974, Geography and faunal provinces in the Tremadoc epoch: Society of Economic Palaeontologists and Mineralogists Special Publication 21, p. 203–218.

Manuscript Accepted by the Society September 4, 1998

Printed in U.S.A.